Mathematics: Theory & Applications

Editors

Richard V. Kadison
Isidore M. Singer

I. M. Gelfand
M. M. Kapranov
A. V. Zelevinsky

Discriminants, Resultants, and Multidimensional Determinants

Birkhäuser
Boston • Basel • Berlin
[redacted]

Israel M. Gelfand
Department of Mathematics
Rutgers University
New Brunswick, N.J. 08903

Mikhail M. Kapranov
Department of Mathematics
Northwestern University
Evanston, IL 60208

Andrei V. Zelevinsky
Department of Mathematics
Northeastern University
Boston, MA 02115

Library of Congress Cataloging In-Publication Data

Gelfand, I. M. (Israel Moiseevich)
Discriminants, resultants, and multidimensional determinants /
I. M. Gelfand, M. M. Kapranov, A. V. Zelevinsky.
p. cm. -- (Mathematics : theory & applications)
Includes bibliographical references and index.
ISBN 0-8176-3660-9
1. Determinants. 2. Discriminant analysis. I. Kapranov, M. M.
(Mikhail M.), 1962- . II. Zelevinsky, Andrei V., 1953- .
III. Title. IV. Series: Mathematics (Boston, Mass.)
QA191.G45 1994 93-46733
516.3'5--dc20 CIP

Printed on acid-free paper
 Birkhäuser

ISBN 0-8176-3660-9
ISBN 3-7643-3660-9
Typeset by the Authors in TEX.
Printed and bound by Quinn-Woodbine, Woodbine, NJ.
Printed in the U.S.A.

9 8 7 6 5 4 3 2 [redacted]

Contents

II. A-DISCRIMINANTS AND A-RESULTANTS

CHAPTER 5. Toric Varieties

CHAPTER 6. Newton Polytopes and Chow Polytopes

CHAPTER 7. Triangulations and Secondary Polytopes

CHAPTER 8. A-Resultants and Chow Polytopes of Toric Varieties

CHAPTER 9. A-Discriminants

CHAPTER 10. Principal A-Determinants

Preface

This book has expanded from our attempt to construct a general theory of hypergeometric functions and can be regarded as a first step towards its systematic exposition. However, this step turned out to be so interesting and important, and the whole program so overwhelming, that we decided to present it as a separate work. Moreover, in the process of writing we discovered a beautiful area which had been nearly forgotten so that our work can be regarded as a natural continuation of the classical developments in algebra during the 19th century.

We found that Cayley and other mathematicians of the period understood many of the concepts which today are commonly thought of as modern and quite recent. Thus, in an 1848 note on the resultant, Cayley in fact laid out the foundations of modern homological algebra. We were happy to enter into spiritual contact with this great mathematician.

The place of discriminants in the general theory of hypergeometric functions is similar to the place of quasi-classical approximation in quantum mechanics. More precisely, in [GGZ] [GKZ2] [GZK1] a general class of special functions was introduced and studied, the so-called A-hypergeometric functions. These functions satisfy a certain holonomic system of linear partial differential equations (the A-hypergeometric equations). The A-discriminant, which is one of our main objects of study, describes singularities of A-hypergeometric functions. According to the general principles of the theory of linear differential equations, these singularities are governed by the vanishing of the highest symbols of A-hypergeometric equations. The relation between differential operators and their highest symbols is the mathematical counterpart of the relation between quantum and classical mechanics; so we can say that hypergeometric functions provide a "quantization" of discriminants.

In our work on hypergeometric functions we found connections with many questions in algebra and combinatorics. We hope that this book brings to light some of these connections. One of the algebraic concepts which seems to us particularly important is that of hyperdeterminants (analogs of determinants for multi-dimensional "matrices.") After rediscovering hyperdeterminants in connection with hypergeometric functions, we found that they too, had been introduced by Cayley in the 1840s. Unfortunately, later on, the study of hyperdeterminants was largely abandoned in favor of another, more straightforward definition (cf. [P]). The only other work on hyperdeterminants of which we are aware is an important paper by Schläfli [Schl]. In this volume we give a detailed treatment of hyperdeterminants with the hope of attracting the attention of other mathematicians to this subject.

We would like to thank S.I. Gelfand, M.I. Graev and V.A. Vassiliev, who, through discussions and collaboration, have much influenced our understanding of the vast and beautiful field of hypergeometric functions.

Introduction

I

In this book we study discriminants and resultants of polynomials in several variables. The most familiar example is the discriminant of a quadratic polynomial $f(x) = ax^2 + bx + c$. This is

$$\Delta(f) = b^2 - 4ac, \tag{1}$$

which vanishes when $f(x)$ has a double root.

More generally, we can consider a polynomial $f(x_1, \ldots, x_k)$ of degree $\leq d$ in k variables. An analog of a multiple root for f is a point where f vanishes together with all its first partial derivatives $\partial f/\partial x_i$. The *discriminant* $\Delta(f)$ is a polynomial function in the coefficients of f which vanishes whenever f has such a "multiple root." The existence of Δ is not quite trivial; however, it can be shown that $\Delta(f)$ exists and is unique up to sign if we require it to be irreducible and to have relatively prime integer coefficients. For instance, the discriminant of a cubic polynomial in one variable ($k = 1$, $d = 3$) is given by

$$\Delta(a_0+a_1x+a_2x^2+a_3x^3) = a_1^2a_2^2-4a_1^3a_3-4a_0a_2^3-27a_0^2a_3^2+18a_0a_1a_2a_3. \tag{2}$$

There is a subtle point in the definition of $\Delta(f)$: that is, $\Delta(f)$ depends not only on f but also on the choice of a degree bound d. For instance, the formula (2) applied to a quadratic polynomial gives a different expression from (1). With this in mind, we introduce the following more general version of a discriminant. Let A be a finite set of monomials in k variables, and let $\mathbf{C}^A$ denote the space of all polynomials with complex coefficients all of whose monomials belong to A. The *A-discriminant* $\Delta_A(f)$ is an irreducible polynomial in the coefficients of $f \in \mathbf{C}^A$ which vanishes whenever f has a multiple root $(x_1, \ldots, x_k)$ with all $x_i \neq 0$ (the last condition is added to be able to ignore trivial multiple roots which can appear if all monomials from A have high degree). The A-discriminant will be one of our main objects of study.

The notion of the A-discriminant includes as special cases several fundamental algebraic concepts. If we take $A = \{1, x, \ldots, x^m, y, yx, \ldots, yx^n\}$, for example, then a typical polynomial from $\mathbf{C}^A$ has the form $f(x) + yg(x)$. Its A-discriminant is the *resultant* of f and g: it vanishes whenever f and g have a common root.

More generally, the resultant of $k + 1$ polynomials $f_0, \ldots, f_k$ in k variables is defined as an irreducible polynomial in the coefficients of $f_0, \ldots, f_k$, which vanishes whenever these polynomials have a common root. The resultant can be treated as a special case of the A-discriminant of an auxiliary polynomial $f_0(x) + \sum_{i=1}^k y_i f_i(x)$, $x = (x_1, \dots, x_k)$.

Another important example occurs when A consists of n^2 monomials $x_i y_j$, $i, j = 1, \dots, n$. A typical polynomial from $\mathbf{C}^A$ is now a bilinear form $f(x, y) = \sum a_{ij} x_i y_j$ whose A-discriminant is the determinant of the matrix $\|a_{ij}\|$.

The last example has a natural generalization: we can take A as the set of all multilinear monomials in three or more groups of variables. An element $f \in \mathbf{C}^A$ (i.e., a multilinear form) is represented by a higher-dimensional "matrix" $\|a_{i_1 \dots i_r}\|$. Thus the A-discriminant Δ_A in this case is a polynomial function of a "matrix" which extends the notion of a determinant. Following Cayley [Ca1], we call this Δ_A the *hyperdeterminant* of $\|a_{i_1 \dots i_r}\|$. For example, the hyperdeterminant of a $2 \times 2 \times 2$ matrix $\|a_{ijk}\|$, $i, j, k = 0, 1$, is given by

$$(a_{000}^2 a_{111}^2 + a_{001}^2 a_{110}^2 + a_{010}^2 a_{101}^2 + a_{011}^2 a_{100}^2)$$

$$-2(a_{000}a_{001}a_{110}a_{111} + a_{000}a_{010}a_{101}a_{111} + a_{000}a_{011}a_{100}a_{111} + a_{001}a_{010}a_{101}a_{110}$$

$$+a_{001}a_{011}a_{110}a_{100} + a_{010}a_{011}a_{101}a_{100}) + 4(a_{000}a_{011}a_{101}a_{110} + a_{001}a_{010}a_{100}a_{111}).$$

The study of hyperdeterminants was initiated by Cayley [Ca1] and Schläfli [Schl] but then was largely abandoned for 150 years. We present a treatment of hyperdeterminants in Chapter 14.

II

Let $\nabla_A = \{f \in \mathbf{C}^A : \Delta_A(f) = 0\}$ be the hypersurface in the space of polynomials consisting of polynomials with vanishing A-discriminant. We shall be mainly concerned with the following two closely related problems:

(a) the study of the geometric properties of the hypersurface ∇_A;

(b) finding an explicit algebraic expression of the discriminant Δ_A.

To illustrate the importance of problem (a), consider the special case when A consists of all monomials in $x_1, \dots, x_k$ of a given degree d. Every $f \in \mathbf{C}^A$ (i.e., a homogeneous form of degree d) defines a hypersurface $\{f = 0\}$ in the projective space P^{k-1}. It is easy to see that ∇_A consists exactly of those f for which the hypersurface $\{f = 0\}$ is *singular*. Therefore the complement $\mathbf{C}^A - \nabla_A$ parametrizes all smooth hypersurfaces of a given degree in the projective space. To understand the geometric structure of $\mathbf{C}^A - \nabla_A$ is an important instance of the general moduli problem in algebraic geometry.

Equally important is the situation over the real numbers. Hilbert's 16th problem (classifying isotopy types of smooth real hypersurfaces of given degree d)

amounts to the study of connected components of $\mathbf{R}^A - \nabla_A$, the space of real polynomials with a non-vanishing discriminant.

Problem (b) has a long and glorious history. Explicit formulas for discriminants and resultants were the focus of several remarkable mathematicians in the last century. Many ingenious formulas were found by Cayley, Sylvester and their followers. However, we are still very far from a complete understanding of discriminants. For instance, an explicit polynomial expression for Δ_A is known only in a very limited number of special cases. Such formulas would be of great importance for the problem of finding explicit solutions of systems of polynomial equations. Problems of this kind are of interest not only for theoretical reasons, but are encountered more and more on a practical level because of the progress in computer technology.

III

We will use three main approaches in our study of discriminants and resultants:

- a geometric approach via projective duality and associated hypersurfaces;
- an algebraic approach via homological algebra and determinants of complexes (Whitehead torsion);
- a combinatorial approach via Newton polytopes and triangulations.

The geometric approach to discriminants is based on the observation that the discriminantal variety ∇_A is projectively dual to a certain variety X_A defined by a simple parametric representation. For example, if A consists of all monomials of degree d in k variables then X_A is the projective space P^{k-1} in its Veronese embedding. In the general case, X_A is the projective *toric* variety associated with A. The notion of the projectively dual variety $X^\vee$ makes sense for an arbitrary projective variety $X \subset P^{n-1}$: it is the closure of the set of all hyperplanes in P^{n-1} which are tangent to X at some smooth point. Thus the problem of finding the discriminant is a particular case of a more general geometric problem: find the equation(s) of $X^\vee$. We call this equation (in the case where $X^\vee$ is a hypersurface) the *X-discriminant.*

Although the resultants can be formally treated as discriminants of a special kind (see above), they have their own interesting geometric meaning. As for discriminants, we can associate the resultant to any projective variety $X \subset P^{n-1}$. Instead of $X^\vee$, we now consider the *associated hypersurface* $\mathcal{Z}(X)$ of X. If $\dim X = k - 1$ then $\mathcal{Z}(X)$ is the locus of all codimension k projective subspaces in P^{n-1} which meet X. The equation of $\mathcal{Z}(X)$ in the appropriate Grassmannian is the classical *Chow form* of X. This can be represented as a polynomial in the coefficients of k linear forms defining a subspace from $\mathcal{Z}(X)$. We call this polynomial the X-*resultant* (the classical resultant of polynomials in several variables is a special case of this construction).

In Part I of this book we examine X-discriminants and X-resultants (or, in other words, projective duality and associated hypersurfaces) in the general context of projective geometry.

IV

The algebraic approach to discriminants and resultants which we use here goes back to Cayley. In his breathtaking 1848 note [Ca4] * he outlined a general method of writing down the resultant of several polynomials in several variables. We were very surprised to find that Cayley introduced in this note several fundamental concepts of homological algebra: complexes, exactness, Koszul complexes, and even the invariant now sometimes called the Whitehead torsion or Reidemeister-Franz torsion of an exact complex. The latter invariant is a natural generalization of the determinant of a square matrix (which itself was a rather recent discovery back in 1848!), so we prefer to call it the determinant of a complex. Using this terminology, Cayley's main result is that the resultant is the determinant of the Koszul complex.

Cayley's method is very general: without much effort it can be adapted to the study of X-discriminants and X-resultants associated as above to an arbitrary projective variety X. To get more detailed information, we complement Cayley's method with more recent tools such as coherent sheaves, perverse sheaves, microlocal geometry and D-modules.

V

Under a combinatorial approach we treat polynomials in the most naive way: as sums of monomials. To the best of our knowledge, there were no attempts in the classical literature to understand discriminants and resultants from this point of view, i.e., to describe which monomials can appear in them and with which coefficients. This is probably because the number of occurring monomials is usually very large. For example, the discriminant of a cubic form in three variables contains 2040 monomials (we are obliged to S. Duzhin who first showed it to us some years ago). At first glance, there seems to be no structure at all in these monomials and their coefficients. However, such a structure exists! The "magic crystal" that brings it to light is the concept of a *Newton polytope*.

Every monomial $x_1^{\omega_1} \cdots x_n^{\omega_n}$ in n variables can be visualized as a lattice point $(\omega_1, \ldots, \omega_n)$ in $\mathbf{R}^n$. The Newton polytope $N(F)$ of a polynomial $F(x_1, \ldots, x_n)$ is

* This note is reproduced as an appendix in this book

the convex hull in $\mathbf{R}^n$ of all lattice points representing monomials occurring in F. The structure of this polytope is deeply related to the geometry of the hypersurface $\{F = 0\}$. In fact, the asymptotic behavior of this hypersurface "at infinity" is controlled by the *extreme monomials* of F which correspond to the vertices of $N(F)$.

The notion of a Newton polytope goes back to Newton, and made some isolated appearances in the 19th century, cf. [Br 2]. More recently, some spectacular applications of Newton polytopes to classical algebraic problems (the number of solutions of systems of polynomial equations) have been found by A. Kouchnirenko, D. Bernstein, A. Khovansky [Ber], [Kou], [Kh]. We make use of these results in Part II.

It was a very surprising discovery for us when we realized that the Newton polytopes of A-discriminants admit a very nice combinatorial description. We recall that A is a finite set of monomials in k variables. As before, we represent the monomials from A as lattice points in $\mathbf{R}^k$. Hence we can consider the convex hull $Q \in \mathbf{R}^k$ of the set A. Our main result (which is the central point of Part II) is a description of the Newton polytope $N(\Delta_A)$ in terms of Q and A. Roughly speaking, it turns out that vertices of $N(\Delta_A)$ (i.e., extreme monomials in the A-discriminant) correspond to some *triangulations* of Q into simplices all of whose vertices lie in A.

The extreme monomial in Δ_A corresponding to a triangulation T of Q is determined explicitly once we know all the simplices in T and their volumes. The coefficient of this monomial is the product of numbers of the form $V_i^{V_i}$ where the V_i are the volumes of the simplices of T under suitable normalization. This provides an explanation of such coefficients as $4 = 2^2$ or $27 = 3^3$ in the formulas (1) and (2) above. The expression $\prod V_i^{V_i}$ (or, rather, its logarithm $\sum V_i \log(V_i)$) brings to mind the entropy of a probability distribution. It would be interesting to find a "probabilistic" reason for its appearance in discriminants. Even more intriguing is the fact that this appearance is not isolated—entropy-like expressions enter the formula for the rational uniformization of the variety ∇_A (see Chapter 9).

To illustrate the above description, consider the simplest A-discriminant of a quadratic polynomial $ax^2 + bx + c$ given by (1). Here A consists of $0, 1, 2 \in \mathbf{Z}$, the polytope Q is the segment $[0, 2]$, with its two "triangulations". The first one consists of just one 1-dimensional "simplex" $[0, 2]$ of length 2, corresponding to the term $-4ac$ in (1). The second "triangulation" consists of two "simplices": $[0, 1]$ and $[1, 2]$, corresponding to the term b^2. Similarly, for the case of a cubic polynomial in one variable, we have $A = \{0, 1, 2, 3\}$ and $Q = [0, 3]$. There are now 4 triangulations of Q which correspond to the first four terms in the discriminant (2). Our final example is the determinant of a 2×2 matrix given by a familiar formula $\Delta = ad - bc$. We have already seen that this is also a special case of an A-discriminant. The set A now consists of the vertices of a square Q; the terms ad and $-bc$ correspond to two triangulations of Q by means of one of its diagonals.

The description of the Newton polytope of Δ_A leads to a purely geometric notion of the *secondary polytope* $\Sigma(A)$ of a point configuration A. This is a polytope whose vertices correspond to the so-called *coherent* triangulations of the convex hull Q of A. Secondary polytopes and their generalizations (*fiber polytopes* introduced and studied by Billera and Sturmfels [BS1], [BS2]) are quite interesting by themselves. A triangulation of a polytope Q can be viewed as a discrete analog of a Riemannian metric on Q. So $\Sigma(A)$ can be seen as a kind of combinatorial Teichmüller space parametrizing such metrics. This reminds us of the work of Penner [Pen] who constructed a combinatorial model for the Teichmüller space of a Riemann surface in terms of its curvilinear triangulations.

VI

As mentioned in the Preface, our interest in the subject arose from the theory of hypergeometric functions [Ge] [GGZ] [GKZ2] [GZK1]. Although this theory is not formally present in the book, its influence is felt in several places. In a sense, one can say that hypergeometric functions provide a "quantization" of the discriminants. More precisely, to a finite set of monomials A, we associate a certain holonomic system of differential equations on the space $\mathbf{C}^A$ whose solutions are the so-called *A-hypergeometric functions*. The highest symbols of the equations of this system define, in the cotangent bundle of $\mathbf{C}^A$, the *characteristic variety* of the system. One of the components of this variety, when projected back to $\mathbf{C}^A$ is the discriminantal hypersurface ∇_A and the projections of other components are similar hypersurfaces associated to subsets of A.

The notion of a coherent triangulation, which plays such an essential part in our combinatorial approach to discriminants, was first brought to our attention by the analysis of A-hypergeometric functions. In fact, every coherent triangulation of the convex hull Q of A produces an explicit basis in the space of A-hypergeometric functions. This basis consists of a finite number of power series whose coefficients are products of the values of the Euler Γ-function.

VII

The book is subdivided into three parts. The first part is devoted to discriminants and resultants associated with arbitrary projective subvarieties. Most of the results here are classical but, to the best of our knowledge, have been never systematically treated in a book. Chapter 1 discusses projective duality. Chapter

2 introduces the Cayley method of expressing the discriminant as the determinant of a complex. Chapter 3 presents a parallel treatment of the resultants. Finally, Chapter 4 gives an exposition of the theory of Chow varieties (parameter spaces for projective subvarieties of given dimension and degree).

In Part II we consider A-discriminants and A-resultants. Geometrically, this corresponds to the specialization of the setting of Part I to projective toric varieties. We review toric varieties in Chapter 5 and the work of Bernstein and Kouchnirenko on Newton polytopes in Chapter 6. In Chapter 7 we present our main combinatorial-geometric construction: the secondary polytopes. In Chapters 8 – 11 this construction is related to Newton polytopes of A-discriminants and A-resultants. The main link between discriminants and triangulations is the so-called *principal A-determinant*. This is a certain product of discriminants whose Newton polytope is precisely the secondary polytope $\Sigma(A)$. For discriminants themselves, the correspondence between triangulations and the vertices of the Newton polytope is, in general, many-to-one.

Finally, Part III is devoted to the most classical examples of discriminants and resultants. The case of polynomials in one variable is treated in Chapter 12. Surprisingly, the point of view of Newton polytopes leads to new results even in this case. We treat the case of forms in several variables in Chapter 13, and hyperdeterminants in Chapter 14.

Geometrically, all of these examples correspond to varieties which are products of projective spaces $P^{l_1} \times \cdots \times P^{l_r}$ in a suitable projective embedding.

VIII

We did not attempt in this volume to collect all that is known about discriminants and resultants. The choice of material reflects both personal interests and the expertise of the authors.

Let us give a brief overview of some of the developments not included here but closely related to our subject. The following list is by no means complete.

First of all, an old tradition going back to Cayley and Sylvester, includes discriminants and resultants in the general context of the invariant theory of the group $GL(n)$. This approach involves expressing discriminants and resultants using the *symbolic method* (see e.g., [Go]). Our combinatorial approach focuses on the monomials, and thus is based on the action of the algebraic torus $(\mathbf{C}^*)^n$, not on the action of the whole group $GL(n)$.

Second, the study of discriminants and resultants constitutes only a part of Elimination Theory. There are other aspects of this theory which we do not discuss. Among those, we can mention the study of certain resultant ideals using commutative algebra, which goes back to Macaulay [Macaul2]. A more modern treatment of these questions was undertaken by Jouanolou [Jo].

Macaulay made another intriguing contribution to the theory by giving an ingenious refinement of the Cayley method [Macaul1]. It would be interesting to put his approach in the general framework of this book.

We did not discuss at all computational applications of explicit formulas for discriminants and resultants. An interesting work in this direction was recently done by J. Canny [Can].

The material we present gives rise to many natural questions. Some of them are currently under active investigation. We did not try to give the most up-to-date account of these developments. Let us just mention some directions of current research.

In the geometric direction, the structures related to Chow forms and X-resultants have led to the notion of the so-called *Chow quotient* for an action of an algebraic group [KSZ]. These quotients, which are different from the quotients provided by the geometric invariant theory, already have some promising applications. For example, in [Ka2], [Ka3] they were related to Grothendieck-Knudsen moduli spaces of stable curves.

In the algebraic direction, there is the work concerning explicit polynomial expressions for discriminants and resultants. In some cases it is possible to classify all such expressions which can be obtained by the Cayley method [SZ], [WZ].

There are other kinds of formulas for discriminants and resultants, in terms of products of some irrational factors (such as the values of one polynomial at the roots of another one). A rather general version of such a formula was given in [PS].

As for the combinatorial direction, we should mention further studies of the properties and applications of secondary and fiber polytopes [BFS], [BGS], [BS1], [BS2]. The Newton polytopes of classical resultants made an unexpected appearance in the axiomatics of monoidal 2-categories [KV]. The study of the Chow forms associated to determinantal varieties and related combinatorial structures has been initiated in [SZ2], [BZ]. Finally, a different approach to the combinatorics of A-resultants was developed in [Stu1].

IX

We are grateful to L.J. Billera, A.G. Khovansky, A.G. Kouchnirenko, S.Yu. Orevkov, B. Sturmfels and J. Weyman for helpful discussions on various questions related to discriminants and resultants. The actual writing of the book began in 1990/91 when two of the authors (M.K., A.Z.) were visiting Cornell University and were partially supported by that institution. In addition, the research of the authors was partially supported by NSF grants. The second author was partially supported by an A. P. Sloan Research fellowship. We would like to thank Birkhäuser and especially Ann Kostant for continuing help during the editorial process.

Conventions

We always work over the field **C** of complex numbers, unless otherwise specified. In many cases it is straightforward to extend the definitions and results to an arbitrary ground field, but we usually leave this to the reader.

Topological terminology (the closure, open and closed sets, etc.) usually refers to Zariski topology.

A few words about the organization of the material. The book is divided into 14 chapters. Each chapter consists of several sections, and each section is divided into subsections numbered by the letters A, B, C, etc. The numeration of sections starts anew in every chapter, and the numeration of all statements (theorems, definitions, examples etc.) starts anew in every section. Thus, the reference to Theorem 4.2 without specifying a chapter, means the second statement in Section 4 of the current chapter.

PART I

General Discriminants and Resultants

CHAPTER 1

Projective Dual Varieties and General Discriminants

1. Definitions and basic examples

A. Projective duality

We denote by P^n the standard complex projective space of dimension n. Thus a point of P^n is given by $(n+1)$ homogeneous coordinates $(x_0 : \ldots : x_n)$, $x_i \in \mathbf{C}$, which are not all equal to 0 and are regarded modulo simultaneous multiplication by a non-zero number. More generally, if V is a finite-dimensional complex vector space, then we denote by $P(V)$ the projectivization of V, i.e., the set of 1-dimensional vector subspaces in V. Thus $P^n = P(\mathbf{C}^{n+1})$.

If $W \subset V$ is a vector subspace, then $P(W)$ is naturally a subset in $P(V)$. Subsets of this form are called projective subspaces. As usual, projective subspaces of dimension 1,2 or of codimension 1 are called lines, planes, and hyperplanes.

Consider the projective space $P = P(V)$, where V is as before. Hyperplanes in P form another projective space P^* which is the projectivization of the dual vector space V^*. Conversely, to every point p in P, we can associate a hyperplane $p^\vee$ in P^*, namely the set of all hyperplanes in P containing p, see Figure 1. Thus $(P^*)^*$ is naturally identified with P. If $L \subset P$ is a projective subspace then we can consider the projective subspace $L^\vee \subset P^*$, the intersection of all hyperplanes $p^\vee$, $p \in L$. Clearly, $\operatorname{codim} L^\vee = \dim L + 1$, and $(L^\vee)^\vee = L$. We say that $L^\vee$ is *projectively dual* to L. If L is the projectivization of a linear subspace $K \subset V$ then $L^\vee$ is the projectivization of the orthogonal complement $K^\perp \subset V^*$.

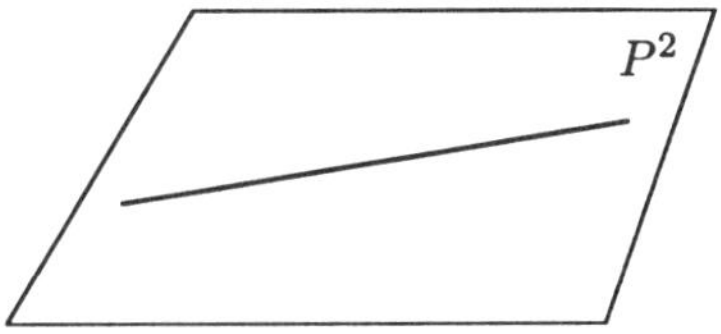

Figure 1a. A point in $(P^2)^*$

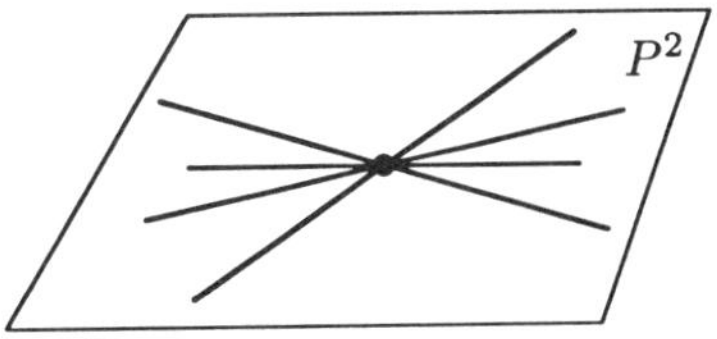

Figure 1b. A line in $(P^2)^*$

Remarkably, projective duality can be extended to an involutive correspondence between non-linear algebraic subvarieties in P and P^*. More precisely, let

$X \subset P$ be a closed irreducible algebraic subvariety. A hyperplane $H \subset P$ is said to be *tangent* to X if there exists a smooth point $x \in X$ such that $x \in H$ and the tangent space to H at x contains the tangent space to X at x. Denote by $X^\vee \subset P^*$ the closure of the set of all hyperplanes tangent to X. The variety $X^\vee$ is called *projectively dual* to X.

In the case when X is smooth and does not lie in any hyperplane, the projectively dual variety $X^\vee$ has the following geometric interpretation: a hyperplane H belongs to $X^\vee$ if and only if the intersection $H \cap X$ is singular. Here $H \cap X$ is regarded as a scheme, and "singular" means "not a smooth algebraic variety."

It is often helpful to consider a projective variety $X \subset P(V)$ together with the associated conic variety (or simply the cone) $Y \subset V$ formed by 0 and all vectors whose projectivization lies in X. The cone $Y^\vee \in V^*$ associated to $X^\vee$ will be called *projectively dual* to Y. In particular, each linear subspace $K \subset V$ is a cone, and its projectively dual cone is $K^\perp \subset V^*$.

The name "dual variety" is justified by the following fundamental fact.

Theorem 1.1. (Biduality Theorem). *For any projective variety $X \subset P$, we have $(X^\vee)^\vee = X$. Moreover, if z is a smooth point of X and H is a smooth point of $X^\vee$, then H is tangent to X at z if and only if z, regarded as a hyperplane in P^*, is tangent to $X^\vee$ at H.*

We postpone the proof of Theorem 1.1 until Section 3.

Let us mention a corollary here which shows that "typically" the projective dual variety is a hypersurface. We shall say that an algebraic variety $X \subset P$ is *ruled in projective spaces* of dimension r, if there is a Zariski open subset $U \subset X$ which is the union of r-dimensional projective subspaces lying on U (in this case it is easy to show, by a closedness argument, that one can in fact take $U = X$; but we do not need this). For example, a quadratic cone in the projective 3-space is ruled in projective lines.

Corollary 1.2. *If $X^\vee$ is not a hypersurface, say,* codim $X^\vee = r + 1$, *then X is a variety ruled in projective spaces of dimension r.*

Proof. By the biduality theorem, the statement is equivalent to the following: if codim $X = r + 1$, then $X^\vee$ is ruled in projective spaces of dimension r. But this is obvious. Indeed, the condition for a hyperplane H to be tangent to X at a smooth point x is that $T_x H$ contains the tangent space $T_x X$. For a given x, all H with this property form a projective space of dimension r.

B. General discriminants

As before, let X be a closed algebraic subvariety in a projective space P, and let

$X^\vee \subset P^*$ be the projectively dual variety.

Proposition 1.3. *If X is irreducible then $X^\vee$ is irreducible.*

Proof. Let $X_{sm} \subset X$ be the smooth locus of X. Consider the set $W_0 \subset P \times P^*$ of pairs (x, H) such that $x \in X_{sm}$, and H is the hyperplane tangent to X at x. Let W be the Zariski closure of W_0. By definition, $X^\vee = pr_2(W)$, the projection of W on P^*. On the other hand, the first projection $pr_1 : W_0 \to X_{sm}$ makes W_0 into a bundle over X_{sm} whose fibers are projective spaces. If X is irreducible we conclude that each of the varieties X_{sm}, W_0, W, and $X^\vee$ is also irreducible, thus proving the proposition.

Now we suppose that X is an irreducible variety. It will be convenient to think of P as $P(V^*)$, the projectivization of the vector space V^* dual to a finite dimensional vector space V. Then $X^\vee$ is an irreducible subvariety in $P(V)$.

Suppose that $X^\vee$ is a hypersurface in $P(V)$. We shall call the *X-discriminant* and denote by Δ_X the defining polynomial of $X^\vee$. This is an irreducible homogeneous polynomial function on V such that $X^\vee$ is given by the equation $\Delta_X = 0$. Note that Δ_X is defined only up to a non-zero constant multiple. When codim $X^\vee > 1$, we set $\Delta_X = 1$.

The definition of Δ_X can be reformulated as follows. Let $Y \subset V^*$ be a cone over X, and suppose $x_1, \ldots, x_k$ are local coordinates on Y. Every $f \in V$ is a linear form on V^*. After restricting f to Y it becomes an algebraic function in $x_1, \ldots, x_k$, which will also be denoted by f. The X-discriminant Δ_X can be defined as an irreducible polynomial, which vanishes at $f \in V$ whenever the function $f(x_1, \ldots, x_k)$ has a "multiple root", i.e., it vanishes at some $y \in Y - \{0\}$ together with all of its first derivatives $\partial f/\partial x_i$.

Example 1.4. Consider the d-dimensional projective space $P^d = P(V^*)$ with homogeneous coordinates $z_0, \ldots, z_d$, and let $X \subset P^d$ be the curve formed by points

$$(x^d : x^{d-1}y : x^{d-2}y^2 : \ldots : xy^{d-1} : y^d), \quad x, y \in \mathbf{C}, \quad (x, y) \neq (0, 0)$$

(the Veronese curve). The space V consists of linear forms in $z_0, \ldots, z_d$. Every linear form $l(z) = \sum a_i z_i$ is uniquely recovered from its restriction to (the cone over) X which is a binary form $f(x, y) = \sum_{i=0}^{d} a_i x^{d-i} y^i$. We can also write f in a non-homogeneous form: $f(x) = \sum_{i=0}^{d} a_i x^{d-i}$. The condition that $l \in X^\vee$ means that the corresponding polynomial $f(x)$ has a multiple root. Hence Δ_X is an irreducible polynomial in the coefficients $a_0, a_1, \ldots, a_d$, vanishing whenever $f(x) = \sum a_i x^{d-i}$ has a multiple root. We see that Δ_X is the classical discriminant of a polynomial in one variable.

Returning to the general case, suppose that $X \subset P(V^*)$ does not lie in a hyperplane, and is smooth, and $X^\vee$ is a hypersurface. Then a vector $f \in V$ satisfies $\Delta_X(f) = 0$ if and only if the hyperplane section $\{f = 0\}$ of X is singular. Our next result shows that the information about Δ_X allows us to find the singular point of this section.

Theorem 1.5. *Suppose, $X \subset P^{n-1}$ is smooth, and $X^\vee \subset (P^{n-1})^*$ is a hypersurface. Let $z_1, \dots, z_n$ be homogeneous coordinates on P^{n-1}, and $a_1, \dots, a_n$ the dual homogeneous coordinates on $(P^{n-1})^*$. Suppose $f = (a_1 : \dots : a_n)$ is a smooth point of $X^\vee$. Then the hyperplane section $\{f = 0\}$ of X has a unique singular point z, and the coordinates of z are given by $(\frac{\partial \Delta_X}{\partial a_1}(f) : \dots : \frac{\partial \Delta_X}{\partial a_n}(f))$.*

Proof. Let $H \subset P^{n-1}$ be the hyperplane corresponding to f. By the Biduality Theorem 1.1, H is tangent to X at z if and only if the hyperplane in $(P^{n-1})^*$ corresponding to z is tangent to $X^\vee$ at f. Since, by our assumption, f is a smooth point of $X^\vee$, such a point z is unique and is given by $z_i = \frac{\partial \Delta_X}{\partial a_i}(f)$.

2. Duality for plane curves

To become more intuitive about projectively dual varieties, we look at the case of plane curves.

A. Parametric representation of the dual

For any irreducible curve $X \subset P^2$ which is not a line, the dual variety $X^\vee \subset (P^2)^*$ is also an irreducible curve, according to Proposition 1.3. Let x, y, z be homogeneous coordinates on P^2, and p, q, r the dual homogeneous coordinates on $(P^2)^*$. The duality associates to every irreducible homogeneous polynomial $f(x, y, z)$ of degree ≥ 2 (the equation of a curve X) an irreducible homogeneous polynomial $F(p, q, r) = \Delta_X$ (the equation of $X^\vee$). The polynomial F is defined by f uniquely up to scalar multiple.

It is not so easy to find F from f. However, it is quite easy to write down a *parametric representation* of $X^\vee$ given a parametric representation of X. We shall do it using appropriate affine coordinates in P^2 and $(P^2)^*$. We choose an affine chart $\mathbf{C}^2 = \{z \neq 0\} \subset P^2$ with affine coordinates x, y (so the third homogeneous coordinate z is set to be 1). The dual chart $\mathbf{C}^{2*} \subset (P^2)^*$ with coordinates p, q is obtained by setting the third homogeneous coordinate r in $(P^2)^*$ to be -1. Geometrically, $\mathbf{C}^{2*}$ consists of lines in P^2 not passing through the point $(0, 0) \in \mathbf{C}^2 \subset P^2$. Every such line that meets $\mathbf{C}^2$ is given by the affine equation $px + qy = 1$, and the coordinates p, q of the line are the coefficients in this equation. Note that the line in $(P^2)^*$ with coordinates $p = q = 0$ does not meet $\mathbf{C}^2$, i.e., it is the line "at infinity."

A local parametric equation of X has the form $x = x(t)$, $y = y(t)$, where t is a local coordinate on X, and $x(t)$, $y(t)$ are analytic functions. By definition, the dual curve $X^\vee$ has the parametrization $p = p(t)$, $q = q(t)$, where $p(t)x+q(t)y+1 = 0$ is the affine equation of the tangent line to X at $(x(t), y(t))$. Hence, the parametric representation of $X^\vee$ has the form

$$p(t) = \frac{-y'(t)}{x'(t)y(t) - x(t)y'(t)}, \quad q(t) = \frac{x'(t)}{x'(t)y(t) - x(t)y'(t)}. \tag{2.1}$$

This formula readily implies biduality theorem 1.1 for plane curves. Indeed, applying (2.1) once again, we find that the parametric representation of $X^{\vee\vee}$ is $x = u(t)$, $y = v(t)$, where

$$u(t) = \frac{-q'(t)}{p'(t)q(t) - p(t)q'(t)}, \quad v(t) = \frac{p'(t)}{p'(t)q(t) - p(t)q'(t)}.$$

Substituting here the values of $p(t)$ and $q(t)$ from (2.1), we find that $u(t) = x(t)$, $v(t) = y(t)$, hence $X^{\vee\vee} = X$.

B. Biduality and caustics

To give an intuitive sense of the biduality theorem in P^2, we express the notion of tangency in the dual projective plane $(P^2)^*$ in terms of the original plane P^2. By definition, a tangent line to a curve at some point is the line which contains this point and which is infinitesimally close to the curve near this point.

In our situation, a point of $(P^2)^*$ is a line $l \subset P^2$. A curve C in $(P^2)^*$ is a 1-parameter family of lines in P^2; see Figure 2. A line in $(P^2)^*$ is a pencil $x^\vee$ of all lines in P^2 passing through a given point $x \in P^2$; see Figure 1b. The condition that $x^\vee$ is tangent to C at l means that the line $l \in (P^2)^*$ is a member of the family C, a point x lies on l, and other lines from C near l are infinitesimally close to the pencil $x^\vee$. This is usually expressed by saying that x is a *caustic point* for the family of lines C.

One can imagine (by making a negative photograph of Figure 2) that a light ray of some fixed intensity is coming along each line of C. Then the total brightness of the ongoing light in an arbitrary small neighborhood of a caustic point x will be infinite, though there may be only one ray (line from C) meeting the point x itself. The set of all caustic points of the family of lines C is usually called the *caustic* of C. This is nothing more than the projectively dual curve $C^\vee \subset P^2$. Now the biduality theorem means that every curve coincides with the caustic for the family of its tangent lines. This is intuitively obvious. The "dual" form of this theorem is less obvious: it means that *every* 1-parameter family C of lines in P^2 consists of tangent lines to some curve in P^2, and this curve is the caustic for C. An example of a 1-parameter family of lines which does not a priori come as tangent lines to some curve is given by reflecting a parallel beam of light in a curved mirror.

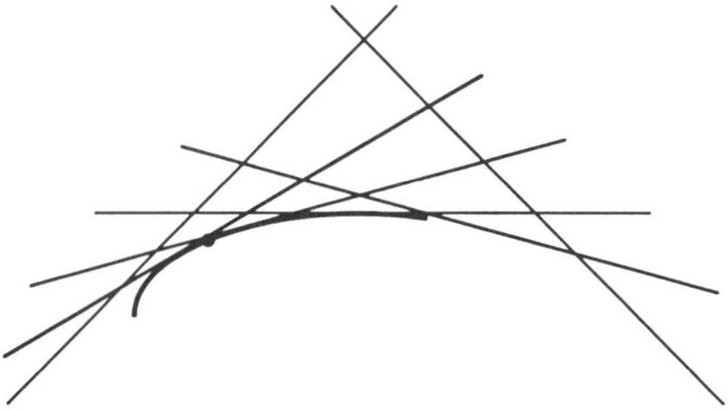

Figure 2. A family of lines and its caustic point

C. First examples

Example 2.1. Let $X \subset P^2$ be a smooth conic. In homogeneous coordinates, which we now prefer to denote by x_1, x_2, x_3, the curve X is given by

$$(Ax, x) = \sum_{i,j=1}^{3} a_{ij} x_i x_j = 0,$$

where $A = \|a_{ij}\|$ is a non-degenerate symmetric 3×3 matrix. We claim that $X^\vee \subset (P^2)^*$ is also a conic curve defined by the inverse matrix A^{-1}. It is clear that the tangent line to X at a point $x^{(0)} \in X$ is given by $(Ax^{(0)}, x) = 0$. Hence the point $\xi \in (P^2)^*$ corresponding to this tangent line has homogeneous coordinates $Ax^{(0)}$, which implies $(A^{-1}\xi, \xi) = 0$.

We see that the duality operation $f \to F$ on homogeneous polynomials can be regarded as a generalization of the matrix inversion. The analog of the set of non-degenerate matrices (where the matrix inversion behaves continuously) is the set of polynomials defining non-singular curves. The condition of non-singularity is again a certain discriminantal condition, as we shall see later (Example 4.15).

Example 2.2. Let $X \subset P^2$ be the cubic parabola, whose equation in affine coordinates (x, y) (in $A^2 \subset P^2$) is $y = x^3$. At $(0, 0)$, the curve X has an *inflection point* (or a *flex*): the tangent line l to X at this point actually has the second order of tangency. Let us look at the behavior of the dual curve $X^\vee$ in the neighborhood of l.

The line l is not covered by the affine coordinate system described in subsection A since it contains the origin. So we introduce a new affine chart in $(P^2)^*$

consisting of lines with affine equations $y = \xi x - \eta$. We can treat the equation of X as parametric: $x = t$, $y = t^3$. By calculating the tangent line we find the parametric representation of $X^\vee$ in our coordinates to be $\xi(t) = 3t^2$, $\eta(t) = 2t^3$. Hence $X^\vee$ is given by $4\xi^3 = 27\eta^2$. Thus $X^\vee$ is a semi-cubic parabola (see Figure 3).

We see that the singular point (cusp) of $X^\vee$ corresponds to the inflection point $(0, 0)$ of X. Considered in the entire projective plane, X itself has a cusp at infinity, as can be seen by choosing the appropriate coordinates. The dual curve $X^\vee$ has a cusp at (0,0) and an inflection point at infinity.

Similarly we can consider the curve $y = x^k$ which has an inflection point at $(0, 0)$ of order $(k - 1)$. Its dual curve has the equation $(k-1)^{k-1}\xi^k = k^k\eta^{k-1}$ and has a complicated cusp.

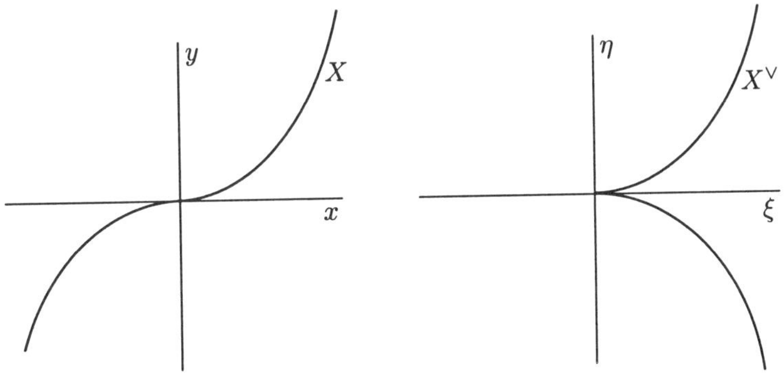

Figure 3. A flex and a cusp

Now consider arbitrary curves. Apart from flexes, there is another source of singularities for the dual curve $X^\vee$: the double tangents to X (see Figure 4). Such a tangent will be a point of self-intersection of $X^\vee$, typically of the type called a *node* (see Figure 4b).

The notion of cusps and nodes for arbitrary plane curves is defined as follows. Suppose that a curve is given by an affine equation $f(x, y) = 0$ and the point in question is $(0, 0)$ so $f(0, 0) = 0$. Let f_x, f_y, f_{xx} and so on denote the partial derivatives of f. The point $(0, 0)$ is said to be a node if $f_x(0, 0) = f_y(0, 0) = 0$, and the quadratic form, given by second derivatives

$$\psi(p, q) = f_{xx}(0, 0)p^2 + 2f_{xy}(0, 0)pq + f_{yy}(0, 0)q^2,$$

has two distinct non-zero linear factors. So a node is a point of intersection of two smooth branches with distinct tangents. The point $(0, 0)$ is said to be a (simple) cusp if $f_x(0, 0) = f_y(0, 0) = 0$, the form $\psi(p, q)$ is non-zero and is a square of a linear form.

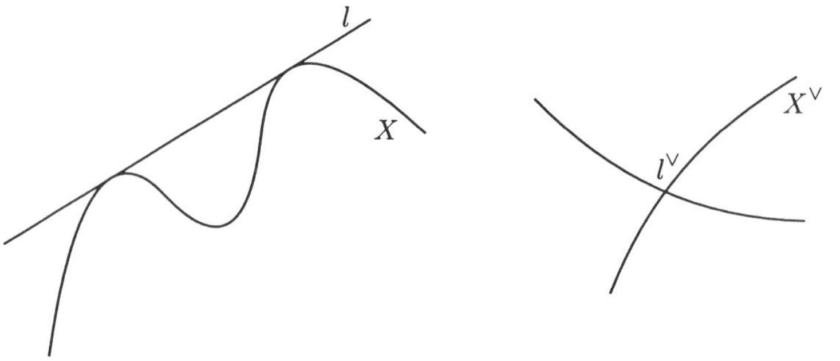

Figure 4. A double tangent and a node

In general, for a smooth curve X, cusps of $X^\vee$ correspond to the simple inflection points of X, i.e., such points where the tangent line has tangency to X of exactly the second order and is not tangent to X anywhere else. Similarly, nodes of $X^\vee$ correspond to simple bitangents of X (i.e., such that both tangencies are of first order and there are no other points of tangency). Both these statements can be easily seen in local coordinates.

Example 2.3. Consider the curve X_0 in $\mathbf{C}^2$ given by

$$x^a + y^a = 1, \quad a \geq 2. \tag{2.2}$$

We regard $\mathbf{C}^2$ as an affine chart of the projective plane P^2 with homogeneous coordinates x, y, z. The closure of X_0 in P^2 (denoted by X) has the homogeneous equation $x^a + y^a = z^a$. This curve is called the *Fermat curve* of order a. We are going to find the projective dual curve.

The equation (2.2) can be written in parametric form:

$$x = t, \quad y = \sqrt[a]{1 - t^a}.$$

Using (2.1) we get the parametric representation of the dual curve in coordinates p, q as follows:

$$p(t) = t^{a-1}, \quad q(t) = (1 - t^a)^{\frac{a-1}{a}}. \tag{2.3}$$

The relation between $p(t)$ and $q(t)$ has the form

$$p^{\frac{a}{a-1}} + q^{\frac{a}{a-1}} = 1. \tag{2.4}$$

Hence $X^\vee$ is given by an equation similar to (2.2) but with the new exponent $b = a/(a-1)$ which satisfies the relation $\frac{1}{a} + \frac{1}{b} = 1$. This fact is the algebraic version of the duality between Banach spaces L_p, which is well-known in analysis; see, e.g., [Ru], Theorem 6.16.

If $a \geq 3$, the equation (2.4) involves multivalued fractional power functions. We can put this equation into polynomial form by constructing an equation of degree $a(a-1)$ for $X^\vee$ as follows. Consider the quantities $u = p^{\frac{a}{a-1}}$, $v = q^{\frac{a}{a-1}}$. Then on $X^\vee$ we have $u + v = 1$. We have the obvious equalities

$$\begin{cases} u^{a-1} + v^{a-1} = p^a + q^a, \\ u^{2(a-1)} + v^{2(a-1)} = p^{2a} + q^{2a}, \\ \vdots \qquad\qquad \vdots \\ u^{(a-1)^2} + v^{(a-1)^2} = p^{(a-1)a} + q^{(a-1)a}. \end{cases} \tag{2.5}$$

The left hand side of each of these equalities is a symmetric polynomial in u and v, so it can be expressed as a polynomial in $u + v$ and uv. More precisely, for any d, we have

$$u^d + v^d = (-1)^d d \sum_{i+2j=d} (-1)^{i+j} \frac{(i+j-1)!}{i!j!} (u+v)^i (uv)^j. \tag{2.6}$$

This is the classical Girard formula (see e.g., [MS]). We shall have more occasions later to use this formula and its generalizations.

Note that $u + v = 1$ on $X^\vee$ and $(uv)^{a-1} = p^a q^a$ is a monomial in p and q. Hence any monomial in $(u + v)$ and uv has the form $\varphi(p, q)(uv)^j$ where φ is a monomial in p and q and $0 \leq j \leq a - 2$. Thus, after substituting the Girard formula on the left hand side of the i-th equality in (2.5), we can write it in the form

$$\varphi_{i0}(p, q) + \varphi_{i1}(p, q) \cdot uv + \varphi_{i2}(p, q) \cdot (uv)^2 + \cdots + \varphi_{i,a-2}(p, q) \cdot (uv)^{a-2} = 0, \tag{2.7}$$

where φ_{ij} are polynomials in p, q. (For example, $p^{ia} + q^{ia}$ on the right hand side of the i-th equation will contribute to φ_{i0}.) Thus the vector $(1, uv, (uv)^2, \ldots, (uv)^{a-2})$ is a solution of the system of $a - 1$ linear equations (2.7) on $a - 1$ unknowns. The condition of compatibility of this system, i.e., the condition

$$\det \| \varphi_{ij}(p, q) \|_{i=1,\ldots,a-1, j=0,\ldots,a-2} = 0$$

is a polynomial equation of degree $a(a-1)$ equivalent to $p^{\frac{a}{a-1}} + q^{\frac{a}{a-1}} = 1$.

For example, let us transform equation

$$p^{\frac{3}{2}} + q^{\frac{3}{2}} = 1, \tag{2.8}$$

defining the curve dual to $x^3+y^3=1$. We have, by setting $u=p^{\frac{3}{2}},\ v=q^{\frac{3}{2}}$, the following:

$$u^2+v^2=(u+v)^2-2uv=1-2uv,$$

$$u^4+v^4=(u+v)^4-4(u+v)^2uv+2(uv)^2=1-4uv+2(uv)^2=(1+2p^3q^3)-4uv.$$

So, from (2.5), we get a linear system in $(1, uv)$:

$$\begin{cases}(1-p^3-q^3)-2uv=0,\\(1+2p^3q^3-p^6-q^6)-4uv=0\end{cases},$$

and the equation of the dual curve can be written

$$\begin{aligned}F(p,q)&=-\frac{1}{2}\begin{vmatrix}1-p^3-q^3 & -2\\ 1+2p^3q^3-p^6-q^6 & -4\end{vmatrix}\\&=p^6+q^6-2p^3q^3-2p^3-2q^3+1=0.\end{aligned}\tag{2.9}$$

The same answer can be obtained by squaring (2.8), expressing $2p^{\frac{3}{2}}q^{\frac{3}{2}}=1-p^3-q^3$ and then squaring again.

D. Plücker formulas

Typically, the dual of a smooth plane curve will be singular. The nature and number of typical singularities are given by the Plücker formulas.

Proposition 2.4. *Let $X\subset P^2$ be a smooth curve of degree d. Then $X^\vee$ has degree $d(d-1)$. For a generic smooth curve of degree d, the curve $X^\vee$ contains $3d(d-2)$ cusps and $(1/2)d(d-2)(d^2-9)$ nodes. In other words, a generic smooth plane curve of degree d has $3d(d-2)$ inflection points and $(1/2)d(d-2)(d^2-9)$ double tangents.*

The proof can be found in [GH], Chapter 2, Section 4.

Note that although the degree of $X^\vee$ (denoted by $d^\vee$) equals $d(d-1)$ we cannot conclude by biduality that $d=(d^\vee)(d^\vee-1)$ because $X^\vee$ is singular and Proposition 2.4 is not applicable. Also let us note that if we take an arbitrary smooth plane curve then $X^\vee$ can have more complicated singularities than just cusps and nodes.

For example, if X is a smooth cubic, then $X^\vee$ is a curve of degree 6 with exactly 9 simple cusps (corresponding to 9 inflection points of X) and no other singularities. Here the assumption of genericity is not essential. Indeed, any inflectional tangent l to X cannot have an order of tangency more than 2 (since then the multiplicity of intersection $l\cap X$ would be at least 4). So all cusps of $X^\vee$ are simple. Similarly, any double tangent to X would have intersection multiplicity

≥ 4 which is impossible. So $X^\vee$ is a very special curve of degree 6. The dual of a generic curve of degree 6 will have degree 30.

There is a more general and symmetric version of the Plücker formulas in which both X and $X^\vee$ are allowed to have singularities.

Proposition 2.5. *Let $X \subset P^2$ be an irreducible curve of degree d having, as its only singularities, κ cusps and ν nodes and such that $X^\vee$ also has only cusps and nodes as its only singularities. Denote by $d^\vee, \kappa^\vee, \nu^\vee$ the degree, the numbers of cusps and of nodes of $X^\vee$. Then*

$$d^\vee = d(d-1) - 3\kappa - 2\nu,$$
$$\kappa^\vee = 3d(d-2) - 8\kappa - 6\nu.$$

The number $\nu^\vee$ can be found from the equation $d = d^\vee(d^\vee - 1) - 2\kappa^\vee - 3\nu^\vee$ obtained by applying Proposition 2.5 to $X^\vee$.

For the proof of this proposition, we again refer to [GH], Chapter 2, Section 4.

E. Schläfli's formula for the dual of a smooth plane cubic.

Let $X \subset P^2$ be a smooth cubic curve with homogeneous equation $f(x_0, x_1, x_2) = 0$. Denote by p_0, p_1, p_2 the dual coordinates in P^{2*} and by $F(p_0, p_1, p_2) = 0$ the homogeneous equation of the dual curve $X^\vee$. From the Plücker formula we find that $\deg(F) = 6$. There is an explicit formula for F in terms of f (L. Schläfli [Schl]).

Denote by f_i, f_{ij} etc. the partial derivatives $\partial f/\partial x_i$, $\partial^2 f/\partial x_i \partial x_j$ etc. Consider the polynomial

$$V(p, x) = \begin{vmatrix} 0 & p_0 & p_1 & p_2 \\ p_0 & f_{00}(x) & f_{01}(x) & f_{02}(x) \\ p_1 & f_{10}(x) & f_{11}(x) & f_{12}(x) \\ p_2 & f_{20}(x) & f_{21}(x) & f_{22}(x) \end{vmatrix}.$$

Schläfli's formula is as follows.

Theorem 2.6. *The equation $F(p_0, p_1, p_2)$ of $X^\vee$ is equal, up to a non-zero constant factor, to the polynomial*

$$G(p_0, p_1, p_2) = \begin{vmatrix} 0 & p_0 & p_1 & p_2 \\ p_0 & \frac{\partial^2 V}{\partial x_0^2}(p) & \frac{\partial^2 V}{\partial x_0 \partial x_1}(p) & \frac{\partial^2 V}{\partial x_0 \partial x_2}(p) \\ p_1 & \frac{\partial^2 V}{\partial x_1 \partial x_0}(p) & \frac{\partial^2 V}{\partial x_1^2}(p) & \frac{\partial^2 V}{\partial x_1 \partial x_2}(p) \\ p_2 & \frac{\partial^2 V}{\partial x_2 \partial x_0}(p) & \frac{\partial^2 V}{\partial x_2 \partial x_1}(p) & \frac{\partial^2 V}{\partial x_2^2}(p) \end{vmatrix}. \quad (2.10)$$

Example 2.7. Consider the case of the Fermat cubic, defined by

$$f(x_0, x_1, x_2) = -x_0^3 + x_1^3 + x_2^3 = 0.$$

In affine coordinates $x = x_1/x_0,\ y = x_2/x_0$, the equation will be $x^3 + y^3 = 1$, the kind considered in Example 2.3. Then

$$-V(p, x) = 36(-p_0^2 x_1 x_2 - p_1^2 x_0 x_2 + p_2^2 x_0 x_1)$$

and the equation of the dual curve is (after dividing by a constant)

$$F(p_0, p_1, p_2) = p_0^6 + p_1^6 + p_2^6 - 2p_0^3 p_1^3 - 2p_1^3 p_2^3 - 2p_0^3 p_2^3$$

which is the homogeneous version of (2.9).

The proof of Theorem 2.6, which will occupy the rest of this section, uses some classical concepts of projective geometry. Consider the *total polarization* of the cubic form f, i.e., the symmetric trilinear form $A(x, y, z)$, $x, y, z \in \mathbf{C}^3$, given by

$$A(x, y, z) = \frac{1}{6} \sum_{i,j,k=0}^{2} f_{ijk} x_i y_j z_k, \tag{2.11}$$

where f_{ijk} are the third partial derivatives of f (they are constant since $\deg f = 3$). Then $f(x) = A(x, x, x)$. For any $x = (x_0 : x_1 : x_2) \in P^2$, we consider the conic

$$K_x = \{x' \in P^2 :\ A(x, x', x') = 0\} = \left\{x' : \sum_{i,j=0}^{2} f_{ij}(x) x_i' x_j' = 0\right\} \tag{2.12}$$

known as the *first polar* of X at x. (We shall have another occasion to use polars later in Chapter 4, Section 2.)

For $p = (p_0 : p_1 : p_2) \in P^{2*}$, we denote by l_p the corresponding line in P^2, i.e., the line $\sum p_i x_i = 0$. Let us give a geometric interpretation of the expressions in Theorem 2.6.

Lemma 2.8. *We have $V(p, x) = 0$ if and only if the line l_p is tangent to the conic K_x.*

Proof. The polynomial $V(p, x)$ is a quadratic form in p_i whose coefficients are the cofactors of the Hessian matrix $\|f_{ij}(x)\|$. So this quadratic form is, for any given x, a scalar multiple of the form given by the inverse matrix $\|f_{ij}(x)\|^{-1}$. Hence, by Example 2.1, the vanishing of this form is equivalent to the tangency of l_p and the conic K_x given by (2.12). The lemma is proved.

For a given $p \in P^{2*}$, let $C_p \subset P^2$ be the locus of x such that $V(p, x) = 0$, i.e., K_x is tangent to l_p. Since $V(p, x)$ has degree 2 in x_i, we find that C_p is also a conic.

Lemma 2.9. *We have $G(p) = 0$ (where G is defined by (2.10)) if and only if the line l_p is tangent to C_p.*

The proof is similar to that of Lemma 2.8.

Let us now note that G is homogeneous of degree 6. Thus to prove Theorem 2.6, we need to show that G is not identically zero and G vanishes on $X^\vee$. The second of these assertions is implied by the following lemma.

Lemma 2.10. *Let $p \in P^{2*}$ be such that the line l_p is tangent to X at a point x which is not an inflection point. Then the conic C_p contains x and is tangent to X at x.*

Proof. We first prove that $x \in C_p$, i.e., K_x contains x and is tangent to X at x. We use the trilinear form A given by (2.11). If $x \in X$ then $A(x, x, x) = 0$, so $x \in K_x$. Moreover, the tangent line to both X and K_x at x consists of y such that $A(x, x, y) = 0$. So K_x is tangent to X at x, i.e., $x \in C_p$.

Let us now show that C_p is tangent to X at x. Let us choose some parametric representation of the curve C_p near x, say, $t \mapsto z(t) = (x + tx' + ...)$ where t is the local parameter. Since the coefficients p_i of the equation of the tangent line l_p are proportional to the partial derivatives f_i, we obtain the value of $V(p, z(t))$ to be

$$\begin{vmatrix} 0 & f_0(x) & f_1(x) & f_2(x) \\ f_0(x) & f_{00}(x + tx' + ...) & f_{01}(x + tx' + ...) & f_{02}(x + tx' + ...) \\ f_1(x) & f_{10}(x + tx' + ...) & f_{11}(x + tx' + ...) & f_{12}(x + tx' + ...) \\ f_2(x) & f_{20}(x + tx' + ...) & f_{21}(x + tx' + ...) & f_{22}(x + tx' + ...) \end{vmatrix}.$$

Let us multiply the second, third and fourth columns of the above matrix with x_0, x_1, x_2 respectively, and subtract half their sum from the first column. Now perform the same operation with respect to rows. Then, by virtue of the Euler identity for the homogeneous functions f_i and the linearity of f_{ij}, we get the matrix (denoted by $W(t)$) of the following form:

$$\begin{pmatrix} t(\sum x_i' f_i) + O(t^2) & O(t) & O(t) & O(t) \\ O(t) & f_{00}(x + tx' + ...) & f_{01}(x + tx' + ...) & f_{02}(x + tx' + ...) \\ O(t) & f_{10}(x + tx' + ...) & f_{11}(x + tx' + ...) & f_{12}(x + tx' + ...) \\ O(t) & f_{20}(x + tx' + ...) & f_{21}(x + tx' + ...) & f_{22}(x + tx' + ...) \end{pmatrix}$$

where, by $O(t)$, we have denoted the functions having at least the first order of vanishing at $t = 0$. It follows that

$$V(p, z(t)) = \det W(t) = t \cdot \left(\sum_{i=0}^{2} x_i' f_i + O(t^2) \right) \cdot H_f(z(t)),$$

where $H_f = \det\|f_{ij}\|$ is the Hessian determinant of f. We have assumed that our chosen point x is not an inflection point of X so $H_f(x) \neq 0$. Therefore, since $V(p, z(t)) = 0$ identically in t, we obtain that $\sum x_i' f_i(x) = 0$. This precisely means that C_p is tangent to X. Lemma 2.10 is proved.

To establish Theorem 2.6, it remains to show that the polynomial $G(p)$ is not identically zero. By Lemma 2.9 this means that, for a generic $p \in P^{2*}$, the line $l_p \subset P^2$ intersects C_p at two distinct points. By definition of C_p, this means that there are two distinct points $z, z' \in l_p$ such that the conics K_z and $K_{z'}$ are tangent to l_p.

Note that the conics $K_x, x \in P^2$, form a linear system of curves (i.e., the coefficients of the equation of K_x are linear functions of x). Note also that there are no points common to all K_x. (Indeed, if z is such a point, then $A(x, z, z) = 0$ for any x, so z should be a singular point of X.) This implies that, for generic $x, y \in P^2$, the intersection $K_x \cap K_y$ consists of four distinct points. We take such points x, y and take l_p to be the line $< x, y >$ joining x and y. We can and will assume, moreover, that K_x and K_y are nonsingular and l_p is not tangent to X. We claim that l_p satisfies the conditions of the previous paragraph (and so $G(p) \neq 0$).

Let $K_x \cap K_y = \{q_1, \dots, q_4\}$. Since K_x, K_y are conics, no three of the q_i are collinear. All the conics through $q_1, \dots, q_4$ form a 1-dimensional linear system (pencil) $\mathcal{L} \cong P^1$. The conics $K_z, z \in l_p$, contain $q_1, \dots, q_4$ and also form a pencil. Hence $K_z, z \in l_p$ are all the conics through $q_1, \dots, q_4$. Note that l_p does not contain any of the q_i. Indeed, otherwise we would have

$$A(x, q_i, q_i) = A(y, q_i, q_i) = A(q_i, q_i, q_i) = 0$$

for some i. This means that $q_i \in X$ and the line $< x, y > = l_p$ is tangent to X at q_i, contrary to our assumption. Now the nonvanishing of G is a consequence of the next lemma.

Lemma 2.11. *Let $q_1, \dots, q_4$ be four points in P^2 of which no three are collinear and let $l \subset P^2$ be a line that does not meet any of the q_i. Then there are two distinct conics through $q_1, \dots, q_4$ tangent to l.*

Proof. By choosing the appropriate homogeneous coordinates x_0, x_1, x_2 in P^2, we can assume that $q_1 = (1:0:0), q_2 = (0:1:0), q_3 = (0:0:1), q_4 = (1:1:1)$. Consider the transformation (Cremona inversion)

$$\Psi : P^2 \longrightarrow P^2, \qquad (x_0 : x_1 : x_2) \longmapsto \left(\frac{1}{x_0} : \frac{1}{x_1} : \frac{1}{x_2} \right).$$

This transformation takes the system of conics through $q_1, \ldots, q_4$ into the system of straight lines through $q_4 = (1:1:1)$. The line l is taken into a conic $\Psi(l)$ not containing q_4. Clearly there are two tangent lines to $\Psi(l)$ through q_4. Applying the inverse transformation $\Psi^{-1} = \Psi$, we get two conics through $q_1, \ldots, q_4$ tangent to l. This concludes the proof of Lemma 2.11 and Theorem 2.6.

3. The incidence variety and the proof of the biduality theorem

A. The incidence variety and the conormal bundle

We shall use the following standard terminology and notation. If M is a smooth algebraic variety, then TM denotes the tangent bundle of M. If $Z \subset M$ is a smooth (not necessarily closed) algebraic subvariety, then TZ is a subbundle in $TM|_Z$, the restriction of TM to Z. The quotient by this subbundle is called the *normal bundle* of Z in M and denoted by $T_Z M$. By taking the duals to TM and $T_Z M$, we get the *cotangent bundle* T^*M of M and the *conormal bundle* $T^*_Z M$ of Z in M. Note that $T^*_Z M$ can be naturally regarded as a subvariety in T^*M.

Let $X \subset P^n$ be a projective variety and let $X^\vee \subset P^{n*}$ be its projective dual. We denote the set of smooth points of X by X_{sm}. Let $W_X^0 \subset P^n \times P^{n*}$ be the set of pairs (x, H) where $x \in X_{sm}$ and H is a hyperplane in P^n tangent to X at x. Denote the Zariski closure of W_X^0 by W_X. We shall call W_X the *incidence variety* corresponding to X.

Let us denote by $\mathrm{pr}_1, \mathrm{pr}_2$ the projections of $P^n \times P^{n*}$ to the first and second factors. We have the following.

(1) The variety $X^\vee$ coincides with $\mathrm{pr}_2(W_X)$.

(2) The projection $\mathrm{pr}_1 : W_X^0 \to X_{sm}$ is a projective bundle.

More precisely, consider the conormal bundle $T^*_{X_{sm}} P^n$. Then pr_1 identifies W_X^0 with the projectivization $P(T^*_{X_{sm}} P^n)$ of this bundle. Indeed, the choice of a hyperplane $H \subset P^n$ tangent to X at x is equivalent to the choice of a hyperplane $T_x H$ in the tangent space $T_x P^n$ which contains $T_x X$. The equation of such a hyperplane is a covector in the conormal space to X at x.

The biduality theorem can be reformulated by saying that

$$W_X = W_{X^\vee}. \tag{3.1}$$

It will be convenient for us to prove (3.1) by working in vector spaces instead of projective ones.

We assume that our P^n is $P(V)$ and P^{n*} is $P(V^*)$. Let $Y \subset V$ be the affine cone over X, i.e., Y consists of 0 and all vectors $v \in V$ whose projectivization is in X. Let $Y^\vee \subset V^*$ be the similar cone over $X^\vee$ so that, in the terminology of

Section 1A, Y and $Y^\vee$ are dual cones. Let Y_{sm} be the smooth part of Y. Denote by $\mathrm{Con}(Y)$ the closure in the cotangent bundle T^*V of the conormal bundle $T^*_{Y_{sm}}V$.

The space T^*V is canonically identified with $V \times V^*$. Denote by pr_i, $i = 1, 2$, the projections of this product to the first and second factors. Then $Y^\vee$ coincides with $\mathrm{pr}_2(\mathrm{Con}(Y))$. An equivalent reformulation of (3.1) is

$$\mathrm{Con}(Y) = \mathrm{Con}(Y^\vee) \tag{3.2}$$

in the natural identification of T^*V and T^*V^* with $V \times V^*$. We shall prove (3.2). For this we need some background.

B. Lagrangian varieties

Let M be any smooth algebraic variety, $m = \dim M$. Recall [Ar] that the cotangent bundle T^*M carries a canonical *symplectic structure*, i.e., a differential 2-form ω with the following two properties.

(1) $d\omega = 0$, i.e., ω is closed.

(2) If (x, ξ) is any point of T^*M then ω defines a non-degenerate skew product on the tangent space $T_{(x,\xi)}(T^*M)$.

The form ω admits the following description in terms of a local coordinate system $(x_1, \dots, x_m)$ in M. Let ξ_i be the fiberwise linear function on T^*M given by the pairing with the vector field $\partial/\partial x_i$. Then $(x_1, \dots, x_m, \xi_1, \dots, \xi_m)$ forms a local coordinate system in T^*M. The form ω is defined by

$$\omega = \sum_{i=1}^{m} dx_i \wedge d\xi_i. \tag{3.3}$$

Although (3.3) uses a choice of coordinates in M, the form ω does not depend on this choice.

An irreducible closed subvariety $\Lambda \subset T^*M$ is called *Lagrangian* if $\dim \Lambda = \dim M = m$ and the restriction of ω to the smooth part Λ_{sm} of Λ vanishes as a 2-form on Λ_{sm}.

An important example of a Lagrangian variety is obtained as follows. Let $Z \subset M$ be any irreducible subvariety, Z_{sm} the smooth part of Z and let

$$\mathrm{Con}(Z) = \overline{T^*_{Z_{sm}}}(M) \tag{3.4}$$

be the closure of the conormal bundle of Z_{sm}. We shall call $\mathrm{Con}(Z)$ the *conormal variety* of Z (since it is in general not a bundle). We claim that $\mathrm{Con}(Z)$ is Lagrangian. Indeed, the statement about the dimension is obvious. To see that $\omega|_{\mathrm{Con}(Z)} = 0$, it suffices to work over the smooth part of Z. Let $x_1, \dots, x_m$ be a local coordinate system on M such that $x_1 = \cdots = x_r = 0$ is the local system of equations of Z. Then the fibers of the conormal bundle over points of Z are generated by 1-forms $dx_{r+1}, \ldots, dx_m$. Hence $\xi_{r+1} = \cdots = \xi_m = 0$ on $T^*_{Z_{sm}} M$ and by (3.3) we find that $\omega = 0$ on $T^*_{Z_{sm}} M$. So Con(Z) is Lagrangian.

Note that the variety Con(Z) $\subset T^*M$ is *conic*, i.e., it is invariant under dilations of fibers of T^*M.

The converse statement is crucial for us.

Proposition 3.1. *Any conic Lagrangian subvariety* $\Lambda \subset T^*M$ *has the form* $\mathrm{Con}(Z)$ *for some irreducible subvariety* $Z \subset M$.

Proof. Let $\mathrm{pr} : T^*M \to M$ be the projection. Define $Z = \mathrm{pr}(\Lambda)$. This is an irreducible subvariety in M. Let z be any smooth point of Z. We claim that the fiber $\mathrm{pr}^{-1}(z) \cap \Lambda$ is contained in the conormal space $(T^*_Z M)_z$. Indeed, let ξ be any covector in $\mathrm{pr}^{-1}(z) \cap \Lambda$. Since $T^*_z M$ is a vector space, we can regard ξ as a tangent vector to T^*M at a point $z \in M \subset T^*M$ (here we identify M with the zero section of the projection $\mathrm{pr} : T^*M \to M$). Since Λ is Lagrangian, ξ is orthogonal with respect to ω to any tangent vector $v \subset T_z Z$. But the scalar product with respect to ω of any such vertical vector (coming from an element ξ of $T^*_z M$) and any horizontal tangent vector (coming from a vector $v \in T_z M$) is equal to $\xi(v)$, the standard pairing of a vector and a covector. This can be seen at once from (3.3). Hence $\mathrm{pr}^{-1}(z) \cap \Lambda \subset (T^*_Z M)_z$.

We have proved that $\Lambda \subset \mathrm{Con}(Z)$. Since Λ and $\mathrm{Con}(Z)$ are irreducible varieties of the same dimension, we conclude that $\Lambda = \mathrm{Con}(Z)$. Proposition 3.1 is proved.

C. Proof of the biduality theorem

We shall prove the equality (3.2). We specialize the considerations of subsection B to the case of the smooth variety $M = V$ and the subvariety $Y \subset V$. The conormal space $\mathrm{Con}(Y)$ is a Lagrangian subvariety in T^*V.

The identification $T^*V = V \times V^* = T^*V^*$ takes the canonical symplectic form on T^*V to minus the canonical symplectic form on T^*V^*. Indeed, choosing a linear system of coordinates $x_1, \ldots, x_m$ in V and the dual coordinates $\xi_1, \ldots, \xi_m$ in V^*, the form on T^*V can be written as $\sum dx_i \wedge d\xi_i$, and the form on T^*V^* as $\sum d\xi_i \wedge dx_i$. Hence $\mathrm{Con}(Y)$ regarded as a subvariety in T^*V^* will still be Lagrangian. Moreover, $Y \subset V$ being the cone over X, is invariant under dilations of V. Hence $\mathrm{Con}(Y) \subset V \times V^*$ is invariant under dilations of V as well as under dilations of V^*. This means that $\mathrm{Con}(Y)$ regarded as a subvariety in T^*V^* will be conic (i.e., invariant under dilations of the fibers of $T^*V^* \to V^*$). Thus, by Proposition 3.1, $\mathrm{Con}(Y) = \mathrm{Con}(Z)$ where $Z \subset V^*$ is the projection of $\mathrm{Con}(Y)$.

But this projection coincides with $Y^\vee$. Hence $\mathrm{Con}(Y) = \mathrm{Con}(Y^\vee)$. This concludes the proof of the biduality theorem.

D. The incidence variety as the desingularization of the dual

We return to the situation of subsection A in which we have the projective variety $X \subset P^n$ and the dual variety $X^\vee \subset P^{n*}$. Consider once again the diagram of projections

$$X \xleftarrow{\mathrm{pr}_1} W_X \xrightarrow{\mathrm{pr}_2} X^\vee \tag{3.5}$$

where W_X is the incidence variety. We have proved that W_X coincides with $W_{X^\vee}$. Hence, over the smooth locus of $X^\vee$, the map pr_2 is a projective bundle. A similar statement holds, of course, for pr_1.

Proposition 3.2.
(a) If X is smooth then W_X is smooth.
(b) If $X^\vee$ is a hypersurface then $\mathrm{pr}_2 : W_X \to X^\vee$ is a birational isomorphism.

Proof. (a) This follows from the fact that W_X is a projective bundle over $X_{sm} = X$.

(b) If $X^\vee$ is a hypersurface then $\dim X^\vee = \dim W_X = n - 1$. Since pr_2 is generically a projective bundle, its generic fiber is a 0-dimensional projective space, i.e., a point. In other words, pr_2 is birational.

Thus, in the most interesting case when X is smooth and $X^\vee$ is a hypersurface, the variety W gives a resolution of singularities of $X^\vee$.

4. Further examples and properties of projective duality

A. Duality and projections

We start with the study of the behavior of projective duality under *projection*. Let P be an n-dimensional projective space and $L \subset P$ a projective subspace of some dimension $k > 0$. The *quotient projective space* P/L has, as points, $(k+1)$-dimensional projective subspaces in P containing L.

The projection with center L (or *from* L) is the map $\pi_L : P - L \to P/L$ which takes any point $x \in P - L$ to the $(k+1)$-dimensional subspace spanned by x and L. The space P/L and the projection π_L can be visualized inside the initial projective space P. To do this, we choose another, $(n-k)$-dimensional projective subspace $H \subset P$ not intersecting L. Then any $(k+1)$-plane containing L intersects H at exactly one point, so P/L becomes identified with H (see Figure 5).

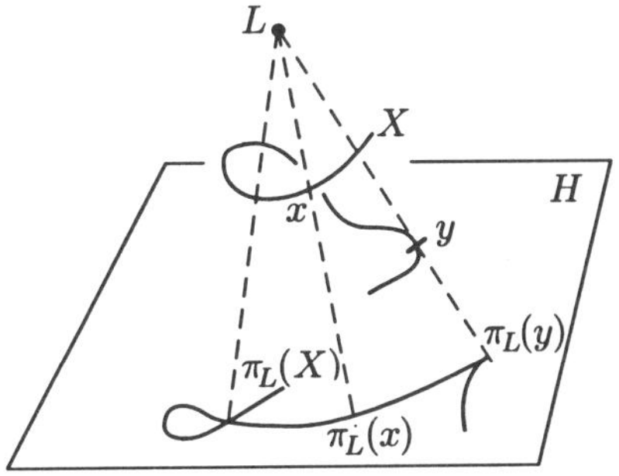

Figure 5. A space curve and its projection from a point

By definition, the dual projective space $(P/L)^*$ is embedded into P^* as a projective subspace. Namely, $(P/L)^*$ is identified with the space of hyperplanes in P containing L.

Proposition 4.1. *Let $X \subset P$ be an algebraic subvariety not intersecting the projective subspace L and such that* $\dim(X) < \dim(P/L)$. *Then*

$$(\pi_L(X))^\vee \quad \subset \quad L^\vee \cap X^\vee. \tag{4.1}$$

If, moreover, $\pi_L : X \to \pi_L(X)$ is an isomorphism of algebraic varieties, then the inclusion in (4.1) is an equality, i.e.,

$$(\pi_L(X))^\vee \quad = \quad L^\vee \cap X^\vee. \tag{4.2}$$

Before proving Proposition 4.1, let us reformulate it in terms of general discriminants, i.e., equations of dual varieties. To do this, we represent all of the projective spaces above as projectivizations of vector spaces. Let

$$0 \to K \xrightarrow{\alpha} V \xrightarrow{\beta} W \to 0$$

be an exact sequence of vector spaces. Consider the projective spaces $P = P(V^*)$, $L = P(W^*)$. Then $P/L = P(K^*)$. The map π_L is induced by $\alpha^* : V^* \to K^*$.

If $X \subset P(V^*)$ is a subvariety then the X-discriminant Δ_X is a polynomial function on V. The discriminant $\Delta_{\pi_L(X)}$ is a polynomial function on K.

Corollary 4.2. *In the situation described above, let $X \subset P = P(V^*)$ be a projective subvariety not intersecting $L = P(W^*)$ and such that* $\dim X < \dim P(K^*)$.

Then $\Delta_{\pi_L(X)}$ is a factor of the restriction to K of Δ_X. If, in addition, $\pi_L : X \to \pi_L(X)$ is an isomorphism then

$$\Delta_{\pi_L(X)} = \Delta_X|_K. \tag{4.3}$$

Proof of Proposition 4.1. A hyperplane in P/L is just a hyperplane in P containing L. It follows from our assumption that, if H regarded as a hyperplane in P/L, is tangent to $\pi_L(X)$ at some smooth point $y = \pi_L(x)$ (where $x \in X$ is smooth), then H, regarded as a hyperplane in P, is tangent to X at x. This proves the inclusion (4.1).

Suppose now that $\pi_L : X \to \pi_L(x)$ is an isomorphism. Let us denote by $X_0^\vee$ the open part of $X^\vee$ consisting of hyperplanes that are tangent to X at a smooth point. So $X^\vee$ is a closure of $X_0^\vee$. Similarly, introduce the open part $\pi_L(X)_0^\vee$ of $\pi_L(X)^\vee$. Since π_L induces an isomorphism of smooth loci $X_{sm} \to \pi_L(X)_{sm}$, we have

$$\pi_L(X)_0^\vee = L^\vee \cap X_0^\vee.$$

Now $\pi_L(X)^\vee$ is the closure of $\pi_L(X)_0^\vee$. Thus, to prove (4.2) it suffices to show that $L^\vee \cap X^\vee$ coincides with the closure of $L^\vee \cap X_0^\vee$. Since $X^\vee = \overline{X_0^\vee}$, the assertion we need can be stated as follows.

Lemma 4.3. *Let $H(t)$, $t \in \mathbf{C}$, $|t| < 1$, be a 1-parameter family of hyperplanes in P (depending analytically on t) such that for $t \neq 0$ the hyperplane $H(t)$ is tangent to X at some smooth point $x(t)$ and the hyperplane $H(0)$ contains L. Then there is another 1-parameter family of hyperplanes $H'(t)$ such that:*
(a) *$H'(0) = H(0)$,*
(b) *For $t \neq 0$ the hyperplane $H'(t)$ is tangent to X at $x(t)$ and in addition contains L.*

Proof of the lemma. Consider the point $x(0) = \lim_{t\to 0} x(t) \in X$. This limit exists since P is a compact manifold. However, $x(0)$ may be a singular point of X. Let $T_{x(0)}X$ be the Zariski tangent space to X at $x(0)$. By a slight abuse of notation we regard it as a projective subspace in P containing $x(0)$. The isomorphism $\pi_L : X \to \pi_L(X)$ induces an isomorphism of Zariski tangent spaces at any point. This implies that $T_{x(0)}X$, regarded as a projective subspace in P, does not intersect L.

For any $t \neq 0$, consider $T_{x(t)}X$ as a projective subspace in P. We denote this subspace by $T(t)$. Let $T(0)$ be the limit position of these subspaces as $t \to 0$. Then $T(0)$ is contained in $T_{x(0)}X$. Hence $T(0)$ does not intersect L either. We get a 1-parameter family $T(t)$ of projective subspaces in P of the same dimension, none of which intersects L. The hyperplane $H(t)$ for $t \neq 0$ contains $T(t)$ by definition of tangency. Hence $H(0)$ contains $T(0)$. Denote by $< L, T(t) >$ the projective subspace spanned by L and $T(t)$. This space has the same dimension for all t, namely $\dim L + \dim X + 1$. For $t = 0$ we have that $H(0)$ contains $< L, T(0) >$. Hence we can construct a 1-parameter family of hyperplanes $H'(t)$ such that $H'(0) = H(0)$ and $H'(t)$ contains $< L, T(t) >$. This is the required family.

Lemma 4.3 and Proposition 4.1 are proved.

Note that the condition that $\pi_L : X \to \pi_L(X)$ be an isomorphism cannot be dropped. Indeed, critical points of $\pi_L : X \to \pi_L(X)$ will produce extra components of $L^\vee \cap X^\vee$, as can be seen in Figure 5: any 2-plane containing the line $< L, y >$ belongs to $X^\vee$. Therefore $L^\vee \cap X^\vee$ is the union of the curve $(\pi_L(X))^\vee$ and the line $\pi_L(y)^\vee$.

Now consider the situation when X is a projective variety embedded into a projective space P and P is embedded, as a projective subspace, into another projective space M. Then we can regard X as a subvariety in either P or M. Correspondingly, X has two dual varieties, $(X^\vee)_P \subset P^*$ and $(X^\vee)_M \subset M^*$. To relate them, note that P^* is the quotient projective space $M^*/(P^\vee)_M$ and therefore we have the projection

$$\pi : M^* - (P^\vee)_M \longrightarrow P^*$$

with the center $(P^\vee)_M$. In particular, if $Z \subset P^*$ is any subvariety then we can form the cone over Z with apex $(P^\vee)_M$. This is a subvariety in M^* defined as the union of $(P^\vee)_M$ and all the fibers $\pi^{-1}(z)$, $z \in Z$.

Proposition 4.4. *In the above notation the variety $(X^\vee)_M$ is the cone over $(X^\vee)_P$ with apex $(P^\vee)_M$.*

The proof is obvious.

Let us reformulate this proposition in terms of general discriminants. Consider a surjection $\pi : E \to V$ of vector spaces and put $P = P(V^*)$, $M = P(E^*)$. Then we have an embedding $j : P \hookrightarrow M$. If X is a subvariety in P then Δ_X is a polynomial function on V. If we regard X as a subvariety in M, i.e., consider $j(X) \subset M$, then $\Delta_{j(X)}$ is a polynomial function on E.

Corollary 4.5. *In the described situation we have, for any $f \in E$,*

$$\Delta_{j(X)}(f) = \Delta_X(\pi(f)).$$

In other words, $\Delta_{j(X)}$ does not depend on some of the arguments and forgetting these arguments yields Δ_X.

B. Linear normality and invertible sheaves

Definition 4.6. A projective variety $X \subset P^n$ is called *non-degenerate* if it does not lie in a hyperplane. A variety X is called *linearly normal* if it is non-degenerate and cannot be represented as an isomorphic projection of a non-degenerate variety from a projective space of higher dimension.

The embedding of a linearly normal variety can be described intrinsically, in terms of the variety itself and a certain invertible sheaf on it. Let us recall this correspondence between invertible sheaves and projective embeddings [GH] [Hart].

By an invertible sheaf on an algebraic variety, we mean the sheaf of sections of some algebraic line bundle. We shall not distinguish notationally a line bundle from the corresponding invertible sheaf.

Let X be a projective variety and let $\mathcal{L}$ be an invertible sheaf on X. Let $V = H^0(X, \mathcal{L})$ be the space of global sections of $\mathcal{L}$. It is well-known that $\dim V < \infty$. To any point $x \in X$ we associate a linear subspace $\Gamma(x) \subset V$ consisting of all sections vanishing at x. Clearly, the codimension of $\Gamma(x)$ can be only 0 or 1. If $\operatorname{codim}(\Gamma(x)) = 0$, the point x is called a *base point* for $\mathcal{L}$. If there are no base points we obtain a regular map

$$\Gamma = \Gamma_{\mathcal{L}} : X \to P(V^*). \tag{4.4}$$

The sheaf $\mathcal{L}$ is called *very ample* if $\Gamma_{\mathcal{L}}$ is an embedding and it is called *ample* if some tensor power $\mathcal{L}^{\otimes m}$, $m \geq 1$, is very ample.

Before going further, let us recall the construction of invertible sheaves on projective spaces. For any $d \in \mathbf{Z}$, we denote by $\mathcal{O}(d)$ or $\mathcal{O}_{P(V^*)}(d)$ the sheaf of degree d homogeneous functions on $P(V^*)$. More precisely, let $\pi : V^* - \{0\} \to P(V^*)$ be the canonical projection. If $U \subset P(V^*)$ is a Zariski open set then the sections of $\mathcal{O}(d)$ over U are, by definition, regular functions f on $\pi^{-1}(U) \subset V^*$ which are homogeneous of degree d, i.e., such that $f(\lambda x) = \lambda^d f(x)$, $\lambda \in \mathbf{C}^*$. It is well known [Hart] that $\mathcal{O}(d)$ corresponds to a line bundle on $P(V^*)$, also denoted by $\mathcal{O}(d)$. For example, the fiber of $\mathcal{O}(-1)$, as a line bundle, over a point of $P(V^*)$ represented by a 1-dimensional subspace $l \subset V^*$ is l itself.

Now let X be a smooth irreducible variety embedded into $P(V^*)$ as a closed subvariety. By $\mathcal{O}_X(d)$ we shall denote the restriction of the line bundle $\mathcal{O}_{P(V^*)}(d)$ to X, as well as the corresponding sheaf of sections. Such sections can be described in a similar way to the above description for the whole $P(V^*)$ by considering regular homogeneous functions on $\pi^{-1}(U)$ where $U \subset X$ is Zariski open.

Proposition 4.7. *A projective variety $X \subset P(V^*)$ is linearly normal if and only* *if it is projectively isomorphic to the image of $\Gamma_{\mathcal{L}}$ for some very ample invertible sheaf $\mathcal{L}$ on X.*

The proof of Proposition 4.7 is well-known, so we shall only sketch it. Consider the projective space $P = P(V^*)$. If $d \geq 0$ then any homogeneous polynomial $f \in S^d(V)$ on V^* defines a global section of $\mathcal{O}_P(d)$. It is well-known [GH] [Hart] that the resulting map $S^d(V) \to H^0(P, \mathcal{O}(d))$ is an isomorphism.

Restricting our sheaf on X and taking $d = 1$, we obtain the restriction morphism

$$\mathrm{res} : V \to H^0(X, \mathcal{O}(1)). \tag{4.5}$$

Proposition 4.7 is a consequence of the following.

Proposition 4.8. *A variety X is linearly normal if and only if the map* res *is an isomorphism.*

Proof. The fact that res is not injective means that there are non-zero linear forms vanishing on X, i.e., X, lies in a hyperplane. If res is not surjective then the map $\Gamma_{\mathcal{O}(1)}$ gives a non-degenerate embedding of X to a projective space $P(H^0(X, \mathcal{O}(1))^*)$ of higher dimension. The initial embedding is obtained from this one by the projection with center $P(\mathrm{Im}(\mathrm{res})^{\perp})$. Conversely, if X can be obtained as an isomorphic projection of a non-degenerate subvariety $\tilde{X}$ in a larger projective space $P(W^*)$ then we have a proper embedding $V \hookrightarrow W$. The map $W \to H^0(X, \mathcal{O}(1))$, given by restricting linear functions from $P(W^*)$ to $\tilde{X}$, is an injection, since any non-zero vector from its kernel defines a hyperplane in $P(W^*)$ containing $\tilde{X}$, which contradicts the assumption that $\tilde{X}$ is non-degenerate. Thus the map res, which is the composition of the two described maps, is not surjective.

Note that every projective variety X is an isomorphic projection of a linearly normal variety: namely of the image of X in the embedding given by $\mathcal{O}(1)$. Thus Corollaries 4.2 and 4.5 imply that, in finding general discriminants, we can restrict our attention to the class of linearly normal varieties.

C. More examples of dual varieties

Example 4.9. Quadrics.

(a) Let $X \subset P(V)$ be a smooth quadric hypersurface, defined by $(Ax, x) = 0$, where $A : V \to V^*$ is a non-degenerate symmetric bilinear form. Then $X^\vee \in P(V^*)$ is a quadric hypersurface defined by the condition $(A^{-1}\xi, \xi) = 0$ where $A^{-1} : V^* \to V$ is the bilinear form on V^* inverse to A.

The proof is entirely similar to that of Example 2.1, so we omit it.

(b) Let $X \subset P(V)$ be a singular quadric hypersurface defined by the equation

$(Ax, x) = 0$ where the bilinear form A is degenerate. Let $K \subset V$ be the kernel of A. Then A induces a non-degenerate bilinear form $\bar{A}$ on V/K. The dual vector space to V/K is $K^{\perp} \subset V^*$. Hence we obtain the inverse form $\bar{A}^{-1}$ on $K^{\perp}$. The subvariety projectively dual to X will be the quadric hypersurface in $P(K^{\perp}) \subset P(V^*)$, defined by the non-degenerate form $\bar{A}^{-1}$.

To show this we note X is a cone with apex $P(K)$ so the result follows from the dual form of Proposition 4.4.

Example 4.10. Determinantal varieties. We fix natural numbers $m \leq n$ and let V be the vector space of m by n matrices $A = \|a_{ij}\|$ with complex entries. Let $r < m$. Consider the variety $Y_r \subset V$ consisting of matrices A such that $\mathrm{rk}A \leq r$. Let $X_r \subset P(V)$ be the projectivization of Y_r. We identify the dual space V^* with V using the bilinear form

$$(A, B) = \sum_{i,j} a_{ij} b_{ij}. \tag{4.6}$$

After this identification we can regard the dual variety $X_r^{\vee} \subset P(V^*)$ as an embedding into $P(V)$.

Proposition 4.11. *The variety $X_r^{\vee}$ coincides with X_{m-r}.*

Proof. We shall explicitly calculate tangent spaces to Y_r at smooth points. Note that every matrix of rank exactly r is a smooth point of Y_r. Indeed, they form an open subset and the group $GL(m)$ acts on them transitively (in fact, all the other points of Y_r are singular, see [ACGH] [Harr]; we do not need this). So we shall restrict ourselves to considering only these points.

We shall proceed in two steps. First we consider the particular case $m = n$ and $r = n - 1$, i.e., the subvariety of square matrices with determinant zero.

Let $A \in Y_{n-1}$ be an n by n matrix of rank $n - 1$. Denote by $\tilde{A}$ the matrix of cofactors of A, i.e.,

$$(\tilde{A})_{ij} = (-1)^{i+j} \det \|a_{pq}\|_{p \neq i, q \neq j}.$$

In other words, if A is the matrix of an operator $\mathbf{A} : \mathbf{C}^n \to \mathbf{C}^n$ then $\tilde{A}$ is the matrix of the exterior power $\bigwedge^{n-1}(\mathbf{A})$.

Lemma 4.12. *The tangent hyperplane $T_A Y_{n-1}$ consists of matrices $B \in V$ such that $(\tilde{A}, B) = 0$, where $(\tilde{A}, B)$ is defined by (4.6).*

Proof of the lemma. A matrix B belongs to $T_A Y_{n-1}$ if and only if

$$\left.\frac{d}{dt}\right|_{t=0} \det(A + tB) = 0.$$

An explicit expansion of the determinant gives

$$\frac{d}{dt}\bigg|_{t=0} \det(A + tB) = (\tilde{A}, B),$$

from which the statement follows.

By the lemma, the dual cone $Y_{n-1}^{\vee}$ is the closure of the set of matrices $\tilde{A}$ corresponding to all A of rank $n-1$. Note that $\tilde{A}$ has rank 1 if rk $A = n-1$, and $\tilde{A} = 0$ if rk $A < n-1$. Hence $Y_{n-1}^{\vee} = Y_1$. This proves Proposition 4.11 in the particular case when $m = n, r = n-1$.

Now let us consider the general case. The variety Y_r is defined by the vanishing of all the minors of order $r+1$. For any $(r+1)$-element subsets $I \subset \{1, \dots, m\}$, $J \subset \{1, \dots, n\}$ and any $(m \times n)$ - matrix S, we denote by S_{IJ} the submatrix of S with rows from I and columns from J. Let $A \in Y_r$. It follows from Lemma 4.12 above that the matrix B lies in $T_A Y_r$ if and only if $(B_{IJ}, \tilde{A_{IJ}}) = 0$ for any $r+1$-element subsets I, J of rows and columns. Denote by $[\tilde{A_{IJ}}]$ the m by n matrix given by

$$[\tilde{A_{IJ}}]_{pq} = \begin{cases} \pm\det\|a_{ij}\|_{i\in I-\{p\}, j\in J-\{q\}} & \text{if } p \in I, q \in J; \\ 0 & \text{otherwise.} \end{cases}$$

Then the orthogonal complement to $T_A Y_r$ is spanned, as a vector space, by matrices $[\tilde{A_{IJ}}]$.

We shall show that for one special matrix A of rank r, the space $(T_A Y_r)^{\perp}$ consists of matrices of rank $\leq m-r$ and contains some matrices of exactly rank $m-r$. Since all the matrices of a given rank are equivalent under the action of $GL(n)$, Proposition 4.11 will follow.

We take A such that $A_{ij} = 1$ if $i = j \leq r$ and $A_{ij} = 0$ otherwise. For any $i \in \{1, \dots, m\}, j \in \{1, \dots, n\}$ denote by $E(i, j)$ the corresponding matrix unit: $E(i, j)_{p,q} = 1$ if $p = i, q = j$ and $E(i, j)_{p,q} = 0$ otherwise. An explicit computation shows that $[\tilde{A_{IJ}}] = 0$ unless both I and J contain $[r] = \{1, 2, \dots, r\}$; furthermore, if $I = [r] \cup \{i\}$, $J = [r] \cup \{j\}$ then $[\tilde{A_{IJ}}] = \pm E(i, j)$. Clearly, each linear combination of such matrices has rank $\leq m-r$ and some of them have rank $m-r$. Proposition 4.11 is proved.

Example 4.13. Consider the particular case of matrices of rank 1 in Example 4.10. Each such matrix can be seen as a product of a vector and a covector. Therefore the variety X_1 is the product $P^{m-1} \times P^{n-1}$ of two projective spaces. The resulting embedding of this product is known as the *Segre embedding* [Harr] [Sh]. We can write the space V of m by n matrices as the tensor product $E \otimes F$ where $\dim E = m$, $\dim F = n$. In this notation the Segre embedding has the form

$$P(E) \times P(F) \hookrightarrow P(E \otimes F), \quad (e, f) \mapsto e \otimes f.$$

The dual variety to $P^{k-1} \times P^{m-1}$ consists, therefore, of matrices that are not of full rank. For $k = m$, this is the determinantal variety. In this case the X-discriminant is the determinant of a square matrix.

Another instructive case from Example 4.10 is $m = 2$, $n > 2$ and $r = 1$. The projectivization X_1 of Y_1 is the product $P^1 \times P^{n-1}$ and hence is smooth. The dual variety is again isomorphic to X_1. So both of these varieties are smooth and have codimension n in their ambient projective spaces. In particular, this provides an example of a *smooth* subvariety whose dual is not a hypersurface.

Example 4.14. Let X be the product of three projective spaces in its Segre embedding:

$$X = P(E) \times P(F) \times P(G) \hookrightarrow P(E \otimes F \otimes G), \quad (e, f, g) \mapsto e \otimes f \otimes g.$$

By choosing bases in E, F, G, we can identify the space $E \otimes F \otimes G$ as well as its dual, with the space of 3-dimensional matrices (arrays) $A = \|a_{ijk}\|$ where $i = 1, \dots, \dim E$, $j = 1, \dots., \dim F$, $k = 1, \dots, \dim G$. The X-discriminant is a polynomial in variables a_{ijk}. In view of the previous example it is natural to call Δ_X the *hyperdeterminant* of the 3-dimensional matrix $\|a_{ijk}\|$. In a similar way, we define the hyperdeterminant of an r-dimensional matrix $\|a_{i_1,\dots,i_r}\|$. A detailed study of hyperdeterminants will be undertaken in Chapter 14.

Example 4.15. Let X be the projective space P^{k-1} in its *Veronese embedding* of degree d. In other words (see [Harr] [Sh]), $X = P(E)$ where E is a k-dimensional vector space and the embedding has the form

$$X = P(E) \hookrightarrow P(S^d E), \quad e \mapsto e^d.$$

The space $(S^d E)^* = S^d E^*$ is identified with the space of homogeneous polynomials (forms) of degree d in k variables. The dual variety $X^\vee \subset P(S^d E^*)$ consists of polynomials f for which the hypersurface $\{f = 0\} \subset P(E)$ is singular. The X-discriminant in this case is the classical discriminant of a form of degree d in k variables. It was introduced in this generality in 1842 by G. Boole (see [Ca1]), who found that the degree of the discriminant is kd^{k-1}. Beyond this, little is known about the discriminant Δ_X (in this example). Except in the case of a binary or quadratic form, it is never written in an expanded form, in part because it is so cumbersome. For example, the discriminant of the ternary cubic form $\sum_{i+j+k=3} a_{ijk} x^i y^j z^k$ has degree 12 as a polynomial in (a_{ijk}) (and it is well-known in the theory of modular forms). The expanded form of this polynomial contains 2040 terms (we would like to thank S. V. Duzhin for making available to us a computer-generated formula for $\Delta(\{a_{ijk}\})$).

Example 4.16. Let $a \geq 2$ be a natural number. Consider the hypersurface $X \subset P^m$ with the affine equation

$$x_1^a + \cdots + x_m^a = 1$$

(or the homogeneous equation $x_1^a + \cdots + x_m^a = x_0^a$). Introduce an affine chart in P^{m*} consisting of hyperplanes with affine equations of the form $\sum_{i=1}^m p_i x_i = 1$. So $p_1, \ldots, p_m$ are coordinates in this chart. An explicit calculation similar to that of Example 2.3 shows that the hypersurface dual to X can be defined, in coordinates $p_1, \ldots, p_n$, by

$$p_1^{\frac{a}{a-1}} + \cdots + p_m^{\frac{a}{a-1}} = 1. \tag{4.7}$$

Like the procedure described in Example 2.3, the irrational equation (4.7) can be replaced by a polynomial equation of degree $a(a-1)^{m-1}$.

Example 4.17. Our last example is a "mixture" of Segre and Veronese embeddings giving a common generalization of Examples 4.14 and 4.15. Let X be the product of several projective spaces $P(E_1) \times P(E_2) \times \cdots \times P(E_r)$ taken in the embedding into $P(S^{d_1}(E_1) \otimes S^{d_2}(E_2) \otimes \cdots \otimes S^{d_r}(E_r))$, where $d_1, \ldots, d_r$ are some positive integers. Let $V = S^{d_1}(E_1) \otimes S^{d_2}(E_2) \otimes \cdots \otimes S^{d_r}(E_r)$, then the dual vector space V^* is identified with the space of multihomogeneous forms $f(x_1, \ldots, x_r)$ ($x_i \in E_i$) having degree d_i in each vector variable x_i. We call the X-discriminant the *multigraded discriminant*. The hyperdeterminant in Example 4.14 is a special case of the multigraded discriminant when all degrees d_i are equal to 1. We shall give a criterion for the non-triviality of the multigraded discriminant (i.e., determine for which dimensions and degrees the dual variety $X^\vee$ is a hypersurface) in the next section. Additional properties of multigraded discriminants will be considered in Chapter 13.

5. The Katz dimension formula and its applications

A. Katz dimension formula

Let $X \subset P^n$ be a k-dimensional projective variety, and $X^\vee \subset (P^n)^\vee$ be the projectively dual variety of X. The following result due to N. Katz [Kat] expresses the codimension of $X^\vee$ in terms of the rank of certain Hessian matrix.

We write the ambient projective space P^n containing X as $P(V)$, where V is a vector space of dimension $n+1$. Then $X^\vee \subset P(V^*)$. The elements of V^* are homogeneous functions of degree 1 on the cone $Y \subset V$ over X. For $x \in P(V)$ let $x^\perp \in V^*$ denote the orthogonal complement to x, i.e., the codimension 1 subspace of functions vanishing at x. Now let x_0 be a smooth point of X. Then one can choose linear functionals $T_0 \in V^* - x_0^\perp$ and $T_1, \ldots, T_k \in x_0^\perp$ so that the functions

$t_1 = T_1/T_0, t_2 = T_2/T_0, \ldots, t_k = T_k/T_0$ are local coordinates on X near x_0. For every $U \in x^\perp$ the function $u = U/T_0$ on a neighborhood of x_0 in X is an analytic function of $t_1, \ldots, t_k$ such that $u(0, 0, \ldots, 0) = 0$. Consider the Hessian matrix

$$\operatorname{Hes}(u) = \operatorname{Hes}(U; T_0, T_1, \ldots, T_k; x_0) = \left\| \frac{\partial^2 u}{\partial t_i \partial t_j}(0, 0, \ldots, 0) \right\|_{1 \le i, j \le k}. \tag{5.1}$$

Theorem 5.1. *We have*

$$\operatorname{codim} X^\vee = 1 + \min \operatorname{corank} \operatorname{Hes}(u), \tag{5.2}$$

the minimum over all possible choices of x_0 and U.

Proof. We start with the following obvious remark. If $\pi : Z \to S$ is any regular map of algebraic varieties then the dimension of the closure of $\pi(Z)$ is equal to the maximum value of the rank of the Jacobian matrix of π at smooth points of Z.

To apply this in our situation, let $W_0 \subset P(V) \times P(V^*)$ be the smooth part of the incidence variety consisting of pairs (x, H) such that x is a smooth point of X, and H is a hyperplane in $P(V)$ tangent to X at x (cf. Section 3). Let $\pi : W_0 \to P(V^*)$ be the second projection. Then $X^\vee$ is the closure of $\pi(W_0)$. To prove our theorem, we compute the Jacobian matrix of π in appropriate local coordinates and relate it to the Hessian matrices appearing in the theorem.

Let $(x_0, H_0) \in W$. Choose $T_0 \in V^* - x_0^\perp$ and $T_1, \ldots, T_k \in x_0^\perp$ as above so that the functions $t_1 = T_1/T_0, t_2 = T_2/T_0, \ldots, t_k = T_k/T_0$ are local coordinates on X near x_0. Extend $T_1, \ldots, T_k$ to a basis $\{T_1, \ldots, T_k, U_1, U_2, \ldots, U_{n-k}\}$ of $x_0^\perp$. Then each of the functions $u_i = U_i/T_0$ $(i = 1, \ldots, n-k)$ on X is an analytic function of $(t_1, \ldots, t_k)$ in a neighborhood of $(0, 0, \ldots, 0)$, and we have $u_i(0, 0, \ldots, 0) = 0$.

Let $x \in X$ be a point close to x_0 with local coordinates $(t_1, \ldots, t_k)$ (so in homogeneous coordinates $x = (1 : t_1 : \ldots : t_k : u_1 : \ldots : u_{n-k})$). Let H be the hyperplane in $P(V)$, defined by a linear form $(\sum_{j=0}^k \tau_j T_j + \sum_{i=1}^{n-k} \eta_i U_i) \in V^*$. Then H is tangent to X at x if and only if the function

$$\tau_0 + \sum_{j=1}^{k} \tau_j t_j + \sum_{i=1}^{n-k} \eta_i u_i$$

vanishes at x together with all its first derivatives. We see that for a given x, the hyperplanes tangent to X at x form a projective space of dimension $n - k - 1$ with homogeneous coordinates $(\eta_1 : \eta_2 : \ldots : \eta_{n-k})$, and the remaining coordinates τ_j of H are given by

$$\tau_j = -\sum_{i=1}^{n-k} \eta_i \partial u_i / \partial t_j \quad (j = 1, \ldots, k), \tag{5.3}$$

$$\tau_0 = -(\sum_{j=1}^{k} \tau_j t_j + \sum_{i=1}^{n-k} \eta_i u_i) = -\sum_{i=1}^{n-k} \eta_i u_i + \sum_{j=1}^{k} \sum_{i=1}^{n-k} \eta_i t_j \partial u_i / \partial t_j. \quad (5.4)$$

Without loss of generality, we can assume that the hyperplane H_0 has coordinates $(\eta_1 : \eta_2 : \ldots : \eta_{n-k}) = (0 : 0 : \ldots : 1)$. It follows that we can set $\eta_{n-k} = 1$ in the above formulas, and use $(t_1, \ldots, t_k, \eta_1, \ldots, \eta_{n-k-1})$ as local coordinates on W near (x_0, H_0) (so that (x_0, H_0) has all the coordinates equal to 0).

In these coordinates the projection $\pi : W_0 \to P(V^*)$ takes the form

$$(t_1, \ldots, t_k, \eta_1, \ldots, \eta_{n-k-1}) \mapsto (\tau_0, \tau_1, \ldots, \tau_k, \eta_1, \ldots, \eta_{n-k-1}),$$

where $\tau_0, \tau_1, \ldots, \tau_k$ are given by (5.3) and (5.4) (with $\eta_{n-k} = 1$). An obvious calculation shows that the Jacobian matrix of π at the origin is the following $n \times (n-1)$ matrix:

$$\frac{\partial(\tau_0, \tau_1, \ldots, \tau_k, \eta_1, \ldots, \eta_{n-k-1})}{\partial(t_1, \ldots, t_k, \eta_1, \ldots, \eta_{n-k-1})}(0, \ldots, 0)$$

$$= \begin{pmatrix} 0 & 0 \\ -\mathrm{Hes}\,(u_{n-k}) & -\left(\frac{\partial(u_1, \ldots, u_{n-k-1})}{\partial(t_1, \ldots, t_k)}(0, \ldots, 0)\right)^t \\ 0 & I_{n-k-1} \end{pmatrix}. \quad (5.5)$$

Here horizontal sizes of the blocks are $1, k, n-k-1$, and vertical sizes are $k, n-k-1$; the $k \times k$ matrix $\mathrm{Hes}\,(u_{n-k})$ is defined by (5.1), and I_{n-k-1} is the identity matrix of order $n-k-1$.

Clearly, the Jacobian matrix (5.5) has rank

$$n - k - 1 + \mathrm{rk}\,\mathrm{Hes}\,(u_{n-k}).$$

It follows that

$$\dim X^\vee = n - k - 1 + \max \mathrm{rk}\,\mathrm{Hes}\,(u), \quad (5.6)$$

the maximum over the same choices as in Theorem 5.1. The equality (5.6) is equivalent to (5.2), and the theorem is proved.

Theorem 5.1 implies the following criterion for the non-triviality of the X-discriminant.

Corollary 5.2. *The dual variety $X^\vee$ is a hypersurface if and only if there exist a smooth point $x_0 \in X$ and linear forms $U, T_0, T_1, \ldots, T_k$ as in Theorem 5.1 such that the Hessian matrix* $\mathrm{Hes}\,(u)$ *is invertible.*

Using this criterion, Knop and Menzel [Knop–Me] classified all irreducible representations of simple Lie algebras such that the dual of the highest vector orbit is a hypersurface.

B. The dimension of the dual to a hypersurface: Segre's theorem

Although the projective duality is an involutive operation, the most typical situation, from the intuitive point of view, is when the projective dual $X^\vee$ is a hypersurface. Indeed, if $X^\vee$ is not a hypersurface, say, $\operatorname{codim} X^\vee = r + 1$ then, by Corollary 1.2, X is covered by the projective spaces of dimension r lying on X, which is certainly atypical behavior. For example, if X is a curve other than a straight line then $X^\vee$ is always a hypersurface.

Therefore by taking projective duals to hypersurfaces, we can get a considerable amount of subvarieties which are not hypersurfaces. This is of interest since it is very easy to define a hypersurface by an equation, whereas the description by equations of subvarieties of higher codimension is much more difficult.

This line of thought was formulated explicitly by B. Segre [Se]. The first step in this approach is to find the codimension of the dual to a hypersurface X. The answer, due to Segre, is as follows.

Theorem 5.3. *Let $f(x_0, \dots, x_n)$ be an irreducible homogeneous polynomial of degree d and let $X \subset P^n$ be the hypersurface with the equation $f = 0$. Let m be the largest number with the following property: any m by m minor of the Hessian matrix $\|\partial^2 f/\partial x_i \partial x_j\|$ (which is a polynomial in x_i of degree $m(d-2)$) is divisible by f. Then* $\dim X^\vee = m - 2$.

Proof. This is very similar to the proof of Theorem 5.1. Let W_0 and $\pi : W_0 \to X^\vee$ have the same meaning as in that proof. Since X is a hypersurface, the other projection $\rho : W_0 \to X$ is birational, so $\dim W_0 = \dim X$. The rational map $\pi \circ \rho^{-1} : X \to X^\vee$ has dense image so $\dim X^\vee$ is equal to the generic rank of the Jacobian matrix of $\pi \circ \rho^{-1}$. Let this rank be r. Note that for a smooth point $x \in X$ we have

$$(\pi \circ \rho^{-1})(x) \quad = \quad \left(\frac{\partial f}{\partial x_0}(x) : \dots : \frac{\partial f}{\partial x_n}(x)\right).$$

Taking partial derivatives, we find that r equals the generic rank of the Hessian matrix of f on X minus one. This is precisely what is claimed in the theorem.

The first non-trivial case is the case of space curves. In this case Theorem 5.3 gives the following.

Corollary 5.4. *An irreducible surface in P^3 with equation $f(x_0, \dots, x_3) = 0$ is projectively dual to a space curve if and only if the Hessian* $\det\|\partial^2 f/\partial x_i \partial x_j\|_{i,j=0,\dots,3}$ *is divisible by f.*

C. Dimension of the dual of the product

Let $X_1 \subset P^{n_1}$ and $X_2 \subset P^{n_2}$ be two irreducible projective varieties. The product

$X_1 \times X_2$ is naturally embedded in $P^{n_1n_2+n_1+n_2}$ via the Segre embedding $P^{n_1} \times P^{n_2} \to P^{n_1n_2+n_1+n_2}$. Let $(X_1 \times X_2)^\vee \subset (P^{n_1n_2+n_1+n_2})^\vee$ be the projectively dual variety. Using Theorem 5.1, we shall calculate $\dim (X_1 \times X_2)^\vee$ in terms of the dimensions of X_1, X_2 and their duals. To present the answer in the most symmetric form, we introduce the following notation. Let $X \subset P^n$ be an irreducible projective variety, and let $X^\vee \subset (P^n)^\vee$ be the projectively dual variety. We set

$$\mu(X) = \dim X + \operatorname{codim} X^\vee - 1. \qquad (5.7)$$

Theorem 5.5. (Product Theorem) *We have*

$$\mu(X_1 \times X_2) = \max (\dim X_1 + \dim X_2, \mu(X_1), \mu(X_2)). \qquad (5.8)$$

Proof. Let $\dim X_1 = k_1$, $\dim X_2 = k_2$. Let $x_0 = (x_{01}, x_{02})$ be a smooth point of $X_1 \times X_2$. We choose local coordinates $t_{11}, t_{21}, \dots, t_{k_1,1}$ on X_1 near x_{01} and $t_{12}, t_{22}, \dots, t_{k_2,2}$ on X_2 near x_{02} as in subsection A. So for $\nu = 1, 2$, a point $x_\nu \in X_\nu$ close to $x_{0\nu}$ has homogeneous coordinates

$$x_\nu = (1 : t_{1\nu} : \ldots : t_{k_\nu,\nu} : u_{1\nu} : \ldots : u_{n_\nu-k_\nu,\nu}),$$

where each $u_{i\nu}$ is an analytic function of $t_{1\nu}, t_{2\nu}, \dots, t_{k_\nu,\nu}$ vanishing at the origin.

By definition of the Segre embedding, a point $(x_1, x_2) \in X_1 \times X_2$ has as homogeneous coordinates all pairwise products of homogeneous coordinates of x_1 and x_2. Therefore, the set of homogeneous coordinates consists of 1 and the following eight sets of variables:

$$t_1 = \{t_{j1}\},\ t_2 = \{t_{j2}\},\ u_1 = \{u_{i1}\},\ u_2 = \{u_{i2}\},$$

$$tt = \{t_{j_1,1}t_{j_2,2}\},\ tu = \{t_{j1}u_{i2}\},\ ut = \{u_{i1}t_{j2}\},\ uu = \{u_{i_1,1}u_{i_2,2}\} \qquad (5.9)$$

(here tt, tu, ut, and uu are simply symbols not to be confused with products). Among these coordinates we choose those in t_1 and t_2 as local coordinates on $X_1 \times X_2$ near x_0. The coordinates from the remaining six sets are analytic functions of these local coordinates vanishing at the origin.

To apply Theorem 5.1, we have to consider the Hessian matrix $\operatorname{Hes}(u)$, where u is a linear combination of all the coordinates in (5.9) regarded as a function of local coordinates from t_1 and t_2. By some abuse of notation we write u as

$$u = t_1 + t_2 + u_1 + u_2 + tt + tu + ut + uu,$$

where each of the summands stands for a linear combination of the variables of the corresponding group. We write Hes (u) as

$$\mathrm{Hes}\,(u) = \begin{pmatrix} A_{11} & A_{12} \\ A_{21} & A_{22} \end{pmatrix},$$

where

$$A_{\alpha\beta} = \left\| \frac{\partial^2 u}{\partial t_{i\alpha} \partial t_{j\beta}}(0, 0, \dots, 0) \right\|_{1 \le i \le k_\alpha, 1 \le j \le k_\beta}.$$

An easy computation shows that $A_{11} = \mathrm{Hes}\,(u_1)$, $A_{22} = \mathrm{Hes}\,(u_2)$, so that

$$\mathrm{Hes}\,(u) = \begin{pmatrix} \mathrm{Hes}\,(u_1) & A \\ {}^tA & \mathrm{Hes}\,(u_2) \end{pmatrix}, \tag{5.10}$$

where A can be an arbitrary $k_1 \times k_2$ matrix.

We need to find the minimal possible value of the corank of Hes (u). Clearly, such a value is attained for generic u_1, u_2, and A. By Theorem 5.1, we have

$$\mathrm{corank}\,\mathrm{Hes}\,(u_1) = \mathrm{codim}\,X_1^\vee - 1, \quad \mathrm{corank}\,\mathrm{Hes}\,(u_2) = \mathrm{codim}\,X_2^\vee - 1. \tag{5.11}$$

To complete the proof of Theorem 5.5 we use the following result from linear algebra.

Proposition 5.6. *Let E be a finite dimensional vector space decomposed into the direct sum $E = E_1 \oplus E_2$ of two subspaces having dimensions k_1 and k_2. For any two integers $c_1 \in [0, k_1]$, $c_2 \in [0, k_2]$, let* Bil (c_1, c_2) *denote the set of all symmetric bilinear forms φ on E such that, for $\nu = 1, 2$, the restriction of φ on V_ν has corank c_ν. Then a generic form $\varphi \in$* Bil (c_1, c_2) *has corank*

$$\mathrm{corank}\,(\varphi) = \max\,(0, c_1 - k_2, c_2 - k_1). \tag{5.12}$$

Combining Proposition 5.6 with Theorem 5.1 and (5.11), we see that

$$\mathrm{codim}\,(X_1 \times X_2)^\vee - 1$$

$$= \max\,(0, \mathrm{codim}\,X_1^\vee - 1 - \dim X_2, \mathrm{codim}\,X_2^\vee - 1 - \dim X_1). \tag{5.13}$$

Adding $\dim X_1 + \dim X_2$ to both sides of (5.13), we obtain (5.8).

Proof of Proposition 5.6. We think of φ as a linear map $\varphi : E_1 \oplus E_2 \to E_1^* \oplus E_2^*$. For $\alpha, \beta = 1, 2$, let $\varphi_{\alpha\beta}$ be the component of φ acting from E_β to E_α^*. The condition that φ is symmetric means that $\varphi_{11}^* = \varphi_{11}$, $\varphi_{22}^* = \varphi_{22}$, $\varphi_{21}^* = \varphi_{12}$. By definition, corank $(\varphi) = \dim \operatorname{Ker} \varphi$, where

$$\operatorname{Ker} \varphi = \{(-x_1, x_2) \in E_1 \oplus E_2 : \varphi_{11}(x_1) = \varphi_{21}^*(x_2),\ \varphi_{21}(x_1) = \varphi_{22}(x_2)\}. \quad (5.14)$$

Without loss of generality we can assume that $k_1 \leq k_2$. For a generic $\varphi \in \operatorname{Bil}(c_1, c_2)$ the map $\varphi_{21} : E_1 \to E_2^*$ is injective. Hence for a given $x_2 \in E_2$ there is at most one x_1 such that $(-x_1, x_2) \in \operatorname{Ker} \varphi$, namely

$$x_1 = \varphi_{21}^{-1} \varphi_{22}(x_2). \quad (5.15)$$

Let $U = \varphi_{22}^{-1}(\operatorname{Im} \varphi_{21}) \subset E_2$, and let $\psi : U \to E_1^*$ be given by

$$\psi = \varphi_{21}^* - \varphi_{11} \varphi_{21}^{-1} \varphi_{22}. \quad (5.16)$$

Using (5.14) and (5.15), we see that $\operatorname{Ker} \varphi$ is isomorphic to $\operatorname{Ker} \psi$.

We have

$$\dim (\operatorname{Im} \varphi_{22}) = k_2 - c_2, \ \dim (\operatorname{Im} \varphi_{21}) = k_1.$$

For φ generic, these two subspaces are in general position in E_2^*. Now consider two cases.

Case 1. $c_2 \geq k_1$. Then $\operatorname{Im} \varphi_{22} \cap \operatorname{Im} \varphi_{21} = (0)$. Therefore, $U = \operatorname{Ker} \varphi_{22}$, and ψ coincides with the restriction of φ_{21}^* to U. It follows that

$$\operatorname{corank} \varphi = \dim (\operatorname{Ker} \varphi_{22} \cap \operatorname{Ker} \varphi_{21}^*).$$

But $\operatorname{Ker} \varphi_{22} \cap \operatorname{Ker} \varphi_{21}^*$ is the orthogonal complement of $\operatorname{Im} \varphi_{22} + \operatorname{Im} \varphi_{21}$, and hence $\operatorname{corank} \varphi = k_2 - (k_2 - c_2 + k_1) = c_2 - k_1$.

Case 2. $c_2 \leq k_1$. Then $\dim (\operatorname{Im} \varphi_{22} \cap \operatorname{Im} \varphi_{21}) = k_1 - c_2$, hence $\dim U = k_1$. Therefore, for generic φ, we have $\operatorname{Ker} \varphi_{21}^* \cap W = (0)$, i.e., φ_{21}^* is an isomorphism between U and E_1^*. Now replace φ_{21} by $\lambda \varphi_{21}$ for a non-zero scalar λ. Then ψ given by (5.16) is replaced by

$$\lambda \varphi_{21}^* - \lambda^{-1} \varphi_{11} \varphi_{21}^{-1} \varphi_{22} = \lambda(\varphi_{21}^* - \lambda^{-2} \varphi_{11} \varphi_{21}^{-1} \varphi_{22}).$$

The latter map is invertible for sufficiently large $|\lambda|$, so $\operatorname{corank} \varphi = \dim (\operatorname{Ker} \psi) = 0$.

Clearly, (5.12) holds in both cases. Proposition 5.6 and Theorem 5.5 are proved.

D. Some corollaries of the Product Theorem

First, Theorem 5.5 immediately generalizes to the product of more than two factors.

Theorem 5.7. *Let $X_1, X_2, \dots, X_r$ be embedded irreducible projective varieties, and let $X_1 \times \cdots \times X_r$ be their product in the Segre embedding. Then*

$$\mu(X_1 \times \cdots \times X_r) = \max(\dim X_1 + \cdots + \dim X_r, \mu(X_1), \dots, \mu(X_r)). \quad (5.17)$$

Theorem 5.7 immediately implies the following criterion.

Corollary 5.8. *The dual variety $(X_1 \times \cdots \times X_r)^\vee$ is a hypersurface if and only if*

$$\mu(X_j) \leq \dim X_1 + \cdots + \dim X_r \text{ for } j = 1, \dots, r. \quad (5.18)$$

If $X = P^k$ is taken in the tautological embedding into itself then $X^\vee = \emptyset$, and we use the convention that $\dim(P^k)^\vee = -1$, hence $\operatorname{codim}(P^k)^\vee = k + 1$, hence $\mu(P^k) = 2k$.

Corollary 5.9. *Let X be an embedded irreducible projective variety. Then the dual variety $(X \times P^l)^\vee$ is a hypersurface if and only if*

$$\operatorname{codim} X^\vee - 1 \leq l \leq \dim X. \quad (5.19)$$

Proof. In view of (5.18), $(X \times P^l)^\vee$ is a hypersurface if and only if $\mu(X) \leq \dim X + l$ and $2l \leq \dim X + l$, which is equivalent to (5.19).

Corollary 5.9 shows that even if the X-discriminant is constant, we can associate a family of discriminants with X corresponding to the products $X \times P^l$ for l running over the string $\{\operatorname{codim} X^\vee - 1, \operatorname{codim} X^\vee, \dots, \dim X\}$. We shall see later that the discriminant in this family corresponding to the maximal value $l = \dim X$ has a nice geometric meaning, namely it can be identified with the Chow form of X (see Chapter 3, Section2D below). The geometric interpretations for other values of l can be found in [WZ 2].

The following result is a special case of Corollary 5.8.

Corollary 5.10. *The dual variety $(P^{k_1} \times \cdots \times P^{k_r})^\vee$ is a hypersurface if and only if*

$$2k_j \leq k_1 + \cdots + k_r \quad (5.20)$$

for $j = 1, \dots, r$.

Without loss of generality we can assume that $k_1 = \max(k_1, k_2, \ldots, k_r)$. Then (5.20) is equivalent to the condition $k_1 \leq k_2 + \cdots + k_r$. According to Example 4.14, this is the criterion for the existence of the hyperdeterminant for the matrix format $(k_1+1) \times \cdots \times (k_r+1)$. In particular, for $r = 2$ we recover an obvious fact that a usual (2-dimensional) matrix possessing the non-trivial determinant must be square. From this perspective, (5.20) can be regarded as the multidimensional analog of the notion of a square matrix. These questions will be treated in more detail in Chapter 14.

The criterion of Corollary 5.10 can be extended to the case when X_j for $j = 1, \ldots, r$ is the projective space P^{k_j} in the Veronese embedding into $P(S^{d_j}(C^{k_j+1}))$, where $d_1, \ldots, d_r$ are any positive integers (in Corollary 5.10, all d_j are equal to 1).

Corollary 5.11. *Suppose X_j for $j = 1, \ldots, r$ is the projective space P^{k_j} in the Veronese embedding into $P(S^{d_j}(\mathbf{C}^{k_j+1}))$. Then the dual variety $(X_1 \times \cdots \times X_r)^\vee$ is a hypersurface if and only if (5.20) holds for all j such that $d_j = 1$ (in particular this is always the case if all $d_j > 1$).*

Proof. We have seen already that $\mu(X_j) = 2k_j$ if $d_j = 1$. It is known that if $d_j > 1$ then $X_j^\vee$ is a hypersurface (cf. Example 4.15), hence $\mu(X_j) = k_j$. Now our statement follows from Corollary 5.10.

Corollary 5.11 gives the criterion of existence of the *multigraded discriminant*, i.e., the discriminant of a multihomogeneous form on $P^{k_1} \times \cdots \times P^{k_r}$ of multidegree $(d_1, \ldots, d_r)$ (cf. Example 4.17).

CHAPTER 2

The Cayley Method for Studying Discriminants

In Chapter 1 we introduced, for any projective variety $X \subset P^n$, the X-discriminant Δ_X which is the equation of the projective dual variety $X^\vee$ (so Δ_X is a constant if $X^\vee$ is not a hypersurface). We now explain the method that allows us, for a smooth X, to write down at least in principle, the polynomial Δ_X. The method goes back to the remarkable paper by Cayley [Ca4] on elimination theory, in which the foundations were laid for what is now called homological algebra. The Cayley method can also be applied to other similar problems, such as finding resultants (see Chapter 3). For discriminants, the Cayley method can be described in modern terminology as consisting of three steps:

Step 1. Interpret the vanishing of $\Delta_X(f)$ (i.e., tangency of X with the hyperplane $f = 0$) as a violation of the exactness of a certain complex of coherent sheaves on X. Usually the violation of exactness occurs at the points of tangency.

Step 2. Interpret the non-exactness of a complex of sheaves as the non-exactness of a certain complex of vector spaces whose terms are fixed and whose differential varies with f.

This is usually done by considering the complex of global sections of our complex of sheaves (possibly after tensor multiplication by a sufficiently ample invertible sheaf to ensure good behavior of the functor H^0). Instead of a complex of vector spaces, we can also construct a spectral sequence taking into account the higher cohomology of the sheaves from the complex.

Step 3. The X-discriminant is the determinant of the complex (spectral sequence) from step 2.

The notion of the determinant of a complex (or spectral sequence) is described in detail in Appendix A. Throughout the present chapter we shall assume familiarity with this notion.

The Cayley method gives an expression of $\Delta_X(f)$ as an alternating product of determinants of certain matrices whose entries depend linearly on f. In some simple instances, the alternating product will reduce to just one determinant.

1. Jet bundles and Koszul complexes

We now will carry out the first step of the Cayley method: interpret the vanishing of $\Delta_X(f)$ as the violation of exactness of a complex of sheaves. The complex in question will be the Koszul complex, obtained from a jet bundle. We start by reviewing some background material on jets. For more details the reader is referred to [Sau].

A. Jets

Let X be a smooth irreducible algebraic variety. By $\mathcal{O}_X$ we denote the sheaf of regular functions on X. For any algebraic vector bundle E on X, we shall denote by the same letter E the sheaf of regular sections of this bundle, which is a locally free sheaf of $\mathcal{O}_X$-modules. The sheaves corresponding to line bundles (vector bundles of rank 1) are known as *invertible sheaves*.

Let $\mathcal{L}$ be an algebraic line bundle on X. We consider the bundle $J(\mathcal{L})$ of *first jets* of sections of $\mathcal{L}$. By definition, the fiber of $J(\mathcal{L})$ at a point $x \in X$ is the quotient of the space of all sections of $\mathcal{L}$ near x by the subspace of sections which vanish at x together with their first derivatives. In other words,

$$J(\mathcal{L})_x = \mathcal{L}/I_x^2\mathcal{L}, \tag{1.1}$$

where $I_x \subset \mathcal{O}_x$ is the ideal of functions vanishing at x. Thus $J(\mathcal{L})$ is a vector bundle on X of rank $\dim X + 1$.

To any section f of $\mathcal{L}$, we associate a section $j(f)$ of $J(\mathcal{L})$ called the *first jet* of f. Namely, the value of $j(f)$ at any $x \in X$ is the class of f modulo $I_x^2\mathcal{L}$ (this is a vector in the fiber $J(\mathcal{L})_x$). The correspondence $f \mapsto j(f)$ is $\mathbf{C}$-linear:

$$j(f_1 + f_2) = j(f_1) + j(f_2), \quad j(\lambda f) = \lambda j(f), \ \lambda \in \mathbf{C}, \tag{1.2}$$

(but this correspondence, being in fact a differential operator, is not $\mathcal{O}_X$-linear). Note that every vector in every fiber of $J(\mathcal{L})$ can be represented as the value of some $j(f)$, but not every section of $J(\mathcal{L})$ has the form $j(f)$.

The relevance of jets to discriminants is as follows. Consider the projective space $P(V^*)$ where V is a vector space. Suppose $X \subset P(V^*)$ is a smooth closed irreducible subvariety and take $\mathcal{L} = \mathcal{O}_X(1)$ (see Section 4B, Chapter 1). Any vector $f \in V$ gives a linear function on V^* and hence can be regarded as a section of $\mathcal{L}$. The following proposition is an immediate consequence of definitions.

Proposition 1.1. *A vector $f \in V$ represents a point in the dual variety $X^\vee \subset P(V)$ if and only if the section $j(f)$ of the bundle $J(\mathcal{L})$ vanishes at some point $x \in X$.*

Let us point out several properties of jet bundles. The first is the isomorphism

$$J(\mathcal{L}) \simeq J(\mathcal{O}_X) \otimes \mathcal{L}. \tag{1.3}$$

This isomorphism is only between sheaves of sections of the two bundles in (1.3) and does not come from a fiberwise isomorphism of the bundles themselves. Its construction is as follows. A section of $J(\mathcal{O}_X)$ over $U \subset X$ is an assignment, to any $x \in U$, of a function $\psi(x, y)$ defined for y close to x and considered modulo I_x^2. Therefore a section of $J(\mathcal{O}_X) \otimes \mathcal{L}$ is an assignment of a similar function $\psi(x, y)$ but taking, for each x, values in the fiber $\mathcal{L}_x$. We can think of ψ as being defined on a small neighborhood of the diagonal in $U \times U$, so for $x \in U$ we have a section $y \mapsto \psi(y, x)$ of $\mathcal{L}$ defined near x. Taking, for each x, the first jet of this section, we get a section of $J(\mathcal{L})$. This defines the required isomorphism.

Similarly to (1.3), we get an isomorphism of sheaves of sections:

$$J(\mathcal{L} \otimes \mathcal{M}) \simeq J(\mathcal{L}) \otimes \mathcal{M} \tag{1.4}$$

for any two line bundles $\mathcal{L}$, $\mathcal{M}$ on X.

For any $i \geq 0$ let Ω_X^i be the sheaf of regular differential i-forms on X. The second property of jets we want to mention is the exact sequence of vector bundles

$$0 \to \Omega_X^1 \otimes \mathcal{L} \xrightarrow{\alpha} J(\mathcal{L}) \xrightarrow{\beta} \mathcal{L} \to 0. \tag{1.5}$$

Here β is the map which takes any vector $j(f)(x) \in J(\mathcal{L})_x$ into $f(x) \in \mathcal{L}_x$.

Consider now the simplest case when X is a projective space.

Proposition 1.2. *The jet bundle $J(\mathcal{O}_{P(V^*)}(1))$ is a trivial bundle naturally identified with $\mathcal{O}_{P(V^*)} \otimes V$.*

Proof. The space V is identified with the space of global sections of $\mathcal{O}(1)$ on $P(V^*)$. For any $x \in P(V^*)$, we get a linear map of vector spaces

$$V \to J(\mathcal{O}_{P(V^*)}(1))_x, \quad f \mapsto j(f)(x).$$

This map is an isomorphism, which may be verified, for instance, using local coordinates.

Proposition 1.2 admits a certain generalization to an arbitrary variety X and a line bundle $\mathcal{L}$ on X. Let us denote by Y the total space of the bundle $\mathcal{L}^*$ from which the zero section is deleted. In the important special case when X lies in a projective space $P(V^*)$ and $\mathcal{L} = \mathcal{O}(1)$, the manifold $Y \subset V^* - \{0\}$ is the punctured cone over X, i.e., the set of all non-zero points in V^* whose projectivization lies in X. Any local section f of $\mathcal{L}$ defines a function $\tilde{f}$ on Y homogeneous of degree 1.

Denote by $\pi : Y \to X$ the standard projection. Denote by ξ the Euler vector field on Y (i.e., the generator of the group of homotheties). The vector field ξ defines the Lie derivation Lie_ξ on tensor fields on Y, in particular, on the sheaves Ω^k_Y of differential k-forms. For any $l \in Z$, let us define a subsheaf $\Omega^1_{l,Y} \subset \Omega^1_Y$ whose sections are 1-forms γ on Y such that $\mathrm{Lie}_\xi(\gamma) = l \cdot \gamma$, i.e., γ is homogeneous of degree l under homotheties. Note that $\Omega^1_{l,Y}$ is not a sheaf of $\mathcal{O}_Y$-modules, in particular, not a coherent sheaf on Y. Let π_* denote the direct image of sheaves under π.

Proposition 1.3. *We have a natural isomorphism of sheaves on X*

$$\psi : J(\mathcal{L}) \to \pi_*(\Omega^1_{1,Y}).$$

In other words, sections of $J(\mathcal{L})$ over a domain U are naturally identified with differential 1*-forms γ over $\pi^{-1}(U)$ such that* $\mathrm{Lie}_\xi(\gamma) = \gamma$. *Moreover, let f be any local section of $\mathcal{L}$, and let $\tilde{f}$ be the corresponding homogeneous (of degree* 1*) function on Y. Then under the isomorphism ψ, the section $j(f)$ goes to the* 1*-form $d\tilde{f}$ on Y.*

Proof. The isomorphism ψ will be constructed fiberwise in order to satisfy the second assertion of the proposition. The fiber of $J(\mathcal{L})$ at a point x is $\mathcal{L}/I_x^2\mathcal{L}$, see (1.1). Consider the correspondence

$$f \mod I_x^2 \longmapsto d(\tilde{f})|_{\pi^{-1}(x)}$$

where on the right stands the restriction of $d(\tilde{f})$, as a section of Ω^1_Y, to the fiber $\pi^{-1}(x)$. It is straightforward to see that this correspondence is well-defined, gives a required isomorphism of fibers and that these isomorphisms for various x glue together into an isomorphism of vector bundles.

B. Koszul complexes

Let X be an irreducible algebraic variety. Let E be an algebraic vector bundle on X of rank r and let s be a global section of E. Consider the complex of sheaves on X

$$\mathcal{K}_+(E,s) = \left\{0 \to \mathcal{O}_X \xrightarrow{s} E \xrightarrow{\wedge s} \bigwedge^2 E \xrightarrow{\wedge s} \cdots \to \bigwedge^r E \to 0\right\}, \tag{1.6}$$

whose differential is given by exterior multiplication with s. We call $\mathcal{K}_+(E,s)$ the (positive) *Koszul complex* associated to E and s. We fix the grading in this complex by assigning the degree j to $\bigwedge^j E$.

It is also convenient to consider the complex dual to (1.6), i.e., the complex

$$\mathcal{K}_-(E,s) = \left\{0 \to \bigwedge^r E^* \to \cdots \to \bigwedge^2 E^* \xrightarrow{i_s} E^* \xrightarrow{i_s} \mathcal{O}_X \to 0\right\}. \quad (1.7)$$

Here the differential is given by contraction with s, i.e., by the map $i_s : \bigwedge^j E^* \to \bigwedge^{j-1} E^*$ dual to the map $\bigwedge^{j-1} E \to \bigwedge^j E$ given by exterior multiplication by s. We normalize the grading of $\mathcal{K}_-(E,s)$ by assigning the degree $(-j)$ to $\bigwedge^j E^*$. Note that we have an isomorphism of complexes

$$\mathcal{K}_-(E,s) \quad \cong \quad \mathcal{K}_+(E,s) \otimes \bigwedge^r E^* \, [r], \quad (1.8)$$

where r in brackets means the shift in the grading of the complex by r.

The reason for the introduction of both of these complexes is that it is often more straightforward to work with $\mathcal{K}_+$, whereas $\mathcal{K}_-$ has simpler cohomology sheaves, as we shall see from the following proposition.

Proposition 1.4.

(a) *The exactness of any of $\mathcal{K}_+(E,s)$, $\mathcal{K}_-(E,s)$ is equivalent to the fact that s does not vanish anywhere on X.*

(b) *Suppose that X is smooth and that s vanishes along a smooth subvariety $Z \subset X$ of codimension exactly $r = \mathrm{rk}\, E$ and is transverse to the zero section. Then $\mathcal{K}_-(E,s)$ has only one non-trivial cohomology sheaf, namely $\mathcal{O}_Z$ (regarded as a sheaf on X) in (highest) degree 0. In this situation $\mathcal{K}_+(E,s)$ has the only non-trivial cohomology sheaf in the highest degree r, and this sheaf is the restriction $\bigwedge^r (E)|_Z$ regarded as a sheaf on X.*

(c) *More generally, suppose that X is not necessarily smooth but possesses a smooth morphism $\pi : X \to S$ to some (possibly singular) variety S. Suppose also that the zero locus Z of s is such that $\pi : Z \to S$ is a smooth morphism,* codim $Z = \mathrm{rk}\, E$ *and, after restricting s to each fiber of π (which is smooth by assumption), s intersects the zero section transversally. Then all the conclusions of part (b) hold.*

For the definition of a smooth morphism in part (c) we refer the reader to [Hart], Chapter 3, Section 10. Informally, the notion of a smooth morphism is a relative version of the notion of a smooth variety over a field. Thus, if $\pi : X \to S$ is a smooth morphism, then both X and S may be singular, but every fiber of π is smooth. An example of a smooth morphism is given by a projection $Z \times S \to S$ where S is any variety (possibly singular) and Z is smooth. In fact, we will not need any other examples in this book.

Proof of Proposition 1.4 (a) If W is a vector space and $w \in W$ is a non-zero vector then the differential in the exterior algebra $\bigwedge^\bullet(W)$, given by the exterior multiplication with w, is exact. This can easily be seen by choosing a basis of W containing w. The dual differential i_w in $\bigwedge^\bullet W^*$ is therefore also exact. Applying this to our situation, we see that if the section s does not vanish anywhere on X then the complexes $\mathcal{K}_\pm(E, s)$ will be fiberwise exact, and hence exact as complexes of sheaves. If, however, s vanishes at some $x \in X$, then the cokernel of the last differential in $\mathcal{K}_\pm(E, s)$ is non-trivial. Indeed, trivializing E near x, we represent s as a collection of functions $(f_1, \dots, f_r)$, and represent the last differential in both complexes as

$$\mathcal{O}^r \to \mathcal{O}, \qquad (u_1, \dots, u_r) \longmapsto \sum u_i f_i.$$

If s vanishes at x, then all f_i also vanish and the above map is not surjective.

(b), (c) Trivializing E near any given point, we identify E locally with $\mathcal{O}^r$ and s with a collection of functions $(f_1, \dots, f_r)$, as above. Our conditions imply that $f_1, \dots, f_r$ form a regular sequence, i.e., every f_i is not a zero divisor in $\mathcal{O}_X/(f_1, \dots, f_{i-1})$. This yields, in a standard way (see, e.g., [GH] Chapter 5, Section 3) that both complexes $\mathcal{K}_\pm(E, s)$ are exact everywhere except in the last term. Now the last cohomology of $\mathcal{K}_-(E, s)$ is $\mathcal{O}_Z$. The statement about $\mathcal{K}_+$ follows from (1.8).

C. Interpretation of $X^\vee$ in terms of Koszul complexes

Let $X \subset P(V^*)$ be a smooth projective variety. We apply the machinery of subsection B to the jet vector bundle $E = J(\mathcal{L})$ where $\mathcal{L} = \mathcal{O}_X(1)$. Propositions 1.1 and 1.4 (a) give the following corollary.

Corollary 1.5. *Let $\mathcal{L} = \mathcal{O}_X(1)$. A vector $f \in V$ represents a point in $X^\vee$ if and only if any of the following complexes of sheaves on X is not exact:*

$$\mathcal{K}_+(J(\mathcal{L}), j(f)) = \left\{0 \to \mathcal{O}_X \xrightarrow{j(f)} J(\mathcal{L}) \xrightarrow{\wedge j(f)} \bigwedge^2 J(\mathcal{L}) \xrightarrow{\wedge j(f)} \cdots\right\} \tag{1.9}$$

$$\mathcal{K}_-(J(\mathcal{L}), j(f)) = \left\{\cdots \to \bigwedge^2 J(\mathcal{L})^* \to J(\mathcal{L})^* \to \mathcal{O}_X \to 0\right\}. \tag{1.10}$$

This concludes the first step of the Cayley method.

2. Discriminantal complexes

Here we perform the second and third steps of the Cayley method: interpret the non-exactness of a complex of coherent sheaves as non-exactness of a certain complex of finite dimensional vector spaces and then consider the determinant of this complex.

A. The general discriminant as the determinant of a complex

Let $X \subset P(V^*)$ be a smooth projective variety. We denote $\mathcal{L} = \mathcal{O}_X(1)$, so that any $f \in V$ can be regarded as a section of $\mathcal{L}$. Let $\mathcal{M}$ be another invertible sheaf (line bundle) on X. We define the *discriminantal complexes* $C^\bullet_+(X, \mathcal{M})$ and $C^\bullet_-(X, \mathcal{M})$ as complexes of global sections of Koszul complexes (1.9) and (1.10) tensored with $\mathcal{M}$. More precisely,

$$C^i_+(X, \mathcal{M}) = H^0\left(X, \bigwedge^i J(\mathcal{L}) \otimes \mathcal{M}\right),$$

$$C^i_-(X, \mathcal{M}) = H^0\left(X, \bigwedge^{-i} J(\mathcal{L})^* \otimes \mathcal{M}\right). \tag{2.1}$$

Thus the terms of these complexes are fixed and the differential, induced by exterior multiplication (or contraction) with $j(f)$ depends on $f \in V$.

We denote the differentials in both complexes by ∂_f. The following statement is an immediate consequence of (1.2) and properties of the exterior product.

Proposition 2.1. *In each of the $C^\bullet_\pm(X, \mathcal{M})$, the differentials ∂_f, $f \in V$, satisfy*

$$\partial_{\lambda f + \mu g} = \lambda \partial_f + \mu \partial_g, \quad \partial_f \partial_g = -\partial_g \partial_f, \quad f, g \in V,\ \lambda, \mu \in \mathbf{C}.$$

In other words, the ∂_f, $f \in V$, make each $C^\bullet_\pm(X, \mathcal{M})$ into a graded module over the exterior algebra $\bigwedge^\bullet(V)$.

It is known that the (non-) exactness of a complex of sheaves does not necessarily imply the (non-) exactness of the corresponding complex of global sections, and the obstructions to this are given by the higher cohomology of the sheaves of the complex. So we shall first consider the case when these obstructions vanish.

We shall say that the discriminantal complex $C^\bullet_+(X, \mathcal{M})$ is *stably twisted* if all of the terms $\bigwedge^i J(\mathcal{L}) \otimes \mathcal{M}$ of the corresponding complex of sheaves have no higher cohomology. Similarly, we shall say that $C^\bullet_-(X, \mathcal{M})$ is stably twisted if all of the terms $\bigwedge^i J(\mathcal{L})^* \otimes \mathcal{M}$ have no higher cohomology.

Let us recall the following theorem due to Serre ([Hart], Chapter 3, Theorem 5.2). Let $\mathcal{M}$ be an ample line bundle (see Chapter 1, Section 4B) and let $\mathcal{F}$ be any coherent sheaf on X. Then for $l \gg 0$ the sheaves $\mathcal{F} \otimes \mathcal{M}^{\otimes l}$ have no higher cohomology. The next proposition is a simple consequence of this theorem.

Proposition 2.2. *Suppose that the line bundle $\mathcal{M}$ on X is ample. Then for $l \gg 0$ the discriminantal complexes $C^\bullet_\pm(X, \mathcal{M}^{\otimes l})$ are stably twisted.*

For a stably twisted discriminantal complex we can perform the second step of the Cayley method.

Proposition 2.3. *Suppose that $C^\bullet_+(X, \mathcal{M})$ (resp. $C^\bullet_-(X, \mathcal{M})$) is stably twisted. Let $f \in V$ be such that its projectivization does not lie in the dual variety $X^\vee \subset P(V)$. Then the complex $(C^\bullet_+(X, \mathcal{M}), \partial_f)$ (resp. $(C^\bullet_-(X, \mathcal{M}), \partial_f)$) is exact.*

Proof. If $f \notin X^\vee$, then the Koszul complexes (1.9) and (1.10) are exact complexes of sheaves. So it suffices to note that exactness is preserved under tensoring with $\mathcal{M}$ and to apply the following general fact.

Lemma 2.4. *If $\mathcal{F}^\bullet$ is a finite exact complex of sheaves on a topological space X and $H^p(X, \mathcal{F}^i) = 0$ for any i and any $p > 0$, then the complex of global sections of $\mathcal{F}^\bullet$ is exact.*

Proof. This is a particular case of the "abstract de Rham theorem" for sheaves, see [GH] Chapter 3, Section 5, or, for an elementary proof, [Hir] no. 2.12. This theorem says that if a complex of sheaves $\mathcal{F}^\bullet$ is a right resolution of a sheaf $\mathcal{G}$ and the $\mathcal{F}^i$ do not have higher cohomology, then the complex of global sections of $\mathcal{F}^\bullet$ calculates the cohomology of $\mathcal{G}$. Our statement corresponds to the case where $\mathcal{G} = 0$.

Suppose now that $C^\bullet_-(X, \mathcal{M})$ is stably twisted. Let us choose a basis in each vector space $C^i_-(X, \mathcal{M})$. Denote the system of bases thus obtained by e. Then, for generic $f \in V$, we can consider the determinant of the based exact complex $(C^\bullet_-(X, \mathcal{M}), \partial_f, e)$ (see Appendix A, Definition 7). The correspondence

$$f \mapsto \det(C^\bullet_-(X, \mathcal{M}), \partial_f, e)$$

is a rational function on the space V as follows from an explicit formula for the determinant of a complex (Appendix A, Theorem 13). Let us abbreviate this function by $\Delta^-_{X,\mathcal{M}}$. A different choice of bases in $C^i_-(X, \mathcal{M})$ results in the multiplication of $\Delta^-_{X,\mathcal{M}}$ by a non-zero constant. We regard $\Delta^-_{X,\mathcal{M}}$ as being defined up to a non-zero constant multiple. With this understanding, the choice of bases does not matter.

Similarly, if $C^\bullet_+(X, \mathcal{M})$ is stably twisted, we introduce the rational function $\Delta^+_{X,\mathcal{M}}$ on V defined up to a non-zero constant multiple.

Theorem 2.5. *If the discriminantal complex $C^\bullet_-(X, \mathcal{M})$ is stably twisted then, up to a non-zero constant factor, we have*

$$\Delta^-_{X,\mathcal{M}}(f) = \Delta_X(f) \tag{2.2}$$

where Δ_X is the X-discriminant. If $C^\bullet_+(X, \mathcal{M})$ is stably twisted then

$$(\Delta^+_{X,\mathcal{M}})(f)^{(-1)^{\dim(X)+1}} = \Delta_X(f). \tag{2.3}$$

In particular, the left hand sides of (2.2) and (2.3) are polynomials. These polynomials are non-constant precisely when $X^\vee$ is a hypersurface.

B. Proof of Theorem 2.5

This proof is slightly more technical than the previous material and makes use of the formalism of derived categories. The reader who is unfamiliar with this formalism may wish to skip this subsection. For a general background on derived categories, see [KS].

First of all, if codim $X^\vee \geq 2$, then the determinant of $C^\bullet_\pm(X, \mathcal{M})$ is a rational function on V which is regular and takes non-zero values outside a subvariety of codimension ≥ 2. Such a function is necessarily constant. So we can and will assume in the rest of the proof that $X^\vee$ is a hypersurface.

Instead of considering complexes $(C^\bullet_\pm(X, \mathcal{M}), \partial_f)$ with varying differentials ∂_f, $f \in V$, we consider the universal complex over the symmetric algebra $S^\bullet(V^*)$ for which all of the above individual complexes are fibers. More precisely, we introduce the complex

$$C^\bullet_+(X, \mathcal{M}) \otimes S^\bullet(V^*)$$
$$= \left\{ C^0_+(X, \mathcal{M}) \otimes S^\bullet(V^*) \xrightarrow{\partial} C^1_+(X, \mathcal{M}) \otimes S^\bullet(V^*) \xrightarrow{\partial} \cdots \right\}.$$

The differential ∂ equals $\sum \partial_{f_i} \otimes \varphi_i$, where $f_0, \ldots, f_n$ and $\varphi_0, \ldots, \varphi_n$ are dual bases of V and V^*, and φ_i also stands for the multiplication operator. Proposition 2.1 implies that $\partial^2 = 0$. In a similar way we define the complex $C^\bullet_-(X, \mathcal{M}) \otimes S^\bullet(V^*)$.

Let $\mathbf{C}(V)$ be the field of rational functions on V, i.e., the field of fractions of $S^\bullet(V^*)$. The complexes $\big(C^\bullet_\pm(X, \mathcal{M}) \otimes S^\bullet(V^*), \partial\big)$ will become, after the extension of scalars from $S^\bullet(V^*)$ to $\mathbf{C}(V)$, exact complexes of finite-dimensional vector spaces over $\mathbf{C}(V)$. Clearly, $\Delta^\pm_{X,\mathcal{M}} \in \mathbf{C}(V)$ are just the determinants of these complexes with respect to the chosen bases. Hence the prime factorization of $\Delta^\pm_{X,\mathcal{M}}$ is given by Theorem 30 from Appendix A. Namely, let $\pi \in S^\bullet(V^*)$ be an irreducible polynomial and let $Z_\pi \subset V$ be the corresponding hypersurface. Then the π-adic order of $\Delta^{\pm}_{X,\mathcal{M}}$ is equal to the alternating sum of multiplicities at Z_π of the cohomology modules of $C^\bullet_\pm(X, \mathcal{M}) \otimes S(V^*)$. Let, as before, $Y^\vee \subset V$ denote the cone over the projective subvariety $X^\vee$. Our assertion is a consequence of the following fact.

Proposition 2.6. *Let $\mathcal{M}$ be such that $C^\bullet_-(X, \mathcal{M})$ (resp. $C^\bullet_+(X, \mathcal{M})$) is stably twisted. Suppose that* $\operatorname{codim} X^\vee = 1$. *Then all of the cohomology modules of $C^\bullet_-(X, \mathcal{M}) \otimes S(V^*)$ (resp. $C^\bullet_+(X, \mathcal{M}) \otimes S(V^*)$) except for the one on the far right have support on subvarieties of codimension ≥ 2. The cohomology module on the far right has only one irreducible component in its support which is a hypersurface. This component is $Y^\vee$ and its multiplicity is* 1.

We shall concentrate on the proof of Proposition 2.6. Recall that, by a theorem of Serre's [Ser], the category of coherent sheaves on $P(V)$ is equivalent to the category of graded finitely generated $S^\bullet(V^*)$-modules considered modulo finite-dimensional modules. We consider the complexes of sheaves on $P(V)$ associated to the complexes $C^\bullet_\pm(X, \mathcal{M}) \otimes S^\bullet(V^*)$ of graded $S^\bullet(V^*)$-modules. We denote these complexes by $\mathcal{O}\big(C^\bullet_\pm(X, \mathcal{M})\big)$. Thus $\mathcal{O}\big(C^\bullet_+(X, \mathcal{M})\big)$ has the form

$$C^0_+(X, \mathcal{M}) \otimes \mathcal{O}_{P(V)} \to C^1_+(X, \mathcal{M}) \otimes \mathcal{O}_{P(V)}(1) \to \cdots \tag{2.4}$$

and $\mathcal{O}(C^\bullet_-(X, \mathcal{M}))$ has the form

$$\cdots \to C^{-1}_-(X, \mathcal{M}) \otimes \mathcal{O}_{P(V)}(-1) \to C^0_-(X, \mathcal{M}) \otimes \mathcal{O}_{P(V)}. \tag{2.5}$$

By Serre's theorem it is enough to prove that all of the cohomology sheaves of (2.4) or (2.5) except for the one on the far right are supported on subvarieties of codimension ≥ 2, the cohomology sheaf on the far right is supported on $X^\vee \subset P(V)$ and the multiplicity equals 1.

We shall interpret complexes (2.4) and (2.5) as direct images in the derived category. Let $W \subset X \times P(V)$ be the incidence variety, i.e., the set of pairs (x, H) where H is a hyperplane tangent to X at x (see Chapter 1, Section 3). Let p_1, p_2 be the projections of W to X and $P(V)$, and let p_X, p_V be the projections of the whole $X \times P(V)$. Since $\mathcal{L}$ is the restriction to X of $\mathcal{O}_{P(V^*)}(1)$, we have the maps

$$V \xrightarrow{\text{res}} H^0(X, \mathcal{L}) \xrightarrow{j} H^0(X, J(\mathcal{L})).$$

Therefore we have a map of $V \otimes V^* = \operatorname{End}(V)$ into the space of sections of the vector bundle $p_X^*(J(\mathcal{L})) \otimes p_V^*(\mathcal{O}(1))$ on $X \times P(V)$. Let

$$\sigma \in H^0\big(X \times P(V), p_X^*\big(J(\mathcal{L})\big) \otimes p_V^*(\mathcal{O}(1))\big)$$

be the image of the identity element in End(V). It can be seen as the "universal section" since the restriction of σ to any fiber $X \times \{\mathbf{C}f\}$ is just the section $j(f)$ of $J(\mathcal{L})$. To simplify further formulas, denote $k = \dim(X) + 1$.

Proposition 2.7. *The subvariety $W \subset X \times P(V)$ is smooth and has codimension k. The section σ vanishes exactly along W and meets the zero section transversally.*

The proof follows from the representation of W as a projective bundle over X (Chapter 1, Section 3).

Now the general properties of Koszul complexes (Proposition 1.4) imply the following:

Corollary 2.8. *The complex of sheaves on $X \times P(V)$:*

$$\cdots \to p_X^*\left(\bigwedge^2 J(\mathcal{L}^*)\right) \otimes p_V^*\mathcal{O}(-2) \to p_X^*\left(J(\mathcal{L})^*\right) \otimes p_V^*\mathcal{O}(-1) \to \mathcal{O}_{X\times P(V)} \tag{2.6}$$

is a left resolution of the structure sheaf $\mathcal{O}_W$, regarded as a sheaf on $X \times P(V)$. The dual complex

$$\cdots \to p_X^*\left(\bigwedge^{k-1} J(\mathcal{L})\right) \otimes p_V^*\mathcal{O}(k-1) \to p_X^*\left(\bigwedge^{k} J(\mathcal{L})\right) \otimes p_V^*\mathcal{O}(k) \tag{2.7}$$

has the only non-trivial cohomology sheaf in the far right term. This cohomology sheaf equals the restriction on W of the line bundle $p_X^*(\Omega_X^{k-1} \otimes \mathcal{L}^{\otimes(k)}) \otimes p_V^*\mathcal{O}(k)$, (this restriction is regarded as a sheaf on $X \times P(V)$).

Denote by $\mathcal{E}$ the line bundle on W in Corollary 2.8 (the one defined as the restriction).

Proposition 2.9. *If the discriminantal complex $C_-^\bullet(X, \mathcal{M})$ is stably twisted, then the corresponding complex $\mathcal{O}(C_-^\bullet(X, \mathcal{M}))$ of sheaves on $P(V)$ is isomorphic in the derived category to the direct image $Rp_{2*}(p_1^*(\mathcal{M}))$. Similarly, if $C_+^\bullet(X, \mathcal{M})$ is stably twisted, then $\mathcal{O}(C_+^\bullet(X, \mathcal{M}))$ is isomorphic to $Rp_{2*}(p_1^*(\mathcal{E} \otimes \mathcal{M}))$.*

Proof. Let p_X, p_V be the projections of $X \times P(V)$ to X and $P(V)$, respectively. Recalling that p_1 and p_2 are the projections of $W \subset X \times P(V)$ to X and $P(V)$, we have the isomorphisms in the derived category

$$Rp_{2*}(p_1^*\mathcal{M}) \cong Rp_{V*}\big(\mathcal{O}_W \otimes p_X^*\mathcal{M}\big) \cong Rp_{V*}\big(\mathcal{K} \otimes p_X^*\mathcal{M}\big)$$

where $\mathcal{K}$ is the complex (2.6). The second of the above isomorphisms follows from Corollary 2.8. Note now that $\mathcal{O}\left(C_-^\bullet(X, \mathcal{M})\right)$ coincides with $p_{V*}(\mathcal{K} \otimes p_X^*\mathcal{M})$, the complex obtained by applying to every term of $\mathcal{K}\otimes p_X^*\mathcal{M}$ the ordinary (not derived) direct image functor p_{V*}. Thus the question is to compare the termwise ordinary direct image with the full derived direct image. If $C_-^\bullet(X,\mathcal{M})$ is stably twisted then for every term $\mathcal{K}^j$ of the complex $\mathcal{K}$, the higher direct images $R^i p_{V*}(\mathcal{K}^j\otimes p_X^*\mathcal{M})$ vanish for $i>0$. So the statement of Proposition 2.9 about $\mathcal{O}\left(C_-^\bullet(X,\mathcal{M})\right)$ follows from the next general lemma (the statement about $\mathcal{O}(C_+^\bullet)$ is similar).

Lemma 2.10. *Let $p: Y\to S$ be a continuous map of topological spaces and let $\mathcal{F}^\bullet$ be a finite complex of sheaves on Y such that $R^i p_*\mathcal{F}^j = 0$ for any $i>0$ and any j. Then the complex $Rp_*\mathcal{F}^\bullet$ is isomorphic in the derived category to $p_*\mathcal{F}^\bullet$, the termwise direct image.*

Proof of the lemma. For any complex of sheaves $\mathcal{F}^\bullet$, there is a canonical morphism $p_*\mathcal{F}^\bullet\to Rp_*\mathcal{F}^\bullet$ of complexes of sheaves on S. Let us show that it is a quasi-isomorphism, i.e., it induces isomorphisms of sheaves of cohomology $\underline{H}^i(p_*\mathcal{F}^\bullet)\to\underline{H}^i(Rp_*\mathcal{F}^\bullet)$ for all i. We have $\underline{H}^i(Rp_*\mathcal{F}^\bullet)=\mathbf{R}^i p_*\mathcal{F}^\bullet$, the hyper-direct images of $\mathcal{F}^\bullet$. As for any derived functor, there is a spectral sequence of sheaves $E_1^{ij}=R^j p_*\mathcal{F}^i\Rightarrow\mathbf{R}^{i+j}p_*\mathcal{F}^\bullet$. So our assumptions imply that this sequence is reduced to one complex $p_*\mathcal{F}^\bullet$. Lemma 2.10 and Proposition 2.9 are proved.

Let us now complete the proof of Proposition 2.6. Since we assume that $X^\vee$ is a hypersurface, the map $p_2: W_X\to P(V)$ is a birational isomorphism of W_X and $X^\vee\subset P(V)$ (Proposition 3.2 of Chapter 1). Hence the complex $Rp_{2*}(p_1^*\mathcal{M})$ of sheaves on $P(V)$ can have only one cohomology sheaf supported on a hypersurface, namely $R^0p_{2*}(p_1^*\mathcal{M})$ and its multiplicity along $X^\vee$ is equal to 1. Hence for any hypersurface $Z\subset P(V)$, we have

$$\sum_i(-1)^i\mathrm{mult}_Z\underline{H}^i\left(\mathcal{O}(C_-^\bullet(X,\mathcal{M}))\right) = \sum_i(-1)^i\mathrm{mult}_Z R^i p_{2*}(p_1^*\mathcal{M}) = \begin{cases}0, & \text{if } Z\neq X^\vee,\\ 1, & \text{if } Z=X^\vee.\end{cases}$$

Here the $\underline{H}^i$ mean, as before, the cohomology sheaves of a complex of sheaves. This, together with Theorem 30 from Appendix A, completes the proof of Proposition 2.6 for $C_-^\bullet(X,\mathcal{M})$. For $C_+^\bullet(X,\mathcal{M})$, the reasoning is similar. Proposition 2.6 and Theorem 2.5 are proved.

C. Examples

Example 2.11. The Sylvester formula. Let X be the projective line $P^1=P(\mathbf{C}^2)$ embedded into $P^d=P(S^d\mathbf{C}^2)$ by the Veronese embedding (Example 1.3, Chapter 1). The space $V=(S^d\mathbf{C}^2)^*$ is the space of binary forms

$$f(x_0,x_1)=a_0x_0^d+a_1x_0^{d-1}x_1+\cdots+a_dx_1^d$$

of degree d, and $\Delta_X(f)$ is the classical discriminant of f.

The line bundle $\mathcal{L}$, which is the restriction to $X = P^1$ of $\mathcal{O}_{P^d}(1)$ is, in terms of the projective line P^1 itself, $\mathcal{O}_{P^1}(d)$. According to Proposition 1.2, the vector bundle $J(\mathcal{L})$ is isomorphic to $\mathcal{O}_{P^1}(d-1) \oplus \mathcal{O}_{P^1}(d-1)$. We take the twisting bundle $\mathcal{M}$ to be $\mathcal{O}_{P^1}(2d-3)$.

Recall the standard facts about the cohomology of sheaves $\mathcal{O}(l)$ on projective spaces [Hart].

Theorem 2.12. *Let V be a vector space of dimension k. If $l \geq 0$ then*

$$H^0(P(V^*), \mathcal{O}(l)) = S^l V$$

and all the other cohomology of $\mathcal{O}(l)$ vanish. If $l \leq -k$, then

$$H^{k-1}(P(V^*), \mathcal{O}(l)) = S^{-k-l}V^* \otimes \bigwedge^k V$$

and all the other cohomology of $\mathcal{O}(l)$ vanish. If $1-k \leq l \leq -1$ then all the cohomology of $\mathcal{O}(l)$ vanish.

Applying this theorem to our situation, we find that the discriminantal complex $C_-^\bullet(X, \mathcal{M})$ is

$$0 \to H^0(\mathcal{O}(d-2) \oplus \mathcal{O}(d-2)) \xrightarrow{\partial_f} H^0(\mathcal{O}(2d-3)) \tag{2.8}$$

(the left term will vanish since the sheaf $\mathcal{O}(-1)$ on P^1 has no sections). We also see that (2.8) is stably twisted. Hence Δ_X equals, up to a constant factor, to the determinant of this complex, i.e., to the determinant of the square matrix ∂_f. Since $H^0(P^1, \mathcal{O}(m))$ is the space $S^m\mathbf{C}^2$ of binary forms of degree m, we can write ∂_f in (2.8) as an operator

$$S^{d-2}\mathbf{C}^2 \oplus S^{d-2}\mathbf{C}^2 \xrightarrow{\partial_f} S^{2d-3}\mathbf{C}^2.$$

An easy calculation shows that ∂_f is given by

$$\partial_f(u, v) = \left(\frac{\partial f}{\partial x_0}\right) \cdot u + \left(\frac{\partial f}{\partial x_1}\right) \cdot v.$$

Thus Theorem 2.5 amounts in our case to the classical *Sylvester formula*

$$\Delta(a_0x_0^d + \cdots + a_dx_1^d)$$
$$= \frac{(-1)^{d-1}}{d^{d-2}} \cdot \begin{vmatrix} a_1 & 2a_2 & \ldots & (d-1)a_{d-1} & da_d & \ldots & 0 \\ 0 & a_1 & \ldots & (d-2)a_{d-2} & (d-1)a_{d-1} & \ldots & 0 \\ \vdots & \vdots & \ddots & \vdots & \vdots & \ddots & \vdots \\ 0 & 0 & \ldots & a_1 & 2a_2 & \ldots & da_d \\ da_0 & (d-1)a_1 & \ldots & 2a_{d-2} & a_{d-1} & \ldots & 0 \\ 0 & da_0 & \ldots & 3a_{d-3} & 2a_{d-2} & \ldots & 0 \\ \vdots & \vdots & \ddots & \vdots & \vdots & \ddots & \vdots \\ 0 & 0 & \ldots & da_0 & (d-1)a_1 & \ldots & a_{d-1} \end{vmatrix}. \tag{2.9}$$

The numerical constant $(-1)^{d-1}/d^{d-2}$ here is, strictly speaking, irrelevant since we consider the discriminant up to a non-zero scalar factor. It becomes relevant if we use the standard normalization of the classical discriminant, see Chapter 12.

Example 2.13. The dual of an algebraic curve. Let $X \subset P(V^*)$ be a smooth algebraic curve of degree d and genus g. Thus the degree (or the first Chern class) of the line bundle $\mathcal{L} = \mathcal{O}_X(1)$ on X is equal to d. The bundle $J(\mathcal{L})$ has rank 2 and, by the exact sequence (1.5), the line bundle $\bigwedge^2 J(\mathcal{L}) = \Omega^1_X \otimes \mathcal{L}^{\otimes 2}$ has degree $2d + 2g - 2$ (we have taken into account that $\deg \Omega^1_X = 2g - 2$). Let us take the twisting bundle $\mathcal{M}$ to be a generic line bundle on X of degree $2d + g - 3$. Then $C^\bullet_-(X, \mathcal{M})$ has the form

$$H^0\left(X, \bigwedge^2 J(\mathcal{L})^* \otimes \mathcal{M}\right) \to H^0(X, J(\mathcal{L})^* \otimes \mathcal{M}) \xrightarrow{\partial_f} H^0(X, \mathcal{M}). \qquad (2.10)$$

Recall the classical Riemann-Roch theorem for curves together with a couple of easy consequences of this theorem ([GH], Chapter 2, Section 3).

Theorem 2.14. *Let X be a smooth irreducible projective curve of genus g. Then*

(a) *for any line bundle $\mathcal{N}$ on X, we have*

$$\dim H^0(X, \mathcal{N}) - \dim H^1(X, \mathcal{N}) = \deg \mathcal{N} + 1 - g;$$

(b) *a generic line bundle on X of degree $g - 1$ has $H^0 = H^1 = 0$;*

(c) *any line bundle $\mathcal{N}$ on X with* $\deg \mathcal{N} > 2g - 2$ *has $H^1(X, \mathcal{N}) = 0$.*

By statement (b) above, the far left term of (2.10) is equal to zero since $\bigwedge^2 J(\mathcal{L})^* \otimes \mathcal{M}$ is a generic line bundle of degree $g - 1$. Moreover, it follows from (c) that $C^\bullet_-(X, \mathcal{M})$ is in this case stably twisted. Hence the X-discriminant $\Delta_X(f)$, $f \in V$, is the determinant of the square matrix ∂_f. The size of the matrix ∂_f, i.e., the dimension of $H^0(X, \mathcal{M})$ equals, by the Riemann-Roch theorem, $2d - 2 + g$. Thus $X^\vee$ is a hypersurface of degree $m = 2d - 2 + g$. Note that the equation of this hypersurface is the determinant of an $m \times m$ matrix of linear forms on V given by ∂_f. This implies the following geometric property of $X^\vee$. Let $\mathbf{C}^{m\times m}$ be the space of all $m \times m$ matrices and $\nabla_m \subset P(\mathbf{C}^{m\times m})$ be the projectivization of the space of matrices with determinant zero. Then $X^\vee$ is isomorphic to a plane section of ∇_m.

3. The degree and the dimension of the dual

Let $X \subset P^n$ be a smooth subvariety. In this section we address the question of how to find the codimension and the degree of the dual variety $X^\vee$ in terms of the invariants of X.

A. The degree of the X-discriminant

We introduce $d^\vee$ to be equal to deg $X^\vee$ in case codim $X^\vee = 1$ and to 0 in case codim $X^\vee > 1$. Obviously, $d^\vee$ is just the degree of the X-discriminant Δ_X (recall that we have set $\Delta_X = 1$ in case codim $X^\vee > 1$).

Theorem 2.5 identifies Δ_X with the determinant of any of the discriminantal complexes $C^\bullet_\pm(X, \mathcal{M})$, provided the complex is stably twisted. This implies a formula for $d^\vee = \deg \Delta_X$.

Theorem 3.1. (a) *If a line bundle $\mathcal{M}$ on X is such that $C^\bullet_-(X, \mathcal{M})$ is stably twisted, then*

$$d^\vee = \sum_{i=-\dim(X)-1}^{0} (-1)^i \cdot i \cdot \dim C^i_-(X, \mathcal{M}).$$

(b) *If $\mathcal{M}$ is such that $C^\bullet_+(X, \mathcal{M})$ is stably twisted, then*

$$d^\vee = (-1)^{\dim(X)+1} \sum_{i=0}^{\dim(X)+1} (-1)^i \cdot i \cdot \dim C^i_+(X, \mathcal{M}).$$

Proof. This follows from Theorem 2.5 and the homogeneity property of the determinant of any based exact complex $(C^\bullet, \partial, e)$:

$$\det(C^\bullet, \lambda\partial, e) = \lambda^r \det(C^\bullet, \partial, e), \quad r = \sum_i (-1)^i \cdot i \cdot \dim C^i, \tag{3.1}$$

(see Appendix A, Corollary 15).

Thus $d^\vee$ is expressed through the dimensions of vector spaces

$$C^i_-(X, \mathcal{M}) = H^0\left(X, \bigwedge^{-i} J(\mathcal{O}_X(1))^* \otimes \mathcal{M}\right),$$

$$C^i_+(X, \mathcal{M}) = H^0\left(X, \bigwedge^{i} J(\mathcal{O}_X(1)) \otimes \mathcal{M}\right).$$

These dimensions (and hence $d^\vee$) can be expressed through simpler quantities associated with X. We shall do this assuming $\mathcal{M} = \mathcal{O}_X(l)$, where $l \gg 0$.

For any coherent sheaf $\mathcal{F}$ on X, we write

$$\mathcal{F}(l) = \mathcal{F} \otimes \mathcal{O}_X(l), \quad h^i(\mathcal{F}) = \dim H^i(X, \mathcal{F}), \quad \chi(\mathcal{F}) = \sum (-1)^i h^i(\mathcal{F}). \tag{3.2}$$

The number $\chi(\mathcal{F})$ is called the *Euler characteristic* of $\mathcal{F}$.

It follows from the Riemann-Roch-Hirzebruch theorem [Hir] that $\chi(\mathcal{F}(l))$ is a polynomial in l. This polynomial is called the *Hilbert polynomial* of $\mathcal{F}$ and denoted by $h_{\mathcal{F}}(l)$. For $l \gg 0$, the higher cohomology of $\mathcal{F}(l)$ vanishes, so we have

$$h_{\mathcal{F}}(l) = \chi(\mathcal{F}(l)) = h^0(\mathcal{F}(l)). \tag{3.3}$$

This is the more familiar definition of Hilbert polynomials, see [Hart], Chapter 1, Theorem 7.5, and also [Ser].

The bundle $\bigwedge^i J(\mathcal{O}_X(1))$ is included in the exact sequence

$$0 \to \Omega_X^i(i) \to \bigwedge^i J(\mathcal{O}_X(1)) \to \Omega_X^{i-1}(i) \to 0, \tag{3.4}$$

which is obtained from (1.5) (with $\mathcal{L} = \mathcal{O}_X(1)$) by taking the exterior power. This implies that, for $l \gg 0$, we have

$$\dim\, C_+^i(X, \mathcal{O}_X(l)) = h^0(\Omega_X^i(i+l)) + h^0(X, \Omega_X^{i-1}(i+l)). \tag{3.5}$$

Let T_X be the tangent bundle of X. By considering the sequence dual to (3.4) and tensoring with $\mathcal{O}_X(l)$ we get

$$\dim\, C_-^i(X, \mathcal{O}(l)) = h^0\left(\bigwedge^{-i} T_X(i+l)\right) + h^0\left(\bigwedge^{-i-1} T_X(i+l)\right). \tag{3.6}$$

We arrive at the following.

Theorem 3.2. *The number* $d^\vee = \deg\, \Delta_X$ *is expressed through Hilbert polynomials of* Ω_X^i *or* $\bigwedge^i T_X$ *as follows: for any* $l \in Z$ *we have*

$$d^\vee = (-1)^{\dim(X)+1} \sum_{i=0}^{\dim(X)+1} (-1)^i \cdot i \cdot \left(h_{\Omega_X^i}(i+l) + h_{\Omega_X^{i-1}}(i+l)\right)$$

$$= \sum_{i=-\dim(X)-1}^{0} (-1)^i \cdot i \cdot \left(h_{\bigwedge^{-i} T_X}(i+l) + h_{\bigwedge^{-i-1} T_X}(i+l)\right). \tag{3.7}$$

Proof. For $l \gg 0$, this follows from Theorem 3.1 and from the equalities (3.5), (3.6) and (3.3). The extension to arbitrary $l \in \mathbf{Z}$ is obtained if we note that any of the sums in (3.7) is a polynomial in l which takes the value $d^\vee$ for any $l \gg 0$, and hence this polynomial is identically equal to $d^\vee$.

B. General formula for codim $X^\vee$ *and* deg $X^\vee$ *in terms of Hilbert polynomials*

For $0 \le i \le \dim X$, we introduce the polynomial

$$p_i(l) = h_{\bigwedge^i T_X}(l). \tag{3.8}$$

For any $l \in \mathbf{Z}$, we introduce the polynomial $f_l(q)$ in a formal variable q by

$$f_l(q) = \sum_{i=0}^{\dim(X)+1} (-1)^i \big(p_i(i+l) + p_{i-1}(i+l)\big) q^i, \tag{3.9}$$

where p_{-1} and $p_{\dim(X)+1}$ are understood to be 0. Clearly $f_l(q)$ is also a polynomial in l.

Note that the representation of $d^\vee$ as the second sum in (3.7) implies that $d^\vee = -f_l'(1)$, minus the derivative of $f_l(q)$ at $q = 1$. Note also that $f_l(1) = 0$ since this is a polynomial in l which for $l \gg 0$ is equal to the Euler characteristic of the generically exact discriminantal complex. Thus, near $q = 1$, we have $f_l(q) = d^\vee(1-q) + O((1-q)^2)$. In fact, the following more general result is true.

Theorem 3.3. *Let $l \in \mathbf{Z}_+$ be any non-negative integer. The codimension of $X^\vee$ equals the order of the zero of $f_l(q)$ at $q = 1$. Let this order be μ and let $f_l(q) = a_\mu(l)(1-q)^\mu + O((1-q)^{\mu+1})$. Then $a_\mu(l) = \deg(X^\vee) \cdot \binom{\mu+l-1}{\mu-1}$.*

Note that for a fixed μ both $a_\mu(l) = (-1)^\mu f_l^{(\mu)}(1)/\mu!$ and the binomial coefficient $\binom{\mu+l-1}{\mu-1}$ depend on l polynomially. So we can extrapolate the equality in Theorem 3.3 for some negative l as well. Namely, if $\mu = \operatorname{codim} X^\vee$ is as above, then for any $l \notin \{-1, -2, \ldots, -\mu+1\}$, we have

$$\deg X^\vee = \frac{(-1)^\mu f_l^{(\mu)}(1)}{\mu \cdot (l+1)(l+2)\ldots(l+\mu-1)} \tag{3.10}$$

and for $l \in \{-1, -2, \ldots, -\mu+1\}$, we have $a_\mu(l) = 0$ so $f_l(q) = O((1-q)^{\mu+1})$.

C. Formula for codim $X^\vee$ *and* deg $X^\vee$ *in terms of Chern classes (Katz-Kleiman-Holme)*

Before giving the proof of Theorem 3.3, we present another formula for the degree of $X^\vee$ due to N. Katz and S. Kleiman [Kat][Kl] in the case where $\operatorname{codim} X^\vee = 1$ and to A. Holme [Hol 1-2] in the general case. We assume that $X \subset P^n$ is a smooth irreducible projective variety of dimension m.

Let $c_i(\Omega^1_X) \in H^{2i}(X, Z)$ be the i-th Chern class of the vector bundle Ω^1_X. Consider the *Chern polynomial* of X with respect to the given projective embedding:

$$c_X(q) = \sum_{i=0}^{m} q^{i+1} \int_{X \cap P^{n-i}} c_{m-i}(\Omega^1_X). \tag{3.11}$$

Here P^{n-i} is any projective subspace in P^n of dimension $n-i$. The "integral" (i.e., the value of the cohomology class on a cycle) does not depend on the choice of P^{n-i}. Note that

$$\int_{X \cap P^{n-i}} c_{m-i}(\Omega^1_X) = \int_X c_{m-i}(\Omega^1_X) c_1(\mathcal{O}_X(1))^i. \tag{3.12}$$

Theorem 3.4. *Let $X \subset P^n$ be a smooth projective variety. Then the codimension of $X^\vee$ equals the order of the zero at $q = 1$ of the polynomial $c_X(q) - c_X(1)$. If this order is μ then* $\deg(X^\vee) = c_X^{(\mu)}(1)/\mu!$.

Example 3.5. Let X be the product of projective spaces $P^1 \times P^3$ embedded into $P^7 = P(\mathbf{C}^2 \otimes \mathbf{C}^4)$ by the Segre embedding. In other words, X is the projectivization of the space of 2×4 matrices of rank 1.

The dual variety $X^\vee$ is again $P^1 \times P^3$ (see Example 4.13 of Chapter 1) and hence has codimension 3. The cohomology ring of any projective space P^l is $Z[t]/t^{l+1}$, the element t being the class of a hyperplane. The total Chern class $\sum c_i(\Omega^1_{P^l})$ equals, as is well-known (see, e.g., [MS], Section 14)

$$\sum c_i(\Omega^1_{P^l}) = (1-t)^{l+1} = 1 - (l+1)t + \binom{l+1}{2} t^2 - \cdots + (-1)^l (l+1) t^l$$

(the term t^{l+1} is dropped since it equals 0 in the cohomology ring). For our variety X, we obtain by the Künneth formula, $H^\bullet(X) = Z[s,t]/(s^2, t^4)$. The fundamental class of X is st^3, and the total Chern class of Ω^1_X is $(1-2s)(1-4t+6t^2-4t^3)$. The first Chern class of the restriction of $\mathcal{O}_{P^7}(1)$ to X is $s+t$. The number $m = \dim(X)$ is equal to 4. Thus $\int_X c_{m-i}(\Omega^1_X) c_1(\mathcal{O}_X))^i$ is equal to the coefficient of st^3 in $(1-2s)(1-4t+6t^2-4t^3)(s+t)^i$. So, by using (3.12), we find the Chern polynomial (3.11) to be

$$c_X(q) = 8q - 16q^2 + 20q^3 - 14q^4 + 4q^5.$$

We have

$$c_X(1) = 4, \quad c'_X(1) = c''_X(1) = 0, \quad c'''_X(1) = 24.$$

This gives $\operatorname{codim}(X^\vee) = 3$, $\deg(X^\vee) = 24/3! = 4$.

D. Proof of Theorem 3.3

We shall prove this theorem by calculations in the *Grothendieck ring* $K(P(V))$ of coherent sheaves on P^n. We recall the main properties of these rings referring to [BGI] [Man] for a detailed exposition.

By definition, for any smooth projective variety Z, the Abelian group $K(Z)$ is generated by the symbols $[\mathcal{F}]$ for coherent sheaves $\mathcal{F}$ on Z; whenever we have a short exact sequence

$$0 \to \mathcal{F} \to \mathcal{G} \to \mathcal{H} \to 0, \tag{3.13}$$

we impose the relation $[\mathcal{G}] = [\mathcal{F}] + [\mathcal{H}]$. The element $[\mathcal{F}] \in K(Z)$ is called the *class* of $\mathcal{F}$. For any finite complex of coherent sheaves $\mathcal{F}^\bullet$, we define its class

$$[\mathcal{F}^\bullet] = \sum (-1)^i [\mathcal{F}^i] \in K(Z). \tag{3.14}$$

It follows that the class of an exact complex equals 0, and hence quasi-isomorphic complexes have the same class. By the Hilbert syzygy theorem ([GH], Chapter 5, Section 4), each coherent sheaf $\mathcal{F}$ on Z has a finite resolution $\mathcal{P}^\bullet$ by locally free sheaves. This permits us to extend to $K(Z)$ several available (and well-behaved) constructions for locally free sheaves. For instance, the ring structure on $K(Z)$ is first defined on classes of locally free sheaves using tensor multiplication. By Hilbert's theorem these classes generate $K(Z)$, so multiplication is extended to the whole $K(Z)$. The function taking a vector bundle to its rank extends to a homomorphism $\mathrm{rk} : K(Z) \to \mathbf{Z}$ called the generic rank. The Euler characteristic χ defines another homomorphism from $K(Z)$ to $\mathbf{Z}$.

The group $K(Z)$ is equipped with the so-called *codimension filtration* $K(Z) = F^0 \supset F^1 \supset \cdots$ where F^i is generated by classes of sheaves with support on a subvariety of codimension $\geq i$. The quotient F^i/F^{i+1} is the *Chow group* of codimension i cycles on Z modulo rational equivalence. More precisely, if $\mathcal{F}$ is a sheaf with support of codimension i, we associate to it the cycle

$$\sum_{\operatorname{codim} C = i} \operatorname{mult}_C(\mathcal{F})\, C,$$

where C runs over all irreducible subvarieties of codimension i and mult denotes the multiplicity of $\mathcal{F}$ along C (see Appendix A).

Let us recall the structure of the Grothendieck ring of a projective space.

Proposition 3.6. *Denote by $q \in K(P^n)$ the class of the invertible sheaf $\mathcal{O}(-1)$. This class generates the ring $K(P(V))$ which is isomorphic to $\mathbf{Z}[q]/(1-q)^n$. The filtration on $K(P^n)$ by the codimension of the support coincides with the filtration by powers of the ideal $(1-q)$.*

Proposition 3.7. *Let $M \subset P^n$ be an irreducible algebraic variety of codimension μ and degree d. Let $\mathcal{F}$ be a coherent sheaf on M of generic rank r. Then the class of $\mathcal{F}$ in $K(P(V))$ has the form $d \cdot r \cdot (1-q)^\mu + O((1-q)^{\mu+1})$.*

Proof. The quotients of the filtration by codimension of the support are the Chow groups of P^n. The correspondence $Z \mapsto \deg(Z)$ establishes an isomorphism between the Chow group of codimension μ cycles in P^n and $\mathbf{Z}$. The class of a sheaf $\mathcal{F}$ in $K(P^n)$ lies in F^μ; its multiplicity along M is r and along any other subvariety is 0. Therefore the image of $\mathcal{F}$ in $F^\mu/F^{\mu+1}$ is r times the image of $\mathcal{O}_M$. To find the image of $\mathcal{O}_M$, note that the class of M in the Chow group equals d times the class of a projective subspace $P^{n-\mu}$ in this group. Such a subspace is given by the vanishing of μ linear functionals $f_1, \ldots, f_\mu$. These functions can be regarded as sections of $\mathcal{O}_{P^n}(1)$ and together form a section $s = (f_1, \ldots, f_\mu)$ of the sheaf $\mathcal{O}_{P^n}(1)^{\oplus\mu}$. Thus the structure sheaf $\mathcal{O}_{P^{n-\mu}}$ (regarded as a sheaf on the whole P^n) has the Koszul resolution

$$\mathcal{K}_-\left(\mathcal{O}_{P^n}(1)^{\oplus\mu}, s\right) = \left\{\cdots \to \mathcal{O}_{P^n}(-2)^{\oplus\binom{\mu}{2}} \to \mathcal{O}_{P^n}(-1)^{\oplus\mu} \to \mathcal{O}_{P^n}\right\}.$$

The class of this resolution in $K(P^n)$ is $(1-q)^\mu$. Proposition 3.7 is proved.

Now we finish the proof of Theorem 3.3. We assume that the ambient space P^n containing X is $P(V^*)$, and therefore $X^\vee$ lies in $P(V)$. For $l \in Z$, we consider the complex $\mathcal{O}(C_-^\bullet(X, \mathcal{O}_X(l))$ of sheaves on $P(V)$ defined by (2.5). Suppose $l \gg 0$. According to Proposition 3.7, the class in $K(P(V))$ of this complex equals the image in $\mathbf{Z}[q]/(1-q)^n$ of the polynomial $\varphi_l(q) = \sum(-1)^i \dim C_-^{-i}(X, \mathcal{O}(l))q^i$. By (3.6), this polynomial equals $f_l(q)$.

We consider the incidence variety $W \subset X \times P(V)$ and its projections p_1, p_2 to X and $P(V)$, as in the proof of Theorem 2.5. Then, by Proposition 2.9, the complex $\mathcal{O}(C_-^\bullet(X, \mathcal{O}_X(l))$ is quasi-isomorphic to the direct image in the derived category of the line bundle $p_1^*\mathcal{O}(l)$ on W. The projection $p_2 : W \to X^\vee$ is generically a bundle whose fibers are projective spaces of dimension $\mu - 1$, where $\mu = \operatorname{codim} X^\vee$. Thus the generic rank of $Rp_{2*}(p_1^*\mathcal{O}(l))$ for $l \gg 0$ equals

$$\dim H^0(P^{\mu-1}, \mathcal{O}(l)) = \dim S^l(\mathbf{C}^\mu) = \binom{\mu+l-1}{\mu-1}.$$

Now Theorem 3.3 follows from Propositions 3.6 and 3.7.

E. Proof of Theorem 3.4

Along with Chern classes we consider the exterior power operations $\bigwedge^i : K(X) \to K(X)$. Their definition, due to Grothendieck, is similar to that of the c_i. Namely, for classes of vector bundles, $\bigwedge^i$ is the usual exterior power and for an exact sequence of vector bundles of type (3.13), we have

$$\left[\bigwedge^p \mathcal{G}\right] = \sum_{i+j=p} \left[\bigwedge^i \mathcal{F}\right] \cdot \left[\bigwedge^j \mathcal{H}\right], \tag{3.15}$$

which follows by considering a suitable filtration in $\bigwedge^p \mathcal{G}$. Introduce, for a vector bundle $\mathcal{F}$, the formal series

$$\lambda_{\mathcal{F}}(t) = \sum_i \bigwedge^i (\mathcal{F}) t^i \in K(X)[[t]]. \tag{3.16}$$

Then we define, for any $\alpha = \sum m_i [\mathcal{F}_i] \in K(X)$, the element $\bigwedge^i(\alpha)$ as the i-th coefficient of $\prod \lambda_{\mathcal{F}_i}(t)^{m_i}$.

As we mentioned earlier, the r-th quotient $gr^r_F K(X)$ of the codimension filtration F on $K(X)$ is the Chow group of codimension r algebraic cycles on X modulo rational equivalence. Denote this group by $CH^r(X)$. There is a natural homomorphism $\mathrm{cl} : CH^r(X) \to H_{2(\dim(X)-r)}(X, \mathbf{Z})$ which takes an algebraic cycle into its homology class. Note that the Poincaré duality on X identifies $H_{2(\dim(X)-r)}(X, \mathbf{Z}) \cong H^{2r}(X, \mathbf{Z})$. Grothendieck has defined Chern classes

$$c_r^{\mathrm{Groth}} : K(X) \longrightarrow CH^r(X) = \mathrm{gr}^r_F K(X)$$

which lift the ordinary Chern classes with values in $H^{2r}(X, \mathbf{Z})$. By definition (cf. [BGI] [Man]), if $\alpha \in K(X)$ is an element of generic rank p, then

$$c_r^{\mathrm{Groth}}(\alpha) = \bigwedge^r (\alpha - p + r - 1) \mod F^{r+1} K(X). \tag{3.17}$$

Here integers are embedded into $K(X)$ as multiples of $[\mathcal{O}_X]$ which is the unit with respect to the ring structure on $K(X)$. The exterior power in (3.17) belongs to $F^r K(X)$. In fact, definitions in [BGI], [Man] give a seemingly different formula for c_r^{Groth} involving the so-called γ-operations. However, this formula is equivalent to (3.17), as can be seen from elementary algebraic manipulations.

Intuitively, (3.17) is very transparent. Namely, the element $\alpha - p + r$ has generic rank r. Suppose for a moment that it is represented as the class of a vector bundle E on X of rank r which has enough global sections. The r-th Chern class of the rank r bundle E is therefore represented by a cycle which is the zero locus of a generic section s of E. Denoting this locus by Z, we form the Koszul complex $\mathcal{K}_+(E, s)$ (see Section 1B) which is a left resolution of the line bundle $\bigwedge^r E|_Z$, regarded as a sheaf on the whole X. The class in $K(X)$ of $\mathcal{K}_+(E, s)$ is

$$\sum (-1)^i \left[\bigwedge^i E\right] = (-1)^r \bigwedge^r ([E] - 1)] = (-1)^r \bigwedge^r (\alpha - p + r - 1).$$

But this class is the same as $(-1)^r$ times the class of a line bundle on Z (the sign comes from the fact that the cohomology of $\mathcal{K}_+(E, s)$ is of degree r), so $\bigwedge^r(\alpha - p + r - 1)$ lies in $F^r K(X)$ and represents Z in $\mathrm{gr}_F^r K(X)$.

Note that, for $r = m = \dim(X)$, the homomorphism $\mathrm{cl} : \mathrm{gr}_F^m K(X) \to H_0(X, \mathbf{Z}) = \mathbf{Z}$ is obtained by taking the Euler characteristic. So we get the following lemma.

Lemma 3.8. *Let* $\dim(X) = m$ *and* $\alpha \in K(X)$ *be an element of generic rank* p. *Then*

$$\int_X c_m(\alpha) = \chi\left(\bigwedge^m(\alpha - p + m - 1)\right).$$

This lemma is all that we need to deduce Theorem 3.4 from Theorem 3.3. Let us write $\mathcal{L} = \mathcal{O}_X(1)$, $\xi = c_1(\mathcal{L}) \in H^2(X, \mathbf{Z})$. Denote also $\Omega = \Omega_X^1$, and let $c(\Omega) = \sum c_i(\Omega) \in H^\bullet(X, \mathbf{Z}) = \bigoplus H^j(X, \mathbf{Z})$ be the total Chern class of Ω. We can write

$$\frac{c_X^{(k)}(1)}{k!} = \int_X \left(\frac{c(\Omega)}{(1-\xi)^{k+1}}\right)_{2m}, \tag{3.18}$$

where the fraction is supposed to be expanded as a power series in ξ and the subscript $2m$ means the homogeneous part of degree $2m$ of a non-homogeneous element of $H^\bullet(X, \mathbf{Z})$. More precisely, to see (3.18), we express both sides as linear combinations of the numbers $\int_X c_{m-i}(\Omega)\xi^i$ and find that the coefficients in these combinations are the same.

Note that $(1-\xi)^{k+1}$ is the total Chern class of the bundle $(k+1)\mathcal{L}^*$, the direct sum of $k+1$ copies of $\mathcal{L}^*$. By the Whitney sum formula for the total Chern class (which is applicable to differences as well as sums of elements of $K(X)$), we have

$$\frac{c(\Omega)}{(1-\xi)^{k+1}} = c\big([\Omega] - (k+1)[\mathcal{L}^*]\big).$$

Applying Lemma 3.8, we write

$$\begin{aligned} \frac{c_X^{(k)}(1)}{k!} &= \int_X \left(\frac{c(\Omega)}{(1-\xi)^{k+1}}\right)_{2m} \\ &= \int_X c_m\big([\Omega] - (k+1)[\mathcal{L}^*]\big) = \chi\left(\bigwedge^m\big([\Omega] - (k+1)[\mathcal{L}^*] + k\big)\right). \end{aligned} \tag{3.19}$$

We can transform the right hand side of (3.19) in the following way. First, we dualize all three summands under the sign of exterior power. The passage from a vector bundle to its dual extends to a homomorphism $* : K(X) \to K(X)$.

For not necessarily locally free sheaves this homomorphism is given by $[\mathcal{F}] \mapsto \sum_i (-1)^i \left[\underline{\mathrm{Ext}}^i_{\mathcal{O}_X}(\mathcal{F}, \mathcal{O}_X)\right]$. It also commutes with exterior powers. Now, the exterior power in (3.19) lies in the lowest term of the codimension filtration, i.e., it can be represented as a linear combination of classes $[\mathcal{F}]$ for sheaves $\mathcal{F}$ with 0-dimensional support (or, even, structure sheaves of individual points in X). For such a sheaf $\mathcal{F}$, we have $\underline{\mathrm{Ext}}^i(\mathcal{F}, \mathcal{O}_X) = 0$ for $i \neq m$ and $\underline{\mathrm{Ext}}^m$ has the same class as $\mathcal{F}$, so $*([\mathcal{F}]) = (-1)^m[\mathcal{F}]$. Thus

$$\chi\left(\bigwedge^m([\Omega] - (k+1)[\mathcal{L}^*] + k)\right) = (-1)^m \chi\left(\bigwedge^m([T_X] - (k+1)[\mathcal{L}] + k)\right). \tag{3.20}$$

Note that we can tensor every summand in the right hand side of (3.20) with $\mathcal{L}^*$ without changing the answer. Indeed, denoting by $\mathcal{G}$ the virtual bundle under the sign $\bigwedge^m$, we have $\bigwedge^m(\mathcal{G} \otimes \mathcal{L}^*) = \bigwedge^m(\mathcal{G}) \otimes (\mathcal{L}^*)^{\otimes m}$, so our procedure amounts to tensoring the whole exterior power with a line bundle. However, for any sheaf $\mathcal{F}$ with 0-dimensional support and any line bundle $\mathcal{M}$, we have $\chi(\mathcal{F} \otimes \mathcal{M}) = \chi(\mathcal{F})$, and our exterior power is a linear combination of classes of such sheaves. Thus we can write

$$\frac{c_X^{(k)}(1)}{k!} = (-1)^m \chi\left(\bigwedge^m([T_X \otimes \mathcal{L}^*] + k[\mathcal{L}^*] - (k+1))\right). \tag{3.21}$$

We now expand (3.21) using the analog of (3.15) for three summands. Note that, for any $c \geq 0$, we have $\bigwedge^c(-(k+1)) = (-1)^c \binom{k+c}{c}$. One way of seeing this is to remark that the formal series $\sum_c (-1)^c \binom{k+c}{c} t^c = 1/(1+t)^{k+1}$ is inverse to $\sum_c \bigwedge^c(+(k+1))t^c = \sum_c \binom{k+1}{c} t^c = (1+t)^{k+1}$. Therefore we can finally write

$$\frac{c_X^{(k)}(1)}{k!} = (-1)^m \sum_{\substack{a+b+c=m \\ a,b,c \geq 0}} (-1)^c \binom{k}{b} \binom{k+c}{c} \chi(T^a \otimes \mathcal{L}^{-a-b}), \tag{3.22}$$

where we have denoted $T^p = \bigwedge^p T_X$. We have thus expressed $c_X^{(k)}(1)/k!$ through the quantities $\chi(T^r \otimes \mathcal{L}^s) = p_r(s)$ used in the definition of the polynomials $f_l(q)$, see (3.9).

Lemma 3.9. *We have*

$$\frac{c_X^{(k)}(1)}{k!} = \sum_{0 \leq i,j \leq k} (-1)^{i+j} \binom{k-1}{i} \binom{k+m-i-j}{m-i} \frac{f_{-i}^{(j)}(1)}{j!}. \tag{3.23}$$

To prove the lemma, we express both sides of (3.23) as linear combinations of the numbers $\chi(T^r \otimes \mathcal{L}^s)$ and compare the coefficients. It turns out that the corresponding coefficients are the same. This comparison is straightforward and amounts to applying the identity

$$\binom{p}{q} = \sum_{j=0}^{r}(-1)^r\binom{r}{j}\binom{p+r-j}{q+r}$$

holding for any $p, q, r \geq 0$. We leave the details to the reader.

Now Theorem 3.4 follows from Theorem 3.3 and Lemma 3.9 by purely formal reasoning. Indeed, let $\mu = \operatorname{codim} X^\vee$. Then, by Theorem 3.3, for any $l \in Z$, we have $f_l^{(j)}(1) = 0$ for $j \leq \mu - 1$ and all l as well as for $j = \mu$ and $l = -1, -2, \ldots, -\mu + 1$. Therefore $c_X^{(k)}(1) = 0$ for $1 \leq k \leq \mu - 1$, since all the summands in the right hand side of (3.23) vanish in this case. Similarly, for $k = \mu$, all summands vanish except for the one with $i = 0$, $j = \mu$ which equals $(-1)^\mu f_0^{(\mu)}(1)/\mu!$. By Theorem 3.3, this number equals $\deg X^\vee$ and we are done.

4. Discriminantal complexes in terms of differential forms

In Section 2 we introduced, for every smooth projective variety $X \subset P(V^*)$ and every invertible sheaf $\mathcal{M}$ on X, two complexes of vector spaces $C^\bullet_\pm(X, \mathcal{M})$. In this section we study one of these complexes, C_+, in the special case when $\mathcal{M}$ has the form $\mathcal{O}_X(l), l \geq 1$. We abbreviate throughout this section $C^\bullet_+(X, \mathcal{O}(l))$ to $C^\bullet(X, l)$.

A. Forms on the cone and discriminantal complexes

Let $X \subset P(V^*)$ be a smooth projective variety. Let $Y \subset V^*$ be the cone over X from which the singular point 0 is deleted. Also, let $\pi : Y \to X$ be the standard projection and ξ the Euler vector field on Y, defining the Lie derivation Lie_ξ on the sheaves Ω^k_Y. For any $l \in \mathbf{Z}$, let us define a subsheaf $\Omega^k_{l,Y} \subset \Omega^k_Y$ whose sections are k-forms ω on Y such that $\mathrm{Lie}_\xi(\omega) = l \cdot \omega$, i.e., ω is homogeneous of degree l under homotheties.

Proposition 4.1. *There is an isomorphism of sheaves on X*

$$\bigwedge^k J(\mathcal{O}(1)) \cong \pi_*\Omega^k_{k,Y}.$$

In other words, sections of $\bigwedge^k J(\mathcal{O}(1))$ over any open set $U \subset X$ are naturally identified with differential k-forms on $\pi^{-1}(U) \subset Y$ homogeneous of degree k.

Proof. This follows from Proposition 1.3 by taking the exterior power.

Corollary 4.2. *For any $k \geq 0$, the term $C^k(X, l)$ of the discriminantal complex is identified with the space of k-forms ω on Y such that* $\mathrm{Lie}_\xi \omega = (k + l)\omega$. *For any $f \in V$, the differential $\partial_f : C^k(X, l) \to C^{k+1}(X, l)$ corresponds, in this identification, to the exterior multiplication by $df \in \Omega^1(Y)$.*

We shall also need an interpretation of forms on X in terms of forms on Y. Let $\Omega^k_X(l)$ denote the sheaf $\Omega^k_X \otimes \mathcal{O}_X(l)$ on X. Let $i_\xi : \Omega^k_Y \to \Omega^{k-1}_Y$ denote the contraction with the Euler vector field ξ. We have the well-known *Cartan formula*

$$d \circ i_\xi + i_\xi \circ d = \mathrm{Lie}_\xi, \tag{4.1}$$

see, e.g., [Ar]. Let $\mathcal{R}_{k,l}$ be the subsheaf of Ω^k_Y consisting of forms ω such that

$$\mathrm{Lie}_\xi \omega = l \cdot \omega, \quad i_\xi \omega = 0. \tag{4.2}$$

Proposition 4.3. *For any k, l, the sheaf $\Omega^k_X(l)$ is naturally identified with $\pi_* \mathcal{R}_{k,l}$. These isomorphisms take the exterior multiplication*

$$\Omega^k_X(l) \otimes \Omega^{k'}_X(l') \to \Omega^{k+k'}_X(l + l') \tag{4.3}$$

into the usual exterior multiplication of forms on Y.

Proof. The statement obviously reduces to the case where $l = 0$. To establish this case, introduce the local coordinates $(x_0, \dots, x_n)$ on Y such that $(x_1, \dots, x_n)$ are coordinates on X, the projection π forgets x_0 and $\xi = \partial/\partial x_0$. The assertion means that a k-form ω on Y does not contain dx_0 and has coefficients independent of x_0 if and only if it is annihilated by $i_{\partial/\partial x_0}$ and $\mathrm{Lie}_{\partial/\partial x_0}$ which is obvious.

B. The second multiplication in the algebra of forms

We introduce on X the bigraded sheaf of algebras $\mathbf{B} = \bigoplus_{k,l \geq 0} \Omega^k_X(l)$. The multiplication is induced by the usual exterior multiplication of differential forms; hence it is associative and supercommutative with respect to the grading by the degree of differential forms (the supercommutativity means that $\omega \wedge \omega' = (-1)^{kk'} \omega' \wedge \omega$ for $\omega \in \Omega^k_X(l)$, $\omega' \in \Omega^{k'}_X(l')$). We want to introduce another multiplication $*$ in $\mathbf{B}$. Namely, let ω, ω' be two local sections of $\Omega^k_X(l)$ and let $\Omega^{k'}_X(l')$ be defined over some domain U. Following Proposition 4.3, let us view ω and ω' as differential forms on Y. Define

$$\omega * \omega' = \begin{cases} \frac{l}{l+l'} \omega \wedge d\omega' - (-1)^k \frac{l'}{l+l'} d\omega \wedge \omega', & \text{if } l, l' \neq 0 \\ 0, & \text{if } l = 0 \text{ or } l' = 0. \end{cases} \tag{4.4}$$

Proposition 4.4.
(a) *The form* $\gamma = \omega * \omega'$ *satisfies the conditions* $\mathrm{Lie}_\xi(\gamma) = (l+l')\gamma,\ i_\xi(\gamma) = 0$; *therefore* $*$ *defines a multiplication*

$$\Omega_X^k(l) \otimes_{\mathbf{C}} \Omega_X^{k'}(l') \to \Omega_X^{k+k'+1}(l+l'). \tag{4.5}$$

(b) *The multiplication* $*$ *is associative and supercommutative with respect to the shifted grading:*

$$\omega * \omega' = (-1)^{(k+1)(k'+1)}\omega' * \omega,\ \ \omega \in \Omega_X^k(l),\ \omega' \in \Omega_X^{k'}(l'). \tag{4.6}$$

Proof. (a) The Lie derivative is a (super) derivation of the algebra of forms which commutes with the exterior derivative d. Therefore $\mathrm{Lie}_\xi(\omega * \omega') = (l+l')\omega * \omega'$. By (4.1), for $l, l' \neq 0$ we have

$$i_\xi(\omega * \omega') = \frac{l}{l+l'} i_\xi\omega \wedge d\omega' + (-1)^k \frac{l}{l+l'} \omega \wedge i_\xi d\omega' -$$

$$-(-1)^k \frac{l'}{l+l'} i_\xi d\omega \wedge \omega' - (-1)^{2k+1} \frac{l}{l+l'} d\omega \wedge i_\xi \omega'.$$

Using (4.1) again and taking into account that ω and ω' are annihilated by i_ξ, we find the right hand side equal to

$$-(-1)^k \frac{l'l}{l+l'} \omega \wedge \omega' + (-1)^k \frac{ll'}{l+l'} \omega \wedge \omega' = 0.$$

(b) This is verified by straightforward checking similar to that in part (a).

Note that in (4.4) it is essential to divide by $l + l'$ to get associativity. The multiplication $*$ reminds us of the multiplication in Deligne-Beilinson cohomology (cf. [RSS], p. 62).

C. The decomposition of the discriminantal complex

We shall identify the discriminantal complex $C^\bullet(X, l)$ with the cone of a chain map between two smaller complexes. In fact, this identification can be established for the complex of sheaves for which the $C^i(X, l)$ are the spaces of global sections.

Let f be any element of the vector space V (recall that X is embedded in $P(V^*)$). We denote the image of f in $H^0(X, \mathcal{O}(1))$ by the same letter f. Since $\mathcal{O}(1)$ is a part of the algebra $\mathbf{B}$, we obtain two operators of multiplication by f:

$$\partial(f) : \Omega_X^k(l) \to \Omega_X^{k+1}(l+1), \quad \partial(f)\omega = (l+1)f * \omega = f d\omega - l df \wedge \omega,$$

$$\tau(f) : \Omega_X^k(l) \to \Omega_X^k(l+1), \quad \tau(f)\omega = f \cdot \omega.$$

It is clear from the supercommutativity of the two multiplications that, for any $f, g \in V$, we have

$$\partial(f)\partial(g) = -\partial(g)\partial(f), \quad \tau(f)\tau(g) = \tau(g)\tau(f).$$

The commutation law between $\partial(f)$ and $\tau(g)$ is given by the following proposition.

Proposition 4.5. *For $f, g \in V$, we have*

$$\partial(f)\tau(g) = \tau(g)\partial(f)(1 + \mathrm{Lie}_\xi^{-1}) - \tau(f)\partial(g)\mathrm{Lie}_\xi^{-1} \tag{4.7}$$

where the operator Lie_ξ *acts on* $\Omega_X^k(l) = \mathbf{B}_{kl}$ *by multiplication by* l. *In particular, we have*

$$\partial(f)\tau(f) = \tau(f)\partial(f). \tag{4.8}$$

Proof. Consider a local section of $\Omega_X^k(l)$. We view it as a form ω on Y satisfying the equations $\mathrm{Lie}_\xi\omega = l\omega,\ i_\xi\omega = 0$. We have

$$\partial(f)\tau(g)\omega = \partial(f)(g\omega) = fdg \wedge \omega + fgd\omega - (l+1)df \wedge g\omega;$$

$$\tau(g)\partial(f)\omega = g(fd\omega - ldf \wedge \omega) = fgd\omega - ldf \wedge g\omega;$$

$$\tau(f)\partial(g)\omega = fgd\omega - ldg \wedge f\omega.$$

Therefore

$$l\partial(f)\tau(g)\omega = (l+1)\tau(g)\partial(g)\omega - \tau(f)\partial(g)\omega,$$

which is equivalent to our assertion.

Consider, for a given $f \in V$, the following diagram of sheaves on X:

$$\begin{array}{ccccccc}
\mathcal{O}_X & \xrightarrow{0} & \Omega^1(1) & \xrightarrow{\partial(f)} & \Omega^2(2) & \xrightarrow{\partial(f)} & \cdots \\
{\scriptstyle\tau(f)}\downarrow & & {\scriptstyle\tau(f)}\downarrow & & {\scriptstyle\tau(f)}\downarrow & & \\
\mathcal{O}(1) & \xrightarrow{\partial(f)} & \Omega^1(2) & \xrightarrow{\partial(f)} & \Omega^2(3) & \xrightarrow{\partial(f)} & \cdots \\
{\scriptstyle\tau(f)}\downarrow & & {\scriptstyle\tau(f)}\downarrow & & {\scriptstyle\tau(f)}\downarrow & & \\
\mathcal{O}(2) & \xrightarrow{\partial(f)} & \Omega^1(3) & \xrightarrow{\partial(f)} & \Omega^2(4) & \xrightarrow{\partial(f)} & \cdots \\
{\scriptstyle\tau(f)}\downarrow & & {\scriptstyle\tau(f)}\downarrow & & {\scriptstyle\tau(f)}\downarrow & & \\
\vdots & & \vdots & & \vdots & &
\end{array} \tag{4.9}$$

Let us number the rows of (4.9) starting from 0. Denote the sequence of sheaves and maps in the l-th row by $\mathcal{B}^\bullet(X,l)$.

Theorem 4.6. *The sequence $\mathcal{B}^\bullet(X,l)$ is a complex of sheaves on X. Vertical arrows define morphisms of complexes*

$$\tau_l : \mathcal{B}^\bullet(X,l) \to \mathcal{B}^\bullet(X,l+1).$$

The twisted Koszul complex

$$\mathcal{O}(l) \stackrel{j(f)\wedge}{\longrightarrow} J(\mathcal{O}(1)) \otimes \mathcal{O}(l) \stackrel{j(f)\wedge}{\longrightarrow} \left(\bigwedge^2 J(\mathcal{O}(1))\right) \otimes \mathcal{O}(l) \to \cdots \tag{4.10}$$

is isomorphic, as a complex of sheaves, to the cone of the morphism τ_l.

Proof. The first two assertions follow from Propositions 4.4 and 4.5. Let us prove the assertion about the cone. By definition of the chain cone (see Appendix A), this assertion means that there are isomorphisms

$$\left(\bigwedge^k J(\mathcal{O}(1))\right) \otimes \mathcal{O}(l) \longrightarrow \Omega^k_X(k+l) \oplus \Omega^{k-1}(k+l) \tag{4.11}$$

which take, for each local section f of $\mathcal{O}(1)$, the operator of exterior multiplication with $j(f)$ into the operator

$$\begin{pmatrix} \partial(f) & 0 \\ (-1)^k\tau(f) & \partial(f) \end{pmatrix}. \tag{4.12}$$

By Proposition 4.1 we can identify sections of $(\bigwedge^k J(\mathcal{O}(1))) \otimes \mathcal{O}(l)$ with differential k-forms on Y homogeneous of degree $k+l$. The sheaf of such forms was denoted by $\Omega^k_{k+l,Y}$. Similarly, we identify sections of $\Omega^k_X(l)$ with k-forms on Y belonging to $\mathcal{R}_{k,l}$, as in Proposition 4.3.

Lemma 4.7. *We have a direct sum decomposition of sheaves on Y:*

$$\Omega^k_{Y,k+l} = \mathcal{R}_{k,k+l} \oplus d\mathcal{R}_{k-1,k+l} \tag{4.13}$$

where d is the exterior derivative on Y.

Proof. We need to show that each local section ω of $\Omega^k_{Y,k+l}$ can be uniquely written in the form

$$\omega = \varphi_1 + d\varphi_2,\ \varphi_1 \in \mathcal{R}_{k,k+l},\ \varphi_2 \in \mathcal{R}_{k-1,k+l}.$$

To show this, we define

$$\varphi_1 = i_\xi(d\omega)/(k+l), \quad \varphi_2 = i_\xi(\omega)/(k+l).$$

Then by (4.6)

$$d\varphi_2 = \mathrm{Lie}_\xi(\omega)/(k+l) - i_\xi(d\omega)/(k+l) = \omega - \varphi_1.$$

Let us prove uniqueness. Suppose that for some $\varphi_1 \in \mathcal{R}_{k,k+l}$, $\varphi_2 \in \mathcal{R}_{k-1,k+l}$, we have $\varphi_1 + d\varphi_2 = 0$. Then

$$0 = i_\xi(\varphi_1 + d\varphi_2) = i_\xi(d\varphi_2) = Lie_\xi \varphi_2 - di_\xi\varphi_2 =$$

$$= \mathrm{Lie}_\xi \varphi_2 = (k+l)\varphi_2,$$

so $\varphi_2 = 0$. Therefore $\varphi_1 = 0$ as well. Lemma 4.7 is proved.

The required isomorphism (4.11) now follows from the decomposition in Lemma 4.7. We only have to show that exterior multiplication by $j(f)$ goes under this isomorphism to the matrix form (4.12). But this follows from Proposition 1.3. Theorem 4.6 is proved.

Note that the isomorphism constructed in Theorem 4.6 is only an isomorphism of complexes of sheaves and is not $\mathcal{O}_X$-linear. However, such an isomorphism still allows us to deduce the decomposition for the complex of global sections.

Let $B_{k,l} = H^0(X, \Omega^k_X(l))$. For any $f \in V$ we get, by taking global sections of rows in (4.9), a complex of vector spaces

$$B^\bullet(l) = \left\{ B_{0,l} \xrightarrow{\partial(f)} B_{1,l+1} \xrightarrow{\partial(f)} B_{2,l+2} \to \cdots \right\} \tag{4.14}$$

and a morphism of complexes

$$\tau_l(f) : B^\bullet(l) \longrightarrow B^\bullet(l+1). \tag{4.15}$$

Theorem 4.6 has the following corollary.

Corollary 4.8. *There are isomorphisms of vector spaces*

$$C^k(X,l) \cong B_{k,k+l} \oplus B_{k-1,k+l}$$

such that for every $f \in V$ the discriminantal complex $(C^\bullet(X,l), \partial_f)$ is identified with the cone of $\tau_l(f)$.

As another application of Theorem 4.6, let us give a criterion for $C^\bullet(X, l)$ to be stably twisted.

Corollary 4.9. *The discriminantal complex $C^\bullet(X, l)$ is stably twisted if and only if the following sheaves on X have no higher cohomology:*

$$\Omega_X^k(k+l), \ \Omega_X^{k-1}(k+l), \quad k = 0, \dots, \dim X + 1.$$

D. Interpretation of the complexes $B^\bullet(l)$

For $l \gg 0$ the discriminantal complex $C^\bullet(X, l) = C_+^\bullet(X, \mathcal{O}(l))$ is stably twisted by Proposition 2.2. Hence, for a generic $f \in V$, the differential ∂_f in this complex is exact. This means that the chain morphism $\tau_l(f) : B^\bullet(l) \to B^\bullet(l+1)$, whose cone is $(C^\bullet(X, L), \partial_f)$, gives an isomorphism on cohomology spaces. So these spaces for $l \gg 0$ are independent on l. We are going to find these spaces. Again, we start with the study of the complex of sheaves $\mathcal{B}^\bullet(X, l)$ and only later do we take the global sections.

Let $Z \subset X$ be any irreducible hypersurface. By $\Omega_X^k(l \cdot Z)$, we denote the sheaf of meromorphic k-forms on X which are regular outside Z and have poles of order $\leq l$ along Z. In other words, if $\varphi = 0$ is a local equation of Z, then a section of $\Omega_X^k(l \cdot Z)$ is a meromorphic k-form ω such that $\varphi^l \cdot \omega$ is regular. The exterior derivative of forms raises the order of poles by one, so we have the complexes

$$\mathcal{O}_X(l \cdot Z) \xrightarrow{d} \Omega_X^1((l+1)Z) \xrightarrow{d} \Omega_X^2((l+2)Z) \xrightarrow{d} \cdots \tag{4.16}$$

Now let $f \in V$ be a non-zero vector and take Z to be the hypersurface $Z_f \subset X$ given by $f = 0$. Denote by $U = U_f = X - Z_f$ the complement to Z_f and let $j : U_f \to X$ stand for the embedding of U_f into X. The complex (4.16) is a subcomplex in the direct image $j_*\Omega_U^\bullet$ of the de Rham complex of U. More precisely, the consideration of orders of poles along Z defines in $j_*\Omega_U^\bullet$ a natural increasing filtration F by subcomplexes such that (4.16) is its l-th part, $F_l(j_*\Omega_X^\bullet)$.

Proposition 4.10. *For any $l \geq 0$ and any non-zero $f \in V$, the complex of sheaves $\mathcal{B}^\bullet(X, l)$ given by the l-th row of (4.9), is isomorphic to the complex $F_l(j_*\Omega_X^\bullet)$ given by (4.16). This system of isomorphisms takes the morphism $\tau_l : \mathcal{B}^\bullet(X, l) \to \mathcal{B}^\bullet(X, l+1)$ of Theorem 4.6 into the canonical embedding $F_l(j_*\Omega_U^\bullet) \hookrightarrow F_{l+1}(j_*\Omega_U^\bullet)$.*

Proof. We have an isomorphism of sheaves

$$\Omega_X^i(l) \to \Omega_X^i(l \cdot Z), \quad \omega \mapsto \omega/f^l. \tag{4.17}$$

Under these isomorphisms the differential $\partial(f)$ is taken to the usual de Rham differential d, since

$$f^{l+1}d(\omega/f^l) = f^{l+1}((d\omega)/f^l - (l\omega \wedge df)/f^{l+1}) = \partial(f)\omega.$$

The assertion about τ_l is equally straightforward.

Proposition 4.10 implies, by taking global sections, an interpretation of $B^\bullet(X,l)$ as part of the algebraic de Rham complex of the affine variety $U_f = X - Z_f$. This leads to the following.

Proposition 4.11. *Let l be such that $C^\bullet(X,l)$ is stably twisted for any $l' \geq l$. For any $f \notin X^\vee$, we have an isomorphism*

$$H^i\big(B^\bullet(X,l), \partial(f)\big) \cong H^i(U_f, \mathbf{C})$$

where the right hand side is the ordinary complex cohomology of the open set $U_f \subset X$ defined by the condition $f \neq 0$.

The proof of Proposition 4.11 is based on two facts. The first is Grothendieck's algebraic de Rham theorem (see [GH] Chapter 3, Section 5) which implies that $H^\bullet(U_f, \mathbf{C})$ can be calculated using the de Rham complex consisting of all regular forms on U_f, i.e., of forms with poles of arbitrary order along Z_f. The other is a local statement saying that, for any $l \geq 1$, the embedding $F_l(j_*\Omega^\bullet_{U_f}) \hookrightarrow F_{l+1}(j_*\Omega^\bullet_{U_f})$ is a quasi-isomorphism of complexes of sheaves (provided Z_f is smooth). This can be easily established in local coordinates (see [De 1], 3.1.10).

Proposition 4.11 is obtained from these two facts by purely formal reasoning involving the hypercohomology of complexes of sheaves. Since Proposition 4.11 will not be used in the sequel, more details are not required here. The formalism of hypercohomology, however, will be discussed below in Section 5.

E. Example: the second multiplication for the canonical curve

Let us illustrate the constructions of subsection B by the example of a *canonical curve*. By definition, this is a smooth algebraic curve X projectively embedded using the line bundle Ω^1_X of 1-forms on X. We shall denote this bundle simply by Ω. It is known [GH] that, for any curve of genus ≥ 3 which is not hyperelliptic, Ω indeed defines an embedding

$$X \hookrightarrow P(H^0(X,\Omega)^*).$$

The coordinate ring of X in this embedding coincides with $\bigoplus H^0(X, \Omega^{\otimes l})$. This *canonical ring* of X, as it is called, will be denoted by Can (X).

So, applying our constructions to $\mathcal{O}_X(1) = \Omega$, we see that the sheaf of graded algebras $\mathbf{B}$ will only have parts

$$\mathbf{B}_{0,l} = \Omega^{\otimes l}, \quad \mathbf{B}_{1,l} = \Omega^{\otimes (l+1)}.$$

Sections of $\Omega^{\otimes i}$ are classically called i-differentials on the curve X.

The only non-trivial part of the multiplication $*$ is the map

$$\mathbf{B}_{0,i} \otimes_{\mathbf{C}} \mathbf{B}_{0,j} \to \mathbf{B}_{1,i+j} \text{ or } \Omega^{\otimes i} \otimes_{\mathbf{C}} \Omega^{\otimes j} \to \Omega^{\otimes(i+j+1)}. \tag{4.19}$$

If x is a local coordinate on X then we can write an i-differential as $f(x)dx^i$. In coordinate form the multiplication is

$$(f(x)dx^i) * (g(x)dx^j) = \left(\frac{i}{i+j}\frac{df}{dx}g(x) - \frac{j}{i+j}f(x)\frac{dg}{dx}\right)dx^{i+j+1}. \tag{4.20}$$

To understand this multiplication, note that the manifold Y in our case is the tangent bundle TX of X without the zero section. It has rank one so there is an isomorphism of (non-linear) bundles with deleted zero section X:

$$T^*X - X \to TX - X, \quad p \mapsto p^{-1}$$

where, for a covector p at $x \in X$, the vector p^{-1} at the same point is defined by $(p, p^{-1}) = 1$.

Thus every i-differential can be regarded as a function on $T^*X - X$ homogeneous of degree $(-i)$. Recall that T^*X is a symplectic manifold and therefore, for any two functions f, g on T^*X, we have their Poisson bracket $\{f, g\}$, see [Ar]. This defines the structure of the Poisson algebra on Can (X), i.e., $\{f, g\}$ satisfies the Jacobi identity and is a derivation in both f and g.

Proposition 4.13. *The multiplication (4.20) coincides with the Poisson bracket of i-differentials considered as functions on T^*X.*

The proof is straightforward.

Note that we cannot speak of the Jacobi identity while remaining in the framework of the algebra $\mathbf{B}$: we need the identification $\mathbf{B}_{1,i+j} \cong \mathbf{B}_{0,i+j+1}$. In fact, the $*$-product of any three elements of $\mathbf{B}$ is zero.

Let us remark that the Poisson algebra Can (X) has an important "quantization," i.e., a non-commutative filtered algebra $E(X)$ whose associated graded algebra is Can (X). This is the algebra of global pseudo-differential (also called micro-differential) operators on X of order ≤ 0, see [Kas].

5. The discriminant as the determinant of a spectral sequence

In Section 2 we expressed the X-discriminant Δ_X as the determinant of discriminantal complexes $C_{\pm}^{\bullet}(X, \mathcal{M})$. We imposed the condition that the complex be stably twisted, i.e., the terms of the corresponding complex of sheaves do not have higher cohomology. In this section we develop an approach that takes into account the higher cohomology as well, and is applicable to any twisting bundle $\mathcal{M}$.

A. Discriminantal spectral sequences

Recall [Bry] [GH] that with every finite complex of sheaves $\mathcal{F}^{\bullet}$ on a topological space X, there are associated the *hypercohomology groups* $\mathbf{H}^i(X, \mathcal{F}^{\bullet})$. To define them, one takes the complexes of Abelian groups $C_j^{\bullet}$ calculating the cohomology of every individual $\mathcal{F}^j$ (for example, the Čech complexes with respect to an appropriate open covering of X). The differentials $\mathcal{F}^j \to \mathcal{F}^{j+1}$ make this collection of complexes into a double complex $C^{\bullet\bullet}$, and the $\mathbf{H}^i(X, \mathcal{F}^{\bullet})$ are the cohomology groups of its total complex. By construction we have the spectral sequence (which is just the spectral sequence of the double complex $C^{\bullet\bullet}$):

$$E_1^{pq} = H^q(X, \mathcal{F}^p) \quad \Rightarrow \quad \mathbf{H}^{p+q}(X, \mathcal{F}^{\bullet}). \tag{5.1}$$

The first differential in (5.1) is induced by the differential in $\mathcal{F}^{\bullet}$. In particular the complex of global sections

$$\cdots \to H^0(X, \mathcal{F}^i) \to H^0(X, \mathcal{F}^{i+1}) \to \cdots$$

is just the bottom row of the term E_1.

The hypercohomology can also be calculated using another spectral sequence

$$'E_2^{pq} = H^q(X, \underline{H}^p(\mathcal{F}^{\bullet})) \quad \Rightarrow \quad \mathbf{H}^{p+q}(X, \mathcal{F}^{\bullet}) \tag{5.2}$$

where $\underline{H}^p(\mathcal{F}^{\bullet})$ is the p-th cohomology sheaf of $\mathcal{F}^{\bullet}$, i.e.,

$$\underline{H}^p(\mathcal{F}^{\bullet})) = \frac{\operatorname{Ker}\{\mathcal{F}^p \xrightarrow{d} \mathcal{F}^{p+1}\}}{\operatorname{Im}\{\mathcal{F}^{p-1} \xrightarrow{d} \mathcal{F}^p\}}. \tag{5.3}$$

In particular, if $\mathcal{F}^{\bullet}$ is exact then $\mathbf{H}^j(X, \mathcal{F}^{\bullet}) = 0$ for all j.

Now let $X \subset P(V^*)$ be a smooth projective variety, and let $\mathcal{M}$ be a line bundle on X. Denote the sheaf $\mathcal{O}_X(1)$ by $\mathcal{L}$. Let $f \in V$. We apply the above formalism to the Koszul complexes (1.9) and (1.10) tensored by $\mathcal{M}$. We define the *discriminantal spectral sequences* $C_{r,+}^{pq}(X, \mathcal{M}, f)$ and $C_{r,-}^{pq}(X, \mathcal{M}, f)$ to be the spectral sequences (5.1) for these complexes. More precisely,

$$C_{1,+}^{pq}(X, \mathcal{M}, f) = H^q\left(X, \bigwedge^{p}(J(\mathcal{L})) \otimes \mathcal{M}\right),$$

$$C_{1,-}^{pq}(X, \mathcal{M}, f) = H^q\left(X, \bigwedge^{-p}(J(\mathcal{L})^*) \otimes \mathcal{M}\right). \tag{5.4}$$

We denote by $\partial_{r,f}$ the differential in the r-th term of any of these spectral sequences. Thus $\partial_{1,f}$ is just the map induced by the differential in the Koszul complex (i.e., by exterior multiplication or contraction with $j(f)$). Higher differentials are found through the standard rules, by chasing the differentials in the double complex $C^{\bullet\bullet}$ above.

So the first terms of the discriminantal spectral sequences (5.4) do not depend on the choice of $f \in V$, but the differential ∂_1 does (and hence do all the subsequent terms).

Proposition 5.1. *Suppose that the projectivization of $f \in V$ does not belong to the dual variety $X^\vee$. Then the discriminantal spectral sequences $C_{r,\pm}^{\bullet\bullet}(X, \mathcal{M}, f)$ are exact, i.e., they converge to zero.*

Proof. Indeed, the Koszul complexes of sheaves (1.9) and (1.10) are exact if $f \notin X^\vee$. The exactness will be preserved after tensoring by $\mathcal{M}$. Hence the hypercohomology of the tensored complex is zero. This hypercohomology is the limit term of the spectral sequences in question.

Let us choose some system of bases e in the components of $C_{1,+}^{pq}(X, \mathcal{M}, f)$. Then, for $f \notin X^\vee$, we have a based exact spectral sequence and hence its determinant is defined (see Appendix A). So we get a rational function

$$f \mapsto \Delta_{X,\mathcal{M}}^{+}(f) = \det(C_{r,+}^{pq}(X, \mathcal{M}, f), e)$$

on V. A different choice of bases changes $\Delta_{X,\mathcal{M}}^{+}$ by a non-zero constant factor.

In a similar way, we define a rational function $\Delta_{X,\mathcal{M}}^{-}$ by taking the determinant of $C_{r,-}^{pq}(X, \mathcal{M}, f)$.

Theorem 5.2. *For any line bundle $\mathcal{M}$ on X, we have*

$$\Delta_X = \Delta_{X,\mathcal{M}}^{-} = (\Delta_{X,\mathcal{M}}^{+})^{(-1)^{\dim(X)+1}},$$

up to a non-zero constant multiple.

Proof. We shall treat Δ^- and the spectral sequence C_-, the other case being similar. Consider the "universal" spectral sequence of sheaves on $P(V)$ incorporating all of the particular spectral sequences $C^{pq}_{r,-}(X, \mathcal{M}, f)$ of vector spaces. To construct this, we start from the complex (2.6) of sheaves on $X \times P(V)$. Denote this complex by $\mathcal{F}^\bullet$ and conserve the other notation of Section 2. In particular, we denote p_X, p_V the projections of $X \times P(V)$ to X and $P(V)$. Our universal spectral sequence has the form

$$E_1^{ij} = R^i p_{V*}(\mathcal{F}^j \otimes p_X^* \mathcal{M}) \Rightarrow \mathbf{R}^{i+j} p_{V*}(\mathcal{F}^\bullet \otimes p_X^* \mathcal{M}) \tag{5.5}$$

where $\mathbf{R}^{i+j} p_{V*}$ are the hyperdirect images. As in the proof of Theorem 2.5, we calculate the π-adic order of $\det(C^{ij}_{r,-}(x, l, f), e)$. By Theorem 36 from Appendix A, this order coincides with the alternating sum of orders along X_π of the sheaves of the limit of (5.5), i.e., of the $\mathbf{R}^i p_{V*}(\mathcal{F}^\bullet \otimes p_X^* \mathcal{M})$. However, the complex $\mathcal{F}^\bullet \otimes p_X^* \mathcal{M}$ is a resolution of an invertible sheaf $p_1^* \mathcal{M}$ on the incidence variety W (Corollary 2.8). Here p_1, p_2 are projections of W to X and $P(V)$. Thus

$$\mathbf{R}^i p_{V*}(\mathcal{F}^\bullet \otimes p_X^* \mathcal{M}) = \mathbf{R}^i p_{V*}(\mathcal{O}_W \otimes p_X^* \mathcal{M}). \tag{5.6}$$

In the case where $\operatorname{codim} X^\vee = 1$, the projection $p_2 : W \to X^\vee$ is a birational isomorphism and the only hypersurface along which these hyperdirect images can be generically non-zero is $X^\vee$. In the case $\operatorname{codim}(X^\vee) > 1$ the right hand side of (5.6) does not have support on any hypersurface. Theorem 5.2 is proved.

Theorem 5.2 generalizes Theorem 2.5 which is valid only in the case of stably twisted complexes. In this case the whole spectral sequence is reduced to a complex of vector spaces given by the bottom row of E_1.

Let us study a little more about the dependence of the discriminantal spectral sequences on $f \in V$.

Proposition 5.3. *For any $\lambda \in \mathbf{C}^*$, we have a natural system of identifications*

$$\varphi^{pq}_{r,\pm} : C^{pq}_{r,\pm}(X, \mathcal{M}, \lambda f) \cong C^{pq}_{r,\pm}(X, \mathcal{M}, f), \quad r \geq 1,$$

with the following properties:

(a) *Under the r-th identification the differential $\partial_{r,f}$ is homogeneous of degree r in f:*

$$\partial_{r,\lambda f} = \lambda^r \partial_{r,f}. \tag{5.7}$$

(b) *The identification φ_r itself is induced by the homogeneity (5.7) of the previous differential $\partial_{r-1,f}$.*

Let us explain the statement (b). Since $\partial_{r-1,f}$ is homogeneous of degree $r-1$, we have $\operatorname{Ker}\partial_{r-1,\lambda f} = \operatorname{Ker}\partial_{r-1,f}$ and $\operatorname{Im}\partial_{r-1,\lambda f} = \operatorname{Im}\partial_{r-1,f}$ for any $\lambda \in \mathbf{C}^*$. So the cohomology of the differentials $\partial_{r-1,\lambda f}$ and $\partial_{r-1,f}$ are canonically identified with each other.

Proof of Proposition 5.3. We treat the C_- spectral sequence; the other case is similar. Denote by $\mathcal{F}^\bullet$ the Koszul complex (1.10) tensored by $\mathcal{M}$, so that

$$\mathcal{F}^j = \bigwedge^{-j}(J(\mathcal{L})^*) \otimes \mathcal{M}.$$

Let ∂_f denote the differential in this complex corresponding to f. Then ∂_f is linear in f, in particular, $\partial_{\lambda f} = \lambda\partial_f$. The spectral sequence $C^{pq}_{r,-}(X, \mathcal{M}, f)$ is just the sequence (5.1) corresponding to $(\mathcal{F}^\bullet, \partial_f)$. To explicitly construct this sequence, we take an affine covering $\{U_\alpha\}$ of X and consider the Čech complex of each $\mathcal{F}^i$:

$$C^{i0} \to C^{i1} \to \cdots,$$

$$C^{ip} = \bigoplus_{\alpha_0,\ldots,\alpha_p} H^0(U_{\alpha_0} \cap \cdots \cap U_{\alpha_p}, \mathcal{F}^i).$$

The vector spaces C^{ip} form a double complex where the vertical (Čech) differential is independent of $f \in V$ and the horizontal (Koszul) differential is linear in f. The discriminantal spectral sequence is just the standard spectral sequence of the double complex $C^{\bullet\bullet}$. So our statement follows from the homogeneity in f of both differentials and from the standard procedure which defines the spectral sequence of a double complex (see, e.g., [GH], Chapter 3, Section 5).

B. Example: the Bezout formula

We consider the case of Example 2.11. Let $X \cong P^1$ be the Veronese curve of degree d in $P^d = P(V^*)$ where $\dim V = 2$. So $\mathcal{L} = \mathcal{O}_{P^d}(1)|_X$ is, in intrinsic terms, $\mathcal{O}_{P^1}(d)$. The space V is the space of binary forms $f(x_0, x_1)$ of degree d.

We have $J(\mathcal{L}) = \mathcal{O}(d-1) \oplus \mathcal{O}(d-1)$. Let us take the twisting line bundle $\mathcal{M}$ to be $\mathcal{O}(d-2)$ and $f \in V$. The discriminantal spectral sequence $C^{pq}_{r,-}(X, \mathcal{M}, f)$ is the spectral sequence (5.1) of the following complex of sheaves on P^1:

$$\mathcal{O}(-d) \xrightarrow{\partial_f} \mathcal{O}(-1) \oplus \mathcal{O}(-1) \xrightarrow{\partial_f} \mathcal{O}(d-2). \tag{5.8}$$

As explained in Example 2.11, (5.8) is the twisted Koszul complex corresponding to the partial derivatives

$$g(x_0, x_1) = \frac{\partial f}{\partial x_0}, \quad h(x_0, x_1) = \frac{\partial f}{\partial x_1},$$

which are forms of degree $d-1$. In other words, the ∂_f are given by formulas $\partial_f(w) = (hw, -gw)$ for w a local section of $\mathcal{O}(-d)$ and $\partial_f(u, v) = gu + hv$ for u, v local sections of $\mathcal{O}(-1)$.

Since $\mathcal{O}(-1)$ has neither H^0 nor H^1 on P^1 (Theorem 2.12), the only non-trivial differential in the spectral sequence will be

$$\partial_{2,f} : H^1(P^1, \mathcal{O}(-d)) \longrightarrow H^0(P^1, \mathcal{O}(d-2)). \tag{5.9}$$

By Theorem 5.2, the X-discriminant (i.e., the classical discriminant of the form f) equals the determinant of $\partial_{2,f}$.

Denote by S^m the space of binary forms of degree m, so $V = S^d$. Standard information about cohomology of $\mathcal{O}(i)$ (see Theorem 2.12) provides that the vector spaces in (5.9) are naturally identified with $(S^{d-2})^*$ and S^{d-2}, respectively. So we can write (5.9) as

$$\partial_{2,f} : (S^{d-2})^* \to S^{d-2}. \tag{5.10}$$

Thus we have a $(d-1) \times (d-1)$ matrix $\partial_{2,f}$ whose entries are, by Proposition 5.3, quadratic forms in coefficients of f and whose determinant is $\Delta_X(f)$.

We are going to find the matrix $\partial_{2,f}$ explicitly. By (5.10), this matrix can be regarded as an element of $S^{d-2} \otimes S^{d-2}$. We regard this as the space of bihomogeneous forms

$$F(x_0, x_1, y_0, y_1) = \sum_{i,j=0}^{d-2} c_{ij} x_0^{d-2-i} x_1^i y_0^{d-2-j} y_1^j \tag{5.11}$$

of bidegree $(d-2, d-2)$.

Proposition 5.4. *The form F corresponding to $\partial_{2,f}$ equals*

$$F(x_0, x_1, y_0, y_1) = \frac{g(x_0, x_1)h(y_0, y_1) - h(x_0, x_1)g(y_0, y_1)}{x_0 y_1 - x_1 y_0} \tag{5.12}$$

where g and h are partial derivatives of f. Therefore $\Delta_X(f)$ coincides, up to a constant multiple, with the determinant of the matrix $\|c_{ij}\|$ of the coefficients of F.

This is the classical *Bezout formula* for the discriminant.

Note that the ratio in (5.12) is in fact a polynomial since the numerator vanishes whenever (x_0, x_1) is proportional to (y_0, y_1). An explicit form for the coefficients c_{ij} of F is as follows. Suppose that

$$g(x_0, x_1) = \frac{\partial f}{\partial x_0} = \sum_{i=0}^{d-1} a_i x_0^{d-1-i} x_1^i, \tag{5.13}$$

$$h(x_0, x_1) = \frac{\partial f}{\partial x_1} = \sum_{i=0}^{d-1} b_i x_0^{d-1-i} x_1^i. \tag{5.14}$$

Then

$$c_{ij} = \sum_{k=0}^{\min(i,j)-1} (a_k b_{i+j-k} - a_{i+j-k} b_k), \tag{5.15}$$

as can be seen directly by division. In (5.15) we assume that $a_\nu = b_\nu = 0$ for $\nu > d-1$.

Proof of Proposition 5.4. We regard x_0, x_1 as homogeneous coordinates on P^1. Consider the affine covering of P^1 consisting of two charts $U_\alpha = \{x_\alpha \neq 0\}$, $\alpha = 0, 1$. Clearly, for $m \in \mathbf{Z}$, the space of sections $\Gamma(U_0 \cap U_1, \mathcal{O}(m))$ consists of all Laurent polynomials

$$\sum_{i+j=m} q_{ij} x_0^i x_1^j, \quad q_{ij} \in \mathbf{C},$$

homogeneous of degree m. Such a polynomial is regular in U_0 (resp. U_1) if $q_{ij} = 0$ for $j < 0$ (resp. for $i < 0$). The Čech cocycles given by the monomials

$$e_i = x_0^{-d+i+1} x_1^{-i-1} \in \Gamma(U_0 \cap U_1, \mathcal{O}(-d)), \quad i = 0, 1, \ldots, d-2 \tag{5.16}$$

form a basis in $H^1(P^1, \mathcal{O}(-d))$. Proposition 5.4 can be reformulated by saying that

$$\partial_{2,f}(e_i) = \sum_{j=0}^{d-2} c_{ij} x_0^{d-2-j} x_1^j \quad \in \quad H^0(P^1, \mathcal{O}(d-2)) \tag{5.17}$$

where the c_{ij} are given by (5.15).

We shall prove (5.17). The discriminantal spectral sequence, i.e., the spectral sequence (5.1) of the complex of sheaves (5.8) can be found from the double complex

$$\begin{array}{ccccc} \Gamma(U_0 \cap U_1, \mathcal{O}(-d)) & \xrightarrow{\partial_f} & \Gamma(U_0 \cap U_1, 2\mathcal{O}(-1)) & \xrightarrow{\partial_f} & \Gamma(U_0 \cap U_1, \mathcal{O}(d-2)) \\ \delta \uparrow & & \delta \uparrow & & \delta \uparrow \\ \begin{matrix} \Gamma(U_0, \mathcal{O}(-d)) \oplus \\ \Gamma(U_1, \mathcal{O}(-d)) \end{matrix} & \xrightarrow{\partial_f} & \begin{matrix} \Gamma(U_0, 2\mathcal{O}(-1)) \oplus \\ \Gamma(U_1, 2\mathcal{O}(-1)) \end{matrix} & \xrightarrow{\partial_f} & \begin{matrix} \Gamma(U_0, \mathcal{O}(d-2)) \oplus \\ \Gamma(U_1, \mathcal{O}(d-2)) \end{matrix} \end{array}.$$

The columns of this double complex are Čech complexes of individual sheaves in (5.8). The middle differential δ is an isomorphism since its kernel and cokernel are H^0 and H^1 of the sheaf $2\mathcal{O}(-1)$ on P^1, which vanish. According to general formulas for differentials in the spectral sequence of a double complex, we have

$$\partial_{2,f}(z) = \partial_f(\delta^{-1}(\partial_f(z))) \quad \text{for} \quad z \in H^1(P^1, \mathcal{O}(-d)) = \frac{\Gamma(U_0 \cap U_1, \mathcal{O}(-d))}{\operatorname{Im} \delta}.$$

Substituting here for z a basis element e_i from (5.16), we first find

$$\partial_f(e_i) = (he_i, -ge_i) = \left(\sum_{j=0}^{d-1} a_j x_0^{-j+i} x_1^{j-i-1}, \quad -\sum_{j=0}^{d-1} b_j x_0^{-j+i} x_1^{j-i-1}\right)$$

$$\in \quad \Gamma(U_0 \cap U_1, \mathcal{O}(-1)) \oplus \Gamma(U_0 \cap U_1, \mathcal{O}(-1)).$$

For any $s = \sum_{i+j=-1} q_{ij} x_0^i x_1^j$ we shall write

$$s^{(0)} = \sum_{j \geq 0} q_{ij} x_0^i x_1^j, \quad s^{(1)} = \sum_{i \geq 0} q_{ij} x_0^i x_1^j,$$

so the map

$$\delta^{-1} : \Gamma(U_0 \cap U_1, \mathcal{O}(-1)) \to \Gamma(U_0, \mathcal{O}(-1)) \oplus \Gamma(U_1, \mathcal{O}(-1))$$

takes $s \mapsto (s^{(0)}, -s^{(1)})$.

In this notation we find that $\partial_f(\delta^{-1}(\partial_f(e_i)))$ is the global section of $\mathcal{O}(d-2)$ represented by the Laurent polynomial

$$g \cdot (he_i)^{(0)} - h \cdot (ge_i)^{(0)} = g \cdot (he_i)^{(1)} - h \cdot (ge_i)^{(1)}.$$

This polynomial is immediately found to have the form (5.17). Proposition 5.4 is proved.

C. Weyman's bigraded complexes

The calculation of the determinant of a spectral sequence with many non-trivial terms may be very involved. In this subsection we describe a procedure, due to J. Weyman [We 2], which permits us to replace the discriminantal spectral sequence by a complex which incorporates, in a sense, "all the higher differentials at once."

By a *bigraded complex* we shall mean a complex $(C^\bullet, \partial)$ of vector spaces in which each term C^i is equipped with additional grading: $C^i = \bigoplus_{p+q=i} C^{pq}$. We shall sometimes refer to $(C^\bullet, \partial)$ as the *total complex* of the bigraded complex $C^{\bullet\bullet}$.

The differential ∂ in $C^\bullet$ is decomposed into the sum $\partial = \sum_r \partial_r$ where ∂_r is bihomogeneous of bidegree $(r, 1-r)$. Thus ∂_0 has degree $(0, 1)$, i.e., it is "vertical", in a standard visualization of the bigrading; the component ∂_1 (of bidegree $(1, 0)$) is "horizontal" (like the differential in the E_1-term of a spectral sequence), ∂_2 has bidegree $(2, -1)$ (as the differential in E_2) etc.

We consider only bigraded complexes having $\partial_r = 0$ for $r < 0$.

Let now $X \subset P(V)$ be a smooth projective variety, $\mathcal{L} = \mathcal{O}_X(1)$ and let $\mathcal{M}$ be an arbitrary line bundle on X. We shall describe Weyman's construction only for the case of the C_--spectral sequence from subsection A; the other one is similar.

Theorem 5.5. *There exists a bigraded complex* $(C^{\bullet\bullet}, \partial_f)$ *in which*

$$C^{pq} = H^q\left(X, \bigwedge^{-p} J(\mathcal{L})^* \otimes \mathcal{M}\right), \tag{5.18}$$

and the differential ∂_f *depends polynomially on* $f \in V$. *This complex has the following properties:*

(a) *The differential* ∂_f *has the form* $\partial_f = \sum_{r \geq 1} \partial_{r,f}$, *where* $\partial_{r,f}$ *has bidegree* $(r, 1-r)$ *with respect to the bigrading in* $C^{\bullet\bullet}$ *and its matrix elements are homogeneous polynomials of degree r in coefficients of f. Moreover,* $\partial_{1,f}$ *is induced by the differential in the Koszul complex (1.10).*

(b) *The determinant of the total complex* $(C^\bullet, \partial_f)$ *of* $C^{\bullet\bullet}$ *equals, up to a non-zero constant factor, the X-discriminant* $\Delta_X(f)$.

Note that the terms of the complex in Theorem 5.5 are completely determined by X and $\mathcal{M}$: they coincide with the components of the first term of the discriminantal spectral sequence C_-. The definition of the differentials $\partial_{r,f}$ for $r \geq 2$ involves some arbitrary choices, as will be seen from the proof.

Proof of the theorem. As in the proof of Theorem 2.5, we consider the incidence variety $W = W_X \subset X \times P(V)$ and the left resolution (2.6) of the structure sheaf $\mathcal{O}_W$. Let p_X, p_V be projections of $X \times P(V)$ to the first and second factor. Denote by $\mathcal{F}^\bullet$ the tensor product of the complex (2.6) and $p_1^*\mathcal{M}$, so

$$\mathcal{F}^i = p_X^*\left(\bigwedge^{-i} J(\mathcal{L})^* \otimes \mathcal{M}\right) \otimes p_V^*\mathcal{O}(i), \quad i = 0, -1, \ldots, -\dim(X) - 1. \tag{5.19}$$

Then the direct image $Rp_{V*}\mathcal{F}^\bullet$ is quasi-isomorphic to $Rp_{V*}(p_X^*\mathcal{M} \otimes \mathcal{O}_W)$. We first find the direct images of the individual $\mathcal{F}^i$.

Lemma 5.6. *For any* $i \leq 0$, *the complex* $Rp_{V*}\mathcal{F}^i$ *is quasi-isomorphic to the complex*

$$H^0\left(X, \bigwedge^{-i} J(\mathcal{L})^* \otimes \mathcal{M}\right) \otimes \mathcal{O}(i) \xrightarrow{0} H^1\left(X, \bigwedge^{-i} J(\mathcal{L})^* \otimes \mathcal{M}\right) \otimes \mathcal{O}(i) \xrightarrow{0} \cdots \tag{5.20}$$

with zero differential.

Proof of the lemma. To calculate the direct image with respect to $p_V : X \times P(V) \to P(V)$, we should take an affine covering $\{U_\alpha\}$ of X and then take the relative Čech complex of $\mathcal{F}^i$ with respect to this covering. Since $\mathcal{F}^i$ is a tensor product of factors pulled back from X and $P(V)$, this relative Čech complex will have the form $C^\bullet(\mathcal{F}^i) \otimes \mathcal{O}(i)$ where $C^\bullet(\mathcal{F}^i)$ is the usual Čech complex of vector spaces calculating the cohomology of $\bigwedge^{-i} J(\mathcal{L})^* \otimes \mathcal{M}$. But any complex of vector spaces is quasi-isomorphic to the direct sum of its cohomology spaces with zero differential. Lemma 5.6 is proved.

Let us denote by $\mathcal{F}^{\geq i}$ the subcomplex in $\mathcal{F}^\bullet$ obtained by deleting the terms $\mathcal{F}^j$ with $j < i$. For any i, we have an obvious exact triangle in the derived category of coherent sheaves on $X \times P(V)$:

$$\cdots \to \mathcal{F}^{\geq i+1} \to \mathcal{F}^{\geq i} \to \mathcal{F}^i[-i] \to \cdots, \tag{5.21}$$

where $\mathcal{F}^i[-i]$ is the complex consisting of one sheaf $\mathcal{F}^i$ situated in the degree i.

Since the functor Rp_{V*} is exact, we get an exact triangle in the derived category of sheaves on $P(V)$:

$$\cdots \to Rp_{V*}(\mathcal{F}^{\geq i+1}) \to Rp_{V*}(\mathcal{F}^{\geq i}) \to Rp_{V*}(\mathcal{F}^i)[-i] \to \cdots. \tag{5.22}$$

Note that we can replace $Rp_{V*}(\mathcal{F}^i)$ by the complex with the zero differential given by (5.20). We use this to construct inductively a nice complex representing the whole direct image $Rp_{V*}(\mathcal{F}^\bullet)$.

More precisely, we shall construct, for any $i = 0, -1, \ldots, -\dim(X) - 1$ a complex $K_i^\bullet$ of sheaves on $P(V)$ such that:

(a)

$$K_i^j = \bigoplus_{\substack{p+q=j \\ p \geq i}} H^q\left(X, \bigwedge^{-i} J(\mathcal{L})^* \otimes \mathcal{M}\right) \otimes \mathcal{O}(p).$$

(b) The natural embedding of graded sheaves $K_{i+1} \subset K_i$ is compatible with the differentials and the quotient K_i/K_{i+1} is isomorphic to the complex (5.20).

(c) $K_i^\bullet$ is quasi-isomorphic to $Rp_{V*}(\mathcal{F}^{\geq i})$.

We start by defining $K_0^\bullet$ to be the complex (5.20) with $i = 0$, i.e., $K_0^\bullet = H^\bullet(X, \mathcal{M}) \otimes \mathcal{O}_{P(V)}$.

Suppose we have constructed K_{i+1} and want to construct K_i. We have a morphism in the derived category

$$Rp_{V*}(\mathcal{F}^i) = H^\bullet\left(X, \bigwedge^{-i} J(\mathcal{L})^* \otimes \mathcal{M}\right) \otimes \mathcal{O}(i) \xrightarrow{\alpha} K_{i+1}$$

and $Rp_{V*}(\mathcal{F}^{\geq i})$ is the third term of the exact triangle containing α.

Note that each term of K_{i+1} is a sum of some summands of the form $\mathcal{O}(j)$, $j \geq i+1$. Hence the higher Ext-groups between any term of $Rp_{2*}(\mathcal{F}^i)$ and any term of $K^\bullet_{i+1}$ are zero. We use the following general lemma.

Lemma 5.7. *Let $\mathcal{G}^\bullet, \mathcal{H}^\bullet$ be two finite complexes of coherent sheaves on X. Suppose that $\mathrm{Ext}^q_{\mathcal{O}_X}(\mathcal{G}^a, \mathcal{H}^b) = 0$ for any a, b and any $q > 0$. Then $\mathrm{Hom}\,(\mathcal{F}^\bullet, \mathcal{G}^\bullet)$ in the derived category of coherent sheaves on X coincides with the space of homotopy classes of genuine morphisms of complexes $\mathcal{G}^\bullet \to \mathcal{H}^\bullet$.*

For the sake of continuity of the exposition, we give the proof of the lemma later. Applying Lemma 5.7, we realize α as a genuine morphism of complexes (note that there is an arbitrary choice of a morphism in a homotopy class) and define $K^\bullet_i$ as the cone of this morphism. This concludes the inductive construction of $K^\bullet_i$.

Taking $i = -\dim(X) - 1$, we get a model $K_{-\dim(X)\,-1}$ for $Rp_{V*}(\mathcal{F}^\bullet)$ satisfying the conditions (a) to (c) above. Note that a complex of sheaves on $P(V)$ of such a form is the same as a family of bigraded complexes $(C^{\bullet\bullet}, \partial_f)$ with ∂_f polynomially depending on $f \in V$ and satisfying condition (a) of Theorem 5.5. So $(C^{\bullet\bullet}, \partial_f)$ is constructed.

It remains to show that the determinant of this complex equals the X-discriminant. But this is deduced from the quasi-isomorphisms

$$K_{-\dim(X)\,-1} \cong Rp_{V*}(\mathcal{F}^\bullet) \cong Rp_{V*}(p_X^*\mathcal{M} \otimes \mathcal{O}_W),$$

exactly as in the proof of Theorem 2.5. Theorem 5.5 is proved.

Let us note once again that the construction of the differentials in the bigraded complex $C^{\bullet\bullet}$ involves arbitrary choices (coming from the choice of a morphism of complexes in a given homotopy class).

Proof of Lemma 5.7. There is a spectral sequence

$$E_1^{pq} = \bigoplus_{b-a=p} \mathrm{Ext}^q_{\mathcal{O}_X}(\mathcal{G}^a, \mathcal{H}^b) \Rightarrow \mathrm{Hom}_{\mathrm{Der}}(\mathcal{G}^\bullet, \mathcal{H}^\bullet[p+q]), \tag{5.23}$$

where $\mathrm{Hom}_{\mathrm{Der}}$ stands for the morphisms in the derived category. To obtain this sequence, we take a resolution of each $\mathcal{G}^a$ by a complex $I^\bullet_a$ of injective sheaves of $\mathcal{O}_X$-modules, so that these complexes for different a fit together into a double complex $I^{\bullet\bullet}$. Similarly we get a double complex $J^{\bullet\bullet}$ composed of resolutions of the $\mathcal{H}^b$. Denote by $I^\bullet$ and $J^\bullet$ the corresponding total complexes. Then $\mathrm{Hom}_{\mathrm{Der}}(\mathcal{G}^\bullet, \mathcal{H}^\bullet[r])$ is the r-th cohomology of the Hom-complex $\mathrm{Hom}(I^\bullet, J^\bullet)$. Our spectral sequence is obtained by taking a natural filtration in this complex.

Now if $\mathrm{Ext}^q_{\mathcal{O}_X}(\mathcal{G}^a, \mathcal{H}^b) = 0$ for any $q > 0$ and any a, b, then e (5.23) degenerates in the term E_1 which has only the bottom row non-trivial. The cycles with respect to d_1 in this bottom row are elements of $\bigoplus_{b-a=p} \mathrm{Hom}(\mathcal{G}^a, \mathcal{H}^b)$ which are morphisms of complexes (of degree p), and the boundaries are morphisms homotopic to zero. Lemma 5.7 is proved.

CHAPTER 3

Associated Varieties and General Resultants

1. Grassmannians. Preliminary material

The Grassmann variety (or Grassmannian) $G(k, n)$ is the set of all k-dimensional vector subspaces in $\mathbf{C}^n$. For $k = 1$, this is the projective space P^{n-1}. Since vector subspaces in $\mathbf{C}^n$ correspond to projective subspaces in P^{n-1}, we see that $G(k, n)$ parametrizes $(k-1)$-dimensional projective subspaces in P^{n-1}. In a more invariant fashion, we can start from any finite-dimensional vector space V and construct the Grassmannian $G(k, V)$ of k-dimensional vector subspaces in V.

The first example of a Grassmannian other than a projective space is $G(2, 4)$, the variety of all straight lines in the projective space P^3. We shall use this example hereafter. A line in 3-space is easily seen to be dependent on 4 parameters. For example, we may choose some fixed 2-plane (screen) and associate to a generic line the point of intersection with the screen (2 parameters) and the direction at the point of intersection (another two parameters). Thus $G(2, 4)$ is a 4-dimensional manifold.

We are going to give an overview of the various types of coordinates used to represent the points of a Grassmannian. For points in the projective space P^{n-1}, there are two kinds of coordinates:

(a) Affine coordinates $(x_1, \ldots, x_{n-1})$ in an affine chart $\mathbf{C}^{n-1} \subset P^{n-1}$;

(b) homogeneous coordinates $(x_1 : \ldots : x_n)$ defined up to a common constant multiple.

In Grassmannians these two types of coordinates generalize to three types: the affine coordinates, Stiefel coordinates and Plücker coordinates. The Stiefel and Plücker coordinates are two different analogs of homogeneous coordinates in a projective space. We shall describe all three types of coordinates.

A. Affine coordinates

Choose a decomposition of $\mathbf{C}^n$ into the direct sum $\mathbf{C}^k \oplus \mathbf{C}^{n-k}$ of coordinate subspaces. Let $A : \mathbf{C}^k \to \mathbf{C}^{n-k}$ be any linear operator. Then its graph is a k-dimensional subspace in $\mathbf{C}^n$. The set U of subspaces which can be obtained in this way is an open subset of the Grassmannian isomorphic to the affine space $\mathrm{Hom}(\mathbf{C}^k, \mathbf{C}^{n-k})$. Thus any point of U can be determined by a $k \times (n-k)$ matrix $\|a_{ij}\|$. By choosing various coordinate decompositions of $\mathbf{C}^n$, we can cover $G(k, n)$ with affine spaces. It is easy to see that the transition functions between the affine charts thus constructed will have rational functions as components. Thus we have defined a structure of an algebraic variety on $G(k,n)$. The dimension of $G(k,n)$ coincides with the dimension of the space of $k \times (n-k)$-matrices, i.e., it is equal to $k(n-k)$. For a projective space this gives the usual affine coordinates in $\mathbf{C}^{n-1} \subset P^{n-1}$.

Example 1.1. Consider the Grassmannian $G(2,4)$ as the variety of lines in P^3 and let (x, y, z) be the affine coordinates in $\mathbf{C}^3 \subset P^3$. Using affine coordinates in $G(2,4)$ amounts to describing an affine line in $\mathbf{C}^3$ by two equations, $y = a_{11}x + a_{12}$ and $z = a_{21}x + a_{22}$.

B. Stiefel coordinates

Let $v_1, \ldots, v_k$ be a basis of a k-dimensional vector subspace $L \subset \mathbf{C}^n$. Consider the matrix $(k \times n)$ formed by coordinates of vectors v_i:

$$M = \begin{pmatrix} v_{11} & \ldots & \ldots & v_{1n} \\ \ldots & \ldots & \ldots & \ldots \\ v_{k1} & \ldots & \ldots & v_{kn} \end{pmatrix}. \tag{1.1}$$

This is a matrix of full rank k. Given any $k \times n$ matrix M of full rank, we can associate to it the k-dimensional subspace $L(M)$ generated by the rows of M. The matrix entries v_{ij} will be called the *Stiefel coordinates* of L.

For the case $k = 1$, we get the representation of points of P^{n-1} using homogenous coordinates. As with homogeneous coordinates, Stiefel coordinates are not unique. Indeed, we can change M to gM where g is a non-degenerate $k \times k$ matrix (the analog of a constant factor for usual homogeneous coordinates). This will lead to another choice of basis in L. Let $S(k,n)$ be the *Stiefel variety* of all complex $k \times n$ matrices of full rank. Then the Grassmannian $G(k,n)$ is the quotient

$$G(k,n) = S(k,n)/GL(k), \tag{1.2}$$

thus generalizing the representation of the projective space P^{n-1} as the quotient $(\mathbf{C}^n - \{0\})/\mathbf{C}^*$. Here $\mathbf{C}^n - \{0\}$ is $S(n,1)$ and $\mathbf{C}^*$ is $GL(1)$.

The relation between Stiefel coordinates and affine coordinates is as follows. Let $M \subset S(k,n)$ and let g be the $k \times k$-matrix formed by first k columns of M. If g is non-degenerate then we can transform M by the action of $GL(k)$ to the form

$$\begin{pmatrix} 1 & 0 & \ldots & 0 & a_{11} & \ldots & \ldots & a_{1,n-k} \\ 0 & 1 & \ldots & 0 & a_{21} & \ldots & \ldots & a_{2,n-k} \\ \ldots & \ldots & \ldots & \ldots & \ldots & \ldots & \ldots & \ldots \\ 0 & 0 & \ldots & 1 & a_{k1} & \ldots & \ldots & a_{k,n-k} \end{pmatrix} \tag{1.3}$$

where $\|a_{ij}\|$ is a $k \times (n-k)$ matrix. This latter matrix will be the matrix of affine coordinates of the subspace $L(M)$.

C. Plücker coordinates

The Plücker coordinates on $G(k, n)$ give another analog of homogeneous coordinates on a projective space. They are defined up to a common scalar factor. To construct them, we consider the classical *Plücker embedding*

$$G(k, n) \hookrightarrow P\left(\bigwedge^k \mathbf{C}^n\right) \tag{1.4}$$

which takes a k-dimensional subspace $L \subset \mathbf{C}^n$ to the one-dimensional subspace $\bigwedge^k L \subset \bigwedge^k \mathbf{C}^n$. This gives an explicit projective embedding of $G(k, n)$ which, up to now, we had constructed only as an "abstract" algebraic variety.

The coordinates $x_1, \ldots, x_n$ in $\mathbf{C}^n$ induce the coordinates in $\bigwedge^k \mathbf{C}^n$ denoted as $p_{i_1,\ldots,i_k}$, $i_1 < \cdots < i_k$ and called the Plücker coordinates. Thus, every k-dimensional subspace $L \subset \mathbf{C}^n$ has a collection of Plücker coordinates $p_{i_1,\ldots,i_k}(L)$ defined uniquely up to a common non-zero multiple. The number of Plücker coordinates is $\binom{n}{k}$ which is greater than the dimension of the Grassmannian. This means that, in contrast to the case of homogeneous coordinates in the projective space, the Plücker coordinates on $G(k, n)$ are *not independent*, i.e., they are subject to certain relations. These relations will be discussed in subsection D below.

The relation between Plücker and Stiefel coordinates is as follows. Let L be a k-dimensional vector subspace in $\mathbf{C}^n$ given by its $k \times n$ matrix $\|v_{ij}\|$ of Stiefel coordinates. By definition, the Plücker coordinate $p_{i_1,\ldots,i_k}(L)$ is the maximal minor of $\|v_{ij}\|$ formed by columns $i_1, i_2, \ldots, i_k$. If we replace $\|v_{ij}\|$ by $\|w_{ij}\| = g\|v_{ij}\|$ with $g \in GL(k)$ then each maximal minor for $\|w_{ij}\|$ will be obtained from the corresponding minor in $\|v_{ij}\|$ by multiplication with the scalar $\det(g)$.

Suppose that M has the form (1.3). Then the minor $p_{i_1,\ldots,i_k}(M)$ equals (up to sign) a suitable minor of the "short" matrix $A = \|a_{ij}\|$. More precisely, let $I = \{i_1, \ldots, i_k\}$ be the set of columns of our minor of M. This set is decomposed into a disjoint union $I = J \cup K$ where $J = \{i_1, \ldots, i_l\}$ is the set of columns from I which lie inside the unit matrix in (1.3) (i.e., $J = I \cap \{1, \ldots, k\}$) and K is the set of all the other columns. We set $s(J) = (i_1 - 1) + (i_2 - 2) + \cdots + (i_l - l)$. In this notation we have the following immediate fact.

Proposition 1.2. *The minor $p_{i_1,\ldots,i_k}(M)$ of a matrix M of the form (1.3) equals $(-1)^{s(J)}$ times the minor of the matrix $A = \|a_{ij}\|$ given by the set of rows $\{1, \ldots, k\} - J$ and the set of columns K (the empty minor is defined to be equal to 1).*

In other words, the collection of all the Plücker coordinates gives, when written in terms of the affine coordinates, the collection of all the minors (of all sizes, including the empty minor) of an indeterminate $k \times (n-k)$ matrix.

Let V and V^* be dual vector spaces of dimension n. Then $G(k, V)$ is naturally identified with $G(n-k, V^*)$. The identification takes a vector subspace $L \in G(k,n)$ to its orthogonal complement $L^\perp \in G(n-k, V^*)$, and is, therefore, a particular case of the projective duality described in Chapter 1. Thus we can also view $G(k,n)$ as the variety of vector subspaces of *codimension* k in $\mathbf{C}^n$. Correspondingly, we have another way of associating the Plücker coordinates to points of $G(k,n)$. We denote these new coordinates by $q_{j_1,\dots,j_{n-k}}$. If a subspace L is given by $(n-k)$ linear equations $f_1 = 0, \dots, f_{n-k} = 0$ and $f_i(x_1, \dots, x_n) = \sum_j f_{ij}x_j$ then $q_{j_1,\dots,j_{n-k}}$ is the maximal minor of the matrix

$$\begin{pmatrix} f_{11} & \dots & \dots & f_{1n} \\ \dots & \dots & \dots & \dots \\ f_{n-k,1} & \dots & \dots & f_{n-k,n} \end{pmatrix}, \tag{1.5}$$

corresponding to columns with numbers $j_1, \dots, j_{n-k}$. By elementary calculations with determinants, we see that

$$q_{j_1,\dots,j_{n-k}} = (-1)^{s(i_1,\dots,i_k)} p_{i_1,\dots,i_k}, \tag{1.6}$$

where $i_1, \dots, i_k$ are all the elements of the complement of $\{j_1, \dots, j_{n-k}\}$ taken in increasing order and $s(i_1, \dots, i_k)$ is the sign of the permutation $(i_1, \dots, i_k, j_1, \dots, j_{n-k})$. Thus the two sets of Plücker coordinates on $G(k,n)$ are essentially equivalent. Accordingly, we shall use either one of them.

D. Plücker relations

As we mentioned above, the number of Plücker coordinates $p_{i_1,\dots,i_k}$ is greater than the dimension of $G(k,n)$, and hence the $p_{i_1,\dots,i_k}$ are subject to certain relations. These relations are called *Plücker relations.* To write them down, it is convenient to assume that the $p_{i_1,\dots,i_k}$ are defined for any sequence $(i_1, \dots, i_k)$ of distinct integers between 1 and n so that the transposition of any two indices changes the sign of $p_{i_1,\dots,i_k}$ (thus effectively we have the same number of coordinates). In this notation the relations are as follows.

Theorem 1.3.

(a) *For any two sequences $1 \le i_1 < \dots < i_{k-1} \le n$ and $1 \le j_1 < \dots < j_{k+1} \le n$, the Plücker coordinates on $G(k,n)$ satisfy the relation*

$$\sum_{a=1}^{k+1} (-1)^a p_{i_1,i_2,\dots,i_{k-1},j_a} p_{j_1,j_2,\dots,\hat{j}_a,\dots,j_{k+1}} = 0 \tag{1.7}$$

(here the symbol $\hat{j}_a$ means that the index j_a is omitted). Any vector $(p_{i_1,\dots,i_k}) \in \bigwedge^k \mathbf{C}^n$ satisfying all such relations is a vector of the Plücker coordinates of some vector subspace $L \in G(k, n)$.

(b) *Moreover, the graded ideal of all polynomials in $p_{i_1,\dots,i_k}$ vanishing on the image of $G(k, n)$ is generated by the left hand sides of the Plücker relations (1.7).*

We shall not prove this theorem here. The proof of assertion (a) (stated in a more invariant form recalled below) can be found in [GH]. In addition, in Section 1 of Chapter 4 we shall prove a more general Chow–van der Waerden theorem which describes (at least theoretically) the equations of Chow varieties of which the Grassmannian is a particular case. For the particular case of the Grassmannian, the Chow–van der Waerden theorem gives exactly the Plücker relations, as we shall have the opportunity to see. A still more comprehensive treatment of the subject, including the proof of part (b) of Theorem 1.3, can be found in [HP], Chapter 6, Sections 6–7.

Example 1.4. The Grassmannian $G(2, 4)$, which parametrizes lines in P^3, has dimension 4, and is embedded by the Plücker embedding into P^5. There are six Plücker coordinates p_{ij}, $1 \le i < j \le 4$ which are subject to one relation

$$p_{12}p_{34} - p_{13}p_{24} + p_{14}p_{23} = 0. \tag{1.8}$$

Thus $G(2, 4)$ is a quadric in P^5. In terms of the matrix of affine coordinates $A = \|a_{ij}\|$, $i, j = 1, 2$, the Plücker coordinates are the following polynomials in the a_{ij} (Proposition 1.2):

$$p_{12} = 1, \ \ p_{34} = \det(A), \ \ p_{13} = a_{21}, \ \ p_{24} = -a_{12}, \ \ p_{14} = a_{22}, \ \ p_{23} = -a_{11}.$$

The Plücker relation now becomes the identity $\det(A) = a_{11}a_{22} - a_{12}a_{21}$.

Note that in the general case Theorem 1.3 (a) can be seen, after passing to the affine coordinates, as providing the necessary and sufficient conditions for the collection of the Δ_{IJ} defined for all pairs of subsets $I \subset \{1, \dots, k\}$, $J \subset \{1, \dots, n-k\}$, $\#(I) = \#(J)$ to be represented by minors of a $k \times (n-k)$-matrix. The entries of this matrix should be given by 1×1-minors Δ_{ij} in our collection. Thus an obvious solution would be to require that each Δ_{IJ} equals the determinant of the matrix of Δ_{ij}, $i \in I$, $j \in J$. These equations are not quadratic. The possibility of replacing them by a system of quadratic equations comes from considering not just 1×1 and $k \times k$ minors, but those of all sizes.

We finish the discussion of the Plücker relations by describing them in a more invariant form. Denote $V = \mathbf{C}^n$ and let $e_1, \dots, e_n$ be the standard basis of V. A collection $(p_{i_1,\dots,i_k})$ represents a point L of $G(k,n)$ if and only if the k-vector

$$R = \sum_{i_1<\dots<i_k} p_{i_1,\dots,i_k} e_{i_1} \wedge \dots \wedge e_{i_k} \in \bigwedge^k V$$

is *decomposable*, i.e., it has the form $R = v_1 \wedge \dots \wedge v_k,\ \ v_i \in V$ (in which case the v_i form a basis of L). So the problem of finding the relations among the $p_{i_1,\dots,i_k}$ is equivalent to finding conditions for a k-vector R to be decomposable.

We consider, for any linear form $l \in V^*$, the contraction operator

$$i_l : \bigwedge^j V \longrightarrow \bigwedge^{j-1} V \tag{1.9}$$

(cf. Chapter 2, Section 1B). Clearly, $i_{l_1} i_{l_2} = -i_{l_2} i_{l_1}$ for any $l_1, l_2 \in V^*$. For any $\Omega \in \bigwedge^p V^*$, we define the operator $i_\Omega : \bigwedge^j V \longrightarrow \bigwedge^{j-p} V$ of contraction with Ω, setting

$$i_{l_1 \wedge \dots \wedge l_p} = i_{l_1} \circ \dots \circ i_{l_p} \tag{1.10}$$

for a decomposable $\Omega = l_1 \wedge \dots \wedge l_p$, and extending this by linearity to all Ω. In particular, if $p = j$ then, for $R \in \bigwedge^j V$, the element $i_\Omega R$ is a number denoted by $< \Omega, R >$. This is the canonical pairing between $\bigwedge^j V^*$ and $\bigwedge^j V$. Now a coordinate-free version of the Plücker relations is as follows.

Theorem 1.5. *A k-vector $R \in \bigwedge^k V$ is decomposable if and only if, for any $\Omega \in \bigwedge^{k-1} V^*$, we have*

$$(i_\Omega R) \wedge R = 0 \quad \text{in} \quad \bigwedge^{k+1} V. \tag{1.11}$$

It is immediate to see the equivalence of Theorems 1.3 (a) and 1.5. Indeed, the choice of $i_1, \dots, i_{k-1}$ in Theorem 1.3 (a) amounts to a choice of Ω in the form $e'_{i_1} \wedge \dots \wedge e'_{i_{k-1}}$ where the $e'_i \in V^*$ are elements of the basis dual to $\{e_1, \dots, e_n\}$. A choice of $j_1, \dots, j_{k+1}$ gives a choice of a coordinate function on $\bigwedge^{k+1} V$ corresponding to the basis vector $e_{j_1} \wedge \dots \wedge e_{j_{k+1}}$. Evaluating this coordinate of $(i_\Omega R) \wedge R$ for $R = (p_{j_1,\dots,j_k}) \in \bigwedge^k V$, we obtain the left hand side of (1.7).

E. The coordinate ring of the Grassmannian

Let $\mathcal{B} = \bigoplus_d \mathcal{B}_d$ be the homogeneous coordinate ring of $G(k,n)$ in the Plücker embedding. In other words, $\mathcal{B}$ is the quotient of the polynomial ring $\mathbf{C}[p_{i_1,\dots,i_k}]$ by the ideal generated by the relations (1.7), and $\mathcal{B}_d$ is the homogeneous part of $\mathcal{B}$ of degree d. The generators $p_{i_1,\dots,i_k}$ of $\mathcal{B}$ are often denoted by $[i_1, \dots, i_k]$

and are called *brackets*. Thus any element of $\mathcal{B}$ can be represented as a *bracket polynomial*, i.e., a polynomial in $[i_1, \ldots, i_k]$. Such a representation is, of course, not unique because of the Plücker relations. Let $\tilde{G}(k, n) \subset \bigwedge^k \mathbf{C}^n$ be the cone over the Grassmannian in the Plücker embedding consisting of k-vectors which are decomposable; that is, they have the form $v_1 \wedge \cdots \wedge v_k$. Any element of $\mathcal{B}$ can be regarded as a function on $\tilde{G}(k, n)$.

We have defined the ring $\mathcal{B}$ and its homogeneous components $\mathcal{B}_d$ using the Plücker embedding (and hence the Plücker coordinates). The two other types of coordinates give two different descriptions of $\mathcal{B}$ and $\mathcal{B}_d$.

The following characterization of $\mathcal{B}_d$ in terms of the Stiefel and affine coordinates follows from (1.2) and Proposition 1.2. It is sometimes called *the first fundamental theorem of invariant theory.*

Proposition 1.6.

(a) *The space $\mathcal{B}_d$ is naturally identified with the space of polynomials f in the entries of an indeterminate $(k \times n)$-matrix $M = \|v_{ij}\|$ of Stiefel coordinates satisfying the condition*

$$f(gM) = \det(g)^d f(M) \tag{1.12}$$

for any matrix $g \in GL(k)$. The multiplication in the ring $\mathcal{B}$ corresponds to the usual multiplication of polynomials.

(b) *The space $\mathcal{B}_d$ is naturally identified with the space of polynomials in the entries of an indeterminate $k \times (n-k)$-matrix (of affine coordinates) $A = \|a_{ij}\|$ generated by d-tuple products of minors of A of all sizes (including the empty minor 1).*

The group $GL(n)$ acts on the Grassmannian $G(k, n)$ and hence on its coordinate ring $\mathcal{B}$. It is known [FH] that the representation of $GL(n)$ in each $\mathcal{B}_d$ is irreducible. Under the usual parametrization of irreducible polynomial representations of $GL(n)$ by Young diagrams, this representation corresponds to the diagram with k rows of length d.

2. Associated hypersurfaces

The Grassmannian $G(k, n)$ can be seen as the variety parametrizing $(k-1)$-dimensional projective subspaces in P^{n-1}, i.e., $(k-1)$-dimensional subvarieties of degree 1. We shall discuss in this section (and later in Chapter 4) the problem of constructing varieties of a higher degree that parametrize $(k-1)$-dimensional subvarieties in P^{n-1}. The approach to this problem, initiated by Cayley [Ca5] and carried out in the general case by Chow and van der Waerden [C–vdW], starts from the remark that for subvarieties of codimension 1, the problem is easily solved; if $X \subset P^{n-1}$ is a hypersurface of degree d, then we can consider its equation which is a homogeneous polynomial of degree d, defined uniquely up to a constant factor. In other words, the space of hypersurfaces of degree d can be identified with the projective space which is the projectivization of the space of homogeneous polynomials of degree d.

The one-to-one correspondence between hypersurfaces and their equations holds not only for projective spaces, but essentially for any variety with a proper understanding of what is an "equation." The idea now is to associate to any irreducible subvariety X in P^{n-1} a hypersurface $\mathcal{Z}(X)$ in a certain Grassmannian from which X can be recovered.

We start with a discussion of hypersurfaces in Grassmannians and their equations.

A. Hypersurfaces in Grassmannians

Given a hypersurface in a projective space, its degree can be defined as the number of points where it intersects a generic line. The analog of lines in $G(k, n)$ is given by the following family of embedded P^1. Let $N \subset M \subset P^{n-1}$ be a flag formed by $(k-2)$-dimensional and k-dimensional projective subspaces. All $(k-1)$-dimensional projective subspaces containing N and contained in M form a one-dimensional family (pencil) $P_{NM} \cong P^1$. We define the *degree* of a hypersurface $Z \subset G(k, n)$ to be the intersection number of Z with a generic pencil of the form P_{NM}.

In general, it is not true that a hypersurface in a projective variety can be given by the vanishing of an element of its coordinate ring (take the variety in question to be a plane curve and the hypersurface to be a point). However, this is the case for Grassmannians.

Let $\mathcal{B} = \bigoplus \mathcal{B}_d$ be the coordinate ring of $G(k, n)$ in the Plücker embedding (see Section 1E above).

Proposition 2.1.

(a) *The ring $\mathcal{B}$ is factorial, i.e., each $f \in \mathcal{B}$ has a decomposition into irreducible factors which is unique up to constant multiples and permutations of these factors.*

(b) *Let Z be an irreducible hypersurface in $G(k, n)$ of degree d. Then there is an element $f \in \mathcal{B}_d$ defined uniquely up to a constant factor such that Z is given by the equation $f = 0$.*

Proof. The factoriality of $\mathcal{B}$ follows from Proposition 1.6 and the factoriality of the polynomial ring $\mathbf{C}[v_{ij}]$. Indeed, any factor of a polynomial from $\mathcal{B}$ should lie in $\mathcal{B}$. Similarly, if Z is a hypersurface in $G(k, n)$, we lift it to a hypersurface $\tilde{Z}$ in the space of matrices and take the irreducible polynomial $f(v_{ij})$ defining $\tilde{Z}$.

As before, denote by M the matrix of the variables v_{ij}. Since $\tilde{Z}$ is $GL(k)$-invariant, the polynomial $f(gM)$ for any $g \in GL(k)$ is a non-zero scalar multiple of $f(M)$, i.e., $f(gM) = \chi(g)f(M)$. The function $\chi(g)$ is multiplicative: $\chi(g_1g_2) = \chi(g_1)\chi(g_2)$, i.e., it is a character of $GL(k)$. Any such function is a power of the determinant: $\chi(g) = \det(g)^{d'}$ for some d'. By Proposition 1.6 (a), f defines an irreducible element of $\mathcal{B}_{d'}$.

It remains to prove that d' coincides with d, the degree of Z defined above. To do this, note that the equality $f(gM) = \det(g)^{d'} f(M)$ means that f is a section of the invertible sheaf $\mathcal{O}_{G(k,n)}(d')$ (restricted from the projective space of the Plücker embedding). We claim that the restriction of any $\mathcal{O}_{G(k,n)}(l)$ to any pencil $P_{NM} \cong P^1 \subset G(k, n)$ is isomorphic to $\mathcal{O}_{P^1}(l)$. This will imply that $d' = d = \deg Z$ since the section f of $\mathcal{O}_{G(k,n)}(d')$ will have exactly d' zeros on P_{NM}. To justify our claim, it is enough to consider the case $l = 1$, since $\mathcal{O}_{P^1}(l + l') \cong \mathcal{O}_{P^1}(l) \otimes \mathcal{O}_{P^1}(l')$, and similarly for $\mathcal{O}_{G(k,n)}(l + l')$. To treat the case $l = 1$, it suffices to take a non-zero section $\varphi \in H^0(G(k, n), \mathcal{O}(1))$ and show that it has exactly one zero on a generic pencil P_{NM}. We take φ to be a single bracket in $\mathcal{B}$, say $[1, \ldots, k]$. The corresponding hypersurface consists of $(k - 1)$-dimensional projective subspaces which intersect the projective span of basis vectors $e_{k+1}, \ldots, e_n$. A generic pencil P_{NM} contains precisely one $(k - 1)$-dimensional projective subspace belonging to this hypersurface. This completes the proof.

B. Associated hypersurfaces and Chow forms

Let $X \subset P^{n-1}$ be an irreducible subvariety of dimension $k - 1$ and degree d. Consider the set $\mathcal{Z}(X)$ of all $(n - k - 1)$-dimensional projective subspaces L in P^{n-1} that intersect X. This is a subvariety in the Grassmannian $G(n - k, n)$ parametrizing all the $(n - k - 1)$-dimensional projective subspaces in P^n.

Proposition 2.2. *The subvariety $\mathcal{Z}(X)$ is an irreducible hypersurface of degree d in $G(n - k, n)$.*

Proof. The variety $\mathcal{Z}(X)$ is included in the diagram (double fibration)

$$\mathcal{Z}(X) \xleftarrow{q} B(X) \xrightarrow{p} X, \tag{2.1}$$

where $B(X)$ is the variety of pairs (x, L) such that $x \in X$, $L \in G(n - k, n)$ and $x \in L$. The projections q and p are given by forgetting x or L. For dimension reasons, a generic $(n-k-1)$-dimensional projective subspace intersecting X meets

X at only one point. Therefore q is a birational isomorphism. The projection p is a Grassmannian fibration: the fiber of p at $x \in X$ is isomorphic to the Grassmannian $G(n-k-1, n-1)$. Thus if X is irreducible then so is $B(X)$ and hence $\mathcal{Z}(X)$. Furthermore, we have

$$\dim \mathcal{Z}(X) = \dim B(X) = (n-k-1)k + (k-1) = k(n-k) - 1$$
$$= \dim G(n-k, n) - 1.$$

To find the degree of $\mathcal{Z}(X) \subset G(n-k, n)$, we should, according to the definition in subsection A, choose a generic flag $N \subset M \subset P^{n-1}$ of projective subspaces such that $\dim N = n-k-2$, $\dim M = n-k$, and count the number of $n-k-1$-dimensional subspaces $L \in \mathcal{Z}(X)$ satisfying $N \subset L \subset M$. But since $\deg X = d$, the intersection $M \cap X$ will typically consist of d points, say, $x_1, \ldots, x_d$. The subspaces $L \in \mathcal{Z}(X)$ containing N and contained in M will be the projective spans of N and x_i. So their number is equal to d, as required. This proves Proposition 2.2.

We shall call $\mathcal{Z}(X)$ the *associated hypersurface* of X. The construction of $\mathcal{Z}(X)$ can be regarded as an analog of the construction of the projective dual variety in Chapter 1.

Let $\mathcal{B} = \bigoplus \mathcal{B}_m$ be the coordinate ring of the Grassmannian $G(n-k, n)$. By Proposition 2.1, $\mathcal{Z}(X)$ is defined by the vanishing of some element $R_X \in \mathcal{B}_d$ which is unique up to a constant factor. This element will be called the *Chow form* of X. Choosing some basis in $\mathcal{B}_d$, we associate to X the collection of coordinates of R_X in this basis, defined up to a common constant factor. These coordinates will be called the *Chow coordinates* of X. We shall see in subsection C below that X can be recovered from its Chow coordinates, i.e., from the vector R_X.

The three kinds of coordinates on Grassmannians give three different ways of writing elements of the coordinate ring and, in particular, the Chow form of a subvariety.

The approach with the Plücker coordinates leads to writing R_X as a bracket polynomial. The Plücker coordinates in $G(n-k, n)$ are labeled by $(n-k)$-element subsets of $\{1, \ldots, n\}$ and denoted by brackets $[i_1, \ldots, i_{n-k}]$. Sometimes it is convenient to use the dual Plücker coordinates $[j_1, \ldots, j_k]$ (see subsection 1C) where $\{j_1, \ldots, j_k\}$ is the complement of $\{i_1, \ldots, i_{n-k}\}$ in $\{1, \ldots, n\}$.

The Stiefel coordinates represent an $(n-k)$-dimensional subspace in $\mathbf{C}^n$ as the zero set of k linear functionals $l_1(x), \ldots, l_k(x)$, where $l_i(x) = \sum_{j=1}^{n} c_{ij} x_j$. The space of such k-tuples of functionals can be viewed as the space $\mathrm{Mat}(k, n)$ of matrices $\|c_{ij}\|$. The open subset $S(k, n) \subset \mathrm{Mat}(k, n)$, consisting of matrices of full rank k, is fibered over the Grassmannian $G(n-k, n)$, see formula (1.2) above.

Consider the lifting of the associated hypersurface $\mathcal{Z}(X)$ to $S(k, n)$ and its closure $\tilde{\mathcal{Z}}(X)$ in the affine space Mat (k, n). Thus $\tilde{\mathcal{Z}}(X)$ is the subvariety of $(f_1, \ldots, f_k)$ such that

$$X \cap \{f_1 = \cdots = f_k = 0\} \neq \emptyset. \tag{2.2}$$

Since $\mathcal{Z}(X)$ is a hypersurface, it follows that $\tilde{\mathcal{Z}}(X)$ is also a hypersurface. The defining polynomial of $\tilde{\mathcal{Z}}(X)$ is a homogeneous polynomial $\tilde{R}_X(f_1, \ldots, f_k)$. We call $\tilde{R}_X$ the *X-resultant* since its vanishing expresses the compatibility of the system (2.2). It is clear that $\tilde{R}_X$ is just another way of writing the Chow form R_X. More precisely, $\tilde{R}_X$ is the polynomial in c_{ij} corresponding to $R_X \in \mathcal{B}_d$ under the isomorphism of Proposition 1.6 (a). Thus to obtain $\tilde{R}_X$, we need to replace each bracket $[j_1, \ldots, j_k]$ in the (dual) bracket representation of R_X by a polynomial $\det \|c_{i,j_\nu}\|$, $i, \nu = 1, \ldots, k$. The X-resultant $\tilde{R}_X$ has the advantage of being an explicit polynomial in variables c_{ij}, not just an element of some "abstract" vector space.

To write the Chow form (or X-resultant) in terms of affine coordinates, we assume that the coefficient matrix $\|c_{ij}\|$ of a system of linear forms $f_1, \ldots, f_k$ defining a subspace, has the form (1.3). In other words, we set $c_{ij} = \delta_{ij}$ for $1 \leq i, j \leq k$. In this way we associate to a subvariety X a non-homogeneous polynomial in the entries of an indeterminate $k \times (n-k)$-matrix. Such an expression will be useful in Chapter 4 when we shall try to determine which elements of $\mathcal{B}_d$ can arise as Chow forms.

We now give some examples of associated hypersurfaces and Chow forms.

Examples 2.3. (a) Let X be a curve in P^3. Its associated hypersurface is the variety of all lines which intersect X.

(b) When X itself is a hypersurface in P^{n-1}, the Grassmannian $G(n-k, n)$ coincides with P^{n-1} and the associated hypersurface $\mathcal{Z}(X)$ coincides with X.

(c) When X is a point p, then $G(n-k, n)$ is the dual projective space $(P^{n-1})^*$ and $\mathcal{Z}(X)$ is the hyperplane dual to p.

(d) When X is a projective subspace, the variety $\mathcal{Z}(X)$ is known as the *Schubert divisor* in $G(n-k, n)$. We can suppose (after a linear change of coordinates) that the linear equations of X are $x_1 = 0, \ldots, x_{n-k} = 0$ where $x_1, \ldots, x_n$ are coordinate functions. Then the associated variety is given by the vanishing of the bracket (Plücker coordinate) $[1, \ldots, n-k] \in \mathcal{B}_1$.

Example 2.4. Let X be the projective subspace $P^{k-1} = P(\mathbf{C}^k)$, embedded into $P(S^d\mathbf{C}^k) = P^{n-1}$ by the Veronese embedding. Here $n = \binom{k+d-1}{d}$ is the number of all monomials in k variables of degree d. A linear form on P^{n-1} when restricted on P^{k-1} gives a homogeneous polynomial of degree d. Thus, the X-resultant $\tilde{R}_X$

is a function of coefficients of k indeterminate polynomials $f_1(t), \ldots, f_k(t)$ of degree d, where $t = (t_1, \ldots, t_k)$. The vanishing of $\tilde{R}_X$ means that the polynomials f_i have a non-trivial common zero. Therefore $\tilde{R}_X$ coincides with the classical resultant of k homogeneous polynomials of degree d in k variables. This explains our terminology.

C. Recovery of X from $\mathcal{Z}(X)$

Proposition 2.5. *A $(k-1)$-dimensional irreducible subvariety $X \subset P^{n-1}$ is uniquely determined by its associated hypersurface $\mathcal{Z}(X)$. More precisely, a point $p \in P^{n-1}$ lies in X if and only if any $(n-k-1)$-dimensional plane containing p belongs to $\mathcal{Z}(X)$.*

The proof is obvious.

This proposition is an analog of the Biduality Theorem of Chapter 1. As in the case of projective dual varieties, the variety X is recovered as a sort of "caustic variety" of $\mathcal{Z}(X)$. The above proposition makes it clear that X can be recovered from its Chow form or, equivalently, from the coefficients of the X-resultant $\tilde{R}_X(f_1, \ldots, f_k)$. This can be done explicitly as follows. Let $x \in P^{n-1}$ be a point and let $\mathbf{x} \in \mathbf{C}^n$ be some vector whose projectivization is x. Let us represent hyperplanes through x as projectivizations of orthogonal complements

$$\mathbf{x}_S^{\perp} = \{\mathbf{y} : S(\mathbf{x}, \mathbf{y}) = 0\}$$

where $S(\mathbf{x}, \mathbf{y})$ is a skew-symmetric form on $\mathbf{C}^n$. We shall denote the linear form $\mathbf{y} \mapsto S(\mathbf{x}, \mathbf{y})$ by $i_{\mathbf{x}}S$.

Corollary 2.6. *Let $X^{k-1} \subset P^{n-1}$ be an irreducible subvariety and $\tilde{R}_X(f_1, \ldots, f_k)$ be the X-resultant. Let us now consider k indeterminate skew-symmetric forms $S_1(\mathbf{x}, \mathbf{y}), \ldots, S_k(\mathbf{x}, \mathbf{y})$ defined by $S_i(\mathbf{x}, \mathbf{y}) = \sum_{j,r} s_{jr}^{(i)} x_j y_r$, where $\|s_{jr}^{(i)}\|$ for each i is a skew-symmetric matrix of (otherwise) independent variables. For any $\mathbf{x} \in \mathbf{C}^n$, consider the following polynomial in coefficients $(s_{jr}^{(i)})$ of all forms S_i:*

$$P(\mathbf{x}, (s_{jr}^{(i)})) = \tilde{R}_X(i_{\mathbf{x}}(S_1), \ldots, i_{\mathbf{x}}(S_k)).$$

Then the coefficients of P (with respect to the variables $s_{jr}^{(i)}$) are polynomials in $\mathbf{x}$ which form a system of equations for the variety X. In other words, the vanishing of these polynomials defines X set-theoretically.

This corollary (due to Chow and van der Waerden) gives a canonical system of equations for each irreducible variety X of degree d. All these equations have degree d, with a number usually much greater than the codimension of X. It was recently shown by F. Catanese [Cat] that, for smooth X, this canonical system of equations defines X not just set-theoretically, but scheme-theoretically as well. In other words, the *subscheme in* P^{n-1}, defined by these equations, coincides with X (regarded as a subscheme in a standard way). Algebraically, this means that the homogeneous ideal in $\mathbf{C}[x_1, \ldots, x_n]$, formed by the homogeneous polynomials vanishing on X, coincides with the homogeneous ideal generated by the canonical equations in all but a finite number of graded components.

D. The Cayley trick

Let us now describe an important relationship between associated hypersurfaces and projective duality. It turns out that any associated hypersurface can be essentially identified with some dual variety. More precisely, let X be an irreducible closed subvariety of dimension $k-1$ in P^{n-1}. Consider the product $\tilde{X} = X \times P^{k-1}$ as a subvariety of $P((\mathbf{C}^{n\times k})^*) := P(\mathbf{C}^n \otimes (\mathbf{C}^k)^*)$ via the Segre embedding. Identify $\mathbf{C}^n \otimes (\mathbf{C}^k)^*$ with the space Mat (k, n) of $k \times n$-matrices and consider the projection

$$\mathrm{Mat}\,(k, n) \supset S(k, n) \xrightarrow{p} G(n-k, n).$$

Theorem 2.7. *Let X be an irreducible closed subvariety of dimension $k-1$ in P^{n-1} and $\mathcal{Z}(X) \subset G(n-k, \mathbf{C}^n)$ its associated hypersurface. Then the projectively dual variety $\tilde{X}^\vee$ of $\tilde{X}$ equals the closure $\overline{p^{-1}(\mathcal{Z}(X))}$.*

Proof. By definition, $\tilde{X}^\vee$ is the closure of the locus of all $f \in P(\mathbf{C}^{n\times k})$ such that the hyperplane $\{f = 0\} \subset P((\mathbf{C}^{n\times k})^*)$ is tangent to $\tilde{X} = X \times P^{k-1}$ at some smooth point (x, y). This can be reformulated as follows. Let $\mathbf{x} \in \mathbf{C}^n$, $\mathbf{y} \in (\mathbf{C}^k)^*$, $\mathbf{f} \in \mathbf{C}^{n\times k}$ be vectors with projectivizations x, y and f, and let $U_\mathbf{x} \subset \mathbf{C}^n$ be the tangent space at $\mathbf{x}$ to the cone over X. The tangent space at $(\mathbf{x}, \mathbf{y})$ to the cone over $\tilde{X}$ equals $(U_\mathbf{x} \otimes \mathbf{y}) + (\mathbf{x} \otimes (\mathbf{C}^k)^*) \subset (\mathbf{C}^{n\times k})^*$. Therefore f lies in $\tilde{X}^\vee$ if and only if $\mathbf{f}$, considered as a linear form on $(\mathbf{C}^{n\times k})^*$, is orthogonal to a subspace of the form $(U_\mathbf{x} \otimes \mathbf{y}) + (\mathbf{x} \otimes (\mathbf{C}^k)^*)$ for some smooth $x \in X$ and some $y \in P^{k-1}$.

On the other hand, the definitions imply that f lies in $\overline{p^{-1}(\mathcal{Z}(X))}$ if and only if $\mathbf{f}$ is orthogonal to $\mathbf{x} \otimes (\mathbf{C}^k)^*$ for some $x \in X$. Those f, for which this x is smooth, form an open dense subset in $\overline{p^{-1}(\mathcal{Z}(X))}$. Therefore it remains to prove that, if $\mathbf{f}$ is orthogonal to $x \otimes (\mathbf{C}^k)^*$, then there exists a non-zero $\mathbf{y} \in (\mathbf{C}^k)^*$ such that $\mathbf{f}$ is orthogonal to $U_\mathbf{x} \otimes \mathbf{y}$.

Now think of $\mathbf{f}$ as a family $(f_1, \ldots, f_k)$ of linear forms on $\mathbf{C}^n$. Since $\mathbf{f}$ is orthogonal to $\mathbf{x} \otimes (\mathbf{C}^k)^*$, then $f_1(x) = \cdots = f_k(x) = 0$. It follows that

$\{(f_1(u), \ldots, f_k(u)) \ : \ u \in U_{\mathbf{x}}\}$ is a proper subspace of $\mathbf{C}^k$ since $U_{\mathbf{x}}$ is a k-dimensional vector space containing $\mathbf{x}$, and so there is a non-zero $\mathbf{y} \in (\mathbf{C}^k)^*$ orthogonal to this subspace. But this exactly means that $\mathbf{f}$ is orthogonal to $U_{\mathbf{x}} \otimes \mathbf{y}$, and Theorem 2.7 is proved.

Corollary 2.8. *For every irreducible closed subvariety X of dimension $k-1$ in P^{n-1}, we have $\tilde{R}_X = \Delta_{\tilde{X}}$, i.e., the X-resultant equals the $\tilde{X}$-discriminant.*

Example 2.9. We have already seen that the classical resultant of k-homogeneous forms $f_1, \ldots, f_k$ of the same degree d in k variables $x_1, \ldots, x_k$ is the X-resultant for the Veronese subvariety $X = P^{k-1} \subset P(S^d(\mathbf{C}^k))$. In this case Corollary 2.8 shows that the resultant $\tilde{R}_X(f_1, \ldots, f_k)$ is equal to the discriminant of the polynomial

$$\tilde{f}(x_1, \ldots, x_k, y_1, \ldots, y_k) = y_1 f_1(x_1, \ldots, x_k) + \cdots + y_k f_k(x_1, \ldots, x_k),$$

where $y_1, \ldots, y_k$ are new variables. This observation was used by Cayley. Following [GZK3], we call the general result of Corollary 2.8 the *Cayley trick.*

E. Higher associated hypersurfaces

The construction associating to a $(k-1)$-dimensional variety $X \subset P^{n-1}$ its associated hypersurface in $G(n-k, n)$ can be generalized as follows. Let $X \subset P^{n-1}$ be an irreducible subvariety of dimension $m-1 \geq k-1$. First we suppose that X is smooth. Define its $(m-k)$-th associated subvariety $\mathcal{Z}_{m-k}(X) \subset G(n-k, n)$ to be the set of $(n-k-1)$-dimensional subspaces $L \subset P^{n-1}$ such that $L \cap X \neq \emptyset$ and $\dim (L \cap ET_x X) \geq m-k$ for some $x \in L \cap X$. Here $ET_x X$ is the *embedded tangent space* to X at x, i.e., the projective subspace in P^{n-1} of dimension $m-1 = \dim X$ which contains x and has the same tangent space at x as X.

When X is singular, we first consider the subspaces L for which there exists a smooth point x with the above property, and then take the Zariski closure of the set of these subspaces in $G(n-k, n)$. It is clear that the 0-th associated variety of a $(k-1)$-dimensional subvariety is the one previously considered.

Example 2.10. Let $S \subset P^3$ be a surface. Then the first associated variety $\mathcal{Z}_1(S) \subset G(2, 4)$ is the set of all lines in P^3 tangent to S.

Proposition 2.11. *Let X be an irreducible $(m-1)$-dimensional subvariety in P^{n-1} and $k \leq m$. Then $\mathcal{Z}_{m-k}(X)$ is a hypersurface in $G(n-k, n)$.*

The proof is similar to that of Proposition 2.2 and we leave it to the reader.

The "higher associated hypersurfaces" just defined are not necessary for the problem of coordinatizing subvarieties; the usual construction suffices. But we shall see in Chapter 4, Section 3 that they share the key differential-geometric property of ordinary associated hypersurfaces.

3. Mixed resultants

The resultants under study so far were generalizations of the classical resultant of k homogeneous polynomials in k variables *of the same degree d*. This is obviously not the most general setting since we can consider polynomials of different degrees. Of course, such a resultant cannot be interpreted in terms of the one Veronese embedding, since different degrees require different embeddings. In this section we shall abandon the Chow form point of view of resultants and study "mixed" resultants in their own context.

A. The $(\mathcal{L}_1, \ldots, \mathcal{L}_k)$-resultant and its first properties

Let X be a compact algebraic variety of dimension $k-1$. We assume that X is irreducible. Instead of polynomials, for general X we must consider sections of invertible sheaves. Let $\mathcal{L}_1, \ldots, \mathcal{L}_k$ be k invertible sheaves on X. We suppose that each $\mathcal{L}_i$ is very ample, i.e., it defines a projective embedding $\varphi_i : X \to P(V_i^*)$ where $V_i = H^0(X, \mathcal{L}_i)$. A section of $\mathcal{L}_i$ can be seen as the restriction of a linear form from $P(V_i)$ to X. For such a section f_i, denote by $X(f_i) \subset X$ the hypersurface defined by $f_i = 0$. A collection $(f_1, \ldots, f_k) \in \prod_i V_i$ is called *degenerate* if $X(f_1) \cap \cdots \cap X(f_k) \neq \emptyset$. We denote by

$$\nabla_{\mathcal{L}_1, \ldots, \mathcal{L}_k} \subset \prod_{i=1}^{k} V_i$$

the locus of all degenerate tuples of sections. We call this locus the *resultant variety* (associated to $\mathcal{L}_1, \ldots, \mathcal{L}_k$).

Proposition 3.1. *The variety $\nabla_{\mathcal{L}_1, \ldots, \mathcal{L}_k}$ is an irreducible hypersurface in $\prod_i V_i$ invariant under dilations of each V_i.*

Proof. As in the proof of Proposition 2.2, we consider the diagram

$$X \xleftarrow{p_1} B_{\mathcal{L}_1, \ldots, \mathcal{L}_k} \xrightarrow{p_2} \prod V_i, \tag{3.1}$$

where $B_{\mathcal{L}_1, \ldots, \mathcal{L}_k} \subset X \times \prod V_i$ is the set of $(x, f_1, \ldots, f_k)$ such that $f_1(x) = \cdots = f_k(x) = 0$. The maps p_1, p_2 in (3.1) are the projections to X and $\prod V_i$, respectively. Clearly,

$$\nabla_{\mathcal{L}_1, \ldots, \mathcal{L}_k} = p_2\left(B_{\mathcal{L}_1, \ldots, \mathcal{L}_k}\right).$$

So the variety $\nabla_{\mathcal{L}_1,\dots,\mathcal{L}_k}$, being the cone over the image of a projective variety under a regular map, is Zariski closed. Let $n_i = \dim V_i$. For any $x \in X$ the fiber $p_1^{-1}(x)$ is the product of hyperplanes

$$\{f_i \in V_i = H^0(X, \mathcal{L}_i) : f_i(x) = 0\},$$

and hence $\dim p_1^{-1}(x) = \left(\sum n_i\right) - k$. Moreover, since p_2 is the projection of a vector bundle, the variety $B_{\mathcal{L}_1,\dots,\mathcal{L}_k}$ is irreducible. Hence $\nabla_{\mathcal{L}_1,\dots,\mathcal{L}_k}$ is irreducible.

Let us now look at the fibers of p_2. Clearly, $p_2^{-1}(f_1, \dots, f_k)$ is the set of the common zeros of $f_1, \dots, f_k$. Since each $\mathcal{L}_i$ is very ample, we conclude that, for generic $(f_1, \dots, f_k) \in \nabla_{\mathcal{L}_1,\dots,\mathcal{L}_k}$, this set of common zeros has just one point so that the map

$$p_2 : B_{\mathcal{L}_1,\dots,\mathcal{L}_k} \longrightarrow \nabla_{\mathcal{L}_1,\dots,\mathcal{L}_k} \tag{3.2}$$

is a birational isomorphism. Hence

$$\dim \nabla_{\mathcal{L}_1,\dots,\mathcal{L}_k} = \dim B_{\mathcal{L}_1,\dots,\mathcal{L}_k} = \dim X + \left(\sum n_i\right) - k = \left(\sum n_i\right) - 1.$$

Thus $\nabla_{\mathcal{L}_1,\dots,\mathcal{L}_k}$ is a hypersurface. Proposition 3.1 is proved.

We call the irreducible equation of $\nabla_{\mathcal{L}_1,\dots,\mathcal{L}_k}$ the $(\mathcal{L}_1, \dots, \mathcal{L}_k)$*-resultant* and denote it by $R_{\mathcal{L}_1,\dots,\mathcal{L}_k}(f_1, \dots, f_k)$. This is a polynomial function in k sections $f_i \in H^0(X, \mathcal{L}_i)$ and is defined up to a constant factor. When X is given in a particular embedding into a projective space $P^n = P(V^*)$ and $\mathcal{L}_1 = \dots = \mathcal{L}_k = \mathcal{O}_X(1)$, any linear form $f \in V$ gives a section of $\mathcal{O}_X(1)$ and $R_{\mathcal{L}_1,\dots,\mathcal{L}_k}(f_1, \dots, f_k)$ coincides with the X-resultant $\tilde{R}_X(f_1, \dots, f_k)$ from Section 2.

Example 3.2. Let X be the projective space P^{k-1}. Let the invertible sheaves be $\mathcal{L}_i = \mathcal{O}(d_i)$, $i = 1, \dots, k$. The space V_i of sections of $\mathcal{O}(d_i)$ is the space of homogeneous polynomials in k variables of degree d_i. The degeneracy of a tuple of polynomials $(f_1, \dots, f_k)$, $\deg(f_i) = d_i$ means that the f_i have a common non-trivial root. Therefore the $(\mathcal{L}_1, \dots, \mathcal{L}_k)$-resultant coincides in this case with the classical resultant of k forms in k variables.

Proposition 3.3. *The $(\mathcal{L}_1, \dots, \mathcal{L}_k)$-resultant $R_{\mathcal{L}_1,\dots,\mathcal{L}_k}(f_1, \dots, f_k)$ is homogeneous with respect to each f_i. The degree with respect to f_i equals the intersection index*

$$\left(\mathcal{L}_1, \dots, \mathcal{L}_{i-1}, \mathcal{L}_{i+1}, \dots, \mathcal{L}_k\right) = \int_X \prod_{j \neq i} c_1(\mathcal{L}_j)$$

of all $\mathcal{L}_j$, $j \neq i$.

Proof. If we multiply each component f_i of a tuple $(f_1, \dots, f_k)$ by its own constant factor, the degeneracy or non-degeneracy of this tuple will be unchanged. Hence $R_{\mathcal{L}_1,\dots,\mathcal{L}_k}$ is homogeneous with respect to each f_i. To calculate the degree of this homogeneity for a given i, we fix the forms f_j, $j \neq i$ and f_i', f_i'' and consider the resultant of $f_j (j \neq i)$ and $f_i' + \lambda f_i''$ as a polynomial in a scalar variable λ. This polynomial vanishes at some λ_0 if and only if $f_i'(x_0) + \lambda_0 f''(x_0) = 0$ for some $x_0 \in \bigcap_{j \neq i} X(f_j)$. Since the cardinality of $\bigcap_{j\neq i} X(f_j)$ is the intersection index $D = (\mathcal{L}_1, \dots, \mathcal{L}_{i-1}, \mathcal{L}_{i+1}, \dots, \mathcal{L}_k)$, we conclude that the resultant as a polynomial in λ has D roots and hence is of degree D. This proves our proposition.

B. The Cayley trick for mixed resultants

The Cayley trick (the expression of a resultant as a discriminant), described in Section 2D for the X-resultant, can be extended to our mixed case as well. Denote

$$\varphi_i : X \hookrightarrow P(V_i^*), \quad V_i = H^0(X, \mathcal{L}_i) \tag{3.3}$$

to be the projective embedding defined by $\mathcal{L}_i$. In the space $P\left(\prod V_i^*\right)$ the projectivizations of individual factors V_i^* are non-intersecting projective subspaces. For $x \in X$, let $\Pi(x) \subset P\left(\prod V_i^*\right)$ be the $(k-1)$-dimensional projective subspace spanned by $\varphi_1(x), \dots, \varphi_k(x)$ (see Figure 6).

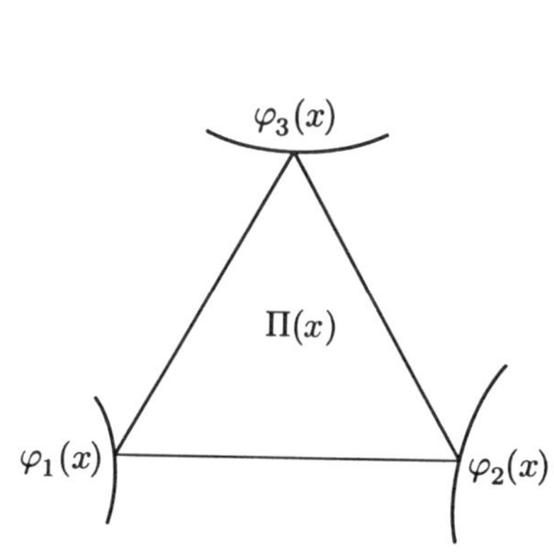

Figure 6.

We define $\tilde{X} \subset P\left(\prod V_i^*\right)$ to be the union of the spaces $\Pi(x)$ for all $x \in X$. As an algebraic variety, $\tilde{X}$ is the projectivization of the vector bundle $\bigoplus \mathcal{L}_i^*$ on X.

Proposition 3.4. *The projectivization*

$$P\left(\nabla_{\mathcal{L}_1,\dots,\mathcal{L}_k}\right) \subset P\left(\prod V_i\right)$$

is projectively dual to $\tilde{X} \subset P\left(\prod V_i^*\right)$.

Proof. This is similar to the proof of Theorem 2.7. We stress only the main points. Let $H \subset P\left(\prod V_i^*\right)$ be a hyperplane corresponding to a non-zero vector (tuple of sections) $(f_1, \dots, f_k) \in \prod V_i$. If H is tangent to $\tilde{X}$ at some point y, and $x \in X$ is such that $y \in \Pi(x)$, then $f_1(x) = \dots = f_k(x) = 0$. Conversely, if $x \in X$ is such that $f_1(x) = \dots = f_k(x) = 0$, then we can find $y \in \Pi(x) = P\left(\bigoplus \mathcal{L}_{i,x}^*\right)$ such that H is tangent to $\tilde{X}$ at y. To do this, we consider the differentials

$$d_x f_i \in \mathrm{Hom}\,(T_x X, \mathcal{L}_{i,x})$$

(they make sense since $f_i(x) = 0$). Each $d_x f_i$ lies in (its own) $(k-1)$-dimensional space $\mathrm{Hom}\,(T_x X, \mathcal{L}_{i,x})$, and there are k of them. Hence there is a non-trivial linear dependence of the form

$$\sum \lambda_i (d_x f_i) = 0, \quad \lambda_i \in \mathcal{L}_{i,x}^*,$$

where the equality is understood in $\mathrm{Hom}\,(T_x X, \mathbf{C})$. The vector $(\lambda_1, \dots, \lambda_k) \in \bigoplus \mathcal{L}_{i,x}^*$ represents a point $y \in \Pi(x)$ where H is tangent to X. We leave to the reader to check details.

Corollary 3.5. *The* $(\mathcal{L}_1, \dots, \mathcal{L}_k)$*-resultant is equal to the* $\tilde{X}$*-discriminant.*

Example 3.6. Let $X = P^1$ be the projective line and $\mathcal{L}_i = \mathcal{O}(d_i),\ i = 1, 2$, where $d_i \geq 1$. The variety $\tilde{X} \subset P^{d_1+d_2+1}$ is a surface known as the *rational normal scroll* of type (d_1, d_2), see [GH] [Harr]. The $(\mathcal{L}_1, \mathcal{L}_2)$-resultant is the classical resultant $R(f_1, f_2)$ of two binary forms of degrees d_1, d_2 (or, equivalently, of two polynomials of degrees d_1, d_2 in one variable). By Proposition 3.4, the zero locus of the classical resultant, when regarded in a projective space, is projectively dual to the rational normal scroll.

Another application of Proposition 3.4 is the formula which "finds" the common root of $(f_1, \dots, f_k)$ by taking partial derivatives of the resultant at $(f_1, \dots, f_k)$. We shall say that a point $x \in X$ is a *simple* common root of $f_1, \dots, f_k$ if the sequence

$$\bigoplus \mathcal{L}_i^* \xrightarrow{(f_1,\dots,f_k)} \mathcal{O}_X \to \mathbf{C}_x \to 0$$

is exact. Here $\mathbf{C}_x$ is the structure sheaf of the point x ("skyscraper sheaf"), i.e., $\mathbf{C}_x = \mathcal{O}_X / I_x$ where I_x is the sheaf of functions vanishing at x.

To put it simply, the condition for x to be a simple common root means that, first of all, there are no common roots other than x, and, second, the intersection of the hypersurfaces $\{f_i = 0\}$ at x is "as generic as possible" (so that the picture near

x cannot be obtained as the limit of pictures with several common roots merging together).

Corollary 3.7. *Let X and $\mathcal{L}_i$ be as before. Suppose that $f_1, \ldots, f_k$ have a simple common root $x \in X$. For each $i = 1, \ldots, k$, consider the differential of $R_{\mathcal{L}_1,\ldots,\mathcal{L}_k}$ at $(f_1, \ldots, f_k)$ with respect to the i-th argument (this is a vector in V_i^*). This vector is non-zero and the corresponding point in the projective space $P(V_i^*)$ coincides with $\varphi_i(x)$, where φ_i is the projective embedding corresponding to $\mathcal{L}_i$.*

Proof. This follows from Corollary 3.5 and Theorem 1.5 of Chapter 1 which find the point of tangency in terms of the discriminants.

Example 3.8. Suppose we are in the situation of Example 3.6, that is, we are dealing with the resultant $R(f_1, f_2)$ of two polynomials f_1, f_2 in one variable of degrees $\deg f_i \leq d_i$. We write

$$f_i(x) = a_{i0} + a_{i1}x + \cdots + a_{id_i}x^{d_i}.$$

The resultant $R(f_1, f_2)$ is a polynomial in the coefficients a_{ij} of f_1, f_2 whose vanishing for particular f_1, f_2 implies that these polynomials have at least one common root α (or else $\deg f_1 < d_1$ and $\deg f_2 < d_2$ which means that the common root is at infinity).

In this situation Corollary 3.7 has the following meaning. Suppose we know the coefficients a_{ij} of f_1 and f_2 and know that f_1 and f_2 have exactly only one simple common root α. Then α can be found explicitly by the formulas:

$$(1 : \alpha : \alpha^2 : \ldots : \alpha^{d_1}) = \left(\frac{\partial R}{\partial a_{10}}(f_1, f_2) : \frac{\partial R}{\partial a_{11}}(f_1, f_2) : \ldots : \frac{\partial R}{\partial a_{1d_1}}(f_1, f_2)\right),$$
$$(1 : \alpha : \alpha^2 : \ldots : \alpha^{d_2}) = \left(\frac{\partial R}{\partial a_{20}}(f_1, f_2) : \frac{\partial R}{\partial a_{21}}(f_1, f_2) : \ldots : \frac{\partial R}{\partial a_{2d_2}}(f_1, f_2)\right). \tag{3.4}$$

Each of these formulas is an equality of two points in the projective space represented by their homogeneous coordinates. We can find α by using just two first components of the vector on the right:

$$\alpha = \frac{(\partial R/\partial a_{11})(f, g)}{(\partial R/\partial a_{10})(f, g)} = \frac{(\partial R/\partial a_{21})(f, g)}{(\partial R/\partial a_{20})(f, g)}. \tag{3.5}$$

It may seem strange that we are actually able to find the common root from the coefficients of the polynomials by using only polynomial operations (in particular, the root belongs to $\mathbf{Q}$ if the coefficients a_{ij} belong to $\mathbf{Q}$). But a little reflection shows that this does not contradict Galois theory: if we assume that α is the only common root then it should be fixed under the Galois group and hence it is rationally known.

C. Resultants associated to vector bundles

The constructions of both the X-discriminant in Chapter 1 and the $(\mathcal{L}_1, \dots, \mathcal{L}_k)$-resultant are particular cases of the following more general construction.

Let X be a $(k-1)$-dimensional irreducible projective variety and let E be a vector bundle on X of rank k. Set $V = H^0(X, E)$. We shall assume that E is *very ample*. This means that the two following conditions hold:

(1) For any $x \in X$, the subspace

$$\gamma(x) = \{s \in H^0(X, E) : \ s(x) = 0\} \subset H^0(X, E) = V$$

has exactly codimension k.

(2) The correspondence $x \mapsto \gamma(x)$ defines a regular embedding of X into the Grassmannian of codimension k subspaces of V.

We define the *E-resultant variety* $\nabla_E \subset V$ as the set of sections which vanish at some point $x \in X$. As in Proposition 3.1, we show that ∇_E is an irreducible hypersurface in V. We define the *E-resultant* R_E to be the irreducible equation of ∇_E. This is a polynomial function on $V = H^0(X, E)$ defined up to a constant factor.

Example 3.9. When $E = \mathcal{L}_1 \oplus \cdots \oplus \mathcal{L}_k$ is a sum of line bundles then the E-resultant coincides with the $(\mathcal{L}_1, \dots, \mathcal{L}_k)$-resultant defined above.

We shall not seriously study E-resultants in their full generality, but merely formulate their simple properties.

Theorem 3.10. *R_E is a homogeneous polynomial function on $V = H^0(X, E)$ of degree equal to $\int_X c_{k-1}(E)$.*

Sketch of the proof. Homogeneity is obvious. To find the degree, we should take two generic sections $s_1, s_2 \in V$ and find the number of values $\lambda \in \mathbf{C}$ such that $s_1 + \lambda s_2$ has a zero. The classical geometric definition of the Chern class $c_{k-1}(E)$ of a rank k bundle E (see [MS]) is as follows. This class (lying in $H^{2k-2}(X, \mathbf{Z})$) is Poincaré dual to the cycle Z of complex codimension $k-1$ consisting of all $x \in X$ such that $s_1(x)$ and $s_2(x)$ are linearly dependent, i.e., proportional. Note that in our situation $\dim Z = 0$. For any $x \in Z$, we can choose λ such that $(s_1 + \lambda s_2)(x) = 0$. Conversely, for the existence of such λ, $s_1(x)$ and $s_2(x)$ must be proportional. So the number of values λ as above is equal to the number of x such that $s_1(x)$ and $s_2(x)$ are proportional, i.e., to $\int_X c_{k-1}(E)$, as claimed.

Next, we describe the Cayley trick for E-resultants. We consider the variety $\tilde{X} = P(E^*)$, the projectivization of the bundle E^*. There is a projection $p : \tilde{X} \to X$ whose fibers are projectivizations of fibers of E, and a natural projection $\pi : E^* - X \to \tilde{X}$, where X is embedded into the total space of E^* as the zero section. For any $l \in \mathbf{Z}$, we denote by $\mathcal{O}(l)_{\text{rel}}$ the invertible sheaf on $\tilde{X}$ defined as follows. For open $U \subset \tilde{X}$, a section of $\mathcal{O}(l)_{\text{rel}}$ over U is a regular function on $\pi^{-1}(U)$ which is homogeneous of degree l with respect to dilations of E^*. We call $\mathcal{O}(l)_{\text{rel}}$ the relative sheaf $\mathcal{O}(l)$ on $\tilde{X} = P(E^*)$. Its restriction to every fiber $p^{-1}(x) = P(E_x^*)$ is the standard sheaf $\mathcal{O}(l)$ of the projective space $P(E_x^*)$ (see Section 4B, Chapter 1).

Note that the direct image $p_*\mathcal{O}(1)_{\text{rel}}$ is identified with E. This follows from the natural identification $H^0(P(W^*), \mathcal{O}(1)) \cong W$ holding for any vector space W (Theorem 2.12, Chapter 2). We conclude that

$$H^0(\tilde{X}, \mathcal{O}(1)_{\text{rel}}) = H^0(X, E) = V.$$

Now the Cayley trick goes as follows.

Theorem 3.11. *The image of $\tilde{X}$ in $P(V^*)$ in the embedding given by $\mathcal{O}(1)_{\text{rel}}$, and the projectivization $P(\nabla_E) \subset P(V)$ are projectively dual to each other.*

The proof is similar to that of Proposition 3.4.

Example 3.12. Suppose that X is a smooth variety and take $E = J(\mathcal{L})$ to be the first jet bundle of a very ample line bundle $\mathcal{L}$. Let $V_0 = H^0(X, \mathcal{L})$ and $V = H^0(X, E)$. The correspondence $f \mapsto j(f)$ (the jet of a section) defines an embedding $V_0 \subset V$. The variety X is embedded into $P(V_0^*)$.

If the projective dual variety $X^\vee \subset P(V_0)$ is a hypersurface then the bundle $E = J(\mathcal{L})$ is very ample, and for any $f \in V_0$, we have $\Delta_X(f) = R_E(j(f))$ where Δ_X is the X-discriminant (i.e., the equation of $X^\vee$). Indeed, $f \in X^\vee$ if and only if $j(f) \in \nabla_E$. Thus the restriction of R_E to $V_0 \subset V$ has the same zero locus as Δ_X. This implies that this restriction is a power of Δ_X. However, the degree of Δ_X is given by Theorem 3.4 of Chapter 2 by

$$\deg \Delta_X = \sum_{i=0}^{k-1} (i+1) \int_X c_{k-1-i}(\Omega_X^1) c_1(\mathcal{L})^i.$$

We claim that

$$\sum_{i=0}^{k-1} (i+1) \int_X c_{k-1-i}(\Omega_X^1) c_1(\mathcal{L})^i = \int_X c_{k-1}(E). \tag{3.6}$$

To see this, consider the exact sequence

$$0 \to \Omega^1_X \otimes \mathcal{L} \to E \to \mathcal{L} \to 0 \tag{3.7}$$

(this is the sequence (1.5) of Chapter 2). For any vector bundle F and any line bundle $\mathcal{L}$, we have (cf. [OSS], Chapter 1)

$$c_p(F \otimes \mathcal{L}) = \sum_{i=0}^{p} \binom{r-i}{p-i} c_i(F) c_1(\mathcal{L})^{k-1-i}, \quad r = \operatorname{rk} F.$$

Applying this formula to $F = \Omega^1_X$, and also applying the Whitney sum formula to (3.7), we deduce the equality (3.6).

4. The Cayley method for the study of resultants

In Chapter 2 we discussed a method for finding discriminants as determinants of certain complexes. This method applies equally well to other elimination problems, in particular to the problem of finding the resultant. In fact, the problem of resultants was the original context in which Cayley introduced his method [Ca4].

A. Koszul complexes and resultant complexes

Let X be an irreducible projective variety of dimension $k-1$ and let $\mathcal{L}_1, \ldots, \mathcal{L}_k$ be invertible sheaves (line bundles) on X. We assume that each $\mathcal{L}_i$ is very ample. Consider the vector bundle $E = \bigoplus \mathcal{L}_i$. Suppose we have fixed sections $f_i \in H^0(X, \mathcal{L}_i)$. Then the collection $(f_1, \ldots, f_k)$ defines a section s of E. Hence we can form two Koszul complexes $\mathcal{K}_\pm(E, s)$ of sheaves on X (the complexes (1.6) and (1.7) of Section 1B, Chapter 2).

Let $\mathcal{M}$ be another line bundle on X. We consider the twisted Koszul complexes $\mathcal{K}_\pm(E, s) \otimes \mathcal{M}$. The complexes of global sections of these complexes of sheaves will be denoted by $C^\bullet_\pm(\mathcal{L}_1, \ldots, \mathcal{L}_k | \mathcal{M})$ and called the *resultant complexes* (associated with $\mathcal{L}_1, \ldots, \mathcal{L}_k$ and $\mathcal{M}$). Note that

$$\bigwedge^p (\mathcal{L}_1 \oplus \cdots \oplus \mathcal{L}_k) = \bigoplus_{1 \le i_1 < \cdots < i_p \le k} \mathcal{L}_{i_1} \otimes \cdots \otimes \mathcal{L}_{i_p}.$$

Therefore the resultant complexes have the form

$$C^\bullet_+(\mathcal{L}_1, \ldots, \mathcal{L}_k | \mathcal{M})$$

$$= \left\{ H^0(X, \mathcal{M}) \to \bigoplus_i H^0(X, \mathcal{L}_i \otimes \mathcal{M}) \to \bigoplus_{i<j} H^0(X, \mathcal{L}_i \otimes \mathcal{L}_j \otimes \mathcal{M}) \to \cdots \right\}, \tag{4.1}$$

$$C_-^\bullet(\mathcal{L}_1, \dots, \mathcal{L}_k|\mathcal{M})$$
$$= \left\{ \cdots \to \bigoplus_{i<j} H^0(X, \mathcal{L}_i^* \otimes \mathcal{L}_j^* \otimes \mathcal{M}) \to \bigoplus_i H^0(X, \mathcal{L}_i^* \otimes \mathcal{M}) \to H^0(X, \mathcal{M}) \right\}. \tag{4.2}$$

The differential in any of these complexes will be denoted by $\partial_{f_1,\dots,f_k}$. It is not strictly necessary to consider both types of complexes since obviously

$$C_+^\bullet(\mathcal{L}_1, \dots, \mathcal{L}_k|\mathcal{M}) = C_-^\bullet\left(\mathcal{L}_1, \dots, \mathcal{L}_k|\mathcal{M} \otimes (\bigotimes_i \mathcal{L}_i)\right)[k], \tag{4.3}$$

where the number k in brackets means the shift of grading of the complex. However, it will be convenient for us to keep the distinction between $C_+^\bullet$ and $C_-^\bullet$.

We shall say that each of the complexes $C_\pm^\bullet(\mathcal{L}_1, \dots, \mathcal{L}_k|\mathcal{M})$ is *stably twisted* if all the terms of the above complex $\mathcal{K}_\pm^\bullet(E, s) \otimes \mathcal{M}$, (of which it is the complex of global sections) have no higher cohomology.

Proposition 4.1. *If the resultant complex $C_+^\bullet(\mathcal{L}_1, \dots, \mathcal{L}_k|\mathcal{M})$ (resp. $C_-^\bullet(\mathcal{L}_1, \dots, \mathcal{L}_k|\mathcal{M})$) is stably twisted, then the differential $\partial_{f_1,\dots,f_k}$ in this complex is exact for any $(f_1, \dots, f_k)$, such that $R_{\mathcal{L}_1,\dots,\mathcal{L}_k}(f_1, \dots, f_k) \neq 0$.*

Proof. Similar to Proposition 2.3 of Chapter 2.

As in Section 2A of Chapter 2, we define the determinants of the resultant complexes

$$R^\pm_{\mathcal{L}_1,\dots,\mathcal{L}_k|\mathcal{M}}(f_1, \dots, f_k) = \det\left(C_\pm^\bullet(\mathcal{L}_1, \dots, \mathcal{L}_k|\mathcal{M}), \partial_{f_1,\dots,f_k}, e\right), \tag{4.4}$$

where e is some system of bases in the terms of the complex. These are rational functions on the space $\prod V_i$ where $V_i = H^0(X, \mathcal{L}_i)$.

Theorem 4.2.

(a) *If the resultant complex $C_+^\bullet(\mathcal{L}_1, \dots, \mathcal{L}_k|\mathcal{M})$ is stably twisted then*

$$R^+_{\mathcal{L}_1,\dots,\mathcal{L}_k|\mathcal{M}} = R^{(-1)^k}_{\mathcal{L}_1,\dots,\mathcal{L}_k},$$

where $R_{\mathcal{L}_1,\dots,\mathcal{L}_k}$ is the $(\mathcal{L}_1, \dots, \mathcal{L}_k)$-resultant.

(b) *If $C_-^\bullet(\mathcal{L}_1, \dots, \mathcal{L}_k|\mathcal{M})$ is stably twisted then*

$$R^-_{\mathcal{L}_1,\dots,\mathcal{L}_k|\mathcal{M}} = R_{\mathcal{L}_1,\dots,\mathcal{L}_k}.$$

Proof. Since this is very similar to the proof of Theorem 2.5 of Chapter 2, we give only an outline here.

We consider the incidence variety $W \subset X \times P(\prod V_i)$ of all tuples $(x, f_1, \ldots, f_k)$ such that $f_1(x) = \cdots = f_k(x) = 0$ (so W is the projectivization of the variety $B_{\mathcal{L}_1,\ldots,\mathcal{L}_k}$ in (3.1)). Let

$$X \overset{p_1}{\longleftarrow} W \overset{p_2}{\longrightarrow} P\left(\prod V_i\right) \tag{4.5}$$

be the natural projections. Let us abbreviate the resultant variety $\nabla_{\mathcal{L}_1,\ldots,\mathcal{L}_k}$ by ∇. Then $P(\nabla) \subset P(\prod V_i)$ coincides with $p_2(W)$ and $p_2 : W \to P(\nabla)$ is a birational isomorphism. Also let

$$p_X : X \times P\left(\prod V_i\right) \to X, \qquad p_V : X \times P\left(\prod V_i\right) \to P\left(\prod V_i\right)$$

be the projections of the product to the factors.

We consider the vector bundle $p_X^*\left(\bigoplus \mathcal{L}_i\right) \otimes p_V^*\mathcal{O}(1)$ on $X \times P\left(\prod V_i\right)$. The space of sections of this bundle is naturally identified with $\mathrm{End}\left(\prod V_i\right)$. The section σ corresponding to the identity endomorphism vanishes exactly along W. Moreover, since $p_1 : W \to X$ is a projective bundle, we can apply Proposition 1.4 (c) of Chapter 2 and get the Koszul resolution of $\mathcal{O}_W$:

$$\cdots \to p_X^*\left(\bigoplus_{i<j}(\mathcal{L}_i^* \otimes \mathcal{L}_j^*)\right) \otimes p_V^*\mathcal{O}(-2) \to p_X^*\left(\bigoplus_i \mathcal{L}_i^*\right) \otimes p_V^*\mathcal{O}(-1) \\ \to \mathcal{O}_{X \times P\left(\prod V_i\right)}. \tag{4.6}$$

Denote this complex by $\mathcal{K}^\bullet$. As in the proof of Theorem 2.5 of Chapter 2, we combine the complexes $\left(C_-^\bullet(\mathcal{L}_1, \ldots, \mathcal{L}_k|\mathcal{M}), \partial_{f_1,\ldots,f_k}\right)$ for various $(f_1, \ldots, f_k) \in \prod V_i$ into a complex of sheaves

$$\mathcal{O}(C_-^\bullet(\mathcal{L}_1, \ldots, \mathcal{L}_k|\mathcal{M}))$$

$$= \left\{\cdots \to C_-^{-1}(\mathcal{L}_1, \ldots, \mathcal{L}_k|\mathcal{M}) \otimes \mathcal{O}(-1) \to C_-^0(\mathcal{L}_1, \ldots, \mathcal{L}_k|\mathcal{M}) \otimes \mathcal{O}\right\} \tag{4.7}$$

on the projective space $P\left(\prod V_i\right)$. This complex is nothing more than the direct image $p_{V*}(\mathcal{K}^\bullet \otimes p_X^*\mathcal{M})$. To complete the proof of Theorem 4.2 (b), we proceed as in the proof of Theorem 2.5 of Chapter 2; we use the fact that $\mathcal{K}^\bullet$ is a resolution of $\mathcal{O}_W$ and that $p_2 : W \to P(\nabla)$ is birational. Similarly for the proof of part (a). We leave the remaining details to the reader.

B. Examples

Example 4.3. The Sylvester formula. Let $X = P^1$ and $\mathcal{L}_i = \mathcal{O}(d_i),\ \ i = 1, 2$. For any d, we denote by S^d the space of homogeneous polynomials of degree d in two variables x_0, x_1. Thus $V_i = H^0(X, \mathcal{L}_i)$ is identified with S^{d_i}.

As we have seen in Example 3.2, the $(\mathcal{L}_1, \mathcal{L}_2)$-resultant of $f_1 \in S^{d_1}$, $f_2 \in S^{d_2}$ is the classical resultant $R(f_1, f_2)$ of two polynomials. To show the meaning of Theorem 4.2 in this example, let us choose the twisting line bundle $\mathcal{M}$ to be $\mathcal{O}(d_1 + d_2 - 1)$. The resultant complex $C^\bullet_-(\mathcal{L}_1, \mathcal{L}_2|\mathcal{M})$ is stably twisted: this follows from general information about the cohomology of sheaves $\mathcal{O}(l)$ on P^1 (Theorem 2.12, Chapter 2). This complex has the form

$$S^{d_2-1} \oplus S^{d_1-1} \xrightarrow{\partial_{f_1, f_2}} S^{d_1+d_2-1}, \tag{4.8}$$

where $\partial_{f_1, f_2}(u, v) = f_1 u + f_2 v$. Our theorem implies that $R(f_1, f_2)$ is, up to a constant factor, the determinant of the matrix ∂_{f_1, f_2} with respect to some fixed bases in S^{d_1-1}, S^{d_2-1} and $S^{d_1+d_2-1}$. We suppose that

$$f_1(x) = a_0 x_0^{d_1} + a_1 x_0^{d_1-1} x_1 + \cdots + a_{d_1} x_1^{d_1},$$

$$f_2(x) = b_0 x_0^{d_2} + b_1 x_0^{d_2-1} x_1 + \cdots + b_{d_2} x_1^{d_2}$$

and choose in each of the spaces S^m in (4.8) the basis of monomials $x_0^{m-i} x_1^i$. Then Theorem 4.2 gives the classical Sylvester formula

$$R(f_1, f_2) = \begin{vmatrix} a_0 & a_1 & a_2 & \dots & a_{d_1-1} & a_{d_1} & 0 & 0 & \dots & 0 \\ 0 & a_0 & a_1 & \dots & a_{d_1-2} & a_{d_1-1} & a_{d_1} & 0 & \dots & 0 \\ \vdots & \vdots & \ddots & \ddots & \vdots & \vdots & \vdots & \ddots & \ddots & \vdots \\ 0 & 0 & 0 & \dots & a_0 & a_1 & a_2 & a_3 & \dots & a_{d_1} \\ b_0 & b_1 & \dots & b_{d_2} & 0 & 0 & \dots & \dots & \dots & 0 \\ 0 & b_0 & b_1 & \dots & b_{d_2} & 0 & 0 & \dots & \dots & 0 \\ \vdots & \vdots & \ddots & \ddots & \ddots & \ddots & \ddots & \ddots & \ddots & \vdots \\ \vdots & \vdots & \ddots & \ddots & \ddots & \ddots & \ddots & \ddots & \ddots & \vdots \\ 0 & 0 & 0 & \dots & \dots & 0 & b_0 & b_1 & \dots & b_{d_2} \end{vmatrix}. \tag{4.9}$$

Example 4.4. Chow forms. Note that Theorem 4.2 gives a way to write down, at least in principle, the Chow form R_X of any irreducible $(k-1)$-dimensional subvariety $X \subset P^{n-1}$. Recall that R_X can be represented as a polynomial function in the coefficients of k indeterminate linear forms $f_1, \dots, f_k$ on $\mathbf{C}^n$. This polynomial function is the same as the $(\mathcal{L}_1, \dots, \mathcal{L}_k)$-resultant where all $\mathcal{L}_i$ are $\mathcal{O}_X(1)$. Thus choosing the twisting sheaf $\mathcal{M}$ to be ample enough (say $\mathcal{M} = \mathcal{O}_X(l),\ l \gg 0$) we can represent R_X as the determinant of the complex $C^\bullet_-(\mathcal{O}(1), \dots, \mathcal{O}(1)|\mathcal{M})$. We can now write $R_X(f_1, \dots, f_k)$ as an alternating product of determinants of some matrices whose entries linearly depend on the coefficients of the f_i.

C. Resultant spectral sequences

Let $X, \mathcal{L}_1, \dots, \mathcal{L}_k$ be as before and let $\mathcal{M}$ be any line bundle on X. As in Section 5A of Chapter 2, we define the *resultant spectral sequences* $C^{pq}_{r,\pm}(\mathcal{L}_1, \dots, \mathcal{L}_k|\mathcal{M})$ to be the spectral sequences of the twisted Koszul complexes

$$\mathcal{K}_\pm(\mathcal{L}_1 \oplus \dots \oplus \mathcal{L}_k, (f_1, \dots, f_k)) \otimes \mathcal{M}.$$

Thus the first terms of these sequences do not depend on $(f_1, \dots, f_k) \in \prod V_i$ and are respectively

$$C^{pq}_{1,+} = \bigoplus_{1 \le i_1 < \dots < i_p \le k} H^q(X, \mathcal{L}_{i_1} \otimes \dots \otimes \mathcal{L}_{i_p} \otimes \mathcal{M}), \quad p \ge 0, \tag{4.10}$$

$$C^{-p,q}_{1,-} = \bigoplus_{1 \le i_1 < \dots < i_p \le k} H^q(X, \mathcal{L}^*_{i_1} \otimes \dots \otimes \mathcal{L}^*_{i_p} \otimes \mathcal{M}), \quad p \ge 0. \tag{4.11}$$

The differentials in the r-th term of any of these spectral sequences will be denoted by $\partial_{r, f_1, \dots, f_k}$. The higher terms of the spectral sequences depend on $f_1, \dots, f_k$ so that in case we wish to emphasize this dependence we will use the notation

$$C^{pq}_{r,\pm}(\mathcal{L}_1, \dots, \mathcal{L}_k|\mathcal{M}, f_1, \dots, f_k).$$

The differentials $\partial_{f_1, \dots, f_k}$ have homogeneity properties similar to those of differentials in the discriminantal spectral sequences (cf. Proposition 5.3, Chapter 2).

Proposition 4.5. *For any $\lambda_1, \dots, \lambda_k \in \mathbf{C}^*$, we have a natural system of identifications*

$$\psi^{pq}_{r\pm} : C^{pq}_{r,\pm}(\mathcal{L}_1, \dots, \mathcal{L}_k|\mathcal{M}, \lambda_1 f_1, \dots, \lambda_k f_k) \to C^{pq}_{r,\pm}(\mathcal{L}_1, \dots, \mathcal{L}_k|\mathcal{M}, f_1, \dots, f_k)$$

with the following properties:

(a) *Under the r-th identification the differential $\partial_{r, f_1, \dots, f_k}$ is multi-homogeneous of multi-degree $(r, \dots, r)$ in $(f_1, \dots, f_k)$:*

$$\partial_{r, \lambda_1 f_1, \dots, \lambda_k f_k} = \lambda_1^r \cdots \lambda_k^r \partial_{r, f_1, \dots, f_k}. \tag{4.12}$$

(b) The identification ψ_r itself is induced by the homogeneity (4.12) of the previous differential $\partial_{r-1,f_1,\ldots,f_k}$.

The meaning of part (b) is similar to that in Proposition 5.3 of Chapter 2. So is the proof, which we leave to the reader.

Proposition 4.6. *If a collection $(f_1, \ldots, f_k) \in \prod V_i$ ($V_i = H^0(X, \mathcal{L}_i)$) is such that $R_{\mathcal{L}_1,\ldots,\mathcal{L}_k}(f_1, \ldots, f_k) \neq 0$ then both spectral sequences*

$$C^{pq}_{r,\pm}(\mathcal{L}_1, \ldots, \mathcal{L}_k|\mathcal{M}, f_1, \ldots, f_k)$$

are exact, that is, they converge to 0.

Proof. Similar to that of Proposition 5.1, Chapter 2.

By Proposition 4.6 we can define the determinant of any of the spectral sequences $C^{pq}_{r,\pm}(\mathcal{L}_1, \ldots, \mathcal{L}_k|\mathcal{M})$. In each case the determinant is a rational function on $\prod V_i$ defined up to a constant factor (see Section 5A of Chapter 2 for details).

Theorem 4.7. *The determinant of the spectral sequence $C^{pq}_{r,-}(\mathcal{L}_1, \ldots, \mathcal{L}_k|\mathcal{M})$ coincides, up to a constant factor, with the $(\mathcal{L}_1, \ldots, \mathcal{L}_k)$-resultant $R_{\mathcal{L}_1,\ldots,\mathcal{L}_k}$. The determinant of $C^{pq}_{r,+}(\mathcal{L}_1, \ldots, \mathcal{L}_k|\mathcal{M})$ coincides with $R^{(-1)^k}_{\mathcal{L}_1,\ldots,\mathcal{L}_k}$.*

The proof is similar to that of Theorem 5.2, Chapter 2 and we do not go into details.

Example 4.8. Bezout formula for the classical resultant. Let $X = P^1$ and $\mathcal{L}_1 = \mathcal{L}_2 = \mathcal{O}(d)$. The space $V = H^0(P^1, \mathcal{O}(d))$ is identified with S^d, the space of binary forms of degree d. Let g, h be two such forms. Then $R_{\mathcal{L}_1,\mathcal{L}_2}$ is the classical resultant of g and h. We calculate it by using the resultant spectral sequence and to this end we take the twisting sheaf $\mathcal{M}$ to be $\mathcal{O}(d-1)$. The resultant spectral sequence $C^{pq}_{r,-}(\mathcal{L}_1, \mathcal{L}_2|\mathcal{M})$ is the spectral sequence of the complex of sheaves

$$\mathcal{O}(-d+1) \xrightarrow{\partial_{g,h}} \mathcal{O}(-1) \oplus \mathcal{O}(-1) \xrightarrow{\partial_{g,h}} \mathcal{O}(d-1). \tag{4.13}$$

The only non-trivial differential in the spectral sequence is

$$\partial_{2,g,h} : H^1(P^1, \mathcal{O}(-d+1)) = (S^{d-1})^* \longrightarrow S^{d-1} = H^0(P^1, \mathcal{O}(d-1)).$$

By Theorem 4.7, we have $R(g, h) = \det(\partial_{2,g,h})$. The method of Section 5, Chapter 2 can be used for the explicit evaluation of $\partial_{2,g,h}$. That is, $\partial_{2,g,h}$ can be regarded as an element of $S^{d-1} \otimes S^{d-1}$, i.e., as a bihomogeneous polynomial $F(x_0, x_1, y_0, y_1)$ in two groups of variables x_i, y_i having the bidegree $(d-1, d-1)$. We have

$$F(x_0, x_1, y_0, y_1) = \frac{g(x_0, x_1)h(y_0, y_1) - h(x_0, x_1)g(y_0, y_1)}{x_0y_1 - y_0x_1},$$

which is the classical Bezout formula for the resultant of two forms of the same degree. The formula for the discriminant (Section 5B Chapter 2) is a particular case corresponding to the case when g and h are partial derivatives of some form $f(x_0, x_1)$.

D. An example: the Sylvester formula for the resultant of three ternary forms

Let $X = P^2$, $\mathcal{L}_1 = \mathcal{L}_2 = \mathcal{L}_3 = \mathcal{O}(d)$. The space $V = H^0(P^2, \mathcal{O}(d))$ is equal to $S^d\mathbf{C}^3$, the space of forms of degree d in three variables x_1, x_2, x_3. There is a remarkable formula for the resultant $R_{\mathcal{L}_1,\mathcal{L}_2,\mathcal{L}_3}$ due to Sylvester (see e.g., [Sal] or [Net], Bd.2, Section 451.) We first describe the formula and then indicate how it can be deduced from Theorem 4.7.

Let k be an integer equal to $d-2$ or $d-1$. (The whole procedure below will be valid for any of these two choices of k). Let $f_1, f_2, f_3 \in S^d\mathbf{C}^3$ be three polynomials. Following Sylvester, we first consider the operator

$$T_{f_1,f_2,f_3} : S^{2d-3-k}\mathbf{C}^3 \oplus S^{2d-3-k}\mathbf{C}^3 \oplus S^{2d-3-k}\mathbf{C}^3 \to S^{3d-3-k}\mathbf{C}^3, \tag{4.14}$$

which takes $(u_1, u_2, u_3) \mapsto u_1 f_1 + u_2 f_2 + u_3 f_3$. This is just a homogeneous part of the last differential in the Koszul complex associated with f_1, f_2, f_3.

For any three non-negative integers α, β, γ such that $\alpha + \beta + \gamma = k$, let us write

$$\begin{cases} f_1 = x_1^{\alpha+1} P^{(1)}_{\alpha\beta\gamma} + x_2^{\beta+1} Q^{(1)}_{\alpha\beta\gamma} + x_3^{\gamma+1} R^{(1)}_{\alpha\beta\gamma} \\ f_2 = x_1^{\alpha+1} P^{(2)}_{\alpha\beta\gamma} + x_2^{\beta+1} Q^{(2)}_{\alpha\beta\gamma} + x_3^{\gamma+1} R^{(2)}_{\alpha\beta\gamma} \\ f_3 = x_1^{\alpha+1} P^{(3)}_{\alpha\beta\gamma} + x_2^{\beta+1} Q^{(3)}_{\alpha\beta\gamma} + x_3^{\gamma+1} R^{(3)}_{\alpha\beta\gamma} \end{cases} \tag{4.15}$$

where $P^{(i)}_{\alpha\beta\gamma}, Q^{(i)}_{\alpha\beta\gamma}, R^{(i)}_{\alpha\beta\gamma}$ are some homogeneous polynomials of degrees respectively $d-\alpha-1, d-\beta-1, d-\gamma-1$. Such a representation of the f_i is clearly possible although not unique (since $\alpha+\beta+\gamma = k < d$, every monomial of degree d is divisible by at least one of $x_1^{\alpha+1}, x_2^{\beta+1}$ or $x_3^{\gamma+1}$). Having chosen such representations, we define the polynomial

$$D_{\alpha\beta\gamma} = \det \begin{vmatrix} P^{(1)}_{\alpha\beta\gamma} & Q^{(1)}_{\alpha\beta\gamma} & R^{(1)}_{\alpha\beta\gamma} \\ P^{(2)}_{\alpha\beta\gamma} & Q^{(2)}_{\alpha\beta\gamma} & R^{(2)}_{\alpha\beta\gamma} \\ P^{(3)}_{\alpha\beta\gamma} & Q^{(3)}_{\alpha\beta\gamma} & R^{(3)}_{\alpha\beta\gamma} \end{vmatrix} \tag{4.16}$$

of degree $3d-3-k$. Define the linear map $D : (S^k\mathbf{C}^3)^* \to S^{3d-3-k}\mathbf{C}^3$ by sending

$$\delta_{\alpha\beta\gamma} \mapsto D_{\alpha\beta\gamma}, \tag{4.17}$$

where $\{\delta_{\alpha\beta\gamma}\}$ is the basis dual to the monomial basis in $S^k\mathbf{C}^3$. This map depends not only on the f_i but on the choices of representations of the f_i in the form (4.15). This dependence, however, is easy to control.

Lemma 4.9. *A different system of representations (4.15) changes any $D_{\alpha\beta\gamma}$ by adding a polynomial from the image of T_{f_1,f_2,f_3}.*

Proof. Suppose that we choose a different representation for, say, f_1, i.e., we replace

$$P^{(1)}_{\alpha\beta\gamma} \mapsto P^{(1)}_{\alpha\beta\gamma} + P, \quad Q^{(1)}_{\alpha\beta\gamma} \mapsto Q^{(1)}_{\alpha\beta\gamma} + Q, \quad R^{(1)}_{\alpha\beta\gamma} \mapsto R^{(1)}_{\alpha\beta\gamma} + R,$$

where $x_1^{\alpha+1}P + x_2^{\beta+1}Q + x_3^{\gamma+1}R = 0$. Then the difference between the new and the old value of $D_{\alpha\beta\gamma}$ will be equal to

$$\det \begin{vmatrix} P & Q & R \\ P^{(2)}_{\alpha\beta\gamma} & Q^{(2)}_{\alpha\beta\gamma} & R^{(2)}_{\alpha\beta\gamma} \\ P^{(3)}_{\alpha\beta\gamma} & Q^{(3)}_{\alpha\beta\gamma} & R^{(3)}_{\alpha\beta\gamma} \end{vmatrix}. \tag{4.18}$$

Since the Koszul complex associated to $x_1^{\alpha+1}, x_2^{\beta+1}, x_3^{\gamma+1}$ is exact, the space of triples (P, Q, R) such that $x_1^{\alpha+1}P + x_2^{\beta+1}Q + x_3^{\gamma+1}R = 0$ is linearly generated by triples of three types:

$$(Ux_2^{\beta+1}, -Ux_1^{\alpha+1}, 0), \quad (Vx_3^{\gamma+1}, 0, -Vx_1^{\alpha+1}), \quad (0, Wx_3^{\gamma+1}, -Wx_2^{\beta+1}) \tag{4.19}$$

where U, V, W are some homogeneous polynomials. In the following formulas let us write for simplicity $P^{(i)}$ instead of $P^{(i)}_{\alpha\beta\gamma}$ and the same for $Q^{(i)}, R^{(i)}$.

If we take (P, Q, R) in, say, the first form in (4.19), we get the determinant (4.18) to be

$$\det \begin{vmatrix} Ux_2^{\beta+1} & -Ux_1^{\alpha+1} & 0 \\ P^{(2)} & Q^{(2)} & R^{(2)} \\ P^{(3)} & Q^{(3)} & R^{(3)} \end{vmatrix}. \tag{4.20}$$

Expanding this determinant in the first row and using (4.15) we obtain that it is equal to $U(R^{(3)}f_2 - R^{(2)}f_3)$ and hence belongs to the image of T_{f_1,f_2,f_3}. The other types of (P, Q, R) given in (4.19) are considered similarly. Lemma 4.9 is proved.

Consider now the linear operator

$$T_{f_1,f_2,f_3} + D : S^{2d-3-k}\mathbf{C}^3 \oplus S^{2d-3-k}\mathbf{C}^3 \oplus S^{2d-3-k}\mathbf{C}^3 \oplus (S^k\mathbf{C}^3)^* \\ \to S^{3d-3-k}\mathbf{C}^3. \tag{4.21}$$

An elementary check shows that, for $k = d-2$ or $d-1$, the dimensions of the source and the target spaces of this operator coincide. By Lemma 4.9, the determinant of $T_{f_1,f_2,f_3} + D$ (with respect to some fixed bases) is independent of the arbitrary choices made in the definitions of D. Now the result of Sylvester is as follows.

Theorem 4.10. *The resultant* $R(f_1, f_2, f_3)$ *is equal to the determinant of the matrix of* $T_{f_1,f_2,f_3} + D$.

In the book of Netto [Net] it is said of the Sylvester formula that it, "...an Strenge der Beweisführung zu wünschen lässt."* No wonder since this formula (published in 1852) involves in fact the determinant of a spectral sequence!

More precisely, let us consider the resultant spectral sequence

$$C_{r,-}^{pq}\big(\mathcal{O}(d), \mathcal{O}(d), \mathcal{O}(d), \mathcal{O}(3d-3-k)\big),$$

i.e., the spectral sequence of the twisted Koszul complex

$$\mathcal{O}(-k-3) \to \mathcal{O}(d-3-k)^{\oplus 3} \to \mathcal{O}(2d-3-k)^{\oplus 3} \to \mathcal{O}(3d-3-k) \quad (4.22)$$

of sheaves on P^2. Note that we have assumed $k = d-2$ or $k = d-1$. Therefore the sheaf $\mathcal{O}(d-3-k)$ is either $\mathcal{O}(-1)$ or $\mathcal{O}(-2)$ and does not have any cohomology. The whole spectral sequence reduces to the form depicted in Figure 7.

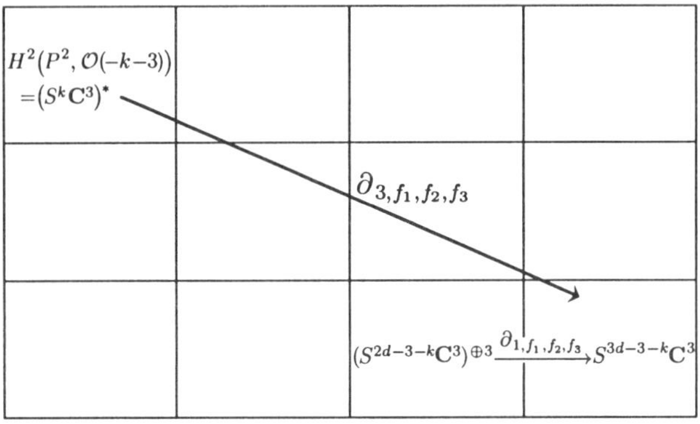

Figure 7.

The differential ∂_{1,f_1,f_2,f_3} is just the operator T_{f_1,f_2,f_3} from (4.14). The transgression ∂_{3,f_1,f_2,f_3} (whose values are defined only modulo the image of ∂_1) is nothing other than the map D from (4.17).

To see that D indeed coincides with ∂_3 (and thus to give a rigorous proof of Theorem 4.10), we have to perform an explicit calculation by using Čech resolutions similar to one used in the proof of Proposition 5.4, Chapter 2. We leave this as an exercise for the reader.

---

* lacks rigor, as far as the proof is concerned.

E. Weyman's complexes

Let $X, \mathcal{L}_1, \ldots, \mathcal{L}_k, \mathcal{M}$ and the resultant spectral sequences $C_{r,\pm}^{pq}(\mathcal{L}_1, \ldots, \mathcal{L}_k|\mathcal{M})$ have the same meaning as in subsection C above. Similar to the procedure described in Section 5C of Chapter 2, each of these spectral sequences can be replaced (non-canonically) by a bigraded complex whose (p, q)-term is the same as $C_{1,\pm}^{pq}$, the corresponding C_1-term. We refer the reader to Section 5C of Chapter 2 for the terminology related to bigraded complexes.

We shall deal with the C_--spectral sequence, the case of the C_+-sequence being analogous.

Theorem 4.11. *There is a bigraded complex $(C^{\bullet\bullet}, \partial_{f_1,\ldots,f_k})$ in which*

$$C^{-p,q} = H^q\left(X, \bigoplus_{1\le i_1<\cdots<i_p\le k} \mathcal{L}_{i_1}^* \otimes \cdots \otimes \mathcal{L}_{i_p}^* \otimes \mathcal{M}\right)$$

and the differential $\partial_{f_1,\ldots,f_k}$ depends polynomially on $f_1, \ldots, f_k$. This complex has the following properties:

(a) The differential $\partial_{f_1,\ldots,f_k}$ has the form $\partial_{f_1,\ldots,f_k} = \sum_{r\ge1} \partial_{r,f_1,\ldots,f_k}$ where $\partial_{r,f_1,\ldots,f_k}$ has bidegree $(r, 1-r)$ with respect to the bigrading in $C^{\bullet\bullet}$ and its matrix elements are polynomials homogeneous of degree r in the coefficients of each f_i. Moreover, $\partial_{1,f_1,\ldots,f_r}$ is induced by the differential in the Koszul complex $\mathcal{K}_-(\mathcal{L}_1 \oplus \cdots \oplus \mathcal{L}_k, (f_1, \ldots, f_k))\otimes\mathcal{M}$.

(b) The determinant of the total complex $(C^\bullet, \partial_{f_1,\ldots,f_k})$ of $C^{\bullet\bullet}$ equals, up to a non-zero constant factor, the $(\mathcal{L}_1, \ldots, \mathcal{L}_k)$-resultant $R_{\mathcal{L}_1,\ldots,\mathcal{L}_k}(f_1, \ldots, f_k)$.

The proof is similar to that of Theorem 5.5 of Chapter 2 and we omit it.

As for the discriminantal case, the resultant spectral sequence has the advantage of being canonical (i.e., it is determined by $f_1, \ldots, f_k$ uniquely), but the calculation of its determinant may be rather involved. On the other hand, Theorem 4.11 realizes $R_{\mathcal{L}_1,\ldots,\mathcal{L}_k}$ in a more convenient form (as the determinant of a bigraded complex) but the differentials in this complex are not canonically defined. We refer the reader to Section 5 of Chapter 2 for a more detailed discussion of the discriminantal case.

CHAPTER 4

Chow Varieties

1. Definitions and main properties

The Grassmann variety $G(k, n)$ parametrizes $(k-1)$-dimensional projective subspaces in P^{n-1}. Projective subspaces are just algebraic subvarieties of degree 1. It is natural to look for parameter spaces parametrizing subvarieties of a given degree $d \geq 1$. Here, however, we encounter some new phenomena. Namely, an irreducible variety can degenerate into a reducible one (e.g., a curve can degenerate into a collection of straight lines). Moreover, consider a reducible variety, say, a union of two distinct lines. Such a variety can degenerate into one line, which apparently has a smaller degree. Of course, in this case it is natural to count the limiting line with multiplicity 2. To take into account all of these possibilities, we need the notion of an algebraic cycle.

A. Algebraic cycles

By a $(k-1)$-dimensional algebraic cycle in P^{n-1}, we mean a formal finite linear combination $X = \sum m_i X_i$ with non-negative integer coeffcients, where the X_i are $(k-1)$-dimensional irreducible closed subvarieties in P^{n-1}. The degree of such a cycle X is defined as $\deg(X) = \sum m_i \deg(X_i)$.

Let $G(k, d, n)$ be the set of all $(k-1)$-dimensional algebraic cycles in P^{n-1} of degree d. The notation is chosen to suggest an analogy with the Grassmannians. Indeed, $G(k, 1, n)$ coincides with the Grassmannian $G(k, n)$ because any algebraic cycle of degree 1 is always given by a projective subspace with multiplicity one. We want to introduce the structure of an algebraic variety on $G(k, n, d)$ generalizing the case of the Grassmannians. Of the three types of coordinates existing for Grassmannians (see Section 1, Chapter 3), only one extends to general $G(k, d, n)$, namely the Plücker coordinates. This is done as follows.

Recall that, for an irreducible $(k-1)$-dimensional subvariety $X \subset P^{n-1}$, we defined (Section 2B Chapter 3) its Chow form (or X-resultant) R_X. This is a polynomial $R_X(f_1, \ldots, f_k)$ in coefficients of k indeterminate linear forms on C^n which vanishes whenever the projective subspace $\{f_1 = \cdots = f_k = 0\}$ of P^{n-1} intersects X. The polynomial R_X has an obvious homogeneity property: for any matrix $g = \|g_{ij}\| \in GL(k)$, we have

$$R_X(g_{11}f_1 + \ldots + g_{1k}f_k, \ldots, g_{k1}f_1 + \cdots + g_{kk}f_k) = \det(g)^d R_X(f_1, \ldots, f_k), \tag{1.1}$$

where $d = \deg X$. The space of polynomials with this property was denoted by $\mathcal{B}_d$ (Section 1 Chapter 3).

Let now $X = \sum m_i X_i$ be a $(k-1)$-dimensional algebraic cycle in P^{n-1} of degree d. We define the Chow form of X as

$$R_X = \prod R_{X_i}^{m_i} \in \mathcal{B}_d. \tag{1.2}$$

The coordinates of the vector R_X are called *Chow coordinates* of X. Their importance is seen from the following theorem of Chow and van der Waerden.

Theorem 1.1. *The map $X \mapsto R_X$ defines an embedding of $G(k, d, n)$ into the projective space $P(\mathcal{B}_d)$ as a closed algebraic variety.*

The variety $G(k, d, n)$ with the algebraic structure defined by the above embedding is called the *Chow variety* and its embedding into $P(\mathcal{B}_d)$ is called the *Chow embedding*. Note, in particular, that Theorem 1.1 implies that the Chow variety is compact so that any (analytic) one-parameter family $X(t)$, $t \neq 0$ of irreducible $(k-1)$-dimensional subvarieties in P^{n-1} has the limit $X(0)$ which is an algebraic cycle. We shall prove Theorem 1.1 later in this section. But first let us discuss some examples.

B. Examples

Example 1.2. *(a)* The Chow variety $G(k, 1, n)$ is the Grassmannian $G(k, n)$ and its Chow embedding coincides with the Plücker embedding, as follows from Example 2.3 (c) of Chapter 3.

(b) Consider the Chow variety $G(n-1, d, n)$, parametrizing cycles of degree d and codimension 1 in P^{n-1}, that is, hypersurfaces. We have seen in Example 2.3 (a) Chapter 3 that the Chow form of an irreducible hypersurface is just its equation which is an irreducible homogeneous polynomial of degree d in n variables. Algebraic cycles of codimension 1 correspond to all nonzero homogeneous polynomials, irreducible or not, of degree d. Thus the Chow variety $G(n-1, d, n)$ is the projective space of such polynomials:

$$G(n-1, d, n) = P(S^d C^n).$$

Example 1.3. Consider the Chow variety $G(2, 2, 4)$ parametrizing 1-dimensional algebraic cycles of degree 2 in P^3. Such a cycle is given by either a curve of degree 2 or by two lines. A curve of degree 2 is just a planar quadric*. Therefore $G(2, 2, 4)$

---

* Indeed, taking three non-collinear points x, y, z on an irreducible curve X of degree 2, we find that the plane Π spanned by x, y, z intersects X in at least 3 points. Since $3 > 2 = \deg(X)$, it follows that $X \subset \Pi$.

has two irreducible components C and D corresponding to planar quadrics and pairs of lines. These components intersect since a quadric can degenerate into a pair of lines. However, the two lines representing the limit of a family of planar quadrics are coplanar, whereas D is formed by all pairs of lines. Conversely, any pair of coplanar lines can be obtained as the limit of a family of planar quadrics. Therefore C and D intersect along the locus of pairs of coplanar lines.

The variety D has dimension 8 since one line in P^3 depends on 4 parameters. It is interesting that the dimension of C is also 8. Indeed, to define a quadric curve X in P^3, we must first define a 2-plane in which lies X. This gives 3 parameters. If a plane $\Pi \subset P^3$ is fixed then all quadrics inside $\Pi \cong P^2$ depend on 5 parameters. Thus the number of parameters adds up to 8.

Example 1.4. Consider the variety $G(2, 3, 4)$ parametrizing 1-dimensional cycles in P^3 of degree 3. There are several possibilities for such a cycle. It may have:
(1) an irreducible curve of degree 3;
(2) a line and a planar quadric curve;
(3) or three lines.

Let C, D, E be subvarieties in $G(2, 3, 4)$ parametrizing cycles of types (1), (2), (3). A cubic curve in P^3 can be of two types (see [Hart], Chapter IV, Section 6): either a planar cubic curve or a so-called *twisted cubic* (rational normal curve). By definition, a twisted cubic is a curve which can be brought by a projective transformation of P^3 to the standard Veronese curve

$$\left\{(x_0^3 : x_0^2 x_1 : x_0 x_1^2 : x_1^3), \quad (x_0 : x_1) \in P^1\right\}. \tag{1.3}$$

Thus C is the union of two components $C_1 \cup C_2$ where C_1 parametrizes planar cubics and C_2 parametrizes twisted cubics. So

$$G(2, 3, 4) = C_1 \cup C_2 \cup D \cup E.$$

The dimension of E (the component parametrizing triples of lines) is 12 (since one line depends on 4 parameters). The dimension of D (the component parametrizing cycles consisting of a line and a planar quadric curve) is also 12, in view of the previous example. It turns out that the dimension of C_2 (the component parametrizing twisted cubics) is also 12. Indeed, all twisted cubics are images of one particular twisted cubic (1.3) under projective transformations. The stabilizer of the curve (1.3) is the group $PGL(2)$ of projective transformations of P^1 embedded into $PGL(4)$ (the group of projective transformations of P^3) via the map

$$GL(2) = GL(\mathbf{C}^2) \hookrightarrow GL(4) = GL(S^3\mathbf{C}^2), \quad g \mapsto S^3 g.$$

Hence $C_2 = PGL(4)/PGL(2)$, and its dimension is equal to $15 - 3 = 12$.

Let us now look at the remaining component C_1 (the space of planar cubics). Its dimension is equal to 3 (the number of parameters defining a plane) plus 9 (the dimension of the space of cubics in a given plane), once again 12.

At this point it becomes tempting to make a naive conjecture that all the components of the variety $G(2, d, 4)$ of all 1-dimensional cycles in P^3 of degree d have the same dimension $4d$ (there is certainly a component of dimension $4d$ parametrizing cycles split into d lines). However, the next example shows that this is not true.

Example 1.5. Consider the Chow variety $G(2, 4, 4)$ of 1-dimensional cycles in P^3 of degree 4. This variety has many components corresponding to the various possibilities that occur for a cycle of degree 4, such as:

(1) an irreducible curve of degree 4;
(2) a cubic curve and a line;
(3) two quadric curves;
(4) a quadric curve ant two lines;
(5) or four lines.

It is clear from the previous examples that all of the components in the cases (2)–(5) have dimension 16. So we concentrate on the subvariety $C \subset G(2, 4, 4)$ parametrizing irreducible curves of degree 4. This variety is also reducible since irreducible curves of degree 4 can be of three different types (again see [Hart], Chapter IV, Section 6), namely,

(1.1) a planar quartic;
(1.2) a rational curve of degree 4;
(1.3) a spatial elliptic curve of degree 4 (intersection of two quadric surfaces).

Let C_1, C_2, C_3 denote the components corresponding to (1.1), (1.2) and (1.3). It is not hard to check that both C_2 and C_3 have dimension 16 (we leave this to the reader). However, the dimension of C_1 is 3 (the number of parameters defining a plane) plus 14 (the dimension of the space of quartics in a given plane), i.e., $\dim C_1 = 17$. Thus the above naive conjecture is not true.

The phenomenon that so many components of $G(2, d, 4)$ have dimension $4d$ can be explained as follows. Let X be a curve in P^3 and for simplicity assume it to be smooth. Let d be the degree of X and let C_X be the component of $G(2, d, 4)$ containing X. To find the dimension of C_X, it is practical to find $T_X C_X$, the (Zariski) tangent space to C_X at X. Vectors from $T_X C_X$ are what is traditionally called *infinitesimal deformations* of X inside P^3, i.e., germs of 1-parameter families of curves $X(t) \subset P^3$, $X(0) = X$, considered "up to terms of second order of vanishing in t," see, e.g., [Hart], Chapter III, Exercise 9.8 for a precise definition. The Kodaira-Spencer deformation theory provides (see *loc. cit.*) that the space of such infinitesimal deformations, i.e., $T_X C_X$, is naturally identified with $H^0(X, \mathcal{N})$ where $\mathcal{N} = T_X P^3$ is the normal bundle of X. Informally, a deformation, i.e., a way of moving X inside P^3, produces a vector of velocity at every $x \in X$. Since we are interested in the movement of X as a whole and not as individual points, the vectors of velocity which are tangent to X should be eliminated. This gives us a section of $\mathcal{N}$. Now we have the following.

Theorem 1.6. *Let $X \subset P^3$ be a smooth curve of degree d (possibly reducible, with arbitrary genus of each component). Then*

$$\chi(X, \mathcal{N}) = \dim H^0(X, \mathcal{N}) - \dim H^1(X, \mathcal{N})$$

equals $4d$. *It follows that* $\dim C_X \geq 4d$, *with the equality if and only if* $H^1(X, \mathcal{N})$ *vanishes.*

Proof of Theorem 1.6. We can assume X to be irreducible. Let g be the genus of X. By the Riemann-Roch theorem for vector bundles on a curve [Hir], we have

$$\chi(X, \mathcal{N}) = c_1(\mathcal{N}) + (\text{rank } \mathcal{N})(1 - g) = c_1(\mathcal{N}) + 2(1 - g),$$

where c_1 stands fot the first Chern class (in our case c_1 is just an integer since it lies in $H^2(X, \mathbf{Z}) = \mathbf{Z}$). We have the exact sequence

$$0 \to TX \to TP^3|_X \to \mathcal{N} \to 0,$$

where TX and TP^3 are tangent bundles of X and P^3. It is known ([Hart], Chapter II Section 8) that, for any n, the line bundle $\bigwedge^n TP^n$ on P^n is isomorphic to $\mathcal{O}_{P^n}(n+1)$. So in our case

$$c_1(TP^3|_X) = c_1\left(\bigwedge^3 TP^3|_X\right) = c_1(\mathcal{O}_{P^3}(4)|_X) = 4d.$$

Taking into account the equality $c_1(TX) = 2 - 2g$, we find $\chi(X, \mathcal{N}) = 4d$.

The cancellation that occurred in the above proof is a specific property of curves in P^3: for, say, curves in P^4, the situation will be quite different.

C. Proof of the Chow–van der Waerden theorem

In this subsection we prove Theorem 1.1, following the original argument of Chow and van der Waerden. We need the following general remarks.

Theorem 1.7 (Abstract theorem on elimination). *Let $f : Y \to Z$ be a regular morphism of quasi-projective algebraic varieties. Suppose that f is projective,* *i.e., it can be factored in the form $X \overset{i}{\hookrightarrow} Z \times P^n \xrightarrow{\pi} Z$, where i is a closed embedding and π is the projection to the second factor. Then, for every Zariski closed subset $R \subset Y$, its image $f(R)$ is Zariski closed in Z.*

For example, if Y itself is a projective variety then every morphism $f : Y \to Z$ is projective and the theorem is applicable.

For the proof of Theorem 1.7, we refer the reader to [Sh]. This proof in principle produces a system of equations for $f(R)$, but this system is impractical. As a typical application, let us mention the following fact.

Proposition 1.8. *The set of polynomials of degree d in n variables which can be decomposed into a product of linear factors is a closed algebraic subvariety in $S^d\mathbf{C}^n$.*

Proof. This follows by considering the map

$$h : (P^{n-1})^d \to P(S^d\mathbf{C}^n) \tag{1.4}$$

which takes (the projectivizations of) linear forms $f_1, \dots, f_d$ into the polynomial $f_1 \cdots f_d$.

Let us now start to prove Theorem 1.1. First, we note that the map $X \mapsto R_X$ is a set-theoretic embedding of $G(k, d, n)$ into $P(\mathcal{B}_d)$. In other words, a cycle is uniquely determined by its Chow form. By decomposing the form into irreducible factors, we reduce the question to the case when a cycle is in fact an irreducible variety. In this case our statement follows from Corollary 2.6 Chapter 3.

It remains only to prove that the image of the Chow embedding is a *closed* algebraic variety in $P(\mathcal{B}_d)$. Any element $F \in \mathcal{B}_d$ is a polynomial in indeterminate linear forms $f_1, \dots, f_k$ or, more precisely, in the coefficients c_{ij} of $f_i = \sum c_{ij}x_j$. What we need to show is that the condition for F to be the Chow form of an algebraic cycle can be expressed by polynomial equations on the coefficients of F.

Proposition 1.9. *A polynomial $F(f_1, \dots, f_k)$ of degree $d \cdot k$ is the Chow form of some cycle from $G(k, d, n)$ if and only if it satisfies the following conditions (a) to (d):*

(a) *F lies in $\mathcal{B}_d$, i.e., a linear transformation $f_i \mapsto \sum_j g_{ij} f_j$ by a non-degenerate matrix $g = \|g_{ij}\|$ multiplies F by $\det(g)^d$.*

(b) *For any fixed $f_1, \dots, f_{k-1} \in (\mathbf{C}^n)^*$, the polynomial $F(f_1, \dots, f_{k-1}, -)$, taking $f_k \mapsto F(f_1, \dots, f_k)$, decomposes into d linear factors $(f_k, \mathbf{x}^1) \cdots (f_k, \mathbf{x}^d)$; here the $\mathbf{x}^\nu$ are some points in $\mathbf{C}^n$. Furthermore, if $F(f_1, \dots, f_{k-1}, -)$ is not identically equal to 0, then the points $\mathbf{x}^\nu = \mathbf{x}^\nu(f_1, \dots, f_{k-1})$ (in this case* *defined uniquely up to rescaling* $\mathbf{x}^\nu \mapsto \lambda_\nu \mathbf{x}^\nu$, $\lambda_\nu \in \mathbf{C}^*$, $\prod \lambda_\nu = 1$*), satisfy the following two conditions:*

(c) *The* $\mathbf{x}^\nu(f_1, \dots, f_{k-1})$ *are annihilated by* $f_1, \dots, f_{k-1}$.

(d) *Let* $S_1(\mathbf{x}, \mathbf{y}), \dots, S_k(\mathbf{x}, \mathbf{y})$ *be* k *indeterminate skew-symmetric forms on* $\mathbf{C}^n$ *(see the context of Corollary 2.6, Chapter 3). Then* $F(i_{\mathbf{x}^\nu} S_1, \dots, i_{\mathbf{x}^\nu} S_k) = 0$ *for all* ν.

Proof. First we show that the Chow form of a cycle $X \in G(k, d, n)$ indeed satisfies the above conditions. The condition (a) is clear. To check the rest, it is enough to assume that X is an irreducible variety. Let $f_1, \dots, f_{k-1}$ be linear forms on $\mathbf{C}^n$ and let $\Pi \subset P^{n-1}$ be the projective subspace defined by $\{f_1 = \cdots = f_{k-1} = 0\}$. Suppose that Π intersects X in finitely many points $x^1, \dots, x^l$ (in particular, its codimension is exactly $k-1$). Then, for any non-zero f_k, we have that $F(f_1, \dots, f_k) = 0$ if and only if the hyperplane $\{f_k = 0\}$ contains at least one of the x^ν. Thus, taking some vector $\mathbf{x}^\nu \in \mathbf{C}^n$ with the projectivization x^ν, we find that for some $\varepsilon \in \mathbf{C}^*$ and any f_k, we have

$$F(f_1, \dots, f_k) = \varepsilon \cdot \prod_\nu (f_k, \mathbf{x}^\nu)^{m_\nu}, \quad m_\nu \in \mathbf{Z}_+, \sum m_\nu = d.$$

So rescaling one of the $\mathbf{x}^\nu$, we get (b). Moreover, the statements (c) and (d) follow in our case by the construction of $\mathbf{x}^\nu$.

Next, if Π intersects X along a subvariety of positive dimension, then the hyperplane $\{f_k = 0\}$ will always meet $\Pi \cap X$ so that the polynomial $F(f_1, \dots, f_{k-1}, -)$ will be identically zero. This polynomial satisfies (a) and (b) and the conditions (c) and (d) are in this case meaningless.

Now let us prove the converse statement: any $F(f_1, \dots, f_k)$, satisfying (a) to (d), is the Chow form of some cycle. First of all note that the validity of these conditions for F is equivalent to that for each irreducible factor of F. Thus we can assume that F is irreducible. Notice that the condition (d) is equivalent to:

(d′) If $f_1', \dots, f_k'$ are linear forms on $\mathbf{C}^n$ vanishing at some $\mathbf{x}^\nu(f_1, \dots, f_{k-1})$ then $F(f_1', \dots, f_k') = 0$.

We can reformulate the conditions (a) to (d′) in terms of the irreducible hypersurface $Z = \{F = 0\} \subset G(n-k, n)$. These geometric conditions are:

(1) Let M be an $(n-k)$-dimensional projective subspace in P^{n-1}. Then the intersection of Z with the projective space M^* of $(n-k-1)$-dimensional subspaces contained in M is either the whole M^* or a union of hyperplanes in M^* orthogonal to some points $x^\nu(M) \in M$.

(2) If $Z \cap M^* \neq M^*$, then any $(n-k-1)$-dimensional projective subspace in P^{n-1}, passing through any point $x^\nu(M)$ for some M in (1), belongs to Z.

Indeed, (1) incorporates the conditions (a), (b) and (c): geometrically, (a) just means that we have a hypersurface in the $G(n-k,n)$; the condition (b) means that $Z \cap M^*$ is either M^* or a union of hyperplanes and (c) means that $x^\nu(M) \in M$. Finally, (2) is a geometric translation of (d′).

Suppose now that Z satisfies (1) and (2). Let $X_0 \subset P^{n-1}$ be the set of all points $x^\nu(M)$ for all $(n-k)$-dimensional projective subspaces $M \subset P^{n-1}$ such that $Z \cap M^* \neq M^*$. Let X be the Zariski closure of X_0. Any $(n-k-1)$-dimensional projective subspace in P^{n-1} intersecting X_0 belongs to Z. Thus the locus of all P^{n-k-1}'s meeting X (denote this locus $\mathcal{Z}(X)$), is contained in Z. Note that, for generic M as above, we have $Z \cap M^* \neq M^*$, and therefore $\mathcal{Z}(X)$ has the same dimension as Z. Thus $\mathcal{Z}(X) = Z$; in particular, $\mathcal{Z}(X)$ is an irreducible hypersurface. This implies that $\dim(X) = k-1$ and $Z = \mathcal{Z}(X)$ is the associated hypersurface of X. Proposition 1.9 is proved.

To prove Theorem 1.1, it remains to show that the conditions (a) to (d) in Proposition 1.9 can be expressed by algebraic equations on the coefficients of F. For (a) this is obvious, and for (b) this follows from Proposition 1.8. To treat the rest of the conditions, we need an auxiliary lemma. In this lemma, for any non-zero vector $\mathbf{x} \in \mathbf{C}^n$, we denote its projectivization by $P(\mathbf{x}) \in P^{n-1}$.

Lemma 1.10. *Let $Y \subset S^d\mathbf{C}^n$ be the variety of polynomials which can be decomposed into a product of linear factors. Let T be any quasi-projective variety and let $\Gamma \subset P^{n-1} \times T$ be any Zariski closed subset. Consider the set $W_\Gamma \subset Y \times T$ which is the union of $\{0\} \times T$ and the subset*

$$\left\{(\mathbf{x}^1 \cdots \mathbf{x}^d,\ t) \in (Y - \{0\}) \times T \ :\ (P(\mathbf{x}^\nu), t) \in \Gamma,\ \forall \nu = 1, \dots, d\right\}. \tag{1.5}$$

Then W_Γ is Zariski closed.

Proof. Let $P(W_\Gamma) \subset P(S^d\mathbf{C}^n) \times T$ be the projectivization of W_Γ, i.e., the quotient of (1.5) by dilations in the first factor. It is enough to show that $P(W_\Gamma)$ is Zariski closed. Let

$$\Gamma_T^d = \left\{(x^1, \dots, x^d,\ t) \in (P^{n-1})^d \times T :\ (x^\nu, t) \in \Gamma,\ \forall \nu\right\}$$

be the d-fold fiber product of Γ with itself over T. It is obviously Zariski closed in $(P^{n-1})^d \times T$. Now $P(W_\Gamma)$ is the image of the composite morphism

$$\Gamma_T^d \hookrightarrow (P^{n-1})^d \times T \xrightarrow{h \times \mathrm{Id}} P(S^d\mathbf{C}^n) \times T,$$

where h is the same as in (1.4). This morphism is projective and the lemma follows from Theorem 1.7.

Now let us prove that the condition (c) of Proposition 1.9, taken together with (a) and (b), is expressed by algebraic equations on the coefficients of the polynomial $F(f_1, \dots, f_k)$. Let Φ be the variety of polynomials satisfying (a) and (b), and let T be the vector space of all $(k-1)$-tuples $(f_1, \dots, f_{k-1})$ of linear forms on $\mathbf{C}^n$. Also, let $Y \subset S^d\mathbf{C}^n$ be as in Lemma 1.10. Thus, for any $F \in \Phi$ and $(f_1, \dots, f_{k-1}) \in T$, we have a polynomial $F(f_1, \dots, f_{k-1}, -) \in Y$. Consider the Zariski closed subset

$$\Gamma \subset P^{n-1} \times T, \quad \Gamma = \{(x, (f_1, \dots, f_{k-1})) : f_1(x) = \cdots = f_{k-1}(x) = 0\}.$$

By Lemma 1.10, the subset $W_\Gamma \subset Y \times T$ is Zariski closed. Now, a polynomial $F \in \Phi$ satisfies the condition (c) if and only if, for any $(f_1, \dots, f_{k-1}) \in T$, the point

$$\big(F(f_1, \dots, f_{k-1}, -), (f_1, \dots, f_{k-1})\big) \in Y \times T$$

belongs to W_Γ. In other words, the subset in Φ, defined by the condition (c), is the intersection

$$\bigcap_{f_1, \dots, f_{k-1}} p_{f_1, \dots, f_{k-1}}^{-1}(W_\Gamma),$$

where $p_{f_1, \dots, f_{k-1}} : \Phi \to Y \times T$ is the regular morphism taking

$$F \quad \longmapsto \quad \big(F(f_1, \dots, f_{k-1}, -), (f_1, \dots, f_{k-1})\big).$$

So the combined conditions (a), (b) and (c) of Proposition 1.9 can be expressed by algebraic equations.

Similarly, let us show that the combined conditions (a), (b) and (d) are given by algebraic equations. Let Φ, T and Y have the same meaning as before, and let Σ be the space of k-tuples $(S_1, \dots, S_k)$ of skew-symmetric bilinear forms on $\mathbf{C}^n$.

Lemma 1.11. *The subset Q in $\Phi \times Y \times \Sigma$ which is the union of $\Phi \times \{0\} \times \Sigma$ and the set*

$$\left\{(F, \mathbf{x}^1 \cdots \mathbf{x}^d, S_1, \dots, S_k) \in \Phi \times (Y - \{0\}) \times \Sigma : F(i_{\mathbf{x}^\nu} S_1, \dots, i_{\mathbf{x}^\nu} S_k) = 0, \ \forall \nu\right\}, \tag{1.6}$$

is Zariski closed.

Proof. As in the proof of Lemma 1.10, it is enough to prove that the projectivization of (1.6), which is a subset in $\Phi \times P(S^d\mathbf{C}^n) \times \Sigma$, is Zariski closed there. This projectivization is the image of the closed subvariety $R \subset \Phi \times (P^{n-1})^d \times \Sigma$ given by

$$R = \left\{(F, x^1, \dots, x^d, S_1, \dots, S_k) : F(i_{\mathbf{x}^\nu} S_1, \dots, i_{\mathbf{x}^\nu} S_k) = 0, \ \forall \nu\right\};$$

(here the $\mathbf{x}^\nu \in \mathbf{C}^n$ are non-zero vectors representing the $x^\nu \in P^{n-1}$) under the projective morphism

$$\mathrm{Id} \times h \times \mathrm{Id} : \quad \Phi \times (P^{n-1})^d \times \Sigma \longrightarrow \Phi \times P(S^d\mathbf{C}^n) \times \Sigma,$$

where h is as in (1.4). The lemma is proved.

Now, a polynomial $F \in \Phi$ satisfies the condition (d) if and only if for any $(S_1, \dots, S_k) \in \Sigma$ and $(f_1, \dots, f_{k-1}) \in T$, the point

$$\big(F,\ F(f_1, \dots, f_{k-1}, -),\ (S_1, \dots, S_k)\big) \in \Phi \times Y \times \Sigma$$

belongs to the subset Q, defined in Lemma 1.11. Thus the set of $F \in \Phi$ satisfying (d) is the intersection

$$\bigcap_{\substack{f_1, \dots, f_{k-1} \\ S_1, \dots, S_k}} q^{-1}_{f_1,\dots,f_{k-1},S_1,\dots,S_k}(Q),$$

where $q_{f_1,\dots,f_{k-1},S_1,\dots,S_k} : \Phi \to \Phi \times Y \times \Sigma$ is the regular map taking

$$F \quad \longmapsto \quad \big(F,\ F(f_1, \dots, f_{k-1}, -),\ (S_1, \dots, S_k)\big).$$

So the combination of conditions (a), (b) and (d) is also algebraic. This concludes the proof of the Chow–van der Waerden theorem.

In the case when $d = 1$, the Chow variety $G(k, 1, n)$ is the Grassmann variety $G(k, n)$ and its Chow embedding into $P(\mathcal{B}_1)$ coincides with the Plücker embedding into $P\left(\bigwedge^k \mathbf{C}^n\right)$ (Example 1.2 (a)). In this case the conditions (a)–(d) of Proposition 1.9 are equivalent to the Plücker relations. More precisely, there is a natural identification of $\mathcal{B}_1$ with $\bigwedge^k \mathbf{C}^n$, so (a) just means that the ambient spaces for the Chow and Plücker embeddings are the same. Conditions (b) and (c) are fulfilled automatically, while (d) (or the equivalent condition (d′), see above) can be seen to be equivalent to the Plücker relations. We leave this as an exercise to the reader.

For $d > 1$, the proof of the Chow–van der Waerden theorem does not give a manageable system of explicit equations for $G(k, d, n)$. We shall discuss this subject in the rest of this chapter.

2. 0-cycles, factorizable forms and symmetric products

In this section we shall study in more detail the Chow variety $G(1, d, n)$ of 0-cycles of degree d in P^{n-1}.

A. Relation to symmetric products

A 0-cycle of degree d is just an unordered collection $\{x_1, \ldots, x_d\}$ of d points (not necessarily distinct) in P^{n-1}. Thus, as a set $G(1, d, n)$ is identified with $\mathrm{Sym}^d(P^{n-1})$, the d-fold symmetric product of P^{n-1}. So we start with a comparison of the definitions of $G(1, d, n)$ and $\mathrm{Sym}^d(P^{n-1})$.

Suppose that our projective space P^{n-1} is $P(V)$ where V is an n-dimensional vector space. The Chow form of a point $x \in P(V)$ is the linear function l_x on V^* given by the scalar product with x:

$$l_x(\xi) = (x, \xi).$$

If $X = \sum m_i x_i$ is a 0-cycle in $P(V)$ then, by our convention, the Chow form R_X is the polynomial $\xi \mapsto \prod l_{x_i}^{m_i}(\xi)$. We arrive at the following.

Proposition 2.1. *The Chow variety $G(1, d, n)$ of 0-cycles in P^{n-1} of degree d is the projectivization of the space of homogeneous polynomials of degree d in n variables which are products of linear forms.*

The set Y of decomposable (into linear factors) polynomials of degree d was already used several times in the course of proving the Chow–van der Waerden theorem. Note that this set has, as its "odd" analog, the set of polyvectors from $\bigwedge^d \mathbf{C}^n$ which are decomposable into wedge products of d vectors. The projectivization of the set of decomposable polyvectors is, as we have seen in Section 1, Chapter 3, nothing more than the Grassmannian $G(d, n)$ in its Plücker embedding. So the variety of 0-cycles is the "even" analog of the Grassmannian.

Recall now the definition of symmetric products. Let X be a quasi-projective algebraic variety. The symmetric product $\mathrm{Sym}^d(X)$ is the quotient of the Cartesian product X^d by the action of the symmetric group S_d permuting the factors. A more precise definition is as follows.

Suppose first that X is an affine variety and R is its coordinate ring. So $R^{\otimes d} = R \otimes \cdots \otimes R$ is the coordinate ring of X^d. The coordinate ring of $\mathrm{Sym}^d(X)$ is, by definition, the subring of S_d-invariants in $R^{\otimes d}$. In other words, this is the ring of regular functions $f(\mathbf{x}_1, \ldots, \mathbf{x}_d)$ of d variables $\mathbf{x}_i \in X$ which are symmetric, i.e., unchanged under any permutation of the $\mathbf{x}_i$.

If X is an arbitrary, not necessarily affine, quasi-projective variety then the symmetric product $\mathrm{Sym}^d(X)$ is defined by gluing affine varieties $\mathrm{Sym}^d(U)$ for various affine open subsets $U \subset X$.

It follows from these definitions that we have a regular morphism of algebraic varieties

$$\gamma : \mathrm{Sym}^d(P^{n-1}) \longrightarrow G(1, d, n), \quad \{\mathbf{x}_1, \ldots, \mathbf{x}_d\} \mapsto \sum \mathbf{x}_i, \tag{2.1}$$

which is set-theoretically a bijection. Note that this does not automatically imply that γ is an isomorphism of algebraic varieties: the morphism from the affine line A^1 to the semicubic parabola $y^2 = x^3$, given by $x(t) = t^2$, $y(t) = t^3$, is bijective but not an isomorphism. So the following fact requires a proof.

Theorem 2.2. *The morphism* $\gamma : \mathrm{Sym}^d(P^{n-1}) \longrightarrow G(1, d, n)$ *is an isomorphism of algebraic varieties (over the field of complex numbers).*

Let us note that over a field of finite characteristic the statement is no longer true [Nee].

B. Symmetric polynomials in vector variables and the proof of Theorem 2.2

Let $x_1, \ldots, x_n$ be homogeneous coordinates in P^{n-1}. We consider the affine space $\mathbf{C}^{n-1}$ inside P^{n-1}, given by the condition $x_n \neq 0$. We can assume that $x_n = 1$ in $\mathbf{C}^{n-1}$ and regard the remaining coordinates $x_1, \ldots, x_{n-1}$ as affine coordinates in $\mathbf{C}^{n-1}$. We are going to compare the symmetric product $\mathrm{Sym}^d(\mathbf{C}^{n-1})$ and its image in $G(1, d, n)$ under the morphism γ in (2.1).

By definition, the coordinate ring of $\mathrm{Sym}^d(\mathbf{C}^{n-1})$ is the ring of symmetric polynomials $f(\mathbf{x}_1, \ldots, \mathbf{x}_d)$ in d vector variables $\mathbf{x}_i \in \mathbf{C}^{n-1}$. We shall denote this ring by $S(d, n-1)$. In our coordinates each $\mathbf{x}_i$ is a vector $(x_{i1}, \ldots, x_{i,n-1})$.

For d scalar variables $x_i \in \mathbf{C}$, the structure of symmetric polynomials is well-known: every symmetric polynomial can be uniquely written as a polynomial in elementary symmetric polynomials

$$e_k(x_1, \ldots, x_d) = \sum_{1 \leq i_1 < \cdots < i_k \leq d} x_{i_1} \cdots x_{i_k}. \tag{2.2}$$

The structure of symmetric polynomials in d vector variables is less trivial. Still, we can define the analogs of elementary symmetric polynomials. To arrive at this definition, note that for the scalar case

$$1 + \sum_{k \geq 1} e_k(x_1, \ldots, x_d) t^k = \prod_i (1 + x_i t).$$

Consider now d vectors $\mathbf{x}_i = (x_{i1}, \ldots, x_{i,n-1}) \in \mathbf{C}^{n-1}$, $i = 1, \ldots, d$ and consider the following polynomial in $n-1$ variables $t_1, \ldots, t_{n-1}$:

$$\prod_i (1 + x_{i1} t_1 + \cdots + x_{i,n-1} t_{n-1}). \tag{2.3}$$

The coefficient of every monomial $t_1^{k_1} \cdots t_{n-1}^{k_{n-1}}$ in this polynomial is obviously a symmetric polynomial in $\mathbf{x}_1, \ldots, \mathbf{x}_d$. We call it an *elementary symmetric polynomial* and denote it by $e_{k_1, \ldots, k_{n-1}}(\mathbf{x}_1, \ldots, \mathbf{x}_d)$.

Thus we have

$$\prod_{i=1}^{d}(1 + x_{i1}t_1 + \cdots + x_{i,n-1}t_{n-1})$$
$$= 1 + \sum_{k_1,\ldots,k_{n-1}} e_{k_1,\ldots,k_{n-1}}(\mathbf{x}_1, \ldots, \mathbf{x}_d)t_1^{k_1} \cdots t_{n-1}^{k_{n-1}}, \tag{2.4}$$

the sum over all multi-indices $(k_1, \ldots, k_{n-1})$ with $1 \le \sum_j k_j \le d$. Note that, for any d vectors $\mathbf{x}_1, \ldots, \mathbf{x}_d \in \mathbf{C}^{n-1} \subset P^{n-1}$, the values $e_{k_1,\ldots,k_{n-1}}(\mathbf{x}_1, \ldots, \mathbf{x}_d)$ are the Chow coordinates of the cycle $X = \sum \mathbf{x}_i \in G(1, d, n)$, i.e., the coefficients of the Chow form R_X. Indeed, the homogeneous coordinates of $\mathbf{x}_i$ are $(x_{i1}, \ldots, x_{i,n-1}, 1)$ so R_X is the homogeneous polynomial in n variables $t_1, \ldots, t_n$ given by

$$R_X(t_1, \ldots, t_n) = \prod_{i=1}^{d}(x_{i1}t_1 + \ldots + x_{i,n-1}t_{n-1} + 1 \cdot t_n).$$

The coefficients of this homogeneous polynomial are obviously the same as the coefficients of the non-homogeneous polynomial (2.3).

Thus we get the following.

Proposition 2.3. *Let $Z^d(\mathbf{C}^{n-1})$ be the open subset in the Chow variety $G(1, d, n)$ consisting of 0-cycles $X = \sum \mathbf{x}_i$ with $\mathbf{x}_i \in \mathbf{C}^{n-1} \subset P^{n-1}$. The ring of regular functions on $Z^d(\mathbf{C}^{n-1})$ is the subring in the ring $S(d, n-1)$ of symmetric polynomials in d vector variables generated by elementary symmetric polynomials. The map $\gamma : \mathrm{Sym}^d(\mathbf{C}^{n-1}) \to Z^d(\mathbf{C}^{n-1})$ is induced by the embedding of the above subring into $S(d, n-1)$.*

So Theorem 2.2 is a consequence of the following.

Theorem 2.4. *Any symmetric polynomial in d vector variables $\mathbf{x}_1, \ldots, \mathbf{x}_d$, $\mathbf{x}_i \in \mathbf{C}^{n-1}$ can be expressed (not necessarily uniquely) as a polynomial in elementary symmetric polynomials $e_{k_1,\ldots,k_{n-1}}(\mathbf{x}_1, \ldots, \mathbf{x}_d)$.*

C. "Fundamental theorem" for symmetric polynomials in vector variables

In this subsection we shall prove Theorem 2.4 and hence Theorem 2.2. Recall that $S(d, n-1)$ denotes the ring of symmetric polynomials in d vector variables $\mathbf{x}_i \in \mathbf{C}^{n-1}$. A convenient way to represent d vector variables $\mathbf{x} = (\mathbf{x}_1, \ldots, \mathbf{x}_d)$ is to write $\mathbf{x}$ as a $d \times (n-1)$ matrix (x_{ij}) with rows $\mathbf{x}_1, \ldots, \mathbf{x}_d$. Let $\mathbf{Z}_+^{d\times(n-1)}$ denote the set of all $d \times (n-1)$ non-negative integer matrices $\omega = (\omega_{ij})$. Each $\omega \in \mathbf{Z}_+^{d\times(n-1)}$ determines a monomial $\mathbf{x}^\omega = \prod_{i,j} x_{ij}^{\omega_{ij}}$. We can also write the monomial $\mathbf{x}^\omega$ in the form

$$\mathbf{x}^\omega = \mathbf{x}_1^{\omega^{(1)}} \cdot \ldots \cdot \mathbf{x}_d^{\omega^{(d)}},$$

where $\omega^{(1)}, \ldots, \omega^{(d)} \in \mathbf{Z}_+^{n-1}$ are the rows of ω.

It is clear that $S(d, n-1)$ as a vector space has a basis consisting of symmetrizations of all possible monomials $\mathbf{x}^\omega$. More precisely, for each $\lambda \in \mathbf{Z}_+^{d\times(n-1)}$, we denote

$$m_\lambda = \sum_{\omega \in S_d \lambda} \mathbf{x}^\omega, \tag{2.5}$$

to be the sum over the orbit of λ under the symmetric group S_d (acting by permutations of rows of λ). Then the functions m_λ, for λ running over some system of representatives of the orbit set $\mathbf{Z}_+^{d\times(n-1)}/S_d$, form a vector space basis of $S(d, n-1)$. In the scalar case ($n = 2$) when $\lambda = (\lambda^{(1)}, \ldots, \lambda^{(d)})$ with all $\lambda^{(i)}$ just non-negative integers, a natural choice of representatives for S_d-orbits are vectors λ satisfying $\lambda^{(1)} \geq \cdots \geq \lambda^{(d)}$. For $n > 2$, there is no such obvious choice.

Notice that each elementary symmetric polynomial $e_{k_1,\ldots,k_{n-1}}(\mathbf{x}_1, \ldots, \mathbf{x}_d)$ is of the form m_λ, where each row of the matrix λ is one of the standard basis vectors $\varepsilon_1, \varepsilon_2, \ldots, \varepsilon_{n-1}$ of $\mathbf{Z}_+^{(n-1)}$. More precisely, λ has k_1 rows equal to ε_1, k_2 rows equal to $\varepsilon_2, \ldots, k_{n-1}$ rows equal to ε_{n-1}, and $d - k_1 - \cdots - k_{n-1}$ zero rows. To prove Theorem 2.4, we introduce another special family of functions m_λ which, in the scalar case $n = 2$, reduces to the *power sums* $x_1^r + x_2^r + \cdots + x_d^r$. This family corresponds to λ having only one non-zero row. For each $\rho \in \mathbf{Z}_+^{(n-1)}$, we denote by p_ρ the function m_λ, where λ has one row equal to ρ and all other rows equal to 0. Thus by definition

$$p_\rho(\mathbf{x}_1, \ldots, \mathbf{x}_d) = \mathbf{x}_1^\rho + \mathbf{x}_2^\rho + \cdots + \mathbf{x}_d^\rho. \tag{2.6}$$

Theorem 2.4 is an immediate consequence of the following two statements.

Proposition 2.5. *Polynomials $p_\rho(\mathbf{x}_1, \ldots, \mathbf{x}_d)$ generate the ring $S(d, n-1)$, i.e., each m_λ can be expressed as a polynomial in the p_ρ's.*

Proposition 2.6. *Every p_ρ can be expressed as a polynomial in the elementary symmetric polynomials $e_{k_1,\ldots,k_{n-1}}$.*

We start with Proposition 2.5. For each matrix $\lambda \in \mathbf{Z}_+^{d\times(n-1)}$ with rows $\lambda^{(1)}, \ldots, \lambda^{(d)}$, we set

$$p_\lambda = p_{\lambda^{(1)}} p_{\lambda^{(2)}} \cdots p_{\lambda^{(d)}}. \tag{2.7}$$

We shall prove the following more precise version of Proposition 2.5.

Proposition 2.5′. *The functions p_λ, for λ running over some system of representatives of the orbits of the S_d-action on $\mathbf{Z}_+^{d\times(n-1)}$, form a vector space basis in $S(d, n-1)$.*

For each $\lambda \in \mathbf{Z}_+^{d\times(n-1)}$, we define the *length* $l(\lambda)$ to be the number of non-zero rows in λ. Expanding the product $p_{\lambda^{(1)}} p_{\lambda^{(2)}} \cdots p_{\lambda^{(d)}}$ in (2.7), and collecting similar terms, we see that p_λ has the form

$$p_\lambda = c_\lambda m_\lambda + \text{ (linear combination of } m_\mu \text{ with } l(\mu) < l(\lambda)), \qquad (2.8)$$

where c_λ is some positive integer. It follows that under a suitable ordering of the index set $\mathbf{Z}_+^{d\times(n-1)}$, the transition matrix from the basis (m_λ) to the family (p_λ) becomes triangular with non-zero diagonal entries. This immediately implies Proposition 2.5′ and hence Proposition 2.5.

Proof of Proposition 2.6. For $\rho = (\rho_1, \ldots, \rho_{n-1}) \in \mathbf{Z}_+^{n-1}$, let us write t^ρ for $t_1^{\rho_1} \cdots t_{n-1}^{\rho_{n-1}}$ and $|\rho|$ for $\sum \rho_i$. Taking the logarithm of the left hand side of (2.4), we obtain

$$\log \prod_{i=1}^{d} (1 + x_{i1}t_1 + \cdots + x_{i,n-1}t_{n-1}) = \sum_{i=1}^{d} \log(1 + x_{i1}t_1 + \cdots + x_{i,n-1}t_{n-1})$$

$$= \sum_{0\neq\rho\in\mathbf{Z}_+^{n-1}} \frac{(-1)^{|\rho|-1}}{|\rho|} \frac{|\rho|!}{\rho_1! \cdots \rho_{n-1}!} p_\rho(\mathbf{x}_1, \ldots, \mathbf{x}_d) t^\rho. \qquad (2.9)$$

Hence the p_ρ are obtained, up to rational factors, as coefficients in the Taylor expansion of

$$\log\left(1 + \sum_{k_1,\ldots,k_{n-1}} e_{k_1,\ldots,k_{n-1}}(\mathbf{x}_1, \ldots, \mathbf{x}_d) t_1^{k_1} \cdots t_{n-1}^{k_{n-1}}\right),$$

i.e., as some polynomials in the $e_{k_1,\ldots,k_{n-1}}$. Proposition 2.6 is proved.

The expression of power sums in terms of elementary symmetric polynomials will be also used later in the classical case of symmetric polynomials in scalar variables.

D. Symmetric products are rational

Recall that by $S(d, n-1)$ we have denoted the ring of symmetric polynomials in d vector variables $\mathbf{x}_1, \ldots, \mathbf{x}_d$, $\mathbf{x}_i \in \mathbf{C}^{n-1}$. By the classical theorem on symmetric polynomials, $S(d, 1)$ is the polynomial ring in elementary symmetric polynomials $e_1, \ldots, e_d$. In other words, $\mathrm{Sym}^d(\mathbf{C})$ is just the affine space $\mathbf{C}^d$. A slightly different form of this result is as follows.

Proposition 2.7. *The symmetric product* $\mathrm{Sym}^d(P^1) = G(1, d, 2)$ *is isomorphic to* P^d.

Proof. For any n, the Chow embedding realizes $G(1, d, n) = \mathrm{Sym}^d(P^{n-1})$ as the subvariety in $P(S^d\mathbf{C}^n)$ whose elements are (projectivizations of) homogeneous polynomials of degree d in n variables that split into linear factors. But every homogeneous polynomial in two variables splits in just such a way. So we get

$$G(1, d, 2) = P(S^d\mathbf{C}^2) = P^d.$$

It is remarkable that the above result admits a "rational" version for symmetric functions in vector variables.

Theorem 2.8. *For any d and n, the variety $\mathrm{Sym}^d(P^{n-1}) = G(1, d, n)$ is rational, i.e., it is birationally isomorphic to the projective space $P^{d(n-1)}$.*

This can be reformulated in terms of fields of functions. Namely, denote by $RS(d, n-1)$ the field of fractions of $S(d, n-1)$, i.e., the field of symmetric rational functions in d vector variables $\mathbf{x}_i \in \mathbf{C}^{n-1}$. Since in Theorem 2.8 we can replace the variety P^{n-1} with its Zariski open subset $\mathbf{C}^{n-1}$, the theorem is a consequence of the following more precise statement.

Theorem 2.8′. *The field $RS(d, n-1)$ is isomorphic to the field of rational functions in $d(n-1)$ variables. An explicit system of (algebraically independent) rational generators of this field is provided by the following elementary symmetric polynomials:*

(a) *d polynomials $e_{k,0,\ldots,0}(\mathbf{x}_1, \ldots, \mathbf{x}_d)$, $k = 1, \ldots, d$.*

(b) *$d(n-2)$ polynomials of the form $e_{k,0,\ldots,0,1,0,\ldots,0}$ where $k = 0, 1, \ldots, d-1$ and 1 may be in any position from the second to the $(n-1)$.*

This theorem was known at the turn of the century (see [Net], Bd.2, §383) but then forgotten and rediscovered again by A. Mattuck [Mat].

Note that elementary symmetric polynomials are just coordinate functions in the Chow embedding of $\mathrm{Sym}^d(P^{n-1})$. Hence Theorem 2.8′ means that a birational isomorphism $\mathrm{Sym}^d(P^{n-1}) \cong P^{d(n-1)}$ can be obtained by a linear projection of $\mathrm{Sym}^d(P^{n-1}) \subset P(S^d\mathbf{C}^n)$ from a certain coordinate subspace. This subspace is spanned by basis vectors corresponding to elementary symmetric polynomials which are not listed in Theorem 2.8′.

Proof of Theorem 2.8′. Let $x_1, \ldots, x_{n-1}$ be standard coordinates in $\mathbf{C}^{n-1}$. To an unordered tuple of points $\{\mathbf{x}_1, \ldots, \mathbf{x}_d\}$, $\mathbf{x}_i \in \mathbf{C}^{n-1}$, we first associate the unordered tuple of their first coordinates $\{x_{1,1}, \ldots, x_{d,1}\}$. Note that the elementary symmetric polynomial $e_{k,0,\ldots,0}(\mathbf{x}_1, \ldots, \mathbf{x}_d)$ is the usual elementary symmetric polynomial $e_k(x_{1,1}, \ldots, x_{d,1})$ in the first coordinates of the $\mathbf{x}_i$. Hence the tuple $\{x_{1,1}, \ldots, x_{d,1}\}$ is completely determined by the values of $e_{k,0,\ldots,0}(\mathbf{x}_1, \ldots, \mathbf{x}_d)$.

Now let $\mathbf{x}_1, \ldots, \mathbf{x}_d$ be given. For any $j \in \{2, \ldots, n-1\}$, consider the polynomial

$$f_j(t) = \sum_{i=1}^{d} x_{i,j} \prod_{\nu \neq i} (t - x_{\nu,1}). \tag{2.10}$$

The coefficients of this polynomial at various powers of t are, up to sign, the elementary symmetric polynomials

$$e_{k,0,\ldots,0,1,0,\ldots,0}(\mathbf{x}_1, \ldots, \mathbf{x}_d), \quad k = 0, \ldots, d-1$$

where 1 in the subscript is in the j-th position. The polynomial $f_j(t)$ satisfies the property:

$$f_j(x_{i,1}) = x_{i,j} \prod_{\nu \neq i} (x_{i,1} - x_{\nu,1}). \tag{2.11}$$

If we assume that the numbers $x_{1,1}, \ldots, x_{d,1}$ are distinct (which is generically the case), then (2.11) gives us the components $x_{1,j}, \ldots, x_{d,j}$ once we know the polynomial f_j and the collection $\{x_{1,1}, \ldots, x_{d,1}\}$. Therefore, under our genericity assumption, an unordered tuple $\{\mathbf{x}_1, \ldots, \mathbf{x}_d\}$ is *uniquely* determined by the values of all the functions $e_{k,0,\ldots,0}(\mathbf{x}_1, \ldots, \mathbf{x}_d)$ and $e_{k,0,\ldots,0,1,0,\ldots,0}(\mathbf{x}_1, \ldots, \mathbf{x}_d)$. To finish the proof, one may use this uniqueness to argue, by Galois theory, that an arbitrary symmetric rational function of $\mathbf{x}_1, \ldots, \mathbf{x}_d$ is uniquely rationally expressible through the functions listed above. We prefer to do essentially the same, but using a more transparent geometric language.

Let $U_1 \subset \mathrm{Sym}^d(\mathbf{C}^{n-1})$ be the open set consisting of unordered d-tuples of points in $\mathbf{C}^{n-1}$ with pairwise distinct first coordinates. Let $U_2 \subset (\mathbf{C}^{n-1})^d = \mathbf{C}^d \times (\mathbf{C}^{n-2}))^d$ be the open set of pairs (a, b), $a = (a_1, \ldots, a_d) \in \mathbf{C}^d$, $b \in (\mathbf{C}^{n-2})^d$ such that all the roots of the polynomial $1 + a_1 x + \cdots + a_d x^d$ are distinct. Consider the correspondence

$$\mathbf{x} = (\mathbf{x}_1, \ldots, \mathbf{x}_d) \mapsto \big((e_{1,0,\ldots,0}(\mathbf{x}), \ldots, e_{d,0,\ldots,0}(\mathbf{x})); \ (e_{k,0,\ldots,0,1,0,\ldots,0}(\mathbf{x}))\big)$$

where, in the last component, k varies between 0 and $d-1$ and 1 can be on any place from 2 to $n-1$. This correspondence gives a regular map $\varphi : U_1 \to U_2$. The above reasoning implies that φ is a bijection. Hence it is an isomorphism since both U_1 and U_2 are smooth. Thus $\mathrm{Sym}^d(\mathbf{C}^{n-1})$ and $(\mathbf{C}^{n-1})^d$ have isomorphic open subsets so they are birationally isomorphic. Since the symmetric functions in Theorem 2.8′ provide a system of coordinates in $(\mathbf{C}^{n-1})^d$, they form a system of rational generators for $RS(d, n-1)$. Theorem 2.8′ is proved.

Before going further, let us make two remarks on symmetric products.

(a) If Y is a smooth manifold of dimension greater than 1, the symmetric products $\mathrm{Sym}^d(Y)$ are singular. The singular points of $\mathrm{Sym}^d(Y)$ correspond to tuples of points of Y some of which coincide. For example, let $d = 2$ and consider $\mathrm{Sym}^2(\mathbf{C}^n)$. This is the quotient of $\mathbf{C}^n \times \mathbf{C}^n$ by the involution $(x, y) \mapsto (y, x)$. Introduce new vector variables $u = (x + y)/2$, $v = (x - y)/2$. The involution in these variables is $(u, v) \mapsto (u, -v)$. In other words, $\mathrm{Sym}^2(\mathbf{C}^n)$ is the product of $\mathbf{C}^n$ and the quotient $\mathbf{C}^n/\pm$ of $\mathbf{C}^n$ by the sign involution $v \mapsto (-v)$. To find out what $\mathbf{C}^n/\pm$ is, note that regular functions on $\mathbf{C}^n/\pm$ are precisely the even regular functions on $\mathbf{C}^n$. Every even polynomial in n variables can be represented as a polynomial in monomials of degree 2, and it means that $\mathbf{C}^n/\pm$ is the affine cone over P^{n-1} in its Veronese embedding defined by monomials of degree 2. This is a singular variety. In the particular case when $n = 2$, this variety $\mathbf{C}^2/\pm$ is the quadratic cone in $\mathbf{C}^3$. Indeed, let s, t be coordinates in $\mathbf{C}^2$. There are three monomials of degree 2: $p = s^2, q = t^2, r = st$. They are subject to one relation $r^2 = pq$ which defines a quadratic cone.

Since the coincidence of only two points is the most generic pattern of degeneration of a tuple, any $\mathrm{Sym}^d(\mathbf{C}^n)$ will behave near such tuples as a product of an affine space and a variety of type $\mathbf{C}^n/\pm$. When more points coincide, the singularity becomes more complicated.

(b) When Y is a smooth surface, the singular variety $\mathrm{Sym}^d(Y)$ admits a canonical resolution of singularities: the so-called Hilbert scheme parametrizing sheaves of ideals in $\mathcal{O}_Y$ of codimension d, see [I]. When $\dim(Y) > 2$, no good resolution of singularities of $\mathrm{Sym}^d(Y)$ is known.

The rest of this section will be devoted to explicit equations for the Chow variety $\mathrm{Sym}^d(P^{n-1}) = G(1, d, n)$. In view of the nature of the Chow embedding (see subsection A), the question can be reformulated as follows.

Given a homogeneous polynomial $f(\mathbf{y}) = f(y_1, \ldots, y_n) = \sum_\omega a_\omega \mathbf{y}^\omega$ of degree d, determine from the coefficients of f whether it is a product of linear factors.

Before proceeding to the general case, we consider an example.

Example 2.9. Let $f(y_1, \ldots, y_n) = \sum a_{ij} y_i y_j$ be a quadratic form defined by a symmetric matrix $A = \|a_{ij}\|$. The form f splits into a product of two linear forms if and only if $\mathrm{rk}\, A \leq 2$. This is equivalent to the fact that all 3×3 minors of A vanish. We see that $\mathrm{Sym}^2(P^{n-1})$ is defined in $P(S^2(\mathbf{C}^n))$ by a system of *cubic* equations. Quadratic equations obviously do not suffice. This is in contrast with the Plücker relations (which determine when a polynomial in anticommuting variables $\xi_1, \ldots, \xi_n$ splits into linear factors).

The conditions for factorization of a general homogeneous polynomial of degree d into linear factors were first derived by Brill [Br1]. It turns out that the equations have degree $d+1$ (in accord with the previous example, where, for d=2, the equations have degree 3).

Brill's answer was given in terms of the symbolic method of invariant theory. We shall translate it into representation-theoretic language. We first explain three main ingredients of Brill's approach: polars, vertical Young multiplication and Newton's power sum symmetric polynomials.

E. Polars

Let $f(x_1, \ldots, x_n)$ be a form (homogeneous polynomial) of degree d. Then for every two vectors x and y, the function $t \mapsto f(tx+y)$ is a polynomial of degree d in t, which can be written as

$$f(tx+y) = \sum_{k=0}^{d} \binom{d}{k} f_{x^k}(x, y) t^k. \tag{2.12}$$

The coefficient $f_{x^k}(x, y)$ is called the k-th *polar* of f; it is a bihomogeneous form of degree k in x and $d-k$ in y. Explicitly, we have

$$f_{x^k}(x, y) = \frac{(d-k)!}{d!} \left(\sum_i x_i \frac{\partial}{\partial y_i} \right)^k f(y). \tag{2.13}$$

To see this it is enough to differentiate (2.12) k times at $t=0$.

Using (2.12) or (2.13), we see that if $f = lh$ is the product of a linear form l and a form h of degree $d-1$, then

$$f_{x^k}(x, y) = \left(1 - \frac{k}{d}\right) l(y) h_{x^k}(x, y) + \frac{k}{d} l(x) h_{x^{k-1}}(x, y). \tag{2.14}$$

In particular, if $f = l^d$ is the d-th power of a linear form l, then

$$l^d_{x^k}(x, y) = l^k(x) l^{d-k}(y). \tag{2.15}$$

F. Power sums

Let $e_k(x_1, \ldots, x_d)$ be the k-th elementary symmetric polynomial in d scalar variables (see (2.2)). For $k \geq 1$ we have the *power sum* symmetric polynomial

$$p_k(x_1, \ldots, x_d) = x_1^k + \cdots + x_d^k,$$

which can be expressed as a universal polynomial with rational coefficients in symmetric polynomials $e_1, \dots, e_d$. The explicit form of this polynomial is obtained, as in subsection C, from the equality

$$\sum_{k=1}^{\infty} \frac{(-1)^{k-1} p_k(x_1, \dots, x_d)}{k} t^k = \log\left(1 + \sum_{i=1}^{d} e_i(x_1, \dots, x_d) t^i\right), \tag{2.16}$$

a particular case of (2.9). By taking the logarithm explicitly, we get the *Girard formula*

$$\begin{aligned} & p_k(x_1, \dots, x_d) \\ & \quad = (-1)^k k \sum_{i_1+2i_2+\dots+di_d=k} (-1)^{i_1+\dots+i_d} \frac{(i_1+\dots+i_d-1)!}{i_1!\cdots i_d!} e_1^{i_1}\cdots e_d^{i_d}. \end{aligned} \tag{2.17}$$

The following two features of this formula will be most important for us:
(1) In the grading such that $\deg(e_i) = i$, the polynomial p_k is homogeneous of degree k.
(2) The monomial e_1^k occurs in p_k with non-zero coefficient.

G. The vertical Young multiplication

Let V be an n-dimensional vector space. It is well-known that the decomposition of the tensor product $S^d(V) \otimes S^d(V)$ as a $GL(V)$-module has the form

$$S^d(V) \otimes S^d(V) = S^{(d,d)}(V) \oplus S^{(d+1,d-1)}(V) \oplus \dots \oplus S^{2d}(V),$$

where $S^{(d+k,d-k)}(V)$ is the Schur functor corresponding to a partition $(d+k, d-k)$ (see e.g., [FH] [Macd]). We shall identify $S^d(V) \otimes S^d(V)$ with the space of bihomogeneous forms $F(x, y)$ of bidegree (d, d) via the map $f \otimes g \mapsto f(x)g(y)$. For $f, g \in S^d(V)$, we denote by $f \odot g$ the component of $f(x)g(y)$ lying in $S^{(d,d)}(V)$. The form $f \odot g$ is called the *vertical (Young) product* of f and g. It is also known as the *apolar covariant* of f and g. When $f \odot g = 0$, the forms f and g are called *apolar* to each other.

An explicit formula for $f \odot g$ is given by the following proposition.

Proposition 2.10. *For $f, g \in S^d(V)$, we have*

$$(f \odot g)(x, y) = \frac{1}{d+1} \sum_{k=0}^{d} (-1)^k \binom{d}{k} f_{y^k}(y, x) g_{x^k}(x, y). \tag{2.18}$$

The proof of Proposition 2.10 will be given a little later. In fact, it would be enough for our purposes to *define* $f \odot g$ using (2.18).

The following property of the vertical product is crucial for Brill's method.

Proposition 2.11. *If $f \in S^d(V)$ and l is a non-zero linear form on V^*, then f is divisible by l if and only if $f \odot l^d = 0$.*

Proof. Substituting $g(y) = l^d(y)$ into (2.18) and using (2.15), we obtain

$$(f \odot l^d)(x, y) = \frac{1}{d+1} \sum_{k=0}^{d} (-1)^k \binom{d}{k} f_{y^k}(y, x) l^k(x) l^{d-k}(y). \tag{2.19}$$

We see that all the terms with $k \neq 0$ in the right hand side of (2.19) are divisible by $l(x)$. Therefore, if $(f \odot l^d)(x, y) = 0$, then $f(x)$ must be divisible by $l(x)$.

Conversely, let $f(x) = l(x)h(x)$ for some form $h(x)$ of degree $d - 1$. Then the polars of f are given by (2.14). Substituting them into (2.19) and regrouping terms, we see that $(f \odot l^d)(x, y) = 0$.

Proof of Proposition 2.10. In our realization, the $GL(V)$-module $S^d(V) \otimes S^d(V)$ is a submodule of the module $S^{2d}(V \oplus V)$ of homogeneous forms $F(x, y)$ of total degree $2d$, where $x, y \in V^*$. Identifying $V \oplus V$ with $V \otimes \mathbf{C}^2$, we see that $S^{2d}(V \oplus V)$ is a representation of the group $GL(V) \times GL(2)$. Consider the subgroup $SL(2) \subset GL(2)$ and its Lie algebra $sl(2)$ with standard generators E_+, E_-, H. These generators act on the forms $F(x, y)$ by the formulas

$$E_+ = \sum_{i=1}^{d} x_i \frac{\partial}{\partial y_i}, \quad E_- = \sum_{i=1}^{d} y_i \frac{\partial}{\partial x_i}, \quad H = \sum_{i=1}^{d} \left(x_i \frac{\partial}{\partial x_i} - y_i \frac{\partial}{\partial y_i} \right).$$

These formulas imply that $S^d(V) \otimes S^d(V)$ is a zero weight subspace of the $sl(2)$-module $S^{2d}(V \otimes \mathbf{C}^2)$:

$$S^d(V) \otimes S^d(V) = \left\{ F \in S^{2d}(V \otimes \mathbf{C}^2) : HF = 0 \right\}.$$

We claim that the submodule $S^{(d,d)}(V) \subset S^d(V) \otimes S^d(V)$ consists of all $sl(2)$-invariant elements; in more concrete terms,

$$S^{(d,d)}(V) = \left\{ F \in S^{2d}(V \otimes \mathbf{C}^2) : E_+ F = E_- F = HF = 0 \right\}.$$

This follows at once from the well-known decomposition of $S^{2d}(V \otimes \mathbf{C}^2)$ as a module over $GL(V) \times GL_2$:

$$S^{2d}(V \otimes \mathbf{C}^2) = \bigoplus_{k=0}^{d} S^{(d+k,d-k)}(V) \otimes S^{(d+k,d-k)}(\mathbf{C}^2),$$

see [FH] [Macd].

Let $\pi : S^d(V) \otimes S^d(V) \to S^{(d,d)}(V)$ be the $GL(V)$-equivariant projection. It is given by the formula

$$\pi(F) = \sum_{k=0}^{d} \frac{(-1)^k}{k!(k+1)!} E_-^k E_+^k(F). \tag{2.20}$$

To justify (2.20), denote by π' the operator in its right hand side. It is enough to show that:

(a) π' is $GL(V)$-equivariant;

(b) $(\pi')^2 = \pi'$ and $\mathrm{Im}(\pi')$ is annihilated by $sl(2)$;

(c) $\pi' \neq 0$ on $S^d(V) \otimes S^d(V)$.

Statement (a) is obvious by construction. Statement (b) is verified by explicit calculations, using the commutation relations in $sl(2)$. Statement (c) follows by computing $\pi'(x_1^d y_2^d)$ and showing that it is non-zero. Since we are going to perform more general computations of that kind below, we omit the details.

By definition, $f \odot g = \pi(f(x)g(y))$. We apply (2.20) for $F = f(x)g(y)$:

$$\begin{aligned}(f \odot g)(x, y) &= \sum_{m=0}^{d} \frac{(-1)^m}{m!(m+1)!} E_-^m \left(f(x) E_+^m g(y)\right) \\ &= \sum_{m=0}^{d} \frac{(-1)^m}{m!(m+1)!} \sum_{k=0}^{m} \binom{m}{k} (E_-^k f(x))(E_-^{m-k} E_+^m g(y)).\end{aligned}$$

By (2.13), $E_-^k f(x) = \frac{d!}{(d-k)!} f_{y^k}(y, x)$. Using the commutation relation $[E_+, E_-] = H$, we obtain after straightforward computations that

$$E_-^{m-k} E_+^m g(y) = \frac{m!(d-k)!}{k!(d-m)!} E_+^k g(y) = \frac{m!d!}{k!(d-m)!} g_{x^k}(x, y).$$

It follows that

$$(f \odot g)(x, y) = \sum_{k=0}^{d} (-1)^k \binom{d}{k}^2 b_{k,d} f_{y^k}(y, x) g_{x^k}(x, y), \tag{2.21}$$

where

$$b_{k,d} = \sum_{m=k}^{d} \frac{(-1)^{m-k}(d-k)!}{(m+1)(m-k)!(d-m)!}.$$

It remains to calculate the coefficients $b_{k,d}$. To do this we represent $b_{k,d}$ as an integral

$$b_{k,d} = \int_0^1 t^k (1-t)^{d-k} dt \tag{2.22}$$

(to prove (2.22) it suffices to expand the integrand in powers of t and integrate it term by term). But this integral is Euler's Beta-function $B(k+1, d-k+1)$, so we get

$$b_{k,d} = \frac{k!(d-k)!}{(d+1)!}.$$

Substituting this expression into (2.21) yields (2.18).

H. Brill's equations

Now we have all the tools for constructing Brill's equations. Let $f \in S^d(V)$ and $x, z \in V^*$. Consider the function

$$E(t) = E_{f,x,z}(t) = \frac{f\big(tf(z)x + z\big)}{f(z)}.$$

Using (2.12), we can write $E(t)$ as

$$E(t) = \sum_{k=0}^{d} e_k t^k,$$

where

$$e_k = \binom{d}{k} f_{x^k}(x, z) f(z)^{k-1}. \qquad (2.23)$$

Clearly, each e_k is a homogeneous form in x, z and f which has degree k in x, $k(d-1)$ in z, and k in coefficients of f. Consider the d-th power sum $p_d(E)$; we will regard it as a function of x depending on f and z as parameters, and denote it by $p_d(E) = P_{f,z}(x)$. By the property (1) of power sums, $P_{f,z}(x)$ is a form of degree d in x whose coefficients are homogeneous polynomials of degree d in f and degree $d(d-1)$ in z. Let

$$B_f(x, y, z) = (f \odot P_{f,z})(x, y); \qquad (2.24)$$

this is a homogeneous form in (x, y, z) of multidegree $(d, d, d(d-1))$ whose coefficients are forms of degree $d+1$ in the coefficients of f. Here is the main result of Brill.

Theorem 2.12. *A form $f(x)$ is a product of linear forms if and only if the polynomial $B_f(x, y, z)$ is identically equal to* 0.

Proof. Let $f(x) = l_1(x) \cdots l_d(x)$ be a factorizable form. Then $E(t)$ can be written as

$$E(t) = \prod_{i=1}^{d} \left(1 + t\frac{f(z)}{l_i(z)} l_i(x)\right).$$

Therefore,

$$P_{f,z}(x) = \sum_{i=1}^{d} \frac{f(z)^d}{l_i(z)^d} l_i(x)^d.$$

It follows that

$$f \odot P_{f,z} = \sum_{i=1}^{d} \frac{f(z)^d}{l_i(z)^d} f \odot l_i^d.$$

Since f is divisible by each l_i, we have $B_f(x, y, z) = f \odot P_{f,z} = 0$ according to Proposition 2.11.

Conversely, let f be such that $f \odot P_{f,z} = 0$ for all z. Using properties (1) and (2) of power sums (see subsection F above), we see that $P_{f,z}(x)$ has the form

$$const \cdot f_x(x, z)^d + \text{ terms divisible by } f(z).$$

Let z be a smooth point of the hypersurface $\{f = 0\}$. For such a point we have $P_{f,z}(x) = const \cdot f_x(x, z)^d$, so $f \odot f_x(x, z)^d = 0$. By Proposition 2.11, $f(x)$ is divisible by the linear form $f_x(x, z)$. But $f_x(x, z) = \frac{1}{d} \sum_{i=1}^{d} \frac{\partial f(z)}{\partial z_i} x_i$ is the equation of the tangent hyperplane to $\{f = 0\}$ at z. It follows that the hypersurface $\{f = 0\}$ coincides with its tangent hyperplane in the neighborhood of z. Hence $\{f = 0\}$ is a union of hyperplanes, implying that f is factorizable. This completes the proof.

Example 2.13. Let $d = 2$, i.e., f is a quadratic form. Let $\varphi(x, y) = f_x(x, y)$ be the symmetric bilinear form such that $f(x) = \varphi(x, x)$. A straightforward computation shows that up to a non-zero multiple

$$B_f(x, y, z) = \det \begin{pmatrix} \varphi(x, x) & \varphi(x, y) & \varphi(x, z) \\ \varphi(y, x) & \varphi(y, y) & \varphi(y, z) \\ \varphi(z, x) & \varphi(z, y) & \varphi(z, z) \end{pmatrix}. \tag{2.25}$$

The coefficients of $B_f(x, y, z)$ are cubic forms in the coefficients of f; by Theorem 2.12, f is a product of two linear forms if and only if all these cubic forms vanish at f. In more invariant terms, the vanishing of (2.25) means that the restriction of φ on every 3-dimensional subspace of V^* is degenerate, i.e., that $\text{rk}(\varphi) \leq 2$. Thus we have recovered the result in Example 2.9.

Theorem 2.12 gives a system of equations which define the symmetric product $\text{Sym}^d(P^{n-1})$ in $P(S^d\mathbf{C}^n)$ *set - theoretically*. J. Weyman has recently shown that in general these equations do not define $\text{Sym}^d(P^{n-1})$ scheme-theoretically, i.e., the subscheme in $P(S^d\mathbf{C}^n)$ defined by these equations is not reduced. In other words, the homogeneous ideal in the polynomial ring of coefficients of indeterminate $f \in S^d\mathbf{C}^n$ differs from the ideal generated by the Brill equations in infinitely many graded components.

3. Cayley-Green-Morrison equations of Chow varieties

As we have already remarked, the proof of the Chow–van der Waerden theorem, given in Section 1C, does not provide a manageable system of equations defining the Chow variety $G(k, d, n)$. In Section 2, we presented an explicit system of equations for $G(1, d, n)$, the Chow variety of 0-cycles. In this section we shall consider cycles of positive dimension. The problem is to recognize associated hypersurfaces (and Chow forms) among all the hypersurfaces in the Grassmannians and their equations. A method to accomplish this was proposed by Cayley [Ca5] and further developed by M. Green and I. Morrison [GrM]. We shall examine the basic ideas of this method.

A. Differential-geometric structure in the Grassmannians

Projective spaces are homogeneous and isotropic, i.e., by a projective transformation, we can not only transform every point to every other point but also any tangent vector at the first point to any tangent vector at the second point. This is no longer the case for general Grassmannians. We shall see that Grassmannians possess a peculiar differential-geometric structure given by a natural stratification of their tangent spaces.

Consider the Grassmannian $G = G(n-k, n)$ of $(n-k)$-dimensional vector subspaces in $\mathbf{C}^n$. Let S be the $(n-k)$-dimensional vector bundle over $G(n-k, n)$ whose fiber over a point represented by a subspace $L \subset \mathbf{C}^n$ is L. This bundle is called *tautological*. Clearly we have the embedding $S \subset \tilde{\mathbf{C}}^n$, where by $\tilde{\mathbf{C}}^n$, we denote the trivial bundle over $G(n-k, n)$ with fiber $\mathbf{C}^n$. Also let TG be the tangent bundle of $G(n-k, n)$.

Proposition 3.1. *There is an isomorphism of vector bundles*

$$TG = \operatorname{Hom}(S, \tilde{\mathbf{C}}^n/S) \tag{3.1}$$

on $G(n-k, n)$ which is equivariant with respect to the action of $GL(n)$.

Proof. The fact is well known, see, e.g., [Harr], Lect. 16 for a detailed discussion. So we give only an intuitive explanation (cf. Section 1B above) here.

Let L be a vector subspace in $\mathbf{C}^n$ of dimension $n-k$. A tangent vector to $G(n-k, n)$ at L is an infinitesimal movement of L. Such a movement is defined by specifying the velocities of individual points. These velocities should form a linear vector field $x \mapsto v(x)$ defined on L. If $v(x) \in L$ for all $x \in L$, then our movement preserves L, that is, it represents a zero vector of $T_L G(n-k, n)$. Therefore $T_L G(n-k, n) = \operatorname{Hom}(L, \mathbf{C}^n/L)$ as claimed.

If a vector space V is represented as $\operatorname{Hom}(B, C)$ where B, C are two other vector spaces, then we can speak about the *rank* of a vector from V, i.e., the dimension of the image of the corresponding operator $B \to C$. If we choose bases in B and C, then V will be identified with the space of matrices of size $\dim(B) \times \dim(C)$ and the rank of a vector is just the rank of the corresponding matrix.

Consider again the Grassmannian $G(n-k,n)$. Let $m = \min(k, n-k)$ and $r < m$. By Proposition 3.1, each tangent space $T_L G$ contains a cone $T_L^r(G)$ formed by vectors of rank $\leq r$. The family of cones $T_L^r(G)$ for a given r is invariant under the action of $GL(n)$ on $G(n-k,n)$. In fact, this family for each r can be recovered from either the smallest $T_L^1(G)$ or the largest cones $T_L^{m-1}(G)$, because of the following well-known proposition (see [Harr], Examples 20.5 and 9.2).

Proposition 3.2. *For any $r < \min(k, n-k)$, denote by Y_r the cone of $k \times (n-k)$ matrices of rank $\leq r$. Then*

(a) *for each r, the variety Y_{r-1} coincides with the set of singular points of Y_r;*

(b) *a matrix A lies in Y_r if and only if there are r matrices $A_1, \ldots, A_r \in Y_1$ such that $A = A_1 + \cdots + A_r$.*

We shall consider the family of largest cones $T_L^{r-1} G(n-k,n)$ in the tangent spaces as the cornerstone of our differential-geometric structure. These cones will be denoted simply by K_L.

Example 3.3. Consider the Grassmannian $G(2,4)$ which is a quadric in P^5 in the Plücker embedding. Each cone K_L is isomorphic to the cone in the space of 2×2 matrices formed by the matrices with zero determinant. Since, for 2×2 matrices, the determinant has degree 2, here the cone is quadratic. In other words, $G(2,4)$ acquires the *conformal structure*, i.e., in each tangent space we have a quadratic form defined up to a constant factor. For other Grassmannians, the cones K_L are no longer quadratic.

A quadratic cone in $\mathbf{C}^4$ contains two families of 2-dimensional vector subspaces (corresponding to the two families of lines on a quadric in P^3). It turns out that, in general, each K_L contains two distinguished families of vector subspaces, which are defined as follows.

Definition 3.4. Let $V = \mathrm{Hom}(B, C)$ be the space of linear operators. We call *α-subspaces* in V vector subspaces of the form

$$E_\alpha(\mathbf{x}) = \{A : B \to C \mid A(\mathbf{x}) = 0\}, \tag{3.2}$$

where $\mathbf{x} \in B$ is a non-zero vector. We call *β-subspaces* the subspaces of the form

$$E_\beta(M) = \{A : B \to C \mid \mathrm{Im}\,(A) \subset M\}, \tag{3.3}$$

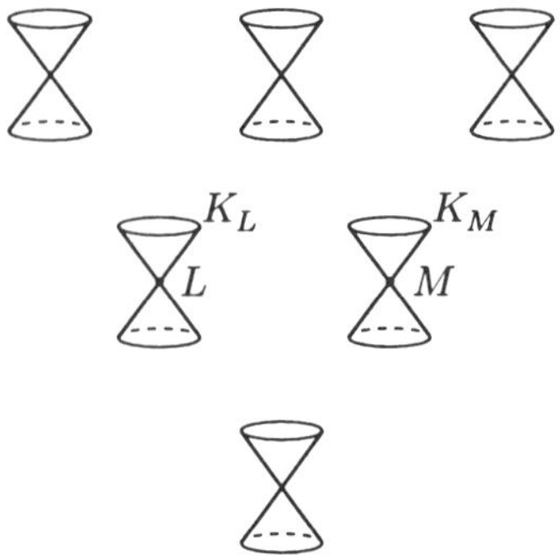

Figure 8. A conformal structure.

where $M \subset C$ is a hyperplane (passing through the origin).

Since the tangent space $T_L G(n-k, n)$ of $G(n-k, n)$ at any point L is identified with $\mathrm{Hom}\,(L, \mathbf{C}^n/L)$, we can speak of α and β subspaces in $T_L G(n-k, n)$. Clearly every such subspace is contained in the cone K_L. We shall also need certain sub-Grassmannians in $G(n-k, n)$, all of whose tangent spaces are α- or β-subspaces.

Let $x \in P^{n-1}$ be any point. We define the corresponding *α-variety* $G_\alpha(x) \subset G(n-k, n)$ to be the locus of all $(n-k-1)$-subspaces in P^{n-1} which contain x. Similarly, for any hyperplane $\pi \subset P^{n-1}$, we define the *β-variety* $G_\beta(\pi) \subset G(n-k, n)$ to be the locus of all $(n-k-1)$-subspaces contained in π. As an algebraic variety, each $G_\alpha(x)$ is isomorphic to the Grassmannian $G(n-k-1, n-1)$ and $G_\beta(\pi)$ is isomorphic to $G(n-k, n-1)$. By definition, the associated hypersurface $\mathcal{Z}(X)$ for a subvariety $X \subset P^{n-1}$ is the union of α-varieties $G_\alpha(x)$ for all $x \in X$.

Proposition 3.5. *Let G be an α-variety in $G(n-k, n)$ and let $L \in G$ be any point. Then the tangent subspace*

$$T_L G \subset T_L G(n-k, n) = \mathrm{Hom}\,(L, \mathbf{C}^n/L)$$

is an α-subspace. Conversely, given $L \in G(n-k, n)$ and an α-subspace $E \subset T_L G(n-k, n)$, there is a unique α-variety G in $G(n-k, n)$ containing L and such that $T_L G = E$. Similar assertions hold for β-varieties and β-subspaces.

Proof. Suppose $G = G_\alpha(x)$ for some $x \in P^{n-1}$. If $\mathbf{x} \in \mathbf{C}^n$ is a vector whose projectivization is x, then $\mathbf{x} \in L$ since $L \in G$. The definitions readily imply that $T_L G = E_\alpha(\mathbf{x})$; (this is intuitively obvious since an infinitesimal movement of L in G can be chosen so that $\mathbf{x}$ will remain fixed). Our proposition is proved in the α case. The β case is similar.

B. Coisotropic hypersurfaces

Definition 3.6. Let B, C be finite-dimensional vector spaces, $V = \mathrm{Hom}(B, C)$ be the space of operators, and $K \subset V$ the cone of operators that are not of full rank. A hyperplane $H \subset V$ is called *coisotropic* if it is tangent to K, i.e., lies in the projectively dual conic variety $K^\vee \subset V^*$ (see Section 1A Chapter 1).

In the case of the above definition we identify the space V^* with $\mathrm{Hom}(C, B)$ using the form $\mathrm{tr}(u \cdot v)$. Then every hyperplane has a linear equation which can be thought of as an operator from C to B. According to Example 4.10 of Chapter 1, the projective dual of the variety of matrices of non-maximal rank is the variety of matrices of rank ≤ 1. Therefore H is coisotropic if and only if its equation has rank 1.

Example 3.7. Let B and C have dimension 2, so the space V consists of 2×2 matrices. The cone K is defined by one equation $\det(A) = 0$. The determinant in this case is a quadratic form. Denote by $<, >$ the corresponding scalar product on V and by $H^\perp$ the orthogonal complement to H. A hyperplane H is coisotropic in our sense if and only if $H^\perp \subset H$. This is the usual coisotropy condition for subspaces in a vector space with a quadratic form.

Proposition 3.8. *Let $H \subset \mathrm{Hom}(B, C)$ be a coisotropic hyperplane. Then H contains a unique α- and a unique β-subspace. Conversely, given an α-subspace and a β-subspace, their linear span is a coisotropic hyperplane.*

Proof. Let $u : C \to B$ be a linear operator of rank 1 which generates the orthogonal complement to H with respect to the pairing $\mathrm{tr}(u \cdot v)$. Its kernel is a hyperplane $M \subset C$, and its image is a 1-dimensional subspace generated by some $\mathbf{x} \in B$. A straightforward check shows that the only α-subspace contained in H is $E_\alpha(\mathbf{x})$, and the only β-subspace contained in H is $E_\beta(M)$.

For the last statement, it remains to prove that every α-subspace E and every β-subspace E' linearly span a hyperplane. This can be shown by an easy dimension count: if $\dim(B) = b$, $\dim(C) = c$, then $\dim(E) = (b-1)c$, $\dim(E') = b(c-1)$, $\dim(E \cap E') = (b-1)(c-1)$, hence $\dim(E + E') = bc - 1$, as required.

Definition 3.9. A hypersurface Z in the Grassmannian $G(n-k, n)$ is called *coisotropic* if, for each smooth point $L \in Z$, the tangent hyperplane $T_L Z$ is coisotropic in $T_L G(n-k, n)$.

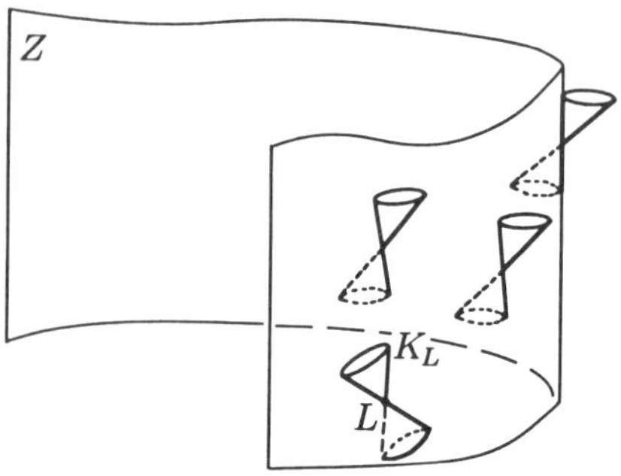

Figure 9. A coisotropic hypersurface.

Proposition 3.10. *Let $X \subset P^{n-1}$ be a $(k-1)$-dimensional irreducible subvariety. Then its associated hypersurface $\mathcal{Z}(X)$ is coisotropic.*

Proof. Let $\mathbf{L} \in \mathcal{Z}(X)$ be an $(n-k)$-dimensional vector subspace with projectivization L. By the definition of associated hypersurface, L intersects X at some point p. We can assume that p is a smooth point of X. Let $\mathbf{p}$ be a non-zero vector in $\mathbf{C}^n$ with projectivization p, and let $Y \subset \mathbf{C}^n$ be the cone over X. Choosing L generically, we can assume that $\mathbf{L} \cap T_{\mathbf{p}}Y = \mathbf{C} \cdot \mathbf{p}$. Let $M \subset \mathbf{C}^n/\mathbf{L}$ be the image of $T_{\mathbf{p}}Y$ under the projection $\mathbf{C}^n \to \mathbf{C}^n/\mathbf{L}$. Clearly, M is a hyperplane in $\mathbf{C}^n/\mathbf{L}$. Identifying, as before, $T_L G(n-k, n) = \mathrm{Hom}\,(\mathbf{L}, \mathbf{C}^n/\mathbf{L})$, we see that $T_L\mathcal{Z}(X)$ consists of operators $A : \mathbf{L} \to \mathbf{C}^n/\mathbf{L}$ such that $A(\mathbf{p}) \in M$. We see that $T_L\mathcal{Z}(X)$ is the sum of the α-subspace $E_\alpha(\mathbf{p})$ and the β-subspace $E_\beta(M)$; hence it is coisotropic, according to Proposition 3.8.

Example 3.11. Consider the Grassmannian $G(2, 4)$ of lines in P^3. As we have seen, its differential structure consists of a family of quadratic cones in tangent spaces which gives rise to a conformal structure. An associated hypersurface in $G(2, 4)$ is the set $\mathcal{Z}(X)$ of all lines meeting a given space curve X. From Proposition 3.10, we see that tangent spaces to any such $\mathcal{Z}(X)$ are coisotropic in the usual sense of quadratic forms.

Let us write the coisotropy condition in the coordinates. We consider the affine coordinates $A = \|a_{ij}\|$, $i = 1, \dots, n-k$, $j = 1, \dots, k$, in an affine chart $\mathbf{C}^{(n-k)\times k} \subset G(n-k, n)$. Recall (Section 1A, Chapter 3) that this chart consists of vector subspaces in $\mathbf{C}^{n-k} \oplus \mathbf{C}^k$ which are graphs of linear operators $A = \|a_{ij}\| : \mathbf{C}^{n-k} \to \mathbf{C}^k$. A hypersurface in the Grassmannian after intersection with the chart $\mathbf{C}^{(n-k)\times k}$ is determined by its equation which is a (non-homogeneous) polynomial f in $(n-k)k$ variables a_{ij}, defined uniquely up to a constant factor. This polynomial may be reducible (if the hypersurface itself contains several irreducible components) but is not allowed to be divisible by squares of irreducible polynomials.

Proposition 3.12. *Let $Z \subset G(n-k, n)$ be a hypersurface. Consider its intersection Z' with the affine chart $\mathbf{C}^{(n-k)\times k}$ with matrix coordinates $\|a_{ij}\|$, $i = 1, \dots, n-k$, $j = 1, \dots, k$. Let $f(a_{ij})$ be the equation of Z'. Then Z is coisotropic if and only if the matrix $\|\partial f/\partial a_{ij}\|$ has rank 1 everywhere on Z', i.e., each minor*

$$\begin{vmatrix} \partial f/\partial a_{ij} & \partial f/\partial a_{il} \\ \partial f/\partial a_{rj} & \partial f/\partial a_{rl} \end{vmatrix}$$

is divisible by f.

Proof. In coordinates $\|a_{ij}\|$ the cones K_L in all the tangent spaces $T_L G(n-k, n)$ are parallel translations of the cone of matrices $\|a_{ij}\|$ that are not of full rank. The cone of matrices of rank ≤ 1 is projectively dual to the latter cone, which implies out statement.

A natural guess would be that associated hypersurfaces can be characterized by the coisotropy condition alone. However, this is not the case as shown by the following example.

Example 3.13. Let S be a surface in P^3 and let $Z \subset G(2, 4)$ be the variety of all lines in P^3 tangent to S at some point (more precisely, Z is the Zariski closure of the set of lines tangent to S at its smooth points). We claim that Z is coisotropic. This is shown by exactly the same argument as in the proof of Proposition 3.10 (we leave the details to the reader).

Generalizing this example, we shall show that the coisotropy property characterizes the higher associated hypersurfaces, see Section 2E, Chapter 3. Recall that ET means the embedded tangent space.

Theorem 3.14.

(a) *Let $k \leq m < n$ and let $X \subset P^{n-1}$ be an irreducible $(m-1)$-dimensional variety. Then its $(m-k)$-th associated hypersurface*

$$\mathcal{Z}_{m-k}(X) = \Big\{ L \in G(n-k, n) \ : \ \dim(L \cap ET_x X) \geq m-k \text{ for some } x \in L \cap X \Big\} \tag{3.4}$$

is coisotropic.

(b) *Any irreducible coisotropic hypersurface $Z \subset G(n-k,n)$ is the higher associated hypersurface $\mathcal{Z}_{m-k}(X)$ for some (uniquely defined) $m \geq k$ and an $(m-1)$-dimensional irreducible subvariety $X \subset P^{n-1}$.*

C. Lagrangian interpretation of higher associated hypersurfaces and the proof of Theorem 3.14

We have seen in Section 3B, Chapter 1 that the cotangent bundle T^*M of any smooth variety M has a natural symplectic structure. To any irreducible subvariety $Z \subset M$, we associated its conormal variety (the closure of the conormal bundle to the smooth locus of Z)

$$\mathrm{Con}\,(Z) = \overline{T^*_{Z_{sm}}(M)} \subset T^*M. \tag{3.5}$$

This is a Lagrangian subvariety in T^*M. In particular, $\dim \mathrm{Con}\,(Z) = \dim (M)$ regardless of the dimension of Z.

The point of the "Lagrangian philosophy" is to work whenever possible with Lagrangian subvarieties $\mathrm{Con}(Z) \subset T^*M$ instead of subvarieties $Z \subset M$. One instance of the usefulness of this approach was seen in Section 3, Chapter 1 in the discussion of projective duality. It turns out that the construction of higher associated hypersurfaces becomes much more transparent in the Lagrangian approach.

More precisely, let $P = P^{n-1}$ be our projective space, $X \subset P$ be an irreducible subvariety of dimension $m-1 \geq k-1$ and $G = G(n-k,n)$ be the Grassmannian. The $(m-k)$-th associated hypersurface $\mathcal{Z}_{m-k}(X)$ is a hypersurface in G. We replace X and $\mathcal{Z}_{m-k}(X)$ by their conormal varieties $\mathrm{Con}\,(X) \subset T^*P$ and $\mathrm{Con}\,(\mathcal{Z}_{m-k}(X)) \subset T^*G$. We shall show that $\mathrm{Con}\,(\mathcal{Z}_{m-k}(X))$ can be obtained from $\mathrm{Con}\,(X)$ by a very simple construction.

Let $F \subset P \times G$ be the flag variety (or incidence variety), i.e.,

$$F = \{(\lambda, L) \in P \times G : \lambda \in L\}. \tag{3.6}$$

In accordance with our general approach, we consider the conormal bundle of F in $P \times G$ which we denote by

$$\Xi = T^*_F(P \times G) \subset T^*(P \times G) = T^*P \times T^*G. \tag{3.7}$$

We have a natural diagram of projections (the Lagrangian correspondence)

$$\begin{array}{ccc} & \Xi & \\ {}^{q_P}\swarrow & & \searrow^{q_G} \\ T^*P & & T^*G \end{array}, \tag{3.8}$$

where, by q_P and q_G, we denote the projections of Ξ to the first and second factors in (3.7). The map q_P is a locally trivial fibration. Therefore if $\Lambda \subset T^*P$ is any irreducible Lagrangian variety, then $q_G(q_P^{-1}(\Lambda)) \subset T^*G$ is also an irreducible variety. It is this simple construction that is relevant to our problem.

Proposition 3.15. *In the above notation, we have*

$$\mathrm{Con}\,(\mathcal{Z}_{m-k}(X)) = q_G(q_P^{-1}(\mathrm{Con}\,(X))). \tag{3.9}$$

The proof is based on a lemma which describes $\Xi = T_F^*(P \times G)$ in intrinsic terms. Let $(\lambda, \xi) \in T^*P$. In other words, $\lambda \in P = P^{n-1}$ is a 1-dimensional subspace in $\mathbf{C}^n$ and $\xi \in T_\lambda^* P$ is a covector at λ. Let also $(\mathbf{L}, \eta) \in T^*G$. We think of $\mathbf{L}$ as an $(n-k)$-dimensional vector subspace in $\mathbf{C}^n$, and denote by L the corresponding projective subspace in P. A covector $\eta \in T_{\mathbf{L}}^*G$ can be regarded as an element of $\mathrm{Hom}\,(\mathbf{C}^n/\mathbf{L}, \mathbf{L})$. Now the question which we want to answer is: when does the tuple $(\lambda, \xi, \mathbf{L}, \eta)$ belong to Ξ?

Lemma 3.16. *Let $(\lambda, \xi, \mathbf{L}, \eta) \in T^*P \times T^*G$ be a point with non-zero ξ, η. If this point belongs to Ξ then the following conditions hold:*
(1) *$\lambda \in L$ (i.e., $(\lambda, L) \in F$).*
(2) *The projectivization L of $\mathbf{L}$ lies in the hyperplane $\pi \subset P$, which is tangent to the vector subspace $\{\xi = 0\} \subset T_\lambda P$.*
(3) *The hyperplane $\{\eta = 0\} \subset T_{\mathbf{L}}G$ is the sum*

$$T_{\mathbf{L}}G_\alpha(\lambda) + T_{\mathbf{L}}G_\beta(\pi),$$

where G_α, G_β are α- and β-varieties (see subsection A). Conversely, if these conditions hold then there is $c \in \mathbf{C}^*$ such that $(\lambda, \xi, \mathbf{L}, c\eta) \in \Xi$.

Proof of the lemma. Let us define the following vector bundle E on $P \times G$. By definition, the fiber of E at a point $(\lambda, \mathbf{L})$ is set to be $\mathrm{Hom}\,(\lambda, \mathbf{C}^n/\mathbf{L})$. Let $s \in H^0(P \times G, E)$ be the section whose value at $(\lambda, \mathbf{L})$ is the natural composite map $\lambda \hookrightarrow \mathbf{C}^n \to \mathbf{C}^n/\mathbf{L}$. This section vanishes whenever $\lambda \subset \mathbf{L}$, i.e., it vanishes on F. The differential ds gives an identification

$$T_F^*(P \times G) \cong E^*|_F. \tag{3.10}$$

So any conormal space $T_F^*(P \times G)_{(\lambda,\mathbf{L})}$ is identified with $\lambda \otimes \mathbf{L}^\perp$ where $\mathbf{L}^\perp \subset \mathbf{C}^{n*}$ is the orthogonal complement to $\mathbf{L}$. The projections

$$T_F^*(P \times G)_{(\lambda,\mathbf{L})} = \lambda \otimes \mathbf{L}^\perp \quad \longrightarrow \quad T_\lambda^* P = \lambda \otimes \lambda^\perp,$$

$$T_F^*(P \times G)_{(\lambda,\mathbf{L})} = \lambda \otimes \mathbf{L}^\perp \quad \longrightarrow \quad T_\mathbf{L}^* G = \mathbf{L} \otimes \mathbf{L}^\perp$$

are induced respectively by the inclusions $\mathbf{L}^\perp \subset \lambda^\perp$, $\lambda \subset \mathbf{L}$ valid for $(\lambda, \mathbf{L}) \in F$. The condition $(\lambda, \xi, \mathbf{L}, \eta) \in \Xi$ means simply that $\xi \in \lambda \otimes \lambda^\perp$ and $\eta \in \mathbf{L} \otimes \mathbf{L}^\perp$ both come from the same vector $\zeta \in \lambda \otimes \mathbf{L}^\perp$.

Now let us suppose that $(\lambda, \xi, \mathbf{L}, \eta) \in \Xi$ and show that the conditions (1)–(3) are satisfied. The fact that ξ comes from $\lambda \otimes \mathbf{L}^\perp$ means that the hyperplane $\pi = \operatorname{Ker} \xi$ contains L. This is the condition (2) of the lemma. Let us write $\xi = \zeta = x \otimes g$ where $x \in \lambda$, $g \in \mathbf{L}^\perp$ (note that λ is 1-dimensional!). Let us remark that the hyperplane π is given by the vanishing of the linear form g. Since $\eta \in \mathbf{L} \otimes \mathbf{L}^\perp$ also comes from $\zeta = x \otimes g$, it is implied that $\operatorname{Ker} \eta \subset \mathbf{L}^* \otimes (\mathbf{L}^\perp)^*$ has the form

$$\operatorname{Ker}\ \eta = (x^\perp \otimes (\mathbf{L}^\perp)^*) + (L^* \otimes g^\perp).$$

But $x^\perp \otimes (\mathbf{L}^\perp)^* = T_\mathbf{L} G_\alpha(\lambda)$ and $\mathbf{L}^* \otimes g^\perp = T_\mathbf{L} G_\beta(\pi)$, so we get the condition (3). This proves the lemma in one direction.

Conversely, suppose that the conditions (1)–(3) hold. As above, (2) implies that $\xi \in \lambda \otimes \mathbf{L}^\perp$, say $\xi = x \otimes g$ where $x \in \lambda$, $g \in \mathbf{L}^\perp$, and that π is given by the vanishing of g. The condition (3) gives that η and $x \otimes g$ have the same kernel and thus are proportional. The lemma is proved.

Proof of Proposition 3.15. As in Section 2E Chapter 3, we use the notation $ET_\lambda X$ for the embedded tangent space to X at a smooth point $\lambda \in X$. Note that both sides of the proposed equality (3.9) are irreducible subvarieties in G. Hence it is enough to prove that for some Zariski open subset $V \subset q_P^{-1}(\operatorname{Con}(X))$ of "generic points," the set $q_G(V)$ is a Zariski dense subset in $\operatorname{Con}(\mathcal{Z}_{m-k}(X))$. We define V as the set of all

$$(\lambda, \xi, \mathbf{L}, \eta) \in q_P^{-1}(\operatorname{Con}(X)) \subset T^*P \times T^*G$$

such that

(a) λ is a smooth point of X.
(b) The projectivization L of $\mathbf{L}$ (which contains the point λ) together with the embedded tangent space $ET_\lambda X$ span the hyperplane $\pi \subset P$ tangent to $\{\xi = 0\}$ at λ.

Note that, under the assumption (b), the covector ξ is determined, up to a scalar factor, by λ and L. Moreover, since both L and $ET_\lambda X$ are contained in the same hyperplane, we have $\dim(L \cap ET_\lambda X) \geq m - k$, i.e., $L \in \mathcal{Z}_{m-k}(X)$.

We consider the open subset $V_1 \subset V$ consisting of $(\lambda, \xi, \mathbf{L}, \eta)$ for which, in addition, $\mathbf{L}$ is a smooth point of $\mathcal{Z}_{m-k}(X)$. We claim that $q_G(V_1)$ is a Zariski dense subset in $\operatorname{Con}(\mathcal{Z}_{m-k}(X))$. Indeed, it remains only to show that if $(\lambda, \xi, \mathbf{L}, \eta) \in V_1$

then $\eta \in (T^*_{\mathcal{Z}_{m-k}(X)}G)_{\mathbf{L}}$ and, moreover, any conormal vector to $\mathcal{Z}_{m-k}(X)$ at $\mathbf{L}$ is thus obtained. By the condition (3) of Lemma 3.16, this is equivalent to the statement that

$$T_{\mathbf{L}}G_\alpha(\lambda) + T_{\mathbf{L}}G_\beta(\pi) = T_{\mathbf{L}}\mathcal{Z}_{m-k}(X). \tag{3.11}$$

However, since X is smooth at λ and $\mathcal{Z}_{m-k}(X)$ is smooth at $\mathbf{L}$, we can replace $X \subset P$ by its first order approximation at λ, i.e., by the embedded tangent space $ET_\lambda X \subset P$ so that

$$T_{\mathbf{L}}\mathcal{Z}_{m-k}(X) = T_{\mathbf{L}}\mathcal{Z}_{m-k}(ET_\lambda X). \tag{3.12}$$

The equality

$$T_{\mathbf{L}}G_\alpha(\lambda) + T_{\mathbf{L}}G_\beta(\pi) = T_{\mathbf{L}}\mathcal{Z}_{m-k}(ET_\lambda X) \tag{3.13}$$

is easy to establish. Indeed, both sides of it are hyperplanes in $T_{\mathbf{L}}G$. The summand $T_{\mathbf{L}}G_\alpha(\lambda)$ corresponds to infinitesimal displacements of $\mathbf{L}$ containing λ. Such displacements lie on the right hand side of (3.13) by definition. The summand $T_{\mathbf{L}}G_\beta(\pi)$ corresponds to infinitesimal displacements of L (the projectivization of $\mathbf{L}$) inside π. Any such displacement intersects $ET_\lambda X$ along a subspace of dimension $\geq m-k$. So the second summand is also contained on the right hand side. The equality (3.13) follows for dimension reasons. Proposition 3.15 is proved.

Having Proposition 3.15 at our disposal, it is very easy to explain why higher associated hypersurfaces are coisotropic. Let $Y \subset T^*G$ be the space of *isotropic covectors*, i.e., covectors defining coisotropic hyperplanes. More precisely, we call a covector $\eta \in T^*_{\mathbf{L}}G = \mathrm{Hom}(\mathbf{C}^n/\mathbf{L}, \mathbf{L})$ coisotropic if the corresponding operator $\mathbf{C}^n/\mathbf{L} \to \mathbf{L}$ has rank ≤ 1. Clearly a hypersurface $Z \subset G$ is coisotropic if and only if its conormal variety $\mathrm{Con}\,(Z) \subset T^*G$ lies in Y. Now part (a) of Theorem 3.14 (higher associated varieties are coisotropic) follows from Proposition 3.15 and the next simple fact.

Proposition 3.17. *In the notation from (3.8) the variety $Y \subset T^*G$ coincides with $q_G(\Xi)$.*

Proof. Our statement is an immediate consequence of the condition (3) in Lemma 3.16 and the fact that coisotropic hyperplanes are precisely sums of an α-subspace and a β-subspace (Proposition 3.8).

Let us now turn to part (b) of Theorem 3.14 (any coisotropic hypersurface is higher associated). Let $Z \subset G$ be an irreducible coisotropic hypersurface. Consider the variety

$$\Lambda = q_P(q_G^{-1}(\mathrm{Con}(Z))) \subset T^*P, \tag{3.14}$$

where q_P, q_G are defined in (3.8).

Lemma 3.18. *Λ is an irreducible conic Lagrangian variety.*

Proof. Λ is conic (i.e., invariant under dilations) since both $\mathrm{Con}(Z) \subset T^*G$ and $\Xi \subset T^*P \times T^*G$ are. The statement that Λ is Lagrangian means that, first, $\dim \Lambda = \dim P = n - 1$ and, second, the restriction to (the smooth locus of) Λ of the symplectic form ω on T^*P vanishes identically.

To see that $\dim \Lambda = n - 1$, note that $\dim Y = k(n-k) + n - 1$, since Y is a fibration over G with the fiber being the cone over the Segre variety $P^{k-1} \times P^{n-k-1}$. So the dimension of fibers of q_G over non-zero covectors from Y is equal to

$$\dim \Xi - \dim Y = \dim P + \dim G - \dim Y$$

$$= (n-1) + k(n-k) - (k(n-k) + n - 1) = 0.$$

In other words, any non-zero covector $\eta \in Y$ lifts uniquely to Ξ. (This can be also seen from the reasoning in the proof of Lemma 3.16.) Since Z is coisotropic, we have that $\mathrm{Con}(Z) \subset Y$ and so $\dim q_G^{-1}\mathrm{Con}(Z) = k(n-k)$. Now the fibers of q_P have dimension

$$\dim \Xi - \dim T^*P \; = \; k(n-k) - (n-1).$$

Therefore

$$\dim \Lambda \; = \; \dim q_P(q_G^{-1}(\mathrm{Con}(Z))) \; \geq \; n - 1.$$

This inequality will suffice if we show that the symplectic form ω vanishes on Λ.

To see this, we introduce local coordinates x_i on P and let $\xi_i = dx_i$ so that the x_i and ξ_i combined are local coordinates on T^*P. Similarly, let (y_j, η_j), $\eta_j = dy_j$ be local coordinates on T^*G. Since $\Xi \subset T^*P \times T^*G$ is Lagrangian, we have

$$\left(\sum_i dx_i \wedge d\xi_i + \sum_j dy_j \wedge d\eta_j \right) \Big|_{\Xi} = 0. \tag{3.15}$$

Since $\mathrm{Con}(Z) \subset T^*G$ is Lagrangian, we get $\left(\sum_j dy_j \wedge d\eta_j \right) \Big|_{\mathrm{Con}(Z)} = 0$. We regard this as an equality on $T^*P \times T^*G$:

$$\left(\sum_j dy_j \wedge d\eta_j \right) \Big|_{\tilde{q}_G^{-1}(\mathrm{Con}(Z))} = 0 \tag{3.16}$$

where $\tilde{q}_G : T^*P \times T^*G \to T^*G$ is the projection. The equations (3.15) and (3.16) imply that, on $T^*P \times T^*G$, we have

$$\left(\sum_i dx_i \wedge d\xi_i \right) \Big|_{\tilde{q}_G^{-1}(\mathrm{Con}(Z)) \cap \Xi} = 0$$

from which, on T^*P, we get $\left(\sum_i dx_i \wedge d\xi_i\right)\Big|_\Lambda = 0$. Lemma 3.18 is proved.

Now let us apply Proposition 3.1 of Chapter 1 which says that any irreducible conic Lagrangian variety $\Lambda \subset T^*P$ has the form $\Lambda = \mathrm{Con}(X)$ for some subvariety $X \subset P$. By our construction (3.14) of Λ we have

$$\mathrm{Con}(Z) = q_G(q_P^{-1}(\Lambda)) = q_G(q_P^{-1}(\mathrm{Con}(X)).$$

It is easy to see from the fact that Z is a hypersurface that $\dim X = m - 1 \geq k - 1$. Proposition 3.15 implies that $Z = \mathcal{Z}_{m-k}(X)$.

This completes the proof of Theorem 3.14.

D. α-distributions and the characterization of associated hypersurfaces

We have established (Theorem 3.14) that coisotropic hypersurfaces in $G(n-k, n)$ are precisely higher associated hypersurfaces of subvarieties $X \subset P^{n-1}$ with $\dim X \geq k - 1$. The question now is how to characterize genuine associated hypersurfaces among all coisotropic ones.

Recall (Proposition 3.8) that any coisotropic hyperplane in Hom (B, C) contains a unique α-subspace and a unique β-subspace. Therefore, if $Z \subset G(n-k, n)$ is a coisotropic hypersurface then for any smooth point $L \in Z$ we have, in the tangent space $T_L Z$, a uniquely defined α-subspace $E(L)$ and a uniquely defined β-subspace $E'(L)$. Clearly these subspaces vary holomorphically with $L \in Z$. So we obtain two *distributions* (or *Pfaff systems*) on Z. We call them the α- and β-distributions of Z and denote them by $\mathcal{E}_{\alpha,Z}$ and $\mathcal{E}_{\beta,Z}$. They are defined on the smooth locus of Z.

It is the α-distribution $\mathcal{E}_{\alpha,Z}$ which will be of special importance for us. Note that if Z is the (genuine) associated hypersurface of some $X \subset P^{n-1}$, $\dim X = k - 1$ then Z is fibered (generically) over X since a generic $(n - k - 1)$-plane intersecting X does so in only one point. Fibers of the arising projection $\pi : Z \to X$ are α-varieties $G_\alpha(x)$, $x \in X$. So in this case the α-distribution $\mathcal{E}_{\alpha,Z}$ consists of planes tangent to fibers of π.

In general, a distribution of p-dimensional subspaces in tangent spaces of a manifold Y is called *integrable* if it possesses a p-dimensional integral submanifold through each point of Y.

Theorem 3.19. *A hypersurface $Z \subset G(n - k, n)$ is the associated hypersurface of some $(k - 1)$-dimensional $X \subset P^{n-1}$ if and only if the next two condition hold:*

(a) *Z is coisotropic;*

(b) *The α-distribution $\mathcal{E}_{\alpha,Z}$ of Z is integrable.*

In this case integral manifolds of $\mathcal{E}_{\alpha,Z}$ are α-varieties $G_\alpha(x)$, and the variety X can be recovered as the set of points x parametrizing these integral manifolds.

Proof. If Z is associated to some X then, as we have just seen, $\mathcal{E}_{\alpha,Z}$ consists of tangent planes to fibers of the (generically defined) projection $Z \to X$, so it is integrable. The converse statement is based on the following.

Lemma 3.20. *Let $\Gamma \subset G(n,k,n)$ be an irreducible variety whose tangent space at every smooth point $L \in \Gamma$ is an α-subspace in $T_L G(n-k,n)$. Then Γ is an α-variety $G_\alpha(x)$ for some $x \in P^{n-1}$.*

Proof. At any smooth point $L \in \Gamma$, we have, by our assumption,

$$T_L\Gamma = T_L G_\alpha(x(L)) \tag{3.17}$$

where $x(L) \in P^{n-1}$ is some point contained in L; clearly, $x(L)$ depends on $L \in \Gamma$ in an analytic way. In other words, (3.17) means that any infinitesimal displacement of L in Γ fixes $x(L)$. Let $M \subset P^{n-1}$ be the closure of the set of all points $x(L)$ for smooth $L \in \Gamma$. We claim that M is a point. Indeed, we have a map

$$\varphi : \Gamma_{sm} \to M, \quad L \mapsto x(L)$$

where Γ_{sm} is the smooth locus of Γ. By definition $\operatorname{Im}\varphi$ is dense in M. On the other hand, the differential of φ at any point

$$d_L\varphi : T_L\Gamma \to T_{x(L)}M$$

is the zero map. To see this, let $v \in T_L\Gamma$ be an infinitesimal movement of L in Γ. By our assumption this infinitesimal movement preserves $x(L)$. This means that the infinitesimal movement of $x(L)$ corresponding to v, is zero, i.e., $(d_L\varphi)(v) = 0$. So M is indeed a point and the lemma is proved.

Having Lemma 3.20, it is easy to finish the proof of Theorem 3.19. Indeed, let $Z \subset G(n-k,n)$ be a coisotropic hypersurface such that the distribution $\mathcal{E}_{\alpha,Z}$ is integrable. By Lemma 3.20, each integral variety of $\mathcal{E}_{\alpha,Z}$ is an α-variety $G_\alpha(x)$ for some $x \in P^{n-1}$. Let $X \subset P^{n-1}$ be the closure of the set of x obtained in this way. Then Z consists precisely of $(n-k-1)$-dimensional projective subspaces intersecting X. Since Z is a hypersurface, it follows that $\dim X = k-1$, and we are done.

E. α-integrability in coordinates

We first recall the Frobenius theorem on the integrability of distributions (Pfaff systems). Let Z be a complex analytic manifold of dimension m and $\alpha_1, \dots, \alpha_r$ be holomorphic 1-forms on Z linearly independent at every point. For any $x \in Z$, the vanishing of linear forms $\alpha_1(x), \dots, \alpha_r(x) : T_x Z \to \mathbf{C}$ defines a vector subspace

$E(x) \subset T_x Z$ of codimension r. We denote by $\mathcal{E}$ the distribution (Pfaff system) formed by these subspaces. The Frobenius theorem is as follows [BCG[3]].

Theorem 3.21. *The distribution $\mathcal{E}$ is integrable if and only if for each i we have*

$$d\alpha_i \wedge \alpha_1 \wedge \cdots \wedge \alpha_r = 0. \tag{3.18}$$

We are going now to implement this theorem for the particular situation when Z is an irreducible coisotropic hypersurface in the Grassmannian $G(n-k, n)$ and $\mathcal{E}$ is the α-distribution $\mathcal{E}_{\alpha,Z}$ of Z. Of course Z may be singular so $\mathcal{E}_{\alpha,Z}$ is defined only at smooth points of Z.

We abbreviate $G = G(n-k, n)$. We shall work with the affine coordinates on G, see Section 1, Chapter 3. Thus we consider the affine chart $\mathbf{C}^{(n-k)\times k} \subset G$ consisting of subspaces which are graphs of linear operators $A = \|a_{ij}\| : \mathbf{C}^{n-k} \to \mathbf{C}^k$. The entries a_{ij}, $i = 1, \dots, n-k$, $j = 1, \dots, k$ are coordinates in this chart.

Let $f(a_{ij}) \in \mathbf{C}[a_{ij}]$ be the equation of the hypersurface Z. This is a polynomial which may be reducible (in case Z is reducible) but not divisible by squares of irreducible polynomials. As we have seen in Proposition 3.12, the coisotropy of Z is equivalent to the fact that the matrix $\|\partial f/\partial a_{ij}\|$ has rank 1 everywhere on Z. In other words, for any $z \in Z$, all rows of the matrix $\|\partial f/\partial a_{ij}\|(z)$ are vectors of length k proportional to each other. Let us assume that, say, the first row, i.e., the vector $(\partial f/\partial a_{11}, \dots, \partial f/\partial a_{1k})(z)$ is non-zero in the neighborhood of a given point of Z. Then this vector generates the 1-dimensional subspace $\operatorname{Im} \|\partial f/\partial a_{ij}\|(z)$. By Proposition 3.8, the α-subspace $E_\alpha(z) \subset T_z Z$ consists of those matrices $B = \|b_{ij}\| \in T_z G = \mathbf{C}^{(n-k)\times k}$ for which

$$(\partial f/\partial a_{11}, \dots, \partial f/\partial a_{1k})(z) \cdot B = 0. \tag{3.19}$$

Thus the distribution $\mathcal{E}_{\alpha,Z}$ is given (on the part of Z where the first row of $\|\partial f/\partial a_{ij}\|$ is not zero) by the vanishing of the following 1-forms:

$$\begin{cases} \alpha_1 = \frac{\partial f}{\partial a_{11}} da_{11} + \cdots + \frac{\partial f}{\partial a_{1k}} da_{1k} \\ \vdots \quad\quad \vdots \\ \alpha_{n-k} = \frac{\partial f}{\partial a_{11}} da_{n-k,1} + \cdots + \frac{\partial f}{\partial a_{1k}} da_{n-k,k} \end{cases} . \tag{3.20}$$

If, by any chance, the row $(\partial f/\partial a_{11}, \dots, \partial f/\partial a_{1k})(z)$ vanishes at some point $z \in Z$, any other non-vanishing row $(\partial f/\partial a_{i1}, \dots, \partial f/\partial a_{ik})(z)$ would do the job. More precisely, we introduce the forms

$$\alpha_j^i = \frac{\partial f}{\partial a_{i1}} da_{j1} + \cdots + \frac{\partial f}{\partial a_{ik}} da_{jk}, i, j = 1, \dots, n-k \tag{3.21}$$

Then, for any smooth point $z \in Z$, there is at least one i such that the 1-forms $\alpha_1^i, \ldots, \alpha_{n-k}^i$ define the α-subspace $E_\alpha(z)$.

Note that the forms (3.20), as well as the forms $\alpha_1^i, \ldots, \alpha_{n-k}^i$ for any i, are linearly dependent on Z. Indeed, on Z we have

$$0 = df = \sum_{i,j} \frac{\partial f}{\partial a_{ij}} da_{ij} = \alpha_1^1 + \ldots + \alpha_{n-k}^{n-k}. \tag{3.22}$$

If, for a given i, the i - th row of the matrix $\|\partial f/\partial a_{ij}\|$ is zero then all the forms α_j^i, $j = 1, \ldots, n-k$ are certainly zero. If this row is a non-zero vector then all the other rows are its multiples, so there are coefficients λ_l such that $\alpha_j^l = \lambda_l \alpha_j^i$ for any j, so (3.22) implies$\sum_l \lambda_l \alpha_l^i = 0$. In other words, to define the distribution $\mathcal{E}_{\alpha,Z}$, we need to take an appropriate subset of the forms $\alpha_1^i, \ldots, \alpha_{n-k}^i$ of cardinality $n-k-1$. This leads to the following conclusion.

Theorem 3.22. *Let $f \in \mathbf{C}[a_{ij}]$ be a square-free polynomial in entries of an $(n-k) \times k$-matrix $\|a_{ij}\|$ such that the hypersurface $Z = \{f = 0\}$ in $\mathbf{C}^{(n-k)\times k} \subset G(n-k, n)$ is coisotropic. The α-distribution $\mathcal{E}_{\alpha,Z}$ is integrable if and only if for each $i_\nu, j_\nu \in 1, \ldots, n-k$, $\nu = 1, \ldots, n-k$ the $(n-k+2)$-form*

$$df \wedge d\alpha_{j_1}^{i_1} \wedge \alpha_{j_2}^{i_2} \wedge \cdots \wedge \alpha_{j_{n-k}}^{i_{n-k}} \tag{3.23}$$

vanishes at every point of Z. In other words, it is required that any component of any form (3.23) (which is a polynomial in a_{ij}) be divisible by the polynomial f.

Proof. This follows from the above analysis and from the Frobenius theorem 3.21 once we observe the following obvious fact. For any differential r-form Ω on $\mathbf{C}^{(n-k)\times k}$, the restriction of Ω on Z vanishes (as an r-form on Z) if and only if $df \wedge \Omega$ vanishes at every point of Z.

Remark 3.23. *(a)* The conditions for a hypersurface $\{f = 0\}$ to be an associated hypersurface, given in Proposition 3.12 and Theorem 3.22, have the form $D_i(f) \equiv 0 \pmod f$ where the D_i are some non-linear differential operators. It is possible in principle to interpret these conditions as algebraic equations on coefficients of f: we need to "eliminate" the indeterminate factor g in $D_i(f) = fg$ which can be done by some linear algebra*. However, the resulting equations will have a degree much higher that the degree of homogeneity of $D_i(f)$ in f. For example,

* The existence of g such that $D_i(f) = fg$ means that $D_i(f)$, regarded as a vector in an appropriate space of polynomials, lies in the image of the operator given by multiplication with f. So we have the condition that some linear system is compatible. This can be reformulated as the vanishing of certain minors.

the conditions of coisotropy (Proposition 3.12) look like quadratic equations in coefficients of f, but in fact they are not.

(b) The approach of [Ca5] and [GrM] used not affine but Plücker coordinates in the Grassmannian. Let us illustrate this for the case of $G(2,4)$. We represent a hypersurface Z by a homogeneous polynomial F in Plücker coordinates p_{ij}, $1 \leq i < j \leq 4$. The polynomial F in now defined not uniquely but modulo the Plücker relation $R = p_{12}p_{34} - p_{13}p_{24} + p_{14}p_{23}$. The coisotropy condition in Plücker coordinates has the form

$$\frac{\partial F}{\partial p_{12}} \cdot \frac{\partial F}{\partial p_{34}} - \frac{\partial F}{\partial p_{13}} \cdot \frac{\partial F}{\partial p_{24}} + \frac{\partial F}{\partial p_{14}} \cdot \frac{\partial F}{\partial p_{23}} \equiv 0 \ (\text{mod}\, F, R). \tag{3.24}$$

It was first proved by Cayley [Ca5] that any hypersurface in $G(2,4)$ whose equation in Plücker coordinates satisfies (3.24) consists either of all lines meeting some curve or of all lines tangent to some surface.

PART II

A-Discriminants and A-Resultants

CHAPTER 5

Toric Varieties

In Part I we studied discriminants and resultants in the general context of projective geometry: our setup was that of an arbitrary projective variety $X \subset P^{n-1}$. We now want to move into a more combinatorial setting, which is closer to the classical concept of discriminants and resultants for *polynomials*. This setting corresponds to the situation when $X \subset P^{n-1}$ is a toric variety. In the present chapter, we have adapted the theory of toric varieties for our purposes. Since there are several references available on the subject [D] [Fu 2] [O], we did not attempt to be exhaustive or self-contained. Our exposition is organized "from the special to the general" so that the general description of toric varieties in terms of fans appears at the very end of the chapter.

1. Projectively embedded toric varieties

A. Monomials and the A-philosophy

A monomial in $k-1$ variables $x_1, \ldots, x_{k-1}$ is a function

$$x^{\omega} = x_1^{\omega_1} \cdots x_{k-1}^{\omega_{k-1}}, \tag{1.1}$$

where $\omega = (\omega_1, \ldots, \omega_{k-1}) \in \mathbf{Z}_+^{k-1}$ is the exponent vector. It will be convenient to allow the ω_i to be arbitrary integers, possibly negative. In this case we get *Laurent monomials.* A Laurent monomial is a well-defined function

$$x \mapsto x^{\omega} : (\mathbf{C}^*)^{k-1} \to \mathbf{C}^*. \tag{1.2}$$

A variety of the form $(\mathbf{C}^*)^{k-1}$ is known as an (algebraic) *torus.* It is a group under component-wise multiplication. A Laurent monomial is nothing more than a character of the torus. By a Laurent polynomial, we mean a finite linear combination of Laurent monomials. Any usual polynomial can be considered as a Laurent polynomial.

Let $A \subset \mathbf{Z}^{k-1}$ be a finite set of integer vectors which can be identified with corresponding monomials. By $\mathbf{C}^A$, we denote the space of Laurent polynomials with monomials from A, i.e., of polynomials of the form

$$f(x) = f(x_1, \ldots, x_{k-1}) = \sum_{\omega \in A} a_{\omega} x^{\omega}. \tag{1.3}$$

The study of many problems involving polynomials becomes more transparent if we consider not individual polynomials but polynomials with indeterminate coefficients. For this, it is useful to fix $A \subset \mathbf{Z}^{k-1}$ and consider all polynomials from $\mathbf{C}^A$ at once. In the subsequent chapters we shall apply this "A-philosophy" to the problem of the discriminant (find whether a given polynomial defines a singular hypersurface) and the resultant (find whether k given polynomials in $k-1$ variables have a common root). As an example outside the scope of the present book, let us mention the general theory of hypergeometric functions [GKZ2]. The study of integrals such as

$$I(f) = \int f(x_1, \ldots, x_{k-1})^{\alpha_0} x_1^{\alpha_1} \cdots x_{k-1}^{\alpha_{k-1}} dx_1 \ldots dx_{k-1}$$

for particular Laurent polynomials f is usually quite difficult. On the other hand, considering $I(f)$ as a function of an *indeterminate* $f \in \mathbf{C}^A$ leads to a meaningful theory. In particular, it is possible to express $I(f)$ in terms of power series of a simple form (the so-called hypergeometric series).

B. The varieties X_A and Y_A

To set the study of Laurent polynomials in the geometric context of Part I, we associate a projective variety to a finite set $A \subset \mathbf{Z}^{k-1}$. Suppose, $A = \{\omega^{(1)}, \ldots, \omega^{(n)}\}$. Define the variety $X_A \subset P^{n-1}$ to be the closure of the set

$$X_A^0 = \{(x^{\omega^{(1)}} : \ldots : x^{\omega^{(n)}}) \ : \ x = (x_1, \ldots, x_{k-1}) \in (\mathbf{C}^*)^{k-1}\}. \tag{1.4}$$

Varieties of the form X_A will now be the main objects of study and we shall now apply the formalism of Part I to X_A.

Along with $X_A \subset P^{n-1}$, we consider the affine variety $Y_A \subset \mathbf{C}^n$, namely the cone over X_A. This can be defined as the closure in $\mathbf{C}^n$ of the set

$$Y_A^0 = \{(x_k \cdot x^{\omega^{(1)}}, \ldots, x_k \cdot x^{\omega^{(n)}}) \ : x = (x_1, \ldots, x_{k-1}) \in (\mathbf{C}^*)^{k-1}, \ x_k \in \mathbf{C}^*\}. \tag{1.5}$$

In a more invariant setting, we can identify the ambient vector space $\mathbf{C}^n$ for Y_A as $(\mathbf{C}^A)^*$, the dual of the space of Laurent polynomials associated with A. Thus, a linear form $\sum a_i z_i$ on $\mathbf{C}^n$ will be written, when convenient, as the Laurent polynomial

$$f(x) = \sum a_i x^{\omega^{(i)}} \in \mathbf{C}^A.$$

Geometrically, this corresponds to the restriction of linear forms to X_A^0. In other words, the hyperplane sections of X_A are certain compactifications of zero loci of polynomials from $\mathbf{C}^A$.

Let us give some important examples of sets A and varieties X_A.

Examples 1.1. *(a)* Let A consist of all monomials in $x_1, \ldots, x_{k-1}$ of degree $\leq d$ that do not contain negative powers of any x_i. Let $\tilde{A}$ consist of all monomials in k variables $x_1, \ldots, x_k$, homogeneous of degree exactly d (not containing negative powers). The sets A and $\tilde{A}$ obviously lead to the same variety $X_A = X_{\tilde{A}}$. This is the projective space P^{k-1} in its Veronese embedding

$$P^{k-1} = P(\mathbf{C}^k) \hookrightarrow P(S^d\mathbf{C}^k), \quad v \mapsto v^d.$$

In particular, for $k = 2$, we get the Veronese curve $C_d \subset P^d$, i.e., the rational normal curve in P^d of degree d.

(b) Let A consist of bilinear monomials $x_i \cdot y_j$, where the x_i and y_j ($i = 1, \ldots, m$; $j = 1, \ldots, n$) are two sets of variables. The variety X_A is the product of two projective spaces $P^{m-1} \times P^{n-1}$ in its Segre embedding

$$P^{m-1} \times P^{n-1} = P(\mathbf{C}^m) \times P(\mathbf{C}^n) \hookrightarrow P(\mathbf{C}^m \otimes \mathbf{C}^n), \quad (v, w) \mapsto v \otimes w.$$

(c) Let A consist of the monomials of the form $x_i \cdot y_j \cdot z_k$, where the x_i ($i = 1, \ldots, m_1$), y_j ($j = 1, \ldots, m_2$), and z_k ($k = 1, \ldots, m_3$) are three sets of variables. Then X_A is the triple product $P^{m_1-1} \times P^{m_2-1} \times P^{m_3-1}$ in its Segre embedding similar to above.

(d) Let A consist of the following monomials in two variables:

$$1, x, x^2, \ldots, x^p, \; y, yx, yx^2, \ldots, yx^q.$$

The variety X_A is the rational normal scroll which we have already encountered in Example 3.6 Chapter 3.

Proposition 1.2. *The varieties X_A, Y_A depend only on the affine geometry of the set $A \subset \mathbf{Z}^{k-1}$. In other words, let $A \subset \mathbf{Z}^{k-1}$, $B \subset \mathbf{Z}^{m-1}$ and $T : \mathbf{Z}^{k-1} \to \mathbf{Z}^{m-1}$ be an integer affine transformation which is injective and such that $T(A) = B$. Then X_A is naturally identified with X_B, and Y_A with Y_B.*

Proof. Let $A = \{\omega^{(1)}, \ldots, \omega^{(n)}\}$ and $B = \{\eta^{(1)}, \ldots, \eta^{(n)}\}$ so $\eta^{(i)} = T(\omega^{(i)})$. Write T in an explicit form:

$$T(a_1, \ldots, a_{k-1}) = \left(c_{1k} + \sum_{j=1}^{k-1} c_{1j}a_j, \ldots, c_{m-1,k} + \sum_{j=1}^{k-1} c_{m-1,j}a_j\right),$$

where $c_{ij} \in \mathbf{Z}$. Consider the map $T^* : (\mathbf{C}^*)^{m-1} \to (\mathbf{C}^*)^{k-1}$, defined by

$$T^*(s_1, \ldots, s_{m-1}) = (t_1, \ldots, t_{k-1}) \quad \text{where} \quad t_j = s_1^{c_{1j}} \cdots s_{m-1}^{c_{m-1,j}}.$$

Then, for any $s = (s_1, \ldots, s_{m-1})$, the points

$$(s^{\eta^{(1)}} : \ldots : s^{\eta^{(n)}}) \in X_B^0 \subset P^{n-1} \quad \text{and} \quad (t^{\omega^{(1)}} : \ldots : t^{\omega^{(n)}}) \in X_A^0 \subset P^{n-1}$$

will be the same since the corresponding vectors in $\mathbf{C}^n$ will be proportional. Since T is injective, T^* is surjective from which $X_A^0 = X_B^0$ and $X_A = X_B$. Similarly, $Y_A = Y_B$.

A particular instance of Proposition 1.2 was already mentioned in Example 1.1 (a) (the sets A and $\tilde{A}$ give the same variety).

Proposition 1.2 implies that in the construction of X_A, for $A \subset \mathbf{Z}^{k-1}$, we can, if we wish, shrink $\mathbf{Z}^{k-1}$ to the smallest affine sublattice in $\mathbf{Z}^{k-1}$ containing A. We introduce special notation for it for later use.

Definition 1.3. If $A \subset \mathbf{Z}^{k-1}$ is a finite subset, then we denote by

$$\mathrm{Aff}_{\mathbf{Z}}(A) = \left\{ \sum_{\omega \in A} n_\omega \cdot \omega \ : \ n_\omega \in \mathbf{Z}, \ \sum n_\omega = 1 \right\}, \tag{1.6}$$

the affine sublattice in $\mathbf{Z}^{k-1}$ generated by A.

C. A general notion of a toric variety

The varieties X_A and Y_A introduced above belong to the following general class of algebraic varieties.

Definition 1.4. A *toric variety* is an irreducible complex algebraic variety X equipped with an action of the algebraic torus $(\mathbf{C}^*)^n$ having an open dense orbit.

This definition is slightly more general than the one usually given in the literature: we do not require normality of X (see Section 2 below for a discussion of this issue).

There is an obvious way of constructing toric varieties. Suppose that we have an action of a torus $H = (\mathbf{C}^*)^m$ on some algebraic variety Z. Let $z \in Z$ be any point. Consider the orbit closure $\overline{Hz}$. By definition, this is a toric variety.

In particular, the variety X_A is obtained by considering the action of the torus $(\mathbf{C}^*)^{k-1}$ on the projective space P^{n-1} given by the formula

$$x \cdot (z_1 : \ldots : z_n) = \left(x^{\omega^{(1)}} z_1 : \ldots : x^{\omega^{(n)}} z_n\right).$$

The variety X_A is the closure of the orbit of the point $(1 : \ldots : 1)$ under this action, so it is a toric variety. Similarly, the variety Y_A is the closure of the orbit of the point $(1, \ldots, 1) \in \mathbf{C}^n$ under the obvious action of the torus $(\mathbf{C}^*)^k = (\mathbf{C}^*)^{k-1} \times \mathbf{C}^*$, see (1.5).

Note that the toric variety $X_A \subset P^{n-1}$ is equivariantly embedded: the action of the torus on X_A extends to the whole P^{n-1}. The following converse statement is almost obvious.

Proposition 1.5. *Let $X \subset P^{m-1}$ be a projective toric variety (with $(\mathbf{C}^*)^{k-1}$ acting on X) in an equivariant embedding. Let $< X >$ be the minimal projective subspace in P^{m-1} containing X. If* $\dim < X > = n-1$, *then there exists a subset $A \subset \mathbf{Z}^{k-1}$ containing n elements and an isomorphism $X_A \to X$, equivariant under the torus and extending to an equivariant projective isomorphism $P^{n-1} \to < X >$ of the ambient projective spaces.*

Proof. An action of $(\mathbf{C}^*)^{k-1}$ on P^{m-1} by projective transformations can always be lifted to a linear action on $\mathbf{C}^m$. Any such action is diagonalizable, so in suitable coordinates it is given by a collection of characters

$$x \in (\mathbf{C}^*)^{k-1} \longmapsto \operatorname{diag}\left(x^{\omega^{(1)}}, \ldots, x^{\omega^{(m)}}\right),$$

for some integer vectors $\omega^{(i)} \in \mathbf{Z}^{k-1}$. Let $z = (z_1 : \ldots : z_m) \in X$ be a point lying on the open orbit of the torus. Some of the z_i may equal 0. Let $A \subset \mathbf{Z}^{k-1}$ be the collection of those $\omega^{(i)}$ for which $z_i \neq 0$. Then we have an isomorphism $X_A \to X$ as required. The rest of the proof is obvious.

Remark 1.6. There are projective toric varieties in the sense of Definition 1.4 that do not have an equivariant projective embedding. For example, consider a nodal cubic curve X in P^2 with homogeneous equation

$$z_0 z_2^2 = z_1^3 - z_0 z_1^2.$$

In the affine coordinates $x = z_1/z_0$, $y = z_2/z_0$, the equation is $y^2 = x^2(x-1)$. The point $p = \{x = y = 0\}$ is the only singular point of X and $X - \{p\} = \mathbf{C}^*$. So there is a $\mathbf{C}^*$-action on X with the open orbit $X - \{p\}$ and thus X is a toric variety. Suppose that X admits an equivariant projective embedding $\varphi : X \hookrightarrow P^{n-1}$. Then the line bundle $\mathcal{L} = \varphi^* \mathcal{O}(1)$ is equivariant with respect to the torus action. Denote by d the degree of $\mathcal{L}$, i.e., the degree of X in our embedding, and let $\operatorname{Pic}_d(X)$ be the moduli space of all line bundles of degree d on X. The torus action on X induces a natural action on $\operatorname{Pic}_d(X)$ under which the point $\mathcal{L}$ is, by our assumption, invariant. But it is easy to see that $\operatorname{Pic}_d(X)$ is isomorphic to $\mathbf{C}^*$ and that the torus action on it is transitive. This contradiction shows that X does not have equivariant projective embeddings.

D. Weight polytopes and torus orbits

We want to recall some statements from the proof of Proposition 1.5 using more invariant terminology.

Let $H = (\mathbf{C}^*)^{k-1}$ be an algebraic torus acting algebraically on a vector space V. Consider the associated action on the projectivization $P(V)$. Take any non-zero $v \in V$, and consider $\overline{P(H \cdot v)}$, the closure in $P(V)$ of the H-orbit of the point corresponding to v. This is a projective toric variety and Proposition 1.5 says that it has the form X_A, where $A = A(v) \subset \mathbf{Z}^{k-1}$ is defined as follows.

According to the standard theorem of the theory of algebraic groups, V can be decomposed into the weight subspaces corresponding to characters of H. A character $\chi : H \to \mathbf{C}^*$ is just a Laurent monomial $\chi(t_1, \ldots, t_{k-1}) = t_1^{\omega_1} \cdots t_{k-1}^{\omega_{k-1}}$. Thus the lattice of characters of H is identified with $\mathbf{Z}^{k-1}$. The weight subspace corresponding to a character χ is defined as

$$V_\chi = \{v \in V : t \cdot v = \chi(t)v \text{ for any } t \in H\}, \tag{1.7}$$

and we have the *weight decomposition*

$$V = \bigoplus_{\chi \in \mathbf{Z}^{k-1}} V_\chi.$$

Correspondingly, for any vector $v \in V$, we shall denote by v_χ its component of weight χ, i.e., the projection of v to V_χ along all the other weight spaces.

Set $A(v) = \{\chi \in \mathbf{Z}^{k-1} : v_\chi \neq 0\}$. Then $\overline{P(H \cdot v)} \cong X_{A(v)}$.

Definition 1.7. Let $H = (\mathbf{C}^*)^{k-1}$ be an algebraic torus acting linearly on a vector space V. Let $v \in V$ be any vector. The *weight polytope* $\mathrm{Wt}(v)$ of v is the convex hull in $\mathbf{R}^{k-1}$ of the set $A(v) = \{\chi \in \mathbf{Z}^{k-1} : v_\chi \neq 0\}$.

This is the first appearance of convex polytopes in this book. Convex geometry will play a very important role in the chapters to follow. Right now we shall describe the correspondence between torus orbits on $\overline{P(H \cdot v)}$ and faces of the polytope $\mathrm{Wt}(v)$.

Let X be a toric variety (whose torus we denote by H). Let $x, x' \in X$ be two points. We shall say that x' is a *toric specialization* of x if

$$x' \in \overline{H \cdot x}.$$

Clearly being a toric specialization is a partial order relation.

Proposition 1.8. *Let H and V be as before. Let $v \in V$ be a non-zero vector and $X = \overline{P(H \cdot v)}$. Then H-orbits on X are in bijection with faces of the polytope* $\mathrm{Wt}(v)$. *More precisely, for any $v' \in V$ representing a point of X, the polytope* $\mathrm{Wt}(v')$ *is a face of* $\mathrm{Wt}(v)$. *If $v', v'' \in V$ are two vectors representing points x', x'' of X then x'' is a toric specialization of x' if and only if* $\mathrm{Wt}(v'') \subset \mathrm{Wt}(v')$.

We reformulate and prove Proposition 1.8 in the particular case of the toric variety $X_A \subset P^{n-1}$ where $A \subset \mathbf{Z}^{k-1}$ is a finite set. By Proposition 1.5, this will suffice to establish Proposition 1.8 in general.

We consider $X_A \subset P^{n-1}$. As in subsection B above, we think of the ambient projective space as $P((\mathbf{C}^A)^*)$ and denote its homogeneous coordinates by z_ω, $\omega \in A$. Consider the polytope $Q \subset \mathbf{R}^{k-1}$, the convex hull of A. This is precisely the weight polytope of the vector $v = (1, \ldots, 1)$ whose orbit closure is X_A. Proposition 1.8 can be reformulated in our situation as follows.

Proposition 1.9. *The torus orbits in X_A are in bijection with non-empty faces of the polytope Q: the orbit $X^0(\Gamma)$ corresponding to a face $\Gamma \subset Q$, is specified inside X_A by conditions*

$$z_\omega = 0 \text{ for } \omega \notin \Gamma, \; z_\omega \neq 0 \text{ for } \omega \in \Gamma.$$

Denote by $X(\Gamma)$ the closure of the orbit $X^0(\Gamma)$. Then $X(\Gamma)$ is isomorphic to $X_{A \cap \Gamma}$. If Γ and Δ are two faces of Q then $X(\Gamma) \subset X(\Delta)$ if and only if $\Gamma \subset \Delta$.

Proof. Let $\Gamma \subset Q$ be a face. Denote by e_Γ the point in P^{n-1} with homogeneous coordinates (z_ω) where $z_\omega = 1$ for $\omega \in \Gamma$ and $z_\omega = 0$ for $\omega \notin \Gamma$. By definition, the points of X_A are limits $\lim_{\tau \to 0} \alpha(\tau) \cdot e_Q$ for all analytic maps α of a punctured disk $\{\tau \in \mathbf{C} : |\tau| < \varepsilon\}$ to $(\mathbf{C}^*)^{k-1}$. Suppose that

$$\alpha(\tau) = (c_1 \tau^{a_1} + \cdots, \ldots, c_{k-1}\tau^{a_{k-1}} + \cdots), \; c_i \in \mathbf{C}^*, \; a_i \in \mathbf{Z}$$

where the dots mean terms of higher order in τ. Let $a = (a_1, \ldots, a_{k-1})$. Consider the linear functional φ_a on $\mathbf{R}^{k-1}$ given by $\varphi_a(b_1, \ldots, b_{k-1}) = \sum a_i b_i$. Let $\Gamma(a) \subset Q$ be the supporting face of φ_a, i.e., the set of all points of Q where φ_a achieves its maximum. Then it is immediate to see that

$$\lim_{\tau \to 0} \alpha(\tau) e_Q = (c_1, \ldots, c_{k-1}) \cdot e_\Gamma.$$

Hence every point of X_A is equivalent under the action of the torus to some e_Γ. Since the closure of the orbit of e_Γ is $X_{A \cap \Gamma}$, all the assertions of the proposition now follow.

We conclude the discussion of weight polytopes with a modification of Proposition 1.8. Let V be a vector space with a linear action of the torus $H = (\mathbf{C}^*)^{k-1}$;

let $v \in V$ be a non-zero vector and $\mathrm{Wt}(v) \subset \mathbf{R}^{k-1}$ its weight polytope. Consider a 1-parameter subgroup in H of the form

$$\tau \longmapsto \tau^a = (\tau^{a_1}, \dots, \tau^{a_{k-1}}).$$

For any covector $w \in V^*$, we consider the Laurent polynomial in τ given by the scalar product

$$\tau \mapsto (\tau^a v, w). \tag{1.8}$$

For any Laurent polynomial $q(\tau) = \sum c_i \tau^i$, we call the term $c_i \tau^i$ with i the largest such that $c_i \neq 0$, the *leading term* of q.

Let $\mathrm{Wt}(v)^a \subset \mathrm{Wt}(v)$ be the face where the linear form $\varphi_a(b_1, \dots, b_{k-1}) = \sum a_i b_i$ achieves its maximum. Suppose the weight decomposition of v has the form

$$v = \sum_{\chi \in \mathbf{Z}^{k-1}} v_\chi.$$

We define

$$v_a = \sum_{\chi \in \mathrm{Wt}(v)^a} v_\chi. \tag{1.9}$$

Then we have the following statement.

Proposition 1.10. *The leading term of the Laurent polynomial (1.8) is equal to*

$$(v_a, w) \cdot \tau^{\varphi_a(\mathrm{Wt}(v)^a)}.$$

The proof is straightforward and left to the reader.

We shall use Proposition 1.10 as a way to recover the weight polytope $\mathrm{Wt}\,(v)$ from the asymptotics of $\tau^a v$ for various 1-parameter subgroups $\tau \longmapsto \tau^a$.

2. Affine toric varieties and semigroups

A. Classification of affine toric varieties

Let S be a commutative semigroup with 0; its *semigroup algebra* $\mathbf{C}[S]$ consists of finite formal sums $\sum_{\gamma \in S} a_\gamma t^\gamma$ where the $a_\gamma \in \mathbf{C}$ are zero for almost all γ and t^γ is a symbol associated to γ. Multiplication in $\mathbf{C}[S]$ is given by the rule

$$t^\gamma \cdot t^{\gamma'} = t^{\gamma + \gamma'}.$$

If S is embedded into a free Abelian group $\mathbf{Z}^k$ then we can view the symbol t^γ, $\gamma = (\gamma_1, \dots, \gamma_k) \in S$ as a Laurent monomial in k variables

$$t^\gamma = t_1^{\gamma_1} \cdots t_k^{\gamma_k}.$$

In particular, $\mathbf{C}[\mathbf{Z}^k]$ is the algebra $\mathbf{C}[t_1, t_1^{-1}, \ldots, t_k, t_k^{-1}]$ of all Laurent polynomials.

Proposition 2.1. *Let $S \subset \mathbf{Z}^k$ be a finitely generated semigroup (with 0). Then* $\operatorname{Spec} \mathbf{C}[S]$ *is an (affine) toric variety.*

Proof. We can assume that S generates $\mathbf{Z}^k$ as an Abelian group. Then $\operatorname{Spec} \mathbf{C}[S]$ has dimension k. Consider the action of the torus $(\mathbf{C}^*)^k$ on $\mathbf{C}[S]$ given by

$$x \cdot t^\gamma = x^\gamma t^\gamma, \; x = (x_1, \ldots, x_k) \in (\mathbf{C}^*)^k, \; \gamma \in S \subset \mathbf{Z}^k. \tag{2.1}$$

The embedding of semigroups $S \subset \mathbf{Z}^k$ gives an embedding

$$(\mathbf{C}^*)^k = \operatorname{Spec} \mathbf{C}[\mathbf{Z}^k] \hookrightarrow \operatorname{Spec} \mathbf{C}[S]$$

of the torus into our variety. The image of this embedding is an open orbit of the action (2.1).

Example 2.2. The semi-cubic parabola Y in $\mathbf{C}^2$ with the equation $y^2 = x^3$ is an affine toric variety. The torus $\mathbf{C}^*$ acts on Y via $t(x, y) = (t^2 x, t^3 y)$. The variety Y can be obtained as $\operatorname{Spec} \mathbf{C}[S]$ where $S \subset \mathbf{Z}_+$ is the semigroup consisting of 0, 2, 3, 4, ... (all non-negative integers except 1).

As another example, consider a finite subset $A \subset \mathbf{Z}^{k-1}$ of size n, and let $X_A \subset P^{n-1}$, $Y_A \subset \mathbf{C}^n$ be toric varieties introduced in Section 1B. Being the cone over X_A, the variety Y_A is affine and can be represented as $\operatorname{Spec} \mathbf{C}[S]$, where $S = S_A$ is the semigroup defined as follows.

Let us embed $\mathbf{Z}^{k-1}$ into $\mathbf{Z}^k = \mathbf{Z}^{k-1} \oplus \mathbf{Z}$ as the lattice of vectors with the last coordinate 1. For $\omega \in \mathbf{Z}^{k-1}$, let $\tilde{\omega} = (\omega, 1)$ be the corresponding vector in $\mathbf{Z}^k$. Let $S_A \subset \mathbf{Z}^k$ be the semigroup generated by $\tilde{\omega}$, $\omega \in A$.

Proposition 2.3. *The variety Y_A coincides with* $\operatorname{Spec} \mathbf{C}[S_A]$.

Proof. If $a = (a_1, \ldots, a_{k-1}) \in \mathbf{Z}^{k-1}$ then the Laurent monomial $t^{\tilde{a}}$ in k variables $t = (t_1, \ldots, t_k)$ equals $t_k \cdot t_1^{a_1} \cdots t_{k-1}^{a_{k-1}}$. So Y_A is the closure in $\mathbf{C}^n$ of the set

$$Y_A^0 = \{(t^{\tilde{\omega}^{(1)}}, \ldots, t^{\tilde{\omega}^{(n)}}), \; t \in (\mathbf{C}^*)^k\},$$

where the $\omega^{(i)}$ are all elements of A, see (1.5).

The coordinate ring $\mathbf{C}[Y_A]$ is generated by restrictions to Y_A of coordinate functions on $\mathbf{C}^n$. These functions, after being restricted to Y_A^0, become Laurent monomials $t^{\tilde{\omega}^{(i)}}$. Hence $\mathbf{C}[Y_A]$ equals the semigroup algebra $\mathbf{C}[S_A]$.

Proposition 2.4. *Any affine toric variety Y has the form* $\operatorname{Spec} \mathbf{C}[S]$ *for some finitely generated semigroup $S \subset \mathbf{Z}^k$, $k \geq 0$.*

Proof. Let $\mathbf{C}[Y]$ be the coordinate ring of Y. Let H be the open orbit of a torus, say T, acting on Y. Then H can be identified with the quotient of T by the stabilizer of some chosen point of H. So we can regard H itself as a torus acting on Y. Let $H \cong (\mathbf{C}^*)^k$. Since $H \subset Y$ is open and Y is irreducible, we have the embedding $\mathbf{C}[Y] \subset \mathbf{C}[H] = \mathbf{C}[\mathbf{Z}^k]$. The H-action on itself gives rise to the action on its coordinate ring, see (2.1). Since the H-action extends to Y, the subring $\mathbf{C}[Y] \subset \mathbf{C}[H]$ is H-invariant. As any linear representation of the algebraic torus, the space $\mathbf{C}[Y]$ decomposes into weight subspaces. But all weight subspaces of $\mathbf{C}[H]$ are 1-dimensional and generated by monomials. Hence $\mathbf{C}[Y]$ itself is generated, as a vector space, by monomials, i.e., it is a semigroup algebra.

Now let $S \subset \mathbf{Z}^k$ be a finitely generated semigroup. Let $K(S) \subset \mathbf{R}^k$ be the polyhedral cone defined as the convex hull of S. Let $Y = \operatorname{Spec} \mathbf{C}[S]$ be the toric variety associated to S.

Proposition 2.5. *Torus orbits on Y are in bijection with faces of the cone $K(S)$. More precisely, the orbit $Y^0(\Gamma)$ corresponding to a face $\Gamma \subset K$ is specified inside Y by the conditions $t^\gamma = 0$ for $\gamma \notin S \cap \Gamma$ and $t^\gamma \neq 0$ for $\gamma \in S \cap \Gamma$. Denote by $Y(\Gamma)$ the closure of the orbit $Y^0(\Gamma)$. Then $Y(\Gamma)$ is isomorphic to* $\operatorname{Spec} \mathbf{C}[S \cap \Gamma]$. *If Γ, Δ are two faces of K then $Y(\Gamma) \subset Y(\Delta)$ if and only if $\Gamma \subset \Delta$.*

Proof. Choose a system of generators $B \subset S$. Consider the affine space $\mathbf{C}^B$ with coordinates z_ω, $\omega \in B$. The variety Y is embedded into $\mathbf{C}^B$ by the map $z_\omega = t^\omega$. The rest of the proof is similar to that of Proposition 1.9 and is therefore omitted.

B. Normality and normalization

Usually the definition of a toric variety includes normality. Let us recall the definition of a normal variety.

Definition 2.6. Let Y be an irreducible affine algebraic variety, $\mathbf{C}[Y]$ be the ring of regular functions on Y and let $\mathbf{C}(Y)$ be the field of rational functions on Y, i.e., the quotient field of $\mathbf{C}[Y]$. The variety Y is called *normal* if $\mathbf{C}[Y]$ is integrally closed in $\mathbf{C}(Y)$, i.e., any rational function $f \in \mathbf{C}(Y)$, satisfying an equation of the form

$$f^m + g_1 f^{m-1} + \cdots + g_m = 0, \quad g_i \in \mathbf{C}[Y],$$

lies in $\mathbf{C}[Y]$. A quasi-projective variety is called *normal* if it can be covered by normal affine varieties.

For a general (not necessarily normal) affine variety Y, the functions $f \in \mathbf{C}(Y)$, satisfying the equations of the form given in Definition 2.6, form a ring $\tilde{\mathbf{C}}[Y]$ called the *integral closure* of $\mathbf{C}[Y]$ in $\mathbf{C}(Y)$. The spectrum $\tilde{Y}$ of this ring is called the *normalization* of Y. The inclusion $\mathbf{C}[Y] \subset \tilde{\mathbf{C}}[Y]$ induces a surjective morphism $\pi : \tilde{Y} \to Y$ called the *normalization morphism* of Y.

If X is a general quasi-projective variety then the normalization morphism of X is defined using an affine covering $X = \bigcup U_i$. More precisely, denote by $\pi_i : \tilde{U}_i \to U_i$ the normalization morphism of U_i. Then it can be shown (see [Sh] for details) that the π_i are compatible with intersections, i.e., the normalization of $U_i \cap U_j$ is the fiber product of $\tilde{U}_i$ and $\tilde{U}_j$ over X. Then we define the normalization $\tilde{X}$ of X to be glued from the $\tilde{U}_i$.

Examples 2.7. *(a)* Consider the semi-cubic parabola Y in $\mathbf{C}^2$, given by the equation $y^2 = x^3$. This is a toric variety (see Example 2.2), however, it is not normal. Indeed, the coordinate ring $\mathbf{C}[Y]$ can be identified with the subring in the polynomial ring $\mathbf{C}[t]$ generated by monomials $x = t^2$ and $y = t^3$. The quotient field $\mathbf{C}(Y)$ of this ring is the full field of rational functions $\mathbf{C}(t)$. The element $f = t \in \mathbf{C}(t)$ lies in the integral closure of $\mathbf{C}[Y]$ since it satisfies the equation $f^4 - (t^2)^2 = 0$. It is easy to see that the integral closure of $\mathbf{C}[Y]$ coincides with $\mathbf{C}[t]$. Thus the normalization of Y is the affine line $\mathbf{C}$, and the normalization morphism $\mathbf{C} \to Y$ acts by $t \mapsto (t^2, t^3)$. Geometrically, Y has a cusp at $(0, 0)$. The normalization morphism straightens this cusp. Note that set-theoretically this morphism is a bijection.

(b) If near some point $y \in Y$ the variety Y is the union of several branches then the normalization morphism separates these branches (see Figure 10).

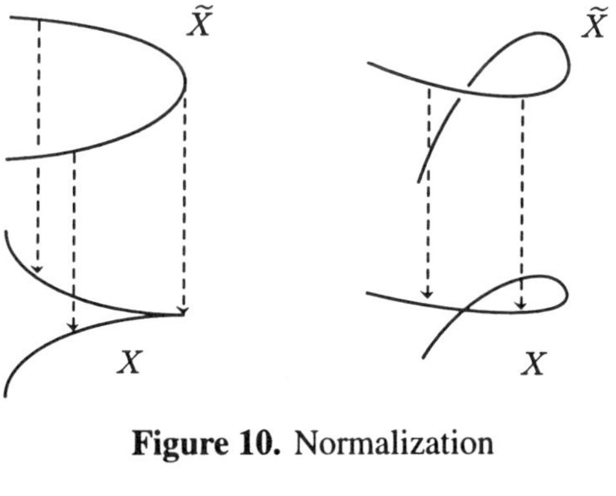

Figure 10. Normalization

The following proposition, due to M. Hochster [Hoch], gives a feeling of what non-normality means for affine toric varieties.

Proposition 2.8.

(a) *Let $S \subset \mathbf{Z}^k$ be a finitely generated semigroup with* 0. *Denote by $K(S) \subset \mathbf{R}^k$, the convex cone which is the convex hull of S and by* $\mathrm{Lin}_{\mathbf{Z}}(S) \subset \mathbf{Z}^k$ *the abelian group generated by S. The integral closure of* $\mathbf{C}[S]$ *is the semigroup algebra of the semigroup* $K(S) \cap \mathrm{Lin}_{\mathbf{Z}}(S)$. *In particular,* $\mathbf{C}[S]$ *is integrally closed (and the corresponding toric variety is normal) if and only if* $S = K(S) \cap \mathrm{Lin}_{\mathbf{Z}}(S)$.

(b) *Any normal affine toric variety has the form* Spec $\mathbf{C}[K \cap \mathbf{Z}^k]$ *for some convex polyhedral cone* $K \subset \mathbf{R}^k$ *given by inequalities with rational coefficients.*

We shall not give a proof here. Let us only remark that in one direction part (a) of the proposition is quite obvious. That is, if an integral vector $b \subset \mathbf{Z}^k$ lies in $K(S) \cap \mathrm{Lin}_{\mathbf{Z}}(S)$ then the corresponding monomial t^b lies in the integral closure of $\mathbf{C}[S]$. Indeed, in this case we have a representation $b = \sum_{a \in S} m_a \cdot a$ where the coefficients $m_a \in \mathbf{Q}$ are nonnegative and almost all equal to zero. Multiplying b with a suitable positive integer r, we find that $rb \in S$. Hence the monomial $f = t^b$ satisfies the equation $f^r - g = 0, \; g \in \mathbf{C}[S]$ and is therefore integral over $\mathbf{C}[S]$.

Remark 2.9. Our basic examples of toric varieties are the varieties X_A, Y_A associated to a finite set $A \subset \mathbf{Z}^{k-1}$. The variety Y_A is the spectrum of the semigroup $S_A \subset \mathbf{Z}^k$ generated by A, see Proposition 2.3. The condition of normality of Y_A given by Proposition 2.8 is quite restrictive. For example, the choice of A, as in Figure 11, gives a non-normal Y_A (and also X_A, as we shall see later).

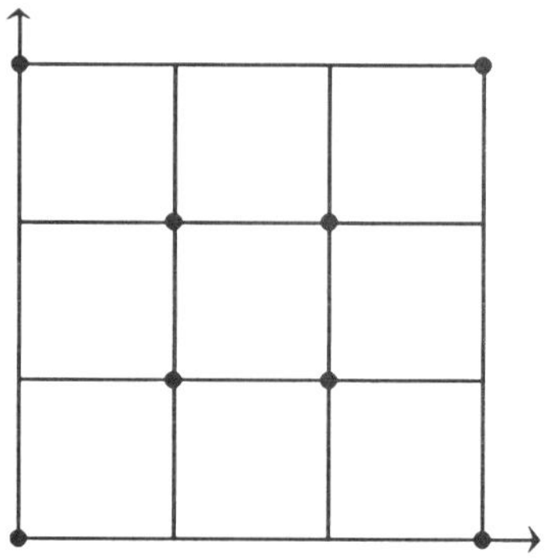

Figure 11.

For the applications we have in mind, it is natural to be able to consider an arbitrary set A; so we do not exclude non-normal toric varieties.

C. Smoothness and quasi-smoothness for affine toric varieties

Proposition 2.10. *The only smooth affine toric varieties are products* $(\mathbf{C}^*)^p \times \mathbf{C}^q$. *In other words, if* $S \subset \mathbf{Z}^m$ *is a finitely generated semigroup, then* $\operatorname{Spec} \mathbf{C}[S]$ *is smooth if and only if* S *is isomorphic to* $\mathbf{Z}^p \times \mathbf{Z}_+^q$ *for some* k, l.

Proof. This is easy (see e.g., [D], subsection 3.3).

Along with the notion of smoothness for affine toric varieties, we consider a weaker notion, the so-called *quasi-smoothness*. Namely, let $S \subset \mathbf{Z}^m$ be a finitely generated semigroup and let Y be the corresponding affine toric variety. We say that Y is quasi-smooth if two conditions are satisfied:

(1) The cone $K(S) \subset \mathbf{R}^m$ (the convex hull of S) is linearly isomorphic to a cone of the form $\mathbf{R}^p \times \mathbf{R}_+^q$.

(2) The normalization morphism $\tilde{Y} \to Y$ is bijective.

By Proposition 2.10, any smooth variety is quasi-smooth. Here are some examples of quasi-smooth but not smooth varieties.

Examples 2.11. *(a)* The semi-cubic parabola $y^2 = x^3$ is a quasi-smooth toric variety. Indeed, the corresponding semigroup $S \subset \mathbf{Z}$ consists of $0, 2, 3, 4, \ldots$. The convex hull $K(S)$ is $\mathbf{R}_+$. The normalization morphism of this parabola is bijective according to Example 2.7 (a).

(b) The quadratic cone $Y \subset \mathbf{C}^3$ with equation $y^2 = xz$ is the spectrum of the semigroup algebra $\mathbf{C}[S]$ where $S \subset \mathbf{Z}^2$ is the semigroup generated by $p = (1, 2), q = (1, 1), r = (1, 0)$. Clearly, $K(S) \subset \mathbf{R}^2$ is the plane angle $\{(a, b) \in \mathbf{R}^2 : 0 \leq b \leq 2a\}$ (see Figure 12); hence it is isomorphic to $\mathbf{R}_+^2$. Furthermore, $S = K(S) \cap \mathbf{Z}^2$. So Y is a normal quasi-smooth variety.

We shall see later that quasi-smooth toric varieties share some of the nice features of smooth ones.

3. Local structure of toric varieties

A. Local structure of X_A near an orbit

Let $A \subset \mathbf{Z}^{k-1}$ be a finite set of lattice points assumed to generate $\mathbf{Z}^{k-1}$ as an affine lattice. Consider the toric variety X_A introduced in Section 1B. By Proposition 1.9, the structure of torus orbits on X_A is governed by the convex polytope $Q \subset \mathbf{R}^{k-1}$, the convex hull of A. Namely, p-dimensional orbits are in bijection with p-dimensional faces $\Gamma \subset Q$. Denote by $X^0(\Gamma)$ the orbit associated to a face Γ. We are going to study the structure of X_A near a point on $X^0(\Gamma)$.

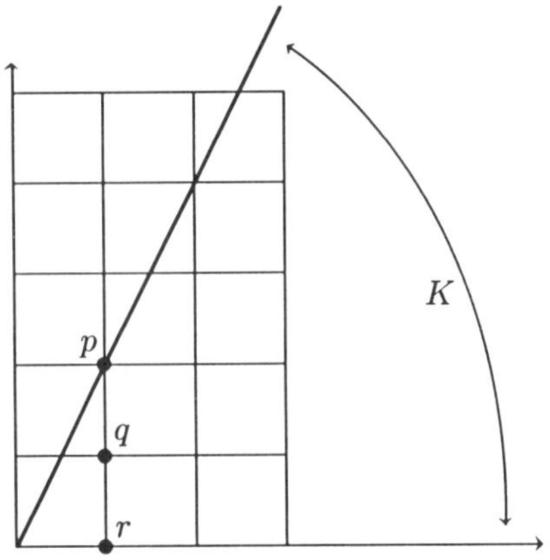

Figure 12.

As in Section 2A, let us embed the affine lattice $\mathbf{Z}^{k-1}$ into the Abelian group $\Xi = \mathbf{Z}^k$ as the subset $\{(a_1, \ldots, a_k) \in \mathbf{Z}^k : a_k = 1\}$. Let S be the subsemigroup in Ξ generated by A and 0. As before, let $K \subset \Xi_{\mathbf{R}} = \mathbf{R}^k$ be the convex hull of S, i.e., the cone with apex 0 and base Q.

For any face $\Gamma \subset Q$, let $\mathrm{Lin}_{\mathbf{R}}(\Gamma) \subset \Xi_{\mathbf{R}}$ be the vector subspace over $\mathbf{R}$ spanned by Γ, so that $\dim \mathrm{Lin}_{\mathbf{R}}(\Gamma) = \dim(\Gamma) + 1$. Consider the quotient lattice

$$\Xi/\Gamma \;:=\; \Xi/(\Xi \cap \mathrm{Lin}_{\mathbf{R}}(\Gamma)).$$

Denote the image of the semigroup S in Ξ/Γ by S/Γ. Clearly S/Γ is a subsemigroup in the free Abelian group Ξ/Γ. Also $(\Xi/\Gamma)_{\mathbf{R}} = \Xi_{\mathbf{R}}/\mathrm{Lin}_{\mathbf{R}}(\Gamma)$.

We introduce the number $i(\Gamma, A)$ as follows. The linear space $\mathrm{Lin}_{\mathbf{R}}(\Gamma)$ has two lattices inside it, both of maximal rank. The first is $\Xi \cap \mathrm{Lin}_{\mathbf{R}}(\Gamma)$. The second, contained in the first, is $\mathrm{Lin}_{\mathbf{Z}}(A \cap \Gamma)$, the Abelian subgroup generated by $A \cap \Gamma$. We define $i(\Gamma, A)$ as the index

$$i(\Gamma, A) = [\Xi \cap \mathrm{Lin}_{\mathbf{R}}(\Gamma) : \mathrm{Lin}_{\mathbf{Z}}(A \cap \Gamma)]. \tag{3.1}$$

This number can also be defined as the index of the affine lattices

$$[\mathbf{Z}^{k-1} \cap \mathrm{Aff}_{\mathbf{R}}(\Gamma) : \mathrm{Aff}_{\mathbf{Z}}(A \cap \Gamma)],$$

see Definition 1.3 for the meaning of $\mathrm{Aff}_{\mathbf{Z}}$. For example, if $A \subset \mathbf{Z}^2$ is the set given in Figure 11, then Q is a square and for any side $\Gamma \subset Q$ the number $i(\Gamma, A)$ equals 3.

Theorem 3.1. *Let $A \subset \mathbf{Z}^{k-1}$ be as before, let Q be the convex hull of A, Γ a non-empty face of Q, and let $X^0(\Gamma)$ be the torus orbit in the toric variety X_A corresponding to Γ. Let $x \in X^0(\Gamma)$ be any point. Then in a neighborhood of x the variety X_A is a union of $i(\Gamma, A)$ branches, each of which is (locally) isomorphic to* Spec $\mathbf{C}[S/\Gamma] \times X^0(\Gamma)$. *These branches are glued together along the common subvariety $\{0\} \times X^0(\Gamma)$, where $0 \in$ Spec $\mathbf{C}[S/\Gamma]$ is the unique 0-dimensional torus orbit.*

Proof. Consider the affine variety $Y_A =$ Spec $\mathbf{C}[S]$ which is the cone over X_A. Let $Y^0(\Gamma) \subset Y_A$ be the cone over the orbit $X^0(\Gamma)$. The local structure of Y_A along $Y^0(\Gamma)$ is the same as that of X_A along $X^0(\Gamma)$ so we shall study Y_A.

Let $Y(\Gamma) \subset Y_A$ be the closure of $Y^0(\Gamma)$. Consider the open subvariety

$$Z = Y_A - \bigcup_{\Sigma:\Gamma\subset\Sigma,\ \Gamma\neq\Sigma} Y(\Sigma).$$

We have $Z \cap Y(\Gamma) = Y_0(\Gamma)$. Let $S(\Gamma) \subset S$ be the intersection $S \cap \mathbf{R}_+\Gamma$, i.e., the subsemigroup generated by $A \cap \Gamma$.

Consider the semigroup

$$\mathrm{Lin}_{\mathbf{Z}}(A \cap \Gamma) + S = \{a + b : a \in \mathrm{Lin}_{\mathbf{Z}}(A \cap \Gamma),\ b \in S\}.$$

Then

$$Z = \mathrm{Spec}\,\mathbf{C}[\mathrm{Lin}_{\mathbf{Z}}(A \cap \Gamma) + S].$$

We have a morphism $p : Z \to Y_0(\Gamma)$, corresponding to the embedding of semigroups

$$\mathrm{Lin}_{\mathbf{Z}}(A \cap \Gamma) \hookrightarrow \mathrm{Lin}_{\mathbf{Z}}(A \cap \Gamma) + S.$$

Let $y_0 \in Y_0(\Gamma)$ be the point at which all of the monomials t^ω, $\omega \in \mathrm{Lin}_{\mathbf{Z}}(A \cap \Gamma)$ are equal to 1. (Recall that Y_A is embedded into the affine space with coordinates z_ω, $\omega \in A$; the point y_0 in this embedding has the following coordinates: $z_\omega = 0$, $\omega \notin \Gamma$ and $z_\omega = 1$, $\omega \in \Gamma$.) Then we have

$$p^{-1}(y_0) = \mathrm{Spec}\ \mathbf{C}[\mathrm{Lin}_{\mathbf{Z}}(A \cap \Gamma) + S)/\mathrm{Lin}_{\mathbf{Z}}(A \cap \Gamma)]. \tag{3.2}$$

Indeed, taking $p^{-1}(y_0)$ amounts to taking the quotient of $\mathbf{C}[\mathrm{Lin}_{\mathbf{Z}}(A \cap \Gamma) + S]$, the coordinate ring of Z, by the relations $t^\omega = 1$, $\omega \in \mathrm{Lin}_{\mathbf{Z}}(A \cap \Gamma)$, defining y_0. This quotient is just the semigroup ring of the quotient semigroup on the right hand side of (3.2).

Denote the above quotient semigroup in (3.2) by Σ. This contains a finite Abelian group $G = (\Xi \cap \mathrm{Lin}_{\mathbf{R}}(\Gamma))/\mathrm{Lin}_{\mathbf{Z}}(A \cap \Gamma)$ of order $i(\Gamma, A)$. The spectrum of the group ring $\mathbf{C}[G]$ is the disjoint union of $i(\Gamma, A)$ points, which will be the set labeling the branches. The quotient Σ/G equals

$$(\mathrm{Lin}_{\mathbf{Z}}(A \cap \Gamma) + S)/(\Xi \cap \mathrm{Lin}_{\mathbf{R}}(\Gamma)) = S/\Gamma.$$

Clearly G acts on Spec $\mathbf{C}[\Sigma] = p^{-1}(y_0)$ and the quotient is Spec $\mathbf{C}[\Sigma/G] =$ Spec $\mathbf{C}[S/\Gamma]$. From this we deduce that there are $i(\Gamma, A)$ branches, each isomorphic to Spec $\mathbf{C}[S/\Gamma]$. Theorem 3.1 is proved.

The theorem just proved implies that whenever $i(\Gamma, A) > 1$ for some Γ, the normalization morphism of X_A will not be bijective since it must separate the branches. This, for example, will be the case for the set A in Figure 11.

B. Smoothness conditions

Proposition 2.10 together with Theorem 3.1 imply the following.

Corollary 3.2. *Let $A \subset \mathbf{Z}^{k-1}$ be a finite set generating $\mathbf{Z}^{k-1}$ as an affine lattice. Let $Q \subset \mathbf{R}^{k-1}$ be the convex hull of A. The projective toric variety X_A is smooth if and only if for every non-empty face $\Gamma \subset Q$ the following conditions hold:*

(a) *The semigroup S/Γ (see subsection A) is free, i.e., isomorphic to $\mathbf{Z}_+^m$, $m = \dim(Q) - \dim(\Gamma) + 1$.*

(b) *The lattices $\mathbf{Z}^{k-1} \cap \mathrm{Aff}_{\mathbf{R}}(\Gamma)$ and $\mathrm{Aff}_{\mathbf{Z}}(A \cap \Gamma)$ coincide so that $i(\Gamma, A) = 1$.*

Proposition 3.3. *If the set A is such that $i(\Gamma, A) = 1$ for any face $\Gamma \subset Q$ then the normalization morphisms of X_A and Y_A are bijective.*

Proof. It is enough to consider Y_A. Let $S \subset \mathbf{Z}^k$ be the semigroup generated by A, and let $K \subset \mathbf{R}^k$ be the convex hull of S. Our assumptions imply that $\mathbf{Z}^k \cap K$ differs from S only in a finite number of lattice points. A point of the normalization of Y_A can be regarded as a ring homomorphism $\varphi : \mathbf{C}[\mathbf{Z}^k \cap K] \to \mathbf{C}$. Suppose there are two points of the normalization mapped into the same point of Y_A, that is, two homomorphisms φ, φ' as above which coincide on $\mathbf{C}[S_A]$. For any $\gamma \in \mathbf{Z}^k \cap K$, there is $m_0 \in \mathbf{Z}_+$ such that, for any $m \geq m_0$, we have $m\gamma \in S_A$ (since $(\mathbf{Z}^k \cap K) - S$ is finite). So $\varphi(t^{m\gamma}) = \varphi'(t^{m\gamma})$, i.e., $\varphi(t^\gamma)^m = \varphi'(t^\gamma)^m$ for $m \geq m_0$. This implies that $\varphi(t^\gamma) = \varphi'(t^\gamma)$. Since this holds for any $\gamma \in \mathbf{Z}^k \cap K$, we have $\varphi = \varphi'$, as required.

C. Admissible semigroups

By Theorem 3.1, the study of arbitrary singularities of toric varieties is reduced to that of singular points given by 0-dimensional torus orbits. To study the latter case, it suffices to consider affine toric varieties of the form $Y =$ Spec $\mathbf{C}[S]$ where S is a semigroup. We first specify which semigroups are to be examined.

Let S be a commutative semigroup with 0. There is a universal Abelian group $\Xi(S)$ associated with S called the *group completion* of S. By definition, $\Xi(S)$ is generated by symbols $a - b$ where $a, b \in S$; these symbols are subject to the relations

$$(a-b)+(c-d)=(a+c)-(b+d).$$

There is a canonical semigroup homomorphism $S \to \Xi(S)$ taking $a \mapsto a - 0$.

Definition 3.4. Let S be a finitely generated commutative semigroup with 0. We say that S is *admissible* if the following conditions hold:

(a) The group completion $\Xi(S)$ is a free Abelian group.

(b) The canonical morphism $S \to \Xi(S)$ is an embedding.

(c) There exists a group homomorphism $h : \Xi(S) \to \mathbf{Z}$ such that $h(a) > 0$ for any $a \in S - \{0\}$.

Let S be an admissible semigroup. We abbreviate $\Xi = \Xi(S)$ and $\Xi_{\mathbf{R}} = \Xi \otimes \mathbf{R}$. So Ξ is a lattice in the real vector space $\Xi_{\mathbf{R}}$. Denote the convex hull of S by $K(S) \subset \Xi_{\mathbf{R}}$. Since $0 \in S$, the set $K(S)$ is a polyhedral cone with apex 0. By condition (c) of admissibility, the cone $K(S)$ is strictly convex, i.e., it does not contain straight lines.

Proposition 3.5. *Let S be an admissible semigroup and let $Y = \operatorname{Spec} \mathbf{C}[S]$ be the corresponding toric variety. Then Y contains exactly one 0-dimensional torus orbit.*

Proof. This follows from Proposition 2.5: torus orbits on Y of any given dimension i correspond to faces of $K(S)$ of dimension i. The unique 0-dimensional orbit corresponds to the apex of $K(S)$.

If $A \subset \mathbf{Z}^{k-1}$ is a finite set then the semigroup $S_A \subset \mathbf{Z}^k$ generated by $A \times \{1\}$ is admissible. The considerations of subsections A and B regarding the toric variety $Y_A = \operatorname{Spec} \mathbf{C}[S_A]$ can be generalized without difficulty to the case of the toric variety $\operatorname{Spec} \mathbf{C}[S]$ for an arbitrary admissible semigroup S. Let us describe briefly these generalizations which will be needed in Chapter 11.

Let S be an admissible semigroup, $Y = \operatorname{Spec} \mathbf{C}[S]$, and let $K(S)$, Ξ have the same meaning as before. For any face $\Gamma \subset K(S)$, we denote by S/Γ the image of S in $\Xi_{\mathbf{R}}/\mathrm{Lin}_{\mathbf{R}}(\Gamma)$. We also write

$$i(\Gamma, S) = [\Xi \cap \mathrm{Lin}_{\mathbf{R}}(\Gamma) : \mathrm{Lin}_{\mathbf{Z}}(S \cap \Gamma)].$$

Thus in the case when $S = S_A$ the number $i(\Gamma, S)$ is the same as $i(\Gamma, A)$, defined by (3.1).

Let $Y^0(\Gamma) \subset Y$ be the orbit corresponding to Γ.

Theorem 3.6.

(a) *In a neighborhood of any point* $y \in Y^0(\Gamma)$*, the variety* Y *is the union of* $i(\Gamma, S)$ *branches locally isomorphic to* $\operatorname{Spec} \mathbf{C}[S/\Gamma] \times Y^0(\Gamma)$.

(b) *The variety* $Y = \operatorname{Spec} \mathbf{C}[S]$ *is quasi-smooth if and only if the following two conditions hold:*

(1) *The cone* $K(S)$ *is simplicial;*

(2) *For any face* $\Gamma \subset K(S)$*, we have* $i(\Gamma, S) = 1$*, i.e.,* $\mathrm{Lin}_{\mathbf{Z}}(S \cap \Gamma) = \Xi \cap \mathrm{Lin}_{\mathbf{R}}(\Gamma)$.

Proof. Part (a) is proved in exactly the same way as in Theorem 3.1. To prove (b) recall (see Section 2C) that, by definition, Y is quasi-smooth if and only if $K(S)$ is simplicial and the normalization morphism of Y is bijective. It suffices to show therefore that the condition (2) of the theorem is equivalent to the bijectivity of the normalization morphism. If $i(\Gamma, S) > 1$ for some Γ, then this morphism cannot be bijective since it would separate the branches near $Y^0(\Gamma)$. Conversely, if $i(\Gamma, S) = 1$ for all Γ, then the same reasoning, as in the proof of Proposition 3.3, shows that the normalization morphism of Y is bijective.

We shall be interested in some numerical invariants of admissible semigroups defined in terms of volumes. We start by describing the normalization of volume forms.

D. The volume form induced by a lattice

Let W be a real affine space of dimension m and let V be the corresponding vector space of translations (this means that W is a principal homogeneous space over V). A translation-invariant volume form on W is defined up to a scalar multiple. To normalize a volume form, it suffices to exhibit a body of volume 1.

By an affine lattice in W, we mean a subset $\Xi \subset W$ which is a principal homogeneous space over a discrete Abelian subgroup $\Sigma \subset V$ of rank equal to $m = \dim W = \dim V$.

A choice of an affine lattice $\Xi \subset W$ defines a volume form Vol_Ξ on W as follows. Consider all the m-dimensional simplices in W with vertices in Ξ. Among them there are the so-called *elementary* simplices, i.e., those whose volume is the minimal possible (note that we can compare the volumes of any two bodies in W because of the affine structure). We specify the volume form Vol_Ξ by assigning the volume 1 to elementary simplices. In this normalization the volume of any convex polytope with vertices in Ξ is an integer.

As an example, let us take $W = \mathbf{R}^m$ and $\Xi = \mathbf{Z}^m$. The unit cube

$$I^m = \{(x_1, \ldots, x_m) \in \mathbf{R}^m : 0 \le x_i \le 1\}$$

has $\mathrm{Vol}_\Xi(I^m) = m!$, as can be seen, e.g., by explicitly triangulating I^m into $m!$ elementary simplices.

Although the volume Vol_Ξ of any lattice polytope is always an integer, it may be impossible to decompose such a polytope P into elementary simplices. For example, take $W = \mathbf{R}^3$, $\Xi = \mathbf{Z}^3$ and take P to be the tetrahedron $ABCD$ inscribed into the unit cube as shown in Figure 13. Then $\mathrm{Vol}_\Xi P = 2$, but P contains no other lattice points except its vertices, so it cannot be decomposed into elementary simplices.

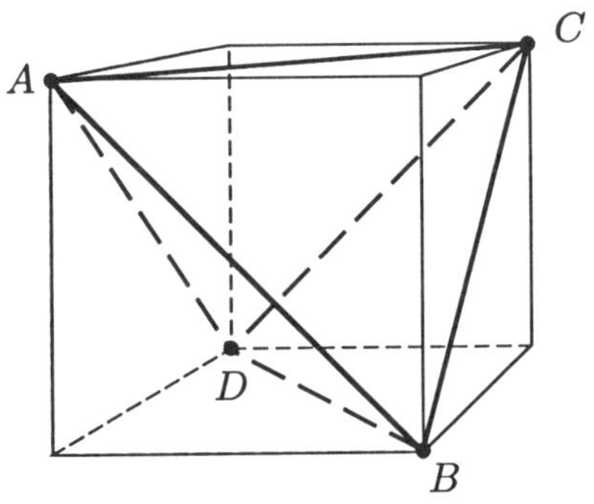

Figure 13.

An obvious asymptotic relation between the volume and the number of lattice points, recalled below, will be used on several occasions in the sequel.

Proposition 3.7. *Let Ξ be an affine lattice in a real affine space W of dimension m. Let $P \subset W$ be a bounded domain with a piecewise smooth boundary. Take some point $O \in W$ and consider the homothetic domains*

$$dP = \{O + d(x - O), x \in P\}$$

for $d > 0$. Then, as $d \to \infty$, the number of lattice points in dP has the asymptotics

$$\#(dP \cap \Xi) = \frac{\mathrm{Vol}_\Xi(P)}{m!} d^m + O(d^{m-1}).$$

Proof. We can assume that $W = \mathbf{R}^m$ and $\Xi = \mathbf{Z}^m$. So we can speak about *lattice cubes* as translations by elements of Ξ of the standard cube $I^m = \{0 \leq x_i \leq 1\}$. The volume of any such cube is $m!$ and the diameter (in the sense of the standard Euclidean distance on $\mathbf{R}^m$) is $\sqrt{m}$. Let $n(d)$ be the number of lattice cubes lying in dP. Associating to any lattice cube $a + I^m$, $a \in \Xi$ its vertex a, we find that $\#(dP \cap \Xi) - n(d)$ does not exceed the number $a(d)$ of lattice points in dP on the distance $\leq \sqrt{m}$ from the boundary. Similarly, $\mathrm{Vol}_\Xi(P)/m! - n(d)$ does not exceed $b(d)$, the volume of the set of points in dP on the distance $\leq \sqrt{m}$ from the boundary. Clearly, $a(d)$ and $b(d)$ both grow for $d \to \infty$ as $const \cdot d^{m-1}$, from which we obtain the statement.

E. The subdiagram volume of an admissible semigroup

Let S be an admissible semigroup, $\Xi = \Xi(S)$ its group completion, and $K(S) \subset \Xi_{\mathbf{R}}$ the convex hull of S. Denote by $K_+(S) \subset K(S)$ the convex hull of the set $S - \{0\}$ and by $K_-(S)$ the closure of the complement $K(S) - K_+(S)$. The set $K_-(S)$ is a bounded (but usually not convex) lattice polyhedron. We call it the *subdiagram part* of S.

Definition 3.8. The *subdiagram volume* of an admissible semigroup S is the number

$$u(S) = \mathrm{Vol}_{\Xi(S)}(K_-(S)),$$

i.e., the volume of the subdiagram part with respect to the volume form induced by $\Xi(S)$, see above. For the trivial semigroup $S = \{0\}$ we set $u(S) = 1$.

Let us emphasize that $u(S)$ depends only on the semigroup S.

Example 3.9. For a free semigroup $S = \mathbf{Z}_+^k$, $k \geq 0$ we have $u(S) = 1$.

Proposition 3.10. *For any admissible semigroup S, the subdiagram volume $u(S)$ is greater than or equal to 1. Moreover, $u(S) = 1$ if and only if S is a free semigroup.*

Proof. The fact that $u(S) \geq 1$ is obvious. Suppose $u(S) = 1$. Then the region $K_-(S)$, being a lattice polyhedron of volume 1, should consist of just one elementary lattice simplex. So the vertices of $K_-(S)$ other than 0 form a $\mathbf{Z}$-basis of $\Xi(S) = \mathbf{Z}^k$. This implies that S consists of non-negative integer combinations of elements of this basis, i.e., $S \cong \mathbf{Z}_+^k$.

Thus $u(S)$ can be seen as the measure of non-freeness of a semigroup S, i.e,. of the singular nature of the toric variety Spec $\mathbf{C}[S]$ (see Corollary 3.2).

F. The multiplicity of a singular point on a (toric) variety

Let Y be an algebraic variety and $y \in Y$ a (possibly singular) point. There is an important numerical invariant $\mathrm{mult}_y Y$, called the *multiplicity* (or local degree) of Y at y, which measures "how singular" a point y is. The most geometrically transparent definition of $\mathrm{mult}_y Y$ can be given in terms of an embedding of Y into a projective space P^m.

Let $Y \subset P^m$, $\dim Y = k$. Choose a generic projective subspace $L \subset P^m$ of codimension k passing through y. Then y is an isolated point of the intersection $Y \cap L$. Now move L away from y, i.e., choose a generic 1-parameter family of projective subspaces $L(t)$, $t \in \mathbf{C}$, $|t| < \varepsilon$ such that $L(0) = L$. Then, for small $t \neq 0$, the isolated point of intersection $y \in L \cap Y$ will split into several intersection points from $L(t) \cap Y$ (see Figure 14).

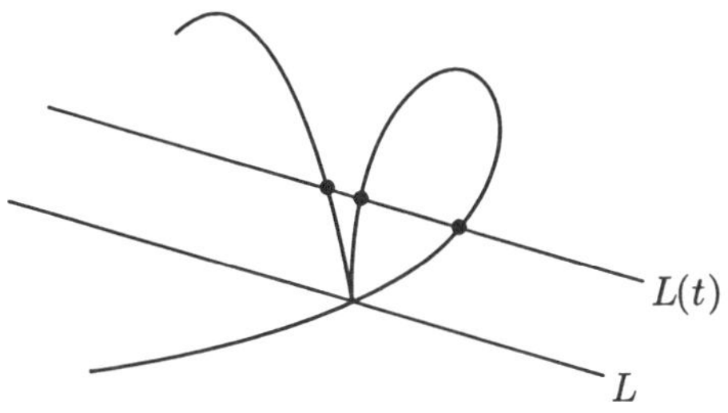

Figure 14.

Definition 3.11. The number of intersection points from $L(t) \cap Y$ arising from the isolated intersection point $y \in L \cap Y$ after a generic deformation $L(t)$ of a generic projective subspace L ($y \in L$, $\operatorname{codim} L = k$) is called *the multiplicity* of Y at y and denoted by $\operatorname{mult}_y Y$.

There are other, more algebraic definitions of multiplicity. First of all, instead of moving L away from y and then counting the intersection points, we can introduce the multiplicity of intersection $i(y; L \cap Y)$ of L and Y at y. There are several approaches to doing this [Mum], [Ful], and we shall not go into detail here. Let us only mention that this leads to the definition of $\operatorname{mult}_y Y$ as

$$\operatorname{mult}_y Y = \min_L \, i(y; L \cap Y), \tag{3.3}$$

where L runs over projective subspaces of codimension $k = \dim(Y)$ containing y and such that y is an isolated point of $L \cap Y$ (see [Mum], Definition 5.9).

A completely intrinsic algebraic definition of multiplicity is obtained as follows. Suppose for simplicity that Y is an affine variety (since we are interested in the local situation near y, this is legitimate). Let $\mathbf{C}[Y]$ be the ring of regular functions on Y and $M_y \subset \mathbf{C}[Y]$ be the ideal of functions vanishing at y. Let M_y^d

be the d-th power of M_y. For $d \gg 0$, the dimension of M_y^d/M_y^{d+1} is known to be given by a polynomial $p_{y,Y}(d)$ in d called the *local Hilbert polynomial* of Y at y. The degree of $p_{y,Y}(t)$ is equal to $k-1$ where $k = \dim(Y)$.

Definition 3.11′. The *multiplicity* $\text{mult}_y Y$ is the number μ such that

$$p_{y,Y}(t) = \frac{\mu}{(k-1)!} t^{k-1} + \text{(terms of lower order.)}$$

The equivalence of Definitions 3.11, 3.11′ and formula (3.3) can be seen by comparing the two definitions of the tangent cone to Y at y: one in terms of a projective embedding and the other in terms of the coordinate ring. We refer the reader to [Mum] and [Harr] for details.

Let us now discuss the main properties of the notion of multiplicity and give some examples .

Proposition 3.12. *The multiplicity* $\text{mult}_y Y$ *is always greater than or equal to* 1. *Moreover,* $\text{mult}_y Y = 1$ *if and only if* y *is a non-singular point of* Y.

Proof. See Corollary 5.15 in [Mum].

Examples 3.13. *(a)* Let $Y \subset \mathbf{C}^m$ be a hypersurface given by $f(x_1, \dots, x_m) = 0$ (here $f(x_1, \dots, x_m)$ may be reducible, but is square-free). Suppose that $f(0) = 0$, i.e., $0 \in Y$. Then $\text{mult}_0 Y$ equals the minimal degree of a non-zero monomial from f. For example, the plane curve with equation $x_1^5 + x_1^3 x_2^3 + x_2^7$ has multiplicity 5 at 0.

(b) If Y_1, Y_2 are two k-dimensional subvarieties in P^m which both contain a point y and do not have irreducible components in common, then

$$\text{mult}_y(Y_1 \cup Y_2) = \text{mult}_y(Y_1) + \text{mult}_y(Y_2).$$

(c) If $Y \subset \mathbf{C}^n$ is the cone over a projective variety $X \subset P^{n-1}$ then $\text{mult}_0 Y = \deg(X)$ is the degree of X.

The last example shows that the notion of multiplicity can be seen as the generalization of the notion of the degree of a projective variety. For toric varieties, multiplicity is calculated as follows.

Theorem 3.14. *Let S be an admissible semigroup,* $Y = \text{Spec}\,\mathbf{C}[S]$ *the corresponding toric variety, and* $0 \in Y$ *the unique* 0*-dimensional torus orbit. Then*

$$\text{mult}_0(Y) = u(S)$$

is the subdiagram volume of S.

Proof. Let Ξ, $\Xi_{\mathbf{R}}$, $K(S)$, $K_+(S)$ and let $K_-(S)$ have the same meaning as in subsection E. Let $k = \text{rank}\, \Xi = \dim Y$. For $d > 0$ let $dK_+(S) = \{da,\ a \in K_+(S)\}$ be the d times dilated polyhedron $K_+(S)$.

For $d \gg 0$, the ideal $M_0^d \subset \mathbf{C}[S]$ is generated as a vector space by monomials t^γ for $\gamma \in dK_+(S) \cap \Xi$. Hence for $d \gg 0$, we have

$$\begin{aligned}\dim(M_0^d / M_0^{d+1}) &= \#((dK_+(S) - (d+1)K_+(S)) \cap \Xi) \\ &= \#(((d+1)\, K_-(S) - dK_-(S)) \cap \Xi)\,.\end{aligned}$$

Applying Proposition 3.7 to $P = K_-(S)$, we conclude that $\dim(M_0^d/M_0^{d+1})$ has the asymptotics

$$\frac{\text{Vol}_\Xi(K_-(S))}{(k-1)!} d^{k-1} + (\text{terms of lower order}).$$

So by Definition 3.11′, we have $\text{mult}_0 Y = \text{Vol}_\Xi(K_-(S)) = u(S)$ as claimed.

Definition 3.15. Let Z be an irreducible subvariety of an algebraic variety Y. The multiplicity $\text{mult}_z Y$ for a generic point $z \in Z$ is called *the multiplicity of Y along Z* and denoted by $\text{mult}_Z Y$.

Theorem 3.16. *Let $A \subset \mathbf{Z}^{k-1}$ be a finite set generating $\mathbf{Z}^{k-1}$ as an affine lattice, and let X_A be the projective toric variety associated to A, see Section 1B. Let $Q \subset \mathbf{R}^{k-1}$ be the convex hull of A. For a face $\Gamma \subset Q$, let $X(\Gamma)$ be the closure in X_A of the torus orbit corresponding to Γ. Then*

$$\text{mult}_{X(\Gamma)}\, X_A = i(\Gamma, A) \cdot u(S/\Gamma)$$

where we have the semigroup S/Γ and the index $i(\Gamma, A)$ as introduced in subsection A.

Proof. This follows from Theorem 3.14 and Theorem 3.1 describing the local structure of X_A near $X(\Gamma)$.

4. Abstract toric varieties and fans

Here we review the language of fans, traditionally used in the description of toric varieties.

A. Classification of abstract toric varieties

Definition 4.1. A *fan* in $\mathbf{R}^k$ is a finite collection $\mathcal{F}$ of convex polyhedral cones such that

(a) Every face of every cone from $\mathcal{F}$ belongs to $\mathcal{F}$.
(b) The intersection of any two cones from $\mathcal{F}$ is a face of both of them.

Note that we do not require that the cones from $\mathcal{F}$ cover the whole space. If this is the case, the fan $\mathcal{F}$ is called *complete*. In general the union of all cones from a fan $\mathcal{F}$ is called the *support* of $\mathcal{F}$ and denoted by $|\mathcal{F}|$. A fan $\mathcal{F}$ is called *rational* if all the cones constituting it are given by inequalities with rational coefficients.

We shall give examples of fans later in subsection B. Right now let us describe the construction of toric varieties from fans. Let $\mathcal{F}$ be a rational fan in $\mathbf{R}^k$ and let $\mathbf{Z}^k \subset \mathbf{R}^k$ be the lattice of vectors with integer coordinates. Let $\Xi \subset (\mathbf{R}^k)^*$ be the dual lattice to $\mathbf{Z}^k$, i.e., the set of linear functionals taking integral values on $\mathbf{Z}^k$. For any cone $K \in \mathcal{F}$, let

$$\check{K} = \{f \in (\mathbf{R}^k)^* : f(x) \geq 0 \text{ for any } \; x \in K\}$$

be the dual cone to K. We associate to K the semigroup $\check{K} \cap \Xi$ and the corresponding semigroup algebra $\mathbf{C}[\check{K} \cap \Xi]$. If $K \subset L$ are two cones from $\mathcal{F}$, then $\check{L} \subset \check{K}$ and hence Spec $\mathbf{C}[\check{K} \cap \Xi]$ is a Zariski open subset in Spec $\mathbf{C}[\check{L} \cap \Xi]$.

Definition 4.2. The *toric variety $X(\mathcal{F})$ associated with a rational fan $\mathcal{F}$ in* $\mathbf{R}^k$ is the result of gluing affine toric varieties $X_K = \operatorname{Spec} \mathbf{C}[\check{K} \cap \Xi]$ by identifying X_L with the corresponding Zariski open subset in X_K whenever $K \subset L$.

By construction, $X(\mathcal{F})$ is a normal variety, since it is glued out of normal affine varieties X_K.

We shall say that a convex cone $K \subset \mathbf{R}^k$ is *simplicial* if it has a simplex as its base. In other words, K is generated, as a convex cone by a part of some basis of $\mathbf{R}^k$. If, moreover, K is generated by a part of some $\mathbf{Z}$-basis of $\mathbf{Z}^k$, then we shall say that K is *strictly simplicial*. In this case $K \cap \mathbf{Z}^k$ and $\check{K} \cap \Xi$ are free semigroups.

The classification theorem for normal toric varieties is as follows.

Theorem 4.3.

(a) *Any normal toric variety is equivariantly isomorphic to a variety of the form $X(\mathcal{F})$ for some rational fan $\mathcal{F}$ in $\mathbf{R}^k$, where k is the dimension of the torus acting on X. This fan is determined uniquely up to a transformation from $GL_k(\mathbf{Z})$.*
(b) *The variety $X(\mathcal{F})$ is compact if and only if the fan $\mathcal{F}$ is complete. The variety $X(\mathcal{F})$ is smooth if and only if every cone from $\mathcal{F}$ is strictly simplicial.*
(c) *The variety $X(\mathcal{F})$ has finitely many orbits of the torus $(\mathbf{C}^*)^k$. These orbits are in bijection with cones from $\mathcal{F}$. The orbit O_K, corresponding to a cone $K \in \mathcal{F}$, has dimension equal to* codim (K). *If K, L are two cones from $\mathcal{F}$ then O_K lies in the closure of O_L if and only if $L \subset K$.*

Proof. See [O], Theorems 1.5, 1.11, 1.10 and Proposition 1.6.

Let X be a (not necessarily normal) toric variety. We shall call the fan of the normalization of X simply the fan of X and denote it by $\mathcal{F}(X)$. The following is a description of $\mathcal{F}(X)$.

Proposition 4.4. *Let X be a toric variety,* $(\mathbf{C}^*)^k \subset X$ *the open torus orbit (identified as above with the torus acting on X),* $e = (1, \ldots, 1) \in (\mathbf{C}^*)^k \subset X$*, and* $\mathcal{F}(X)$ *the fan in* $\mathbf{R}^k$ *corresponding to X. Two integral vectors* $(a_1, \ldots, a_k)$, $(b_1, \ldots, b_k) \in \mathbf{Z}^k$ *belong to the interior of the same cone of* $\mathcal{F}(X)$ *if and only if*

$$\lim_{t\to 0}(t^{a_1}, \ldots, t^{a_k})e = \lim_{t\to 0}(t^{b_1}, \ldots, t^{b_k})e.$$

Proof. For normal toric varieties this is Proposition 1.6 (v) in [O]. The non-normal case follows using normalization, since the fan $\mathcal{F}(X)$ is defined in terms of the normalization.

B. Polytopes and their normal fans

By a (convex) *polytope* we mean a subset in Euclidean space $\mathbf{R}^k$ which is the convex hull of a finite number of points. Polytopes are to be distinguished from convex *polyhedra* by which we mean subsets in $\mathbf{R}^k$ which are the intersections of a finite number of affine half-spaces. Thus polytopes are exactly bounded polyhedra.

Let P be a polytope in $\mathbf{R}^k$ and φ a linear functional on $\mathbf{R}^k$. We call the *supporting face* for φ and denote by P^φ the (maximal) face of P on which φ achieves its maximum. This construction was encountered previously in the proof of Proposition 1.9. Note that φ is constant on P^φ.

Definition 4.5. Let P be a polytope in $\mathbf{R}^k$ and Γ a face of P. The *normal cone* $N_\Gamma P$ is the subset in the dual space $(\mathbf{R}^k)^*$ consisting of linear functionals $\varphi : \mathbf{R}^k \to \mathbf{R}$ such that $P^\varphi = \Gamma$.

Clearly $N_\Gamma P$ is a closed convex cone in $(\mathbf{R}^k)^*$. Moreover, it is immediate to see that the collection of cones $N_\Gamma P$ for various faces $\Gamma \subset P$ forms a complete fan. This fan is called the *normal fan* of P and denoted by $N(P)$.

In a similar way we define the normal fan $N(P)$ for a (possibly unbounded) convex polyhedron P. In this case $N(P)$ may not be complete. Its support is the set of linear functionals which achieve a maximum on P.

The concept of the normal fan provides a way of classifying polytopes.

Definition 4.6. Two polytopes P, $P' \subset \mathbf{R}^k$ are called *normally equivalent* if their normal fans become equal after a linear isomorphism of $\mathbf{R}^k$.

For example, a square and a rectangle are normally equivalent polygons.

If $\mathcal{F}_1, \mathcal{F}_2$ are two fans in $\mathbf{R}^n$, we say that $\mathcal{F}_2$ is a *refinement* of $\mathcal{F}_1$, if each cone from $\mathcal{F}_1$ is a union of cones from $\mathcal{F}_2$. In other words, $\mathcal{F}_2$ is obtained from $\mathcal{F}_1$ by further subdivision.

If P is a polytope and $N(P)$ is its normal fan, a refinement of $N(P)$ can be obtained by cutting out (as with a knife) some vertex, or edge, or any face of codimension at least 2 of P, see Figure 15.

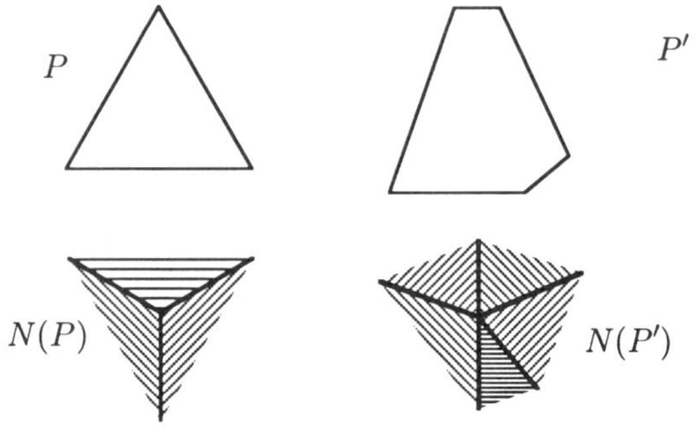

Figure 15.

This notion of "refinement of polytopes" (or, rather, their normal fans) has an interesting interpretation going back to Minkowski.

Definition 4.7. Let P, Q be two polytopes in $\mathbf{R}^k$. Their *Minkowski sum* $P + Q$ is the set of all vector sums $p + q$, $p \in P$, $q \in Q$.

It is clear that $P + Q$ is again a convex polytope. Indeed, consider the direct product $P \times Q \subset \mathbf{R}^k \times \mathbf{R}^k$ and the linear map $\mathbf{R}^k \times \mathbf{R}^k \to \mathbf{R}^k$ given by $(u, v) \mapsto u + v$. The Minkowski sum $P + Q$ is the image of the convex polytope $P \times Q$ under this linear map.

Theorem 4.8.

(a) *Let P, Q be polytopes in $\mathbf{R}^k$. Then the normal fan $N(P+Q)$ of $P+Q$ is the smallest common refinement of $N(P)$ and $N(Q)$, i.e., the collection of cones of the form $K \cap L$, $K \in N(P)$, $L \in N(Q)$.*

(b) *If P, R are polytopes in $\mathbf{R}^k$ and $N(R)$ is a refinement of $N(P)$ then there is $\lambda \in \mathbf{R}$ and a polytope Q such that R is affinely isomorphic to the Minkowski sum $\lambda P + Q$.*

Proof. See [Sm].

C. Fans corresponding to projectively embedded and affine toric varieties

First we consider the situation of Section 1D. Let $H = (\mathbf{C}^*)^{k-1}$ be a torus acting linearly on a vector space V. Consider the associated action of H on the projective space $P(V)$. For any $v \in V$, let $\overline{P(Hv)}$ be the closure in $P(V)$ of the H-orbit of the point corresponding to v. This is a projective (not necessarily normal) toric variety. Recall that we have associated to $v \in V$ its weight polytope $\mathrm{Wt}(v) \subset \mathbf{R}^{k-1}$.

Proposition 4.9. *The fan of the toric variety $\overline{P(Hv)}$ equals the normal fan of* $\mathrm{Wt}(v)$. *In particular,* $\dim(\overline{P(Hv)}) = \dim(\mathrm{Wt}(v))$.

Proof. We have $\overline{P(Hv)} \cong X_{A(v)}$ (see Section 1D). By taking into account the description of $\mathcal{F}(X)$ given in Proposition 4.4, we note that, for $X_{A(v)}$, the required statement was already proved in the proof of Proposition 1.9.

Now let Y be an affine toric variety so $Y = \mathrm{Spec}\ \mathbf{C}[S]$ where S is a semigroup embedded into some $\mathbf{Z}^k$. We can assume that S generates $\mathbf{Z}^k$ as an Abelian group. Let $K \subset \mathbf{R}^k$ be the convex hull of S. The normalization of Y is Spec $\mathbf{C}[K \cap \mathbf{Z}^k]$. Therefore the fan of Y consists of the dual cone $\check{K} \subset \mathbf{R}^{n*}$ together with all its faces. In particular, if K contains a non-trivial vector subspace then $\check{K}$ has dimension strictly less than k.

D. Quasi-smoothness

Definition 4.10. A (not necessarily normal) toric variety X is called *quasi-smooth* if every cone of the fan $\mathcal{F}(X)$ is simplicial and the normalization morphism $\tilde{X} \to X$ is bijective.

Clearly smooth varieties are quasi-smooth. For affine toric varieties this notion of quasi-smoothness coincides with the one given in Section 2C. Indeed, if $X = \mathrm{Spec}\,\mathbf{C}[S]$ and K is the convex hull of S, then the fan corresponding to X consists of faces of the dual cone $\check{K}$. The fact that K has the form $\mathbf{R}^p \times \mathbf{R}_+^q$ is equivalent to the fact that $\check{K}$ is simplicial (it will have smaller dimension, if $p > 0$).

Let us give a reformulation of the notion of quasi-smoothness for varieties of the form X_A. This involves geometric and arithmetic conditions. The geometric condition is as follows.

Definition 4.11. A convex polytope $P \subset \mathbf{R}^{k-1}$ is called *simple* if, for any face $\Gamma \subset P$, the normal cone $N_\Gamma P$ is simplicial.

This definition is equivalent to the following more transparent geometric one: P is simple if every vertex of P has exactly $\dim P$ edges passing through it.

Proposition 4.12. *Let $A \subset \mathbf{Z}^{k-1}$ be a finite set, and $Q \subset \mathbf{R}^{k-1}$ the convex hull of A. The variety X_A is quasi-smooth if and only if Q is a simple polytope and, for any face $\Gamma \subset Q$, the index $i(\Gamma, A)$ (see Section 3A) equals 1.*

Proof. This is an easy consequence of Theorem 3.1.

Since the dual of a simplicial cone is also simplicial, a local model for a quasi-smooth toric variety is Spec $\mathbf{C}[K \cap \mathbf{Z}^k]$ where $K \in \mathbf{R}^k$ is a rational simplicial cone.

CHAPTER 6

Newton Polytopes and Chow Polytopes

1. Polynomials and their Newton polytopes

A. Newton polytopes

Suppose we have a complicated (Laurent) polynomial $f(x_1, \ldots, x_k)$ in k variables. Let A be the set of monomials in f with non-zero coefficients. As we have seen in Chapter 5, to understand the structure of f, it is natural to consider it as a member of the space $\mathbf{C}^A$ of all polynomials whose monomials belong to A. Geometrically, each Laurent monomial $x^\omega = x_1^{\omega_1} \cdots x_k^{\omega_k}$ is represented by a lattice point $\omega = (\omega_1, \ldots, \omega_k) \in \mathbf{Z}^k$. The most important characteristic of f is its *Newton polytope*, defined as follows.

Definition 1.1. Let $f(x_1, \ldots, x_k) = \sum_{\omega \in \mathbf{Z}^k} a_\omega x^\omega$ be a Laurent polynomial in k variables. Its *Newton polytope* $N(f)$ is the convex hull in $\mathbf{R}^k$ of the set $\{\omega : a_\omega \neq 0\}$.

The Newton polytope is a particular case of the weight polytope introduced in Definition 1.7, Chapter 5. This case corresponds to the situation when V is the space of all polynomials in $x_1, \ldots, x_k$ with the action of $H = (\mathbf{C}^*)^k$ given by the scaling of variables: $(t_1, \ldots, t_k) \cdot (x_1, \ldots, x_k) = (t_1 x_1, \ldots, t_k x_k)$.

The following proposition is obvious.

Proposition 1.2.

(a) *The Newton polytope of $f(x_1, \ldots, x_k)$ lies in the hyperplane $\{\omega \in \mathbf{R}^k : (\omega, \varphi) = a\}$ for some $\varphi = (\varphi_1, \ldots, \varphi_k) \in \mathbf{Z}^k, a \in \mathbf{Z}$ if and only if the polynomial f is quasi-homogeneous of weight φ, i.e., $f(t^{\varphi_1} x_1, \ldots, t^{\varphi_k} x_k) = t^a f(x_1, \ldots, x_k)$.*

(b) *The Newton polytope of the product of two polynomials equals the Minkowski sum of the Newton polytopes of factors, see Definition 4.7, Chapter 5.*

If $f(x_1, \ldots, x_k) = \sum_{\omega \in \mathbf{Z}^k} a_\omega x^\omega$ is a Laurent polynomial and Γ is a face of its Newton polytope, then we shall call the *coefficient restriction* of f to Γ and denote by $f\|_\Gamma$ the polynomial $\sum_{\omega \in \Gamma} a_\omega x^\omega$. One use of this notion is given in the next proposition which is a particular case of Proposition 1.10, Chapter 5.

We recall that the leading term of a Laurent polynomial in one variable $q(\tau) = \sum a_i \tau^i$ is the monomial $a_i \tau^i$ where i is the largest such that $a_i \neq 0$, see Section 1D, Chapter 5.

Proposition 1.3. *Let $f(x_1, \ldots, x_k)$ be a Laurent polynomial and $Q \subset \mathbf{R}^k$ be its Newton polytope. Let $\varphi(\omega) = \sum \varphi_i \omega_i$ be a linear functional on $\mathbf{R}^k$ with all φ_i integral. Let $Q^\varphi \subset Q$ be the supporting face of φ, see Section 4B Chapter 5. Then, for any $x_1, \ldots, x_k \in \mathbf{C}^*$, the leading term of the Laurent polynomial $q(t) = f(t^{\varphi_1}x_1, \ldots, t^{\varphi_k}x_k)$ equals $t^{\varphi(Q^\varphi)} f\|_{Q^\varphi}(x_1, \ldots, x_k)$.*

B. Logarithmic maps and amoebas

The notion of the Newton polytope of a polynomial might seem artificial from a geometric point of view. However, this is not so. In this subsection we relate $N(f)$ to the structure of the hypersurface defined by the equation $f = 0$.

Denote by $\log : (\mathbf{C}^*)^k \to \mathbf{R}^k$ the map

$$(x_1, \ldots, x_k) \to (\log|x_1|, \ldots, \log|x_k|). \tag{1.1}$$

For a Laurent polynomial $f(x_1, \ldots, x_k)$, denote by Z_f the hypersurface in $(\mathbf{C}^*)^k$, defined by the equation $f(x_1, \ldots, x_k) = 0$.

Definition 1.4. The *amoeba* of a Laurent polynomial f is the subset $\log(Z_f) \subset \mathbf{R}^k$.

This name is motivated by the following typical shape of $\log(Z_f)$ in two dimensions (see Figure 16).

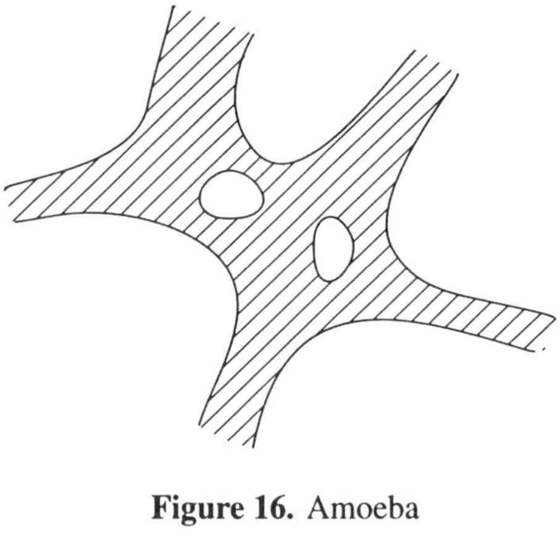

Figure 16. Amoeba

This shape is peculiar because of the thin "tentacles" going off to infinity. A bit later we shall give rigorous statements showing that the behavior of $\log(Z_f)$ is indeed typical. But first we relate the amoeba to the problem of finding Laurent series expansions for the rational function $1/f(x)$. Recall the general properties of Laurent series in several variables and their regions of convergence, see e.g., [Kr].

Proposition 1.5.

(a) *Let* $F(x) = \sum_{\omega \in \mathbf{Z}^k} c_\omega x^\omega$ *be a (formal) Laurent series in* $x_1, \ldots, x_k$ *with complex coefficients* c_ω *(which may be non-zero for all* ω*). Then the domain of convergence of* $F(x)$ *in* $(\mathbf{C}^*)^k$ *has the form* $\log^{-1}(B)$*, where* $B \subset \mathbf{R}^k$ *is a convex subset.*

(b) *If* $\varphi(x)$ *is a holomorphic function in a domain of the form* $\log^{-1}(B)$*, where* $B \subset \mathbf{R}^k$ *is a convex open subset, then there is a unique Laurent series converging to* $\varphi(x)$ *in this domain.*

Applying this proposition to the rational function $1/f$, we deduce the following.

Corollary 1.6. *Let* $f(x)$ *be a Laurent polynomial. All the components of the complement* $\mathbf{R}^k - \log(Z_f)$ *to the amoeba of* f *are convex subsets in* $\mathbf{R}^k$*. They are in bijective correspondence with Laurent series expansions of the rational function* $1/f(x)$*.*

Let $Q \subset \mathbf{R}^k$ be the Newton polytope of a Laurent polynomial $f(x_1, \ldots, x_k) = \sum a_\omega x^\omega$, and let $\gamma \in Q$ be any vertex. Then we can write

$$f(x) = a_\gamma x^\gamma \left(1 + \sum_{\omega \neq \gamma} \frac{a_\omega}{a_\gamma} x^{\omega - \gamma}\right) = a_\gamma x^\gamma (1 + g(x))$$

and construct the Laurent expansion

$$R_\gamma(x) = \frac{1}{f(x)} = a_\gamma^{-1} x^{-\gamma} (1 - g(x) + g(x)^2 - \cdots) \tag{1.2}$$

by using the geometric series. This is a well-defined Laurent series whose exponents lie in the affine cone $-\gamma + \mathbf{R}_+ \cdot (Q - \gamma)$ in $\mathbf{R}^k$; this cone is obtained by drawing half-lines from γ through all points of Q and then translating the result by (-2γ) (see Figure 17).

Let us describe the domain of convergence of R_γ. Let $N_\gamma(Q)$ denote the normal cone of Q at the vertex γ, see Definition 4.5, Chapter 5.

Proposition 1.7. *There is a vector* $b \in N_\gamma(Q)$ *such that the Laurent series* $R_\gamma(x)$ *converges absolutely for any* $x = (x_1, \ldots, x_k) \in (\mathbf{C}^*)^k$ *for which the vector* $\log(x) = (\log|x_1|, \ldots, \log|x_k|)$ *lies in the affine cone* $b + N_\gamma(Q)$*. In particular, for such* x*, we have* $f(x) \neq 0$*.*

Proof. For any $x = (x_1, \ldots, x_k) \in (\mathbf{C}^*)^k$ such that $|g(x)| < 1$, we have by construction (1.2) that $R_\gamma(x)$ converges absolutely. Choose $b \in N_\gamma(Q)$ so that

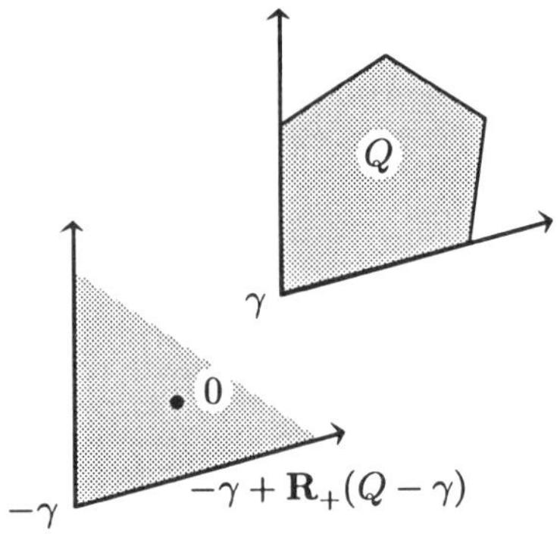

Figure 17. Laurent expansion of $1/f$

$(b, \omega - \gamma) \ll 0$ for all integral points $\omega \in Q$ different from γ. An easy calculation which is left to the reader shows that $|g(x)| < 1$ whenever $\log(x) \in b + N_\gamma(Q)$. This proves our proposition.

Corollary 1.8. *Vertices of the Newton polytope $Q = N(f)$ are in bijection with those connected components of the complement $\mathbf{R}^k - \log(Z_f)$ which contain an affine convex cone with non-empty interior.*

Proof. Let C_γ be the component containing $N_\gamma(Q)+b$, where b is as in Proposition 1.7. Since the normal cones $N_\gamma Q$ cover the whole logarithmic space $\mathbf{R}^k$, only the components of the type C_γ can contain an affine convex cone with a non-empty interior. It remains to show that the components C_γ are all distinct. Suppose $C_\gamma = C_\delta$ for two distinct vertices $\gamma, \delta \in Q$. Let K be the cone which is the convex hull of the normal cones $N_\gamma(Q)$, $N_\delta(Q)$. Since any component of $\mathbf{R}^k - \log(Z_f)$ is a convex set, the component $C_\gamma = C_\delta$ contains some translation of the cone K, say, $c + K$. However, suppose there is $a \in K$ not lying in $N_\gamma(Q)$. Then there is a vertex $\beta \in Q$ which is joined to γ by an edge and such that $(a, \gamma) < (a, \beta)$. This implies that, for $t \gg 0$ and $x \in (\mathbf{C}^*)^k$ such that $\log(x) = c + ta$, the series $R_\gamma(x)$ will not absolutely converge. Indeed, for such x the subseries of (1.2) consisting of terms whose exponents lie on the half-line $\gamma + \mathbf{R}_+ \cdot (\beta - \gamma)$ will have terms not tending to zero. Thus $x \notin C_\gamma$, a contradiction. We see that $N_\gamma(Q) = K = N_\delta(Q)$, hence $\gamma = \delta$. The proposition is proved.

Since the normal cones $N_\gamma Q$ cover the logarithmic space $\mathbf{R}^k$, we see that the amoeba is situated in thin spaces between walls of the translated normal cones (see Figure 18). It follows that the combinatorial structure of the Newton polytope

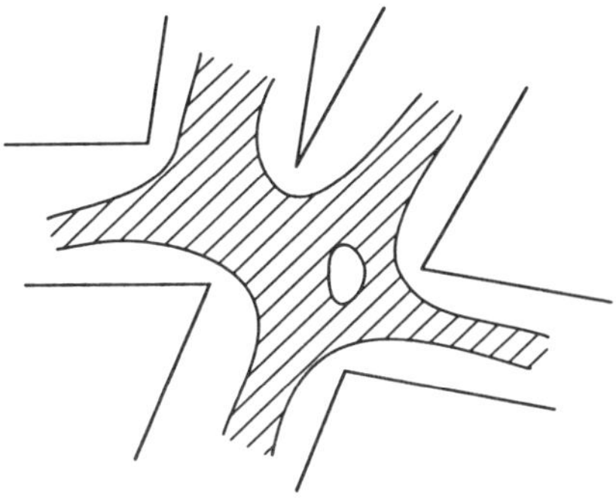

Figure 18. The amoeba between walls

$N(f)$ can be read from the geometry of the hypersurface $Z_f = \{f = 0\}$.

For a real number $M \geq 0$, let $S^{k-1}(M) \subset \mathbf{R}^k$ be the sphere with center 0 and radius M. Consider the intersection $S^{k-1}(M) \cap \log(Z_f)$.

Proposition 1.9. *Suppose that the Newton polytope $N(f)$ has full dimension k. Then the limit subset*

$$\lim_{M\to\infty} \frac{1}{M} \cdot (S^{k-1}(M) \cap \log(Z_f))$$

in the unit sphere exists, and it is the $(k-2)$-skeleton of the cell decomposition of the sphere which is dual to the decomposition of the boundary of the polytope $N(f)$ by its faces.

The limit here can be understood, for example, in the sense of the Hausdorff distance between closed subsets of the sphere, see [Hau].

Proof. By Corollary 1.8, the open set $\mathbf{R}^k - \log(Z_f)$ is the union of some non-intersecting neighborhoods C_γ of translated normal cones $N_\gamma(Q) + b_\gamma$ and, possibly, some other open subsets that neither intersect the C_γ nor contain a convex affine cone with non-empty interior. Thus $\log(Z_f) \subset \mathbf{R}^k - \bigcup C_\gamma$ and the boundary of each C_γ is a part of the boundary of Z_f. It follows that

$$\lim_{M\to\infty} \frac{1}{M} \cdot \left(S^{k-1}(M) \cap \log(Z_f)\right) = \lim_{M\to\infty} \frac{1}{M} \cdot \left(S^{k-1}(M) \cap \left(\mathbf{R}^k - \bigcup C_\gamma\right)\right)$$

$$= \lim_{M\to\infty} \frac{1}{M} \cdot \left(S^{k-1}(M) \cap \left(\mathbf{R}^k - \bigcup (N_\gamma(Q) + b_\gamma)\right)\right)$$

and for the last limit the assertion is obvious.

Remark 1.10. To find, for a given polynomial $f(x)$, all the components of the complement to its amoeba (and hence all the Laurent series for $1/f$) is a difficult and interesting problem. For a polynomial in one variable this requires the separation of roots of f (which are not rationally known). For higher dimensions the amoeba can have "holes" (i.e., components in the complement that do not correspond to vertices of the Newton polytope, e.g., bounded components). The presence of such components is probably governed by the relative sizes of the coefficients, which are ignored in the definition of $N(f)$.

C. Amoebas and the moment map

In subsection B we considered hypersurfaces in a torus. One can consider hypersurfaces in toric varieties as well.

Let $A \subset \mathbf{Z}^k$ be a finite subset generating $\mathbf{Z}^k$ as an affine lattice, and let X_A be the corresponding projective toric variety, see Section 1B, Chapter 5. This variety contains the torus $(\mathbf{C}^*)^k$ as an open subset. Denote, as usual, by $\mathbf{C}^A$ the space of Laurent polynomials with monomials in A, i.e., of polynomials of the form

$$f(x_1, \dots, x_k) = \sum_{\omega \in A} a_\omega x^\omega.$$

Let Y_A be the cone over X_A. Any polynomial from $\mathbf{C}^A$ defines a homogeneous function on Y_A and hence defines a hypersurface in X_A, to be denoted by $\bar{Z}_f$. It is obviously the closure (in the Zariski topology as well as in the ordinary topology) of the previously considered affine hypersurface $Z_f \subset (\mathbf{C}^*)^k \subset X_A$.

Let $Q \subset \mathbf{R}^k$ be the convex hull of A. Then Q is the Newton polytope of a generic $f \in \mathbf{C}^A$. The analog of the logarithmic map (1.1) in the present "compact" situation is given by the *moment map* ([At 1-2], see also [O], p. 94):

$$\mu = \mu_A : X_A \longrightarrow Q, \quad \mu(x) = \frac{\sum_{\omega \in A} |x^\omega| \cdot \omega}{\sum_{\omega \in A} |x^\omega|}. \tag{1.3}$$

Here each x^ω is regarded as a homogeneous function on the cone Y_A over X_A; because of the denominator the expression in (1.3) is homogeneous of degree 0, i.e., it is a well-defined function on X_A. The notation x^ω can be understood in the usual sense, as a Laurent monomial, on the open set $(\mathbf{C}^*)^k \subset X_A$

The properties of the moment map are summarized in the following theorem ([At 2], Th. 2). We omit the proof.

Theorem 1.11.

(a) *The map μ is surjective. For any face $\Gamma \subset Q$ the inverse image $\mu^{-1}(\Gamma)$ coincides with $X(\Gamma)$, the closure of the orbit corresponding to Γ. The orbit itself is the inverse image of the interior of Γ.*

(b) *Let* $(S^1)^k$ *be the subgroup in* $(\mathbf{C}^*)^k$ *consisting of* $(x_1, \dots, x_k)$ *such that* $|x_i| = 1$ *for all* i. *Each fiber* $\mu^{-1}(q)$, $q \in Q$, *is an orbit of* $(S^1)^k$. *If* q *lies inside a* j-*dimensional face of* Q, *then* $\mu^{-1}(q)$ *is isomorphic to* $(S^1)^j$. *In particular,* $\mu^{-1}(q)$ *consists of one point if and only if* q *is a vertex of* Q.

Let $f \in \mathbf{C}^A$. The *compactified amoeba* of f is the image $\mu(\bar{Z}_f) \subset Q$. This notion is really useful only when all coefficients of f are non-zero. We shall denote the space of such polynomials by $(\mathbf{C}^*)^A$ and suppose, in the rest of this section, that $f \in (\mathbf{C}^*)^A$.

A typical pattern of behavior of a compactified amoeba is depicted in Figure 19 and formalized in the following theorem.

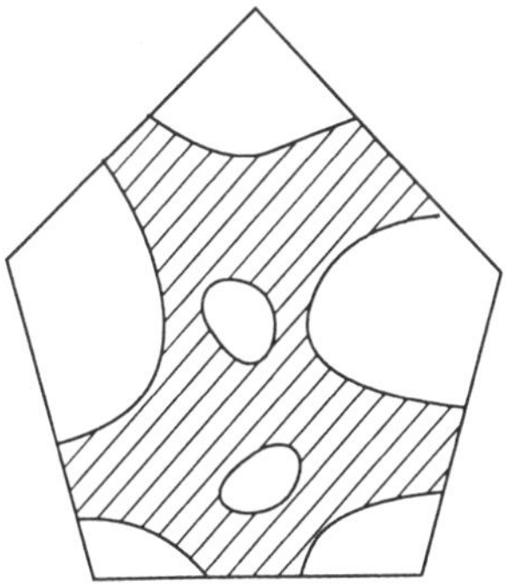

Figure 19. Compactified amoeba

Theorem 1.12. *Let* $f \in (\mathbf{C}^*)^A$ *and let* Q *be the convex hull of* A. *Then*

(a) *for any vertex* $\gamma \in Q$, *there is a neighborhood of* γ *which does not intersect the compactified amoeba* $\mu(\bar{Z}_f)$;

(b) *denote by* $K(\gamma)$ *the component of the complement* $Q - \mu(\bar{Z}_f)$ *containing (the neighborhood of) a vertex* $\gamma \in Q$. *All of these components* $K(\gamma)$ *are distinct.*

Proof. The moment map μ maps the torus $(\mathbf{C}^*)^k \subset X_A$ onto the interior of Q. Moreover, we have the commutative diagram

$$\begin{array}{ccc} (\mathbf{C}^*)^k & \xrightarrow{\log} & \mathbf{R}^k \\ & \searrow^{\mu} & \downarrow \alpha \\ & & \mathrm{Int}(Q) \end{array} \tag{1.4}$$

where α is the map whose value at $\lambda = (\lambda_1, \dots, \lambda_k) \in \mathbf{R}^k$ is given by

$$\alpha(\lambda) = \frac{\sum_{\omega \in A} e^{(\omega,\lambda)} \cdot \omega}{\sum_{\omega \in A} e^{(\omega,\lambda)}} \tag{1.5}$$

and (ω, λ) stands for the standard scalar product $\sum \omega_i \lambda_i$.

Taking into account Corollary 1.8, it remains to show that, for any vertex $\gamma \in Q$, the map α takes a translated normal cone $b + N_\gamma(Q)$ onto a neighborhood of γ in $\mathrm{Int}(Q)$.

If $\lambda \in N_\gamma(Q)$ then $(\lambda, \gamma) > (\lambda, \omega)$ for $\omega \in A - \{\gamma\}$. Hence when λ goes to infinity in the cone $b + N_\gamma(Q)$, in the linear combination $\sum_{\omega \in A} e^{(\omega,\lambda)} \cdot \omega$ the summand corresponding to γ will outweigh all the others. This means that after dividing by $\sum_{\omega \in A} e^{(\omega,\lambda)}$, we shall approach the point γ. Conversely, if $\alpha(\lambda)$ is close to γ, then the summand corresponding to γ in the numerator of (1.5) dominates all the other summands (this can be seen by applying to the vector sum in (1.5) a linear functional on $\mathbf{R}^k$ achieving maximum exactly at γ; we leave the routine calculation to the reader). So λ lies deep inside $N_\gamma(Q)$.

Let us mention another property of compactified amoebas which makes them in some instances easier to handle than those considered in subsection B.

Proposition 1.13. *Let $f(x) \in (\mathbf{C}^*)^A$ be as before; let $\Gamma \subset Q$ be a face and let $f\|_\Gamma$ be the coefficient restriction of f to Γ, see subsection A. Then the compactified amoeba of $f\|_\Gamma$ coincides with the intersection with Γ of the compactified amoeba of f.*

Proof. Let $(z_\omega,\ \omega \in A)$ be the standard homogeneous coordinates in the ambient projective space P^{n-1} of the projective variety X_A (here n is the cardinality of A). The orbit closure $X(\Gamma)$ is given inside X_A by equations $z_\omega = 0$ for $\omega \notin \Gamma$ (Proposition 2.5, Chapter 5). Since the restriction of z_ω to (the cone over) X_A becomes the Laurent monomial x^ω, our statement follows directly by the comparison of moment maps for A and $A \cap \Gamma$.

2. Theorems of Kouchnirenko and Bernstein on the number of solutions of a system of equations

Historically, one of the first applications of Newton polytopes was to the problem of finding the number of solutions of a system of polynomial equations in several variables. In this section we examine these developments.

A. The setting of the problem

Suppose we have $k-1$ Laurent polynomials in $k-1$ variables:

$$f_1(x_1, \dots, x_{k-1}), \dots, f_{k-1}(x_1, \dots, x_{k-1}). \tag{2.1}$$

These polynomials define functions on the algebraic torus $(\mathbf{C}^*)^{k-1}$. We would like to find the number of their common roots *in this torus.* Of course, we must count isolated common roots with appropriate multiplicities.

Example 2.1. Consider a Laurent polynomial in one variable $f(x) = a_r x^r + a_{r+1}x^{r+1} + \cdots + a_s x^s$. The number of its non-zero roots is equal to $s - r$ provided that $a_r, a_s \neq 0$. Note that the Newton polytope of f is the line segment $[r, s]$ so $s - r$ is the length of this segment.

In the general case the number of common roots of $f_1, \ldots, f_{k-1}$ in the torus can vary with the f_i since some roots can go off to the infinity of the torus. Here is one of the possible ways to set the problem about the number of roots precisely.

Let $A \subset \mathbf{Z}^{k-1}$ be a finite set of exponents and $\mathbf{C}^A$ be the vector space of Laurent polynomials

$$f(x) = \sum_{\omega \in A} a_\omega x^\omega$$

with all monomials from A. We assume that A generates $\mathbf{R}^{k-1}$ as an affine space. We shall consider polynomials $f_1, \ldots, f_{k-1}$ from $\mathbf{C}^A$. It is clear that for a *generic* choice of the f_i (i.e., for $(f_1, \ldots, f_{k-1})$ belonging to some dense open subset in $(\mathbf{C}^A)^{k-1}$), the number of common roots of the f_i in the torus is constant. So this number depends only on A and we are interested in finding it.

B. Kouchnirenko theorem

The answer to the problem raised in subsection A was given by Kouchnirenko [Kou] in terms of the polytope $Q \subset \mathbf{R}^{k-1}$, the convex hull of A. Clearly Q is the Newton polytope of a generic $f \in \mathbf{C}^A$. Recall (Section 3D, Chapter 5) that the integral lattice $\mathbf{Z}^{k-1} \subset \mathbf{R}^{k-1}$ induces a volume form $\mathrm{Vol}_{\mathbf{Z}^{k-1}}$ on $\mathbf{R}^{k-1}$ such that the volume of an elementary lattice simplex is 1. In this normalization, the volume of any polytope with vertices on the lattice is an integer. Kouchnirenko's theorem is as follows.

Theorem 2.2. *In the above assumptions the number of common roots of generic polynomials $f_1, \ldots, f_{k-1} \in \mathbf{C}^A$ is equal to the volume* $\mathrm{Vol}_{\mathbf{Z}^{k-1}} Q$.

Before proving the theorem, let us consider an example. Let A consist of all monomials in two variables $x_1^i x_2^j$ with $i, j \geq 0, \quad i + j \leq d$. The polytope Q is the plane triangle with vertices $(0, 0)$, $(d, 0)$, $(0, d)$. Its area with respect to the volume form induced by $\mathbf{Z}^2$ is d^2 (as can be seen, for example, by decomposing it into d^2 elementary lattice triangles). In this case Theorem 2.2 follows from the classical Bezout theorem: the number of common roots of two polynomials of degree $\leq d$ in two variables is d^2. More precisely, the Bezout theorem says that two curves C_1, C_2 in the projective plane P^2 of degree d intersect in d^2 points (if counted with appropriate multiplicities). If C_i is the projective closure of the curve in $\mathbf{C}^2$ given by $f_i(x_1, x_2) = 0$ and f_1, f_2 are generic, then all d^2 points of intersection lie in the torus $(\mathbf{C}^*)^2 \subset \mathbf{C}^2 \subset P^2$.

If, for A, we take some subset of the set of monomials of degree $\leq d$ then the Bezout theorem will be still formally applicable to the curves $C_i \subset P^2$. However, among the d^2 points of the intersection $C_1 \cap C_2$ some may lie outside the torus $(\mathbf{C}^*)^2$ regardless of the choice of the $f_i \in \mathbf{C}^A$. So the number of common roots in the torus can be less than d^2.

To prove the Kouchnirenko theorem, let us reformulate it in more geometric terms. First of all, note that we can reduce the question to the situation when the set A generates $\mathbf{Z}^{k-1}$ as an affine lattice, i.e., $\mathbf{Z}^{k-1} = \mathrm{Aff}_{\mathbf{Z}}(A)$ (see Definition 1.3, Chapter 5). Indeed, let $\{\eta^{(1)}, \dots, \eta^{(k)}\}$ be an affine $\mathbf{Z}$-basis of the lattice $\mathrm{Aff}_{\mathbf{Z}}(A)$, i.e., a subset in this lattice (not necessarily in A) such that $\mathrm{Aff}_{\mathbf{Z}}(A)$ consists of affine linear combinations of $\eta^{(1)}, \dots, \eta^{(k)}$. In particular, any vector $\omega \in A$ can be written in the form $\omega = \sum m_i \eta^{(i)}$ with $m_i \in \mathbf{Z}$, $\sum m_i = 1$. In other words,

$$\omega = \eta^{(k)} + \sum_{i=1}^{k-1} m_i (\eta^{(i)} - \eta^{(k)}).$$

It follows that every polynomial $f \in \mathbf{C}^A$ can be written in the form

$$f(x_1, \dots, x_{k-1}) = x^{\eta^{(k)}} g(y_1, \dots, y_{k-1})$$

where $y_i = x^{\eta^{(i)} - \eta^{(k)}}$. The transformation $(x_1, \dots, x_{k-1}) \mapsto (y_1, \dots, y_{k-1})$ is a finite covering of degree $m = [\mathbf{Z}^{k-1} : \mathrm{Aff}_{\mathbf{Z}}(A)]$. So the number of x-roots is m times the number of y-roots. At the same time

$$\mathrm{Vol}_{\mathbf{Z}^{k-1}}(Q) = m \cdot \mathrm{Vol}_{\mathrm{Aff}_{\mathbf{Z}}(A)}(Q).$$

Thus everything reduces to the lattice $\mathrm{Aff}_{\mathbf{Z}}(A)$ and polynomials in $y_1, \dots, y_{k-1}$.

So we assume that $\mathbf{Z}^{k-1} = \mathrm{Aff}_{\mathbf{Z}}(A)$. We arrange the elements of A in some order: $A = \{\omega^{(1)}, \dots, \omega^{(n)}\}$. Let $X_A \subset P^{n-1}$ be the toric variety introduced in Section 1B Chapter 5.

A generic point $(y_1 : \dots : y_n)$ of X_A has a parametric representation $y_i = x^{\omega^{(i)}}$ for some $x = (x_1, \dots, x_{k-1}) \in (\mathbf{C}^*)^{k-1}$. Therefore a linear form $l(y) = \sum_{i=1}^n a_i y_i$ on P^{n-1} becomes, after restricting to X_A, a Laurent polynomial

$$f(x) = \sum_{i=1}^{n} a_i x^{\omega^{(i)}}.$$

If $l_1(y), \dots, l_{k-1}(y)$ are $k-1$ generic linear forms defining a projective subspace $L \subset P^{n-1}$, then the number of points of $X_A \cap L$ equals the number of solutions of the system of equations

$$f_1(x) = 0, \dots, f_{k-1}(x) = 0, \tag{2.2}$$

where $f_j(x)$ is the polynomial corresponding to $l_j(y)$. Note that it is essential here to assume the l_j generic: otherwise the intersection points may not lie in the image of the parametrization $y_i = x^{\omega^{(i)}}$, $x \in (\mathbf{C}^*)^{k-1}$.

We conclude that the number of common roots of a generic system (2.2) equals the degree of the toric variety X_A. In other words, Kouchnirenko theorem 2.2 is equivalent to the following statement.

Theorem 2.3. *Let $A \subset \mathbf{Z}^{k-1}$ be a finite set of lattice points generating $\mathbf{Z}^{k-1}$ as an affine lattice and let $Q \subset \mathbf{R}^{k-1}$ be the convex hull of A. Then the degree of the toric variety $X_A \subset P^{n-1}$ equals* $\mathrm{Vol}_{\mathbf{Z}^{k-1}} Q$.

This statement is in fact a particular case of Theorem 3.14 from Chapter 5 which describes the multiplicity (or local degree) of any singular point on a toric variety. To specialize to the present situation, we consider the variety Y_A, the affine cone over X_A. The apex of this cone is a singular point of Y_A and the corresponding multiplicity equals $\deg X_A$. The semigroup S_A corresponding to Y_A is generated in $\mathbf{Z}^k = \mathbf{Z}^{k-1} \oplus \mathbf{Z}$ by vectors $(\omega, 1)$, $\omega \in A$. The subdiagram volume $u(S_A)$ which, by the cited theorem, is equal to $\deg (X_A)$, is just the k-dimensional volume of the pyramid P in $\mathbf{R}^k$ with apex 0 and base $Q \times \{1\}$. Obviously $\mathrm{Vol}_{\mathbf{Z}^k}(P) = \mathrm{Vol}_{\mathbf{Z}^{k-1}}(Q)$ and Theorem 2.3 follows.

By analyzing the above arguments it is possible to extract the exact conditions for a system $(f_1, \ldots, f_k)$ to be "generic" in the sense of Theorem 2.2. We leave this to an interested reader.

C. Kouchnirenko theorem for underdetermined systems

In the paper [Kou] Kouchnirenko has proven a more general theorem which will also be important for us. Namely, consider a system of $p \leq k-1$ equations $f_i(x_1, \ldots, x_{k-1}) = 0$, $i = 1, \ldots, p$ where the f_i belong to the space $\mathbf{C}^A$ as before. Let

$$Z_{f_1,\ldots,f_p} = \{x \in (\mathbf{C}^*)^{k-1} : f_1(x) = \cdots = f_p(x) = 0\}$$

be the set of solutions of this system in the torus. This is typically a variety of dimension $k-1-p$ so we cannot speak about the "number" of solutions. However, there is an important numerical invariant of any variety Z replacing the number of points of a finite set, namely the topological Euler characteristic

$$\chi(Z) = \sum_i (-1)^i \dim H^i(Z, \mathbf{C}). \tag{2.3}$$

When Z has dimension 0, the Euler characteristic $\chi(Z)$ is just the cardinality of Z. Now the general theorem of Kouchnirenko is as follows.

Theorem 2.4. *Let $A \in \mathbf{Z}^{k-1}$ be a finite subset affinely spanning $\mathbf{R}^{k-1}$ and let $1 \leq p \leq k-1$. Let also $Q \subset \mathbf{R}^{k-1}$ be the convex hull of A. There is a dense Zariski open subset U in the space $(\mathbf{C}^A)^p$ of p-tuples of polynomials $(f_1, \ldots, f_p)$, $f_i \in \mathbf{C}^A$ such that, for any $(f_1, \ldots, f_p) \in U$, we have*

$$\chi(Z_{f_1,\ldots,f_p}) = (-1)^{k-1-p}\mathrm{Vol}_{\mathbf{Z}^{k-1}}(Q).$$

For the proof of this theorem we refer the reader to the original paper of Kouchnirenko [Kou].

Example 2.5. Let $A = \{1, x_1, \ldots, x_{k-1}\}$ so $\mathbf{C}^A$ consists of affine-linear polynomials $f(x) = a_0 + \sum_{i=1}^{k-1} a_i x_i$. If $a_i \neq 0$ for all i, then the hypersurface in $(\mathbf{C}^*)^{k-1}$ given by the equation $f(x) = 0$ is isomorphic to $\mathbf{C}^{k-2}$ minus the union of $(k-1)$ affine hyperplanes in the general position, i.e., to the variety Z_{k-2} where we denote

$$Z_m = \left\{(t_1, \ldots, t_m) \in \mathbf{C}^m : \ t_i \neq 0, \sum t_i \neq 1\right\}.$$

The fact that $\chi(Z_m) = (-1)^m$ for any m can be easily seen by induction. Indeed, $Z_1 = \mathbf{C} - \{0, 1\}$ has $\chi = -1$. Let $\pi : Z_m \to Z_{m-1}$ be the projection with center $(0, \ldots, 0, 1)$ (see Figure 20). This projection makes Z_m into a fibration over Z_{m-1} with fibers isomorphic to Z_1 so $\chi(Z_m) = \chi(Z_1)\chi(Z_{m-1}) = -\chi(Z_{m-1})$.

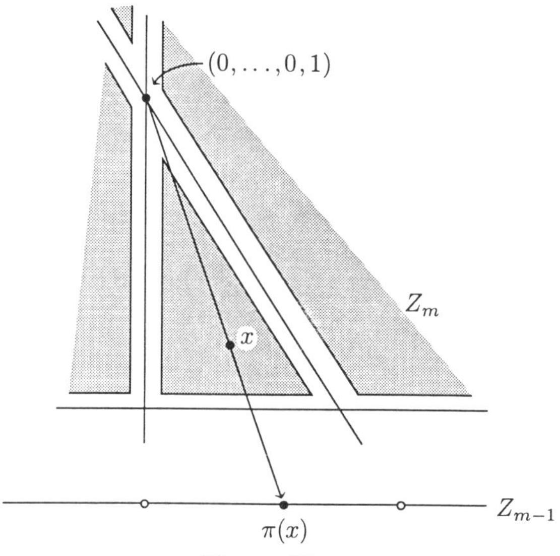

Figure 20.

D. Mixed volumes and the Bernstein theorem

The Bernstein theorem [Ber] extends the Kouchnirenko theorem to the case of systems of equations where each equation has its own set of monomials. Consider $k-1$ sets of monomials $A_1, \dots, A_{k-1} \subset \mathbf{Z}^{k-1}$. We assume that their union generates $\mathbf{Z}^{k-1}$ as an affine lattice. Let $\mathbf{C}^{A_i}$ be the space of all Laurent polynomials with monomials from A_i, i.e., polynomials of the form $\sum_{\omega \in A_i} a_\omega x^\omega$, $x = (x_1, \dots, x_{k-1})$. We are interested in the number of solutions of the system of equations

$$f_1(x) = \cdots = f_{k-1}(x) = 0, \quad x \in (\mathbf{C}^*)^{k-1}$$

for *generic* polynomials $f_i \in \mathbf{C}^{A_i}$. For the case $A_1 = \cdots = A_{k-1} = A$, this number is, by Kouchnirenko's theorem, equal to the volume of the polytope Q, the convex hull of A. In the general case D. Bernstein expressed this number in terms of *mixed volumes*. Let us recall the concept of mixed volumes introduced by Minkowski (see [E] [Liu] [O]).

Let V be a real vector space of dimension m. For any two convex polytopes $P, Q \subset V$, we denote by $P + Q$ their Minkowski sum (Definition 4.7, Chapter 5) and by λP for $\lambda \in \mathbf{R}$ the scaled polytope P, i.e., $\lambda P = \{\lambda p : p \in P\}$. Let Vol be a translation invariant volume form on V. It is known (see the cited literature) that, for any m convex polytopes $P_1, \dots, P_m \subset V$, the expression

$$\mathrm{Vol}(\lambda_1 P_1 + \cdots + \lambda_m P_m) \tag{2.4}$$

is a homogeneous polynomial in $\lambda_1, \dots, \lambda_m$ of degree m.

Definition 2.6. The *mixed volume* $\mathrm{Vol}(P_1, \dots, P_m)$ is the coefficient at the monomial $\lambda_1 \cdots \lambda_m$ in the polynomial (2.4).

More explicitly, we have

$$\mathrm{Vol}(P_1, \dots, P_m) = \frac{1}{m!} \sum_{k=1}^{m} (-1)^{m-k} \sum_{1 \le i_1 < \cdots < i_k \le m} \mathrm{Vol}(P_{i_1} + \cdots + P_{i_k}).$$

The following proposition summarizes the basic properties of mixed volumes.

Proposition 2.7.

(a) $\mathrm{Vol}(P, \dots, P) = \mathrm{Vol}(P)$ *is the usual volume of* P.

(b) *The mixed volume* $\mathrm{Vol}(P_1, \dots, P_m)$ *is invariant under translations of each of the* P_i *and under simultaneous linear transformations of all of the* P_i *with determinant* 1.

Proof. This follows from the homogeneity of the volume under the dilations and from its invariance under translations and linear transformations with determinant 1.

Part (b) above permits us to speak about mixed volumes in an affine space W equipped with a translation invariant volume form Vol. In particular, when W is equipped with an affine lattice Ξ, we have the volume form Vol_Ξ (see Section 3D Chapter 5). We shall write $\mathrm{Vol}_\Xi(P_1, \ldots, P_m)$ for the corresponding mixed volume.

Now the Bernstein theorem is as follows.

Theorem 2.8. *Let $A_1, \ldots, A_{k-1} \subset \mathbf{Z}^{k-1}$ be finite sets such that $A_1 \cup \cdots \cup A_{k-1}$ generates $\mathbf{Z}^{k-1}$ as an affine lattice. Let $Q_i \subset \mathbf{R}^{k-1}$ be the convex hull of A_i, and let $\mathbf{C}^{A_i}$ be the space of Laurent polynomials in $x_1, \ldots, x_{k-1}$ with monomials from A_i. Then there exists a dense Zariski open subset $U \subset \prod \mathbf{C}^{A_i}$ with the following property: for any $(f_1, \ldots, f_{k-1}) \in U$, the number of solutions of the system of equations $f_1(x) = \cdots = f_{k-1}(x) = 0$ in $(\mathbf{C}^*)^{k-1}$ equals the mixed volume* $\mathrm{Vol}_{\mathbf{Z}^{k-1}}(Q_1, \ldots, Q_{k-1})$.

For the proof see [Ber].

Remark 2.9. For the case of two polynomials in two variables, the statement, equivalent to Theorem 2.8, was proved by Minding [Min] in 1841 long before Minkowski introduced mixed volumes.

3. Chow polytopes

In the previous section we defined the Newton polytope of a polynomial. Geometrically, considering a polynomial is equivalent to considering the hypersurface defined by the vanishing of this polynomial. Thus the Newton polytope is a combinatorial invariant of a hypersurface in $(\mathbf{C}^*)^n$. It is natural to try to extend the notion of the Newton polytope from hypersurfaces to algebraic subvarieties of arbitrary dimension. In this section (based on [KSZ2]) we shall discuss such a generalization.

For technical reasons we shall study algebraic subvarieties not in the torus but in the projective space P^{n-1}. This does not seriously restrict generality. For example, a hypersurface in P^{n-1} is given by a homogeneous polynomial in n variables $x_1, \ldots, x_n$ which does not contain negative powers of the x_i. The collection of such polynomials is, of course, less general than the collection of all Laurent polynomials in $x_1, \ldots, x_n$. However, any Laurent polynomial in $n-1$ variables $y_1, \ldots, y_{n-1}$ can be transformed, by a suitable monomial change of variables, to a homogeneous polynomial in n variables (not containing negative powers).

A. Definition of Chow polytopes

In Section 1, Chapter 4 we considered the Chow variety $G(k, n, d)$ of algebraic cycles in P^{n-1} of degree d and dimension $k-1$. This is a projective variety and we recall its projective embedding.

Let $\mathcal{B} = \mathcal{B}(n-k, n) = \bigoplus_d \mathcal{B}_d$ be the homogeneous coordinate ring of the Grassmannian $G(n-k, n)$ (see Section 1E Chapter 3). This ring is generated by the so-called brackets $[i_1, \ldots, i_k]$, $1 \leq i_1 < \cdots < i_k \leq n$. By definition, $[i_1, \ldots, i_k]$ is the polynomial which takes an $(n-k)$-dimensional subspace L with a chosen basis to the Plücker coordinate $p_{j_1,\ldots,j_{n-k}}(L)$, where $\{j_1 < \cdots < j_{n-k}\}$ is the complement to $\{i_1, \ldots, i_k\}$. The brackets are subject to Plücker quadratic relations, see (1.7) of Chapter 3. The Chow variety $G(k, n, d)$ is realized as a Zariski closed subset in the projective space $P(\mathcal{B}_d)$.

This realization associates to a cycle $X = \sum m_i X_i$ its Chow form $R_X = \prod R_{X_i}^{m_i}$, where R_{X_i} is the equation for the locus of $(n-k-1)$-dimensional projective subspaces intersecting X_i. In the case of a hypersurface, the Chow form coincides with the equation of this hypersurface. Thus the notion of the equation is extended to the subvarieties of higher codimension. Informally, the Chow polytope which we are going to define, will be the "Newton polytope of the Chow form."

We fix a system of homogeneous coordinates $x_1, \ldots, x_n$ in P^{n-1}. The algebraic torus $H = (\mathbf{C}^*)^n$ acts on P^{n-1} by independent dilations of the homogeneous coordinates. Therefore, H acts on the space $\mathcal{B}_d$ of (possible) Chow forms of $(k-1)$-dimensional cycles of degree d.

Definition 3.1. Let $X \in G(k, n, d)$ be an algebraic cycle in P^{n-1} of dimension $(k-1)$ and degree d. The *Chow polytope* of X is the weight polytope of the Chow form R_X in the space $\mathcal{B}_d$ (see Definition 1.7 Chapter 5), and is denoted by $\mathrm{Ch}(X)$.

By definition, $\mathrm{Ch}(X)$ lies in the vector space $\mathbf{R}^n = \mathbf{Z}^n \otimes \mathbf{R}$, where $\mathbf{Z}^n$ is identified with the group of characters of the torus H. For any character $\chi \in \mathbf{Z}^n$ of H, as in (1.7) of Chapter 5, we denote by $\mathcal{B}_{d,\chi}$ the weight subspace of $\mathcal{B}_d$ corresponding to χ. We are going to describe the set of possible weights.

Denote by $\Delta(k, n)$ the convex polytope in $\mathbf{R}^n$ which is the convex hull of the $\binom{n}{k}$ points $e_{i_1} + \cdots + e_{i_k}$, $1 \leq i_1 < \cdots < i_k \leq n$, where the e_i are the standard basis vectors. This polytope was introduced in [GM] and called the (k, n)-*hypersimplex*. Clearly, $\Delta(k, n)$ can be seen as the convex hull of the barycenters of the $(k-1)$-dimensional faces of the $(n-1)$-dimensional simplex. In particular, $\Delta(1, n)$ and $\Delta(n-1, n)$ are $(n-1)$-dimensional simplices. The first non-trivial example is provided by the hypersimplex $\Delta(2, 4)$ which is the 3-dimensional octohedron (see Figure 21).

For any $d > 0$, we denote by $d\Delta(k, n)$ the scaled hypersimplex, i.e., the convex hull of the points $d(e_{i_1} + \cdots + e_{i_k})$.

Proposition 3.2. *The weight subspace $\mathcal{B}_{d,\chi}$ is non-zero if and only if $\chi \in d\Delta(k, n)$.*

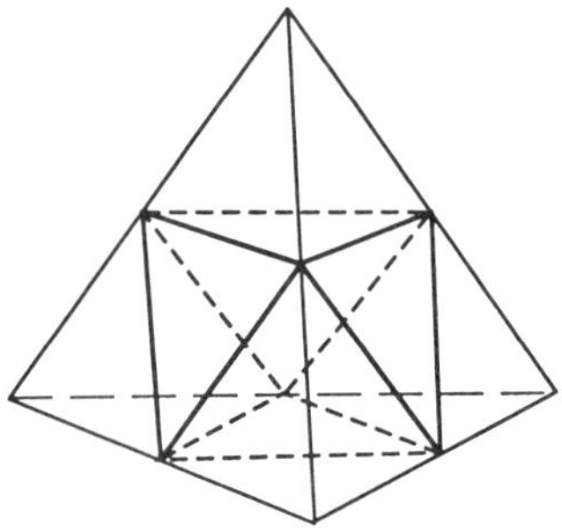

Figure 21. The octohedron as the hypersimplex $\Delta(2, 4)$

Proof. Clearly every element of $\mathcal{B}_d$ is a polynomial of degree d in brackets $[i_1, \ldots, i_k]$, and $[i_1, \ldots, i_k]$ has the weight $e_{i_1} + \cdots + e_{i_k}$. It follows that every χ such that $\mathcal{B}_{d,\chi} \neq 0$, belongs to $d\Delta(k, n)$. Also, the reverse inclusion is not hard to prove by elementary methods. Moreover, we can prove it by observing that the space $\mathcal{B}_d$ is an irreducible representation of the group $GL(n, \mathbf{C})$ with highest weight $d(e_1 + \cdots + e_k)$. Our statement is a special case of the general fact about weight diagrams of irreducible representations of reductive groups: the weights of such a representation with highest weight λ are exactly all weights β such that $\lambda - \beta$ lies in the root lattice, and β lies in the convex hull of the orbit of λ under the action of the Weyl group (see, e.g., [Bou1], Ch. VIII, Section 7, Proposition 5).

The weight subspaces in $\mathcal{B}_d$ can have dimension more than 1. However, the weight subspaces corresponding to the vertices of $d\Delta(k, n)$ are 1-dimensional and generated by the d-th powers $[i_1, \ldots, i_k]^d$.

Now let us discuss some examples of Chow polytopes.

Example 3.3. Let $k = n - 1$, i.e., suppose that X is a hypersurface in P^{n-1} defined by a homogeneous equation $f(x_1, x_2, \ldots, x_n) = 0$. The ring $\mathcal{B}(n - 1, n)$ coincides with the polynomial ring $\mathbf{C}[x_1, \ldots, x_n]$: the variable x_j is identified with the bracket $(-1)^{j-1}[1, 2, \ldots, \hat{j}, \ldots, n]$. This is a particular case of the correspondence between Plücker coordinates in $G(k, n)$ and $G(n - k, n)$, see (1.6) of Chapter 3. The Chow form of X is just its equation $f(x_1, \ldots, x_n)$. Therefore the Chow polytope of X is the same as the Newton polytope of f.

Example 3.4. Let L be a $(k - 1)$-dimensional projective subspace in P^{n-1} and let

$(p_{i_1,\ldots,i_k})$ be its Plücker coordinates. Then

$$R_L = \sum_{1 \le i_1 < \cdots < i_k \le n} p_{i_1,\ldots,i_k}[i_1, \ldots, i_k] \in \mathcal{B}_1$$

is a linear combination of the generators of the algebra $\mathcal{B}$. We say that $\{i_1, \ldots, i_k\}$ is a *base* for L if $p_{i_1,\ldots,i_k} \neq 0$. By definition, the Chow polytope of L is the convex hull of the points $e_{i_1} + \cdots + e_{i_k}$ for all bases $\{i_1, \ldots, i_k\}$ for L.

The definition of bases can be given in several equivalent forms. For example, let $\mathbf{L} \subset \mathbf{C}^n$ be the vector subspace corresponding to L. Then $\{i_1, \ldots, i_k\}$ is a base for L if and only if the restrictions of the coordinate linear forms $x_{i_1}, \ldots, x_{i_k}$ to $\mathbf{L}$ form a basis of $\mathbf{L}^*$.

The polytope $\mathrm{Ch}(L)$ was introduced in [GGMS] and called the *matroid polytope* of L. Note, in particular, that if L is a coordinate subspace spanned by $e_{i_1}, \ldots, e_{i_k}$ then $R_L = [i_1, \ldots, i_k]$ and the Chow polytope $\mathrm{Ch}(L)$ is just a point.

Examples 3.5. *(a)* If X consists of a single point $x = (x_1 : \ldots : x_n)$ then its associated hypersurface $\mathcal{Z}(X)$ is the hyperplane polar to x, and its Chow polytope is the face $\mathrm{Ch}(x)$ with vertices $\{e_i : x_i \neq 0\}$ of the standard $(n-1)$-simplex Δ^{n-1} in $\mathbf{R}^n$.

(b) The Chow form of a cycle $X = \sum_{i=1}^p m_i X_i$ (with all X_i irreducible) is the product $R_X = R_{X_1}^{m_1} R_{X_2}^{m_2} \cdots R_{X_p}^{m_p}$. Thus its Chow polytope is the Minkowski sum of polytopes $\mathrm{Ch}(X) = m_1\mathrm{Ch}(X_1) + m_2\mathrm{Ch}(X_2) + \cdots + m_p\mathrm{Ch}(X_p)$.

(c) If $X = \{x^{(1)}, x^{(2)}, \ldots, x^{(p)}\}$ is a zero-dimensional variety then its associated hypersurface $\mathcal{Z}(X)$ is the hyperplane arrangement polar to X, and, by (a) and (b), its Chow polytope is the Minkowski sum of simplices $\mathrm{Ch}(X) = \sum_{i=1}^p \mathrm{Ch}(x^{(i)})$. It can be quite non-trivial. Consider, for instance, X to be the system of positive roots of type A_{n-1}, i.e., the (projectivization of) the system of vectors $e_i - e_j$, $1 \le i < j \le n$. Then the Chow polytope of X is the so-called *permutohedron* (or general hypersimplex, see [Milg] [GS]). By definition, this is the convex hull of points $(s(1), \ldots, s(n)) \in \mathbf{R}^n$ where s runs over all the $n!$ permutations of numbers $1, \ldots, n$. We shall encounter this polytope in the next chapter.

Definition 3.6. Let X, Y be two cycles in $G(k, n, d)$. We say that Y is a *toric specialization* of X if Y lies in the closure of the torus orbit HX in $G(k, n, d)$.

For points of a toric variety, the notion of toric specialization was already considered in Section 1D, Chapter 5. As in that case, the relation "Y is a toric specialization of X" is a partial order relation on cycles. In particular, the set of all toric specializations of a given cycle X is a partially ordered set (or *poset*).

Corollary 3.7. *Let X be a cycle in P^{n-1} of dimension $k-1$ and degree d. Then the* *face poset of its Chow polytope* Ch(X) *(with the order relation given by inclusion) is isomorphic to the poset of toric specializations of* X *in* $G(k, d, n)$.

This follows from Proposition 1.8, Chapter 5, since the closure of HX is a toric subvariety in $G(k, d, n)$.

B. Vertices of Chow polytopes

For any subset $\sigma \subset \{1, 2, \ldots, n\}$, let $\mathbf{L}_\sigma$ denote the coordinate subspace of $\mathbf{C}^n$ spanned by $\{e_i : i \in \sigma\}$ and L_σ the projectivization of $\mathbf{L}_\sigma$. The torus $H = (\mathbf{C}^*)^n$ leaves L_σ invariant, and so we have a linear representation $\rho_\sigma : H \to GL(\mathbf{L}_\sigma)$ for each σ. For any subvariety $X \subset P^{n-1}$, we abbreviate $H_X = \{t \in H : tX = X\}$.

Proposition 3.8. *Let* X *be an irreducible subvariety in* P^{n-1}, *and let* $\sigma = \sigma(X)$ *denote the minimal subset of* $\{1, 2, \ldots, n\}$ *such that* $X \subset L_\sigma$.

(a) *For a generic point* $x = (x_1 : \ldots : x_n)$ *in* X, *all the coordinates* x_i, $i \in \sigma$, *are non-zero.*

(b) *The dimension of the Chow polytope* Ch(X) *equals* $\#(\sigma) - \dim \rho_\sigma(H_X)$. *If* $x \in X$ *is a generic point, then* $\dim \mathrm{Ch}(X) = \mathrm{codim}_{L_\sigma}\left(\overline{H_X\, x}\right)$.

Proof. (a) Suppose on the contrary that every point $x = (x_1 : \ldots : x_n) \in X$ lies in a proper subspace L_τ of L_σ. Then $X = \bigcup_\tau (X \cap L_\tau)$ is a union of a finite number of its proper closed subvarieties. But this is impossible, since X is assumed to be irreducible.

(b) We have

$$\dim \mathrm{Ch}(X) = \dim \mathrm{Wt}(R_X) = \dim \overline{H \cdot R_X} = \dim H - \dim H_X.$$

It remains to note that the subgroup $H_\sigma = \mathrm{Ker}(\rho_\sigma) \subset H$ lies in H_X and is equal to the stabilizer in H_X of a vector in $\mathbf{C}^n$ whose projectivization is a generic point $x \in X$. Therefore,

$$\begin{aligned} \dim H - \dim H_X &= \dim \rho_\sigma(H) - \dim \rho_\sigma(H_X) \\ &= \#(\sigma) - \dim \rho_\sigma(H_X) = \mathrm{codim}_{L_\sigma}\left(\overline{H_X\, x}\right) \end{aligned}$$

as claimed.

Corollary 3.9. *The coordinate projective subspaces are the only irreducible subvarieties* $X \subset P^{n-1}$ *with* $\dim \mathrm{Ch}(X) = 0$.

The next theorem follows at once from Corollaries 3.7, 3.9 and Example 3.5.

Theorem 3.10. *For any cycle $X \in G(k,d,n)$, the vertices of the Chow polytope* $\mathrm{Ch}(X)$ *are in a bijective correspondence with toric specializations of X of the form* $\sum_{\#(\sigma)=k} m_\sigma L_\sigma$. *The vertex of* $\mathrm{Ch}(X)$ *corresponding to such a cycle equals*

$$\omega = \sum_\sigma m_\sigma \sum_{i\in\sigma} e_i \in d\Delta(k,n).$$

The corresponding weight component of the Chow form R_X is the bracket monomial

$$R_{X,\omega} = const \cdot \prod_{\#(\sigma)=k} [\sigma]^{m_\sigma}.$$

C. Possible edges of Chow polytopes

We have already remarked (Example 3.4) that, for a projective subspace $L \subset P^{n-1}$, (i.e., for a cycle of degree 1) the Chow polytope is the matroid polytope of L, and the vertices are some of the vertices of the hypersimplex $\Delta(k,n)$. The following was proved in [GGMS].

Theorem 3.11. *If $L \subset P^n$ is a $(k-1)$-dimensional projective subspace, then every edge of its matroid polytope is an edge of the hypersimplex $\Delta(k,n)$. Every edge of $\Delta(k,n)$ is parallel to some vector of the form $e_i - e_j$ where $e_1, \ldots, e_n$ is the standard basis of* $\mathbf{R}^n$.

In fact, it was proved in [GGMS] that subpolytopes in $\Delta(k,n)$ whose vertices and edges are among those of $\Delta(k,n)$ correspond to *matroids* (combinatorial structures axiomatizing properties of linear dependence of a collection of vectors in a vector space). An example of such a polytope is given by a square pyramid forming a half of the octohedron $\Delta(2,4)$.

We are now going to generalize Theorem 3.11 to arbitrary Chow polytopes.

A non-zero vector $a = (a_1, \ldots, a_n) \in \mathbf{Z}^n$ is called *admissible* if $g.c.d.(a_1, \ldots, a_n) = 1$ and $a_1 + a_2 + \cdots + a_n = 0$. For such a vector, put $\sigma_+(a) = \{i : a_i > 0\}$ and $\sigma_-(a) = \{i : a_i < 0\}$. We say that a has *level* $\mathrm{lev}(a) := \#(\sigma_+(a)) + \#(\sigma_-(a)) - 2$ and *degree* $\deg(a) := \sum_{i\in\sigma_+(a)} a_i$ $(= -\sum_{i\in\sigma_-(a)} a_i)$.

The promised generalization of Theorem 3.11 is as follows.

Theorem 3.12. *Each edge of the Chow polytope of any cycle $X \in G(k,d,n)$ is parallel to some admissible vector $a \in \mathbf{Z}^n$ with* $\mathrm{lev}\,(a) \le k-1$ *and* $\deg\,(a) \le d$.

We denote by $A(k,d,n)$ the set of admissible $a \in \mathbf{Z}^n$ with $\mathrm{lev}\,(a) \le k-1$ and $\deg\,(a) \le d$. Before proving the theorem let us give some examples of these sets.

Examples 3.13. *(a)* The set $A(k, 1, n)$ consists of vectors $e_i - e_j$ $(1 \leq i, j \leq n,\ i \neq j)$, i.e., it is the root system of type A_{n-1}.

(b) For $k \geq 2$, the set $A(k, 2, n)$ consists of the following vectors: the roots $e_i - e_j$ $(1 \leq i, j \leq n,\ i \neq j)$ and the vectors $\pm(2e_i - e_j - e_p)$ for distinct $1 \leq i, j, p \leq n$.

D. Proof of Theorem 3.12

With an admissible vector a, we associate the $(n-1)$-dimensional subtorus $H[a] := \{(x_1, \dots, x_n) \in H = (\mathbf{C}^*)^n : \prod_{i=1}^n x_i^{a_i} = 1\}$. Clearly, $H[a]$ is irreducible and contains the one-parameter subgroup $\mathbf{C}^* = \{(t, \dots, t)\} \subset H$. Conversely, each irreducible algebraic subtorus of codimension 1 in H containing $\mathbf{C}^*$ is of this form.

Proposition 3.14. *Let X be a $(k-1)$-dimensional irreducible subvariety in P^{n-1} of degree d such that* $\mathrm{Ch}(X)$ *is an interval. Then X is a toric variety $\overline{H[a]x}$ of codimension 1 in some coordinate k-subspace, where $a \in \mathbf{Z}^n$ is an admissible vector with* $\mathrm{lev}(a) \leq k-1$ *and* $\deg(a) = d$. *The Chow polytope* $\mathrm{Ch}(X)$ *equals the interval $[b, b+a]$ for some $b \in \mathbf{Z}^n$.*

Proof. Let $\sigma = \sigma(X)$ be as in Proposition 3.8, and let x be a generic point of X, so $x_i \neq 0$ for $i \in \sigma$ by Proposition 3.8 (a). By Proposition 3.8 (b), the subgroup H_X is of codimension 1 in H. Consider the irreducible component of the identity $H_X^0 \subset H_X$. Then $H_X^0 = H[a]$ for some admissible $a \in \mathbf{Z}^n$. Since $X \subset L_\sigma$, it follows that $\sigma_+(a) \cup \sigma_-(a) \subset \sigma$, and

$$\dim \overline{H[a]x} = \dim L_\sigma - 1 = \#(\sigma) - 2.$$

Since X is a proper irreducible subvariety of L_σ containing $\overline{H[a]x}$, it follows that $X = \overline{H[a]x}$. Therefore, $\#(\sigma) = k+1$, say $\sigma = \{i_1 < i_2 < \cdots < i_{k+1}\}$, and we see that X is a hypersurface in L_σ. This hypersurface X is defined by the equation

$$\prod_{i \in \sigma_+(a)} x_i^{a_i} - c \cdot \prod_{i \in \sigma_-(a)} x_i^{-a_i} = 0 \tag{3.1}$$

for some $c \in \mathbf{C}^*$. These considerations imply $\mathrm{lev}(a) \leq k-1$ and $\deg(a) = d$. By Example 3.3, the Chow form R_X is obtained from (3.1) by the substitution $x_{i_j} \mapsto (-1)^{j-1}[i_1, i_2, \dots, \widehat{i_j}, \dots, i_{k+1}]$. Thus R_X is the sum of two bracket monomials with weights

$$b_+ = d(e_{i_1} + \cdots + e_{i_{k+1}}) - \sum_{i \in \sigma_+(a)} a_i e_i,$$

$$b_- = d(e_{i_1} + \cdots + e_{i_{k+1}}) + \sum_{i \in \sigma_-(a)} a_i e_i,$$

and the Chow polytope of X is the interval $[b_+, b_-]$. We see that $b_- = b_+ + a$, thereby concluding the proof of Proposition 3.14.

Corollary 3.15. *Let* $X = \sum m_i X_i$ *be a cycle in* $G(k, d, n)$ *such that* $\mathrm{Ch}(X)$ *is an interval. Then this interval has the form* $[b, b+a]$ *for some admissible* $a \in \mathbf{Z}^n$ *with* $\mathrm{lev}(a) = k - 1$ *and* $\deg(a) \leq d$. *Moreover, each irreducible component* X_i *of* X *is either a coordinate* $(k-1)$*-subspace or a toric hypersurface as in (3.1).*

Corollary 3.15 is a direct consequence of Proposition 3.14 and Example 3.5 (b). This corollary immediately implies Theorem 3.12. We also have obtained the following additional information about edges of any Chow polytope $\mathrm{Ch}(X)$.

Corollary 3.16. *The toric specialization of a cycle* $X \in G(k, d, n)$ *corresponding to any edge of* $\mathrm{Ch}(X)$ *is a linear combination of coordinate* $(k-1)$*-subspaces (possibly none) and toric hypersurfaces as in (3.1) (at least one).*

CHAPTER 7

Triangulations and Secondary Polytopes

In this chapter we discuss a combinatorial framework for discriminants and resultants related to toric varieties. The main construction introduces a certain class of polytopes, called *secondary polytopes*, whose vertices correspond to certain triangulations of a given convex polytope. These polytopes will play a crucial role later in the study of the Newton polytopes of discriminants and resultants. The constructions in this chapter are quite elementary.

1. Triangulations and secondary polytopes

A. Triangulations

A *triangulation* of a convex polytope $Q \subset \mathbf{R}^{k-1}$ is a decomposition of Q into a finite number of simplices such that the intersection of any two of these simplices is a common face of them both (maybe empty).

In Figure 22 we give examples of a triangulation and a pattern which we do not allow in our notion of triangulation. Notationally, we regard a triangulation as a collection of its simplices of maximal dimension. All the lower-dimensional simplices are just faces of the maximal ones.

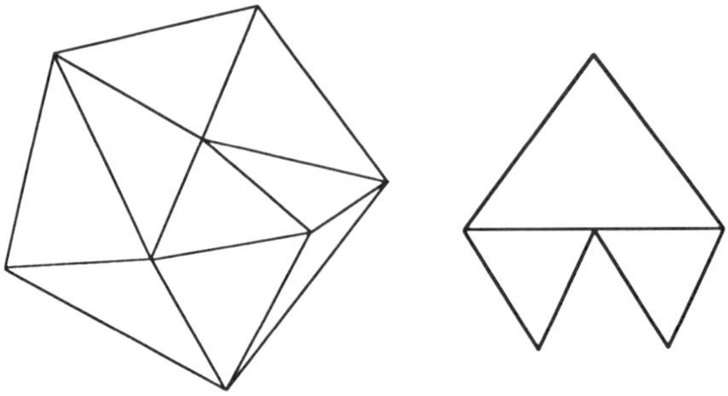

Figure 22.

Throughout this chapter we shall use the notation $\mathrm{Conv}(A)$ for the convex hull of a set A in an affine space.

We are interested in triangulations whose vertices belong to a fixed finite set. Let A be a finite subset of a polytope Q containing all the vertices (i.e., such that $Q = \mathrm{Conv}(A)$). By a *triangulation of* (Q, A), we mean simply a triangulation of

Q into simplices with vertices in A. Note that we do not require every element of A to appear as a vertex of a simplex.

Example 1.1. Here is a simple way to construct a triangulation of (Q, A). Let us suppose that $Q \subset \mathbf{R}^{k-1}$ has full dimension $k - 1$. Take any function $\psi : A \to \mathbf{R}$ and consider, in the space $\mathbf{R}^k = \mathbf{R}^{k-1} \times \mathbf{R}$, the union of vertical half-lines

$$\{(\omega, y) : y \leq \psi(\omega),\ \omega \in A,\ y \in \mathbf{R}\}.$$

Let G_ψ be the convex hull of all these half-lines (Figure 23). This is an unbounded polyhedron projecting onto Q. The faces of G_ψ which do not contain vertical half-lines (i.e., are bounded) form the bounded part of the boundary of G_ψ, which we call the *upper boundary* of G_ψ. Clearly, the upper boundary projects bijectively onto Q.

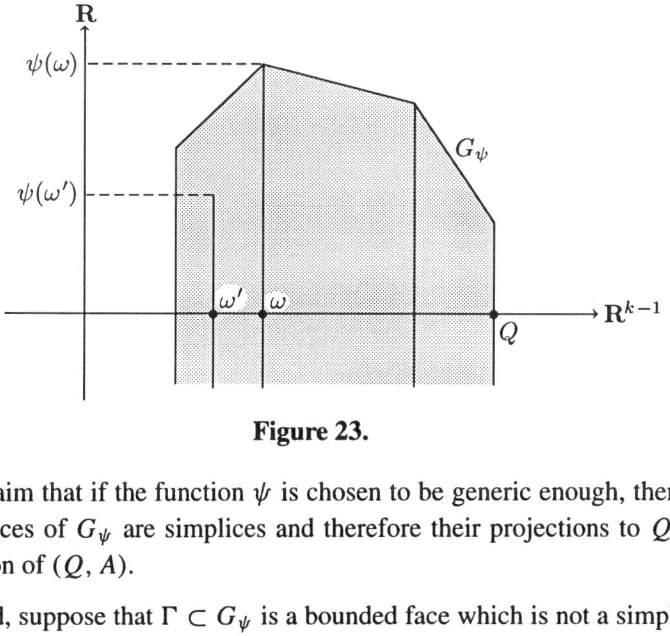

Figure 23.

We claim that if the function ψ is chosen to be generic enough, then all the bounded faces of G_ψ are simplices and therefore their projections to Q form a triangulation of (Q, A).

Indeed, suppose that $\Gamma \subset G_\psi$ is a bounded face which is not a simplex, i.e., Γ contains at least $k + 1$ vertices, say

$$(\omega^{(1)}, \psi(\omega^{(1)})),\ \ldots,\ (\omega^{(k+1)}, \psi(\omega^{(k+1)})) \tag{1.1}$$

where $\omega^{(i)} = (\omega_1^{(i)}, \ldots, \omega_{k-1}^{(i)}) \in \mathbf{R}^{k-1}$. Since the points (1.1) lie on an affine hyperplane in $\mathbf{R}^k$, the determinant

$$\begin{vmatrix} 1 & \omega_1^{(1)} & \ldots & \omega_{k-1}^{(1)} & \psi(\omega^{(1)}) \\ \vdots & \vdots & \vdots & \vdots & \vdots \\ 1 & \omega_1^{(k+1)} & \ldots & \omega_{k-1}^{(k+1)} & \psi(\omega^{(k+1)}) \end{vmatrix} \tag{1.2}$$

vanishes. This determinant is a non-zero linear function in $\psi(\omega^{(i)})$. The vanishing of this function defines a hyperplane in the space $\mathbf{R}^A$ of all functions $A \to \mathbf{R}$. Thus, if we delete the hyperplanes corresponding to all $(k+1)$-element subsets of full rank in A from $\mathbf{R}^A$, then any ψ from the complement will define a triangulation.

This construction will play an important role in the sequel. Note in particular, that we get a simple proof of the following fact: any polytope Q can be triangulated into simplices whose vertices are among the vertices of Q. This fact is not immediately obvious for polytopes of high dimension.

B. Example: triangulations of a circuit

A collection Z of points in an affine space $\mathbf{R}^{k-1}$ is said to form a *circuit* if any proper subset $Z' \subset Z$ is affinely independent (i.e., is the set of vertices of a simplex of some dimension) but Z itself is affinely dependent. (This terminology comes from matroid theory [GGMS] [GS]). Thus, a circuit is obtained by adding just one point to the set of vertices of a simplex (Figure 24).

Figure 24. Circuits

The notion of a circuit uses only affine dependence and hence can be defined in an affine space over any field. Over the field of real numbers, which is the case we consider, a circuit has, besides its cardinality, another invariant called *signature* defined as follows.

Note that up to a real multiple, there is only one real affine relation between elements of Z:

$$\sum_{\omega \in Z} c_\omega \cdot \omega = 0, \quad \sum c_\omega = 0. \tag{1.3}$$

The numbers c_ω are non-zero since any proper subset of Z is affinely independent. Let Z_+ and Z_- be subsets of Z consisting of ω such that c_ω is positive (resp. negative). Clearly, the decomposition of Z into two parts $Z_+ \cup Z_-$ is defined by Z uniquely, up to interchanging of Z_+ and Z_- (which corresponds to the change of signs of all c_ω in (1.3)). The signature of Z is the pair (p, q) where $p = \#(Z_+)$, $q = \#(Z_-)$; it is defined up to interchanging of p and q. We have $p + q = \#(Z)$. This notion is analogous to that of the signature of a real quadratic form.

Proposition 1.2. *Let $Z \subset \mathbf{R}^{k-1}$ be any circuit and $Q = \mathrm{Conv}(Z)$. Then Q has exactly two triangulations with vertices in Z: the triangulation T_+ with simplices $\mathrm{Conv}(Z - \{\omega\})$, $\omega \in Z_+$, and the triangulation T_- with simplices $\mathrm{Conv}(Z - \{\omega\})$, $\omega \in Z_-$.*

Proof. First we show that the $T_\pm$ are indeed triangulations. A point $x \in Q$ can be written as a convex combination

$$x = \sum_{\gamma \in Z} \lambda_\gamma \cdot \gamma, \quad \lambda_\gamma \geq 0, \quad \sum_{\gamma \in Z} \lambda_\gamma = 1.$$

By using (1.3) we can write, for any fixed $\omega \in Z$

$$x = \frac{1}{c_\omega} \sum_{\gamma \neq \omega} (\lambda_\gamma c_\omega - \lambda_\omega c_\gamma)\gamma.$$

It follows that if $\omega \in Z_+$ then the condition that $x \in \mathrm{Conv}(Z - \{\omega\})$ takes the form $\frac{\lambda_\gamma}{c_\gamma} \geq \frac{\lambda_\omega}{c_\omega}$ for all $\gamma \in Z_+$. So each x belongs to at least one simplex $\mathrm{Conv}(Z - \{\omega\})$, $\omega \in Z_+$ and the intersection of any two of these simplices is a common face. In other words, T_+ is a triangulation. For T_- the reasoning is similar.

To see that there are no other triangulations of (Q, A) except $T_\pm$, we observe that every maximal simplex from T_+ and every maximal simplex from T_- have a common point which is interior for both simplices. The proposition is proved.

The triangulations $T_\pm$ for the case of circuits with ≤ 4 vertices are shown in Figure 25.

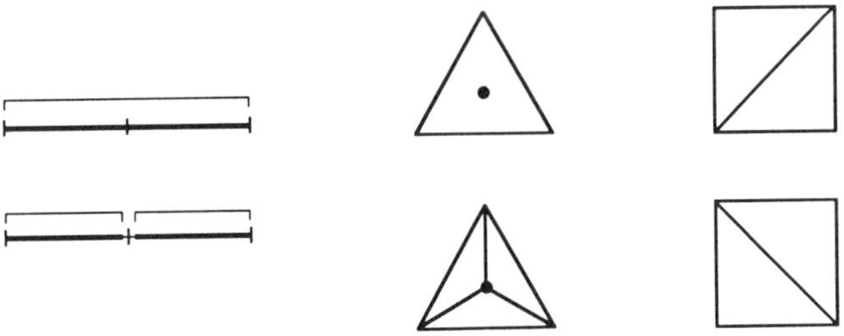

Figure 25. Pairs of triangulations coming from circuits

Proposition 1.2 can be easily extended to the case of a spanning subset $A \subset \mathbf{R}^{k-1}$ of cardinality $k + 1$. Such a subset has the form $A = Z \cup B$ where $Z \subset A$ is a unique circuit and B is such that any point $b \in B$ does not lie in the affine subspace spanned by Z and the remaining points $b' \in B, b' \neq b$ (see Figure 26). Indeed, up to a scalar multiple, there is exactly one affine relation between the elements of A, and Z consists of the points $\omega \in A$ with a non-zero coefficient in this relation .

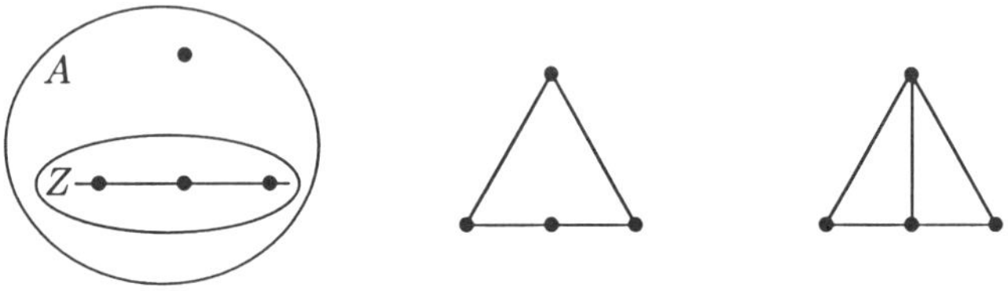

Figure 26. A spanning $(k+1)$-element subset $A \subset \mathbf{R}^{k-1}$ and two triangulations of $\mathrm{Conv}(A)$

The same argument, as in the proof of Proposition 1.2, shows that there are exactly two triangulations of $(\mathrm{Conv}(A), A)$: T_+ with simplices $\mathrm{Conv}(A - \{\omega\})$, $\omega \in Z_+$, and T_- with simplices $\mathrm{Conv}(A - \{\omega\})$, $\omega \in Z_-$. They are obtained from the corresponding two triangulations of $\mathrm{Conv}(Z)$ by taking iterated cones over their simplices.

C. Coherent triangulations

Let A and Q be as above and let T be a triangulation of (Q, A). A continuous function $g : Q \to \mathbf{R}$ will be called *T-piecewise-linear* if it is affine-linear on every simplex of T. Clearly, T-piecewise-linear functions form a vector space.

We call a continuous function $g : Q \to \mathbf{R}$ *concave*, if for any $x, y \in Q$, we have $g(tx+(1-t)y) \geq tg(x)+(1-t)g(y)$, $0 \leq t \leq 1$. By a *domain of linearity* of a concave function $g : Q \to \mathbf{R}$, we mean a subset U in Q such that $g|_U$ is given by some affine-linear function, and which is maximal with this property.

Definition 1.3. A triangulation T of (Q, A) is called *coherent* if there exists a concave T-piecewise-linear function whose domains of linearity are precisely (maximal) simplices of T.

We shall mostly be interested in coherent triangulations*. An example of two non-coherent triangulations is given in Figure 27.

---

* In [GZK3] they were called regular.

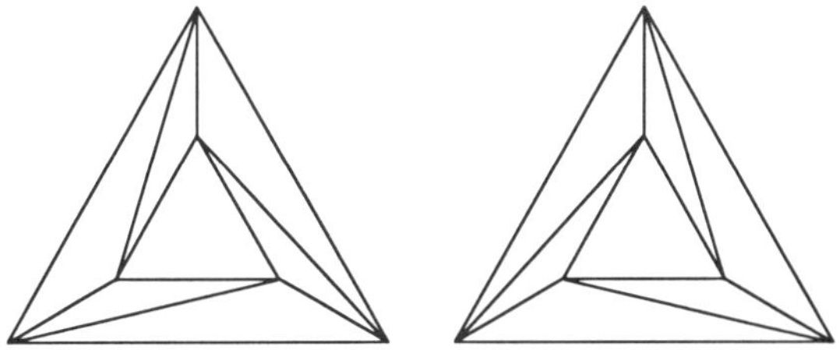

Figure 27. Non-coherent triangulations

The fact that these triangulations are indeed non-coherent can be checked by a straightforward calculation. A proof avoiding calculations may be obtained by using the theory of secondary polytopes developed below (see the remark following Theorem 1.7).

Let T be an arbitrary triangulation of (Q, A), and let $\psi : A \to \mathbf{R}$ be any function. Then there is a unique T-piecewise-linear function $g_{\psi,T} : Q \to \mathbf{R}$ such that, for each $\omega \in A$, which is a vertex of some simplex of T, we have $g_{\psi,T}(\omega) = \psi(\omega)$. The function $g_{\psi,T}$ is obtained by affinely interpolating ψ inside each simplex. Note that the values of ψ at elements of A which are not vertices of any simplex of T do not affect the function $g_{\psi,T}$.

Denote by $\mathbf{R}^A$ the space of all functions $A \to \mathbf{R}$. The correspondence $\psi \mapsto g_{\psi,T}$ defines a surjective linear map from $\mathbf{R}^A$ to the space of T-piecewise-linear functions on Q.

Definition 1.4. Let T be a triangulation of (Q, A). We shall denote by $C(T)$ the cone in $\mathbf{R}^A$ consisting of functions $\psi : A \to \mathbf{R}$ with the following two properties:

(a) The function $g_{\psi,T} : Q \to \mathbf{R}$ is concave.

(b) For any $\omega \in A$ which is not a vertex of any simplex from T, we have $g_{\psi,T}(\omega) \geq \psi(\omega)$.

Clearly, a triangulation T is coherent if and only if the interior of $C(T)$ is non-empty. Moreover, ψ lies in the interior of $C(T)$ if and only if T can be obtained from ψ by the construction of Example 1.1. Thus coherent triangulations are precisely those which can be obtained by this construction.

Proposition 1.5. *Let A and $Q = \text{Conv}(A)$ be fixed. The cones $C(T)$ for all the coherent triangulations of (Q, A) together with all faces of these cones form a complete fan (see Section 4A Chapter 5) in $\mathbf{R}^A$.*

This fan will be called the *secondary fan* of A. This is an example of a fan that does not *a priori* come as the normal fan of some polytope. However, in the next subsection, we shall represent the secondary fan in this form.

Proof of Proposition 1.5. Clearly, the $C(T)$ cover the whole space. So it suffices to show that, for any two coherent triangulations T, T' of (Q, A), the intersection $C(T) \cap C(T')$ is a face of both $C(T)$ and $C(T')$. To see this, note that if $\psi \in \mathbf{R}^A$ belongs to this intersection then the functions $g_{\psi,T}$ and $g_{\psi,T'}$ are both concave. Hence if $\sigma \in T$, $\sigma' \in T'$ are two maximal simplices whose interiors intersect, then $g_{\psi,T}$ and $g_{\psi,T'}$ should be given over $\sigma \cup \sigma'$ by the same affine function. This property is expressed by a set of linear conditions on ψ. This means that $C(T) \cap C(T')$ is the intersection of $C(T)$ with a vector subspace in $\mathbf{R}^A$ as well as the intersection of $C(T')$ with the same vector subspace. Hence it is a face of $C(T)$ and $C(T')$.

The structure of the faces of cones $C(T)$ will be studied in Section 2.

D. Secondary polytopes

As above, let $A \subset \mathbf{R}^{k-1}$ be a finite subset, and $Q = \text{Conv}(A)$. We assume that $\dim(Q) = k - 1$. Fix a translation invariant volume form Vol on $\mathbf{R}^{k-1}$. Let T be a triangulation of (Q, A). By the *characteristic function* of T we shall mean the function $\varphi_T : A \to \mathbf{R}$ defined as follows:

$$\varphi_T(\omega) = \sum_{\sigma:\, \omega \in \text{Vert}(\sigma)} \text{Vol}(\sigma), \tag{1.4}$$

where the summation is over all (maximal) simplices of T for which ω is a vertex. In particular, $\varphi_T(\omega) = 0$ if ω is not a vertex of any simplex of T.

Now we define the main object of study in this chapter.

Definition 1.6. The *secondary polytope* $\Sigma(A)$ is the convex hull in the space $\mathbf{R}^A$ of the vectors φ_T for all the triangulations T of (Q, A).

Our first task will be to find the dimension of $\Sigma(A)$ and to describe which of the functions φ_T will actually be the vertices of $\Sigma(A)$.

To do this, we describe the normal cone to this polytope at every φ_T. Recall that the *normal cone* $N_{\varphi_T}\Sigma(A)$ consists of all linear forms ψ on $\mathbf{R}^A$ such that

$$\psi(\varphi_T) = \max_{\varphi \in \Sigma(A)} \psi(\varphi).$$

The point φ_T is a vertex of $\Sigma(A)$ if and only if the interior of this cone is non-empty. We shall identify $\mathbf{R}^A$ with its dual space by the scalar product

$$(\psi, \varphi) = \sum_{\omega \in A} \psi(\omega)\varphi(\omega).$$

Thus the normal cones lie in $\mathbf{R}^A$.

Theorem 1.7.

(a) *The secondary polytope* $\Sigma(A)$ *has dimension* $n-k$ *where* $n = \#(A)$.

(b) *Vertices of* $\Sigma(A)$ *are precisely the characteristic functions* φ_T *for all coherent triangulations* T *of* (Q, A). *If* T *is a coherent triangulation of* (Q, A) *then* $\varphi_T \neq \varphi_{T'}$ *for any other triangulation* T' *of* (Q, A).

(c) *For any triangulation* T *(coherent or not) the normal cone* $N_{\varphi_T}\Sigma(A)$ *coincides with the cone* $C(T) \subset \mathbf{R}^A$ *introduced in Definition 1.4.*

Note that, for non-coherent triangulations T, the correspondence $T \mapsto \varphi_T$ may not be injective. For example, it is easy to see that $\varphi_T = \varphi_{T'}$ where T, T' are two triangulations in Figure 27. Thus, the fact that they are non-coherent follows from Theorem 1.7 (b).

The proof of Theorem 1.7 will be based on the following two lemmas.

Lemma 1.8. *For every triangulation* T *of* (Q, A) *and every* $\psi \in \mathbf{R}^A$, *we have*

$$(\psi, \varphi_T) = k \int_Q g_{\psi,T}(x)dx. \tag{1.5}$$

Proof. This follows at once from the definitions of φ_T and $g_{\psi,T}$, and the following obvious fact: the integral of an affine-linear function g over a simplex σ is equal to the arithmetic mean of values of g at the vertices of σ times the volume of σ.

Fix $\psi \in \mathbf{R}^A$. We shall evaluate explicitly the maximum $\max_{\varphi\in\Sigma(A)}(\psi, \varphi)$. Namely, as in Example 1.1, let $G_\psi \subset \mathbf{R}^k = \mathbf{R}^{k-1} \times \mathbf{R}$ be the convex hull of all the vertical half-lines $\{(\omega, y) : y \leq \psi(\omega),\ \omega \in A,\ y \in \mathbf{R}\}$. The upper boundary of G_ψ can be regarded as the graph of a piecewise-linear function $g_\psi : Q \to \mathbf{R}$. Explicitly,

$$g_\psi(x) = \max\{y : (x, y) \in G_\psi\}. \tag{1.6}$$

Lemma 1.9.

(a) *The function* g_ψ *is concave.*

(b) *For any triangulation* T *of* (Q, A) *we have* $g_\psi(x) \geq g_{\psi,T}(x)$, $\forall x \in Q$ *where* $g_{\psi,T}$ *was defined in subsection C.*

(c) *We have*

$$\max_{\varphi\in\Sigma(A)}(\psi, \varphi) = k \int_Q g_\psi(x)dx. \tag{1.7}$$

Proof. Part (a) follows by construction. To verify the inequality in (b), it suffices to consider x varying in some fixed simplex σ of T. By definition, $g_{\psi,T}$ is affine-linear over σ and $g_\psi(\omega) \geq \psi(\omega) = g_{\psi,T}(\omega)$ for any vertex $\omega \in \sigma$. So the inequality is valid over σ.

Let us prove (c). The maximum in (1.7) can be taken over the set of the φ_T for all triangulations T of (Q, A), since $\Sigma(A)$ is defined as the convex hull of these φ_T. Hence part (b) together with Lemma 1.8 imply that the left hand side of (1.7) is greater than or equal to the right hand side. To show the equality, it suffices to exhibit a triangulation T for which $g_\psi = g_{\psi,T}$.

To do this, we consider the projections of the bounded faces of the polyhedron G_ψ into Q. These are polytopes with vertices in A. Take a generic ψ' close to ψ. Then the bounded faces of the polyhedron $G_{\psi'}$ (see the construction of Example 1.1) give a triangulation T of (Q, A) which induces a triangulation of each of the above polytopes. Hence g_ψ is T-piecewise-linear and coincides with $g_{\psi,T}$. This proves Lemma 1.9.

Now we can prove Theorem 1.7. Part (c) of this theorem follows at once from Lemmas 1.8 and 1.9. Part (b) follows from part (c) and the fact that the $C(T)$ for coherent T are all distinct and cover the whole $\mathbf{R}^A$ (see Proposition 1.5). To prove part (a), notice that all the cones $C(T) \subset \mathbf{R}^A$ contain the same k-dimensional vector subspace. This subspace consists of functions $\psi : A \to \mathbf{R}$, which are restrictions to A of global affine-linear functions $g : \mathbf{R}^{k-1} \to \mathbf{R}$. Clearly, this is the maximal vector subspace in $\mathbf{R}^A$ common to all $C(T)$. This implies that the codimension of $\Sigma(A)$ in $\mathbf{R}^A$ is equal to k. Theorem 1.7 is completely proved.

Corollary 1.10. *Let $A \subset \mathbf{R}^{k-1}$ be a finite subset affinely spanning $\mathbf{R}^{k-1}$. Then $\Sigma(A)$ has dimension 1, i.e., it is an interval if and only if $\#(A) = k+1$.*

In the situation of Corollary 1.10, the vertices of an interval $\Sigma(A)$ correspond to the triangulations $T_\pm$ constructed in Section B. This shows, in particular, that these two triangulations are always coherent.

Proposition 1.11. *Suppose that A affinely spans $\mathbf{R}^{k-1}$. The affine span of the polytope $\Sigma(A)$ is equal to*

$$\left\{\varphi \in \mathbf{R}^A : \sum_{\omega\in A}\varphi(\omega) = k\cdot \mathrm{Vol}(Q),\ \sum_{\omega\in A}\varphi(\omega)\cdot\omega = k\cdot\int_Q x\,dx\right\}. \tag{1.8}$$

Proof. We know that $\dim(\Sigma(A)) = \#(A) - k$. The equalities (1.8) define an affine subspace of the same dimension. Hence it is enough to prove that these equalities are satisfied on $\Sigma(A)$. In the proof we can assume that $\varphi = \varphi_T$ for a triangulation T of (Q, A).

Now, the first equality is obvious: the volume of each simplex σ of T will be counted in $\sum_{\omega\in A}\varphi_T(\omega)$ exactly k times, one for each vertex of σ. The second equality follows from Lemma 1.8 applied to each component of the vector-valued linear function $x \mapsto x$, $Q \to \mathbf{R}^{k-1}$.

E. The secondary polytope as the Minkowski integral

Now we shall discuss an "integral representation" of the secondary polytope $\Sigma(A)$ (due to Billera and Sturmfels [BS]).

Let us number the elements of A, say $A = \{\omega^{(1)}, \dots, \omega^{(n)}\}$. Consider the standard $(n-1)$-dimensional simplex $\Delta = \Delta^{n-1} \subset \mathbf{R}^A$, i.e., the convex hull of the basis vectors $e_1, \dots, e_n$. Vertices of this simplex correspond, via our numeration, to elements of A. Let $\pi : \Delta \to Q$ denote the affine projection sending each vertex e_i of Δ to the corresponding element $\omega^{(i)} \in A$.

If a polytope Q is given, then a set A such that $Q = \mathrm{Conv}(A)$ can be regarded as a "set of generators" of Q. The simplex Δ, whose vertices are in bijection with elements of A, can be regarded as the "free polytope" on A. From this point of view it is natural to regard the fibers of π as "polytopes of relations" between elements of A which measure the difference between Q and Δ.

For any $x \in Q$, consider the fiber $\pi^{-1}(x)$ of the projection π. If x is an interior point of Q then $\pi^{-1}(x)$ is a convex polytope of dimension $n-k$, i.e., of the same dimension as $\Sigma(A)$. For different points x the structure of $\pi^{-1}(x)$ can be quite different so it is not clear which particular fiber to choose as the "polytope of relations."

It turns out that $\Sigma(A)$ can be defined as a certain average of all of these fibers. More precisely, we call a *section* of π a map $\gamma : Q \to \Delta$ such that $\gamma(x) \in \pi^{-1}(x)$ for $x \in Q$. Consider the set $\int_Q \pi^{-1}(x)dx \subset \mathbf{R}^n$ of all vector integrals $\int_Q \gamma(x)dx$, where γ runs over all Borel measurable sections of π. This set can be called the *Minkowski integral* of the family of polytopes $\pi^{-1}(x)$, $x \in Q$ in complete analogy with the notion of the Minkowski sum of a finite number of polytopes. It follows from the general results of [Au] that $\int_Q \pi^{-1}(x)dx$ is a compact convex subset in $\mathbf{R}^n$.

Theorem 1.12. *Let us identify $\mathbf{R}^A$ with $\mathbf{R}^n$ by the chosen numeration of A. Then $\Sigma(A) = k \cdot \int_Q \pi^{-1}(x)dx$, i.e., the secondary polytope is equal to k times dilated the Minkowski integral of all the fibers of $\pi : \Delta \to Q$.*

Before proving Theorem 1.12, let us exhibit some particular points of $\int_Q \pi^{-1}(x)dx$, i.e., integrals of some particular sections. Let T be any triangulation of (Q, A). For any simplex σ of T consider the corresponding sub-simplex $\tilde{\sigma} \subset \Delta$ whose vertices are the points e_i such that the corresponding $\omega^{(i)} \in A$ are vertices of σ. Clearly, $\pi : \tilde{\sigma} \to \sigma$ is a bijection. Let $\tilde{T} \subset \Delta$ be the simplicial subcomplex in Δ defined as the union of $\tilde{\sigma}$, $\sigma \in T$. This subcomplex bijectively projects onto Q, i.e., it is the graph of some continuous section $\gamma_T : Q \to \Delta$.

Lemma 1.13. *For any T as before, we have the equality of vectors in $\mathbf{R}^A = \mathbf{R}^n$*

$$\varphi_T = k \cdot \int_Q \gamma_T(x)dx,$$

where φ_T is defined by (1.4).

Proof. This is a reformulation of Lemma 1.8. More precisely, we apply an arbitrary linear functional ψ to both parts of the required equality and then use Lemma 1.8.

Proof of Theorem 1.12. The set $\int_Q \pi^{-1}(x)dx$, being a compact convex subset of $\mathbf{R}^n$, is uniquely determined by its *support function*

$$S(\psi) = \max_{\varphi \in \int_Q \pi^{-1}(x)dx} \psi(\varphi).$$

Taking into account (1.7), we have only to show that

$$S(\psi) = \int_Q g_\psi(x)dx \tag{1.9}$$

(we recall that g_ψ is defined by (1.6)).

Let G'_ψ be the convex hull in $\mathbf{R}^k = \mathbf{R}^{k-1} \times \mathbf{R}$ of the set $\{(\omega, \psi(\omega)),\ \omega \in A\}$. This is a convex polytope and its upper boundary coincides with the upper boundary of the polyhedron G_ψ used to define g_ψ. In other words, $g_\psi(x) = \max\{y : (x, y) \in G'_\psi\}$.

In the next lemma we regard $\psi \in \mathbf{R}^n$ as a linear functional on $\mathbf{R}^n$ using the standard scalar product.

Lemma 1.14. Let $x \in Q$ and $\psi \in \mathbf{R}^n$. A real number y belongs to $\psi(\pi^{-1}(x))$ if and only if $(x, y) \in G'_\psi$.

Proof of Lemma 1.14. By definition, the fiber $\pi^{-1}(x)$ is a convex subpolytope of Δ whose points $(\lambda_1, \dots, \lambda_n)$ correspond to various convex representations $x = \sum_i \lambda_i \omega^{(i)}$, where $\lambda_i \geq 0$, $\sum_i \lambda_i = 1$. Therefore, if $y = \psi(\lambda) = \sum_i \psi_i \lambda_i$ then $(x, y) = \sum_i \lambda_i (\omega^{(i)}, \psi_i)$ is the corresponding convex combination of the points $(\omega^{(i)}, \psi_i)$. This implies our statement.

Theorem 1.12 is an immediate consequence of Lemmas 1.13 and 1.14.

Note that the Minkowski integral of fibers makes sense in a more general situation, for an arbitrary projection $\pi : P \to Q$ of two convex polytopes (in the above construction P was a simplex). This leads to a concept of *fiber polytopes* introduced and studied by Billera and Sturmfels in [BS]. This will not be discussed here because the secondary polytopes provide all the combinatorial framework for this book.

F. Secondary polytopes and the Gale transform

Following [BFS], we shall describe the coherent triangulations of (Q, A) and hence vertices of $\Sigma(A)$ in terms of the so-called Gale transform of A. Consider the space $L_A \subset \mathbf{R}^A$ of affine relations between elements of A, i.e.,

$$L_A = \{(\lambda_\omega) \in \mathbf{R}^A : \sum_{\omega \in A} \lambda_\omega \cdot \omega = 0, \ \sum_{\omega \in A} \lambda_\omega = 0\}. \tag{1.10}$$

By Proposition 1.11, $\Sigma(A)$ lies in an affine subspace parallel to L_A. Let the $b_\omega \in L_A^*$ be the coordinate linear forms on L_A. The family of vectors $B = \{b_\omega : \omega \in A\}$ in L_A^* is called the *Gale transform* of A. The following property of the Gale transform is crucial.

Lemma 1.15. *A linear combination $\sum_{\omega \in A} \psi_\omega \cdot b_\omega$ is equal to 0 if and only if there is an affine-linear function g on $\mathbf{R}^{k-1}$ such that $\psi_\omega = g(\omega)$ for all $\omega \in A$.*

Proof. Let the e_ω be standard basis vectors in $\mathbf{R}^A$. Let $\pi : \mathbf{R}^A \to \mathbf{R}^k = \mathbf{R}^{k-1} \times \mathbf{R}$ be the linear map sending e_ω to $(\omega, 1)$. By definition, $\sum_{\omega \in A} \psi_\omega \cdot b_\omega = 0$ if and only if the linear form ψ on $\mathbf{R}^A$ is orthogonal to $\operatorname{Ker} \pi$. Since π is onto, this is equivalent to the fact that $\psi = \tilde{g} \circ \pi$ for some linear form $\tilde{g}$ on $\mathbf{R}^k$. Setting $g(\omega) = \tilde{g}(\omega, 1)$ we obtain our statement.

Lemma 1.15 implies the following lemma whose proof we leave to the reader.

Lemma 1.16.

(a) *Let I be a subset of A. The forms b_ω for $\omega \in I$ form a basis in L_A^* if and only if the set $A - I$ is affinely independent, i.e., it consists of vertices of some $(k-1)$-dimensional simplex in Q.*

(b) *The convex hull of B contains the origin $0 \in L_A^*$ in its interior.*

Now let $\sigma \subset Q$ be any $(k-1)$-dimensional simplex with vertices in A. Denote by C_σ the convex cone (octant) in L_A^* spanned by the b_ω for all $\omega \in A$ which are not among the vertices of σ. We say that $v \in L_A^*$ is *generic* if v does not belong to the boundary of any of the cones C_σ. We call the connected components of the set of all generic vectors $v \in L_A^*$ *dual chambers*. Clearly, every dual chamber Γ is an open cone in L_A^*.

Denote by $p : \mathbf{R}^A \to L_A^*$ the projection dual to the embedding $L_A \subset \mathbf{R}^A$ with respect to the standard scalar product on $\mathbf{R}^A$.

Theorem 1.17. *There is a bijective correspondence between dual chambers and coherent triangulations of (Q, A) defined as follows. For any coherent triangulation T the corresponding dual chamber has the form $(-1) \cdot p(C^0(T))$ where $C^0(T)$ is the interior of $C(T) \subset \mathbf{R}^A$ introduced in Definition 1.4. This chamber coincides with the intersection $\bigcap_{\sigma \in T} C_\sigma$.*

Proof. Let σ be a $(k-1)$- simplex with vertices in A. Let us describe the inverse image $p^{-1}C_\sigma \subset \mathbf{R}^A$. For $\psi \in \mathbf{R}^A$ we consider the affine-linear function $g_{\psi,\sigma} : \mathbf{R}^{k-1} \to \mathbf{R}$ defined by the condition that $g_{\psi,\sigma}(\omega) = \psi(\omega)$ for any vertex ω of σ.

Lemma 1.18. *A vector $\psi \in \mathbf{R}^A$ belongs to $p^{-1}C_\sigma$ if and only if we have $g_{\psi,\sigma}(\omega) \leq \psi(\omega)$ for any $\omega \in A$ which is not a vertex of σ.*

Proof. Every $\omega \in A$ which is not a vertex of σ is uniquely represented as an affine combination

$$\omega = \sum_{\gamma \in \sigma} \mu_\gamma \cdot \gamma.$$

This means that the vector $\lambda \in \mathbf{R}^A$ with components $\lambda_\omega = 1$, $\lambda_\gamma = -\mu_\gamma$ for $\gamma \in \sigma$, and all other components equal to 0 belongs to L_A. Tracing the definitions, we see that the condition that $p(\psi) \in C_\sigma$ is equivalent to $(\psi, \lambda) \geq 0$ for all λ as above. But the inequality $(\psi, \lambda) \geq 0$ can be rewritten as

$$\psi(\omega) \geq \sum_{\gamma \in \sigma} \mu_\gamma \psi(\gamma) = g_{\psi,\sigma}(\omega),$$

and we are done.

Now the proof of Theorem 1.17 is completed as follows. By Definition 1.4, $\psi \in C(T)$ if and only if the piecewise-linear function $g_{\psi,T}$ is concave and $g_{\psi,T}(\omega) \geq \psi(\omega)$ for all $\omega \in A$ which are not vertices of simplices from T. But $g_{\psi,T}$ restricted to any simplex $\sigma \in T$ coincides with the globally affine function $g_{\psi,\sigma}$ so our statement follows from the definition of concavity and Lemma 1.18. (The minus sign in Theorem 1.17 appears because we have to reverse the inequality in Lemma 1.18).

Remark 1.19. Suppose that A lies in $\mathbf{Z}^{k-1}$. The representation of coherent triangulations T and the corresponding cones $C(T)$ by the dual cones C_σ was used in [GZK1] to construct a complete system of solutions of the *A-hypergeometric system*. This is a certain system of linear differential equations on the space $\mathbf{C}^A$. For any simplex σ with vertices in A, there are associated as many as $\mathrm{Vol}(\sigma)$ solutions of the system given by explicit power series in variables a_ω, $\omega \in A$ (here the volume is normalized as in Section 3C Chapter 5). These series converge as long as the vector $(\log|a_\omega|) \in \mathbf{R}^A$ belongs to a certain translation of the cone $p^{-1}C_\sigma$, where $p : \mathbf{R}^A \to L_A^*$ is as before. It was shown in [GZK1] that when σ runs over simplices of some coherent triangulation of (Q, A), all the corresponding series have a common domain of convergence.

2. Faces of the secondary polytope

In Section 1 we introduced, for any finite subset $A \subset \mathbf{R}^{k-1}$, the secondary polytope $\Sigma(A)$. Vertices of $\Sigma(A)$ correspond to the so-called coherent triangulations of the polytope $Q = \mathrm{Conv}(A)$. In this section we describe the faces of $\Sigma(A)$ of arbitrary dimension. They correspond to certain subdivisions of Q into convex polytopes.

A. Polyhedral subdivisions

For technical reasons it is convenient to introduce the notion of a marked polytope. By definition, a *marked polytope* is a pair (Q, A) where $Q \subset \mathbf{R}^{k-1}$ is a convex polytope and $A \subset Q$ is a finite subset containing all the vertices, so $Q = \mathrm{Conv}(A)$. Although Q is determined by A, it is convenient to keep it as a piece of notation.

Definition 2.1. Let (Q, A) be a marked polytope, $Q \subset \mathbf{R}^{k-1}$, $\dim(Q) = k-1$. A *(polyhedral) subdivision* of (Q, A) is a family $S = \{(Q_i, A_i) : i \in I\}$ of marked polytopes such that

(a) each A_i is a subset of A and $\dim(Q_i) = k - 1$;

(b) any intersection $Q_i \cap Q_j$ is a face (possibly empty) of both Q_i and Q_j, and

$$A_i \cap (Q_i \cap Q_j) = A_j \cap (Q_i \cap Q_j);$$

(c) the union of all Q_i coincides with Q.

For example, a triangulation of (Q, A) is a particular case of a subdivision: we take the Q_i to be the simplices of the triangulation and A_i to be the set of vertices of Q_i.

We shall reserve the name triangulation only for subdivisions of the form described above. So, if in a subdivision $S = (Q_i, A_i)$ each Q_i is a simplex but, for some i, the set A_i contains other elements besides vertices of Q_i then S is not referred to as a triangulation.

Let S, S' be two subdivisions of (Q, A), say $S = \{(Q_i, A_i)\}$, $S' = \{(Q'_j, A'_j)\}$. We shall say that S *refines* S' if, for each j, the collection of (Q_i, A_i) such that $Q_i \subset Q'_j$ forms a subdivision of (Q'_j, A'_j). This makes the set of all subdivisions of (Q, A) into a poset. Triangulations are precisely minimal elements of this poset. The maximal element is the subdivision consisting of (Q, A) itself.

Example 2.2. Here is a simple way to construct subdivisions of (Q, A) generalizing the construction of triangulations in Example 1.1. Take any function $\psi : A \to \mathbf{R}$ and let

$$G_\psi = \mathrm{Conv}\{(\omega, y) : y \le \psi(\omega),\ \omega \in A \subset \mathbf{R}^{k-1},\ y \in \mathbf{R}\}$$

be the polyhedron in $\mathbf{R}^k = \mathbf{R}^{k-1} \times \mathbf{R}$ introduced in the cited example. Consider the upper boundary of G_ψ as the graph of a piecewise-linear function $g_\psi : Q \to \mathbf{R}$, so

$$g_\psi(x) = \max\{y : (x, y) \in G_\psi\}.$$

Let the $Q_i \subset Q$, $i \in I$ be the projections of the bounded faces of G_ψ of codimension 1, i.e., the domains of linearity of g_ψ. Let A_i consist of all $\omega \in A \cap Q_i$ such that $g_\psi(\omega) = \psi(\omega)$ (i.e., the point $(\omega, \psi(\omega))$ lies on the boundary of G_ψ). Then $\{(Q_i, A_i)\}$, $i \in I$ forms a subdivision of (Q, A). Denote this subdivision by $S(\psi)$. If ψ is generic, as in Example 1.1, then $S(\psi)$ is a triangulation.

Definition 2.3. A subdivision S of (Q, A) is called *coherent* if it has the form $S(\psi)$ for some $\psi \in \mathbf{R}^A$.

Note that this definition agrees with Definition 1.3 of coherent triangulations (see the remark following Definition 1.4).

Now we have all the ingredients to describe the faces of $\Sigma(A)$. For any coherent subdivision S of (Q, A), denote by $F(S)$ the convex hull in $\mathbf{R}^A$ of the characteristic functions φ_T (see (1.4)) for all triangulations T refining S. Denote also by $C(S)$ the convex cone in $\mathbf{R}^A$ consisting of ψ such that S is a refinement of $S(\psi)$ (again, this agrees with Definition 1.4 when S is a triangulation).

Theorem 2.4. *The faces of $\Sigma(A)$ are precisely the subpolytopes $F(S)$ for all coherent subdivisions S of (Q, A). In addition, $F(S) \subset F(S')$ if and only if S refines S'. The normal cone to $F(S)$ coincides with $C(S)$.*

Proof. Let $\psi \in \mathbf{R}^A$. As usual, we regard ψ as a linear form on $\mathbf{R}^A$ by the scalar product $\sum_{\omega \in A} \psi(\omega)\varphi(\omega)$. Let $\Sigma(A)^\psi$ be the supporting face of $\Sigma(A)$ (see Section 4B, Chapter 5) consisting of all points where ψ attains its maximal value. Clearly all faces of $\Sigma(A)$ have the form $\Sigma(A)^\psi$ for some ψ. By Lemma 1.9 and by definition of $S(\psi)$, we have

$$\Sigma(A)^\psi = F(S(\psi)). \tag{2.1}$$

This implies that the correspondence $S \mapsto F(S)$ is a bijection between coherent subdivisions and the faces of $\Sigma(A)$.

Let us prove the statement about the normal cone. Fix some coherent subdivision S. Let $C(S)^0 = \{\psi : S = S(\psi)\}$. By (2.1), the normal cone $N_{F(S)}\Sigma(A)$ coincides with the closure of $C(S)^0$. It remains to show that $\overline{C(S)^0} = C(S)$. It is easy to see that $C(S)$ is a closed set. Indeed, for ψ to lie in $C(S)$, it is necessary and sufficient that the following conditions hold. First, the restriction of the function g_ψ to any Q_i is affine linear. Second, for any i and any $\omega \in A_i$, we have

$g_\psi(\omega) = \psi(\omega)$. Third, the function g_ψ is concave. Fourth, for $\omega \notin \bigcup A_i$ we have $g_\psi(\omega) \geq \psi(\omega)$. Each of these conditions obviously specifies a closed set, hence $C(S)$ is closed. Thus $\overline{C(S)^0} \subset C(S)$. To prove the reverse inclusion it is enough to show that for generic $\psi \in C(S)$ we have $S(\psi) = S$. This follows by the definition of coherence.

It remains to show that $F(S) \subset F(S')$ if and only if S refines S'. But this follows from the description of the normal cones in (2.1). Theorem 2.4 is proved.

We illustrate Theorem 2.4 by the following example.

Example 2.5. Let Q be a square in the plane, and A be the set of five points consisting of four vertices and the center of Q. The secondary polytope $\Sigma(A)$ is the triangle depicted in Figure 28. The coherent subdivisions of (Q, A) are shown near the corresponding faces of $\Sigma(A)$, the bold points are the points belonging to the subsets A_i of a subdivision.

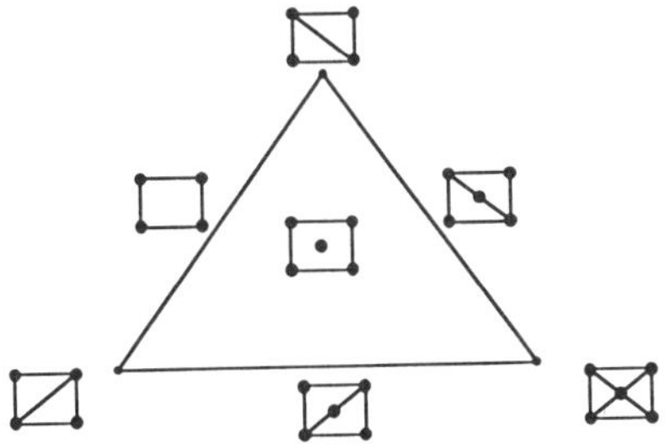

Figure 28. Secondary polytope of a 5 point configuration

Let us find the dimensions of the faces of $\Sigma(A)$. This is tantamount to finding the dimensions of the normal cones $C(S)$ to the faces. Let $S = \{(Q_i, A_i)\}$ be a subdivision of (Q, A). Let $\Lambda(S)$ be the linear subspace in $\mathbf{R}^A$ consisting of functions $\psi : A \to \mathbf{R}$ with the following property:

> There exists a continuous function $g : Q \to \mathbf{R}$ which is affine-linear on each Q_i, and $g(\omega) = \psi(\omega)$ for any ω in any A_i.

For any S, the space $\Lambda(S)$ contains the k-dimensional subspace $\Lambda \subset \mathbf{R}^A$ given by the restrictions to A of globally affine-linear functions. This and the definition of the $C(S)$ readily implies the following.

Corollary 2.6. *The affine span of $C(S)$ coincides with $\Lambda(S)$. Hence the codimension of the face $F(S)$ equals* $\dim(\Lambda(S)) - k$.

This statement can be put into the dual form. Consider again the space L_A of affine relations among elements of A, introduced in (1.10). Let $S = \{(Q_i, A_i)\}$ be a subdivision of (Q, A). For any i, let

$$L_{A_i} = \{\lambda = (\lambda_\omega) \in L_A : \lambda_\omega = 0 \text{ for } \omega \notin A_i\}$$

be the space of affine relations among elements of A_i. Let L_S be the sum $\sum_i L_{A_i}$. The description of normal cones in Theorem 2.4 implies the following refinement of Proposition 1.11.

Corollary 2.7. *For any coherent subdivision S of (Q, A), the affine span of the corresponding face $F(S) \subset \Sigma(A)$ is a parallel translation of L_S, and hence* $\dim(F(S)) = \dim(L_S)$.

B. Facets of $\Sigma(A)$

According to Theorem 2.4, the facets (= faces of codimension 1) of $\Sigma(A)$ correspond to maximal nontrivial coherent subdivisions of (Q, A). We call these subdivisions *coarse*. It seems that they do not have a good general description. We shall describe two special classes of coarse subdivisions.

(1) Suppose $\omega \in A$ is not a vertex of Q. Then the collection S consisting of one marked polytope $(Q, A - \{\omega\})$ is a coarse subdivision of (Q, A). The corresponding facet $F(S)$ of $\Sigma(A)$ is the intersection of $\Sigma(A)$ with the hyperplane in $\mathbf{R}^A$

$$\{\varphi : A \to \mathbf{R} : \varphi(\omega) = 0\}. \tag{2.2}$$

(2) Choose an affine-linear function g on Q. Let

$$Q_+ = \{x \in Q : g(x) \geq 0\}, \quad Q_- = \{x \in Q : g(x) \leq 0\}.$$

Suppose that Q_+ and Q_- are polytopes of full dimension $k - 1$ having all vertices in A. Then taking S to consist of two marked polytopes $(Q_\pm, A \cap Q_\pm)$, we get a coarse subdivision of (Q, A). The corresponding facet $F(S)$ of $\Sigma(A)$ is the intersection of $\Sigma(A)$ with the hyperplane in $\mathbf{R}^A$ consisting of all $\varphi : A \to \mathbf{R}$ such that

$$\sum_{\omega \in Q_+} g(\omega)\varphi(\omega) = k \int_{Q_+} g(x)dx. \tag{2.3}$$

In general, these two types of facets do not exhaust all possible facets of the secondary polytope. In Section 3 we shall give an example of a coarse subdivision of the product of two simplices $\Delta^{m-1} \times \Delta^{m-1}$ which does not belong to either of the above two categories.

C. Edges of $\Sigma(A)$

The vertices of $\Sigma(A)$ correspond to (coherent) triangulations of (Q, A). It is natural to ask when two vertices $\varphi_T, \varphi_{T'}$ corresponding to triangulations T, T' are joined by an edge. Intuitively, one would expect T and T' to differ "as little as possible." A complete description of such "neighboring pairs" of triangulations will be given a bit later. We start with a discussion of the most typical situation.

As above, we assume that $Q = \text{Conv}(A) \subset \mathbf{R}^{k-1}$ is a polytope of full dimension $k-1$. Let T be a triangulation of (Q, A). Suppose that there is a circuit $Z \subset A$ (see Section 1B) of cardinality $k+1$ such that its convex hull $\text{Conv}(Z)$ is a union of simplices of T with vertices in Z. We have seen that $\text{Conv}(Z)$ has exactly two triangulations with vertices in Z, say, T_+ and T_-. The triangulation induced by T must be one of these two, say, T_+. Let us define a new triangulation T' of (Q, A) by replacing T_+ with T_- and leaving the rest of the simplices of T intact (see Figure 29). This new triangulation T' will be called the *modification* of T along Z and denoted by $s_Z(T)$.

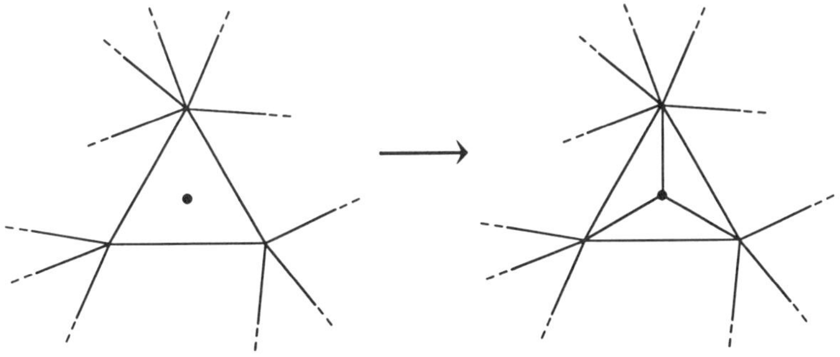

Figure 29. Modification of a triangulation

There is a natural analogy between modifications of triangulations and modifications of smooth manifolds, i.e., gluing handles or surgery [Miln1]. A modification of an n-dimensional manifold consists in taking away an embedded submanifold (with boundary) of the form $S^p \times B^q$ (where S is the sphere, B is the ball, and $p + q = n$) and in its place gluing the manifold $B^{p+1} \times S^{q-1}$. A numerical characteristic of such a modification is the partition of n into the sum of p and q. Similarly, one characteristic of a modification of a triangulation is the partition $\#(Z) = \#(Z_+) + \#(Z_-)$. In Section 4, Chapter 11 we shall establish a direct relation between these two types of modifications.

In general, even when T is coherent, the modified triangulation $s_Z(T)$ may be non-coherent. However, we have the following.

Proposition 2.8. *Let a triangulation T and a circuit Z be as above and let $T' = s_Z(T)$. If both T and T' are coherent then the corresponding vertices $\varphi_T, \varphi_{T'}$ of $\Sigma(A)$ are joined by an edge.*

Proof. Let S be the polyhedral subdivision of (Q, A) consisting of the marked polytope $(\mathrm{Conv}(Z), Z)$ and all the simplices common to T and T'. By our assumption, S is coherent. Let $F(S)$ be the face of $\Sigma(A)$ corresponding to S. By Corollary 2.7,

$$\dim F(S) = \dim L_Z = 1,$$

since there is a unique affine relation among elements of Z.

Now we can proceed to describe all edges of the secondary polytope. In addition to the modifications described above, there are some "degenerate" ones in which the circuit Z has smaller cardinality.

Definition 2.9. Let T be a triangulation of (Q, A), and let $Z \subset A$ be a circuit. We say that T is *supported on* Z if the following conditions hold:

(a) There are no vertices of T inside $\mathrm{Conv}(Z)$ except for the elements of Z itself.
(b) The polytope $\mathrm{Conv}(Z)$ is a union of the faces of the simplices of T.
(c) Let $\mathrm{Conv}(I)$ and $\mathrm{Conv}(I')$ be two simplices (of maximal dimension) of one of the two possible triangulations of $\mathrm{Conv}(Z)$. Then, for every subset $F \subset A - Z$, the simplex $\mathrm{Conv}(I \cup F)$ appears in T if and only if $\mathrm{Conv}(I' \cup F)$ appears.

In the case when $\#(Z) = k + 1$, condition (c) follows from (b). An example of a triangulation supported on a circuit of smaller cardinality is given in Figure 30.

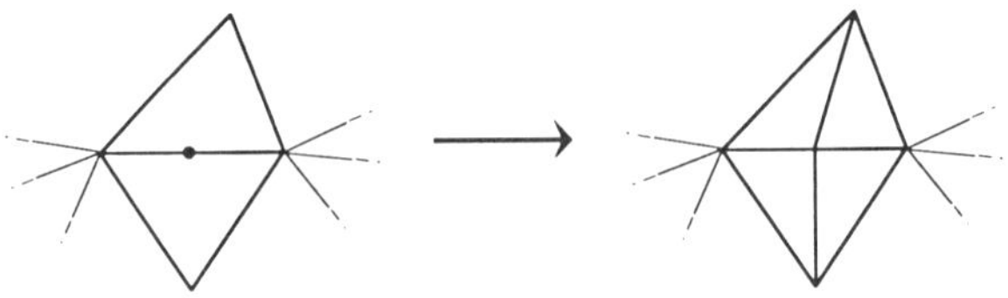

Figure 30. Modification along a circuit of small cardinality

Let T be a triangulation supported on a circuit Z. Then T induces one of two possible triangulations on $\mathrm{Conv}(Z)$, say T_+. As in the special case above, we let $s_Z(T)$ denote the new triangulation of (Q, A) that is obtained from T by taking away all the simplices of the form $\mathrm{Conv}(I \cup F)$ with $\mathrm{Conv}(I) \in T_+$ and adding the simplices of the form $\mathrm{Conv}(I' \cup F)$ with $\mathrm{Conv}(I') \in T_-$ and the same F. We say that $s_Z(T)$ is obtained from T by the modification along Z. It is clear that $s_Z(T)$ is also supported on Z, and $s_Z(s_Z(T)) = T$.

The description of the edges of $\Sigma(A)$ is as follows.

Theorem 2.10. *Let T and T' be two coherent triangulations of (Q, A). The vertices $\varphi_T, \varphi_{T'} \in \Sigma(A)$ are joined by an edge if and only if there is a circuit $Z \subset A$ such that T and T' are both supported on Z and obtained from each other by the modification along Z.*

Theorem 2.10 is proved in the same manner as its special case, namely Proposition 2.8. We have to describe the polyhedral subdivision $S = S(T, T')$ corresponding to the edge joining φ_T and $\varphi_{T'}$. To characterize this subdivision, we introduce the following notion.

Definition 2.11. Suppose that the coherent triangulations T and T' of (Q, A) are obtained from each other by the modification along a circuit $Z \subset A$. We say that a subset $J \subset A - Z$ is *separating* for T and T' if, for some $\omega \in Z$, the set $(Z - \{\omega\}) \cup J$ is the set of vertices of a simplex (of maximal dimension) of T.

If $\#(Z) = k + 1$, then $\emptyset$ is the only separating subset.

Proposition 2.12. *In the situation of Theorem 2.10, the polyhedral subdivision $S(T, T')$ corresponding to the edge $[\varphi_T, \varphi_{T'}]$ consists of the simplices* $(\mathrm{Conv}(I), I)$ *which T and T' have in common and the polyhedra* $(\mathrm{Conv}(Z \cup J), Z \cup J)$ *for all separating subsets $J \subset A - Z$.*

Proposition 2.12 and hence Theorem 2.10 follow readily from Corollary 2.7 which allows us to describe all the subdivisions of (Q, A) corresponding to the edges of $\Sigma(A)$ (i.e., to the faces of dimension 1), in terms of the affine relations. We leave the details to the reader.

3. Examples of secondary polytopes

We now present some examples of the general constructions in Sections 1 and 2. We adopt the notation of Section 1, so throughout this section $A = \{\omega^{(1)}, \dots, \omega^{(n)}\}$ will be the set of points in a real affine space whose convex hull Q has dimension $k - 1$. Then the secondary polytope $\Sigma(A)$ is a convex polytope of dimension $n - k$ in $\mathbf{R}^A = \mathbf{R}^n$. The vertices of $\Sigma(A)$ correspond to coherent triangulations of (Q, A) via (1.4). The affine span of $\Sigma(A)$ is given by Proposition 1.11; it is parallel to the subspace $L_A \subset \mathbf{R}^A$ of affine relations between the elements of A.

A. The case dim $(Q) = 1$*: skew cubes*

We start with the simplest case when $A \subset \mathbf{R}$ consists of points on a line. We can assume that elements of A are increasing real numbers $\omega^{(1)} < \cdots < \omega^{(n)}$ and the "volume form" on $\mathbf{R}$ is the usual length.

The polytope Q is the interval $[\omega^{(1)}, \omega^{(n)}]$ (see Figure 31). A "triangulation" of (Q, A) is just a subdivision of this interval into smaller intervals of the form $[\omega^{(i)}, \omega^{(j)}]$, $i < j$. There are exactly 2^{n-2} such triangulations and they are labeled by subsets of $\{2, 3, \ldots, n-1\}$. More precisely, if I is any such subset, say, $I = \{i_1 < \ldots < i_s\}$ then we define the triangulation $T(I)$ as consisting of intervals

$$[\omega^{(1)}, \omega^{(i_1)}],\quad [\omega^{(i_1)}, \omega^{(i_2)}],\ \ldots,\ [\omega^{(i_s)}, \omega^{(n)}].$$

Clearly all triangulations of (Q, A) are coherent. Therefore $\Sigma(A)$ has 2^{n-2} vertices. There is one well-known polytope with this number of vertices: the cube $[0, 1]^{n-2}$. It turns out that $\Sigma(A)$ is not exactly a cube nor can it be taken into a cube by an affine transformation. However, it is *combinatorially equivalent* to a cube which means that there is an order-preserving bijection of the poset of the faces of $\Sigma(A)$ onto the poset of the faces of the cube.

Figure 31.

Indeed, the subdivisions of (Q, A) (in the sense of Definition 2.1) are naturally labeled by pairs of subsets (I, J) of $\{2, 3, \ldots, n-1\}$ such that $I \cap J = \emptyset$. If $I = \{i_1 < \cdots < i_s\}$ then the corresponding subdivision $S(I, J)$ consists of marked polytopes

$$([\omega^{(1)}, \omega^{(i_1)}],\ \{\omega^{(1)}, \omega^{(2)}, \ldots, \omega^{(i_1)}\} - J),$$
$$([\omega^{(i_1)}, \omega^{(i_2)}],\ \{\omega^{(i_2)}, \omega^{(i_1+1)}, \ldots, \omega^{(i_2)}\} - J),$$
$$\ldots \quad \ldots \quad \ldots$$
$$([\omega^{(i_s)}, \omega^{(n)}],\ \{\omega^{(i_s)}, \omega^{(i_s+1)}, \ldots, \omega^{(n)}\} - J).$$

Clearly, all subdivisions of (Q, A) are coherent, and $S(I, J)$ refines $S(I', J')$ if and only if $I' \subseteq I, J' \subseteq J$. We denote by $F(I, J) = F(S(I, J))$ the face corresponding to a subdivision $S(I, J)$ via Theorem 2.4. In particular, $F(I, \{2, \ldots, n-1\} - I) = \varphi_{T(I)}$ is the vertex of $\Sigma(A)$ corresponding to I. Applying Theorem 2.4, we obtain the following.

Proposition 3.1. *The secondary polytope $\Sigma(A)$ is combinatorially equivalent to an $(n-2)$-dimensional cube $[0, 1]^{n-2} = \{(x_2, \ldots, x_{n-1}) : 0 \le x_i \le 1\}$, the face $F(I, J)$ corresponding to the face of $[0, 1]^{n-2}$ given by the equations $x_i = 1$ for $i \in I$, and $x_i = 0$ for $i \in J$.*

Let us find the precise positions of vertices of $\Sigma(A)$. For $i = 1, \dots, n-1$, let $l_i = \omega^{(i+1)} - \omega^{(i)}$ be the length of $[\omega^{(i)}, \omega^{(i+1)}]$. For a subset $I = \{i_1 < \cdots < i_s\} \subset \{2, \dots, n-1\}$, let $\varphi(I) = \varphi_{T(I)}$ be the vertex of $\Sigma(A)$ corresponding to I. This is a vector in $\mathbf{R}^A$ which is identified with $\mathbf{R}^n$ by our numbering of A. In accordance with this identification, we write any vector φ of $\mathbf{R}^A$ as $(\varphi_1, \dots, \varphi_n)$ where φ_i is the value of $\varphi : A \to \mathbf{R}$ at $\omega^{(i)}$.

According to (1.4), the coordinates of $\varphi(I)$ are given as follows:

$$\varphi(I)_i = \begin{cases} \sum_{j=i_{r-1}}^{i_{r+1}-1} l_j & \text{if } i = i_r \in I \\ \\ 0 & \text{if } i \in \{2, \dots, n-1\} - I \end{cases} \tag{3.1}$$

(with the convention that $i_0 = 1$, $i_{s+1} = n$).

Proposition 1.11 which describes the affine span of $\Sigma(A)$ translates into the following.

Proposition 3.2. *The affine span of $\Sigma(A)$ is an affine subspace of codimension 2 in $\mathbf{R}^n$ given by two equations*

$$\varphi_1 + \cdots + \varphi_n = 2(l_1 + \cdots + l_{n-1}), \tag{3.2}$$

$$l_1\varphi_2 + (l_1 + l_2)\varphi_3 + \cdots + (l_1 + \cdots + l_{n-1})\varphi_n = (l_1 + \cdots + l_{n-1})^2. \tag{3.3}$$

By (3.2) and (3.3), we can express say φ_1 and φ_n in terms of other coordinates, and so use $\varphi_2, \dots, \varphi_{n-1}$ as independent coordinates for $\Sigma(A)$.

By Proposition 3.1, the facets of $\Sigma(A)$ are of the form $F(\emptyset, \{i\})$ and $F(\{i\}, \emptyset)$ for $i = 2, \dots, n-1$. There are exactly the two types of coarse subdivisions described in Section 2C (see (2.2) and (2.3)). So $F(\emptyset, \{i\})$ corresponds to the linear constraint $\varphi_i \geq 0$, and $F(\{i\}, \emptyset)$ corresponds to the linear constraint

$$\sum_{j=i+1}^{n} (\omega^{(j)} - \omega^{(i)})\varphi_j \geq 2 \int_{\omega^{(i)}}^{\omega^{(n)}} (x - \omega^{(i)})dx. \tag{3.4}$$

We can also rewrite (3.4) as

$$\sum_{j=i+1}^{n} (l_i + l_{i+1} + \cdots + l_{j-1})\varphi_j \geq (l_i + l_{i+1} + \cdots + l_{n-1})^2. \tag{3.5}$$

The edges of $\Sigma(A)$ are given by Theorem 2.10. In particular, the vertex $\varphi(\emptyset)$ has the neighboring vertices $\varphi(\{i\})$ for $i \in [2, n-1]$, and

$$\varphi(\{i\}) - \varphi(0) = -(l_i + l_{i+1} + \cdots + l_{n-1})e_1 + (l_1 + l_2 + \cdots + l_{n-1})e_i$$

$$-(l_1 + l_2 + \cdots + l_{i-1})e_n, \tag{3.6}$$

where the e_i are the standard basis vectors in $\mathbf{R}^n$. The opposite vertex $\varphi([2, n-1])$ has the neighboring vertices $\varphi([2, n-1] - \{i\})$ for $i \in [2, n-1]$, and

$$\varphi([2, n-1]) - \varphi([2, n-1] - \{i\}) = -l_i e_{i-1} + (l_{i-1} + l_i)e_i - l_{i-1}e_{i+1} \tag{3.7}$$

(here $[2, n-1]$ stands for $\{2, \ldots, n-1\}$).

Now we consider a special case when $l_1 = l_2 = \cdots = l_{n-1} = 1$. This case will be of particular interest later, in relation to the Newton polytope of the classical discriminant (Chapter 12). Looking at (3.7) we recognize on the right hand side the Cartan matrix

$$\begin{pmatrix} 2 & -1 & 0 & 0 & \cdots & 0 \\ -1 & 2 & -1 & 0 & \cdots & 0 \\ 0 & -1 & 2 & -1 & \cdots & 0 \\ \vdots & \vdots & \vdots & \vdots & \ddots & \vdots \\ 0 & 0 & 0 & 0 & \cdots & 2 \end{pmatrix}$$

of the root system of type A_{n-2}. This suggests the following interpretation of $\Sigma(A)$. Let R be a root system of type A_{n-2} in a real vector space V with the standard choice of simple roots $\alpha_1, \ldots, \alpha_{n-2}$ and the corresponding fundamental weights $\omega_1, \ldots, \omega_{n-2}$. Let $\rho = \omega_1 + \omega_2 + \cdots + \omega_{n-2}$ be the half-sum of all positive roots. Consider the convex polytope $P(2\rho)$ in V which is the intersection of two cones $\sum_i \mathbf{R}_+ \cdot \omega_i$ and $2\rho - \sum_i \mathbf{R}_+ \cdot \alpha_i$.

Proposition 3.3. *If $l_1 = l_2 = \cdots = l_{n-1} = 1$ then the mapping*

$$(\varphi_1, \ldots, \varphi_{n-1}) \mapsto \varphi_2\omega_1 + \varphi_3\omega_2 + \cdots + \varphi_{n-1}\omega_{n-2}$$

is an affine isomorphism between $\Sigma(A)$ and $P(2\rho)$.

Proof. This follows at once from Proposition 3.1 and formulas (3.1), (3.6), and (3.7).

The polytope $P(2\rho)$ plays an important role in the representation theory of the Lie algebra sl_{n-1}. Let λ be a non-negative integral linear combination of fundamental weights $\omega_1, \ldots, \omega_{n-2}$. It was conjectured by Kostant and proved in [BZ] that $\lambda \in P(2\rho)$ if and only if the irreducible sl_{n-1}-module with highest weight λ appears in the decomposition of the exterior algebra of the adjoint representation.

As a concrete example, let us take $A = \{0, 1, 2, 3\}$. By using φ_1 and φ_2 as independent coordinates for $\Sigma(A)$ (see the remark after Proposition 3.2), we find $\Sigma(A)$ to be the quadrangle in the plane depicted in Figure 32.

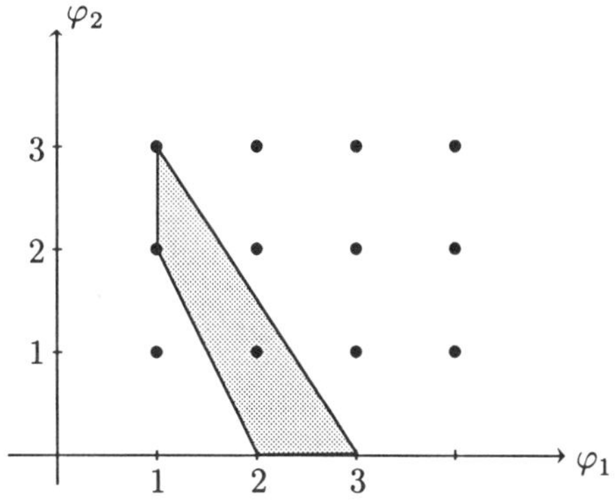

Figure 32.

B. The case $\dim(Q) = 2$*: the associahedron (Stasheff polytope)*

Let $A = \{\omega^{(1)}, \ldots, \omega^{(n)}\} \subset \mathbf{R}^2$ be the set of vertices of a convex n-gon Q in the plane, numbered consecutively around the perimeter (see Figure 33).

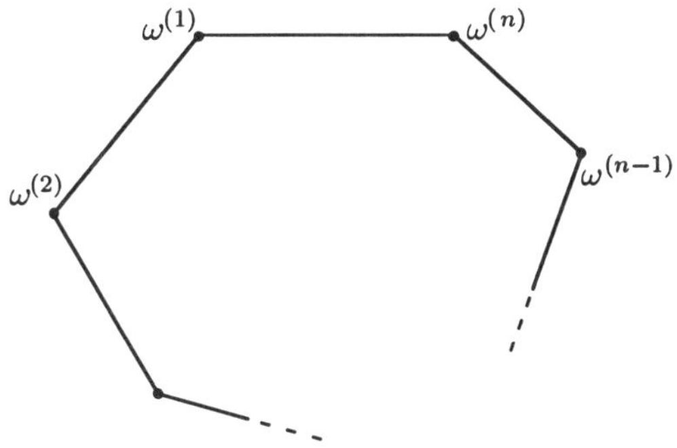

Figure 33.

The corresponding secondary polytope $\Sigma(A)$ is of special interest; we shall see that its vertices are in a natural bijection with various "associativity patterns", i.e., the different ways to insert parentheses in the formal product $x_1 x_2 \cdots x_{n-1}$ of several non-associative variables. For example, two triangulations of a quadrangle correspond to two ways of forming a product of three variables: $(x_1x_2)x_3$ and $x_1(x_2x_3)$. Before describing this correspondence, we want to first see what the general results of Sections 1 and 2 give in this case.

Clearly, every subdivision of (Q, A) is determined by a set of mutually non-crossing diagonals of Q (two diagonals are said to *cross* if they intersect at an interior point of Q). Let S_D denote the subdivision corresponding to a set D of non-crossing diagonals. Clearly, S_D refines $S_{D'}$ if and only if $D' \subset D$.

Proposition 3.4. *Every subdivision of* (Q, A) *is coherent.*

Proof. It is intuitively obvious that one can "fold" Q in a convex way along every set D of non-crossing diagonals. Let us give a formal argument whose idea will also be used in subsection C in a less obvious situation.

Consider the subdivision S_D corresponding to a set of non-crossing diagonals $D = \{d_1, \dots, d_k\}$ of Q. Using induction on k, we can choose an ordering of D so that, for each $j = 1, \dots, k$, the diagonals $d_1, \dots, d_{j-1}$ lie on one side of d_j, and $d_{j+1}, \dots, d_k$ lie on the other side of d_j. Let $\alpha_1, \dots, \alpha_k$ be affine-linear functions on Q defining diagonals $d_1, \dots, d_k$, i.e., each d_j is given by the equation $\alpha_j = 0$. The functions α_j are defined up to scalar multiples. Clearly, we can choose them so that each α_j is non-positive on $d_1, \dots, d_{j-1}$ and non-negative on $d_{j+1}, \dots, d_k$. The subdivision S_D consists of $k+1$ polygons $\mathrm{Conv}(\sigma_0)$, $\mathrm{Conv}(\sigma_1), \dots, \mathrm{Conv}(\sigma_k)$, where each σ_j is the set of vertices of Q satisfying the inequalities $\alpha_1, \alpha_2, \dots, \alpha_j \geq 0$, $\alpha_{j+1}, \dots, \alpha_k \leq 0$.

Now consider the piecewise-linear function g on Q given by

$$g(x) = \alpha_1(x) + \alpha_2(x) + \dots + \alpha_j(x) \text{ for } x \in \mathrm{Conv}(\sigma_j),$$

and define $\psi = (\psi_1, \psi_2, \dots, \psi_n)$ by $\psi_i = g(\omega^{(i)})$. Recalling the definitions of g_ψ in Section 1C and of the coherent subdivision $S(\psi)$ given in Example 2.2, we conclude that $g = g_\psi$, and $S_D = S(\psi)$. The proposition is proved.

The face lattice of $\Sigma(A)$ is given by Theorem 2.4:

Proposition 3.5. *The face lattice of* $\Sigma(A)$ *is anti-isomorphic to the lattice of sets of non-crossing diagonals in* Q.

In particular, $\dim(\Sigma(A)) = n - 3$, and, for each $k = 0, 1, \dots, n - 3$, the k-dimensional faces of $\Sigma(A)$ correspond to sets of $(n-3-k)$ non-crossing diagonals, or equivalently, to subdivisions of (Q, A) into $(n - 2 - k)$ convex subpolygons. Taking $k = 0$, we again recover the fact that the vertices of $\Sigma(A)$ correspond to triangulations of (Q, A), each of which subdivides Q into $n - 2$ triangles.

Example 3.6. For $n = 4$, the polytope $\Sigma(A)$ is an interval; for $n = 5$, it is a plane pentagon and for $n = 6$, it is the 3-dimensional polytope with 14 vertices. These polytopes together with triangulations corresponding to their vertices are depicted in Figure 34.

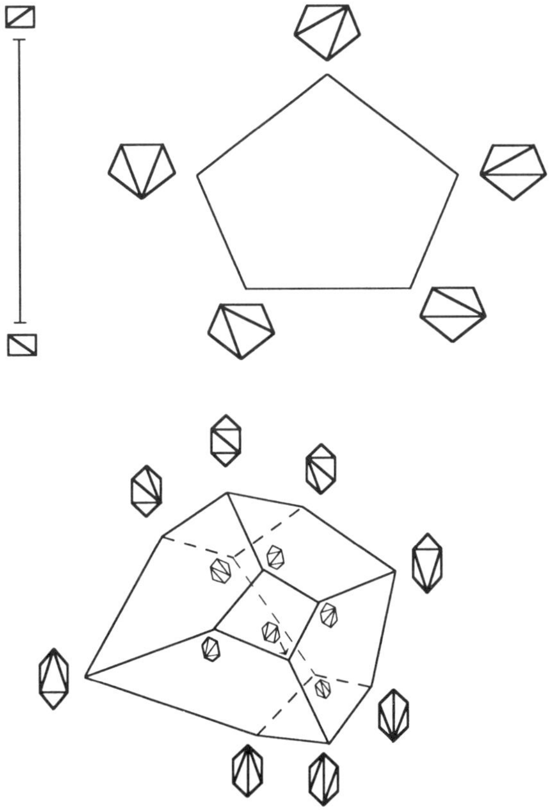

Figure 34. Stasheff polytopes

The problem of describing triangulations and more general subdivisions of a convex polygon is very classical, having attracted the attention of Euler in the 18th century, and Catalan and Cayley in the 19th century. The old work was mainly concentrated around enumeration problems. It was apparently first shown by Euler that the number of triangulations of (Q, A) is the $(n-1)$-st *Catalan number* c_{n-1}, where

$$c_n = \frac{1}{n}\binom{2n-2}{n-1}; \tag{3.8}$$

thus $c_2 = 1,\ c_3 = 2,\ c_4 = 5,\ c_5 = 14,\ c_6 = 42$, etc. The Catalan numbers appear in mathematics under many different disguises. Catalan himself defined c_n as the number of parenthesizings of the formal product $x_1x_2\cdots x_n$. Here to parenthesize the product means to insert a well-formed string of $n-1$ left and $n-1$ right parentheses so that the product can be evaluated in any non-associative algebra.

The correspondence between parenthesizings of $x_1x_2\cdots x_{n-1}$ and triangulations of the convex n-gon Q is established as follows. Let T be any triangulation of (Q, A). The triangles from T have $2n-3$ edges in total: n edges of Q and $n-3$ diagonals. We write every such edge in the form $[a_i, a_j]$ with $i<j$, and we wish to assign to it some parenthesizing of the product $x_ix_{i+1}\cdots x_{j-1}$.

We proceed by induction on $j-i$. For $j=i+1$ there is nothing to do: we simply attach the symbol x_i to the edge $[a_i, a_{i+1}]$ for $i=1,\ldots,n-1$. Now suppose $[a_i, a_k]$ is an edge of T with $k-i>1$. Clearly, there is exactly one triangle of T of the form $\{a_i, a_j, a_k\}$ with $i<j<k$. By induction, we can assume that $[a_i, a_j]$ is already assigned a parenthesizing U of $x_ix_{i+1}\cdots x_{j-1}$, and $[a_j, a_k]$ is already assigned a parenthesizing V of $x_jx_{j+1}\cdots x_{k-1}$. Then we assign to $[a_i, a_k]$ the parenthesizing (UV) of $x_ix_{i+1}\cdots x_{k-1}$. Finally, we associate to T the parenthesizing assigned to the last edge $[a_1, a_n]$ of Q. This procedure is illustrated by Figure 35.

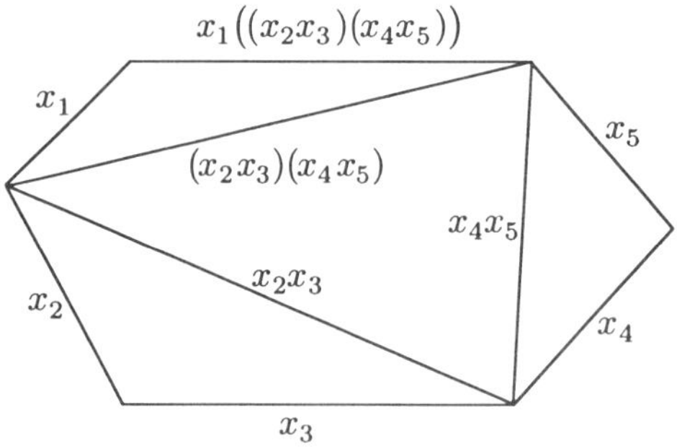

Figure 35. Parenthesizing corresponding to a triangulation

Given the above interpretation, the name *associahedron* is natural for the secondary polytope of the set of vertices of a convex polygon. The faces of the associahedron also can be described in terms of certain parenthesizings.

By a *partial parenthesizing* of $x_1\cdots x_{n-1}$ we mean a string of parentheses obtained by taking away some pairs of corresponding parentheses from a parenthesizing. If π, π' are two partial parenthesizings then we say that π refines π' if π can be obtained from π' by adding new pairs of parentheses. The same construction as above shows that subdivisions of (Q, A) (and hence, the faces of $\Sigma(A)$) are in bijection with partial parenthesizings of $x_1\cdots x_{n-1}$. Clearly, one subdivision refines another if and only if the same is true for corresponding parenthesizings.

The simplicial complex of partial parenthesizings was introduced and studied by Stasheff [St] in relation to the theory of loop spaces (because of that the associahedra are sometimes called Stasheff polytopes). He constructed a geometric realization of the simplicial complex as a convex body but not as a polytope. The question of the existence of a polytopal model of this complex circulated around for some years (see [Lee]). We see that a positive answer to this question is a very special case of the theory of secondary polytopes. Another construction was given by Lee [Lee].

We conclude the discussion of the associahedron by the formula for the number of its faces of any given dimension which generalizes (3.8). For each $k = 0, 1, \ldots, n-3$, let $f(k, n)$ denote the number of sets of k non-crossing diagonals in the n-gon Q, i.e., the number of faces of $\Sigma(A)$ of dimension $(n-3-k)$.

Proposition 3.7. *For $k = 0, 1, \ldots, n-3$, we have*

$$f(k,n) = \frac{1}{n-1}\binom{n-3}{k}\binom{n+k-1}{k+1}. \tag{3.9}$$

As indicated in [Lee], the formula (3.9) was discovered by Kirkman in 1857, with the first complete proof given by Cayley in 1891. The proof below follows the original Cayley one [Ca6].

Proof. We adopt the convention that $f(k, n)$ is defined for all integers k and n but is equal to 0 unless $0 \le k \le n-3$.

Lemma 3.8. *If $k \ge 1$ then $f(k, n)$ satisfies the recursion*

$$f(k,n) = \frac{n}{2k}\sum_{l+m=n+2}\ \sum_{i+j=k-1} f(i,l)f(j,m). \tag{3.10}$$

Proof of Lemma 3.8. Let $\tilde{f}(k, n)$ be the number of *sequences* $d_1, \ldots, d_k$ of non-crossing diagonals in the n-gon Q. Clearly, $\tilde{f}(k,n) = k!f(k,n)$. Now the first diagonal d_1 cuts Q into two pieces, say an l-gon Q' and an m-gon Q'' with $l + m = n + 2$, $l \le m$. For fixed $l < \frac{n+2}{2}$ there are n different choices of d_1, and for $l = \frac{n+2}{2}$ there are $\frac{n}{2}$ such choices. Once we have chosen d_1, the remaining $k-1$ diagonals fall into two groups: say, i of them belong to Q', and $j = k-1-i$ belong to Q''. We arrive at the recursion

$$\tilde{f}(k,n) = n \sum_{l+m=n+2, l\le m}{}' \ \sum_{i+j=k-1} \binom{k-1}{i}\tilde{f}(i,l)\tilde{f}(j,m), \tag{3.11}$$

where $\sum'$ denotes the convention that the terms with $l = m = \frac{n+2}{2}$ are taken with coefficient $\frac{1}{2}$. Clearly, we can replace $\sum'_{l+m=n+2, l\leq m}$ by $\frac{1}{2}\sum_{l+m=n+2}$, and (3.10) becomes a direct consequence of (3.11). The lemma is proved.

Clearly, the $f(k,n)$ are uniquely determined by (3.10) together with the initial conditions

$$f(0,n) = 1 \text{ for } n \geq 3. \tag{3.12}$$

Following Cayley, we introduce two sequences of generating functions:

$$U_k = \sum_n f(k,n)x^n, \quad V_k = 2(k+1)\sum_n \frac{f(k+1,n-2)}{n-2}x^n \quad (k = 0, 1, \ldots). \tag{3.13}$$

They are connected by

$$U_{k+1} = x\frac{d}{dx}\left(\frac{V_k}{2(k+1)x^2}\right). \tag{3.14}$$

Using these functions, (3.10) can be rewritten as

$$\left(\sum_k U_k y^k\right)^2 = \sum_k V_k y^k, \tag{3.15}$$

and (3.12) as

$$U_0 = \frac{x^3}{1-x}. \tag{3.16}$$

Let $X = \frac{x^2}{1-x}$. We claim that the functions U_k and V_k are given by

$$U_k = \frac{x}{(k+1)!}\frac{d^k X^{k+1}}{dx^k}, \quad V_k = \frac{2(k+1)x^2}{(k+2)!}\frac{d^k X^{k+2}}{dx^k}. \tag{3.17}$$

Indeed, it is enough to prove that the functions defined by (3.17) satisfy (3.14), (3.15), and (3.16). The formulas (3.14) and (3.16) are obvious. As for (3.15), Cayley proved that it is satisfied by the functions which are defined by (3.17) not only for our $X = \frac{x^2}{1-x}$ but for an *arbitrary* power series X in x. We leave the proof of this statement as an exercise for the reader (hint: see [Ca6]).

To complete the proof of Proposition 3.7, it remains to expand U_k given by (3.17) into a power series in x using the binomial expansion of $(1-x)^{-(k+1)}$ and differentiating it k times term by term. This gives (3.9).

C. The case of a triangular prism: the permutohedron

Let A be the set of vertices of the triangular prism $Q = \Delta^1 \times \Delta^{m-1}$. Here Δ^1 is the interval $[e_1, e_2]$ in $\mathbf{R}^2$, and Δ^{m-1} is the standard simplex in $\mathbf{R}^m$ with vertices $f_1, f_2, \ldots, f_m$, where the e_i and the f_j stand for the standard basis vectors. So A consists of $2m$ points $\omega_{ij} = e_i + f_j$ $(i = 1, 2,\ j = 1, \ldots, m)$ in $\mathbf{R}^{m+2}$. Using our standard notation we see that $n = \#(A) = 2m$, and $k = \dim(Q) + 1 = m + 1$, hence $\dim(\Sigma(A)) = n - k = m - 1$. We will think of $\Sigma(A)$ as a convex polytope living in the space $\mathbf{R}^{2\times m}$ of real $2 \times m$ matrices.

We shall show that the secondary polytope $\Sigma(A)$ is naturally identified with the *permutohedron* P_m. By definition, P_m is the convex polytope in $\mathbf{R}^m$ with vertices $(s(1), \ldots, s(m))$, where s runs over all permutations of $\{1, 2, \ldots, m\}$. Note that the affine span of P_m is the hyperplane $\big\{(x_1, \ldots, x_m) \in \mathbf{R}^m : \sum x_i = m(m+1)/2\big\}$, so $\dim(P_m) = m - 1$.

Example 3.9. The permutohedron P_2 is an interval, P_3 is a hexagon and P_4 is a 3-dimensional polytope with 24 vertices. These polytopes are depicted in Figure 36. There we use the notation $[j_1, \ldots, j_n]$ for the point with coordinates $(s(1), \ldots, s(n))$ where s is the permutation such that $j_\nu = s^{-1}(\nu)$. Under this notation, two vertices $[j_1, \ldots, j_n]$ and $[k_1, \ldots, k_n]$ are joined by an edge if and only if $(k_1, \ldots, k_n)$ is obtained from $(j_1, \ldots, j_n)$ by permuting two numbers in consecutive positions. A proof of this and other facts about the permutohedron can be found in [Milg] [GS].

Our analysis of $\Sigma(A)$ will be analogous to the case of the associahedron. We start by evaluating ranks of all subsets of A. We write an arbitrary subset $\sigma \subset A$ as $\sigma(J_1, J_2) = \{\omega_{1j}\ (j \in J_1),\ \omega_{2j}\ (j \in J_2)\}$ for some subsets $J_1, J_2 \subset \{1, 2, \ldots, m\}$. By counting the dimension of the space of affine relations L_σ, we readily conclude that the convex hull $\mathrm{Conv}(\sigma)$ has dimension

$$\dim(\mathrm{Conv}(\sigma(J_1, J_2))) = \begin{cases} \#(J_1 \cup J_2) & \text{if } J_1 \cap J_2 \neq \emptyset, \\ \#(J_1 \cup J_2) - 1 & \text{if } J_1 \cap J_2 = \emptyset. \end{cases} \tag{3.18}$$

Now we formulate and prove the analogs of Propositions 3.4 and 3.5. We call a *diagonal section* or simply *diagonal* of Q any hyperplane section d of Q which is an affine span of some vertices of Q but not a facet. It follows from (3.18) that diagonals of Q are in bijection with proper subsets $J \subset \{1, 2, \ldots, m\}$ via

$$J \mapsto d_J = \mathrm{Conv}(\sigma(J, \{1, 2, \ldots, m\} - J)). \tag{3.19}$$

We see that every diagonal of Q is an $(m-1)$-dimensional simplex with vertices in A. As before, two diagonals are said to *cross* if they have a common interior point of Q.

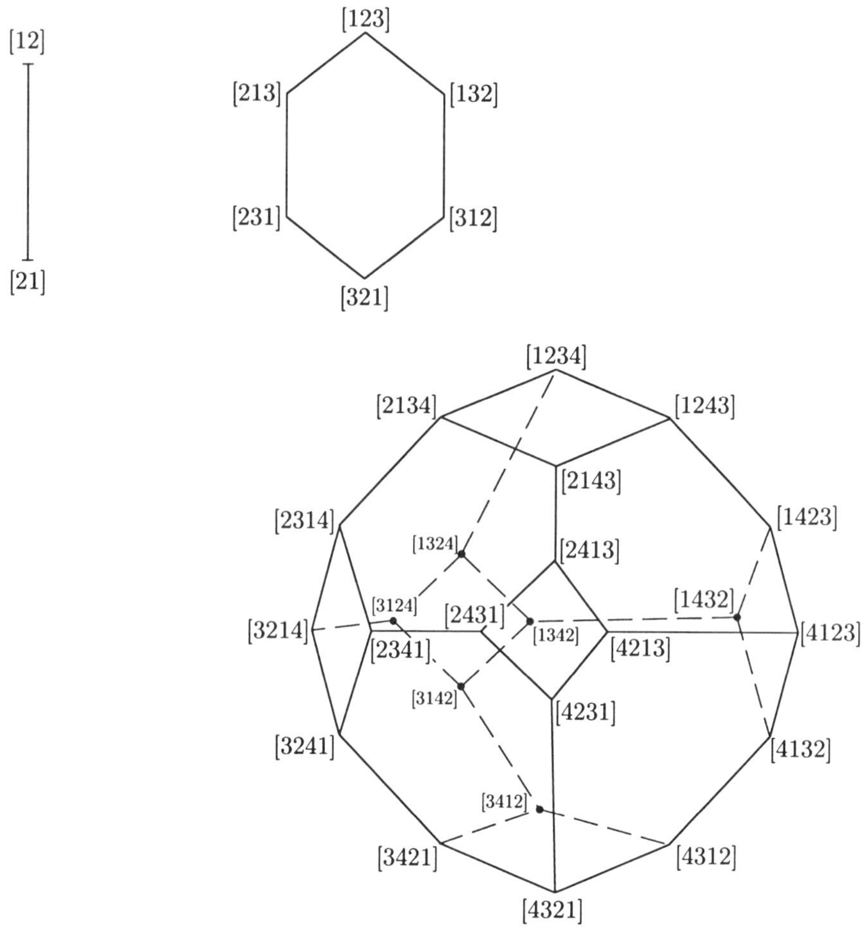

Figure 36. Permutohedra

Clearly, if two diagonals do not cross then their intersection is their common face, which is some simplex of smaller dimension with vertices in A lying in some proper face of Q. It follows that every set of non-crossing diagonals D gives rise to a subdivision S_D of (Q, A). As before, S_D refines $S_{D'}$ if and only if $D' \subset D$.

Proposition 3.10.

(a) *The correspondence $D \mapsto S_D$ is a bijection between all sets of non-crossing diagonals in Q and all subdivisions of (Q, A).*

(b) *Every subdivision of (Q, A) is coherent; hence the face lattice of $\Sigma(A)$ is anti-isomorphic to the lattice of sets of non-crossing diagonals in Q.*

Proof. (a) We have only to show that every subdivision S of (Q, A) has the form

S_D for some set of non-crossing diagonals D. Since A consists of vertices of Q, a subdivision $S = \{(Q_i, A_i)\}$ is uniquely determined by the decomposition $Q = \cup_i Q_i$. Now every interior facet of each Q_i (not lying on the boundary of Q) belongs to some diagonal of Q. But then it must coincide with this diagonal since any diagonal is a simplex with vertices in A so it cannot be further subdivided. This implies our statement.

(b) Let $S = \{(Q_i, A_i)\}$ be a subdivision of (Q, A). It is enough to construct a convex piecewise linear function on Q whose domains of linearity are exactly the polytopes Q_i. This can be done in exactly the same way as in Proposition 3.4., thus proving our proposition.

To describe the face lattice of $\Sigma(A)$ more explicitly, we use the following criterion for the non-crossing of two diagonals.

Proposition 3.11. *Two diagonals d_J and $d_{J'}$ of Q do not cross if and only if one of the sets J, J' contains the other.*

Proof. It is easy to see that Q can be described as

$$Q = \left\{(\alpha, \beta) \in \mathbf{R}^2 \times \mathbf{R}^m : \alpha_i, \beta_j \geq 0, \; \alpha_1 + \alpha_2 = \beta_1 + \cdots + \beta_m = 1\right\}, \quad (3.20)$$

so a point $(\alpha, \beta) \in Q$ lies on the boundary if and only if at least one of the coordinates α_i, β_j is equal to 0. We readily verify that the diagonal d_J is given by

$$d_J = \left\{(\alpha, \beta) \in Q : \alpha_1 = \sum_{j \in J} \beta_j \right\}. \quad (3.21)$$

It follows that if, say $J \subset J'$, then every point $(\alpha, \beta) \in d_J \cap d_{J'}$ has $\beta_j = 0$ for $j \in J' - J$, and hence lies on the boundary of Q. Conversely, suppose that both $J - J'$ and $J' - J$ are non-empty. Then it follows easily from (3.20) and (3.21) that d_J and $d_{J'}$ cross, i.e., they have a common interior point of Q.

Combining Propositions 3.10 and 3.11, we arrive at the following description of the face lattice of $\Sigma(A)$. By a *flag* in $\{1, 2, \ldots, m\}$ we mean a strictly increasing sequence $F = (J_1 \subset J_2 \subset \cdots \subset J_r)$ of proper subsets of $\{1, 2, \ldots, m\}$. A flag F *refines* another flag F' if it contains all subsets from F'.

Corollary 3.12. *The sets of non-crossing diagonals in Q and hence the faces of $\Sigma(A)$ are in refinement-preserving bijection with flags in $\{1, 2, \ldots, m\}$, via*

$$F = (J_1 \subset J_2 \subset \cdots \subset J_r) \mapsto D = \{d_{J_1}, \ldots, d_{J_r}\}.$$

The subdivision $S(F)$ corresponding to a flag F consists of $r+1$ marked polytopes (Q_p, A_p), $p = 1, \ldots, r+1$, where $A_p = \sigma(J_p, \{1, 2, \ldots, m\} - J_{p-1})$ (with the *convention* $J_0 = \emptyset$, $J_{r+1} = \{1, 2, \ldots, m\}$), *and*

$$Q_p = \mathrm{Conv}(A_p) = \left\{(\alpha, \beta) \in Q : \sum_{j \in J_{p-1}} \beta_j \le \alpha_1 \le \sum_{j \in J_p} \beta_j\right\}. \tag{3.22}$$

In particular, we obtain the following description of the vertices of $\Sigma(A)$.

Corollary 3.13. *The vertices of* $\Sigma(A)$ *(i.e., coherent triangulations of* (Q, A)*) are in bijection with complete flags* $F = (J_1 \subset J_2 \subset \cdots \subset J_{m-1})$ *in* $\{1, 2, \ldots, m\}$ *(so* $\#(J_p) = p$ *for* $p = 1, \ldots, m-1$*). The triangulation* $S(F)$ *corresponding to a flag* F *consists of* m *simplices with vertices* $\sigma(J_p, \{1, 2, \ldots, m\} - J_{p-1})$ *for* $p = 1, \ldots, m$.

The symmetric group S_m acts on Δ^{m-1} by permutations of vertices. We extend this action to the action on (Q, A) by affine automorphisms. Then S_m acts on subdivisions and triangulations of (Q, A), and on $\Sigma(A)$. Since S_m acts transitively on the set of complete flags in $\{1, 2, \ldots, m\}$, Corollary 3.13 shows that it also acts transitively on triangulations of (Q, A), or equivalently, on the vertices of $\Sigma(A)$. We can choose for example the flag

$$F = (\{1\} \subset \{1, 2\} \subset \cdots \subset \{1, 2, \ldots, m-1\});$$

then the triangulation $T_0 = S(F)$ consists of m simplices

$$\sigma_p = \mathrm{Conv}\{\omega_{11}, \omega_{12}, \ldots, \omega_{1p}, \omega_{2p}, \omega_{2,p+1}, \ldots, \omega_{2m}\}, \quad p = 1, \ldots, m.$$

Graphically, these simplices correspond to various "staircases" leading from the north-west to the south-east corner of a $2 \times m$ matrix (Figure 38, p.247 below).

It is easy to see that all simplices σ_p have the same volume. We normalize the volume form on Q so that each σ_p has volume 1. This means that each matrix entry φ_{ij} of φ_{T_0} is equal to the number of simplices σ_p having ω_{ij} as a vertex. It follows that

$$\varphi_{T_0} = \begin{pmatrix} m & m-1 & \cdots & 1 \\ 1 & 2 & \cdots & m \end{pmatrix}. \tag{3.23}$$

All the other vertices of $\Sigma(A)$ are obtained from φ_{T_0} by permutations of columns. We conclude that the projection $\mathbf{R}^{2\times m} \to$ g R^m sending each $2 \times m$ matrix to its first row is an affine isomorphism of $\Sigma(A)$ with the permutohedron P_m.

D. The product of two simplices

It would be very interesting to generalize the results in the previous example to the case when A is the set of vertices of $Q = \Delta^{p-1} \times \Delta^{q-1}$, the product of two standard simplices (the triangular prism appears for p=2). Unfortunately, not much is known about the secondary polytope in this case. We do not even know if there exist non-coherent triangulations of (Q, A). Below we give a brief review of some partial results.

First, A consists of pq points $\omega_{ij} = e_i + f_j \in \mathbf{R}^p \times \mathbf{R}^q$, where $e_1, \ldots, e_p$ and $f_1, \ldots, f_q$ are the standard basis vectors and Q is

$$Q = \{(\alpha, \beta) \in \mathbf{R}^p \times \mathbf{R}^q : \alpha_i, \beta_j \geq 0, \ \alpha_1 + \cdots + \alpha_p = \beta_1 + \cdots + \beta_q = 1\}. \quad (3.24)$$

Hence $\dim(\Sigma(A)) = pq - (p+q-1) = (p-1)(q-1)$.

Graphically, it is convenient to represent a subset $B \subset A$ as a *bipartite graph* G with the set of vertices $\{e_1, \ldots, e_p, f_1, \ldots, f_q\}$ and the edges corresponding to pairs (e_i, f_j) for all $\omega_{ij} \in B$ (see Figure 37).

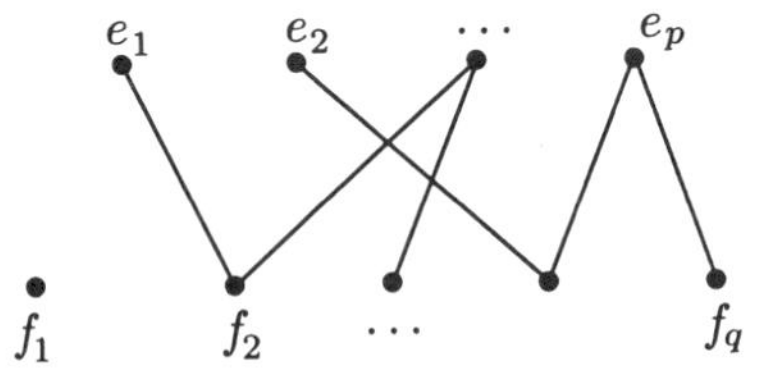

Figure 37. A bipartite graph

The formula (3.18) is generalized as follows:

$$\dim(\mathrm{Conv}(B)) = \sum_{\Gamma} (v(\Gamma) - 1) - 1, \quad (3.25)$$

where the sum is over all connected components Γ of the graph G corresponding to B, and $v(\Gamma)$ is the number of vertices of Γ. In particular, if $\#(B) = p+q-1$ then $\mathrm{Conv}(B)$ is a simplex of full dimension $p+q-2$ if and only if the graph G is connected (and in this case G is a tree with $p+q$ vertices).

Among these simplices we find all "staircases" (or *lattice paths*) leading from the point $(1, 1) \in \mathbf{Z}^2$ to the point (p, q) (see Figure 38).

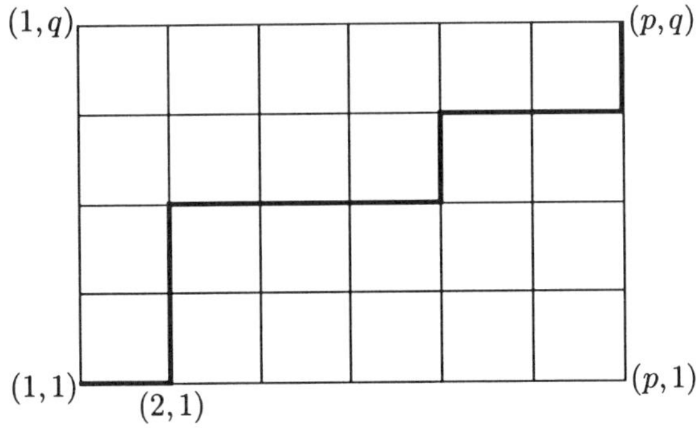

Figure 38. A lattice path

The language of lattice paths is very useful not only for the present example but in many other instances (see e.g. Chapter 12). A lattice path from $(1, 1)$ to (p, q) is a collection L of integral points $(i_1, j_1), (i_2, j_2), \ldots, (i_{p+q-1}, j_{p+q-1})$ such that $(i_1, j_1) = (1, 1), (i_{p+q-1}, j_{p+q-1}) = (p, q)$, and (i_{k+1}, j_{k+1}) is either $(i_k + 1, j_k)$ or $(i_k, j_k + 1)$ for $k = 1, \ldots, p + q - 2$. Graphically, we represent a lattice path by a polygonal line joining successively all the points of L. By a slight abuse of notation, we denote this line as a subset of the plane by the same letter L. It has end-points $(1, 1)$ and (p, q) and consists of several horizontal and vertical intervals. We can also think of L as an actual path from $(1, 1)$ to (p, q) consisting of $(p + q - 2)$ moves by a interval of unit length, each move going either horizontally from left to right, or vertically upwards.

It is convenient to encode such a path by a word $w = w(L) = w_1 w_2 \ldots w_{p+q-2}$ which records our moves in the following way: we write $w_k = A_i$ if the k-th move on our path is the i-th successive horizontal one, and we write $w_k = B_j$ if the k-th move is the j-th successive vertical move. The word $w(L)$ is obtained by a shuffle of two words $A_1 A_2 \ldots A_{p-1}$ and $B_1 B_2 \ldots B_{q-1}$, i.e., by a permutation of these symbols, leaving the A_i's and the B_j's in the same relative order. We call $w(L)$ the *shuffle* of L. Clearly, the correspondence $L \mapsto w(L)$ is a bijection between lattice paths and shuffles of $A_1 A_2 \ldots A_{p-1}$ and $B_1 B_2 \ldots B_{q-1}$. In particular, the number of different lattice paths is equal to the number of shuffles, i.e., to $\binom{p+q-2}{p-1}$.

Returning to our simplices, consider a lattice path L, and let

$$\sigma = \mathrm{Conv}\{\omega_{ij} : (i, j) \in L\}.$$

It is easy to see that σ is defined in $Q = \Delta^{p-1} \times \Delta^{q-1}$ by shuffling two increasing sequences

$$\alpha_1 \leq \alpha_1 + \alpha_2 \leq \cdots \leq \alpha_1 + \cdots + \alpha_{p-1} \text{ and } \beta_1 \leq \beta_1 + \beta_2 \leq \cdots \leq \beta_1 + \cdots + \beta_{q-1}$$

according to the shuffle $w(L)$. More precisely, $\sigma \subset Q$ is defined by the inequalities: $\alpha_1 + \cdots + \alpha_i \leq \beta_1 + \cdots + \beta_j$ whenever A_i precedes B_j in the word w, and $\alpha_1 + \cdots + \alpha_i \geq \beta_1 + \cdots + \beta_j$ whenever B_j precedes A_i.

This description of σ implies that the simplices corresponding to all possible lattice paths from $(1, 1)$ to (p, q) form a triangulation T_0 of (Q, A). This triangulation was known and used in algebraic topology for a long time. It is called *standard* or *canonical*. It is easy to show that T_0 is coherent, i.e., defines a vertex φ_{T_0} of $\Sigma(A)$ (a linear form $\psi \in \mathbf{R}^A$ which attains its maximum on $\Sigma(A)$ exactly at φ_{T_0} can be chosen as $\psi(\omega_{ij}) = ij$). Acting on this vertex by the group $S_p \times S_q$ of all permutations of the index set, we obtain many other vertices.

Thus far, everything follows the same pattern as in the case of a triangular prism. But here is the major difference: if $p, q \geq 3$ then not every coherent triangulation of (Q, A) can be obtained from the standard one by permutations of indices. The case $p = q = 3$ (the product of two triangles) was investigated in detail by A. E. Postnikov. Postnikov proved that all the triangulations in this case are coherent, and up to $S_3 \times S_3$-symmetry, there are 5 different triangulations. They are presented in Figure 39. Here dotted squares represent simplices of a triangulation, with the dots marking the vertices of a simplex. Two such squares are connected by an edge in Fig. 39. when the corresponding simplices have a common facet. We see in particular that $\Sigma(A)$ is a 4-dimensional polytope having $18 + 36 + 6 + 12 + 36 = 108$ vertices.

The vertices of the third type depicted in Figure 39 are especially symmetric. The stabilizer in $S_3 \times S_3$ of such a vertex v has cardinality 6. It can be seen that this stabilizer acts simply transitively on the edges of $\Sigma(A)$ containing v. Thus v is a vertex of a 4-dimensional polytope which has 6 edges passing through it. This means that $\Sigma(A)$ is not a simple polytope (in all the previous examples $\Sigma(A)$ was easily seen to be simple).

Starting with the case $p = 3, q = 4$, the situation becomes even more complicated: not every simplex σ can be transformed into a "staircase" simplex from T_0 by the action of $S_p \times S_q$. An example in the case $p = 3, q = 4$ is provided by

$$\sigma = \mathrm{Conv}\big\{\omega_{11}, \omega_{14}, \omega_{22}, \omega_{24}, \omega_{33}, \omega_{34}\big\}.$$

This simplex can be defined in Q by $\alpha_1 \geq \beta_1, \alpha_2 \geq \beta_2, \alpha_3 \geq \beta_3$.

Example 3.14. Consider the case $p = q$ when $Q = \Delta^{p-1} \times \Delta^{p-1}$ is the product of two simplices of the same dimension. Recall from Section 2B that faces of $\Sigma(A)$ of codimension 1 (= facets) correspond to so-called coarse (= maximal non-trivial) polyhedral subdivisions of (Q, A). We have described in Section 2B two special classes of coarse subdivisions. Now we are going to exhibit a coarse subdivision of (Q, A) which does not belong to any of these classes.

Let $A_0 \subset A$ be the set of vertices $e_i + f_j$ with $i \leq j$, and $Q_0 = \mathrm{Conv}(A_0)$. Denote by ζ the cyclic permutation of indices $1, \ldots, p$ taking each $i \neq p$ to $i + 1$ and p to 1. Let ζ^m be the m-fold iteration of ζ. For each $m = 0, 1, \ldots, p - 1$, we consider the marked polytope (Q_m, A_m), where $A_m = \{e_i + f_j : \zeta^m(i) \leq \zeta^m(j)\}$, and $Q_m = \mathrm{Conv}(A_m)$ (so (Q_m, A_m) is the image of (Q_0, A_0) under the automorphism $(\zeta^m, \zeta^m) \in S_p \times S_p$ of $\Delta^{p-1} \times \Delta^{p-1}$).

Proposition 3.15. *The* (Q_m, A_m), $m = 0, \ldots, p - 1$ *form a coarse coherent subdivision of* $\Delta^{p-1} \times \Delta^{p-1}$.

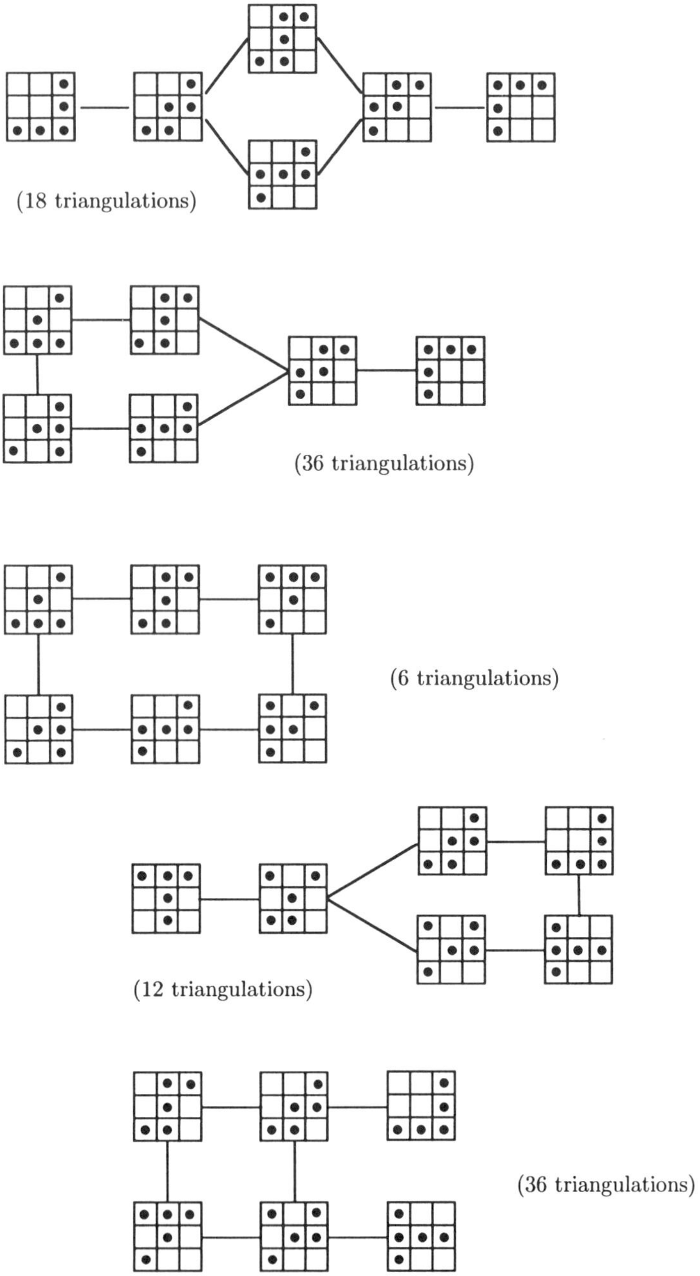

Figure 39. Triangulations of $\Delta^2 \times \Delta^2$ (Postnikov). Only one triangulation in each $S_3 \times S_3$ orbit is depicted; the number of triangulations in each orbit is indicated.

For example, for $p = 2$ we get a subdivision of a square $\Delta^1 \times \Delta^1$ into two triangles.

Proof. A straightforward check shows that, for any $m = 0, \dots, p-1$, we have

$$Q_m = \Big\{(\alpha, \beta) \in Q : \alpha_i + \alpha_{i+1} + \cdots + \alpha_m \leq \beta_i + \beta_{i+1} + \cdots + \beta_m \ (1 \leq i \leq m),$$

$$\alpha_{m+1} + \alpha_{m+2} + \cdots + \alpha_j \geq \beta_{m+1} + \beta_{m+2} + \cdots + \beta_j \ (m+1 \leq j \leq p-1)\Big\}. \quad (3.26)$$

Now the conditions in Definition 2.1 are verified directly, so $\{(Q_m, A_m)\}$ is indeed a polyhedral subdivision of (Q, A) (e.g., to check the condition (b) observe that, for any $l < m$, the intersection $Q_l \cap Q_m$ is a common facet of Q_l and Q_m defined by $\alpha_{l+1} + \cdots + \alpha_m = \beta_{l+1} + \cdots + \beta_m$).

To show that this subdivision is coherent, it is enough to construct a concave piecewise-linear (continuous) function $g : Q \to \mathbf{R}$ whose linearity domains are exactly the polytopes Q_m, cf. Definition 2.3. We define such a g by

$$g|_{Q_m} = (\alpha_1 + \cdots + \alpha_m) - (\beta_1 + \cdots + \beta_m) \quad (3.27)$$

(so $g|_{Q_0} = 0$). The fact that g is well-defined and satisfies all the desired properties follows directly from the above description of the Q_m and their pairwise intersections.

To show that our subdivision is coarse, we use Corollary 2.6. We have only to show that every continuous function $g' : Q \to \mathbf{R}$ whose restriction to each Q_m is affine-linear, has the form $g' = \lambda g + g_0$, where $\lambda \in \mathbf{R}$ and g_0 is a globally affine-linear function on Q. Subtracting, if necessary, from g' a globally affine-linear function, we can assume that $g'|_{Q_0} = 0$. Now for each $m = 1, \dots, p-1$ the affine-linear function $g'|_{Q_m}$ vanishes on the facet $Q_m \cap Q_0$ of Q_m, and hence is divisible by the equation $(\alpha_1 + \cdots + \alpha_m) - (\beta_1 + \cdots + \beta_m)$ of this facet. We conclude that each $g'|_{Q_m}$ is proportional to $g|_{Q_m}$. Since all these restrictions must agree on the intersections $Q_l \cap Q_m$, we conclude that all the coefficients of proportionality are the same, i.e., g' is proportional to g. This completes the proof of our proposition.

Finally let us remark that the volume of $\Delta^{p-1} \times \Delta^{p-1}$ in our usual normalization is equal to the number of simplices in the standard triangulation T_0, i.e., to $\binom{2p-2}{p-1}$. Since the above subdivision has cyclic symmetry, every Q_m has the volume

$$\frac{1}{p}\binom{2p-2}{p-1} = c_p$$

which is the p-th Catalan number (see (3.8)).

CHAPTER 8

A-Resultants and Chow Polytopes of Toric Varieties

We begin, starting with resultants, to apply the general formalism of Part I to discriminants and resultants associated with toric varieties. The treatment of discriminants is left for the next chapter.

1. Mixed $(A_1, \ldots, A_k)$-resultants

A. The definitions

Consider $k-1$ variables $x_1, \ldots, x_{k-1}$. As usual, we associate to any integral vector $\omega \in \mathbf{Z}^{k-1}$ a Laurent monomial x^ω in $x_1, \ldots, x_{k-1}$. For any finite subset $A \subset \mathbf{Z}^{k-1}$, we denote by $\mathbf{C}^A$ the space of Laurent polynomials of the form $\sum_{\omega \in A} a_\omega x^\omega$.

Let $A_1, \ldots, A_k$ be k finite subsets in $\mathbf{Z}^{k-1}$. We shall consider the product $\prod \mathbf{C}^{A_i}$ of the spaces of polynomials corresponding to the A_i. A polynomial from $\mathbf{C}^{A_i}$ will be written as $f_i(x) = \sum_{\omega \in A_i} a_{i,\omega} x^\omega$. Therefore the collection of coefficients $(a_{i,\omega})$ forms a system of coordinates in $\prod \mathbf{C}^{A_i}$. We are going to study conditions under which k polynomials $f_i \in \mathbf{C}^{A_i}$ have a common root. We assume that the following conditions are satisfied:

(1) Each A_i generates $\mathbf{R}^{k-1}$ as an affine space.

(2) All A_i together generate $\mathbf{Z}^{k-1}$ as an affine lattice.

Consider the subvariety $\nabla^0_{A_1,\ldots,A_k} \subset \prod \mathbf{C}^{A_i}$ consisting of k-tuples of polynomials $(f_1, \ldots, f_k)$ for which there is $x \in (\mathbf{C}^*)^{k-1}$ such that all $f_i(x) = 0$. Let $\nabla_{A_1,\ldots,A_k}$ be the closure of $\nabla^0_{A_1,\ldots,A_k}$ (in this case the closure in the Zariski topology coincides with the closure in the usual topology).

Proposition - Definition 1.1. *The variety $\nabla_{A_1,\ldots,A_k}$ is an irreducible hypersurface in $\prod \mathbf{C}^{A_i}$ defined over the rational numbers. Therefore there is a polynomial (unique up to sign) $R_{A_1,\ldots,A_k}(f_1, \ldots, f_k)$ in coefficients $a_{i,\omega}$ of the f_i with the following properties:*

(a) *$R_{A_1,\ldots,A_k}$ has integral coefficients, i.e., it lies in the ring $\mathbf{Z}[(a_{i,\omega})]$;*

(b) *$R_{A_1,\ldots,A_k}$ is an irreducible element of $\mathbf{Z}[(a_{i,\omega})]$ (in particular, the g.c.d. of its coefficients equals 1);*

(c) *If the $f_i \in \mathbf{C}^{A_i}$ are some polynomials which have a common root in $(\mathbf{C}^*)^{k-1}$ then $R_{A_1,\ldots,A_k}(f_1, \ldots, f_k) = 0$.*

The polynomial $R_{A_1,\ldots,A_k}$ is called the $(A_1, \ldots, A_k)$-resultant.

The proof is similar to that of Proposition 3.1 of Chapter 3 and we omit it (a bit later we will show that the $(A_1, \ldots, A_k)$-resultant is in fact a special case of the $(\mathcal{L}_1, \ldots, \mathcal{L}_k)$-resultant in Section 3 of Chapter 3). The only new phenomenon here, as compared to Chapter 3, is the possibility of normalizing the resultant not only up to a non-zero constant factor, but up to sign as well. This is because all the varieties involved are defined over the rational numbers.

Examples 1.2. *(a)* Let $k = 2$, and let A_1 consist of monomials $1, x, \ldots, x^p$, and A_2 consist of monomials $1, x, \ldots, x^q$. Then the (A_1, A_2)-resultant is the classical resultant of two polynomials in one variable.

(b) Let k be arbitrary; let $A_1 = \cdots = A_k$ consist of monomials $1, x_1, \ldots, x_{k-1}$ of degree ≤ 1. Let us write each polynomial f_i, $i = 1, \ldots, k$ as $f_i(x) = a_{i0} + \sum_{j=1}^{k-1} a_{ij} x_j$. Then $R_{A_1,\ldots,A_k}(f_1, \ldots, f_k) = \det\|a_{ij}\|$.

(c) More generally, let $A_1 = \ldots = A_k = A \subset \mathbf{Z}^{k-1}$ be any affinely independent k-element set, i.e., the set of vertices of a $(k-1)$-dimensional simplex in $\mathbf{R}^{k-1}$. Then for a collection $\{f_i(x) = \sum_{\omega \in A} a_{i,\omega} x^\omega : i = 1, \ldots, k\}$, the $(A, \ldots, A)$-resultant $R_{A,\ldots,A}(f_1, \ldots, f_k)$ equals $\det\|a_{i,\omega}\|$.

Now let us relate the $(A_1, \ldots, A_k)$-resultant with the mixed resultant of sections of k invertible sheaves $\mathcal{L}_1, \ldots, \mathcal{L}_k$ on a projective variety X of dimension $(k-1)$ (see Section 3 Chapter 3). To this end we construct a projective variety $X = X_{A_1,\ldots,A_k}$ (which will be a toric variety) and k invertible sheaves $\mathcal{L}_i$, $i = 1, \ldots, k$, in such a way that $\mathbf{C}^{A_i}$ will be embedded into the space of global sections $H^0(X, \mathcal{L}_i)$.

First we recall the construction of the toric variety X_A associated to a set $A \subset \mathbf{Z}^{k-1}$ (see Section 1B Chapter 5). We consider the space $(\mathbf{C}^A)^*$, dual to the space of polynomials. By $(z_\omega)_{\omega \in A}$, we denote the standard coordinates in this space (dual to the coordinates a_ω in $\mathbf{C}^A$). The z_ω serve also as homogeneous coordinates in the projective space $P((\mathbf{C}^A)^*)$. We define X_A as the closure of the image of the map

$$\gamma_A : (\mathbf{C}^*)^{k-1} \to P((\mathbf{C}^A)^*), \quad x \mapsto (z_\omega = x^\omega)_{\omega \in A}. \tag{1.1}$$

Now we generalize this construction.

Definition 1.3. Let $A_1, \ldots, A_k \subset \mathbf{Z}^{k-1}$ be as above. Define $X_{A_1,\ldots,A_k}$ to be the closure of the image of the map

$$\gamma_{A_1} \times \cdots \times \gamma_{A_k} : (\mathbf{C}^*)^{k-1} \to X_{A_1} \times \cdots \times X_{A_k} \subset P((\mathbf{C}^{A_1})^*) \times \cdots \times P((\mathbf{C}^{A_k})^*). \tag{1.2}$$

Proposition 1.4. *The variety $X_{A_1,\ldots,A_k}$ is a toric variety. More precisely, denote by $A_1 + \cdots + A_k$ the set of all pointwise sums $\omega^{(1)} + \cdots + \omega^{(k)}$, $\omega^{(i)} \in A_i$. Then $X_{A_1,\ldots,A_k}$ is isomorphic to $X_{A_1+\cdots+A_k}$. In particular, the convex polytope corresponding to this variety is the Minkowski sum of convex polytopes $Q_i = \mathrm{Conv}(A_i)$.*

Proof. Consider the Segre embedding of $P((\mathbf{C}^{A_1})^*) \times \cdots \times P((\mathbf{C}^{A_k})^*)$ into the projective space $P((\mathbf{C}^{A_1})^* \otimes \cdots \otimes (\mathbf{C}^{A_k})^*)$, the projectivization of the tensor product. Then the weights of the torus $(\mathbf{C}^*)^{k-1}$ on this tensor product are precisely the pointwise sums $\omega^{(1)} + \cdots + \omega^{(k)}$, $\omega^{(i)} \in A_i$, from which we get the assertion.

Denote by $\mathcal{L}_i$ the invertible sheaf on $X_{A_1,\ldots,A_k}$ which is the inverse image of the sheaf $\mathcal{O}(1)$ from the factor $P(\mathbf{C}^{A_i})$. There is a canonical injective map

$$\mathrm{res}_i : \mathbf{C}^{A_i} = H^0(P((\mathbf{C}^{A_i})^*), \mathcal{O}(1)) \to H^0(X_{A_1,\ldots,A_k}, \mathcal{L}_i). \tag{1.3}$$

The map res_i may be not surjective if the variety $X_{A_i} \subset P((\mathbf{C}^{A_i})^*)$ is not linearly normal (see Section 4B Chapter 1).

Proposition 1.5. *Under the maps* res_i *the $(A_1, \ldots, A_k)$-resultant of polynomials corresponds (up to a non-zero constant factor) to the $(\mathcal{L}_1, \ldots, \mathcal{L}_k)$-resultant of sections.*

Proof. We use the Cayley trick for mixed resultants (Section 3B Chapter 3). As in the cited section, we consider the auxiliary variety $\tilde{X} = P\left(\bigoplus \mathcal{L}_i^*\right)$ and its embedding into $P\left(\prod V_i^*\right)$ where $V_i = H^0(X_{A_1,\ldots,A_k}, \mathcal{L}_i)$. Then $R_{\mathcal{L}_1,\ldots,\mathcal{L}_k}$ is the same as the $\tilde{X}$-discriminant $\Delta_{\tilde{X}}$. Let p be the natural projection

$$P\left(\prod V_i^*\right) \to \prod (\mathbf{C}^{A_i})^*$$

induced by $\mathrm{res} = \mathrm{res}_1 \times \cdots \times \mathrm{res}_k$: a hyperplane H in $\prod V_i$ goes to $\mathrm{res}(H)$. Our construction of $X_{A_1,\ldots,A_k}$ implies that p maps the variety $\tilde{X}$ isomorphically to its image. By the same argument, as in the proof of Proposition 3.4 Chapter 3, the $p(\tilde{X})$-discriminant $\Delta_{p(\tilde{X})}$ coincides with $R_{A_1,\ldots,A_k}$. But since $p : \tilde{X} \to p(\tilde{X})$ is an isomorphism, Corollary 4.5 Chapter 1 implies that $\Delta_{p(\tilde{X})}$ coincides with the restriction of $\Delta_{\tilde{X}}$ to $\prod (\mathbf{C}^{A_i})$. This completes the proof.

B. Basic properties

According to Proposition 3.3, Chapter 3, the degree of homogeneity of the $(\mathcal{L}_1, \ldots, \mathcal{L}_k)$-resultant $R_{\mathcal{L}_1,\ldots,\mathcal{L}_k}$ with respect to sections of $\mathcal{L}_i$ equals the intersection index

$$\int_X \prod_{j \neq i} c_1(\mathcal{L}_j) \tag{1.4}$$

of all the invertible sheaves $\mathcal{L}_j$ for $j \neq i$. In our case this intersection index has a transparent interpretation. Namely, this is the number of solutions (in the torus $(\mathbf{C}^*)^{k-1}$) of a generic system of equations $\{f_j(x) = 0 : f_j \in \mathbf{C}^{A_j},\ j \neq i\}$. Indeed, (1.4) is just the number of points of intersection of the zero loci of generic sections $f_j \in H^0(X_{A_1,\ldots,A_k}, \mathcal{L}_j)$, $j \neq i$. If the f_j are generic enough then all the intersection points lie in the open torus orbit $(\mathbf{C}^*)^{k-1} \subset X_{A_1,\ldots,A_k}$. Since every f_j restricts to this orbit as a Laurent polynomial from $\mathbf{C}^{A_j}$, the statement becomes obvious.

By the Bernstein theorem (Theorem 2.8, Chapter 6) the above number of solutions is equal to the mixed volume of the polytopes $Q_1, \ldots, \check{Q}_i, \ldots, Q_k$ (Q_i omitted), where $Q_j = \mathrm{Conv}(A_j)$. Therefore we obtain the following statement.

Proposition 1.6. *The $(A_1, \ldots, A_k)$-resultant $R_{A_1,\ldots,A_k}(f_1, \ldots, f_k)$ is homogeneous with respect to each f_i. The degree of homogeneity with respect to f_i is equal to the mixed volume of all the polytopes Q_j, $j \neq i$ (with respect to the volume form induced by $\mathbf{Z}^{k-1}$).*

The following statement is a direct consequence of Corollary 3.7, Chapter 3 and Proposition 1.5 above.

Proposition 1.7. *Let $(f_1, \ldots, f_k) \in \prod \mathbf{C}^{A_i}$ be a collection of polynomials such that $R_{A_1,\ldots,A_k}(f_1, \ldots, f_k) = 0$ but the differential of $R_{A_1,\ldots,A_k}$ at $(f_1, \ldots, f_k)$ is not zero. Then $f_1, \ldots, f_k$ have at most one common root $x_0 \in (\mathbf{C}^*)^{k-1}$. If such a root exists then for each i the vector in $(\mathbf{C}^{A_i})^*$ with coordinates*

$$\left(z_\omega = \frac{\partial R_{A_1,\ldots,A_k}}{\partial a_{i,\omega}}(f_1, \ldots, f_k) \right)_{\omega \in A_i}$$

is a non-zero constant multiple of the vector with coordinates $(z'_\omega = x_0^{\omega})_{\omega \in A_i}$.

Thus, by knowing the resultant, we can find the common root of a system of polynomials, provided it is unique.

In Chapter 3 we represented the $(\mathcal{L}_1, \ldots, \mathcal{L}_k)$-resultant as the determinant of a certain complex (the resultant complex). Proposition 1.5 implies that the $(A_1, \ldots, A_k)$-resultant can be represented in such a form. However, we shall use the explicit form of the resultant complex only for the case when $A_1 = \ldots = A_k$. This particular case will be the subject of the next section.

2. The A-resultant

We are particularly interested in the $(A_1, \ldots, A_k)$-resultant in the case when all the sets A_i coincide with each other, i.e., $A_1 = \cdots = A_k = A$. We shall call it simply the A-resultant and denote it by R_A. We assume that $A \subset \mathbf{Z}^{k-1}$ is a finite set which affinely generates $\mathbf{Z}^{k-1}$ over $\mathbf{Z}$.

A. The A-resultant as the Chow form

Recall that, to a set $A \subset \mathbf{Z}^{k-1}$ of size n, we associate a toric subvariety $X_A \subset P^{n-1}$ (see Section 1B Chapter 5). The homogeneous coordinates in P^{n-1} correspond naturally to elements of A. We can choose a numbering of these elements: $A = \{\omega^{(1)}, \ldots, \omega^{(n)}\}$; then the coordinates will also be numbered.

The dimension of X_A equals the dimension of the affine span of A in $\mathbf{R}^{k-1}$. Our assumptions imply that this dimension equals $k - 1$.

Let $y_1, \ldots, y_n$ be homogeneous coordinates in P^{n-1}. As noted in Section 1 Chapter 5, any linear form $l(y) = \sum_{j=1}^{n} a_j y_j$, after restriction to the subvariety X_A^0, becomes a Laurent polynomial $f_l(x) = \sum_j a_j x^{\omega^{(j)}}$, $x \in (\mathbf{C}^*)^{k-1}$. Thus we have an identification $l \mapsto f_l$ of the space of linear forms with $\mathbf{C}^A$.

Proposition 2.1. *Let $l_1, \ldots, l_k$ be linear forms on $\mathbf{C}^n$. Then the A-resultant $R_A(f_{l_1}, \ldots, f_{l_k})$ coincides with the resultant $\tilde{R}_{X_A}(l_1, \ldots, l_n)$ corresponding to the toric subvariety X_A.*

The proof is obvious from the definition of the X-resultant, (see Section 2B Chapter 3).

Recall that the X-resultant is very closely related to the Chow form of a projective subvariety X. The Chow form R_X of any $(k-1)$-dimensional subvariety $X \subset P^{n-1}$ of degree d lies in the projective space $P(\mathcal{B}(n-k, n)_d)$ where $\mathcal{B}(n-k, n)$ is the graded coordinate ring of the Grassmannian $G(n-k, n)$. The space $\mathcal{B}(n-k, n)_d$ consists of polynomials in entries of an indeterminate $k \times n$ matrix satisfying the homogeneity condition (Proposition 1.6, Chapter 3). In our case these entries are just the coefficients of the Laurent polynomials f_i. By the Kouchnirenko theorem (Theorem 2.3, Chapter 6), degree d of X_A equals $\mathrm{Vol}(Q)$, where the volume is induced by the lattice $\mathbf{Z}^{k-1}$. So we obtain the following corollary.

Corollary 2.2. *Let $g = \|g_{ij}\|$ be an invertible $k \times k$ matrix. Then, for any $f_1, \ldots, f_k \in \mathbf{C}^A$, denoting $f_i' = \sum_j g_{ij} f_j$, we have*

$$R_A(f_1', \ldots, f_k') = \det(g)^{\mathrm{Vol}(Q)} R_A(f_1, \ldots, f_k).$$

Writing each f_i as $f_i(x) = \sum_{j=1}^{n} a_{ij} x^{\omega^{(j)}}$, we identify the space $(\mathbf{C}^A)^k$ of k-tuples of polynomials with the space of $k \times n$ matrices

$$\begin{pmatrix} a_{11} & a_{12} & \ldots & a_{1n} \\ \vdots & \vdots & \vdots & \vdots \\ a_{k1} & a_{k2} & \ldots & a_{kn} \end{pmatrix}. \tag{2.1}$$

The A-resultant is a polynomial in the matrix entries a_{ij}. Denote by $[i_1, \dots, i_k]$ the polynomial in a_{ij} which is the $k \times k$ minor of the matrix (2.1) on columns with numbers $i_1, \dots, i_k$. The following is another corollary of Proposition 2.1.

Corollary 2.3. *$R_A(f_1, \dots, f_n)$ can be expressed as a polynomial in the brackets $[i_1, \dots, i_k]$.*

Example 2.4. Let $k - 1 = 1$ and A consist of monomials $1, x, \dots, x^d$. Then the A-resultant is the classical resultant of two polynomials in one variable of the same degree d. Let us denote these polynomials as

$$g(x) = a_0 x^d + \cdots + a_d, \quad h(x) = b_0 x^d + \cdots + b_d.$$

The brackets $[ij]$ have the form $[ij] = a_i b_j - a_j b_i$.

We have already mentioned (Example 4.8, Chapter 3) the Bezout formula for the resultant $R(g, h)$. In our present context of non-homogeneous polynomials in one variable (rather than homogeneous polynomials in two variables) this formula is as follows: define the polynomial

$$F(x, y) = \frac{g(x)h(y) - h(x)g(y)}{x - y} = \sum_{i,j=0}^{d-1} c_{ij} x^i y^j.$$

Then $R(g, h) = \det \| c_{ij} \|$. Finding the c_{ij} by explicit division (cf. (5.15), Chapter 2) we get the following expression of $R(g, h)$ as a polynomial in the brackets:

$$R(g,h) = \begin{vmatrix} [01] & [02] & [03] & \dots & [0d] \\ [02] & [03]+[12] & [04]+[13] & \dots & [1d] \\ [03] & [04]+[13] & [05]+[14]+[23] & \dots & [2d] \\ \vdots & \vdots & \vdots & \vdots & \vdots \\ [0d] & [1d] & [2d] & \dots & [d-1,d] \end{vmatrix} \tag{2.2}$$

where the (p, q)-th entry equals $\sum_{i=0}^{\min(p,q)-1} [i, p + q - i]$.

B. The determinantal expression

In Section 4 Chapter 3 we represented any $(\mathcal{L}_1, \dots, \mathcal{L}_k)$-resultant, in particular, the Chow form of any projective subvariety as the determinant of the so-called resultant complex. For the case of the toric variety X_A, this construction can be reformulated in an explicit way.

To describe the graded coordinate ring of X_A, we embed the affine lattice $\mathbf{Z}^{k-1} \supset A$ into $\mathbf{Z}^k = \mathbf{Z}^{k-1} \times \mathbf{Z}$ as the set of integral vectors with the last coordinate 1. By $h : \mathbf{Z}^k \to \mathbf{Z}$, we denote the projection given by this last coordinate.

Let $S = S(A)$ denote the semigroup in $\mathbf{Z}^k$ generated by A. It is graded using the group homomorphism h, and we set

$$S_i := \{\omega \in S(A) : h(\omega) = i\}.$$

The coordinate ring of X_A equals the graded semigroup algebra $\mathbf{C}[S(A)]$. Elements of this semigroup algebra will be written as Laurent polynomials in k variables $\sum_{\omega \in S(A)} c_\omega t^\omega$, where t stands for $(t_1, \ldots, t_k)$ and all but finitely many coefficients c_ω are equal to 0. The component $\mathbf{C}[S(A)]_i \subset \mathbf{C}[S(A)]$ of degree i has, as a vector space, the basis $\{t^\omega : \omega \in S_i\}$.

There is a canonical homomorphism

$$\mathrm{res} : \mathbf{C}[S(A)]_i \to H^0(X_A, \mathcal{O}(i)) \tag{2.3}$$

which is an isomorphism for $i \gg 0$ (this is a general fact valid for the coordinate ring of any projective variety). Note that $\mathbf{C}^A$ is identified with $\mathbf{C}[S(A)]_1$: a polynomial in $(k-1)$ variables $f(x_1, \ldots, x_{k-1}) \in \mathbf{C}^A$ corresponds to $t_k f(t_1, \ldots, t_{k-1}) \in \mathbf{C}[S(A)]_1$.

We consider the resultant complex $C_+^\bullet(\mathcal{O}(1), \ldots, \mathcal{O}(1)|\mathcal{O}(l))$ associated to X_A (see Section 4A Chapter 3) and denote this complex by $C^\bullet(l)$. For $l \gg 0$, we have

$$C^i(l) = \left(\bigoplus_{u \in S_{i+l}} \mathbf{C} \cdot t^u\right) \otimes \bigwedge^i \mathbf{C}^k. \tag{2.4}$$

By a *degree* 1 *form* we mean an element in $\mathbf{C}^A = \mathbf{C}[S(A)]_1$ (see above). For any k-tuple $(f_1, \ldots, f_k)$ of degree 1 forms, the differential $\partial = \partial_{(f_1, \ldots, f_k)} : C^i(l) \to C^{i+1}(l)$ is defined by

$$\partial\big(t^u \otimes (e_{j_1} \wedge \cdots \wedge e_{j_i})\big) := \sum_{j=1}^{k} (t^u \cdot f_j) \otimes (e_j \wedge e_{j_1} \wedge \cdots \wedge e_{j_i}), \tag{2.5}$$

where the e_j are the standard basis vectors in $\mathbf{C}^k$. For the space $C^i(l)$ we choose the basis $\{t^u \otimes (e_{j_1} \wedge \cdots \wedge e_{j_i})\}$. Denote this system of bases for $C^\bullet(l)$ by e.

Theorem 2.5. *For $l \gg 0$, the A-resultant equals*

$$R_A(f_1, \ldots, f_k) = \pm \det\left(C^\bullet(l), \partial_{f_1, \ldots, f_k}, e\right)^{(-1)^k}. \tag{2.6}$$

Proof. The equality up to a constant factor (instead of up to a sign) follows from Theorem 4.2, Chapter 3. To get the equality up to a sign, note that the right hand side of (2.6) is a polynomial in coefficients $a_{i,\omega}$ of the f_i with rational coefficients. Consider its factorization into prime elements of the ring $\mathbf{Z}[a_{i,\omega}]$ of integral polynomials in $a_{i,\omega}$. Among the prime elements of this ring there are the prime *numbers* $p \in \mathbf{Z}$. To show that the equality in (2.6) holds up to sign, we have only to show that no prime numbers are present in the prime factorization of the right hand side.

Indeed, let $p \in \mathbf{Z}$ be a prime number. We consider the modulo p version of the complex $C^\bullet(l)$. Namely, let $C^\bullet_{\mathbf{Z}}$ be the graded Abelian subgroup in $C^\bullet(l)$, integrally spanned by the chosen bases (see above). Let F be the algebraic closure of the field $\mathbf{Z}/p\mathbf{Z}$. We define

$$C^\bullet_F(l) = C^\bullet_{\mathbf{Z}}(l) \otimes_{\mathbf{Z}} F.$$

Let also F^A be the space of polynomials $\sum_{\omega \in A} a_\omega x^\omega$ with coefficients $a_\omega \in F$. For any $f_1, \dots, f_k \in F^A$, we have the complex of F-vector spaces $C^\bullet_F(l)$ with the differential $\partial = \partial_{f_1,\dots,f_k}$ defined similarly to (2.5).

To show that p does not enter into the prime factorization of the right hand side of (2.6) it suffices, by general properties of determinants of complexes (compatibility with base change, see Appendix A, Proposition 24), to show the following.

Lemma 2.6. *For generic polynomials $f_1, \dots, f_k \in F^A$ and $l \gg 0$, the complex of F-vector spaces $(C^\bullet_F(l), \partial_{f_1,\dots,f_k})$ is exact.*

Proof. We consider the algebraic variety $X_A \otimes F$ over the field F and the sheaf $\mathcal{O}(1)$ on this variety. Let $f_1, \dots, f_k \in F^A$ and let $\mathcal{K} = \mathcal{K}_+\big(\mathcal{O}(1)^{\oplus k}, (f_1, \dots, f_k)\big)$ be the Koszul complex of sheaves on $X_A \otimes F$ associated with the f_i (see Section 1B Chapter 2). For the generic f_i their zero loci in $X_A \otimes F$ do not intersect and hence $\mathcal{K}$ is an exact complex of sheaves. On the other hand, $\big(C^\bullet_F(l), \partial_{f_1,\dots,f_k}\big)$ is the complex of global sections of $\mathcal{K} \otimes \mathcal{O}(l)$. So the same argument as in the complex case (see Section 2A Chapter 2) proves the exactness of $C^\bullet_F(l)$ for $l \gg 0$.

Since Lemma 2.6 is proved, the proof of Theorem 2.5 is complete.

3. The Chow polytope of a toric variety and the secondary polytope

A. Some notation and conventions

Let $A \subset \mathbf{Z}^{k-1}$ be a finite subset generating $\mathbf{Z}^{k-1}$ as an affine lattice, and let X_A be the corresponding toric variety. In this section we compute the Chow polytope $\mathrm{Ch}(X_A)$ of X_A. As in Section 1B, Chapter 5, we realize the ambient projective space for X_A as $P((\mathbf{C}^A)^*)$, with the homogeneous coordinates $(z_\omega)_{\omega \in A}$. The torus acting on this space by dilations of homogeneous coordinates is $(\mathbf{C}^*)^A$. Its character lattice is $\mathbf{Z}^A$ and thus $\mathrm{Ch}(X_A)$ is a convex polytope in $\mathbf{R}^A$ (see Section 3, Chapter 6). We denote the standard basis in $\mathbf{Z}^A$ and $\mathbf{R}^A$ by e_ω, $\omega \in A$.

In Section 1D, Chapter 7 we introduced, for any affinely spanning set $A \subset \mathbf{R}^{k-1}$ and any translation invariant volume form Vol on $\mathbf{R}^{k-1}$, the secondary polytope $\Sigma(A) \subset \mathbf{R}^A$. For now we choose the volume form to be $\mathrm{Vol}_{\mathbf{Z}^{k-1}}$, the form induced by $\mathbf{Z}^{k-1}$ (so that the volume of an elementary lattice simplex is 1 (see Section 3D Chapter 5).

Note that both $\Sigma(A)$ and $\mathrm{Ch}(X_A)$ lie in $\mathbf{R}^A$ and have vertices on $\mathbf{Z}^A$.

B. Description of $\mathrm{Ch}(X_A)$*: the statement of results*

Theorem 3.1 [KSZ2]. *The Chow polytope* $\mathrm{Ch}(X_A)$ *coincides with the secondary polytope* $\Sigma(A)$.

The proof of this theorem will occupy the rest of the section. We start with a discussion of the result.

The vertices of $\Sigma(A)$ correspond to coherent triangulations of $Q = \mathrm{Conv}(A)$ with vertices in A (see Theorem 1.7, Chapter 7). The vertices of $\mathrm{Ch}(X_A)$ correspond to extreme toric degenerations of $X_A \subset P((\mathbf{C}^A)^*)$, i.e., toric degenerations into algebraic cycles consisting entirely of coordinate projective subspaces (see Theorem 3.10, Chapter 6). Let us explain how these two classes of objects are connected.

Let $\sigma \subset Q$ be a $(k-1)$-dimensional simplex with vertices in A and let $\mathrm{Vert}(\sigma) \subset A$ be the set of vertices of σ. We denote by L_σ the coordinate projective subspace in $P((\mathbf{C}^A)^*)$ with the coordinates z_ω for $\omega \in \mathrm{Vert}(\sigma)$. We denote by $[\sigma]$ the bracket $[\mathrm{Vert}(\sigma)]$ corresponding to the k-element set $\mathrm{Vert}(\sigma)$ (see Section 3 Chapter 6). Thus, $[\sigma]$ is an element of degree 1 of the coordinate ring $\mathcal{B}(n-k, n)$ of the Grassmannian $G(n-k, n)$. It is defined uniquely up to a sign; to normalize the sign, one needs an ordering of vertices of σ.

Theorem 3.1 is a consequence of the following more refined statement.

Theorem 3.2. *The coherent triangulations of* (Q, A) *correspond bijectively to the extreme toric degenerations of* X_A *via*

$$T \mapsto \sum_{\sigma \in T} \mathrm{Vol}(\sigma)\, L_\sigma .$$

The corresponding weight component of the Chow form R_A *equals*

$$R_{A,\varphi_T} = \pm \prod_{\sigma \in T} [\sigma]^{\mathrm{Vol}(\sigma)} .$$

Theorem 3.2 shows, in particular, that the Chow form of X_A carries the information about all coherent triangulations of (Q, A): all the simplices of a coherent triangulation T together with their volumes can be directly read off the corresponding extreme term of R_A.

Now let us discuss the general idea of the proof of Theorems 3.1 and 3.2. To prove that two polytopes $\mathrm{Ch}(X_A)$ and $\Sigma(A)$ in $\mathbf{R}^A$ coincide, it suffices to show that, for any generic linear function $\lambda : \mathbf{R}^A \to \mathbf{R}$, the vertices

$$\mathrm{Ch}(X_A)^\lambda \in \mathrm{Ch}(X_A), \quad \Sigma(A)^\lambda \in \Sigma(A)$$

where λ achieves its maximum, coincide with each other.

If T is a coherent triangulation of (Q, A) and $\varphi_T \in \Sigma(A)$ is the corresponding vertex then the normal cone $N_{\varphi_T}\Sigma(A)$ (i.e., the cone of linear functions achieving their maximum at φ_T) was described in Theorem 1.7, Chapter 7. Namely, under the standard identification $(\mathbf{R}^A)^* = \mathbf{R}^A$, the cone $N_{\varphi_T}\Sigma(A)$ coincides with the cone $C(T) \subset \mathbf{R}^A$ of "concave T-piecewise-linear functions," (see Definition 1.4, Chapter 7 for details).

On the other hand, by Proposition 1.10, Chapter 5, the information about the normal cones of $\mathrm{Ch}(X_A)$ can be extracted from asymptotics of translations of R_A under the action of various 1-parameter subgroups in the torus $(\mathbf{C}^*)^A$. More specifically, choose an integral vector $\lambda = (\lambda_\omega)_{\omega \in A} \in \mathbf{Z}^A$, and define a 1-parameter subgroup

$$\tau \mapsto \tau^\lambda = \left(\tau^{\lambda_\omega}\right)_{\omega \in A} \in (\mathbf{C}^*)^A. \tag{3.3}$$

We are interested in the action of this subgroup on the space $\mathcal{B}(n-k, n)_d$ that contains R_A. This action is induced from the action on $\mathbf{C}^A$, written as

$$f = \sum_{\omega \in A} a_\omega x^\omega \mapsto \tau^\lambda f = \sum_{\omega \in A} \tau^{\lambda_\omega} a_\omega x^\omega. \tag{3.4}$$

After these preparations, we see that Theorem 3.2 is equivalent to the following.

Theorem 3.3. *Let T be a coherent triangulation of (Q, A), and let $\lambda = (\lambda_\omega)_{\omega \in A} \in \mathbf{Z}^A \cap \mathrm{Int}(C(T))$ (here* Int *stands for the interior). Then, for any $f_1, \ldots, f_k \in \mathbf{C}^A$, the leading term of the Laurent polynomial $\tau \mapsto R_A(\tau^\lambda f_1, \ldots, \tau^\lambda f_k)$ equals*

$$\pm \prod_{\sigma \in T} [\sigma]^{\mathrm{Vol}(\sigma)} \cdot \tau^{\langle \lambda, \varphi_T \rangle}.$$

Recall that by the leading term of a Laurent polynomial $\sum_i c_i \tau^i$, we mean the term $c_i \tau^i$ with the largest i such that $c_i \neq 0$.

The proof of Theorem 3.3 will be given in subsection D below. It will use the representation of R_A as the determinant of a complex (Theorem 2.5) and a certain filtration in this complex induced by λ. To carry this out, we need some technique which we shall also use on several occasions later.

C. Kouchnirenko resolutions

Let $S \subset \mathbf{Z}^k$ be a finitely generated semigroup, and let $K \subset \mathbf{R}^k$ be the convex hull of S. We assume that $\dim(K) = k$. Let $g : K \to \mathbf{R}$ be a concave, continuous piecewise-linear function, homogeneous of degree 1. We assume that the linear functions constituting g have rational coefficients and that g takes integral values on S.

We have an increasing filtration on the semigroup algebra $\mathbf{C}[S]$:

$$F_m\mathbf{C}[S] = \left\{ \sum_{\omega \in S,\ g(\omega) \le m} a_\omega t^\omega \right\}, \quad m \in \mathbf{Z}, \tag{3.5}$$

(see Figure 40).

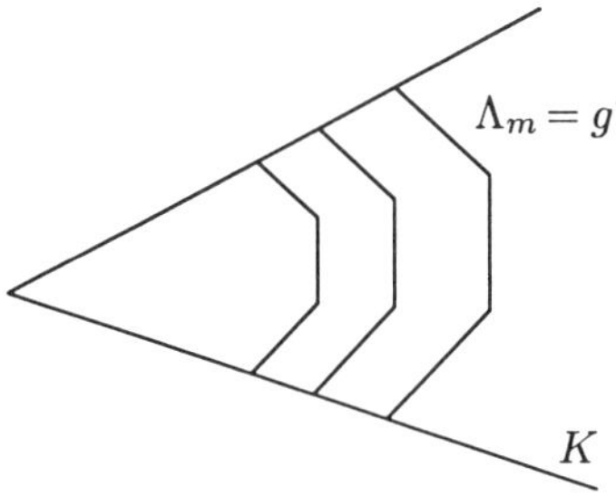

Figure 40. The Kouchnirenko filtration on $\mathbf{C}[S]$

The concavity of g implies that F is a ring filtration: $F_m \cdot F_{m'} \subset F_{m+m'}$. We are interested in the associated graded ring

$$\mathrm{gr}^F_\bullet \mathbf{C}[S] = \bigoplus_m \mathrm{gr}^F_m \mathbf{C}[S], \quad \mathrm{gr}^F_m \mathbf{C}[S] = \frac{F_m\mathbf{C}[S]}{F_{m-1}\mathbf{C}[S]}. \tag{3.6}$$

As a vector space

$$\mathrm{gr}^F_m \mathbf{C}[S] = \bigoplus_{\substack{\omega \in S \\ g(\omega) = m}} \mathbf{C} \cdot t^\omega. \tag{3.7}$$

We are going to describe a resolution of the vector space (3.7) constructed by Kouchnirenko [Kou]. Let $\Lambda = \Lambda_m \subset K$ be the level set $g^{-1}(m)$. This is a polyhedral hypersurface in K. As such, it is the union of its faces of dimension $k-1$. For any face $\Gamma \subset \Lambda$ (not necessarily of dimension $k-1$), we consider the vector space

$$\mathbf{C}[S \cap \Gamma] = \bigoplus_{\omega \in S \cap \Gamma} \mathbf{C} \cdot t^{\omega}. \tag{3.8}$$

Thus we have a surjection

$$\bigoplus_{\substack{\Gamma \subset \Lambda \\ \dim(\Gamma) = k-1}} \mathbf{C}[S \cap \Gamma] \to \mathrm{gr}^F_m \mathbf{C}[S]. \tag{3.9}$$

Kouchnirenko's idea is to extend this surjection to a left resolution of $\mathrm{gr}^F_m \mathbf{C}[S]$ by taking into account the faces of Λ of all dimensions.

Let $\Gamma \subset \Lambda$ be a face of dimension $k-1-r$. We denote by $\mathrm{OR}(\Gamma)$ the *orientation space* of Γ. This is a 1-dimensional vector space which can be described, e.g., as the cohomology space $H^{k-1-r}(\Gamma, \partial\Gamma, \mathbf{C})$ where $\partial\Gamma$ is the boundary of Γ. A choice of orientation of Γ gives an identification $\mathrm{OR}(\Gamma) = \mathbf{C}$. Suppose we fix some orientation of $\mathbf{R}^k$. Then, for any $(k-1)$-dimensional face $\Gamma \subset \Lambda$, the function g defines an orientation of Γ and hence an identification $\mathrm{OR}(\Gamma) = \mathbf{C}$.

If Γ and Θ are two faces of Λ such that $\Gamma \subset \Theta$ and $\dim \Theta = \dim \Gamma + 1$ then there is a canonical map

$$\varepsilon(\Gamma, \Theta) : \mathrm{OR}(\Gamma) \to \mathrm{OR}(\Theta) \tag{3.10}$$

which is just the coboundary homomorphism

$$H^{\dim(\Gamma)}(\Gamma, \partial\Gamma) \to H^{\dim(\Theta)}(\Theta, \partial\Theta).$$

Whenever $\Gamma \subset \Phi$ are two faces of Λ such that $\dim \Phi = \dim \Gamma + 2$, we have

$$\sum_{\Theta : \Gamma \subset \Theta \subset \Phi} \varepsilon(\Theta, \Phi) \circ \varepsilon(\Gamma, \Theta) = 0 \tag{3.11}$$

(the sum has in fact two summands). We shall use the $\varepsilon(\Gamma, \Theta)$ as sign factors to define a complex.

We call a face $\Gamma \subset \Lambda$ *interior* if it does not lie on the boundary $\partial\Lambda$. Define vector spaces

$$U^{-j} = \bigoplus_{\substack{\Gamma \subset \Lambda-\text{interior} \\ \dim(\Gamma) = k-1-j}} \mathbf{C}[S \cap \Gamma] \otimes \mathrm{OR}(\Gamma). \tag{3.12}$$

These spaces fit together into a complex

$$U^\bullet = \{U^{-k+1} \xrightarrow{\delta} \cdots \xrightarrow{\delta} U^{-1} \xrightarrow{\delta} U^0\} \tag{3.13}$$

with the differential δ defined as follows. For $\dim \Gamma = k-1-j$, $\omega \in S \cap \Gamma$ and $\xi \in \mathrm{OR}(\Gamma)$, we set

$$\delta(t^\omega \otimes \xi) = \sum_{\substack{\Theta \supset \Gamma - \text{ interior} \\ \dim(\Theta)=k-j}} t^\omega \otimes \varepsilon(\Gamma, \Theta)(\xi). \tag{3.14}$$

By (3.11), we have $\delta^2 = 0$.

Theorem 3.4. *The complex $U^\bullet$ is a left resolution of* $\mathrm{gr}^F_m \mathbf{C}[S]$.

We call $U^\bullet$ the *Kouchnirenko resolution* of $\mathrm{gr}^F_m \mathbf{C}[S]$.

Proof. The complex $U^\bullet$ is acted on by the torus $(\mathbf{C}^*)^k$. The weight component $U^\bullet_\omega$ of weight $\omega \in \mathbf{Z}^k$ is simply the complex formed by the multiplicity spaces of the monomial t^ω in various terms of $U^\bullet$. This component is zero unless $\omega \in \Lambda$ and in the latter case it is described as follows. Let $\Gamma \subset \Lambda$ be the smallest interior face containing ω. Let $r = \dim \Gamma$. Then $U^\bullet_\omega$ is the augmented cochain complex of the polyhedral $(k-1-r)$-sphere whose p-cells correspond to $(r+p)$-dimensional faces of Λ containing Γ. So this complex is exact everywhere except at the end. The theorem is proved.

We also need a slightly different version of the Kouchnirenko resolutions. Consider the field $\mathbf{C}((\tau^{-1}))$ of the Laurent series of the form $\sum_{j=-\infty}^{p} a_j \tau^j$. This field has an increasing filtration

$$F_m \mathbf{C}((\tau^{-1})) = \left\{ \sum_{j \le m} a_j \tau^j \right\}, \quad m \in \mathbf{Z}. \tag{3.15}$$

The associated graded ring

$$\mathrm{gr}^F_\bullet \mathbf{C}((\tau^{-1})) = \mathbf{C}[\tau, \tau^{-1}] \tag{3.16}$$

is the ring of Laurent polynomials.

Let the semigroup S, the cone K and the function g be as before. Consider the ring $\mathbf{C}[S]((\tau^{-1}))$ with the increasing filtration

$$F_m \mathbf{C}[S]((\tau^{-1})) = \left\{ \sum_{\omega \in S,\ j \in \mathbf{Z}} a_\omega t^\omega \tau^j, \quad j \le g(\omega) + m \right\}. \tag{3.17}$$

The resolution of the associated graded ring $\mathrm{gr}^F_\bullet \mathbf{C}[S]((\tau^{-1}))$ can be constructed as in Theorem 3.4. More precisely, let P be the polyhedral subdivision of K into cones which are domains of linearity of g (so the faces of the polyhedral hypersurface $\Lambda = g^{-1}(m)$ considered above are just the intersections of Λ with cones from P).

A cone $\kappa \in P$ (of any dimension) will be called *interior* if it is not contained in the boundary ∂K. For any $\kappa \in P$, we define the orientation space of κ as

$$\mathrm{OR}(\kappa) = H^{\dim(\kappa)}(\kappa, \partial\kappa \cup \kappa^\infty, \mathbf{C})$$

where κ^∞ is a neighborhood of infinity in κ. For example, we can take

$$\kappa^\infty = \{z \in \kappa : \|z\| \geq 1\},$$

where $\|z\|$ is the standard Euclidean norm on $\mathbf{R}^k$. For the κ of maximal dimension, we have a canonical identification $\mathrm{OR}(\kappa) = \mathbf{C}$ induced by the chosen orientation of $\mathbf{R}^k$.

If $\kappa \subset \rho$ are two cones of P of adjacent dimensions then we have, as before, the map

$$\varepsilon(\kappa, \rho) : \mathrm{OR}(\kappa) \to \mathrm{OR}(\rho)$$

satisfying the analog of (3.11). For any $j = 0, 1, \ldots, k$, we consider the space

$$W^{-j} = \bigoplus_{\substack{\kappa \in P-\text{interior} \\ \dim(\kappa)=k-j}} \mathbf{C}[\tau, \tau^{-1}] \otimes \mathbf{C}[S \cap \kappa] \otimes \mathrm{OR}(\kappa). \tag{3.18}$$

These spaces are arranged into a complex of graded $\mathbf{C}[\tau, \tau^{-1}]$-modules

$$W^\bullet = \{W^{-k} \xrightarrow{\delta} \cdots \xrightarrow{\delta} W^{-1} \xrightarrow{\delta} W^0\} \tag{3.19}$$

whose differential δ is given by a formula similar to (3.14):

$$\delta(t^\omega \otimes \xi) = \sum_{\substack{\rho \supset \kappa-\text{ interior} \\ \dim(\rho)=k-j+1}} t^\omega \otimes \varepsilon(\kappa, \rho)(\xi). \tag{3.20}$$

Theorem 3.5. *The complex $W^\bullet$ is a left resolution of* $\mathrm{gr}^F_\bullet \mathbf{C}[S]((\tau^{-1}))$.

The proof is similar to that of Theorem 3.4 and we omit it.

D. Proof of Theorem 3.3

Now we are in position to prove Theorem 3.3. By Theorem 2.5, we have the determinantal representation for the A-resultant:

$$R_A(f_1, \dots, f_k) = \pm\det(C^\bullet(l), \partial_{f_1,\dots,f_k}, e) \tag{3.21}$$

where $C^\bullet(l)$ is the resultant complex. Choosing of $\lambda = (\lambda_\omega)_{\omega\in A} \in \mathbf{Z}^A$, we get a complex $\tilde{C}^\bullet(l)$ of vector spaces over $\mathbf{C}((\tau^{-1}))$ which is obtained from $(C^\bullet(l), \partial_{\tau^\lambda f_1,\dots,\tau^\lambda f_k})$ by extension of scalars. The determinant of this complex with respect to the system of bases e introduced above equals the Laurent polynomial $R_A(\tau^\lambda f_1, \dots, \tau^\lambda f_k)$.

We now introduce an increasing filtration on $\tilde{C}^\bullet(l)$. Let $g_{T,\lambda} : Q \to \mathbf{R}$ be the T-piecewise-linear function such that $g_{T,\lambda}(\omega) = \lambda_\omega$ for any $\omega \in A$ which is a vertex of a simplex of T (see Section 1C, Chapter 7). The fact that λ lies in $C(T)$ means that $g_{T,\lambda}$ is concave. Recall that the space $\mathbf{R}^{k-1}$ containing Q is embedded into $\mathbf{R}^k = \mathbf{R}^{k-1} \times \mathbf{R}$ as the set of vectors with the last coordinate 1. Let $K = \mathbf{R}_+ Q \subset \mathbf{R}^k$ be the cone generated by Q. We extend the function $g_{T,\lambda} : Q \to \mathbf{R}$ to a homogeneous function $g : K \to \mathbf{R}$ of degree 1. Clearly g is also concave.

We use the notation of Section 2B. In particular, $h : \mathbf{Z}^k \to \mathbf{Z}$ is the last coordinate, $S = S(A) \subset \mathbf{Z}^k$ is the semigroup generated by A and $S_i = \{\omega \in S(A) : h(\omega) = i\}$.

Now we introduce a filtration on $\tilde{C}^\bullet(l)$ by

$$F_m\tilde{C}^i(l) \quad := \quad \bigoplus_{u\in S_{i+l}} \left(F_{m+g(u)}\mathbf{C}((\tau^{-1})) \cdot t^u \right) \otimes \bigwedge^i \mathbf{C}^k, \tag{3.22}$$

where the filtration of $\mathbf{C}((\tau^{-1}))$ is defined by (3.15).

Proposition 3.6. *The differential ∂ of $\tilde{C}^\bullet(l)$ is compatible with the filtration F, i.e., $\partial\big(F_m\tilde{C}^i(l)\big) \subseteq F_m\tilde{C}^{i+1}(l)$.*

Proof. By the concavity of g, we have

$$g(u+\omega) \;\geq\; g(u) + g(\omega) \;=\; g(u) + \lambda_\omega$$

for all $u \in S_{i+l}$ and $\omega \in S_1$. A typical element in $F_m\tilde{C}^i(l)$ is a $\mathbf{C}$-linear combination of elements $\tau^s t^u \otimes (e_{j_1} \wedge \cdots \wedge e_{j_i})$ where $s \leq m + g(u)$. The image of such an element under the differential ∂ is a $\mathbf{C}$-linear combination of elements

$$\tau^{s+\lambda_\omega} t^{u+\omega} \otimes (e_j \wedge e_{j_1} \wedge \cdots \wedge e_{j_i}) \quad \text{where} \quad \omega \in S_1.$$

This element lies in $F_m\tilde{C}^{i+1}(l)$ because

$$s+\lambda_\omega \;\leq\; m+g(u)+\lambda_\omega \;\leq\; m+g(u+\omega).$$

This proves our proposition.

We now pass to associated graded objects. For the field $\mathbf{C}((\tau^{-1}))$ the associated graded ring is the Laurent polynomial ring $\mathbf{C}[\tau,\tau^{-1}]$. For each i, the space $\mathrm{gr}^F\tilde{C}^i(l)$ is a free $\mathbf{C}[\tau,\tau^{-1}]$-module having the same basis $\{t^u\otimes(e_{j_1}\wedge\cdots\wedge e_{j_i})\}$ as above.

By Proposition 3.6, we obtain an associated graded complex $\mathrm{gr}^F\tilde{C}^\bullet(l)$, which is a complex of finitely-generated free $\mathbf{C}[\tau,\tau^{-1}]$-modules. The matrix of its differential $\mathrm{gr}^F\partial$ is the leading term in the expansion of the matrix of ∂ as a Laurent series in τ with matrix coefficients. Therefore (by Theorem 27 from Appendix A) the determinant of the complex $\mathrm{gr}^F\tilde{C}^\bullet(l)$ equals the leading term of the determinant of $\tilde{C}^\bullet(l)$. For integers $l\gg 0$, the determinant of $\mathrm{gr}^F\tilde{C}^\bullet(l)$ equals the leading term of the Laurent polynomial $\tau\mapsto R_A(\tau^\lambda f_1,\ldots,\tau^\lambda f_k)$. For the proof of Theorem 3.3, it now suffices to show that

$$\det\left(\mathrm{gr}^F\tilde{C}^\bullet(l)\,,\;e\right)=\pm\prod_{\sigma\in T}\left(\,[\,\sigma\,]^{\mathrm{Vol}(\sigma)}\,\tau^{\lambda_\sigma\cdot\mathrm{Vol}(\sigma)}\right), \tag{3.23}$$

where $\lambda_\sigma=\sum_{\omega\in\mathrm{Vert}(\sigma)}\lambda_\omega$. To prove (3.23) we use the Kouchnirenko resolution.

Let σ be any simplex (of arbitrary dimension) of the triangulation T. We say that σ is interior if it does not lie on ∂Q. The simplicial cone in $\mathbf{R}^k$ generated by σ will be denoted by $\mathbf{R}_+\sigma$. We abbreviate $S_\sigma := S(A)\cap\mathbf{R}_+\sigma$ and $S_{i,\sigma} := S_i\cap S_\sigma$. We define a complex

$$W_\sigma^\bullet(l) \quad=\quad \left\{\,W_\sigma^0(l)\xrightarrow{\partial_\sigma}W_\sigma^1(l)\xrightarrow{\partial_\sigma}\cdots\xrightarrow{\partial_\sigma}W_\sigma^k(l)\,\right\}$$

of free $\mathbf{C}[\tau,\tau^{-1}]$-modules as follows. Set

$$W_\sigma^i(l)\quad:=\quad\left(\bigoplus_{u\in S_{i+l,\sigma}}\mathbf{C}[\tau,\tau^{-1}]\cdot t^u\right)\otimes\bigwedge^i\mathbf{C}^k \tag{3.24}$$

and

$$\partial\left(t^u\otimes(e_{j_1}\wedge\cdots\wedge e_{j_i})\right):=\sum_{j=1}^k\left(t^u\cdot(\tau^\lambda f_j)|_\sigma\right)\otimes(e_j\wedge e_{j_1}\wedge\cdots\wedge e_{j_i}), \tag{3.25}$$

where $\left(\sum_{\omega\in A} c_\omega x^\omega\right)|_\sigma := \sum_{\omega\in \mathrm{Vert}(\sigma)} c_\omega x^\omega$. For $\nu = 0, 1, \ldots, k-1$, we introduce the direct sum of complexes

$$W^{\bullet,-\nu}(l) := \bigoplus_{\substack{\sigma\in T \\ \mathrm{codim}(\sigma)=\nu}} W^\bullet_\sigma(l) \otimes \mathrm{OR}(\sigma),$$

where $\mathrm{OR}(\sigma) = \mathrm{OR}(\mathbf{R}_+\sigma)$ is the orientation space defined in subsection C above.

Previously, Kouchnirenko's construction provided us with resolutions of some vector spaces. We now want to construct resolutions for not just vector spaces, but complexes. Let $C^\bullet$ be a complex of vector spaces. By a *left resolution* of $C^\bullet$ we mean a double complex of the form

$$0 \to D^{\bullet,-m} \to D^{\bullet,-m+1} \to \ldots \to D^{\bullet,0} \to C^\bullet \to 0 \tag{3.26}$$

with exact rows. More precisely, we think of $C^\bullet$ and each $D^{\bullet,-j}$ as columns of our double complex. These columns are, therefore, complexes and maps in (3.26) are morphisms of complexes. With this understanding, we state our next result.

Proposition 3.7. *The complex* $\mathrm{gr}^F \tilde{C}^\bullet(l)$ *of free* $\mathbf{C}[\tau, \tau^{-1}]$*-modules has a left resolution*

$$0 \to W^{\bullet,-k+1}(l) \to W^{\bullet,-k+2}(l) \to \cdots \to W^{\bullet,0}(l) \to \mathrm{gr}^F \tilde{C}^\bullet(l) \to 0. \tag{3.27}$$

Proof. Let us fix j and define a complex of vector spaces

$$0 \to W^{j,-k+1}(l) \to \cdots \to W^{j,0}(l) \to \mathrm{gr}^F \tilde{C}^j(l) \to 0. \tag{3.28}$$

All the vector spaces in (3.28) are already defined and we have only to define the differential. To do this we note that the graded vector space which is the direct sum of all the terms in (3.28) is naturally identified with the direct sum of $\binom{k}{j}$ copies of the part of Kouchnirenko's complex $W^\bullet$ spanned by monomials from S_{l+j}. So we define the differential in (3.28) to be the direct sum of $\binom{k}{j}$ copies of Kouchnirenko's differential (3.20). The commutativity of these new differentials with the differentials (3.25) (i.e., the fact that we indeed get a double complex) is verified immediately. The exactness of rows follows from Theorem 3.5. The proposition is proved.

Note that each orientation space $\mathrm{OR}(\sigma)$ has a distinguished basis vector defined up to a sign. Tensoring this vector with the bases $\{t^u \otimes (e_{j_1} \wedge \cdots \wedge e_{j_i})\}$ as above, we get a system of bases in $W^\bullet_\sigma(l)$ and hence in $W^{\bullet,-\nu}(l)$ which are defined up to signs and permutations of basis vectors. For an abbreviation of notation we denote all these bases by e.

Proposition 3.8. *We have*

$$\det\left(\mathrm{gr}^F \tilde{C}^\bullet(l)\,,\ e\right) = \pm\prod_{\nu=0}^{k-1} \det\left(W^{\bullet,-\nu}(l)\,,\ e\right)^{(-1)^\nu}$$
$$= \pm\prod_{\sigma\in T} \det\left(W^\bullet_\sigma(l)\,,\ e\right)^{(-1)^{\mathrm{codim}(\sigma)}}.$$

Proof. Our strategy is to replace $\mathrm{gr}^F \tilde{C}^\bullet(l)$ by its resolution given in Proposition 3.7. So we consider the following part of the double complex (3.27):

$$W^{\bullet\bullet}(l) = \left\{W^{\bullet,-k+1}(l) \to \ldots \to W^{\bullet,0}(l)\right\}.$$

Let $W^\bullet(l)$ be the associated simple complex. Consider the increasing filtration G in $W^\bullet(l)$ such that $G_\nu W^\bullet(l)$ is the subcomplex formed by $W^{j,-l}(l)$ with $l \le \nu$. Quotients of this filtration are shifted complexes $W^{\bullet,-\nu}(l)[\nu]$. Thus, by the multiplicativity of determinants of complexes with respect to filtrations (Proposition 17, Appendix A),

$$\det(W^\bullet(l), e) \;=\; \pm\prod_{\nu=0}^{k-1} \det\left(W^{\bullet,-\nu}(l)\,,\ e\right)^{(-1)^\nu}.$$

Now note that every complex (3.28) has determinant ±1. Indeed, let $W^{j,-\nu}_{\mathbf{Z}}(l) \subset W^{j,-\nu}(l)$ be the (free) $\mathbf{Z}[\tau, \tau^{-1}]$-submodule generated by the basis vectors described above. Let $\tilde{C}^\bullet_{\mathbf{Z}}(l)$ be a similar submodule in $\tilde{C}^\bullet(l)$. Note that (3.28) comes from a complex of free $\mathbf{Z}[\tau, \tau^{-1}]$-modules

$$0 \to W^{j,-k+1}_{\mathbf{Z}}(l) \to \cdots \to W^{j,0}_{\mathbf{Z}}(l) \to \mathrm{gr}^F \tilde{C}^j_{\mathbf{Z}}(l) \to 0$$

which is exact as well. This is because Kouchnirenko's complexes, being direct sums of augmented cochain complexes of polyhedral spheres, are defined and exact over $\mathbf{Z}$. Thus the determinant of (3.28) is an invertible element in $\mathbf{Z}[\tau, \tau^{-1}]$, i.e., it has the form $\pm\tau^l$, $l \in \mathbf{Z}$. However, the matrix elements of differentials in (3.28) do not depend on τ. Thus the determinant is equal to ±1, as claimed.

Now consider another filtration in $W^\bullet(l)$ whose quotients are shifted rows of $W^{\bullet\bullet}(l)$. We find, similar to the above, that

$$\det\left(\mathrm{gr}^F \tilde{C}^\bullet(l), e\right) = \det\left(W^\bullet(l), e\right),$$

from which we obtain the assertion of the proposition.

In view of Proposition 3.8, formula (3.23) and hence Theorem 3.3 are implied by the following result.

Proposition 3.9. *For any simplex $\sigma \in T$ and $l \gg 0$ we have*

$$\det(W_\sigma^\bullet(l) , e) = \pm[\sigma]^{\mathrm{Vol}(\sigma)} \tau^{\lambda_\sigma \mathrm{Vol}(\sigma)}.$$

In particular, this determinant is ± 1 if σ does not have the full dimension $k-1$.

Proof. We proceed by analyzing three cases.
Case 1: $\dim(\sigma) = k-1$, $\mathrm{Vol}(\sigma) = 1$. The set σ is a basis of $\mathbf{Z}^k$. All complexes in question are invariant under change of basis in $\mathbf{Z}^k$, so we may assume $\sigma = \{e_1, \ldots, e_k\}$. In this case $W_\sigma^\bullet(l)$ is the usual Koszul resolution of the k linear forms $(\tau^\lambda f_i)|_\sigma$. The determinant of this complex is the resultant of these linear forms, which equals $[\sigma] \cdot \tau^{\lambda_\sigma}$.

Case 2: $\dim(\sigma) = k-1$, $\mathrm{Vol}(\sigma) > 1$. Let $\mathbf{Z}\sigma$ be the subgroup of $\mathbf{Z}^k = \mathbf{Z}A$ generated by σ. By our choice of the volume form, the index of $\mathbf{Z}\sigma$ in $\mathbf{Z}^k$ equals $d := \mathrm{Vol}(\sigma)$. Choose representatives $u_1, u_2, \ldots, u_d$ in $\mathbf{Z}^k$ modulo $\mathbf{Z}\sigma$. For $j = 1, 2, \ldots, d$, define a subspace $W_{\sigma,j}^i(l)$ of $W_\sigma^i(l)$ by taking all the summands in (3.24) indexed by $u \in S_{i+l,\sigma} \cap (u_j + \mathbf{Z}\sigma)$. The differential in (3.25) maps $W_{\sigma,j}^i(l)$ into $W_{\sigma,j}^{i+1}(l)$. Hence we get subcomplexes $W_{\sigma,j}^\bullet(l)$ for $j = 1, 2, \ldots, d$ such that $W_\sigma^\bullet(l)$ is the direct sum of these subcomplexes. Hence

$$\det(W_\sigma^\bullet(l) , e) = \prod_{j=1}^{d} \det(W_{\sigma,j}^\bullet(l) , e).$$

For $l \gg 0$ each complex $W_{\sigma,j}^\bullet(l)$ is isomorphic to the standard Koszul resolution in Case 1, which implies the assertion.

Case 3: $\dim(\sigma) < k-1$. As in the cases 1 and 2, the complex $W_\sigma^\bullet(l)$ is isomorphic to a direct sum of Koszul resolutions of k generic linear forms in $\dim(\sigma)+1$ variables. This complex fails to be exact under a specialization of the coefficients if and only if the forms have a non-zero common root. This imposes a condition of codimension $k - \dim(\sigma) \geq 2$, and hence $\det(W_\sigma^\bullet(l) , e)$ is a non-zero constant. A similar argument as in the proof of Theorem 2.5 (reduction modulo p) shows that $\det(W_\sigma^\bullet(l) , e) = \pm 1$.

Theorems 3.1, 3.2 and 3.3 are completely proved.

CHAPTER 9

A-Discriminants

We now introduce the second main object of study: the A-discriminant Δ_A.

1. Basic definitions and examples

A. Definitions and first examples

Our setup now will be the same as in Section 1, Chapter 5. Namely, we choose a finite subset A in the integral lattice $\mathbf{Z}^{k-1}$ whose elements ω correspond to Laurent monomials $x^\omega = x_1^{\omega_1} \cdots x_{k-1}^{\omega_{k-1}}$ in $k-1$ variables. We consider the space $\mathbf{C}^A$ of Laurent polynomials of the form $f(x) = \sum_{\omega \in A} a_\omega x^\omega$.

We let $\nabla_0 \subset \mathbf{C}^A$ denote the set of all f for which there exists $x^{(0)} \in (\mathbf{C}^*)^{k-1}$ such that

$$f(x^{(0)}) = (\partial f/\partial x_i)(x^{(0)}) = 0 \ for\ all\ i. \tag{1.1}$$

Let ∇_A be the closure of ∇_0. It is not hard to see that ∇_A is an irreducible variety defined over $\mathbf{Q}$. Indeed, let X_A be the toric variety associated to A (Section 1B Chapter 5). Then we have the following fact, which is obvious from the definitions.

Proposition 1.1. *The variety ∇_A is conical, i.e., it is invariant under the multiplication by scalars. Its projectivization $P(\nabla_A)$ is the variety projectively dual to X_A.*

Now we can give the definition of the A-discriminant.

Definition 1.2. If the set $A \subset \mathbf{Z}^{k-1}$ has the property that $\nabla_A \subset \mathbf{C}^A$ is a subvariety of codimension 1, then by the *A-discriminant* we mean an irreducible integral polynomial $\Delta_A(f)$ in the coefficients a_ω, $\omega \in A$ of $f \in \mathbf{C}^A$ which vanishes on ∇_A. Such a polynomial is uniquely determined up to sign. If $\operatorname{codim} \nabla_A > 1$, we set $\Delta_A = 1$.

Thus Δ_A is a particular case of the general discriminants defined in Chapter 1: under the notation of that chapter, we have $\Delta_A = \Delta_{X_A}$. We start with some simple properties and then give basic examples.

Proposition 1.3. *The polynomial Δ_A is homogeneous. In addition, it satisfies $(k-1)$ quasi-homogeneity conditions: for all monomials $\prod a_\omega^{m(\omega)}$ in Δ_A, the vector $\sum m(\omega) \cdot \omega \in \mathbf{Z}^{k-1}$ is the same.*

Proof. If $f \in \nabla_A$ and $\lambda_0, \lambda_1, \ldots, \lambda_{k-1}$ are nonzero numbers, then the polynomial

$$g(x_1, \ldots, x_{k-1}) = \lambda_0 f(\lambda_1 x_1, \ldots, \lambda_{k-1} x_{k-1})$$

is obviously in ∇_A as well. This implies our proposition.

In fact, the variety ∇_A, and hence also the polynomial Δ_A, depend only on the affine geometry of $A \subset \mathbf{Z}^{k-1}$.

Proposition 1.4. *Let $A \subset \mathbf{Z}^{k-1}$, $B \subset \mathbf{Z}^{m-1}$ be two finite subsets and $T : \mathbf{Z}^{k-1} \to \mathbf{Z}^{m-1}$ be an integral affine transformation which is injective and such that $T(A) = B$. Then under the corresponding identification of $\mathbf{C}^A$ and $\mathbf{C}^B$, the variety ∇_A is identified with ∇_B and the polynomial Δ_A is identified with Δ_B.*

Proof. In this case X_A and X_B are naturally identified (Proposition 1.2, Chapter 5). Hence the projectively dual varieties are also identified.

Remark 1.5. The above proposition means, in particular, that we can shrink, if necessary, the affine lattice $\mathbf{Z}^{k-1}$ containing A to the lattice $\mathrm{Aff}_{\mathbf{Z}}(A)$, affinely generated over $\mathbf{Z}$ by A (see Definition 1.3, Chapter 5).

Examples 1.6. We consider the same choices of sets A which were discussed in Examples 1.1, Chapter 5.

(a) Let A consist of all monomials of degree $\leq d$ in $k-1$ variables $x_1, \ldots, x_{k-1}$. The space $\mathbf{C}^A$ consists of all polynomials $f(x_1, \ldots, x_{k-1})$ of degree $\leq d$. Equivalently, let $\tilde{A}$ consist of all homogeneous monomials in k variables $x_1, \ldots, x_k$ of degree exactly d. Then $\mathbf{C}^{\tilde{A}}$ is the space $S^d\mathbf{C}^k$ of forms of degree d in k variables. There is an obvious identification

$$\mathbf{C}^{\tilde{A}} \to \mathbf{C}^A, \quad f(x_1, \ldots, x_k) \mapsto f(x_1, \ldots, x_{k-1}, 1),$$

which takes $\Delta_{\tilde{A}}$ to Δ_A. The polynomial $\Delta_{\tilde{A}}$ is the classical discriminant of a form of degree d in k variables, discussed in Example 4.15 Chapter 1. Recall, in particular, Boole's formula $\deg \Delta = k(d-1)^{k-1}$.

(b) Let A consist of bilinear monomials $x_i \cdot y_j$, where the x_i and y_j ($i = 1, \ldots, m$; $j = 1, \ldots, n$) are two sets of variables. Then $\mathbf{C}^A$ consists of bilinear forms $f(x, y) = \sum a_{ij} x_i y_j$ and is identified with the space of $m \times n$ matrices $\|a_{ij}\|$. The A-discriminant $\Delta_A(f) = \Delta_A(\|a_{ij}\|)$ is identically equal to 1 unless $m = n$, and in this case it is the determinant of the square matrix $\|a_{ij}\|$. The monomials in $\Delta_A(f)$ have an obvious combinatorial significance: they correspond to the permutations of the set of m elements, and the coefficients are the signs of the permutations. This explains our interest in the monomials appearing in the A-discriminant in other cases.

(b′) Let A consist of trilinear monomials

$$x_i \cdot y_j \cdot z_l, \; i = 1, \ldots, m_1, \; j = 1, \ldots, m_2, \; l = 1, \ldots, m_3.$$

The previous example makes it natural to refer to the A-discriminant of a polynomial

$$\sum_{i=1}^{m_1} \sum_{j=1}^{m_2} \sum_{l=1}^{m_3} a_{ijk} x_i y_j z_l \in \mathbf{C}^A$$

as the *hyperdeterminant* of the three-dimensional “matrix” $\|a_{ijl}\|$. This concept (and its obvious generalization to higher-dimensional “matrices”) was introduced by Cayley [Ca1] at almost the same time as that of the determinant of a square matrix. Hyperdeterminants were later studied by Schläfli [Schl] and, after the break of almost 150 years, by the authors [GKZ3]. In view of the previous example, the monomials in the hyperdeterminant form a “higher analog” of the symmetric group. We shall present our treatment of the hyperdeterminants in Chapter 14. (We note that there is another notion of the “determinant” of a multidimensional matrix which is different from ours (see, e.g., [P] [So]). It is based on a direct generalization of the determinant formula for a square matrix and includes a summation over the product of several symmetric groups.)

(c) Let A consists of monomials

$$1, x, x^2, \ldots, x^p, \; y, yx, yx^2, \ldots, yx^q.$$

Then $\mathbf{C}^A$ consists of polynomials $\Phi(x, y) = f(x) + yg(x)$ where f, g are polynomials in one variable x of degrees not greater than p or q, respectively. In this case $\Delta_A(\Phi) = R(f, g)$ is the classical resultant of f and g (see Example 3.6 Chapter 3). This relation holds in a more general context.

Let $A_1, \ldots, A_k \subset \mathbf{Z}^{k-1}$ be finite subsets satisfying the assumptions of Section 1A, Chapter 8. Let $R_{A_1,\ldots,A_k}$ be the $(A_1, \ldots, A_k)$-resultant; it is a function of k polynomials $f_i \in \mathbf{C}^{A_i}$. Let $A \subset \mathbf{Z}^{2k-2} = \mathbf{Z}^{k-1} \times \mathbf{Z}^{k-1}$ be the following set:

$$A = (A_1 \times \{e_1\}) \cup \cdots \cup (A_{k-1} \times \{e_{k-1}\}) \cup (A_k \times \{0\}), \tag{1.2}$$

where the e_i are the standard basis vectors of $\mathbf{Z}^{k-1}$. Then $\mathbf{C}^A$ is the space of polynomials of the form

$$f_k(x) + \sum_{i=1}^{k-1} y_i f_i(x),$$

where $f_i \in \mathbf{C}^{A_i}$. We have the following statement (“Cayley trick”).

Proposition 1.7. *We have*

$$R_{A_1,\ldots,A_k}(f_1,\ldots,f_k) = \Delta_A\left(f_k(x) + \sum_{i=1}^{k-1} y_i f_i(x)\right).$$

Proof. This follows from Corollary 3.5, Chapter 3 about the relation between projectively dual and associated varieties. The proof is so simple that we repeat it here for our particular case. If $x^{(0)}$ is a common root of $(f_1,\ldots,f_k)$, then we can find $y_1^{(0)},\ldots,y_{k-1}^{(0)}$ such that the polynomial $f_k+\sum y_i f_i$ vanishes at the point $y_1^{(0)},\ldots,y_{k-1}^{(0)},x_1^{(0)},\ldots,x_k^{(0)}$ along with its first derivatives. We do this by solving a linear system. Conversely, if $(y^{(0)},x^{(0)})$ is such a point, then

$$f_i(x^{(0)}) = \left.\frac{\partial(f_k+\sum_{i=1}^{k-1} y_i f_i)}{\partial y_i}\right|_{(x^{(0)},y^{(0)})} = 0$$

for $1 \le i \le k-1$, and so $f_k(x^{(0)}) = 0$ as well.

B. The case of a circuit

Let $A \subset \mathbf{Z}^{k-1}$ be a circuit. This means (see Section 1B, Chapter 7) that A is affinely dependent, but any proper subset of A is affinely independent. In this case the A-discriminant Δ_A can be calculated explicitly. We present this formula, following Kouchnirenko [Kou].

We can assume that A generates $\mathbf{Z}^{k-1}$ as an affine lattice. So $\#(A) = k+1$. There is, up to scaling, just one affine relation between elements of A:

$$\sum_{\omega\in A} m_\omega\cdot\omega = 0,\quad \sum_{\omega\in A} m_\omega = 0. \tag{1.3}$$

We normalize such a relation uniquely up to sign by requiring that all m_ω be integers with the greatest common divisor equal to 1. Note that

$$|m_\omega| = \mathrm{Vol}_{\mathbf{Z}^{k-1}}(\mathrm{Conv}(A-\{\omega\})).$$

Let $A_+, A_- \subset A$ be the sets of ω such that m_ω is positive (resp. negative).

Proposition 1.8. *Suppose that $A \subset \mathbf{Z}^{k-1}$ is a circuit which generates $\mathbf{Z}^{k-1}$ as an affine lattice. Let $f = \sum_{\omega\in A} a_\omega x^\omega$ be an indeterminate polynomial from $\mathbf{C}^A$. Then the A-discriminant of f is a non-zero scalar multiple of the polynomial*

$$\left(\prod_{\omega\in A_+} m_\omega^{m_\omega}\right)\prod_{\omega\in A_-} a_\omega^{-m_\omega} - \left(\prod_{\omega\in A_-} m_\omega^{-m_\omega}\right)\prod_{\omega\in A_+} a_\omega^{m_\omega}, \tag{1.4}$$

where the m_ω are defined as above.

Proof. First, we show that the polynomial (1.4) vanishes for $f \in \nabla_0$, i.e., when the system (1.1) has a solution $x^{(0)} \in (\mathbf{C}^*)^{k-1}$. Indeed, (1.1) can be written as

$$\sum_{\omega \in A} a_\omega (x^{(0)})^\omega = 0, \ \sum_{\omega \in A} a_\omega (x^{(0)})^\omega \cdot \omega = 0. \tag{1.5}$$

Comparing (1.5) with (1.3) we conclude that the vectors $(a_\omega (x^{(0)})^\omega)_{\omega \in A}$ and $(m_\omega)_{\omega \in A}$ are proportional to each other. To eliminate $x^{(0)}$, we apply to both vectors the function $(y_\omega)_{\omega \in A} \mapsto \prod_{\omega \in A} y_\omega^{m_\omega}$ (since $\sum_{\omega \in A} m_\omega = 0$, this function takes the same value at proportional vectors). We obtain the equality

$$\prod_{\omega \in A} \left(\frac{a_\omega}{m_\omega}\right)^{m_\omega} = 1; \tag{1.6}$$

its polynomial form is exactly the vanishing of (1.4).

Conversely, suppose (1.4) vanishes at some $f \in \mathbf{C}^A$. We can assume that f is generic, so that all a_ω are non-zero. This implies (1.6), i.e., that the vector with components $(y_\omega = \frac{a_\omega}{m_\omega})_{\omega \in A}$ satisfies the relation

$$\prod_{\omega \in A} y_\omega^{m_\omega} = 1. \tag{1.7}$$

Using the fact that A affinely spans $\mathbf{Z}^{k-1}$, it is easy to see that every solution of (1.7) has the form $(y_\omega = c(x^{(0)})^{-\omega})$ for some non-zero constant c and some $x^{(0)} \in (\mathbf{C}^*)^{k-1}$. This, in turn, implies $f \in \nabla_0$.

The above arguments show that (1.4) defines ∇_A set-theoretically. So, up to a scalar multiple, it must be a power of Δ_A. But, since it is a sum of only two monomials, we conclude that it can be only the first power of Δ_A, which completes the proof.

2. The discriminantal complex

The problem of finding the A-discriminant Δ_A, raised in Section 1, is a special case of a more general problem addressed in Chapter 1: finding the equation of $X^\vee$, the projectively dual variety to a given projective variety $X \subset P^{n-1}$. In our present case X is the toric variety X_A.

Suppose that a projective variety X is smooth. In Theorem 2.5 of Chapter 2 we have represented the equation of $X^\vee$ as the determinant of the so-called discriminantal complex. The terms of this complex consist of some differential forms on the affine variety $Y \subset \mathbf{C}^n$ which is the cone over X (see Section 4,

Chapter 2). In our present situation we can make the complexes more explicit using a description of differential forms on tori and toric varieties going back to Danilov [D], see also [O]. For the convenience of the reader we recall this description from scratch, starting with the case of a torus.

A. Differential forms on a torus

Let $H = (\mathbf{C}^*)^k$ be a torus and let $\Xi = \mathrm{Hom}(H, \mathbf{C}^*) = \mathbf{Z}^k$ be its character lattice. The ring $\mathbf{C}[H]$ of regular functions on H is identified with the group algebra $\mathbf{C}[\Xi]$ of Ξ. This means that we can regard a Laurent polynomial

$$f(x) = \sum_{\omega \in \mathbf{Z}^k} a_\omega x^\omega \in \mathbf{C}[H]$$

as a complex-valued function $\omega \mapsto a_\omega$ with finite support on $\Xi = \mathbf{Z}^k$. In this language the multiplication in $\mathbf{C}[H]$ is given by the convolution product: if (a_ω) and (b_ω) are elements of $\mathbf{C}[H]$ then their product (c_ω) is given by

$$c_\omega = \sum_{\omega' + \omega'' = \omega} a_{\omega'} b_{\omega''}.$$

We want to describe in similar terms the de Rham complex formed by the $\Omega^i(H)$, the spaces of regular differential i-forms on H.

Let $\Xi_{\mathbf{C}} = \Xi \otimes \mathbf{C}$ be the complexification of the free Abelian group Ξ. By a *discrete vector field* on Ξ we shall mean an assignment $\omega \mapsto v_\omega$ which takes any $\omega \in \Xi$ to a vector $v_\omega \in \Xi_{\mathbf{C}}$ such that $v_\omega = 0$ for all but finitely many ω. For $\omega \in \Xi$, $v \in \Xi_{\mathbf{C}}$, we denote by (ω, v) the discrete vector field equal to v at ω and 0 elsewhere.

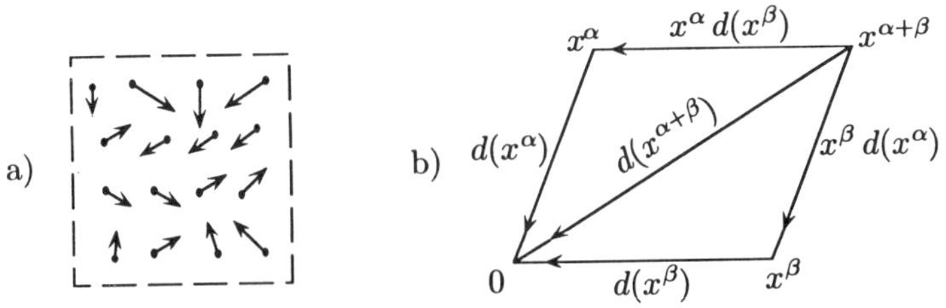

Figure 41. (a) A discrete vector field on $\mathbf{Z}^2$
(b) The proof of the Leibniz rule

Denote by $\mathrm{Vec}^1(\Xi)$ the space of all discrete vector fields on Ξ. Clearly $\mathrm{Vec}^1(\Xi)$ is a free $\mathbf{C}[\Xi]$-module of rank k under the obvious convolution product

$$\mathbf{C}[\Xi] \otimes \mathrm{Vec}^1(\Xi) \to \mathrm{Vec}^1(\Xi), \quad x^\alpha \otimes (\omega, v) \mapsto (\omega + \alpha, v). \tag{2.1}$$

Proposition 2.1. *There is a canonical isomorphism* $\varphi : \Omega^1(H) \to \mathrm{Vec}^1(\Xi)$ *of* $\mathbf{C}[\Xi]$*-modules. This isomorphism takes* $d(x^\gamma) \in \Omega^1(H)$ *into the discrete vector field* $(\gamma, -\gamma)$.

Geometrically, the field $(\gamma, -\gamma)$ is just one vector at the point γ which joins this point with 0 (see Figure 41 b).

Proof. By associating to $x^\alpha d(x^\beta)$ the discrete vector field $(\alpha + \beta, -\beta)$ we get the required isomorphism. The inverse isomorphism is defined as follows:

$$(\omega, v) \mapsto -x^\omega \cdot \frac{d(x^v)}{x^v} = -x^\omega \cdot d\log(x^v), \tag{2.2}$$

where $v \in \Xi$. This correspondence is additive in v and hence extends to any $v \in \Xi_{\mathbf{C}}$ by linearity.

It is instructive to see the validity of the Leibniz rule $d(x^{\alpha+\beta}) = x^\alpha d(x^\beta) + x^\beta d(x^\alpha)$ in the language of discrete vector fields (Figure 41 b).

Let us introduce the space $\mathrm{Vec}^i(\Xi)$ of *discrete i-vector fields* on Ξ whose elements are finitely supported functions $\omega \mapsto \lambda_\omega$ mapping Ξ to $\bigwedge^i(\Xi_{\mathbf{C}})$. For any $\omega \in \Xi$ and $\lambda \in \bigwedge^i(\Xi_{\mathbf{C}})$ we shall denote by (ω, λ) the discrete i-vector field on Ξ equal to λ at ω and 0 elsewhere.

The space $\mathrm{Vec}^0\,\Xi$ is just the group algebra $\mathbf{C}[\Xi]$. We have multiplication

$$\mathrm{Vec}^i\,\Xi \otimes \mathrm{Vec}^j\,\Xi \to \mathrm{Vec}^{i+j}\,\Xi, \quad (\omega, \lambda) \otimes (\eta, \mu) \mapsto (\omega + \eta, \lambda \wedge \mu). \tag{2.3}$$

Proposition 2.2. *The space* $\Omega^i(H)$ *of regular differential i-forms on H is naturally identified (as a* $\mathbf{C}[H]$*-module) with* $\mathrm{Vec}^i(\Xi)$. *Under this identification the exterior product in* $\Omega^\bullet(H)$ *corresponds to the product (2.3) on* $\mathrm{Vec}^\bullet(\Xi)$. *The exterior derivative of a form represented by a discrete i-vector field* $\omega \mapsto \lambda_\omega$ *on* Ξ *is represented by the discrete* $(i+1)$*-vector field* $\omega \mapsto \lambda_\omega \wedge (-\omega)$.

Proof. The isomorphism in question takes

$$x^{\alpha_0} d(x^{\alpha_1}) \wedge \cdots \wedge d(x^{\alpha_i}) \mapsto (-1)^i (\alpha_0 + \cdots + \alpha_i, \alpha_1 \wedge \cdots \wedge \alpha_i).$$

The inverse isomorphism takes

$$(\alpha, \beta_1 \wedge \cdots \wedge \beta_i) \mapsto (-1)^i x^\alpha \frac{d(x^{\beta_1})}{x^{\beta_1}} \wedge \cdots \wedge \frac{d(x^{\beta_i})}{x^{\beta_i}} \tag{2.4}$$

where α, $\beta_j \in \Xi$. The correspondence (2.4) is additive in the β's and extends to the arbitrary $\beta_j \in \Xi_{\mathbf{C}}$ by linearity.

B. Differential forms on an affine toric variety

Let $\Xi = \mathbf{Z}^k$ be a free Abelian group of rank k. Let $S \subset \Xi$ be a finitely generated semigroup containing 0 and generating Ξ as a group. Let Ξ_R be the real vector space $\Xi \otimes R$ and let $K \subset \Xi_R$ be the convex hull of S. This is a convex polyhedral cone with apex 0. For any face $\Gamma \subset K$, we denote by $\mathrm{Lin}_{\mathbf{C}}(\Gamma)$ the smallest $\mathbf{C}$-vector subspace in $\Xi_{\mathbf{C}}$ containing Γ. Clearly, the complex dimension of $\mathrm{Lin}_{\mathbf{C}}(\Gamma)$ equals the dimension of Γ in the usual sense. For example, $\mathrm{Lin}_{\mathbf{C}}(K)$ is the whole $\Xi_{\mathbf{C}}$. For any $\omega \in S$, let $\Gamma(\omega)$ be the smallest face of K containing ω. In particular, if ω lies in the interior of K then $\Gamma(\omega) = K$.

Definition 2.3. Denote by $\mathrm{Vec}^i(S)$ the space of discrete i-vector fields (λ_ω) on Ξ with the properties:

(a) $v_\omega = 0$ if $\omega \notin S$;

(b) for any $\omega \in S$ we have $v_\omega \in \bigwedge^i \mathrm{Lin}_{\mathbf{C}}(\Gamma(\omega))$.

Figure 42 illustrates this definition.

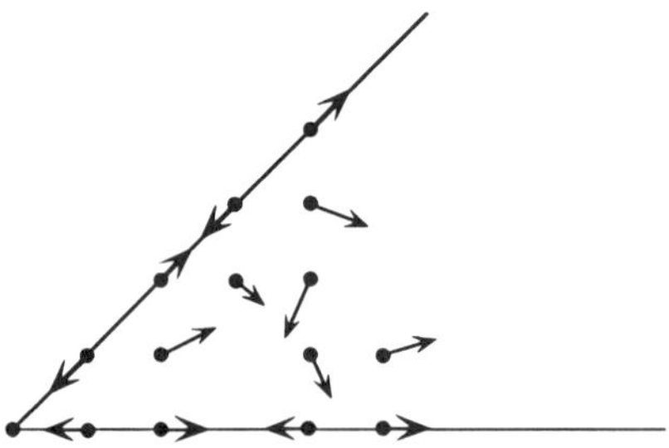

Figure 42. A discrete vector field from $\mathrm{Vec}^1(S)$

As in subsection A, for $\omega \in S$ and $\lambda \in \bigwedge^i \mathrm{Lin}_{\mathbf{C}}(\Gamma(\omega))$, we denote by (ω, λ) the discrete i-vector field equal to λ at ω and 0 elsewhere.

Clearly, $\mathrm{Vec}^0(S)$ is the semigroup algebra of S. The multiplication (2.3) restricts to the multiplication

$$\mathrm{Vec}^i S \otimes \mathrm{Vec}^j S \longrightarrow \mathrm{Vec}^{i+j} S. \tag{2.5}$$

With respect to (2.5), $\mathrm{Vec}^i S$ is a finitely generated module over $\mathbf{C}[S] = \mathrm{Vec}^0 S$. Let us also define the "exterior derivative"

$$d : \mathrm{Vec}^i S \to \mathrm{Vec}^{i+1} S, \quad d((\omega, \lambda)) = (\omega, -\omega \wedge \lambda). \tag{2.6}$$

Proposition 2.4.

(a) *The maps d satisfy $d^2 = 0$ and the Leibniz rule thus making the direct sum* $\mathrm{Vec}^\bullet S = \bigoplus_i \mathrm{Vec}^i S$ *into a supercommutative differential graded algebra.*

(b) *If $S = \mathbf{Z}_+^a \times \mathbf{Z}^b$ then* $\mathrm{Vec}^i S$ *is identified with the space of regular differential i-forms on $\mathbf{C}^a \times (\mathbf{C}^*)^b$, and (2.5) and (2.6) coincide with the usual exterior multiplication and differential.*

Proof. Obvious.

Let $Y = \mathrm{Spec}\,\mathbf{C}[S]$ be the affine toric variety corresponding to S. By Serre's theorem [Hart] any module over the coordinate ring of an affine algebraic variety gives rise to a coherent sheaf on this variety. In our case we consider the $\mathbf{C}[S]$-module $\mathrm{Vec}^i S$. The coherent sheaf on Y corresponding to this module will be denoted by $\tilde{\Omega}_Y^i$ and called the sheaf of *Danilov i-forms* on Y.

Theorem 2.5.

(a) *Let Y_{sm} be the smooth locus of Y. Then the restriction of $\tilde{\Omega}_Y^i$ to Y_{sm} is naturally identified with the sheaf of regular differential i-forms on Y_{sm}.*

(b) *The maps (2.3) and (2.6) extend to morphisms of sheaves on Y*

$$\tilde{\Omega}_Y^i \otimes \tilde{\Omega}_Y^j \to \tilde{\Omega}_Y^{i+j}; \qquad d : \tilde{\Omega}_Y^i \to \tilde{\Omega}_Y^{i+1} \tag{2.7}$$

which, after restriction to Y_{sm}, coincide with the usual exterior multiplication and differential on forms.

Proof. Clearly, Y_{sm} is a union of open subsets of the form $\mathbf{C}^a \times (\mathbf{C}^*)^b$ invariant under the torus action. The coordinate ring of any such subset in Y has the form $\mathbf{C}[S']$ where S' is obtained from S by inverting some elements. Our statement now follows from Proposition 2.4 (b) and the identification

$$\mathrm{Vec}^i(S') = \mathrm{Vec}^i S \otimes_{\mathbf{C}[S]} \mathbf{C}[S'],$$

which can be verified immediately.

C. Combinatorial description of the discriminantal complex

Let $A \subset \mathbf{Z}^{k-1}$ be a finite set of n lattice points (=Laurent monomials) and let $X_A \subset P^{n-1}$ be the corresponding toric variety, see Section 1B, Chapter 5. Let us assume that X_A is smooth and has dimension $k-1$. In this case the formalism of Chapter 2 is applicable and we can represent the A-discriminant, i.e., the equation of the variety projectively dual to X_A, as the determinant of the discriminantal complex

$$\Delta_X(f) = const \cdot \det(C_+^\bullet(X_A, \mathcal{M}), \partial_f, e)^{(-1)^k} \tag{2.8}$$

where $\mathcal{M}$ is a sufficiently ample invertible sheaf on X_A (Theorem 2.5, Chapter 2). We take $\mathcal{M} = \mathcal{O}(l)$, $l \gg 0$. We shall use the results of the previous subsection to describe this complex quite explicitly.

Shrinking, if necessary, the lattice $\mathbf{Z}^{k-1}$, we can assume that it is affinely generated by A. As before, we embed $\mathbf{Z}^{k-1}$ into a free Abelian group $\Xi = \mathbf{Z}^k = \mathbf{Z}^{k-1} \times \mathbf{Z}$ as the set of lattice points with the last component equal to 1. We denote by $h : \Xi \to \mathbf{Z}$ the projection given by this last component.

Let $S \subset \Xi$ be the semigroup generated by A and 0. The semigroup algebra $\mathbf{C}[S]$ is graded by means of h (i.e., the degree of the monomial t^u, $u \in S$, is set to be $h(u)$). Under this grading $\mathbf{C}[S]$ is the homogeneous coordinate ring of the projective toric variety X_A and also the affine coordinate ring of the affine toric variety Y_A (the cone over X_A). The space $\mathbf{C}^A$ is embedded into $\mathbf{C}[S]$ as the graded component of degree 1.

Let l be an integer. Consider the graded vector space $C^\bullet(A, l)$, where

$$C^i(A,l) = \bigoplus_{u \in S, h(u)=l+i} \bigwedge^i \mathrm{Lin}_{\mathbf{C}}\Gamma(u). \tag{2.9}$$

Let us denote a typical element in the u-th summand of $C^i(A,l)$ by (u, λ) where $\lambda \in \bigwedge^i \mathrm{Lin}_{\mathbf{C}}\Gamma(u)$. For any $f(x) = \sum_{\omega \in A} a_\omega x^\omega \in \mathbf{C}^A$, we define the differential $\partial_f : C^i(A,l) \to C^{i+1}(A,l)$ by

$$\partial_f(u, \lambda) = -\sum_{\omega \in A} a_\omega \cdot (\omega + u, \omega \wedge \lambda). \tag{2.10}$$

It is straightforward to see that $\partial_f^2 = 0$.

Theorem 2.6. *Assume that X_A is smooth. For $l \gg 0$ the complex $(C^\bullet(A,l), \partial_f)$ coincides with the discriminantal complex $(C_+^\bullet(X_A, \mathcal{O}(l)), \partial_f)$.*

Proof. By Corollary 4.2, Chapter 2, the space $C_+^i(X_A, \mathcal{O}(l))$ (denoted there by $C^i(X_A, l)$) is identified with the space of differential i-forms on $Y - \{0\}$ homogeneous of degree $i + l$. On the other hand, (2.9) is the $(i+l)$-th graded component of the $\mathbf{C}[S]$-module $\mathrm{Vec}^i S$ with respect to the following grading: $\deg(u, \lambda) = h(u)$. This grading is obviously compatible with the similar grading on $\mathbf{C}[S]$ defined above.

By our assumption X_A is smooth and hence $Y_A - \{0\}$ is the smooth locus of Y_A. So the sheaf $\tilde{\Omega}^i_{Y_A}$ on Y_A corresponding to $\mathrm{Vec}^i S$ is identified, after restriction to $Y_A - \{0\}$, with the sheaf of i-forms $\Omega^i_{Y_A - \{0\}}$. Note that Theorem 2.5 gives a natural homomorphism of vector spaces

$$\Phi : \mathrm{Vec}^i S \longrightarrow \{i\text{-forms on } Y_A - \{0\}\}. \tag{2.11}$$

Let $j : Y_A - \{0\} \hookrightarrow Y_A$ be the embedding. Since the point 0 has codimension at least 2 in Y_A, the direct image under j of any coherent sheaf on $Y_A - \{0\}$, in particular, of the sheaf $\Omega^i_{Y_A-\{0\}}$, is a coherent sheaf on Y. The homomorphism Φ comes (by taking global sections) from a morphism

$$\varphi : \tilde{\Omega}^i_{Y_A} \to j_* \Omega^i_{Y_A-\{0\}}$$

of coherent sheaves on Y_A whose existence follows from Theorem 2.5. Since φ is an isomorphism outside 0, it follows that Ker φ and Coker φ are coherent sheaves on Y supported at 0. Therefore, by taking global sections, we find that Ker Φ and Coker Φ are finite-dimensional vector spaces. Since both spaces in (2.11) are graded, we conclude that, for $l \gg 0$, the induced map of $(i + l)$-th graded components is an isomorphism, i.e.,

$$C^i_+(X_A, \mathcal{O}(l)) \cong C^i(A, l).$$

The fact that the differentials in these complexes agree under this isomorphism follows from Theorem 2.5 (b). Theorem 2.6 is proved.

The above theorem implies that (in the case of smooth X_A) the A-discriminant equals $const \cdot D^{(-1)^k}$ where D is the determinant of $C^\bullet(A, l)$. We shall give a more precise formula, valid up to sign.

Namely, consider the following system of bases in the vector spaces $C^i(A, l)$. For any $u \in S$ the vector space $\mathrm{Lin}_{\mathbf{C}}\Gamma(u)$ contains a $\mathbf{Z}$-lattice $\mathrm{Lin}_{\mathbf{C}}\Gamma(u) \cap \Xi$. We choose any $\mathbf{Z}$-basis in this lattice as a basis of $\mathrm{Lin}_{\mathbf{C}}\Gamma(u)$. Correspondingly, we choose the basis in $\bigwedge^i(\mathrm{Lin}_{\mathbf{C}}\Gamma(u))$ formed by exterior products of basis vectors in $\mathrm{Lin}_{\mathbf{C}}\Gamma(u)$. Finally, we choose as a basis in the direct sum (2.9), the union of the chosen bases in the summands. Let e be the resulting system of bases in the terms $C^i(A, l)$.

Theorem 2.7. *Assume that X_A is smooth. Then for $l \gg 0$ we have*

$$\Delta_A(f) = \pm \det(C^\bullet(A, l), \partial_f, e)^{(-1)^k}.$$

Proof. Up to a constant factor, the statement follows from Theorem 2.5, Chapter 2. To prove it up to a sign, we proceed as in the proof of Theorem 2.5, Chapter 8. Namely, we denote by $C^\bullet_{\mathbf{Z}}(A, l)$ the natural $\mathbf{Z}$-form of $C^\bullet(A, l)$:

$$C^i_{\mathbf{Z}}(A, l) = \bigoplus_{u \in S, h(u) = l + i} \bigwedge\nolimits^i_{\mathbf{Z}} (\mathrm{Lin}_{\mathbf{C}}\Gamma(u) \cap \Xi).$$

Now let p be an arbitrary prime number, let F be the algebraic closure of $\mathbf{Z}/p\mathbf{Z}$ and consider the graded F-vector space

$$C_F^\bullet(A,l) = C_{\mathbf{Z}}^\bullet(A,l) \otimes_{\mathbf{Z}} F.$$

For any $f \in F^A$ we get a differential ∂_f in $C_F^\bullet(A,l)$. As in the proof of Theorem 2.5 Chapter 8, it suffices for our purposes to show that, for generic $f \in F^A$, the complex $(C_F^\bullet(A,l), \partial_f)$ is exact. The description of differential forms on a smooth toric variety given in subsection B, remains valid over a field of any characteristic. So our statement is proved in the same way as generic exactness of the discriminantal complex over $\mathbf{C}$ (Theorem 2.3, Chapter 2). This completes the proof of Theorem 2.7.

D. The degree of the A-discriminant

We continue to assume that X_A is smooth and that A affinely spans $\mathbf{Z}^{k-1}$ over $\mathbf{Z}$. Let $Q \subset \mathbf{R}^{k-1}$ be the convex hull of A. For each face $\Gamma \subset Q$, let $\mathrm{Aff}_{\mathbf{R}}(\Gamma)$ be the smallest real affine subspace containing Γ. This space comes equipped with an affine $\mathbf{Z}$-lattice $\mathrm{Aff}_{\mathbf{Z}}(\Gamma \cap A)$ generated by $\Gamma \cap A$. Since we assume X_A to be smooth, this lattice coincides with $\mathrm{Aff}_{\mathbf{R}}(\Gamma) \cap \mathbf{Z}^{k-1}$ (Corollary 3.2, Chapter 5).

As in Section 4D, Chapter 5, the above lattice gives rise to a volume form on $\mathrm{Aff}_{\mathbf{R}}(\Gamma)$ normalized so that the volume of an elementary lattice simplex equals 1. Let us denote this form as Vol_Γ.

Theorem 2.8. *Suppose X_A is smooth. Then the degree of homogeneity of the A-discriminant equals*

$$\sum_{\Gamma \subset Q} (-1)^{\operatorname{codim} \Gamma} (\dim \Gamma + 1) \cdot \mathrm{Vol}_\Gamma(\Gamma).$$

In particular, this sum is always non-negative; it equals zero if and only if $\Delta_A = 1$.

Proof. We retain the notation of the previous subsections. Thus $S \subset \mathbf{Z}^k$ is the semigroup generated by A and 0; it is graded by $h : S \to \mathbf{Z}_+$. Denote by S_l the graded component $\{u \in S : h(u) = l\}$. We also need the convex hull K of S. This is a polyhedral cone with apex 0 whose base is Q. We extend h to a linear functional (denoted also h) from K to $\mathbf{R}$ and denote $K_l = \{u \in K : h(u) = l\}$. Since X_A is assumed to be smooth and, in particular, normal, the intersection $K \cap \mathbf{Z}^k$ differs from S in finitely many points only.

Let $\Gamma \subset Q$ be a face. Then the cone $\mathbf{R}_+\Gamma$ generated by Γ is a face of K. Let Γ^0 be the interior of Γ. Consider the set $S_l \cap \mathbf{R}_+\Gamma^0$. By the above, for $l \gg 0$ this set coincides with $K_l \cap \mathbf{R}_+\Gamma^0$. Thus, for large l, the number $\#\left(S_l \cap \mathbf{R}_+\Gamma^0\right)$

coincides with the number of integer points in the l times dilated open polytope Γ^0. It is known [D] that for $l \gg 0$ this number is given by a polynomial in l which we denote by $p_\Gamma(l)$ (it is closely related to the so-called Ehrhart polynomial counting the number of integer points in the dilations of Γ, not Γ^0). The leading term of this polynomial is, by Proposition 3.7, Chapter 5,

$$\frac{\mathrm{Vol}_\Gamma(\Gamma)}{(\dim\Gamma)!}l^{\dim\Gamma}.$$

By Corollary 14 from Appendix A, the degree of Δ_A is equal to

$$\deg(\Delta_A) = \sum_{i=0}^{k}(-1)^{k-i}i\cdot\dim C^i(A,l).$$

By representing $C^i(A,l)$ as a direct sum over $u \in S_{i+l}$ (see (2.9)) and separating the u's lying in different faces of K, we find that, for $l \gg 0$,

$$\dim C^i(A,l) = \sum_{\Gamma\subset Q}\binom{\dim\Gamma+1}{i}p_\Gamma(l+i).$$

Substituting this expression into the above formula for $\deg(\Delta_A)$ we obtain, after some easy algebraic transformations, that

$$\deg(\Delta_A) = \sum_{\Gamma\subset Q}(-1)^{\mathrm{codim}\,\Gamma}(\dim\Gamma+1)\sum_{i=0}^{\dim\Gamma}(-1)^{\dim\Gamma-i}\binom{\dim\Gamma}{i}\cdot p_\Gamma(l+1+i).$$

To deduce Theorem 2.8 from this expression, we have only to show that the inner sum is equal to $\mathrm{Vol}_\Gamma(\Gamma)$. We shall use the following elementary lemma.

Lemma 2.9. *Let $p(t) = a_0t^r + \cdots + a_r$ be a polynomial of degree r. Then for any value of t the sum $\sum_{i=0}^{r}(-1)^i\binom{r}{i}p(t-i)$ is equal to $r!a_0$.*

The lemma is well-known; the easiest way to prove it is to observe that the sum in question is the iterated difference $\Delta^r p(t)$, where $\Delta p(t) = p(t) - p(t-1)$.

To complete the proof of Theorem 2.8, it is enough to apply Lemma 2.9 to each polynomial $p(t) = p_\Gamma(t+\dim\Gamma+1)$.

Examples 2.10. Let us illustrate Theorem 2.8 for the sets A in Examples 1.6.

(a) Let A consist of all monomials in $x_1, \ldots, x_{k-1}$ of degree $\le d$ (or, equivalently, of all homogeneous monomials in $x_1, \ldots, x_k$ of degree d). Since $X_A = P^{k-1}$ is smooth, Theorem 2.8 is applicable. The polytope Q is the simplex

$$\left\{(t_1, \ldots, t_{k-1}) \in \mathbf{R}^{k-1} : t_i \ge 0, \sum t_i \le d\right\}$$

of dimension $k-1$. For each $i = 0, \dots, k-1$, this simplex has exactly $\binom{k}{i+1}$ faces of dimension i, and each of them has the normalized volume equal to d^i. By Theorem 2.8,

$$\deg(\Delta_A) = \sum_{i=0}^{k-1} (-1)^{k-1-i}(i+1)\binom{k}{i+1} d^i$$

$$= k\sum_{i=0}^{k-1} (-1)^{k-1-i}\binom{k-1}{i} d^i k(d-1)^{k-1}, \tag{2.12}$$

the last equality being the binomial formula. By Lemma 2.9, $\deg(\Delta_A)$ is equal to $k(d-1)^{k-1}$. This is Boole's formula (see Examples 1.6 above).

(b) Let A consist of bilinear monomials $x_i \cdot y_j$, $i = 1, \dots, m$; $j = 1, \dots, n$. The polytope Q is the product of two simplices $\Delta^{m-1} \times \Delta^{n-1}$. A face of Q is given by a pair of non-empty subsets $I \subset \{1, \dots, m\}$, $J \subset \{1, \dots, n\}$ and is itself a product $\Delta^{i-1} \times \Delta^{j-1}$ where $i = \#(I)$, $j = \#(J)$. The normalized volume of such a face is $\binom{i+j-2}{i-1}$. Since $X_A = P^{m-1} \times P^{n-1}$ is smooth, the degree of Δ_A is given by Theorem 2.8. We obtain

$$\deg(\Delta_A) = \sum_{i=1}^{m}\sum_{j=1}^{n} (-1)^{m+n-i-j}(i+j-1)\binom{m}{i}\binom{n}{j}\binom{i+j-2}{i-1}. \tag{2.13}$$

Without loss of generality, we can assume that $n \le m$. To simplify (2.13) we rewrite it as

$$\sum_{i=0}^{m} (-1)^i \binom{m}{i} p(m-i),$$

where $p(t)$ is a polynomial of degree n given by

$$p(t) = \sum_{j=1}^{n} (-1)^{n-j} \frac{1}{(j-1)!}\binom{n}{j} t(t+1)\cdots(t+j-1).$$

Using Lemma 2.9, we see that $\deg(\Delta_A) = 0$ unless $m = n$, and in the latter case $\deg(\Delta_A) = n$. This is in accordance with the fact that Δ_A is identically 1 for $m \ne n$ and coincides with the determinant of a $n \times n$ matrix if $m = n$.

The above two examples have a common generalization to the case when A consists of all multihomogeneous monomials of a given multidegree in several groups of variables. An application of Theorem 2.8 to this case will be discussed in Section 2, Chapter 13.

(c) Let A consist of monomials

$$1, x, x^2, \dots, x^m, y, yx, yx^2, \dots, yx^n.$$

The polytope Q is the trapezoid depicted in Figure 43.

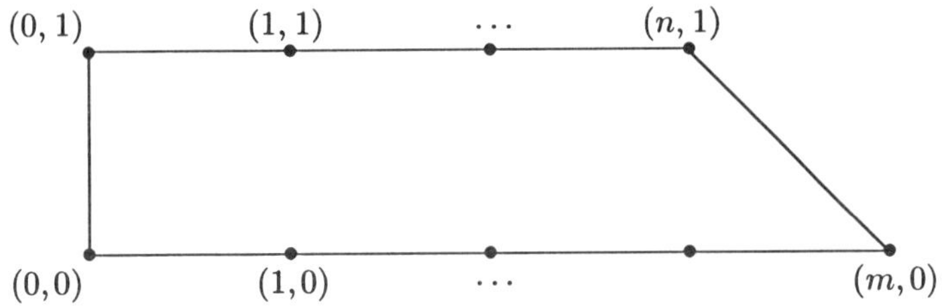

Figure 43.

Its area normalized with respect to $\mathbf{Z}^2$ is $m+n$. Under the normalizations used in Theorem 2.8, the horizontal sides have lengths m and n, and two other sides are of length 1. Four vertices of Q each should be ascribed the "volume" 1. Hence Theorem 2.8 gives

$$\deg(\Delta_A) = 3(m+n) - 2(m+n+2) + 1 \cdot 4 = m+n.$$

This is in accordance with the interpretation of Δ_A as the resultant of two polynomials of degrees m and n in one variable, see Example 1.6 (c).

3. A differential-geometric characterization of A-discriminantal hypersurfaces

In this section, which is based on [Ka1], we exhibit one characteristic property of discriminantal hypersurfaces regarded as hypersurfaces in tori. As we have seen in Section 1, Chapter 6, the geometry of a hypersurface in a torus is closely related to the Newton polytope of the Laurent polynomial defining this hypersurface. The differential-geometric property described in this section can be compared with the description of the Newton polytope of the A-discriminant given in Chapter 11 below.

A. The Gauss map in an algebraic group

Let G be an algebraic group. For each $g \in G$, let $l_g : G \to G$ be the left multiplication by g. Let $\mathfrak{g}$ be the Lie algebra of G. Let $Z \subset G$ be an irreducible algebraic hypersurface (possibly with singularities). The (left) *Gauss map* of Z is the rational map $\gamma_Z : Z \to P(\mathfrak{g}^*)$ which takes a smooth point $z \in Z$ into $d(l_z^{-1})(T_z Z)$, i.e., to the translation to unity of the tangent hyperplane to Z at z. This translation is a hyperplane in $T_e G = \mathfrak{g}$, i.e., a point in $P(\mathfrak{g}^*)$.

Note that both varieties Z and $P(\mathfrak{g}^*)$ have the same dimension. This raises the following natural problem.

Problem. *Classify algebraic hypersurfaces $Z \subset G$ such that $\gamma_Z : Z \to P(\mathfrak{g}^*)$ is a birational isomorphism.*

In what follows we shall consider only the case when $G = (\mathbf{C}^*)^m$ is an algebraic torus. In this case we shall refer to γ_Z as the *logarithmic Gauss map* since explicit formulas for it involve logarithmic derivatives.

It turns out that the above problem for tori can be completely solved and the class of hypersurfaces in question essentially coincides with the class of A-discriminantal hypersurfaces.

B. The reduced A-discriminantal variety

Let $A \subset \mathbf{Z}^{k-1}$ be a finite subset of cardinality n, which generates $\mathbf{Z}^{k-1}$ as an affine lattice. Let $\nabla_A \subset \mathbf{C}^A$ be the corresponding discriminantal variety. We consider the action of $(\mathbf{C}^*)^k$ on $\mathbf{C}^A$ given by

$$(t_1, \dots, t_k) : f(x_1, \dots, x_{k-1}) \mapsto t_k f(t_1 x_1, \dots, t_{k-1} x_{k-1}). \tag{3.1}$$

This action preserves ∇_A. Consider the subset $(\mathbf{C}^*)^A \subset \mathbf{C}^A$ as an algebraic torus acting on $\mathbf{C}^A$ componentwise. A point of $(\mathbf{C}^*)^A$ will be denoted as $(z_\omega)_{\omega \in A}$ where $z_\omega \in \mathbf{C}^*$. Then (3.1) comes from a homomorphism of tori

$$\varphi : (\mathbf{C}^*)^k \to (\mathbf{C}^*)^A, \quad \varphi(t_1, \dots, t_k)_\omega = t_k t_1^{\omega_1} \cdots t_{k-1}^{\omega_{k-1}}.$$

Let $\varphi^* : \mathbf{Z}^A \to \mathbf{Z}^k$ be the dual homomorphism of character lattices. Since we assume that A generates $\mathbf{Z}^{k-1}$ as an affine lattice, φ^* is surjective. Set $L_A = \operatorname{Ker} \varphi^*$. Clearly, L_A consists of all affine relations between elements of A, i.e., of families $(c_\omega)_{\omega \in A}$, $c_\omega \in \mathbf{Z}$ such that

$$\sum_{\omega \in A} c_\omega \cdot \omega = 0 \quad \text{and} \quad \sum_{\omega \in A} c_\omega = 0.$$

Let $H(L_A) = \operatorname{Spec} \mathbf{C}[L_A] = \operatorname{Hom}(L_A, \mathbf{C}^*)$ be the algebraic torus whose character lattice is L_A. Then we have an exact sequence of tori

$$1 \to (\mathbf{C}^*)^k \xrightarrow{\varphi} (\mathbf{C}^*)^A \xrightarrow{p} H(L_A) \to 1, \tag{3.2}$$

where p is the natural projection. Since ∇_A is $(\mathbf{C}^*)^k$-invariant, the intersection $\nabla_A \cap (\mathbf{C}^*)^A$ is the inverse image, under p, of some subvariety $\tilde{\nabla}_A \subset H(L_A)$ which is uniquely defined. Clearly, $\operatorname{codim}(\nabla_A) = \operatorname{codim}(\tilde{\nabla}_A)$. We shall be interested only in the case when ∇_A and $\tilde{\nabla}_A$ are hypersurfaces. In this case we shall call $\tilde{\nabla}_A$ the *reduced A-discriminantal hypersurface*. The defining equation of $\tilde{\nabla}_A$ is obtained from the A-discriminant Δ_A by specializing some k of the variables a_ω to be 1 so as to kill the k quasi-homogeneities.

Now we state our main result.

Theorem 3.1. *Let $G = (\mathbf{C}^*)^m$ be an algebraic torus, and let $Z \subset G$ be an irreducible algebraic hypersurface. The following conditions are equivalent:*
(a) *The logarithmic Gauss map $\gamma_Z : Z \to P^{m-1}$ is birational.*
(b) *There exist $k > 0$, a $(k+m)$-element subset $A \subset \mathbf{Z}^{k-1}$ affinely generating $\mathbf{Z}^{k-1}$, and an isomorphism of algebraic varieties $G \to H(L_A)$ (i.e., a group isomorphism of tori followed by a translation) which takes Z to the reduced A-discriminantal hypersurface $\tilde{\nabla}_A$.*

The proof of this theorem will be given later in this section. First, let us give some examples of hypersurfaces with birational γ_Z provided by Theorem 3.1.

Examples 3.2. *(a)* The hypersurface $Z \subset (\mathbf{C}^*)^{r^2}$ consisting of $r \times r$-matrices $\|a_{ij}\|$ such that

$$\begin{vmatrix} 1 & 1 & \dots & 1 \\ 1 & a_{11} & \dots & a_{1r} \\ \vdots & \vdots & \dots & \vdots \\ 1 & a_{r1} & \dots & a_{rr} \end{vmatrix} = 0,$$

has a birational logarithmic Gauss map.

(b) The hypersurface $Z \subset (\mathbf{C}^*)^{d-1}$, consisting of $(a_1, \dots, a_{d-1})$, such that the polynomial $x^d + x^{d-1} + a_1 x^{d-2} + \dots + a_{d-1}$ has a multiple root, possesses a birational logarithmic Gauss map.

(c) Consider the affine hyperplane Z in $(\mathbf{C}^*)^m$ given by the equation $\sum a_i = 1$. It is straightforward to see that Z possesses a birational logarithmic Gauss map. The same, of course, will hold for any hyperplane $\{\sum \lambda_i a_i = 1\}$ where all $\lambda_i \neq 0$, since it is a torus translation of the above hyperplane.

The hyperplane Z can be realized as a reduced A-discriminantal hypersurface if the set A consists of the following monomials in two variables x, y:

$$A = \{1, x, \dots, x^m, y, yx\}.$$

Then $\mathbf{C}^A$ consists of polynomials of the form $f(x) + yg(x)$ where $f(x) = a_m x^m + \dots + a_0$ has degree $\leq m$ and g is an affine-linear polynomial $b_1 x + b_0$. The A -discriminant $\Delta_A(f, g)$ is the classical resultant $R(f, g)$ which in our case is equal to

$$a_0 b_1^m - a_1 b_1^{m-1} b_0 + \cdots \pm a_m b_0^m.$$

To get the reduced A-discriminantal hypersurface, we should specialize three variables to 1, say b_0, b_1 and a_0. This specialization gives an affine hyperplane as above.

C. The Horn uniformization

The proof of Theorem 3.1 is based on some formulas from the paper [Hor] of Horn devoted to study of hypergeometric series. Crucial for us will be an explicit formula for a rational uniformization of the reduced A-discriminantal variety. This formula appeared in [Hor] although it was not realized at the time that the image of the corresponding map is in fact a discriminantal variety.

Let $A \subset \mathbf{Z}^{k-1}$ be as before. Let us write $A = \{\omega^{(1)}, \ldots, \omega^{(n)}\}$ where each $\omega^{(j)}$ is a vector $(\omega_1^{(j)}, \ldots, \omega_{k-1}^{(j)}) \in \mathbf{Z}^{k-1}$. Thus we identify $\mathbf{Z}^A$ and $\mathbf{C}^A$ with $\mathbf{Z}^n$ and $\mathbf{C}^n$. Put $m = n - k$. Choose a $\mathbf{Z}$-basis $\{a^{(1)}, \ldots, a^{(m)}\}$ of the lattice $L_A \subset \mathbf{Z}^n$, where each $a^{(p)}$ is an integral vector $(a_1^{(p)}, \ldots, a_n^{(p)}) \in \mathbf{Z}^n$. The choice of the basis $\{a^{(p)}\}$ identifies the torus $H(L_A)$ with $(\mathbf{C}^*)^m$.

Let us define a rational map $h : P^{m-1} \to (\mathbf{C}^*)^m$ taking a point with homogeneous coordinates $\lambda = (\lambda_1 : \ldots : \lambda_m)$ to $(\Psi_1(\lambda), \ldots, \Psi_m(\lambda))$ where

$$\Psi_l(\lambda_1, \ldots, \lambda_m) = \prod_{j=1}^{n} \left(\sum_{p=1}^{m} a_j^{(p)} \lambda_p \right)^{a_j^{(l)}}. \tag{3.3}$$

Since for each $a = (a_1, \ldots, a_n) \in L_A$ we have $\sum a_j = 0$, each Ψ_l is homogeneous of degree 0. We shall call h the *Horn uniformization.*

Theorem 3.3.

(a) *Under the identification* $(\mathbf{C}^*)^m \cong H(L_A)$, *defined by the chosen basis of* L_A, *the image of* $h : P^{m-1} \to (\mathbf{C}^*)^m$ *is identified with the reduced A-discriminantal variety* $\tilde{\nabla}_A \subset H(L_A)$.

(b) *If* $\tilde{\nabla}_A$ *is a hypersurface then* $h : P^{m-1} \to \tilde{\nabla}_A$ *is a birational isomorphism. In this case the inverse to h coincides with the logarithmic Gauss map* $\gamma_{\tilde{\nabla}_A}$.

Thus part (b) of Theorem 3.3 proves the implication $(a) \Rightarrow (b)$ of Theorem 3.1.

Proof of Theorem 3.3. Let us define h in more invariant terms. Consider the rational map

$$h' : L_A \otimes \mathbf{C} \longrightarrow \mathrm{Hom}(L_A, \mathbf{C}^*) = H(L_A)$$

such that for $(a_1, \dots, a_n) \in L_A \otimes \mathbf{C}$ and $(b_1, \dots, b_n) \in L_A$, $b_i \in \mathbf{Z}$ we have

$$h'(a_1, \dots, a_n)(b_1, \dots, b_n) = \prod_{j=1}^{n} a_j^{b_j}.$$

The map h' gives rise to a rational map $P(L_A \otimes \mathbf{C}) \to H(L_A)$, also denoted by h', since $\sum b_j = 0$ for each $b \in L_A$.

Lemma 3.4. *Under the identification $L_A \otimes \mathbf{C} \to \mathbf{C}^m$ induced by the choice of basis, the map h' is identified with h.*

The proof is straightforward and left to the reader.

In order to prove part (a) of Theorem 3.3, we recall the standard uniformization of ∇_A arising from the fact that it is a projective dual variety. Denote the space $\mathbf{C}^A = \mathbf{C}^n$ by V and consider the $(\mathbf{C}^*)^k$-action on the dual space V^* which is dual to (3.1). Let $Y = Y_A \subset V^*$ be the affine toric variety corresponding to A (see Section 1B Chapter 5). Recall that Y is the closure of the orbit of the point $(1, \dots, 1)$ with respect to the $(\mathbf{C}^*)^k$-action on V^*. The conic variety $\nabla_A \subset V$ is projectively dual to $Y \subset V^*$. Therefore we have a diagram of rational maps

$$Y^\vee = \nabla_A \xleftarrow{\alpha} P(T_Y^*(V^*)) \xrightarrow{\beta} Y,$$

where $T_Y^*(V^*)$ is the conormal bundle to Y and P stands for the projectivization (see Section 3A Chapter 1). (Since we are working birationally, we are free to delete from any of our varieties any proper algebraic subvariety, in particular, to delete from Y all the singular points where the conormal bundle does not make sense.)

If $Y^\vee$ is a hypersurface then α is birational, as it follows from Proposition 3.2 Chapter 1.

The conormal space to Y at $(1, \dots, 1)$ is naturally identified with $L_A \otimes \mathbf{C}$. Indeed, a linear functional on the ambient vector space V^* of Y can be viewed, after restricting to Y, as a Laurent polynomial in $t_1, \dots, t_k$ of the form $g(t_1, \dots, t_k) = t_k f(t_1, \dots, t_{k-1})$ where $f = \sum_{\omega \in A} a_\omega t^\omega \in \mathbf{C}^A$. The linear functional corresponding to such a g vanishes on the tangent space to Y at $(1, \dots, 1)$ if and only if $t_i \partial g / \partial t_i = 0$ for all $i = 1, \dots, k$. This means precisely that the coefficients a_ω should form an affine relation between the ω's, i.e., $(a_\omega) \in L_A \otimes \mathbf{C}$.

By using the $(\mathbf{C}^*)^k$-action on Y, we get a birational trivialization of the conormal bundle to Y, i.e., a map

$$\varepsilon : (\mathbf{C}^*)^k \times (L_A \otimes \mathbf{C}) \longrightarrow T_Y^*(V^*).$$

Also let $\delta : (\mathbf{C}^*)^k \times (L_A \otimes \mathbf{C}) \to \mathbf{C}^A = V$ be induced by the $(\mathbf{C}^*)^k$-action on $V \supset L_A \otimes \mathbf{C}$. Then, by the above identification of the conormal space to Y and by the fact that Y is the closure of the orbit of $(1, \ldots, 1)$, we find that the image of δ is ∇_A.

Consider the commutative diagram

$$\begin{array}{ccc} L_A \otimes \mathbf{C} & \xrightarrow{h=h'} & H(L_A) = \mathrm{Hom}(L_A, \mathbf{C}^*) \\ \downarrow & & \uparrow \\ \mathbf{C}^n = \mathbf{Z}^n \otimes \mathbf{C} & \xrightarrow{\tilde{h}} & (\mathbf{C}^*)^n = \mathrm{Hom}(\mathbf{Z}^n, \mathbf{C}^*) \end{array} \tag{3.4}$$

where the map $\tilde{h}$ is given by the same formula as h', i.e.,

$$\tilde{h}(a_1, \ldots, a_n)(b_1, \ldots, b_n) = \prod_j a_j^{b_j}$$

for each $(a_1, \ldots, a_n) \in \mathbf{C}^n$, $(b_1, \ldots, b_n) \in \mathbf{Z}^n$. The left vertical map is the natural injection, the right vertical one is the projection p considered in subsection B.

Now part (a) and the first statement of part (b) of Theorem 3.3 will follow from the commutativity of (3.4) and of the following diagram

$$\begin{array}{ccccc} & & \mathbf{C}^A & \xrightarrow{\tilde{h}} & \mathbf{C}^A \\ & \overset{\delta}{\nearrow} & & & \uparrow \\ (\mathbf{C}^*)^k \times (L_A \otimes \mathbf{C}) & & & & \\ & \underset{\varepsilon}{\searrow} & & & \\ & & T_Y^*(V^*) & \xrightarrow{\alpha} & \nabla_A \end{array} \tag{3.5}$$

We leave to the reader to check that (3.5) is commutative.

To prove part (b) of Theorem 3.3, note the identities

$$\frac{\partial \log \Psi_i}{\partial \lambda_j} = \frac{\partial \log \Psi_j}{\partial \lambda_i}, \quad i, j = 1, \ldots, m, \tag{3.6}$$

which follow immediately from (3.3). Now we consider the Jacobian matrix $J(\lambda_1, \ldots, \lambda_m) = \|\partial \log \Psi_i / \partial \lambda_j\|$ and regard it as a matrix-valued rational function in the λ's. Since Ψ_j and hence $\log \Psi_j$ are homogeneous of degree 0, we have the identities

$$\sum_{i=1}^{m} \lambda_i \frac{\partial \log \Psi_j}{\partial \lambda_i} = 0 \quad \text{for } j = 1, \ldots, m. \tag{3.7}$$

In other words, the matrix $J(\lambda_1, \ldots, \lambda_m)$ annihilates the column vector $(\lambda_1, \ldots, \lambda_m)^t$. If the image $\tilde{\nabla}_A$ of h is a hypersurface, then the kernel of $J(\lambda_1, \ldots, \lambda_m)$ for generic $(\lambda_1, \ldots, \lambda_m)$ is one-dimensional and hence generated by $(\lambda_1, \ldots, \lambda_m)^t$.

On the other hand, the matrix $J(\lambda) = J(\lambda_1, \dots, \lambda_m)$ is symmetric by (3.6). Therefore the image $\mathrm{Im}(J(\lambda))$ is the orthogonal complement, with respect to the standard quadratic form $\sum x_i^2$, to $\mathrm{Ker}(J(\lambda)) = \mathbf{C} \cdot (\lambda_1, \dots, \lambda_m)^t$. This means that $\lambda_1, \dots, \lambda_m$ are coefficients of the linear equation of the hyperplane $\mathrm{Im}(J(\lambda))$.

Denote by $\log : (\mathbf{C}^*)^m \to \mathbf{C}^m$ the (multivalued) map which takes $(z_1, \dots, z_m)$ to $(\log(z_1), \dots, \log(z_m))$. In invariant language, log is the logarithmic map from the torus $H(L_A)$ to its Lie algebra. The hyperplane $\mathrm{Im}(J(\lambda)) \subset \mathbf{C}^m$ above is nothing more than the tangent hyperplane to the image $\log(\tilde{\nabla}_A)$ at the point $\log(h(\lambda))$. Hence the fact that the λ_i are coefficients of the equation of this hyperplane means that the point of P^{m-1} represented by λ is the image of $h(\lambda)$ under the logarithmic Gauss map. Theorem 3.3 and hence the implication (a) $\Rightarrow$ (b) of Theorem 3.1, are proved.

D. End of the proof of Theorem 3.1

Let $Z \subset (\mathbf{C}^*)^m$ be a hypersurface such that the logarithmic Gauss map $\gamma_Z : Z \to P^{m-1}$ is birational. Denote by $\Psi : P^{m-1} \to Z$ the inverse rational map, given by m rational functions $\Psi_i(\lambda_1, \dots, \lambda_m)$, $i = 1, \dots, m$, homogeneous of degree 0.

We reverse the arguments in subsection C. As a first step, let us prove that the Ψ_i satisfy (3.6). More precisely, we have the following lemma.

Lemma 3.5. *If $\Psi = (\Psi_1, \dots, \Psi_m)$, $\Psi_i = \Psi_i(\lambda_1, \dots, \lambda_m)$ is the inverse to the logarithmic Gauss map, then*

$$\log \Psi_i = \frac{\partial}{\partial \lambda_i}\left(\sum_{j=1}^{m} \lambda_j \log \Psi_j\right). \tag{3.8}$$

Proof. As in subsection C, let $J(\lambda) = J(\lambda_1, \dots, \lambda_m)$ denote the Jacobian matrix $\|\partial \log \Psi_i / \partial \lambda_j\|$, and $\log : (\mathbf{C}^*)^m \to \mathbf{C}^m$ the componentwise logarithm map. The tangent space to $\log(Z)$ at $z = \Psi(\lambda)$ is the image of $J(\lambda)$. Since each $\Psi_i(\lambda)$ is homogeneous of degree 0, the kernel $\mathrm{Ker}\,(J(\lambda))$ contains the column vector $(\lambda_1, \dots, \lambda_m)^t$.

Let $(\mu_1, \dots, \mu_m)$ be the coefficients of the equation of $\mathrm{Im}(J(\lambda))$, so that $\mu_1\xi_1 + \dots + \mu_m\xi_m = 0$ for $(\xi_1, \dots, \xi_m) \in \mathrm{Im}(J(\lambda))$.

To say that Ψ is inverse to the logarithmic Gauss map of its image $Z = \mathrm{Im}(\Psi)$ is equivalent to saying that, for any $\lambda = (\lambda_1, \dots, \lambda_m)$, the above vector $\mu = (\mu_1, \dots, \mu_m)$ is proportional to λ. But μ generates the 1-dimensional vector space $\mathrm{Ker}(J^t(\lambda))$. Hence we have

$$\sum_{j=1}^{m} \lambda_j \frac{\partial \log \Psi_j}{\partial \lambda_i} = 0, \quad j = 1, \dots, m. \tag{3.9}$$

Now we can prove our lemma. Indeed, we can rewrite (3.9) as

$$\sum_{j \neq i} \lambda_j \frac{\partial \log \Psi_j}{\partial \lambda_i} = -\lambda_i \frac{\partial \log \Psi_i}{\partial \lambda_i}.$$

Adding this equality with the obvious equality

$$\frac{\partial}{\partial \lambda_i}(\lambda_i \log \Psi_i) = \lambda_i \frac{\partial \log \Psi_i}{\partial \lambda_i} + \log \Psi_i,$$

we obtain (3.8).

To complete the proof of Theorem 3.1, it remains to classify rational solutions of (3.6) which are homogeneous of degree 0. This classification is essentially due to Horn [Hor].

Proposition 3.6. *Suppose that rational functions $\Psi_i(\lambda_1, \dots, \lambda_m)$, $i = 1, \dots, m$, satisfy (3.6) and are homogeneous of degree 0. Then there exist $n > 0$, an integral $m \times n$ matrix $\|a_i^{(l)}\|$, $i = 1, \dots, n$, $l = 1, \dots, m$ and non-zero constants $\xi_1, \dots, \xi_m$ such that each $\xi_l \Psi_l$ has the form (3.3).*

Proof. Let $p_1, \dots, p_n$ be all the irreducible polynomials entering into the factorization of the numerator or denominator of at least one Ψ_j. The polynomials p_i are homogeneous. Let $d_i = \deg(p_i)$. Thus we have

$$\Psi_l(\lambda_1, \dots, \lambda_m) = \varepsilon_l \prod_{j=1}^{n} p_j(\lambda)^{a_{jl}}, \quad \varepsilon_l \in \mathbf{C}^*,\ a_{jl} \in \mathbf{Z}.$$

The equality $\partial \log \Psi_l / \partial \lambda_r = \partial \log \Psi_r / \partial \lambda_l$ can be rewritten as

$$\sum_{j=1}^{n} \frac{a_{jl}(\partial p_j / \partial \lambda_r)}{p_j} = \sum_{j=1}^{n} \frac{a_{jr}(\partial p_j / \partial \lambda_l)}{p_j}. \tag{3.10}$$

In each of the fractions appearing in (3.10) the numerator has degree less than that of the denominator. It follows that

$$a_{jl} \frac{\partial p_j}{\partial \lambda_r} = a_{jr} \frac{\partial p_j}{\partial \lambda_l}, \quad l, r = 1, \dots, m,\ j = 1, \dots, n. \tag{3.11}$$

In other words, for each $j = 1, \dots, n$, the $2 \times m$ matrix

$$\begin{pmatrix} a_{j1} & a_{j2} & \cdots & a_{jm} \\ \frac{\partial p_j}{\partial \lambda_1} & \frac{\partial p_j}{\partial \lambda_2} & \cdots & \frac{\partial p_j}{\partial \lambda_m} \end{pmatrix}$$

has rank one. This means that the numerical vector $(a_{j1}, \dots, a_{jm})$ is proportional (over the field $\mathbf{C}(\lambda_1, \dots, \lambda_m)$ of rational functions) to the vector of polynomials $(\frac{\partial p_j}{\partial \lambda_1}, \frac{\partial p_j}{\partial \lambda_2}, \dots, \frac{\partial p_j}{\partial \lambda_m})$. So we have

$$\frac{\partial p_j}{\partial \lambda_r} = a_{jr}\delta_j(\lambda) \tag{3.12}$$

for some polynomials $\delta_j(\lambda)$. Since the p_j are homogeneous, we have

$$\sum_{r=1}^{m} \lambda_r \frac{\partial p_j}{\partial \lambda_r} = d_j p_j.$$

By (3.12), we conclude that

$$\left(\sum_r \lambda_r a_{jr}\right) \delta_j(\lambda) = d_j p_j(\lambda).$$

If $d_j > 1$ then $\deg(\delta_j) > 0$ and this contradicts the irreducibility of p_j. Hence $d_j = 1$ for all j. So the p_j are linear homogeneous functions, say

$$p_j(\lambda_1, \dots, \lambda_m) = \sum_{r=1}^{m} b_{jr}\lambda_r.$$

By (3.12), $b_{jr} = \delta_j \cdot a_{jr}$ where $\delta_j \in \mathbf{C}^*$. Hence

$$\Psi_l(\lambda) = \varepsilon_l \Big(\prod_j \delta_j^{a_{jl}}\Big) \cdot \prod_{j=1}^{n} \left(\sum_{r=1}^{m} a_{jr}\lambda_r\right)^{a_{jl}},$$

as claimed in Proposition 3.6.

Now let us finish the proof of Theorem 3.1. Since Z is a hypersurface, the rank of the matrix $\|a_j^{(l)}\|$, $j = 1, \dots, n$, $l = 1, \dots, m$, given by Proposition 3.6, equals m, i.e., the rows $a^{(l)} \in \mathbf{Z}^n$ are linearly independent. Denote by $L \subset \mathbf{Z}^n$ the Abelian subgroup generated by the $a^{(l)}$. Then L is a primitive lattice, i.e., $L = (L \otimes \mathbf{Q}) \cap \mathbf{Z}^n$ (otherwise the map Ψ given by (3.3) would be a ν-fold cover of its image, where $\nu = [(L \otimes \mathbf{Q}) \cap \mathbf{Z}^n : L]$). Since the Ψ_l are homogeneous, each $a \in L$ has the sum of components equal to 0. Thus our claim is a consequence of the following lemma.

Lemma 3.7. *Let $L \subset \mathbf{Z}^n$ be a primitive subgroup of rank m such that $\sum_{j=1}^n a_j = 0$ for any $a = (a_1, \dots, a_n) \in L$. Then there is an n-element set $A \subset \mathbf{Z}^{n-m-1}$ such* *that L is the lattice of affine relations among elements of A, and A generates $\mathbf{Z}^{n-m-1}$ as an affine lattice.*

Proof. Put $\Xi = \mathbf{Z}^n/L$. This is a free Abelian group of rank $n-m$. Let $\omega^{(j)} \in \Xi$ be the image of the j-th basis vector of $\mathbf{Z}^n$, and let $A = \{\omega^{(1)}, \dots, \omega^{(n)}\}$. Since each $a \in L$ has $\sum a_j = 0$, a homomorphism $\mathbf{Z}^n \to \mathbf{Z},\ a \mapsto \sum a_j$ factors through a homomorphism $\Xi \to \mathbf{Z}$, denoted by h. Then A is contained in the affine lattice $\{u \in \Xi : h(u) = 1\}$ and (affinely) generates it. Identifying this lattice with $\mathbf{Z}^{n-m-1}$, we obtain our statement. Theorem 3.1. is completely proved.

E. Discriminantal varieties and entropy

The formulas of subsections C, D can be rewritten in a nice way. Consider once more the space $\mathbf{C}^m = L_A \otimes \mathbf{C}$. The matrix $\|a_j^{(l)}\|,\ j = 1, \dots, n,\ l = 1, \dots, m$, defines n linear functionals

$$p_j(\lambda) = p_j(\lambda_1, \dots, \lambda_m) = \sum_{l=1}^{m} a_j^{(l)} \lambda_l$$

on this space. The rational functions $\Psi_l(\lambda)$ defined by (3.3) satisfy (3.6), i.e., the differential form $\Omega = \sum_l \log(\Psi_l) d\lambda_l$ is closed. By Lemma 3.5, $\Omega = dS$ where $S = \sum_l \lambda_l \log \Psi_l$.

Proposition 3.8. *The function $S = \sum \lambda_l \log \Psi_l$ can be written as*

$$S(\lambda) = \sum_{j=1}^{n} p_j(\lambda) \log p_j(\lambda).$$

Proof. Using (3.3), we obtain

$$S(\lambda) = \sum_{l=1}^{m} \lambda_l \log \Psi_l(\lambda) = \sum_{l=1}^{m} \lambda_l \sum_{j=1}^{n} a_j^{(l)} \log p_j(\lambda) = \sum_j p_j(\lambda) \log p_j(\lambda),$$

as required.

The function S looks like the entropy of a probability distribution, the linear functionals $p_j(\lambda)$ playing the role of the probabilities of elementary events. Curiously, we have $\sum_j p_j(\lambda) = 0$ (compare with the usual rule $\sum p_j = 1$ for probabilities). This is one of the mysterious appearances of entropy-like expressions in the theory of discriminants (cf. Proposition 1.8).

Recall that the reduced discriminantal variety $\tilde{\nabla}_A \subset (\mathbf{C}^*)^m$ has the parametrization $x_i = \Psi_i(\lambda_1, ..., \lambda_m)$. Lemma 3.5 and Proposition 3.8 imply that $\log(\tilde{\nabla}_A) \subset$

$\mathbf{C}^m$, the image of $\tilde{\nabla}_A$ under the (multivalued) map $\log : (\mathbf{C}^*)^m \to \mathbf{C}^m$, has the parametrization $y_i = \partial S/\partial \lambda_i$ (here the $y_i = \log(x_i)$ are the coordinates in $\mathbf{C}^m$). This can be rephrased by saying that $\log(\tilde{\nabla}_A)$ is projectively dual to the graph of the entropy S in the following sense.

Note that S is homogeneous of degree 1, so its graph (denote it Γ) is a conic analytic subvariety in $\mathbf{C}^{m+1}$. Let $(z, \lambda_1, \ldots, \lambda_m)$ be the coordinates in $\mathbf{C}^{m+1}$ so that Γ is given by $z = S(\lambda_1, \ldots, \lambda_m)$. Let $P(\Gamma) \subset P^m$ be the projectivization of Γ. Consider the dual projective space P^{m*} with homogeneous coordinates $w, y_1, \ldots, y_m$ dual to $z, \lambda_1, \ldots, \lambda_m$. Let $P(\Gamma)^\vee \subset P^{m*}$ be the hypersurface projectively dual to $P(\Gamma)$. Consider the affine chart $\mathbf{C}^m$ in P^{m*} given by $w \neq 0$. We assume that $w = 1$ in this chart and take $y_1, \ldots, y_m$ as affine coordinates. In these coordinates the intersection $P(\Gamma)^\vee \cap \mathbf{C}^m$ is nothing more than $\log(\tilde{\nabla}_A)$, as it follows from the above mentioned parametrization.

Note that the (non-reduced) A-discriminantal hypersurface ∇_A is projectively dual to the toric variety X_A.

F. The relation to hypergeometric functions

The aim of Horn's paper cited above was to study the domains of convergence of hypergeometric series in two (or more) variables. More precisely, Horn called a power series

$$F(x) = \sum_{\nu_1, \ldots, \nu_m} c_{\nu_1, \ldots, \nu_m} x_1^{\nu_1} \cdots x_m^{\nu_m}$$

hypergeometric if the ratios of coefficients

$$R_i(\nu_1, \ldots, \nu_m) = \frac{c_{\nu_1, \ldots, \nu_{i-1}, \nu_i + 1, \nu_{i+1}, \ldots, \nu_m}}{c_{\nu_1, \ldots, \nu_m}}$$

are rational functions in $\nu_1, \ldots, \nu_m$. It is known that a power series converges "up to its first singularity" (see Proposition 1.5 Chapter 6). On the other hand, the knowledge of the functions R_i allows us to determine the growth of the coefficients and hence the convergence radius in any given direction. Thus we can obtain the information about singularities of the analytic function represented by $F(x)$. The most interesting component in the singularity locus is the hypersurface with the parametric presentation

$$x_i = \Psi_i(\lambda_1, \ldots, \lambda_m), \text{ where } \Psi_i(\lambda_1, \ldots, \lambda_m) = \lim_{t \to \infty} R_i(t\lambda_1, \ldots, t\lambda_m) \tag{3.13}$$

(and other components are pull-backs of similar hypersurfaces from the spaces with a smaller number of variables). This is the Horn uniformization*.

However, it appears to be a recent remark [GKZ4] that any component of the singular locus of every hypergeometric series is in fact the reduced A-discriminantal variety for a suitable A.

The fact that the Ψ_i defined by (3.13), satisfy (3.6), follows from the cocycle relations

$$R_i(\nu)R_j(\nu + e_i) = R_j(\nu)R_i(\nu + e_j), \tag{3.14}$$

where the $e_i \in \mathbf{Z}^m$ are standard basis vectors. These relations hold by definition of the R_i. In other words, (3.14) means that the R_i form a 1-cocycle of the group $\mathbf{Z}^m$ with coefficients in the multiplicative group of rational functions $\mathbf{C}(\nu_1, \dots, \nu_m)^*$ on which $\mathbf{Z}^m$ acts by translations of variables.

The classification of solutions of (3.6) given in Proposition 3.6 is a limit case of the classification of solutions of (3.14) (i.e., the description of the corresponding cohomology group) due to Birkeland and Ore [Bir], [Ore].

* In principle, the limits in (3.13) may be identically equal to zero (this would mean that the corresponding series diverges) or infinity. In order for the limits to be other than 0, ∞, the degree of the numerator of R_i should be the same as the degree of the denominator. This is the case for hypergeometric series satisfying differential equations with regular singularities.

CHAPTER 10

Principal A-Determinants

Our aim in this and in the following chapter is to study the Newton polytope of the A-discriminant Δ_A. This will be done through an intermediary object, the so-called principal A-determinant E_A. Like the A-discriminant, $E_A = E_A(f)$ is a polynomial function in coefficients a_ω of an indeterminate polynomial $f \in \mathbf{C}^A$. We shall do the following:

(1) give a complete description of the Newton polytope of E_A. It turns out to coincide with the secondary polytope $\Sigma(A)$ (see Chapter 7);
(2) give a formula (prime factorization) expressing E_A as a product of Δ_A and discriminants corresponding to some subsets of A;
(3) give a formula for the product of values of a polynomial at its critical points in terms of the principal A-determinants.

The prime factorization of E_A will allow us to express Δ_A through the principal determinants corresponding to some subsets of A, thus providing information about the Newton polytope of Δ_A. This will be done in Chapter 11.

1. Statements of main results

In this section we only give definitions and formulate the results. Proofs are given only when they are immediate; More involved proofs are postponed until later sections of this chapter.

A. The principal A-determinant

Let $A \subset \mathbf{Z}^{k-1}$ be a finite subset which affinely generates $\mathbf{Z}^{k-1}$ over $\mathbf{Z}$. As usual, let $\mathbf{C}^A$ be the space of Laurent polynomials $f(x_1, \ldots, x_{k-1}) = \sum_{\omega \in A} a_\omega x^\omega$ with monomials from A. For k polynomials $f_1, \ldots, f_k \in \mathbf{C}^A$, let $R_A(f_1, \ldots, f_k)$ be their A-resultant (see Section 2A Chapter 8). Now for any $f = f(x_1, \ldots, x_{k-1}) \in \mathbf{C}^A$ we define

$$E_A(f) = R_A\left(x_1 \frac{\partial f}{\partial x_1}, \ldots, x_{k-1} \frac{\partial f}{\partial x_{k-1}}, f\right). \tag{1.1}$$

Note that each $x_i(\partial f/\partial x_i)$ belongs to $\mathbf{C}^A$ so (1.1) makes sense. Clearly E_A is a polynomial function in coefficients of f. We call $E_A(f)$ the *principal A-determinant* of f.

This name is explained by the fact that E_A possesses a determinantal representation which is considerably simpler than the representation of Δ_A (see Section 2C Chapter 9). This is just the determinantal representation for R_A (see Section 2B Chapter 8), under the specialization (1.1). Since we shall make use of this construction, let us explain it in more detail.

We again adopt the notation and conventions of Section 2B Chapter 8. So we embed A into the Abelian group $\mathbf{Z}^k = \mathbf{Z}^{k-1} \times \mathbf{Z}$ by $\omega \mapsto \tilde{\omega} = (\omega, 1)$. As before, let $h : \mathbf{Z}^k \to \mathbf{Z}$ be the projection given by the last coordinate, S the subsemigroup in $\mathbf{Z}^k$ generated by A and 0, and for each $l \in \mathbf{Z}$, let S_l be the slice $\{u \in S : h(u) = l\}$.

Let $l \in \mathbf{Z}$ and $f = \sum_{\omega \in A} a_\omega x^\omega \in \mathbf{C}^A$. Define the complex $L^\bullet(A, l)$ with the differential ∂_f by setting

$$L^i(A, l) = \bigoplus_{u \in S_{l+i}} \bigwedge^i (\mathbf{C}^k). \tag{1.2}$$

A typical element of the u-th summand will be denoted by (u, λ), where $\lambda \in \bigwedge^i(\mathbf{C}^k)$. We define $\partial_f : L^i(A, l) \to L^{i+1}(A, l)$ by

$$\partial_f(u, \lambda) \quad = \quad -\sum_{\omega \in A} a_\omega \cdot (u + \omega, \lambda \wedge \omega). \tag{1.3}$$

It is immediate that $\partial_f^2 = 0$.

The spaces $L^i(A, l)$ are equipped with natural $\mathbf{Z}$-lattices

$$L^i_{\mathbf{Z}}(A, l) = \bigoplus_{u \in S_{i+l}} \bigwedge^i_{\mathbf{Z}} (\mathbf{Z}^k).$$

We denote by e some system of $\mathbf{Z}$-bases in these lattices.

Proposition 1.1. *For $l \gg 0$ and generic $f \in \mathbf{C}^A$, the differential ∂_f in $L^\bullet(A, l)$ is exact and*

$$E_A(f) = \pm \det(L^\bullet(A, l), \partial_f, e)^{(-1)^k}.$$

Proof. Comparing the above definitions with formulas (2.4) and (2.5) in Section 2, Chapter 8, we see that $L^\bullet(A, l)$ is obtained by the specialization (1.1) from the resultant complex $C^\bullet(l)$. Thus our statement is a special case of Theorem 2.5, Chapter 8.

Consider the particular case when X_A is a smooth variety. Comparing $L^\bullet(A, l)$ with the discriminantal complex $C^\bullet(A, l)$ (see (2.9), (2.10) in Chapter 9), we see that $C^\bullet(A, l)$ is a subcomplex of $L^\bullet(A, l)$. This subcomplex is more complicated than the complex itself, because each $C^i(A, l)$ is a sum of different summands, while all the summands in $L^i(A, l)$ are the same.

B. The prime factorization of E_A

Let $Q \subset \mathbf{R}^{k-1}$ be the convex hull of A. For any non-empty face $\Gamma \subset Q$, we introduced (see Section 3A Chapter 5) the semigroup S/Γ which is the image of S in the quotient lattice $\mathbf{Z}^k/\mathbf{Z}^k \cap \mathrm{Lin}_{\mathbf{Z}}(\Gamma)$. We also introduced the index

$$i(\Gamma, A) = [\mathbf{Z}^k \cap \mathrm{Lin}_{\mathbf{R}}(\Gamma) : \mathrm{Lin}_{\mathbf{Z}}(A \cap \Gamma)].$$

We denote by $X^0(\Gamma) \subset X_A$ the torus orbit corresponding to Γ and its closure by $X(\Gamma)$. Let $\mathrm{mult}_{X(\Gamma)} X_A$ be the multiplicity of X_A along $X(\Gamma)$ (see Definition 3.11, Chapter 5). By Theorem 3.16 of Chapter 5,

$$\mathrm{mult}_{X(\Gamma)} X_A = i(\Gamma, A) \cdot u(S/\Gamma),$$

where $u(\cdot)$ stands for the subdiagram volume of a semigroup.

For any face $\Gamma \subset Q$ and any $f = \sum_{\omega \in A} a_\omega x^\omega \in \mathbf{C}^A$, we denote by $f\|_\Gamma = \sum_{\omega \in A \cap \Gamma} a_\omega x^\omega$ the coefficient restriction of f to Γ (cf. Section 1, Chapter 6). For $f \in \mathbf{C}^A$ we write $\Delta_{A\cap\Gamma}(f) := \Delta_{A\cap\Gamma}(f\|_\Gamma)$.

Now we describe the prime factorization of the principal A-determinant.

Theorem 1.2. *The principal A-determinant is equal to*

$$E_A(f) = \pm \prod_{\Gamma \subset Q} \Delta_{A\cap\Gamma}(f)^{i(\Gamma, A)\cdot u(S/\Gamma)},$$

the product taken over all non-empty faces $\Gamma \subset Q$ including Q itself. In particular, if X_A is smooth then

$$E_A(f) = \pm \prod_{\Gamma \subset Q} \Delta_{A\cap\Gamma}(f).$$

The proof will be given in Section 2 below.

Examples 1.3. Let us illustrate Theorem 1.2 for the sets A given in Examples 1.6, Chapter 9.

(a) Let A consist of all integral monomials of degree d in $x_1, \ldots, x_k$. The polytope $Q \subset \mathbf{R}^k$ is the $(k-1)$-dimensional simplex

$$\left\{(t_1, \ldots, t_k) \in \mathbf{R}^k : t_i \geq 0, \sum t_i = d\right\}.$$

There are $2^k - 1$ (non-empty) faces of Q, corresponding to non-empty subsets $I \subset \{1, \dots, k\}$. More precisely, the face $\Gamma(I)$ corresponding to a subset I is given by the equations $t_i = 0, i \notin I$. Since X_A is smooth, Theorem 1.2 gives

$$E_A(f) = \prod_{\emptyset \neq I \subset \{1,\dots,k\}} \Delta_{A \cap \Gamma(I)}(f).$$

For example, if $k = 2$, then a polynomial $f \in \mathbf{C}^A$ has the form

$$f(x_1, x_2) = a_0 x_1^d + a_1 x_1^{d-1} x_2 + \cdots + a_d x_2^d$$

and

$$E_A(f) = a_0 a_d \Delta_A(f),$$

where Δ_A is the classical discriminant of a binary form of degree d.

If $k = 3$, then we can write $f \in \mathbf{C}^A$ as

$$f(x_1, x_2, x_3) = \sum_{i+j \leq d} a_{ij} x_1^i x_2^j x_3^{d-i-j},$$

and

$$E_A(f) = \Delta_A(f) \cdot a_{00} a_{0d} a_{d0} \cdot \Delta(a_{0d} x_2^d + a_{0,d-1} x_2^{d-1} x_3 + \cdots + a_{00} x_3^d) \cdot$$

$$\Delta(a_{d0} x_1^d + a_{d-1,0} x_1^{d-1} x_3 + \cdots + a_{00} x_3^d) \cdot \Delta(a_{d0} x_1^d + a_{d-1,1} x_1^{d-1} x^2 + \cdots + a_{0,d} x_2^d).$$

Here three Δ factors are the discriminants of binary forms, and Δ_A is the discriminant of a ternary form of degree d.

(b) Let A consist of bilinear monomials $x_i y_j$, $i = 1, \dots, m$, $j = 1, \dots, n$. A polynomial $f(x, y) = \sum a_{ij} x_i y_j \in \mathbf{C}^A$ is determined by its matrix of coefficients $\|a_{ij}\|$ of size $m \times n$. The polytope Q is the product of two simplices $\Delta^{m-1} \times \Delta^{n-1}$. Its faces are products $\Gamma_1 \times \Gamma_2$ where Γ_1 is a face of Δ^{m-1} and Γ_2 is a face of Δ^{n-1}. In other words, faces correspond to pairs of non-empty subsets $I \subset \{1, \dots, m\}$, $J \subset \{1, \dots, n\}$ of arbitrary cardinalities. Let us denote the face corresponding to a pair (I, J) by $\Gamma(I, J)$. It was shown in Example 1.6 (b), Chapter 9 that the $A \cap \Gamma(I, J)$-discriminant $\Delta_{A \cap \Gamma(I,J)}(f)$ identically equals 1 if I and J have different cardinalities, and if $\#(I) = \#(J)$ then $\Delta_{A \cap \Gamma(I,J)}(f)$ equals the minor $\Delta_{IJ}(\|a_{ij}\|)$ of the matrix $\|a_{ij}\|$ on the rows from I and columns from J. Therefore

$$E_A(f) = E_A(\|a_{ij}\|) = \prod_{I,J} \Delta_{IJ}(\|a_{ij}\|) \tag{1.4}$$

is the product of all the square minors (of all sizes, including 1×1) of the $(m \times n)$-matrix $\|a_{ij}\|$.

For example, if $m = n = 2$ then

$$E_A \begin{pmatrix} a_{11} & a_{12} \\ a_{21} & a_{22} \end{pmatrix} = a_{11}a_{12}a_{21}a_{22}(a_{11}a_{22} - a_{12}a_{21}).$$

(c) Let A consist of $1, x, \ldots, x^m, y, yx, \ldots, yx^n$. The space $\mathbf{C}^A$ consists of polynomials $g(x, y) = f_0(x) + yf_1(x)$ where

$$f_0(x) = a_0 + a_1 x + \cdots + a_m x^m, \quad f_1(x) = b_0 + b_1 x + \cdots + b_n x^n$$

are polynomials in one variable x of degrees m, n. The polytope Q is the trapezoid in Figure 43. Therefore

$$E_A(f_0(x) + yf_1(x)) = a_0 a_m b_0 b_n \Delta_m(f_0) \cdot \Delta_n(f_1) \cdot R_{m,n}(f_0, f_1) \tag{1.5}$$

where Δ_m is the classical discriminant of a polynomial of degree m in one variable and $R_{m,n}$ is the classical resultant of two polynomials of degrees m, n in one variable.

C. The Newton polytope of E_A

Recall that E_A is a polynomial in coefficients a_ω of an indeterminate polynomial $f(x) = \sum_{\omega \in A} a_\omega x^\omega \in \mathbf{C}^A$. Let us write E_A in an expanded form, as a sum of monomials:

$$E_A(f) = \sum_{\varphi: A \to \mathbf{Z}_+} c_\varphi \cdot \prod_{\omega \in A} a_\omega^{\varphi(\omega)} \tag{1.6}$$

where φ runs the exponent vectors of monomials in E_A and $c_\varphi \in \mathbf{Z}$ is the coefficient at the monomial corresponding to φ. The Newton polytope $N(E_A)$ is, by definition, the convex hull in $\mathbf{R}^A$ of all vectors $\varphi : A \to \mathbf{Z}_+$ such that $c_\varphi \neq 0$.

Recall that we defined the secondary polytope $\Sigma(A) \subset \mathbf{R}^A$ (Section 1D Chapter 7). Its vertices are in bijection with coherent triangulations of (Q, A): the vertex corresponding to a triangulation T is the function $\varphi_T : A \to \mathbf{R}$ given by

$$\varphi_T(\omega) = \sum_{\substack{\sigma \in T \\ \omega \in \mathrm{Vert}(\sigma)}} \mathrm{Vol}\,(\sigma) \tag{1.7}$$

where Vol is a fixed translation invariant volume form on $\mathbf{R}^{k-1}$. In the sequel we always assume that the volume form $\mathrm{Vol} = \mathrm{Vol}_{\mathbf{Z}^{k-1}}$ is induced by the lattice $\mathbf{Z}^{k-1}$ (see Section 3D Chapter 5).

Theorem 1.4.

(a) *The Newton polytope of E_A coincides with the secondary polytope $\Sigma(A)$. In particular, vertices of E_A are the functions φ_T for all the coherent triangulations of (Q, A).*

(b) *If T is a coherent triangulation of (Q, A) then the coefficient at the monomial $\prod_{\omega\in A} a_\omega^{\varphi_T(\omega)}$ in E_A equals*

$$c_{\varphi_T} = \pm \prod_{\sigma\in T} \mathrm{Vol}\,(\sigma)^{\mathrm{Vol}\,(\sigma)}.$$

Proof. This is a corollary of Theorems 3.1 and 3.2 of Chapter 8 describing the Chow polytope of the toric variety X_A (i.e., the weight polytope of the resultant R_A) and the extreme monomials in R_A. We have only to show that under specialization (1.1) the bracket monomial $\prod_{\sigma\in T}[\sigma]^{\mathrm{Vol}\,(\sigma)}$ of Theorem 3.2, Chapter 8 becomes the monomial $\pm\prod_{\sigma\in T} \mathrm{Vol}\,(\sigma)^{\mathrm{Vol}\,(\sigma)} \cdot \prod_{\omega\in A} a_\omega^{\varphi_T(\omega)}$ (this means, in particular, that this specialized monomial is non-zero, and so the Newton polytope of E_A coincides with the Chow polytope). Remembering the definition of the brackets $[\sigma]$, we see that under (1.1), each $[\sigma]$ becomes

$$\pm\mathrm{Vol}\,(\sigma) \prod_{\omega\in\sigma} a_\omega. \tag{1.8}$$

This immediately implies our statement.

Examples 1.5. *(a)* Let $A = \{1, x, x^2\}$. The space $\mathbf{C}^A$ consists of quadratic polynomials $f(x) = ax^2 + bx + c$. The polytope Q is the segment $[0, 2]$. The discriminant $\Delta_A(f)$ is equal to $b^2 - 4ac$. The principal A-determinant has the form $E_A(f) = ac(b^2 - 4ac) = ab^2c - 4a^2c^2$; its Newton polytope is the segment in $\mathbf{R}^A = \mathbf{R}^3$ with the end-points $(1, 2, 1)$ and $(2, 0, 2)$. There are exactly two triangulations of (Q, A). The first triangulation T_1 consists of just one "simplex" which is the whole $[0, 2]$. Its "volume" equals 2. The corresponding vertex φ_{T_1} of $\Sigma(A)$ is $(2, 0, 2)$, and the corresponding term in E_A is $-4a^2c^2$. The second triangulation T_2 consists of two "simplices" which are segments $[0, 1]$ and $[1, 2]$, both of volume 1. The vertex $\varphi_{T_2} \in \Sigma(A)$ is $(1, 2, 1)$, and the corresponding term in E_A is ab^2c.

The generalization of this example to polynomials of arbitrary degree in one variable will be discussed later (see Section 2, Chapter 12).

(b) Let A consist of bilinear monomials $x_i y_j$ with $i = 1, 2,\ j = 1, \ldots, n$. Then $\mathbf{C}^A$ is identified with the space of $2\times n$- matrices $\|a_{ij}\|$. By (1.4) above, we

have

$$E_A(\|a_{ij}\|) = \prod_{i=1}^{2}\prod_{j=1}^{n} a_{ij} \cdot \prod_{k<l}(a_{1k}a_{2l} - a_{2k}a_{1l})$$

$$= \prod_{i=1}^{2}\prod_{j=1}^{n} a_{ij} \cdot \begin{vmatrix} a_{11}^{n-1} & a_{12}^{n-1} & \dots & a_{1n}^{n-1} \\ a_{11}^{n-2}a_{21} & a_{12}^{n-2}a_{22} & \dots & a_{1n}^{n-2}a_{2n} \\ \vdots & \vdots & \vdots & \vdots \\ a_{21}^{n-1} & a_{22}^{n-1} & \dots & a_{2n}^{n-1} \end{vmatrix},$$

where the last equality follows from the formula for the Vandermonde determinant. We see that E_A has $n!$ monomials (corresponding to all the summands in the determinant) and the Newton polytope of E_A is the permutohedron P_n, i.e., the convex hull of an orbit of the action of the symmetric group S_n in $\mathbf{R}^n$ (see Section 3C Chapter 7). This is in accord with the fact that the secondary polytope $\Sigma(A)$ is the permutohedron.

Generalizing this example to the case when $\mathbf{C}^A$ is the space of matrices of arbitrary size $m \times n$ we see that the Newton polytope of the product of all minors (of all sizes) of such a matrix coincides with the secondary polytope $\Sigma(\Delta^{m-1} \times \Delta^{n-1})$ of the product of two simplices. The relationship between triangulations of $\Delta^{m-1} \times \Delta^{n-1}$ and the extreme monomials in the product of minors is not yet fully understood. Let us mention only that the standard triangulation T_0 (see Section 3D Chapter 7) corresponds to the product of diagonal terms in all the minors. We leave checking this fact (i.e., that the exponent vector of the product of all diagonal terms coincides with φ_{T_0}), to the reader.

Remark 1.6. The coefficients of the extreme monomials in E_A, given by Theorem 1.4, have the form

$$\prod V_i^{V_i} = e^{\sum V_i \log V_i}.$$

We have already encountered entropy-like expressions $\sum p_i \log p_i$ (see Section 3E Chapter 9). This is the second appearance of such expressions in the theory of discriminants.

Here the role of probabilities p_i is played by the volumes Vol (σ) of the simplices σ in a given coherent triangulation T of (Q, A). Since $\sum_{\sigma \in T} \text{Vol}\,(\sigma) = \text{Vol}\,(Q)$, we can regard the collection of Vol $(\sigma)/$Vol (Q), $\sigma \in T$, as a probability distribution in the following model: we randomly throw points into the polytope Q and look at the probability of the event that a point lands in a given simplex of a triangulation. By Theorem 1.4, the coefficient of $\prod_{\omega \in A} a_\omega^{\varphi_T(\omega)}$ in the principal A-determinant is essentially the exponent of the entropy of this model. It would be nice to find a "probabilistic" explanation of this.

D. Applications to the geometry of discriminantal hypersurfaces

As we have seen (Section 1B Chapter 6), the structure of the Newton polytope of a Laurent polynomial F in n variables is closely related to the geometry of the hypersurface $\{F = 0\}$ in the torus $(\mathbf{C}^*)^n$.

We explore this relation for the polynomial $F = E_A$ in variables a_ω, $\omega \in A$. So we regard E_A as a Laurent polynomial, i.e., as a function on the torus $(\mathbf{C}^*)^A$. In other words, in this subsection we s consider only the polynomials $f \in \mathbf{C}^A$ all of whose coefficients are non-zero.

If T is a coherent triangulation of (Q, A) then the normal cone $C(T)$ to the secondary polytope $\Sigma(A)$ at its vertex φ_T was described in Theorem 1.7, Chapter 7. It consists of those functions $\lambda = (\lambda_\omega) : A \to \mathbf{R}$ whose values at vertices of T extend to a concave T-piecewise-linear function $g_{T,\lambda} : Q \to \mathbf{R}$ and whose values at elements ω of A which are not vertices of T do not exceed the values of $g_{T,\lambda}(\omega)$.

Let $Z \subset (\mathbf{C}^*)^A$ be the zero set of E_A. Since the A-discriminant Δ_A is a divisor of E_A, the variety Z contains the discriminantal variety ∇_A intersected with $(\mathbf{C}^*)^A$.

Corollary 1.8 of Chapter 6 together with Theorem 1.4 imply the following.

Corollary 1.7. *Let T be a coherent triangulation of (Q, A). There exists a vector $b = (b_\omega) \in C(T)$ with the following property: whenever $f(x) = \sum_{\omega\in A} a_\omega x^\omega \in (\mathbf{C}^*)^A$ is such that the vector* $(\log|a_\omega|) \in \mathbf{R}^A$ *lies in the translated cone $C(T) + b$, we have $E_A(f) \neq 0$ (and hence $\Delta_A(f) \neq 0$).*

Plainly, whenever the logarithms of the absolute values of the coefficients of $f \in (\mathbf{C}^*)^A$ are "concave enough" with respect to any given triangulation, then f has a non-zero discriminant. Thus such an f defines a smooth hypersurface in $(\mathbf{C}^*)^{k-1}$ (since the presence of a singular point in the torus is, by the definition of Δ_A, a sufficient condition for the vanishing of Δ_A). If X_A is smooth then the hypersurface in X_A given by f is also smooth (since Δ_A is the equation of the variety projective dual to X_A).

We shall discuss in Chapter 11 the consequences of this for real algebraic geometry.

Remark 1.8. There is another way of proving Corollary 1.7 which can provide, in principle, some bounds for the vector b. This is done by using the theory of hypergeometric functions. Namely, in [GGZ], to a set $A \subset \mathbf{Z}^{k-1}$ there is associated the *A-hypergeometric system* (cf. Remark 1.19 Chapter 7). This is a holonomic system of linear differential equations on a function $\mathbf{C}^A \to \mathbf{C}$. The set of singular points of this system is the hypersurface in $\mathbf{C}^A$ which is the union of the hypersurfaces $\{\Delta_{A\cap\Gamma}(f) = 0\}$ for all the faces $\Gamma \subset Q$. Taking into account the prime factorization of E_A, we can write this hypersurface by one equation

$E_A(f) = 0$.

On the other hand, in [GZK1] we constructed, for any coherent triangulation T of (Q, A), a complete system of solutions of the A-hypergeometric system given by the Laurent power series. These series have a common domain of convergence of exactly the form stated in Corollary 1.7: this domain is determined by some $b \in C(T)$ and consists of $f = \sum a_\omega x^\omega \in (\mathbf{C}^*)^A$ such that $(\log|a_\omega|) \in C(T) + b$; therefore, in this domain the polynomial E_A does not vanish. This method allows us, in principle, to give an effective bound for b by estimating the growth of the coefficients of the A-hypergeometric series.

E. General formalism of principal A-determinants

For future convenience, we need to put the principal A-determinants into a slightly more general context. Suppose we are given a free Abelian group $\Xi \cong \mathbf{Z}^k$ of rank k and a finite subset $A \subset \Xi$ satisfying the following condition: there exists a homomorphism $h : \Xi \to \mathbf{Q}$ equal to 1 on A. Note that we do not require that A generates Ξ, nor that h takes integer values on Ξ; in particular, we allow the situation when the subgroup generated by A has rank $< k$. As usual, let $Q = \mathrm{Conv}(A) \subset \Xi_{\mathbf{R}} = \Xi \otimes \mathbf{R}$.

Suppose also that we have a finitely generated semigroup $S \subset \Xi$ satisfying the following condition: S contains A and the convex hull of S is the minimal convex cone containing A. As before, we denote $S_i = \{u \in S : h(u) = i\}$. Note that we do not require that A generates S as a semigroup.

As usual, we identify points of Ξ with Laurent monomials in k variables. Let $\mathbf{C}^A$ denote the space of Laurent polynomials with monomials from A. Fix some $f \in \mathbf{C}^A$. We now define the complex $L^\bullet(A, l, S, \Xi)$. Its terms are

$$L^i(A, l, S, \Xi) = \bigoplus_{a \in S_{i+l}} \bigwedge^i \Xi_{\mathbf{C}}. \tag{1.9}$$

The differential ∂_f is given by (1.3).

Proposition 1.9. *For $l \gg 0$ the complex $(L^\bullet(A, l, S, \Xi), \partial_f)$ is exact whenever $\Delta_{A\cap\Gamma}(f) \neq 0$ for any face $\Gamma \subset Q$.*

The proof of this proposition as well as of Proposition 1.10 and Theorem 1.11 below will be given later in Section 3.

We introduce the $\mathbf{Z}$-lattice in each term $L^i(A, l, S, \Xi)$ by

$$L^i_{\mathbf{Z}}(A, l, S, \Xi) = \bigoplus_{a \in S_{i+l}} \bigwedge^i \Xi. \tag{1.10}$$

Let e be some system of $\mathbf{Z}$-bases in the lattices (1.10). For any $f \in \mathbf{C}^A$ and $l \gg 0$, we set

$$E_{A,l}(f, S, \Xi) = \det(L^\bullet(A, l, S, \Xi), \partial_f, e)^{(-1)^k}.$$

This gives us a rational function on $\mathbf{C}^A$ defined up to a sign.

Proposition 1.10. *The rational function $E_{A,l}(f, S, \Xi)$ is independent (up to a sign) of the choice of $l \gg 0$.*

So we write $E_A(f, S, \Xi)$ for $E_{A,l}(f, S, \Xi)$ with $l \gg 0$. This function includes the principal A-determinant $E_A(f)$ considered above, as a special case corresponding to the following situation:

$\Xi = \mathbf{Z}^{k-1} \times \mathbf{Z}$, and $h : \Xi \to \mathbf{Z}$ is given by the last coordinate;

A affinely generates $\mathbf{Z}^{k-1} \times \{1\}$;

S is the semigroup generated by A.

In fact, any $E_A(f, S, \Xi)$ can be reduced to this special case, as the following theorem shows.

Theorem 1.11.

(a) *If the linear span* $\mathrm{Lin}_{\mathbf{R}}(A) \subset \Xi_{\mathbf{R}}$ *has dimension less than k then* $E_A(f, S, \Xi) = \pm 1$ *identically.*

(b) *If* $\mathrm{Lin}_{\mathbf{R}}(A)$ *has full dimension k then*

$$E_A(f, S, \Xi) = \pm\big[\Xi : \mathrm{Lin}_{\mathbf{Z}}(A)\big]^{\mathrm{Vol}_\Theta(Q)} \cdot E_A(f)^{[\mathrm{Lin}_{\mathbf{Z}}(S_{\mathbf{Z}}):\mathrm{Lin}_{\mathbf{Z}}(A)]}$$

where $\mathrm{Lin}_{\mathbf{Z}}$ *means the Abelian subgroup generated by a set;* $\Theta = \mathrm{Lin}_{\mathbf{Z}}(S) \cap \{u \in \Xi : h(u) = 1\}$*; and* $S_{\mathbf{Z}} = \{u \in S : h(u) \in \mathbf{Z}\}$.

The appearance of the semigroup $S_{\mathbf{Z}}$ in the theorem is natural since the complexes (1.9) involve the summation over points which always belong to $S_{\mathbf{Z}}$. In most (but not all) applications we will be interested in, we have $S_{\mathbf{Z}} = S$.

Theorem 1.11 shows that the increase of generality in passing from $E_A(f)$ to $E_A(f, S, \Xi)$ is largely nominal. However, many formulas below can be written much more simply in terms of $E_A(f, S, \Xi)$.

F. Coefficient restrictions of E_A

By Theorem 2.4 Chapter 7, the faces of the Newton polytope $N(E_A) = \Sigma(A)$ correspond to coherent polyhedral subdivisions $P = \{(Q_i, A_i)\}$ of (Q, A). Let $F(P) \subset N(E_A)$ be the face corresponding to a subdivision P. We shall describe the coefficient restriction

$$E_A\|_{F(P)} = \sum_{\varphi \in F(P)} c_\varphi \prod_{\omega \in A} a_\omega^{\varphi(\omega)}, \tag{1.11}$$

where the c_φ are coefficients of E_A (see (1.6)). If P is a triangulation then $F(P)$ is a vertex and the coefficient restriction is just the monomial in E_A described in Theorem 1.4. The general answer is formulated in terms of the generalized A-determinants of subsection E. As before we assume that $A \subset \mathbf{Z}^{k-1}$ is embedded into the Abelian group $\mathbf{Z}^k$, and that $S \subset \mathbf{Z}^k$ is the semigroup generated by A.

Theorem 1.12. *Let $P = \{(Q_i, A_i)\}$ be a coherent subdivision of (Q, A). Then*

$$E_A(f)\|_{F(P)} = \prod_i E_{A_i}(f, S^{(i)}, \mathbf{Z}^k)$$

where $S^{(i)} = S \cap \mathbf{R}_+ Q_i$ is the intersection of S with the cone over Q_i.

Here and later on the principal determinant associated to a subset $B \subset A$ is regarded as a function on $\mathbf{C}^A$ via the coefficient restriction $\mathbf{C}^A \to \mathbf{C}^B$ (this applies, in particular, to $B = A_i$ in Theorem 1.12).

By using Theorem 1.11, we can reformulate Theorem 1.12 in terms of the E_{A_i}.

Theorem 1.12′. *In the above assumptions we have*

$$E_A\|_{F(P)} = \prod_i [\mathbf{Z}^k : \mathrm{Lin}_{\mathbf{Z}}(A_i)]^{\mathrm{Vol}_{\mathbf{Z}^{k-1}}(Q_i)} E_{A_i}(f)^{[\mathbf{Z}^k : \mathrm{Lin}_{\mathbf{Z}}(A_i)]}.$$

More precisely, we should note that $\mathrm{Lin}_{\mathbf{Z}}(S^{(i)}) = \mathrm{Lin}_{\mathbf{Z}}(S) = \mathbf{Z}^k$ (this is because $S^{(i)}$ is the intersection of S with a cone having non-empty interior). Thus the lattice Θ in Theorem 1.11 coincides with $\mathbf{Z}^{k-1}$. Also the semigroup $S^{(i)}_{\mathbf{Z}}$ coincides with $S^{(i)}$, since $S^{(i)}$ is generated by A_i.

Let us prove Theorem 1.12. By Proposition 1.3 Chapter 6, the coefficient restriction of any Laurent polynomial $F(x_1, \ldots, x_n)$ to any face of its Newton polytope can be found from leading terms of 1-variable polynomials of the form

$$\tau \mapsto F\big(\tau^{\lambda_1} x_1, \ldots, \tau^{\lambda_n} x_n\big).$$

Hence Theorem 1.12 is a consequence of the following statement.

Theorem 1.13. *Let $P = \{(Q_i, A_i)\}$ be a coherent polyhedral subdivision of (Q, A) and let $C(P) \subset \mathbf{R}^A$ be the corresponding cone (Section 2A, Chapter 5). Let $\lambda = (\lambda_\omega)_{\omega \in A}$ be an integral vector lying in the interior of $C(P)$. Then for a generic $f = \sum a_\omega x^\omega \in \mathbf{C}^A$, the leading term of the Laurent polynomial $\tau \mapsto E_A(\sum_\omega \tau^{\lambda_\omega} a_\omega x^\omega)$ equals*

$$\pm \prod_i E_{A_i}(f, S^{(i)}, \mathbf{Z}^k)\tau^{\lambda(F(P))},$$

where the $S^{(i)}$ are the same as in Theorem 1.12 and $\lambda(F(P))$ is the value of λ (regarded as a linear functional on $\mathbf{R}^A$) at any point of $F(P)$.

The proof of Theorem 1.13 is quite similar to that of Theorem 3.3, Chapter 8. Using Proposition 1.1, we interpret E_A as the determinant of the complex (1.2), then, out of λ, construct a filtered complex over the field $\mathbf{C}((\tau^{-1}))$ and study the associated graded complex using the Kouchnirenko resolution. The first term of this resolution will be the direct sum of complexes calculating the $E_{A_i}(f, S^{(i)}, \mathbf{Z}^k)$ and other terms will have determinant ± 1 by Theorem 1.11 (a). We leave the details to the reader (cf. also [GZK3]).

G. Modifications of triangulations and signs in Theorem 1.4

In Theorem 1.4 we described only the absolute value of the coefficient in the monomial of E_A corresponding to a coherent triangulation T. However, the sign of the coefficient is also of interest. For instance, when A consists of bilinear monomials $x_i y_j$, $i = 1, 2$, $j = 1, \ldots, n$, the polynomial E_A is essentially the Vandermonde determinant (see Example 1.5 (b)) so monomials in E_A correspond to permutations of $\{1, \ldots, n\}$ and the sign of a monomial equals the sign of the corresponding permutation.

Since E_A itself is defined only up to a sign, it is possible to describe the coefficients only up to a simultaneous change of sign. Thus we can compare the signs of any two monomials.

Let T, T' be two coherent triangulations of (Q, A) such that the corresponding vertices φ_T, $\varphi_{T'}$ of the secondary polytope are joined by an edge. Let c_{φ_T}, $c_{\varphi_{T'}}$ be the corresponding coefficients in E_A. Let $p(T, T') \in \{0, 1\}$ be such that $(-1)^{p(T,T')}$ is the sign of the ratio $c_{\varphi_T}/c_{\varphi_{T'}}$.

Since any two vertices of a polytope can be joined by an edge path, knowing all the $p(T, T')$ will allow us to compare the signs of any two coefficients corresponding to vertices of $N(E_A)$. (This, in fact is how the sign of a permutation is defined: we postulate that the sign changes under any transposition and ascribe sign $+1$ to the identity permutation.)

By Theorem 2.10, Chapter 7, the vertices φ_T, $\varphi_{T'}$ are joined by an edge if and only if the coherent triangulations T, T' of (Q, A) are obtained from each other by a modification along some circuit $Z \subset A$. We shall use the terminology of Section 2, Chapter 7.

Theorem 1.14. *Let T and T' be two coherent triangulations of (Q, A) obtained from each other by a modification along a circuit $Z \subset A$. Then*

$$p(T, T') \equiv \left(\sum_{J \text{ separ.}} (\mathrm{Vol}\,(\mathrm{Conv}\,(J \cup Z)) + \left[\mathbf{Z}^{k-1} : \mathrm{Aff}_{\mathbf{Z}}(J \cup Z)\right]) \right) \pmod 2$$

where the summation is over all separating subsets.

Note that if we embed A into $\mathbf{Z}^k$ as above, then $[\mathbf{Z}^{k-1} : \mathrm{Aff}_{\mathbf{Z}}(J \cup Z)] = [\mathbf{Z}^k : \mathrm{Lin}_{\mathbf{Z}}(J \cup Z)]$. In the formulation of Theorem 1.14 we prefer to use the affine notation since it does not involve the embedding into $\mathbf{Z}^k$. Note also that in [Lo] a seemingly different formula for the $p(T, T')$ was stated; however, it becomes identical with the one above once we take into account the normalization of the volume Vol (Conv$(J \cup Z)$).

Proof. We consider the coefficient restriction $E_A\|_{[\varphi_T, \varphi_{T'}]}$ of E_A to the edge $[\varphi_T, \varphi_{T'}]$. This restriction is a polynomial which depends essentially on one variable.

For any Laurent polynomial F in the a_ω with real coefficients whose Newton polytope is a segment, let $p(F) \in \{0, 1\}$ be such that $(-1)^{p(F)}$ is the sign of the ratio of the two extreme coefficients of F. Clearly, the assignment $F \mapsto p(F)$ has the following multiplicative property: for any two polynomials F and G whose Newton polytopes are segments parallel to each other, we have $p(FG) \equiv p(F) + p(G) \pmod 2$.

Using Theorem 1.12 and the description of the polyhedral subdivision corresponding to $[\varphi_T, \varphi_{T'}]$ (Proposition 2.12 Chapter 7), we see that $E_A\|_{[\varphi_T, \varphi_{T'}]}$ equals, up to multiplication by a monomial and a constant, the product

$$\prod_{J \text{ separ.}} E_{J \cup Z}(f, S_J, \mathbf{Z}^k), \text{ where } S_J = S \cap \mathrm{Conv}(\mathbf{R}_+(J \cup Z)).$$

Each of the factors in this product has, as its Newton polytope, a segment parallel to $[\varphi_T, \varphi_{T'}]$. Hence we can write, using the multiplicative property of $F \mapsto p(F)$ and Theorem 1.11,

$$\begin{aligned} p(E_A\|_{[\varphi_T, \varphi_{T'}]}(f)) &\equiv \sum_{J \text{ separ.}} p(E_{J \cup Z}(f, S_J, \mathbf{Z}^k)) \\ &\equiv \sum_{J \text{ separ.}} [\mathbf{Z}^{k-1} : \mathrm{Aff}_{\mathbf{Z}}(J \cup Z)]\; p(E_{J \cup Z}(f)) \pmod 2. \end{aligned} \tag{1.12}$$

The principal determinant $E_{J \cup Z}$ is calculated with respect to the Abelian subgroup in $\mathbf{Z}^k = \mathbf{Z}^{k-1} \times \mathbf{Z}$ generated by $J \cup Z$. To prove Theorem 1.14, we need another lemma.

Call a set $B \subset \mathbf{R}^{k-1}$ *weakly dependent* if it is obtained from an affinely independent set (the set of vertices of a simplex) by adding exactly one point lying in its affine span. Such a subset contains a unique circuit. Note that the subsets $J \cup Z$ above are weakly dependent.

Lemma 1.15. *Let Ξ be an affine lattice and let $B \subset \Xi$ be a weakly dependent subset which affinely generates Ξ over $\mathbf{Z}$. Then*

$$p(E_B) \equiv \mathrm{Vol}_{\,\Xi}(\mathrm{Conv}(B)) + 1 \pmod 2.$$

Let us deduce Theorem 1.14 from Lemma 1.15. We apply this lemma to $B = J \cup Z$ and $\Xi = \mathrm{Aff}_{\mathbf{Z}}(J \cup Z)$. Therefore the summand in (1.12) corresponding to J is equal to

$$[\mathbf{Z}^{k-1} : \mathrm{Aff}_{\mathbf{Z}}(J \cup Z)]\big(\mathrm{Vol}_{\,\mathrm{Aff}_{\mathbf{Z}}(J\cup Z)}(\mathrm{Conv}(J \cup Z)) + 1\big)$$

$$= \mathrm{Vol}_{\,\mathbf{Z}^{k-1}}(\mathrm{Conv}(J \cup Z)) + [\mathbf{Z}^{k-1} : \mathrm{Aff}_{\mathbf{Z}}(J \cup Z)],$$

as claimed.

Proof of Lemma 1.15. Let Z be the circuit contained in B. By Theorem 1.2, the polynomial E_B has the form $\pm\prod_\Gamma \Delta_{B\cap\Gamma}$, where Γ runs through all faces of Conv (B). But if such a face does not coincide with $\mathrm{Conv}(Z)$, then we have $\Delta_{B\cap\Gamma} = \pm 1$ since $B\cap\Gamma$ is independent and hence the corresponding discriminantal subvariety has codimension ≥ 2. In other words, $E_B = \Delta_Z$.

Let $Z = Z_+ \cup Z_-$ be the decomposition of Z into the positive and negative parts (see Section 1B Chapter 7). We set

$$m(\omega) = \mathrm{Vol}_{\,\mathrm{Aff}_{\mathbf{Z}}(Z)}\,\mathrm{Conv}(Z - \{\omega\}) \text{ for } \omega \in Z_+,$$

$$m(\omega) = -\mathrm{Vol}_{\,\mathrm{Aff}_{\mathbf{Z}}(Z)}\,\mathrm{Conv}(Z - \{\omega\}) \text{ for } \omega \in Z_-.$$

Since B affinely generates Ξ over $\mathbf{Z}$, it follows that $m(\omega) = \pm\mathrm{Vol}_{\,\Xi}\,\mathrm{Conv}(B - \{\omega\})$ for $\omega \in Z_\pm$.

Let us use the explicit formula for Δ_Z given in Proposition 1.8, Chapter 9. The first term in this formula has a positive coefficient, and the coefficient in the second term has the sign

$$-\prod_{\omega\in Z_-}(-1)^{m_\omega} = (-1)^{\mathrm{Vol}_{\,\Xi}(\mathrm{Conv}(B))+1}$$

as required.

H. The product of values of a polynomial at its critical points

Suppose we are given a finite set $A \subset \mathbf{Z}^{k-1}$ of Laurent monomials in the variables $x_1, \ldots, x_{k-1}$, as before. For a polynomial $f \in \mathbf{C}^A$, we let

$$\text{Sing } (f) = \{\alpha \in (\mathbf{C}^*)^{k-1} : (\partial f/\partial x_i)(\alpha) = 0 \ \forall i = 1, \ldots, k-1\} \qquad (1.13)$$

denote the set of critical points of f. As before, we suppose that A affinely generates $\mathbf{Z}^{k-1}$ over $\mathbf{Z}$. This implies that, for generic $f \in \mathbf{C}^A$, the set Sing (f) is finite, and all of the critical points are non-degenerate. For such f we define the number

$$\Pi_A(f) = \prod_{\alpha \in \text{Sing } (f)} f(\alpha). \qquad (1.14)$$

Since (1.14) is a symmetric expression in the irrationalities $\alpha \in$ Sing (f), elementary Galois theory shows that Π_A is a rational function on $\mathbf{C}^A$. Unlike the discriminant Δ_A, the function Π_A depends not only on the affine geometry of A, but also on the choice of zero (origin) in $\mathbf{Z}^{k-1}$.

Our purpose now is to find the prime factorization of the rational function Π_A (in the ring $\mathbf{Z}[(a_\omega)_{\omega \in A}]$). It turns out that Π_A has a simple expression in terms of the principal A-determinant and the principal $(A \cap \Gamma)$-determinants for the *facets* $\Gamma \subset Q$ (i.e., faces of codimension 1). This gives us the prime factorization of Π_A since the prime factorization of the principal determinants is given by Theorem 1.2.

As we said, Π_A depends on the choice of the origin of the coordinate system in $\mathbf{Z}^{k-1}$. Let ω_0 denote this origin.

Definition 1.16. A polytope $Q \subset \mathbf{R}^{k-1}$ is said to be *convenient* if the origin ω_0 does not lie in the affine span of any proper face $\Gamma \subset Q$. (cf. Kouchnirenko [Kou]).

It is clear that any polytope can be made into a convenient one by a suitable translation. Thus, we suppose that $Q = \text{Conv}(A)$ is convenient.

We regard $\mathbf{Z}^{k-1}$ as an additive group with the neutral element ω_0. Let $\Gamma \subset Q$ be a facet. We define the group homomorphism $h_\Gamma : \mathbf{Z}^{k-1} \to \mathbf{Q}$ by the conditions $h_\Gamma|_\Gamma = 1$ and $h_\Gamma(\omega_0) = 0$.

We define a number $\rho(\Gamma) \in \mathbf{Z}$, called the *distance* from Γ to ω_0 as follows; its absolute value $|\rho(\Gamma)|$ is given by

$$|\rho(\Gamma)|^{-1} = \min\{a > 0 : a = h_\Gamma(\omega) \text{ for some } \omega \in \mathbf{Z}^{k-1}\}.$$

In addition, we set $\rho(\Gamma) > 0$ if $h_\Gamma|_Q \geq 1$ (i.e., Γ separates ω_0 from Q), and $\rho(\Gamma) < 0$ if $h_\Gamma|_Q \leq 1$. Let $K(\Gamma)$ be the cone with apex ω_0 generated by Γ.

As usual, we embed $\mathbf{Z}^{k-1}$ into $\mathbf{Z}^k$ as the set of points $(M_1, \ldots, m_{k-1}, 1)$. Let $K(A)$ be the cone in $\mathbf{R}^k$ with apex 0 generated by the image of A in this embedding.

Theorem 1.17. *If Q is a convenient polytope, then we have*

$$\Pi_A(f) \\ = \pm E_A(f, K(A)\cap \mathbf{Z}^k), \mathbf{Z}^k) \prod_{\mathrm{codim}(\Gamma)=1} E_{A\cap\Gamma}(f, K(\Gamma)\cap\mathbf{Z}^{k-1}, \mathbf{Z}^{k-1})^{\rho(\Gamma)},$$

where Γ runs through all facets of Q.

Here the factors in the product are the generalized determinants of subsection E. The proof will be given in Section 4 below.

Using Theorem 1.11, we can rewrite Theorem 1.17 in terms of ordinary principal determinants. It is easy to see that

$$[\mathbf{Z}^{k-1} : \mathrm{Lin}_{\mathbf{Z}}(A\cap\Gamma)] = |\rho(\Gamma)| \cdot i(\Gamma, A),$$

where

$$i(\Gamma, A) = [\mathrm{Aff}_{\mathbf{R}}(\Gamma)\cap\mathbf{Z}^{k-1} : \mathrm{Aff}_{\mathbf{Z}}(A\cap\Gamma)]$$

is the affine index encountered earlier. Also the index $[\mathbf{Z}^{k-1} : \mathrm{Lin}_{\mathbf{Z}}(S_{\mathbf{Z}})]$ for the semigroup $S = K(\Gamma)\cap\mathbf{Z}^{k-1}$ is equal to $|\rho(\Gamma)|$. A similar index for $S = K(A)\cap\mathbf{Z}^k$ is equal to 1. Combining this information with Theorem 1.11 we have:

Corollary 1.18. *In the situation of Theorem 1.17 we have*

$$\Pi_A(f) = \pm[\mathbf{Z}^{k-1} : \mathrm{Aff}_{\mathbf{Z}}(A)]^{\mathrm{Vol}_{\mathbf{Z}^{k-1}}(Q)} \cdot E_A(f)^{[\mathbf{Z}^{k-1}:\mathrm{Aff}_{\mathbf{Z}}(A)]} \\ \prod_{\mathrm{codim}(\Gamma)=1} \Big(|\rho(\Gamma)|\cdot i(\Gamma, A)\Big)^{\mathrm{Vol}_{\mathrm{Aff}_{\mathbf{Z}}(A\cap\Gamma)}(\Gamma)\cdot\rho(\Gamma)} \cdot E_{A\cap\Gamma}(f)^{\rho(\Gamma)\cdot i(\Gamma, A)}.$$

Example 1.19. Let $A = \{x, x^2, \ldots, x^d\}$, so $\mathbf{C}^A$ consists of all polynomials in one variable of degree $\leq d$ with constant term 0. The polytope Q is the segment $[1, d]$, so Q is convenient and has two faces $\{1\}$ and $\{d\}$ of codimension 1. We have $\rho(\{1\}) = 1$ and $\rho(\{d\}) = -d$. The indices $i(\Gamma, A)$ are equal to 1. For $f(x) = a_1x + \cdots + a_dx^d$ we have $E_A(f) = a_1a_d\Delta_A(f)$, $E_{\{1\}}(f) = a_1$, $E_{\{d\}}(f) = a_d$. The 0-dimensional volume of a point is 1. Hence Corollary 1.18 gives

$$\Pi_A(f) = E_A(f)\cdot 1^1 E_{\{1\}}(f)\cdot d^{-d}E_{\{d\}}(f)^{-d} = d^{-d}a_1^2a_d^{1-d}\Delta_A(f)$$

which can, of course, also be seen by elementary means.

2. Proof of the prime factorization theorem

In this section we prove Theorem 1.2 which describes the prime factorization of the principal A-determinant E_A. The general line of the proof is as follows. We interpret the exponents of prime factors in E_A in terms of multiplicities of some coherent sheaves on $\mathbf{C}^A \times (\mathbf{C}^A)^*$ along some subvarieties. Then we interpret the numbers $i(\Gamma, A) \cdot u(S/\Gamma)$ as multiplicities along the same subvarieties of some constructible sheaves on the same space. Finally, we compare the two kinds of sheaves using the theory of $\mathcal{D}$-modules and the Riemann-Hilbert correspondence.

A. The logarithmic de Rham complex

We assume the setup of Section 1A. So $A \subset \mathbf{Z}^{k-1}$ is a finite subset affinely generating $\mathbf{Z}^{k-1}$ over $\mathbf{Z}$. We realize $\mathbf{Z}^{k-1}$ as the set of vectors in $\mathbf{Z}^k$ with the last coordinate 1. We denote by $S = S_A$ the semigroup in $\mathbf{Z}^k$ generated by A and by $Y = Y_A = \text{Spec } \mathbf{C}[S]$ the affine toric variety corresponding to A (see Section 1B Chapter 5).

We consider the free $\mathbf{C}[S]$-module

$$L^i(S) = \mathbf{C}[S] \otimes \bigwedge^i \mathbf{C}^k = \bigoplus_{\gamma \in S} \bigwedge^i \mathbf{C}^k. \tag{2.1}$$

Elements of $L^i(S)$ can be regarded as discrete i-vector fields on $\mathbf{Z}^k$ with support in S, in the sense of Section 2A Chapter 9. As explained there, these vector fields represent meromorphic differential i-forms on Y.

Let us denote the union of all torus orbits of codimension ≥ 1 by $Y_1 \subset Y$ and the open orbit by $Y^0 = Y - Y_1$.

We consider the coherent algebraic sheaf on Y corresponding to the $\mathbf{C}[S]$-module $L^i(S)$. We denote this sheaf by $\Omega^i_Y(\log Y_1)$ and call it the sheaf of *logarithmic i-forms*. As in Section 2A Chapter 9 we see that the restriction of $\Omega^i_Y(\log Y_1)$ to Y^0 coincides with $\Omega^i_{Y^0}$.

Example 2.1. If A is the set of vertices of a simplex then $S \cong \mathbf{Z}^k_+$ is a free semigroup, so $Y = \mathbf{C}^k$ and $Y_1 \subset Y$ is the union of all coordinate hyperplanes in $\mathbf{C}^k$. In this case $\Omega^i_Y(\log Y_1)$ is the usual sheaf of i-forms on Y with logarithmic singularities along Y_1, see [De 1]. This can be immediately seen from formula (2.4) of Chapter 9 describing the correspondence between discrete i-vector fields on the lattice and meromorphic forms on the torus.

We denote a typical element from the γ-th summand in (2.1) by (γ, λ) where $\lambda \in \bigwedge^i \mathbf{C}^k$. As in Section 2A Chapter 9, we define the map

$$d : L^i(S) \to L^{i+1}(S), \quad d(\gamma, \lambda) = (\gamma, -\gamma \wedge \lambda) \tag{2.2}$$

which gives rise to a morphism of sheaves

$$d : \Omega^i_Y(\log Y_1) \to \Omega^{i+1}_Y(\log Y_1) \tag{2.3}$$

extending the usual exterior derivative of forms on Y^0. Similarly, by using the identical formula to (2.3), Chapter 9, we define the exterior multiplication

$$\Omega^i_Y(\log Y_1) \otimes \Omega^j_Y(\log Y_1) \to \Omega^{i+j}_Y(\log Y_1). \tag{2.4}$$

Let Y_{an} be the complex analytic space corresponding to the algebraic variety Y. For any coherent algebraic sheaf $\mathcal{F}$ on Y, let $\mathcal{F}_{\mathrm{an}}$ be the corresponding analytic sheaf on Y_{an}. We are interested in the complex of sheaves

$$\Omega^\bullet_{Y,\mathrm{an}}(\log Y_1) = \left\{\Omega^0_{Y,\mathrm{an}}(\log Y_1) \xrightarrow{d} \Omega^1_{Y,\mathrm{an}}(\log Y_1) \xrightarrow{d} \cdots\right\} \tag{2.5}$$

which we call the *logarithmic de Rham complex* of Y.

Let $j : Y^0 \hookrightarrow Y$ be the embedding of the open orbit. Since the restriction of (2.5) to Y^0 is the usual de Rham complex on Y^0, which is a resolution of the constant sheaf $\underline{\mathbf{C}}_{Y^0}$, we get a morphism in the derived category of sheaves on Y:

$$\zeta : \Omega^\bullet_{Y,\mathrm{an}}(\log Y_1) \longrightarrow Rj_*(\underline{\mathbf{C}}_{Y^0}). \tag{2.6}$$

Along with the analytic version of the de Rham complex, we use the formal version, i.e., the one obtained by considering the formal power series as coefficients of our forms. More precisely, let Z be an algebraic variety and let $z \in Z$ be a point. We denote $\mathcal{O}^\wedge_{z,Z} = \lim\operatorname{inv}\,(\mathcal{O}_Z/I_z^m)$ where $I_z \subset \mathcal{O}_Z$ is the ideal of functions vanishing at z. The ring $\mathcal{O}^\wedge_{z,Z}$ is called the *formal completion* of $\mathcal{O}_Z$ at z. If $\mathcal{F}$ is any coherent algebraic sheaf on Z then its formal completion at z is defined as $\mathcal{F}^\wedge_z = \mathcal{F} \otimes_{\mathcal{O}_Z} \mathcal{O}^\wedge_{z,Z}$.

Returning to our situation of a toric variety Y, we have the vector spaces $\Omega^i_Y(\log Y_1)^\wedge_y$ which are connected by the differentials induced by (2.3).

Theorem 2.2.

(a) *The morphism (2.6) is a quasi-isomorphism.*

(b) *For any $y \in Y$ the map*

$$\Omega^\bullet_{Y,\mathrm{an}}(\log Y_1)_y \longrightarrow \Omega^\bullet_{Y,\mathrm{an}}(\log Y_1)^\wedge_y \tag{2.7}$$

from the complex of stalks at y to the formal completion at y, is a quasi-isomorphism of complexes of vector spaces.

Proof. For the case when Y is a normal variety, this follows from the results of Danilov [D]. Let us treat the general case.

The statement (a) means that, for any $y \in Y$, the morphism (2.6) induces isomorphisms on the cohomology of complexes of stalks at y:

$$H^r(\Omega^\bullet_{Y,\mathrm{an}}(\log Y_1)_y) \to H^r(Rj_*(\underline{\mathbf{C}}_{Y^0})_y) = H^r(U \cap Y^0, \mathbf{C}), \tag{2.8}$$

where U is a small neighborhood of y in Y.

To prove (2.8) we look at the torus orbit containing y. Let $K \subset \mathbf{R}^k$ be the convex hull of the semigroup S. This is a cone with base $Q = \mathrm{Conv}(A)$. Torus orbits on Y correspond to faces of K or, equivalently, to all (possibly empty) faces of Q. For a face $\Gamma \subset Q$ let $Y^0(\Gamma) \subset Y$ be the corresponding orbit.

Suppose that $y \in Y^0(\Gamma)$. The local structure of Y near y was described in Section 3A Chapter 5. Namely, near y the variety Y has the structure of the product $Y^0(\Gamma) \times T$ where the transversal slice T has the form

$$T = \mathrm{Spec}\ \mathbf{C}[\Sigma], \quad \Sigma = \big(\mathrm{Lin}_{\mathbf{Z}}(A \cap \Gamma) + S\big)/\mathrm{Lin}_{\mathbf{Z}}(A \cap \Gamma). \tag{2.9}$$

We denote the unique torus fixed point on T simply by 0.

The semigroup Σ contains a finite Abelian group

$$G = (\mathbf{Z}^k \cap \mathrm{Lin}_{\mathbf{R}}(\Gamma))/\mathrm{Lin}_{\mathbf{Z}}(A \cap \Gamma)$$

of order $i(\Gamma, A)$. We have $\Sigma/G = S/\Gamma$, where the semigroup S/Γ is the image of S in $\mathbf{C}^k/\mathrm{Lin}_{\mathbf{C}}(\Gamma)$. Let

$$q : \Sigma \to \Sigma/G = S/\Gamma \tag{2.10}$$

be the natural projection. It follows that the cohomology group on the right hand side of (2.8) is the direct sum of $i(\Gamma, A)$ copies of the r-th cohomology group of the open orbit in Spec $\mathbf{C}[S/\Gamma]$, i.e., it is

$$\left(\bigwedge^r (\mathbf{C}^k/\mathrm{Lin}_{\mathbf{C}}(\Gamma))\right)^{\oplus i(\Gamma, A)} \tag{2.11}$$

As for the complex $\Omega^\bullet_{Y,\mathrm{an}}(\log Y_1)_y$, we find that it splits into the (topological) tensor product of two complexes:

$$\left(\Omega^\bullet_{Y^0(\Gamma),\mathrm{an}}\right)_y \otimes \Omega^\bullet_{T,\mathrm{an}}(\log T_1)_0. \tag{2.12}$$

Here T_1 is the union of torus orbits on T of codimension ≥ 1, and $\Omega^\bullet_{T,\mathrm{an}}(\log T_1)$ is the coherent analytic sheaf on T_{an} corresponding to the $\mathbf{C}[\Sigma]$-module

$$L^i(\Sigma) = \bigoplus_{u \in \Sigma} \bigwedge^i (\mathbf{C}^k/\mathrm{Lin}_{\mathbf{C}}(\Gamma)) \tag{2.13}$$

(the $\mathbf{C}[\Sigma]$-action is defined by (2.10) and the inclusion $S/\Gamma \subset \mathbf{C}^k/\mathrm{Lin}_{\mathbf{C}}(\Gamma)$). As before, an element in the u-th summand of (2.13) will be denoted by (u, λ). The differential in $\Omega^\bullet_{T,\mathrm{an}}(\log T_1)$ comes from the map similar to (2.2)

$$d : L^i(\Sigma) \to L^{i+1}(\Sigma), \quad (u, \lambda) \mapsto (u, -q(u) \wedge \lambda) \tag{2.14}$$

with q from (2.10).

Since $Y^0(\Gamma)$ is a smooth variety, the complex $(\Omega^\bullet_{Y^0(\Gamma,\mathrm{an})})_y$ is a right resolution of $\mathbf{C}$. It remains therefore to prove that the r-th cohomology of the complex $\Omega^\bullet_{T,\mathrm{an}}(\log T_1)_0$ is identified with (2.11). An element of the stalk $\Omega^m_{T,\mathrm{an}}(\log T_1)_0$ is a (possibly infinite) sum

$$\sum_{u\in\Sigma}(u, \lambda_u), \quad \lambda_u \in \bigwedge^m \left(\mathbf{C}^k/\mathrm{Lin}_{\mathbf{C}}(\Gamma)\right) \tag{2.15}$$

where the vectors λ_u grow at most polynomially. If $u \in \Sigma$ is such that $q(u) \neq 0$ then the complex

$$\bigwedge^0\left(\mathbf{C}^k/\mathrm{Lin}_{\mathbf{C}}(\Gamma)\right) \xrightarrow{\wedge q(u)} \bigwedge^1\left(\mathbf{C}^k/\mathrm{Lin}_{\mathbf{C}}(\Gamma)\right) \xrightarrow{\wedge q(u)} \cdots$$

is exact, so every $\lambda_u \in \bigwedge^m\left(\mathbf{C}^k/\mathrm{Lin}_{\mathbf{C}}(\Gamma)\right)$, annihilated by multiplication with $q(u)$, has the form

$$\lambda_u = q(u) \wedge \mu_u, \quad \mu_u \in \bigwedge^{m-1}\left(\mathbf{C}^k/\mathrm{Lin}_{\mathbf{C}}(\Gamma)\right).$$

It is clear (compare [D], Proposition 13.4) that the μ_u can be also chosen to grow at most polynomially. This proves that

$$H^m\left(\Omega^\bullet_{T,\mathrm{an}}(\log T_1)_0\right) = \left\{ \sum_{u\in\Sigma, q(u)=0} (u, \lambda_u) \right\}$$

which is the same as (2.11). Part (a) of Theorem 2.2 is proved. To prove (b) we apply the same reasoning to formal series of the form (2.15) (without any growth conditions on the λ_u).

B. The exponents of prime factors in E_A

We retain the notation of subsection A. Recall in addition that the semigroup $S \subset \mathbf{Z}^k$ is graded by the homomorphism $h : \mathbf{Z}^k \to \mathbf{Z}$ given by the last coordinate. So $\mathbf{C}[S]$ is a graded algebra. We regard the space $\mathbf{C}^A$ as the degree 1 component of $\mathbf{C}[S]$. Therefore, we regard polynomials from $\mathbf{C}^A$ as regular functions on Y which are homogeneous of degree 1 with respect to dilations of Y.

For any i, the space $L^i(S)$ introduced in (2.1) is a graded $\mathbf{C}[S]$-module. The terms $L^i(A,l)$ of the complex calculating E_A (see Section 1A) are just graded components of $L^i(S)$. This implies the following.

Proposition 2.3. *For $l \gg 0$ the space $L^i(A,l)$ is identified with the space of global sections $\omega \in H^0(Y, \Omega^i_Y(\log Y_1))$ which are homogeneous of degree i with respect to dilations of Y. The differential $\partial_f : L^i(A,l) \to L^{i+1}(A,l)$, for $f \in \mathbf{C}^A$, is identified with exterior multiplication by df.*

Proposition 2.4. *Suppose that $f \in \mathbf{C}^A$ is such that, for any face $\Gamma \subset Q$, we have $\Delta_{A\cap\Gamma}(f) \neq 0$. Then for $l \gg 0$ the complex $(L^\bullet(A,l), \partial_f)$ is exact.*

Proof. Consider the complex $(L^\bullet(S), df)$ of graded $\mathbf{C}[S]$-modules with the differential given by the multiplication with df. This is just the direct sum of all the $(L^\bullet(A,l), \partial_f)$. It is enough to prove that under our assumptions all the cohomology spaces of $(L^\bullet(S), df)$ are finite-dimensional (so they will be situated in only finitely many graded components). This can be reformulated by saying that the complex of sheaves on Y

$$\Omega^0_Y(\log Y_1) \xrightarrow{df\wedge} \Omega^1_Y(\log Y_1) \xrightarrow{df\wedge} \cdots \tag{2.16}$$

is exact outside the point 0 (the only 0-dimensional torus orbit).

Let $y \neq 0$ be any point of Y. Suppose that y lies in the torus orbit $Y^0(\Gamma)$ where $\Gamma \subset Q$ is a face (non-empty, since $y \neq 0$). The assumption that $\Delta_{A\cap\Gamma}(f) \neq 0$ means that the restriction $f|_{Y^0(\Gamma)}$ does not have critical points.

Now, as in the proof of Theorem 2.2, we use the identification of a neighborhood of y in Y with a neighborhood of $(y, 0)$ in $Y^0(\Gamma) \times T$ where T is described in (2.9).

Let us prove that the stalk at y of the complex (2.16) is exact. It is enough to show this after the formal completion at y. As in (2.12) this completion, regarded as a graded vector space, splits into the (completed) tensor product

$$\left(\Omega^\bullet_{Y^0(\Gamma)}\right)^\wedge_y \check{\otimes}\, \Omega^\bullet_T(\log T_1)^\wedge_0 . \tag{2.17}$$

Consider the filtration on (2.17) by the degree of the second factor. This filtration is compatible with the differential coming from (2.16) and the quotients are complexes of the form

$$\left(\Omega^\bullet_{Y^0(\Gamma)}, d\big(f|_{Y^0(\Gamma)}\big)\right)^\wedge_y \check{\otimes}\, \Omega^i_T(\log T_1)^\wedge_0 .$$

Since $f|_{Y^0(\Gamma)}$ does not have critical points, the first factor in the above product is exact whereby the completion of (2.16) is also exact. The proposition is proved.

Corollary 2.5. *The prime factorization of E_A in the ring $\mathbf{Z}[(a_\omega)]$ has the form*

$$E_A(f) = c \cdot \prod_{\Gamma \subset Q} \Delta_{A\cap\Gamma}(f)^{m(\Gamma)}, \tag{2.18}$$

where $c \in \mathbf{Q}$ and the $m(\Gamma)$ are some non-negative exponents.

Our aim now is to find the numbers $m(\Gamma)$. Denote the space $\mathbf{C}^A$ by V. The coefficients a_ω of an indeterminate polynomial $f(x) = \sum_{\omega\in A} a_\omega x^\omega$ are affine coordinates in V. Denote by $\mathbf{C}[a]$ the ring of polynomials in a_ω, i.e., of regular functions on V.

Choose $l > 0$. The complex $L^\bullet(A, l)$ gives rise to a complex of free $\mathbf{C}[a]$-modules

$$\mathbf{C}[a] \otimes L^0(A, l) \to \mathbf{C}[a] \otimes L^1(A, l) \to \cdots \tag{2.19}$$

whose differentials have the form $\sum_{\omega\in A} a_\omega \otimes \partial_{x^\omega}$. Denote this complex by $L^\bullet(A, l)[a]$.

Let $\Gamma \subset Q$ be a face. Denote by $Y(\Gamma)^\vee \subset V$ the conic variety projectively dual to the orbit closure $Y(\Gamma) \subset Y$ corresponding to Γ. In the rest of this subsection we shall assume that $Y(\Gamma)^\vee$ has codimension 1, so that $\Delta_{A\cap\Gamma}$ is its equation. Note that only in this case do we need to find the exponent $m(\Gamma)$.

According to Theorem 30 of Appendix A, $m(\Gamma)$ can be expressed as

$$m(\Gamma) = (-1)^k \mathrm{mult}_{Y(\Gamma)^\vee} H^\bullet(L^\bullet(A, l)[a]). \tag{2.20}$$

Here mult stands for the multiplicity of a module along an irreducible component of its support, see the discussion immediately preceding Theorem 30 from Appendix A. As usual, we extend multiplicity to graded modules (like the $H^\bullet$ of a complex) by alternating summation.

Let V^* be the space dual to $V = \mathbf{C}^A$. Let (b_ω) be coordinates in V^* dual to (a_ω), and $\mathbf{C}[b]$ be the ring of polynomials in b_ω, i.e., the ring of regular functions on V^*. The variety Y is embedded into V^*.

We regard $L^i(S) = \bigoplus_l L^i(A, l)$ as a $\mathbf{C}[b]$-module using the homomorphism $\mathbf{C}[b] \to \mathbf{C}[S]$ taking b_ω to the monomial t^ω. Let $\mathbf{C}[a, b] = \mathbf{C}[a] \otimes \mathbf{C}[b]$ be the ring of polynomials in a's and b's, i.e., of regular functions on $V \times V^*$. Consider the complex of $\mathbf{C}[a, b]$-modules

$$\mathbf{C}[a] \otimes L^0(S) \to \mathbf{C}[a] \otimes L^1(S) \to \cdots \tag{2.21}$$

with the differential defined by the same formula $\sum a_\omega \otimes \partial_{x^\omega}$. Let us denote this complex by $L^\bullet(S)[a]$.

Recall (Section 3B Chapter 1) that to any irreducible subvariety Z of a smooth variety M, there corresponds a Lagrangian variety $\mathrm{Con}(Z) \subset T^*M$. In our situation, the product $V \times V^*$ is identified with the cotangent bundles of both V^* and V. Under this identification the conormal variety $\mathrm{Con}(Y(\Gamma)) \subset T^*V^*$ corresponds to the conormal variety $\mathrm{Con}(Y(\Gamma)^\vee) \subset T^*V$ (since $Y(\Gamma)$ and $Y(\Gamma)^\vee$ are projectively dual). Let us denote this variety simply by $\mathrm{Con}(\Gamma)$.

Proposition 2.6. *We have*

$$m(\Gamma) = (-1)^k \mathrm{mult}_{\mathrm{Con}(\Gamma)} H^\bullet(L^\bullet(S)[a]).$$

Proof. Consider in $\mathbf{C}[a, b]$ the grading given by $\deg(a_\omega) = 0$, $\deg(b_\omega) = 1$. Let $\pi_\Gamma : \mathrm{Con}(\Gamma) \to Y(\Gamma)^\vee$ be the natural projection. Since $\dim(\mathrm{Con}(\Gamma)) = \dim(V) = n$, and, by our assumptions, the variety $Y(\Gamma)^\vee \subset V$ has codimension 1, we find that the fibers of π_Γ over generic points have dimension 1. We claim that for $l \gg 0$ we have

$$\mathrm{mult}_{\mathrm{Con}(\Gamma)} H^i L^\bullet(S)[a] = \mathrm{mult}_{Y(\Gamma)^\vee} H^i(L^\bullet(A, l)[a]).$$

Indeed, $\mathrm{Con}(\Gamma)$ is a line bundle over $Y(\Gamma)^\vee$ at its generic point. Thus we have a trivialization $\pi_\Gamma^{-1}(U) \cong U \times \mathbf{C}$ for sufficiently small Zariski open $U \subset Y(\Gamma)^\vee$. Let F be the field of rational functions on $Y(\Gamma)^\vee$. Then the field of rational functions on $\mathrm{Con}(\Gamma)$ is identified, by the above trivialization, with $F(t)$. Let $\mathcal{O}_{Y(\Gamma)^\vee, V}$ be the local ring of the irreducible subvariety $Y(\Gamma)^\vee \subset V$ (see [Hart]). By definition, this ring is obtained from $\mathbf{C}[V] = \mathbf{C}[a]$ by inverting all polynomials which do not vanish identically on $Y(\Gamma)^\vee$, i.e., all polynomials prime to $\Delta_{A \cap \Gamma}$. Let $\mathbf{m}$ be the maximal ideal in $\mathcal{O}_{Y(\Gamma)^\vee, V}$, so that $\mathcal{O}_{Y(\Gamma)^\vee, V}/\mathbf{m} = F$. For a $\mathbf{C}[a]$-module N we have, by definition of multiplicity,

$$\mathrm{mult}_{Y(\Gamma)^\vee} N = \sum_j \dim_F \mathbf{m}^j N / \mathbf{m}^{j+1} N.$$

Now we take $N = H^i L^\bullet(A, l)[a]$ for some i. In this case $\mathbf{m}^j N/\mathbf{m}^{j+1} N$ is the l-th graded component of a certain graded module M over $F[t]$ (the grading on $F[t]$ is given by $\deg(t) = 1$). Namely, let us consider the ring $\mathcal{O}_{Y(\Gamma)^\vee, V}[b]$ with the grading $\deg b_\omega = 1$. Let $I_{\mathrm{Con}(\Gamma)}$ be the ideal in this ring whose elements are functions vanishing on Con (Γ). Then

$$\mathcal{O}_{Y(\Gamma)^\vee, V}[b]/I_{\mathrm{Con}(\Gamma)} = F[t]$$

by virtue of the trivialization $\pi_\Gamma^{-1}(U) = U \times \mathbf{C}$ above. Consider the graded module

$$M = \frac{I^j_{\mathrm{Con}(\Gamma)} H^i L^\bullet(S)[a]}{I^{j+1}_{\mathrm{Con}(\Gamma)} H^i L^\bullet(S)[a]}.$$

Then $\mathbf{m}^j N/\mathbf{m}^{j+1}N$ is the l-th graded component of M.

Our statement follows now from the following obvious fact.

Let F be a field and $M = \bigoplus M_l$ be a finitely generated graded $F[t]$-module. Then for $l \gg 0$ we have

$$\dim_{F(t)} M \otimes_{F[t]} F(t) = \dim_F M_l.$$

C. Constructible sheaves and characteristic cycles

We start by recalling background material on constructible sheaves. For a detailed treatment see [KS].

Let M be a complex algebraic variety. A sheaf $\mathcal{F}$ on M is called *constructible* if there is a Whitney stratification [KS] of M by locally closed algebraic subvarieties M_α such that the restriction of $\mathcal{F}$ to each M_α is a locally constant sheaf of finite rank. A constructible sheaf $\mathcal{F}$ on M defines a function $M \to \mathbf{Z}_+$ which associates to $x \in M$ the rank $\mathrm{rk}(\mathcal{F}_x)$ of the stalk of $\mathcal{F}$ at x.

A complex $\mathcal{F}^\bullet$ of sheaves on M is called constructible if each cohomology sheaf $\underline{H}^i(\mathcal{F}^\bullet)$ is constructible and almost all cohomology sheaves are zero. Every sheaf will be regarded as a one-term complex placed in degree 0, so all of the following discussion is applicable to individual sheaves as well.

A constructible complex $\mathcal{F}^\bullet$ on M defines a function $M \to \mathbf{Z}$ denoted by

$$x \to \chi(\mathcal{F}^\bullet, x) = \sum (-1)^i \mathrm{rk}(\underline{H}^i(\mathcal{F}^\bullet)_x). \tag{2.22}$$

If $\varphi : M \to N$ is a regular map of algebraic varieties and $\mathcal{F}^\bullet$ is a constructible complex on M then the direct image $R\varphi_*(\mathcal{F}^\bullet)$ is a constructible complex on N.

Let φ be a regular function on an algebraic variety M and let $\mathcal{F}^\bullet$ be a constructible complex on M. To φ and $\mathcal{F}^\bullet$, there are associated two new constructible complexes on the subvariety $Z = \varphi^{-1}(0) \subset M$ called the complexes of *nearby cycles* and *vanishing cycles* and denoted respectively $\Psi_\varphi(\mathcal{F}^\bullet)$ and $\Phi_\varphi(\mathcal{F}^\bullet)$ (see [De1] [KS]).

Let us first describe $\Psi_\varphi(\mathcal{F}^\bullet)$. Before giving a formal definition let us look at the underlying idea. Let $x \in \varphi^{-1}(0)$ be any point. Take a small ball B_x around x. Call the space of nearby i-cycles at x for φ with coefficients in $\mathcal{F}^\bullet$ the space of (hyper)cohomology

$$\mathbf{H}^i(B_x \cap \varphi^{-1}(\varepsilon), \mathcal{F}^\bullet) \tag{2.23}$$

where $\varepsilon \neq 0$ is a small complex number (see Figure 44).

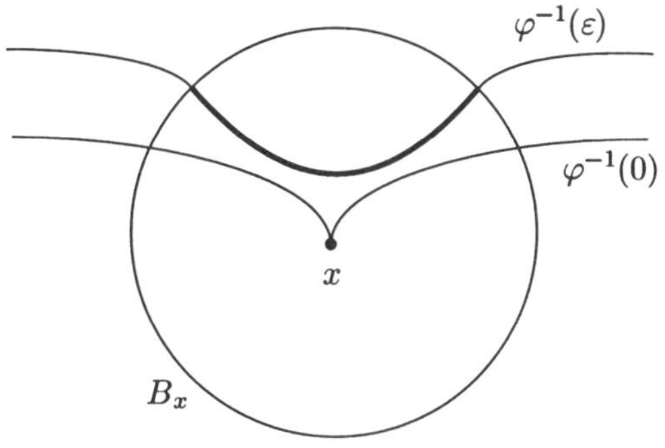

Figure 44. Nearby cycles

We would like to define a constructible complex $\Psi_\varphi(\mathcal{F}^\bullet)$ such that, for any $x \in \varphi^{-1}(0)$, the stalk of $\Psi_\varphi(\mathcal{F}^\bullet)$ at x is identified with (2.23). An obvious subtle point is that we do not have any canonical choice of ε. The formal definition which we are now going to state takes care of this subtlety by considering the universal covering of the space of all possible ε.

Let $e : \mathbf{C} \to \mathbf{C}^*$ be the universal covering (the exponential function). Let $U = M - \varphi^{-1}(0)$, so φ gives a map $U \to \mathbf{C}^*$. Consider the fiber product $\tilde{U}$ of U and $\mathbf{C}$ over $\mathbf{C}^*$, so that we have maps $\pi : \tilde{U} \to U$ (which is a covering with the Galois group $\mathbf{Z}$) and $\tilde{\varphi} : \tilde{U} \to \mathbf{C}$. Let also $i : \varphi^{-1}(0) \hookrightarrow M,\ \ j : U \hookrightarrow M$ be the embeddings.

Definition 2.7. The complex of nearby cycles is a complex on $\varphi^{-1}(0)$ given by

$$\Psi_\varphi(\mathcal{F}^\bullet) = i^* R(j\pi)_*(\pi^*\mathcal{F}^\bullet).$$

In fact, we shall be interested only in numerical invariants of this complex for which the naive point of view of (2.23) is sufficient.

Proposition 2.8. *Let $x \in \varphi^{-1}(0)$ be any point. Then in the notation of (2.22) and (2.23), we have*

$$\chi(\Psi_\varphi(\mathcal{F}^\bullet), x) = \chi(B_x \cap \varphi^{-1}(\varepsilon), \mathcal{F}^\bullet) := \sum_i (-1)^i \dim \mathbf{H}^i(B_x \cap \varphi^{-1}(\epsilon), \mathcal{F}^\bullet).$$

The proof is straightforward and is obtained by unraveling the definitions of the direct and inverse images.

There is a canonical map $can : i^*(\mathcal{F}^\bullet) \to \Psi_\varphi(\mathcal{F}^\bullet)$ of complexes of sheaves on $\varphi^{-1}(0)$ (recall that we denote by $i : \varphi^{-1}(0) \hookrightarrow M$ the embedding).

Definition 2.9. The complex of vanishing cycles $\Phi_\varphi(\mathcal{F}^\bullet)$ is *the chain cone* of the map can.

Thus we have an exact sequence of sheaves

$$\cdots \to i^*\underline{H}^p(\mathcal{F}^\bullet) \to \underline{H}^p(\Psi_\varphi(\mathcal{F}^\bullet)) \to \underline{H}^p(\Phi_\varphi(\mathcal{F}^\bullet)) \to i^*\underline{H}^{p+1}(\mathcal{F}^\bullet) \to \cdots$$

so $\Phi_\varphi(\mathcal{F}^\bullet)$ measures the difference between the cohomology of the generic fiber $\varphi^{-1}(\varepsilon)$ and the special fiber $\varphi^{-1}(0)$.

For example, if $M = \mathbf{C}$, $\mathcal{F} = \underline{\mathbf{C}}$ is the constant sheaf on $\mathbf{C}$ (regarded as a complex placed in degree 0) and $\varphi(z) = z$ then $\Phi_\varphi(\mathcal{F}) = 0$, since there is no difference between stalks of $\mathcal{F}$ at $0 = \varphi^{-1}(0)$ and other points.

Proposition 2.8 implies the following fact.

Proposition 2.10. *The stalk Euler characteristic $\chi(\Phi_\varphi(\mathcal{F}^\bullet), x)$ at any point $x \in \varphi^{-1}(0)$ can be calculated as*

$$\chi(B_x \cap \varphi^{-1}(0), \mathcal{F}^\bullet) - \chi(B_x \cap \varphi^{-1}(\varepsilon), \mathcal{F}^\bullet)$$

where B_x is a small ball around x.

Now let M be a smooth complex algebraic variety with the cotangent bundle T^*M. Let $\mathcal{F}^\bullet$ be a constructible complex on M. The *characteristic cycle* of $\mathcal{F}^\bullet$ is an algebraic cycle in T^*M of the form

$$\mathbf{SS}(\mathcal{F}^\bullet) = \sum_{\Lambda \subset T^*M} c(\Lambda; \mathcal{F}^\bullet) \cdot \Lambda, \tag{2.24}$$

where Λ runs over conic Lagrangian subvarieties in T^*M and the $c(\Lambda; \mathcal{F}^\bullet)$ are some integers defined as follows [Gi] [KS].

As we saw in Proposition 3.1, Chapter 1, every conic Lagrangian subvariety $\Lambda \subset T^*M$ has the form

$$\mathrm{Con}(Z) = \overline{T^*_{Z_{\mathrm{sm}}}(M)},$$

i.e., is the closure of the conormal bundle to the smooth part of some possibly singular closed subvariety $Z \subset M$. Take $\Lambda = \mathrm{Con}(Z)$. Take a generic point $x \in Z$ and a generic function φ in a neighborhood of x such that $\varphi = 0$ on Z and the cotangent vector $d_x\varphi$ lies in $T^*_{Z_{\mathrm{sm}}}(M)$. Then the coefficient $c(\Lambda; \mathcal{F}^\bullet)$ is given by

$$c(\Lambda; \mathcal{F}^\bullet) = (-1)^{\dim(M)-1}\chi(\Phi_\varphi(\mathcal{F}^\bullet), x). \tag{2.25}$$

It is known that the right hand side of (2.25) does not depend on the choice of generic φ and x as above. If $\Lambda = M$ is the zero section of the cotangent bundle, we set $c(\Lambda; \mathcal{F}^\bullet)$ to be the stalk Euler characteristic $\chi(\mathcal{F}^\bullet, x)$ for a generic point $x \in M$.

Informally, $c(\Lambda; \mathcal{F}^\bullet)$ measures the "jump" of $\mathcal{F}^\bullet$ at a generic point of the subvariety $Z \subset M$ corresponding to Λ. For example, if $\mathcal{F}$ is a constant sheaf then $c(\Lambda; \mathcal{F}^\bullet) = 0$ for any $\Lambda \neq M$. Moreover, it is known that if $\{M_\alpha\}$ is a Whitney stratification of M such that the $\underline{H}_i(\mathcal{F}^\bullet)$ are locally constant on each M_α then only the varieties $\mathrm{Con}(\overline{M_\alpha})$ can occur in $\mathbf{SS}(\mathcal{F}^\bullet)$ with non-zero coefficients, see [KS].

D. The case of toric varieties: interpretation of multiplicities

Now we return to the setup of subsections A, B, and the notation there. So $V = \mathbf{C}^A$; $Y = Y_A \subset V^*$ is an affine toric variety associated with A; $Y^0(\Gamma)$ is the torus orbit corresponding to the face $\Gamma \subset Q$, $Q = \mathrm{Conv}(A)$; $Y^0 = Y^0(Q)$ is the open orbit; and $Y(\Gamma)$ is the closure of $Y^0(\Gamma)$.

Let $j : Y^0 \hookrightarrow V^*$ be the embedding. Then $Rj_*\underline{\mathbf{C}}_{Y^0}$ is a constructible complex on V^* whose cohomology sheaves are locally constant (in fact, constant) along each $Y^0(\Gamma)$. For several reasons it will be more convenient to consider the shifted complex $Rj_*\underline{\mathbf{C}}_{Y^0}[n-k]$, where $k = \dim Y$, $n = \#(A) = \dim V$. We are interested in the multiplicity of the Lagrangian variety $\mathrm{Con}(\Gamma) = \mathrm{Con}(Y(\Gamma))$ in the characteristic cycle of this shifted complex (see (2.25)).

Theorem 2.11. *For any face $\Gamma \subset Q$, we have*

$$c(\mathrm{Con}(\Gamma), Rj_*\underline{\mathbf{C}}_{Y^0}[n-k]) = i(\Gamma, A)u(S/\Gamma).$$

Here $i(\Gamma, A)$ and $u(S/\Gamma)$ are the same as in Theorem 1.2.

Of course, the shift amounts to multiplying the multiplicity in question by $(-1)^{n-k}$.

Proof. We use the "transversal slice" to $Y(\Gamma)$ constructed in Theorem 3.1 Chapter 5. Let us use the terminology of Section 3 Chapter 5.

Let S be any admissible semigroup, and let $\Xi = \mathbf{Z}^m$ be its group completion. Choose a system B of generators of S. Let $\mathbf{C}^B$ denote, as usual, the space of Laurent polynomials $f(x_1, \ldots, x_m) = \sum_{\omega \in B} a_\omega x^\omega$. So the (a_ω) form a system of coordinates in $\mathbf{C}^B$. Consider also the dual space $(\mathbf{C}^B)^*$ with coordinates z_ω dual to the a_ω.

Let $Y_S = \mathrm{Spec}\, \mathbf{C}[S]$ be the toric variety corresponding to S. It is embedded into the space $(\mathbf{C}^B)^*$ as the closure of the set

$$Y_S^0 = \{(z_\omega)_{\omega \in B} : z_\omega = x^\omega,\ x \in (\mathbf{C}^*)^m\}$$

which is the open torus orbit. Let $j : Y_S^0 \hookrightarrow (\mathbf{C}^B)^*$ be the embedding. Let $0 \in Y_S$ be the unique zero-dimensional torus orbit.

Theorem 2.11 will be a consequence of Theorem 3.1 Chapter 5 and the following fact.

Theorem 2.12. *Under the above assumptions, the multiplicity*

$$c(\mathrm{Con}(0), Rj_*\underline{\mathbf{C}}_{Y_S^0}[\#(B) - m])$$

equals $u(S)$, the subdiagram volume of the semigroup S.

Proof of Theorem 2.12. Let $K \subset \Xi_{\mathbf{R}} = \mathbf{R}^m$ be the convex hull of S, and let $K_+ = \mathrm{Conv}\,(S - \{0\})$, $K_- = K - K_+$ so $u(S) = \mathrm{Vol}_{\,\Xi}(K_-)$. We assume that the set B of generators of S does not contain 0. Let $P_1 \subset \mathbf{R}^m$ be the convex hull of $B \cup \{0\}$, and P_2 be the convex hull of B. Then $K_- = P_1 - P_2$, so $u(S) = \mathrm{Vol}_{\,\Xi}(P_1) - \mathrm{Vol}_{\Xi}(P_2)$ (see Figure 45).

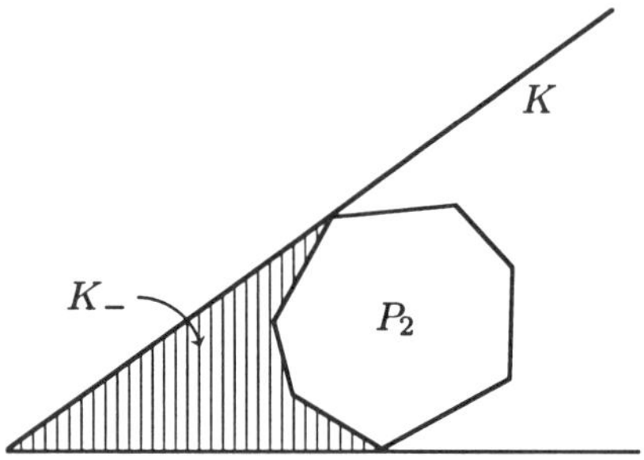

Figure 45.

After these preparations, let us calculate the multiplicity in Theorem 2.12. By definition,

$$c(\mathrm{Con}(0), Rj_*\underline{\mathbf{C}}_{Y_S^0}[\#(B) - m]) = \chi(\Phi_\varphi(Rj_*\underline{\mathbf{C}}_{Y_S^0})[\#(B) - m], 0),$$

where φ is a generic linear functional on the ambient space $(\mathbf{C}^B)^*$. By Proposition 2.10, this number equals

$$(-1)^{m-1}\Big(\chi(\varphi^{-1}(0), Rj_*\underline{\mathbf{C}}_{Y_S^0}) - \chi(\varphi^{-1}(\varepsilon), Rj_*\underline{\mathbf{C}}_{Y_S^0})\Big)$$

(there is no need to intersect with a small ball around 0 since the Euler characteristics of the parts of $\varphi^{-1}(0)$ and $\varphi^{-1}(\varepsilon)$ situated away from 0 will cancel out). After restricting to $Y_S^0 \cong (\mathbf{C}^*)^m$ a generic linear function φ becomes a generic Laurent polynomial

$$f(x) = \sum_{\omega \in B} a_\omega x^\omega \ \in \mathbf{C}^B.$$

So our multiplicity can be written as the difference of the usual Euler characteristics (with coefficients in $\mathbf{C}$):

$$c(\mathrm{Con}(0), Rj_* \underline{\mathbf{C}}_{Y_S^0}) \\ = (-1)^{m-1}\chi\{x \in (\mathbf{C}^*)^m : f(x) = \varepsilon\} - (-1)^{m-1}\chi\{x \in (\mathbf{C}^*)^m : f(x) = 0\}.$$

By Theorem 2.4 of Chapter 6, the first summand equals $\mathrm{Vol}_{\Xi}(P_1)$ (since P_1 is the Newton polytope of the polynomial $f(x) - \varepsilon$); the second one, by the similar reason, is equal to $\mathrm{Vol}_{\Xi}(P_2)$. So the difference equals $u(S) = \mathrm{Vol}_{\Xi}(P_1) - \mathrm{Vol}_{\Xi}(P_2)$. Theorem 2.12 and thus 2.11 are proved.

E. End of the proof: Riemann-Hilbert correspondence

We now prove that for any face $\Gamma \subset Q$

$$m(\Gamma) = i(\Gamma, A) u(S/\Gamma), \tag{2.26}$$

where $m(\Gamma)$ is the exponent of $\Delta_{A \cap \Gamma}$ in E_A. This will establish Theorem 1.2 up to a scalar multiple. Finally, at the end of the subsection we shall explain how to take care of the scalar multiple.

By Proposition 2.6, $m(\Gamma)$ is the alternating sum of multiplicities along Con (Γ) of a certain complex of $\mathbf{C}[a, b]$-modules. By Theorem 2.11, $i(\Gamma, A) \cdot u(S/\Gamma)$ is the multiplicity of Con (Γ) in the characteristic cycle of some constructible complex of sheaves. To relate these objects to each other we use a certain ring of differential operators whose associated graded ring is $\mathbf{C}[a, b]$. We shall use the Riemann-Hilbert correspondence between holonomic regular $\mathcal{D}$-modules and constructible sheaves [KK] [Meb 1-2]. Let us recall the main features of this correspondence.

Let M be a smooth complex algebraic variety and let $\mathcal{D}_M$ be the sheaf of (algebraic) linear differential operators on M. This is a sheaf of filtered rings, the filtration F being defined by the order of the differential operators. The associated graded sheaf of rings is $S^\bullet(T_M)$, the symmetric algebra of the tangent bundle of M.

We are interested in coherent sheaves of $\mathcal{D}_M$-modules which will be called just $\mathcal{D}_M$-modules (left or right). For example, the sheaf $\mathcal{O}_M$ is a left $\mathcal{D}_M$-module. If $\mathcal{M}$ is a $\mathcal{D}_M$-module, then it is known that $\mathcal{M}$ possesses a good filtration, i.e., a filtration $F_\bullet \mathcal{M}$ which is compatible with $F_\bullet \mathcal{D}_M$ and such that the associated graded module $\mathrm{gr}^F_\bullet(\mathcal{M})$ is coherent over $S^\bullet T_M$. The support of $\mathrm{gr}^F_\bullet(\mathcal{M})$ is a conic subvariety in T^*M. It is known to be independent of the choice of a good filtration. This support is denoted Char($\mathcal{M}$) and called the *characteristic variety* of $\mathcal{M}$. If Λ is any irreducible component of Char($\mathcal{M}$) then the multiplicity $\mathrm{mult}_\Lambda \mathrm{gr}^F_\bullet(\mathcal{M})$ is also independent of F. Let us denote this multiplicity by $c(\Lambda; \mathcal{M})$. Therefore we can form the *characteristic cycle* of $\mathcal{M}$:

$$\mathbf{Char}(\mathcal{M}) = \sum_{\Lambda \subset T^*M} c(\Lambda; \mathcal{M}) \cdot \Lambda. \tag{2.27}$$

It is known that dim (Char($\mathcal{M}$)) $\geq$ dim (M). If dim(Char($\mathcal{M}$)) = dim(M), a module $\mathcal{M}$ is called *holonomic*. In this case Char($\mathcal{M}$) is a Lagrangian subvariety [Bj] [Gi] [KS] and hence the characteristic cycle **Char**($\mathcal{M}$) is an object of the same nature as the characteristic cycles of constructible complexes defined in subsection C above.

The *de Rham complex* of a right $\mathcal{D}_M$-module $\mathcal{M}$ is the complex

$$\mathrm{DR}(\mathcal{M}) = \mathcal{M} \otimes^L_{\mathcal{D}_M} \mathcal{O}_{M,\mathrm{an}}, \tag{2.28}$$

where $\otimes^L$ is the left derived functor of the tensor product and $\mathcal{O}_{M,\mathrm{an}}$ is the sheaf of complex analytic functions on M. We consider the functor DR for complexes of right $\mathcal{D}$-modules as well.

The Riemann-Hilbert correspondence is the functor DR from (complexes of) holonomic $\mathcal{D}$-modules to complexes of sheaves of **C**-vector spaces. We need the following fundamental properties of this correspondence [KK], [Meb 1-2].

Theorem 2.13.

(a) *If $\mathcal{M}$ is a holonomic right $\mathcal{D}_M$-module then* DR($\mathcal{M}$) *is a constructible complex.*

(b) *If $\mathcal{M}^\bullet$ is a complex of coherent right $\mathcal{D}_M$-modules such that the cohomology modules $\underline{H}^i(\mathcal{M}^\bullet)$ are holonomic and regular (see* loc. cit.*) then we have the equality of algebraic cycles in T^*M*

$$\sum_i (-1)^i \mathbf{Char}(\underline{H}^i(\mathcal{M}^\bullet)) = (-1)^{\dim M} \mathbf{SS}(\mathrm{DR}(\mathcal{M}^\bullet)).$$

In this theorem the regularity condition is essential. We shall use the following fact [KK], [Meb 1-2] giving a necessary and sufficient condition for regularity.

Theorem 2.14. *Let $\mathcal{M}^\bullet$ be a complex of coherent right $\mathcal{D}_M$-modules such that the cohomology modules $\underline{H}^i(\mathcal{M}^\bullet)$ are holonomic. The condition that all these modules are regular is equivalent to the following condition:*

(*) *Let $x \in M$ be any point and let $\mathcal{O}^{\wedge}_{x,M}$ be the formal completion of $\mathcal{O}_M$ at x. Let*

$$\mathrm{DR}(\mathcal{M}^{\bullet}, x)^{\wedge} = \mathcal{M}^{\bullet} \otimes^{L}_{\mathcal{D}_M} \mathcal{O}^{\wedge}_{x,M}$$

be the formal de Rham complex at x. Then the natural map from the stalk at x of $\mathrm{DR}(\mathcal{M}^{\bullet})$ *to* $\mathrm{DR}(\mathcal{M}^{\bullet}, x)^{\wedge}$ *is a quasi-isomorphism.*

Let us now proceed to the proof of (2.26). By Theorem 2.11, this is equivalent to the following.

Theorem 2.15. *For any face $\Gamma \subset Q$, we have*

$$m(\Gamma) = c(\mathrm{Con}(\Gamma), Rj_{*}\underline{\mathbf{C}}_{Y^0}[n-k]).$$

As in subsection B above, let V^* be the dual space to $V = \mathbf{C}^A$, and (b_ω) the distinguished coordinates in V^*. Let Dif be the Weyl algebra of polynomial differential operators on V^*, generated by the b_ω and the $\partial/\partial b_\omega$. We introduce a filtration F on Dif by letting $F_k\mathrm{Dif}$ consist of the operators of order $\leq k$. The associated graded ring $\mathrm{gr}^F_{\bullet}\mathrm{Dif}$ will be identified with the algebra of polynomials $\mathbf{C}[a, b]$, where we associate the image of $\partial/\partial b_\omega$ to the variable a_ω. The category of finitely generated right modules over the algebra Dif is equivalent to the category of coherent sheaves of right $\mathcal{D}_{V^*}$-modules (see [Bj]).

Now consider the exterior derivative $d : L^k(S) \to L^{k+1}(S)$ defined by (2.2). We also consider the right Dif-modules $L^k(S) \bigotimes_{\mathbf{C}[b]} \mathrm{Dif}$ and the maps $\tilde{\partial}_k : L^k(S) \otimes \mathrm{Dif} \to L^{k+1}(S) \otimes \mathrm{Dif}$, having the form

$$\tilde{\partial}_k(\lambda \otimes p) = d(\lambda) \otimes p + \sum_{\omega \in A} \partial_{x^\omega}(\lambda) \otimes (\partial/\partial b_\omega)p, \tag{2.29}$$

where $\lambda \in L^k(S)$, $p \in \mathrm{Dif}$. We introduce a filtration on $L^k(S) \otimes \mathrm{Dif}$ by setting $F_m(L^k(S) \otimes \mathrm{Dif}) = L^k(S) \otimes F_{k+m}\mathrm{Dif}$. The following proposition is checked directly.

Proposition 2.16.

(a) *The right* Dif*-modules $L^k(S) \otimes_{\mathbf{C}[b]} \mathrm{Dif}$ together with the differentials $\tilde{\partial}_k$ and the above filtration form a filtered complex.*

(b) *The graded complex of* $\mathbf{C}[a, b]$*-modules associated to the filtered complex of* Dif*-modules $L^{\bullet}(S) \bigotimes_{\mathbf{C}[b]} \mathrm{Dif}$ is isomorphic to the complex $L^{\bullet}(S)[a]$ (see(2.21)).*

If M is a finitely generated right Dif-module then, by the de Rham complex of M, the de Rham complex (2.28) of the $\mathcal{D}_{V^*}$-module is

$$\mathcal{M} = M \otimes_{\mathrm{Dif}} \mathcal{D}_{V^*}$$

canonically associated to M. The same for complexes of Dif-modules.

We calculate the de Rham complexes of the complex of right Dif-modules in Proposition 2.16 in terms of the ordinary de Rham complexes of differential forms. More precisely, we introduced in subsection A the logarithmic de Rham complex $\Omega^\bullet_{Y,\mathrm{an}}(\log Y_1)$.

Proposition 2.17. *The complex* $\mathrm{DR}(L^\bullet(S) \otimes \mathrm{Dif})$ *is naturally identified with the direct image to* V^* *of the logarithmic de Rham complex* $(\Omega^\bullet_{Y,\mathrm{an}}(\log Y_1),\ d)$.

Proof. The $\mathcal{D}_{V^*}$-module, corresponding to $L^i(S)\otimes \mathrm{Dif}$ is nothing but $\Omega^i_Y(\log Y_1)\otimes_{\mathcal{O}_{V^*}} \mathcal{D}_{V^*}$. Hence after tensor multiplication over $\mathcal{D}_{V^*}$ with $\mathcal{O}_{V^*}$, we get back $\Omega^i_Y(\log Y_1)$. Similarly we compare the differentials.

Combining Theorem 2.14 and Theorem 2.2 (b) implies that all the cohomology modules of the complex

$$(L^\bullet(S) \otimes \mathrm{Dif}) \otimes_{\mathrm{Dif}} \mathcal{D}_{V^*} \tag{2.30}$$

are regular. Proposition 2.17 together with Theorems 2.13 and 2.2 (a) imply that the characteristic cycle (2.24) of the constructible complex $Rj_*\underline{\mathbf{C}}_{Y^0}$ coincides, up to a sign $(-1)^n$, with the characteristic cycle (2.27) of the complex (2.30) of $\mathcal{D}_{V^*}$-modules. The latter characteristic cycle is defined in terms of multiplicities of associated graded modules. Therefore

$$m(\Gamma) = \mathrm{mult}_{\mathrm{Con}(\Gamma)} H^\bullet(L^\bullet(S)[a]) = (-1)^k(-1)^n c(\mathrm{Con}(\Gamma),\ Rj_*\underline{\mathbf{C}}_{Y^0}).$$

This proves Theorem 2.15 and hence (2.26).

To complete the proof of Theorem 1.2, we need to show that no prime number $p \in \mathbf{Z}$ can enter the prime factorization of E_A in the ring $\mathbf{Z}[(a_\omega)]$. This is done as in the proof of Theorem 2.5 Chapter 8. Namely, we consider the field F which is the algebraic closure of $\mathbf{Z}/p\mathbf{Z}$ and define the F-form of the complex $L^\bullet(l, A)$. This is a complex $L^\bullet_F(A, l)$ of F-vector spaces with the differential ∂_f depending on a polynomial $f \in F^A$ with coefficients in F. As before, it is enough to prove that, for a generic $f \in F^A$, the complex $(L^\bullet_F(A, l), \partial_f)$ is exact. But this follows as in Proposition 2.4 by considering sheaves of logarithmic forms on the F-variety obtained from Y.

Theorem 1.2 is completely proved.

3. Proof of the properties of generalized A-determinants

In this section we keep the notation of Section 1E.

A. Proof of Proposition 1.9

We first consider the case when S generates Ξ. The proof is entirely similar to that of Proposition 2.4: we consider the toric variety $Y = \operatorname{Spec} \mathbf{C}[S]$, and the complex in question will be exact whenever the restriction of f to any torus orbit of positive dimension in Y has no critical points. Since we assume that the convex cone generated by A coincides with $\operatorname{Conv}(S)$, the conditions $\Delta_{A\cap\Gamma}(f) \neq 0$, $\Gamma \subset Q$, guarantee the absense of critical points. We leave the details to the reader.

If $\operatorname{Lin}_{\mathbf{Z}}(S)$ has the same rank as Ξ then we are reduced to the above case since

$$L^\bullet(A, l, S, \Xi) = L^\bullet(A, l, S, \operatorname{Lin}_{\mathbf{Z}}(S))$$

(the choice of the lattice affects only the choice of bases in the terms of these complexes).

Suppose now that $\operatorname{Lin}_{\mathbf{Z}}(\Xi)$ has rank $k - d < k = \operatorname{rk}(\Xi)$. Introduce on $L^\bullet(l, A, S, \Xi)$ an increasing filtration F by

$$F_p L^i(A, l, S, \Xi) = \bigoplus_{u \in S_{i+l}} \operatorname{Im}\left\{ \bigwedge^p \operatorname{Lin}_{\mathbf{C}}(S) \otimes \bigwedge^{i-p} (\Xi_{\mathbf{C}}) \longrightarrow \bigwedge^i \Xi_{\mathbf{C}} \right\}. \qquad (3.1)$$

Clearly, F makes $L^\bullet(A, l, S, \Xi)$ into a filtered complex and the quotients of this filtration are

$$\operatorname{gr}_p^F L^\bullet(A, l, S, \Xi) = L^\bullet(A, l, S, \Xi \cap \operatorname{Lin}_{\mathbf{R}}(A)) \otimes \bigwedge^p \mathbf{C}^d \, [p]. \qquad (3.2)$$

In other words, the p-th quotient is the direct sum of $\binom{d}{p}$ copies of the p times shifted complex corresponding to the lattice of the same rank as $\operatorname{Lin}_{\mathbf{Z}}(S)$. For such complexes we already know the exactness. Since the exactness of the quotients of a filtration implies the exactness of a complex, we are done.

B. Proofs of Proposition 1.10 and Theorem 1.11

The fact that for $l \gg 0$ the rational function $f \mapsto E_{A,l}(f, S, \Xi)$ is independent of l up to a constant is proved as in Proposition 2.6. We leave the details to the reader.

The independence up to a sign is known (Proposition 1.1) for the case when S and Ξ are generated by A as a semigroup and a group, respectively. In the general case this will follow from the next two comparison lemmas, which will also imply Theorem 1.11.

Lemma 3.1.

(a) *If* $\operatorname{rk}\operatorname{Lin}_{\mathbf{Z}}(A) = \operatorname{rk}\Xi$ *then*

$$E_{A,l}(f, S, \Xi) = \pm[\Xi : \operatorname{Lin}_{\mathbf{Z}}(S)]^{\operatorname{Vol}_{\Theta}(Q)} E_{A,l}(f, S, \operatorname{Lin}_{\mathbf{Z}}(S)), \tag{3.3}$$

where $\Theta = \operatorname{Lin}_{\mathbf{Z}}(S) \cap \{u \in \Xi : h(u) = 1\}$.

(b) *If* $\operatorname{rk}\operatorname{Lin}_{\mathbf{Z}}(A) < \operatorname{rk}\Xi$ *then* $E_{A,l}(f, S, \Xi)$ *is identically equal to* ± 1.

Lemma 3.2. *Suppose that* Ξ *is generated by* S *as an Abelian group and let* $S(A)$ *be the subsemigroup in* Ξ *generated by* A. *Then, in the notation of Theorem 1.11, we have*

$$E_{A,l}(f, S, \Xi) = E_{A,l}(f, S(A), \Xi)^{[\operatorname{Lin}_{\mathbf{Z}}(S_{\mathbf{Z}}):\operatorname{Lin}_{\mathbf{Z}}(A)]}. \tag{3.4}$$

It remains therefore to prove these two lemmas.

Proof of Lemma 3.1. (a) The principal determinants on both sides of (3.3) are determinants of the same complex, but with different choices of bases, corresponding to the lattices $L^{\bullet}_{\mathbf{Z}}(A, l, S, \Xi)$ and $L^{\bullet}_{\mathbf{Z}}(A, l, S, \operatorname{Lin}_{\mathbf{Z}}(A))$. Hence, their ratio is the alternating product

$$\frac{E_{A,l}(f, S, \Xi)}{E_{A,l}(f, S, \operatorname{Lin}_{\mathbf{Z}}(S))} = \prod_i \left[L^i_{\mathbf{Z}}(A, l, S, \Xi) : L^i_{\mathbf{Z}}(A, l, S, \operatorname{Lin}_{\mathbf{Z}}(A))\right]^{(-1)^{k-i}}.$$

For typographic reasons we introduce the notation $a \uparrow b$ for the power a^b. Clearly,

$$\left[\bigwedge^i \Xi : \bigwedge^i \operatorname{Lin}_{\mathbf{Z}}(S)\right] = [\Xi : \operatorname{Lin}_{\mathbf{Z}}(S)] \uparrow \binom{k}{i-1},$$

where $\binom{k}{-1}$ is taken to be 0. Hence, our alternating product is equal to

$$[\Xi : \operatorname{Lin}_{\mathbf{Z}}(S)] \Bigg\uparrow \left(\sum_{i=0}^{k} (-1)^{k-i} \binom{k}{i-1} \#(S_{l+i})\right). \tag{3.5}$$

For large l the cardinality $\#(S_l)$ is a certain polynomial $p_S(l)$ in l (cf. the proof of Theorem 2.8 Chapter 9 for similar situation). The easiest way to see this in our case is to note that $\#(S_l)$ is the dimension of the l-th graded component of the semigroup algebra $\mathbf{C}[S]$. It is well known that, for any finitely generated graded commutative algebra $R = \bigoplus R_l$, the dimension $\dim R_l$ is for $l \gg 0$ given by a polynomial in l, the *Hilbert polynomial* of R (see Section 3F Chapter 5 above or [Hart], Chapter I, Section 7).

To find the leading term of $p_S(l)$, we note that, for large l, the numbers $\#(S_l)$ and $\#(lQ \cap \mathrm{Lin}_{\mathbf{Z}}(S))$ differ by the contribution from points situated near the boundary of the cone $K = \mathrm{Conv}(S)$, so this difference has asymptotics $o(l^{k-1})$. Thus, by Proposition 3.7 Chapter 5, we find that

$$p_S(l) = (\mathrm{Vol}_{\Theta}(Q) \cdot l^{k-1}/(k-1)!) + (\text{terms of degree } < k-1).$$

The cardinality of S_{i+l} has, as a function of l, the same leading term. Therefore, by Lemma 2.9 Chapter 9, the exponent in (3.5) is equal to $\mathrm{Vol}_{\Theta}(Q)$, proving (3.3).

(b) We use the filtration F of (3.1). There is a similar filtration on the **Z**-lattices. Using multiplicativity of determinants under filtrations and (3.2), we conclude that

$$E_{A,l}(f, S, \Xi) = E_A(f, S, \mathrm{Lin}_{\mathbf{Z}} A) \uparrow \sum_{p=0}^{d} (-1)^p \binom{d}{p}. \tag{3.6}$$

By the binomial formula, the exponent in (3.6) is equal to 0, from which we get the assertion.

Proof of Lemma 3.2. Let U be any subset in S such that $U + A \subset U$. We define the subcomplex $L^\bullet(A, l, U, \Xi)$ in $L^\bullet(A, l, S, \Xi)$ by putting

$$L^i(A, l, U, , \Xi) = \bigoplus_{u \in S_{i+l} \cap U} \bigwedge^i \Xi_{\mathbf{C}}.$$

The terms of this complex are equipped with obvious **Z**-lattices obtained by summing the lattices $\bigwedge^i \Xi$.

Recall that by $S_{\mathbf{Z}}$, we denoted the semigroup $\{u \in S : h(u) \in \mathbf{Z}\}$. Let $\gamma \subset \mathrm{Lin}_{\mathbf{Z}}(S_{\mathbf{Z}})$ be any coset of $\mathrm{Lin}_{\mathbf{Z}}(A)$. We take $U = S \cap \gamma$. Obviously,

$$L^\bullet(A, l, S, \Xi) = \bigoplus_{\gamma \in \mathrm{Lin}_{\mathbf{Z}}(S_{\mathbf{Z}})/\mathrm{Lin}_{\mathbf{Z}}(A)} L^\bullet(A, l, S \cap \gamma, \Xi). \tag{3.7}$$

So $E_{A,l}(f, S, \Xi)$ is the product of the determinants of the complexes on the right hand side of (3.7) (taken with respect to the above lattices). It remains to show the following lemma.

Lemma 3.3. *Let $S(A) \subset \Xi$ be the semigroup generated by A. Then for $l \gg 0$ and any coset γ as above, we have*

$$\det(L^\bullet(l, A, S \cap \gamma, \Xi), \partial_f) = \pm \det(L^\bullet(l, A, S(A), \Xi), \partial_f).$$

Proof of Lemma 3.3. Let $\xi \in S$ be an element lying in the interior of $\text{Conv}(S)$. Suppose $h(\xi) = m$. For any l the translation by ξ gives rise to an embedding of sets

$$S(A)_{l+i} \to S_{l+m+i} \cap (\xi + \text{Lin}_{\mathbf{Z}}(A))$$

and hence to an embedding of complexes

$$\varphi_{l,\xi} : L^\bullet(l, A, S(A), \Xi) \hookrightarrow L^\bullet(l+m, A, S \cap (\xi + \text{Lin}_{\mathbf{Z}}(A)), \Xi).$$

For $l \gg 0$ and generic $f \in \mathbf{C}^A$, both of these complexes are exact so $\text{Coker}\,\varphi_{l,\xi}$ is exact as well. By definition, the i-th term of $\text{Coker}\,\varphi_{l,\xi}$ is equal to the direct sum of the copies of $\bigwedge^i \Xi_{\mathbf{C}}$ over the points of the set-theoretic difference

$$W_{i,l} = \Big(S_{l+m+i} \cap (\xi + \text{Lin}_{\mathbf{Z}}(A))\Big) - \big(S(A)_{l+i} + \xi\big). \tag{3.8}$$

So this term has a distinguished $\mathbf{Z}$-lattice (the sum over $W_{i,l}$ of the copies of $\bigwedge^i \Xi$) and we can speak of the determinant of $\text{Coker}\,\varphi_{l,\xi}$.

We need to show that for $l \gg 0$ we have $\det(\text{Coker}\,\varphi_{l,\xi}, \partial_f) = \pm 1$. Let us note that the set (3.8) is, for $l \gg 0$, "situated near the boundary" of $K = \text{Conv}(S) \subset \Xi_{\mathbf{R}}$. More precisely, for any face $\Gamma \subset K$ of codimension 1, there are finitely many affine hyperplanes $H_{\Gamma,1}, \ldots, H_{\Gamma,r(\Gamma)}$ parallel to $\text{Lin}_{\mathbf{R}}(\Gamma)$ such that, for every i, l, we have

$$W_{i,l} \subset \bigcup_{\Gamma,j} H_{\Gamma,j}.$$

Let $C^\bullet_{\Gamma,l}$ be the subcomplex of $\text{Coker}\,\varphi_{l,\xi}$ whose i-th term is the sum of $\bigwedge^i \Xi_{\mathbf{C}}$ over the points of $W_{i,l} \cap \bigcup_j H_{\Gamma,j}$. The terms of $C^\bullet_{\Gamma,l}$ are equipped with obvious $\mathbf{Z}$-lattices. We have a surjection of complexes

$$\bigoplus_{\substack{\Gamma \subset K \\ \text{codim}\,\Gamma = 1}} C^\bullet_{\Gamma,l} \longrightarrow \text{Coker}\,\varphi_{\xi,l}. \tag{3.9}$$

The kernel of (3.9) comes from points of $W_{i,l}$ which lie in more than one $H_{\Gamma,j}$. Since $l \gg 0$, we can assume that whenever $H_{\Gamma,j} \cap H_{\Gamma',j'} \cap K \neq \emptyset$, the faces $\Gamma, \Gamma' \subset K$ also intersect. Thus the kernel of (3.9) comes from points of $W_{i,l}$ lying on finitely many translates of subspaces $\text{Lin}_{\mathbf{R}}(\Sigma)$ where $\Sigma \subset K$ is a face of codimension 2. Continuing in this way, we get, for any face $\Gamma \subset K$ of arbitrary codimension, finitely many translates $H_{\Gamma,j}$ of $\mathrm{Lin}_{\mathbf{R}}(\Gamma)$ and a complex $C^{\bullet}_{\Gamma,l}$ similar to the above so that we have a Kouchnirenko-type resolution

$$\cdots \to \bigoplus_{\substack{\Gamma\subset K \\ \mathrm{codim}\,\Gamma=2}} C^{\bullet}_{\Gamma,l} \to \bigoplus_{\substack{\Gamma\subset K \\ \mathrm{codim}\,\Gamma=1}} C^{\bullet}_{\Gamma,l} \to \mathrm{Coker}\,\varphi_{\xi,l} \tag{3.10}$$

(see Section 3C Chapter 8). However, the determinant of every $C^{\bullet}_{\Gamma,l}$ with respect to the natural system of $\mathbf{Z}$-lattices is equal to 1. Indeed, there is a filtration of $C^{\bullet}_{\Gamma,l}$ with quotients of the form $L^{\bullet}(A\cap\Gamma, l, S', \Xi)$ where S' is a subsemigroup in Γ. To get such a filtration it is enough to pick a sufficiently generic vector $v\in K$ and define $F^sC^i_{\Gamma,l} = \bigoplus \bigwedge^i \Xi_{\mathbf{C}}$ where the sum is over $u\in\bigcup_j H_{\Gamma,j}$ such that $u\in sv+K$. This is a filtration labeled by $s\in\mathbf{R}_+$ with jumps occurring for only finitely many values of s, namely for those s for which the boundary of $sv+K$ contains a relatively open part of some $H_{\Gamma,j}$. It is straightforward to see that, for such s, the quotient indeed has the specified form.

Since the determinant of each $L^{\bullet}(A\cap\Gamma, l, S', \Xi)$ is equal to ±1 by Lemma 3.1 (a), we conclude that $\det(\mathrm{Coker}\,\varphi_{l,\xi}) = \pm1$. Lemma 3.3 is proved.

4. The proof of the product formula

Now we prove Theorem 1.17. We shall use the terminology and notation of Section 1H.

A. The Koszul complex on the torus and its filtrations

We start by constructing a complex whose determinant can be expressed explicitly in terms of $\Pi_A(f)$.

Let T denote the torus $(\mathbf{C}^*)^{k-1}$. Let Ω^i_T be the sheaf of regular differential i-forms on T and $\Omega^i(T)$ be the vector space of global sections of this sheaf. For $\alpha\in T$ by $\Omega^i_{T,\alpha}$, we denote the fiber at α of Ω^i_T considered as a vector bundle.

Consider the complex $(\Omega^{\bullet}(T), df)$ whose differential is the exterior multiplication by the 1-form df. If f has only non-degenerate (Morse) critical points, then the only non-zero cohomology of this complex is

$$H^{k-1}(\Omega^{\bullet}(T), df) = \Omega^{k-1}(T)/df\wedge\Omega^{k-2}(T) = \bigoplus_{\alpha\in\mathrm{Sing}\,(f)} \Omega^{k-1}_{T,\alpha} \tag{4.1}$$

(because of the standard properties of the Koszul complex, see Section 1B Chapter 2). So if we consider the endomorphism of $(\Omega^{\bullet}(T), df)$, given by multiplication by f, then it is natural to expect that the determinant of this endomorphism (= the determinant of the complex obtained as the cone of the endomorphism) will be related to $\Pi_A(f)$. However, the terms of $\Omega^\bullet(T)$ are infinite-dimensional. Namely, $\Omega^i(T)$ can be identified with $\mathbf{C}[\mathbf{Z}^{k-1}] \otimes \bigwedge^i \mathbf{C}^{k-1}$ (see Section 2A Chapter 9). So our first step will be to replace $\Omega^\bullet(T)$ by a certain finite-dimensional complex.

The above identification of $\Omega^i(T)$ with the tensor product allows us to write $\Omega^i(T)$ as $\bigoplus_{\gamma \in \mathbf{Z}^{k-1}} \bigwedge^i \mathbf{C}^{k-1}$. Denote a typical element of the γ-th summand by (γ, λ). For a Laurent polynomial $f = \sum_{\omega \in A} a_\omega x^\omega$ the differential given by the exterior multiplication with df can be identified with $\partial_f = \sum_{\omega \in A} a_\omega \partial_\omega$, where

$$\partial_\omega(\gamma, \lambda) = (\omega + \gamma, \ -\omega \wedge \lambda)$$

(the addition is understood relative to the group structure on $\mathbf{Z}^{k-1}$ with ω_0 as zero).

Suppose we are given an integer $l \geq 0$. Consider the subspace

$$V^i(l) = \bigoplus_{\gamma \in (l+i)Q} \bigwedge^i \mathbf{C}^{k-1} \subset \Omega^i(T).$$

Exterior multiplication by df takes $V^i(l)$ to $V^{i+1}(l)$, so that $(V^\bullet(l), df)$ is a subcomplex in $(\Omega^\bullet(T), df)$.

Proposition 4.1. *If $f \in \mathbf{C}^A$ is a generic polynomial, then for $l \gg 0$ the embedding of complexes $(V^\bullet(l), df) \hookrightarrow (\Omega^\bullet(T), df)$ is a quasi-isomorphism.*

Proof. Let $\Lambda = \partial Q$ be the boundary of the polytope Q. It is the union of two subsets: the positive part Λ_+ and the negative part Λ_-, defined as follows (see Figure 46).

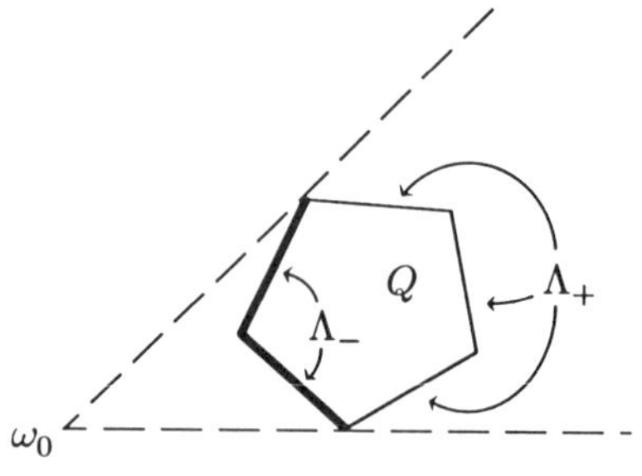

Figure 46.

The set Λ_- consists of all $\xi \in Q$ for which the interval $[\omega_0, \xi]$ contains no other points of Q; we also define $\Lambda_+ = \partial Q - \Lambda_-$. In other words, Λ_+ (resp. Λ_-) is the union of the facets of Q having positive (resp. negative) distance from ω_0.

Let $\Sigma \subset \mathbf{R}^{k-1}$ be the convex cone with apex ω_0 generated by Q (if ω_0 lies in the interior of Q, then $\Sigma = \mathbf{R}^{k-1}$). The intersection $\Sigma \cap \mathbf{Z}^{k-1}$ has a natural semigroup structure with ω_0 as zero.

We define an increasing filtration F on $\Sigma \cap \mathbf{Z}^{k-1}$ by letting $F_m(\Sigma \cap \mathbf{Z}^{k-1})$ consist of all $\xi \in \Sigma \cap \mathbf{Z}^{k-1}$ such that $(1/m)\xi \in \mathrm{Conv}(\{\omega_0\} \cup \Lambda_+)$. Similarly, we define a decreasing filtration G on $\Sigma \cap \mathbf{Z}^{k-1}$ by letting $G^m(\Sigma \cap \mathbf{Z}^{k-1})$ consist of all ξ for which the interval $[\omega_0, (1/m)\xi]$ intersects with Λ_-. Clearly, both filtrations are compatible with the semigroup structure, i.e., $F_k + F_l \subset F_{k+l}$, $G^k + G^l \subset G^{k+l}$. The filtrations F and G on $\Sigma \cap \mathbf{Z}^{k-1}$ give rise to ring filtrations on the semigroup algebra $\mathbf{C}[\Sigma \cap \mathbf{Z}^{k-1}]$, also denoted by F and G.

Let $Z = \mathrm{Spec}\ \mathbf{C}[\Sigma \cap \mathbf{Z}^{k-1}]$ be the toric variety corresponding to the semigroup $\Sigma \cap \mathbf{Z}^{k-1}$. Thus, if ω_0 lies outside Q then Z is an affine normal toric variety with a unique zero-dimensional torus orbit. If ω_0 lies inside Q then $Z = T$ is the torus.

Let $Z_1 \subset Z$ be the union of the torus orbits of codimension ≥ 1 (so Z_1 is empty in the case when ω_0 is inside Q). Consider the sheaf $\Omega^i_Z(\log Z_1)$ of logarithmic forms on Z (see Section 2A). The space of global sections $\Gamma(Z, \Omega^i_Z(\log Z_1))$ will be denoted simply by $\mathbf{\Omega}^i$. This is the direct sum of the copies of $\bigwedge^i \mathbf{C}^{k-1}$ over all the points in $\Sigma \cap \mathbf{Z}^{k-1}$.

Now suppose that $f \in \mathbf{C}^A$ is a Laurent polynomial. We consider the complex of vector spaces $(\mathbf{\Omega}^\bullet, df)$ which is obtained by taking the global sections of the complex of sheaves $(\Omega^\bullet_Z(\log Z_1), df)$ on Z. We introduce filtrations F and G in $(\mathbf{\Omega}^\bullet, df)$ by setting

$$F_m \mathbf{\Omega}^i = \bigoplus_{\gamma \in F_{m+i}(\Sigma \cap \mathbf{Z}^{k-1})} \bigwedge^i \mathbf{C}^{k-1}, \quad G^m \mathbf{\Omega}^i = \bigoplus_{\gamma \in G^{m+i}(\Sigma \cap \mathbf{Z}^{k-1})} \bigwedge^i \mathbf{C}^{k-1}. \tag{4.2}$$

In this notation the complex $V^\bullet(l)$ which is of interests to us is $(F_l \cap G^l)(\mathbf{\Omega}^\bullet, df)$.

Lemma 4.2. *For generic $f \in \mathbf{C}^A$ and for $l \gg 0$ the embedding*

$$G^l(\mathbf{\Omega}^\bullet, df) \hookrightarrow (\Omega^\bullet(T), df)$$

is a quasi-isomorphism.

Proof. The space $G^l \mathbf{\Omega}^i$ is a module over $\mathbf{C}[\Sigma \cap \mathbf{Z}^{k-1}]$, and hence it is the space of global sections of some coherent sheaf on Z which we denote by $G^l \Omega^i_Z(\log Z_1)$. Thus, we have a decreasing filtration of the sheaf $\Omega^i_Z(\log Z_1)$ by subsheaves $G^m \Omega^i_Z(\log Z_1)$ which is invariant under the exterior multiplication by df. Hence, any cohomology space of $G^l(\Omega^\bullet, df)$ is the space of global sections of the corresponding cohomology sheaf of $G^l(\Omega^\bullet_Z(\log Z_1), df)$. In the next lemma we find these cohomology sheaves.

Lemma 4.3. *For generic $f \in \mathbf{C}^A$ and $l \gg 0$, we have*

$$\underline{H}^{k-1}(G^l(\Omega^\bullet, df)) = \bigoplus_{\alpha \in \mathrm{Sing}\,(f)} i_{\alpha *}\Omega^{k-1}_{T,\alpha}, \tag{4.3}$$

where $i_\alpha : \{\alpha\} \hookrightarrow T$ is the embedding of the one-point set. The rest of the cohomology sheaves are zero.

Proof. A generic polynomial $f \in \mathbf{C}^A$, regarded as a function on T, has only Morse critical points. The open T-orbit in Z is identified with T. The restriction of each sheaf $G^l\Omega^i_Z(\log Z_1)$ coincides, under this identification, with Ω^i_T, so that our assertion holds there. It remains to show that the complex $(G^l\Omega^\bullet_Z(\log Z_1), df)$ is acyclic over all points of Z_1.

If ω_0 is inside Q then there is nothing to prove since Z coincides with T (see above). So we suppose that ω_0 is outside Q. Let $z \in Z$ be the unique point fixed under T. The sheaf $G^l\Omega^i_Z(\log Z_1)$ coincides with $\Omega^i_Z(\log Z_1)$ on the complement of $\{z\}$. Hence, to prove our claim for points different from z it suffices to consider the complex $(\Omega^\bullet_Z(\log Z_1), df)$. Let $L \subset Z$ be a torus orbit different from $\{z\}$ and from T. Since Q is a convenient polytope, the restriction $f|_L$ is quasi-homogeneous, and so it vanishes at all of its critical points. Consequently, if $\Delta_{\Gamma\cap A}(f) \neq 0$ for all of the faces Γ of the cone Σ, it follows that $f|_L$ has no critical points. An argument similar to the proof of Proposition 2.4 shows that $(\Omega^\bullet_Z(\log Z_1), df)$ is acyclic over L in this case.

It remains to prove the acyclicity over the point z. We shall use an easy general lemma.

Lemma 4.4. *Suppose that Y is an algebraic variety, $y \in Y$, and $\mathcal{F}^\bullet$ is a finite complex of coherent sheaves on Y. Then $\mathcal{F}^\bullet$ is acyclic at y if and only if the complex of vector spaces $\mathcal{F}^\bullet \otimes \mathbf{C}_y$ is acyclic where $\mathbf{C}_y$ is the skyscraper sheaf at y.*

Proof. Only the "if" part needs a proof. To this end, suppose that $\mathcal{F}^\bullet \otimes \mathbf{C}_y$ is acyclic but $\mathcal{F}^\bullet$ is not acyclic at y, i.e., some cohomology sheaf $\underline{H}^j(\mathcal{F}^\bullet)$ is not zero in a neighborhood of y. Let j be maximal with this property. But since the tensor product functor is left-exact, we have $\underline{H}^j(\mathcal{F}^\bullet) \otimes \mathbf{C}_y = H^j(\mathcal{F}^\bullet \otimes \mathbf{C}_y) = 0$ contradicting the choice of j.

We apply Lemma 4.4 to $Y = Z$, $y = z$ and $\mathcal{F}^\bullet = \left(\Omega^\bullet_Z(\log Z_1), df\right)$. Thus, to prove Lemma 4.3 we need to show the acyclicity of the complex of vector spaces

$G^l\Omega^\bullet_Z(\log Z_1)\otimes\mathbf{C}_z$. For $l \gg 0$ the space $G^l\Omega^i_Z(\log Z_1)\otimes\mathbf{C}_z$ can be identified with the space of functions from $(l+i)\Lambda_- \cap \mathbf{Z}^{k-1}$ to $\Lambda^i\mathbf{C}^{k-1}$; here $(l+i)\Lambda_-$ stands for the image of Λ_- under the dilation with center ω_0 and coefficient $l+i$. Now we can relate, using a version of the Kouchnirenko resolution (Section 3C Chapter 8), the complex $(G^l\Omega^\bullet_Z(\log Z_1)\otimes\mathbf{C}_z, df)$ with the complexes defining principal determinants. In the next lemma, describing this resolution, we use the complexes $L^\bullet$ (see Section 1E).

Lemma 4.5. *There exists a left resolution of the complex $G^l\Omega^\bullet_Z(\log Z_1)\otimes\mathbf{C}_z$, whose p-th term is the direct sum*

$$\bigoplus_{\substack{\Gamma\subset\Lambda_-,\\ \operatorname{codim}\Gamma=p+1}} L^\bullet(A\cap\Gamma, l, K(\Gamma)\cap\mathbf{Z}^{k-1}, \mathbf{Z}^{k-1}). \tag{4.4}$$

Here Γ runs through the faces of Q of codimension $p+1$, contained in Λ_-, and $K(\Gamma)$ is the cone generated by Γ with apex ω_0.

The proof is completely analogous to the proofs of Theorems 3.4 and 3.5 in Chapter 8.

Since the summands in (4.4) are acyclic for generic $f\in\mathbf{C}^A$, we now obtain Lemma 4.3. This in turn implies Lemma 4.2, because we find that $(\Omega^\bullet_T, df)$ has the same cohomology as $(G^l(\Omega^\bullet_Z(\log Z_1)), df)$, namely $\bigoplus_{\alpha\in\mathrm{Sing}(f)} i_{\alpha*}\Omega^{k-1}_{T,\alpha}$.

Now we are in a position to prove Proposition 4.1. We consider the increasing filtration

$$(F_l\cap G^l)(\boldsymbol{\Omega}^\bullet, df)\subset(F_{l+1}\cap G^l)(\boldsymbol{\Omega}^\bullet, df)\subset\cdots\subset G^l(\boldsymbol{\Omega}^\bullet, df).$$

Lemma 4.6. *For generic $f\in\mathbf{C}^A$ and $l\gg 0$ all embeddings*

$$(F_{l+i}\cap G^l)(\boldsymbol{\Omega}^\bullet, df)\subset(F_{l+i+1}\cap G^l)(\boldsymbol{\Omega}^\bullet, df)$$

are quasi-isomorphisms for $i=0,1,2,\ldots$.

Proof. For the quotient complexes of the embeddings in question, we have the Kouchnirenko resolutions in terms of the complexes $L^\bullet(A\cap\Gamma, l, K(\Gamma)\cap\mathbf{Z}^{k-1}, \mathbf{Z}^{k-1})$ corresponding to the faces Γ of Q that lie in Λ_+, by analogy with Lemma 4.5. Hence, the quotient complexes are acyclic for generic f.

Lemma 4.6 implies Proposition 4.1 since $V^\bullet(l)=(F_l\cap G^l)(\boldsymbol{\Omega}^\bullet, df)$.

B. Proof of Theorem 1.17

We have seen (Proposition 4.1 together with formula (4.1)) that, for generic f and $l \gg 0$, the complex $(V^\bullet(l), df)$ has only the highest dimensional cohomology, in degree $k-1$, which is naturally isomorphic to $\bigoplus_{\alpha \in \text{Sing}(f)} \Omega^{k-1}_{T,\alpha}$. Consequently, for such l we have an identification (see Proposition 3 Appendix A)

$$\varepsilon_l : \text{Det}(V^\bullet(l)) \to \left(\bigotimes_{\alpha \in \text{Sing}(f)} \Omega^{k-1}_{T,\alpha} \right)^{\otimes (-1)^{k-1}} \tag{4.5}$$

Here Det means the determinantal vector space of a complex. The bases of monomials in the terms of $V^\bullet(l)$ determine some basis (i.e., vector) $e(l)$ in $\text{Det}(V^\bullet(l))$. This vector is determined uniquely up to sign. Note that for any two non-zero vectors of a 1-dimensional vector space we can speak about their ratio which is a non-zero number.

Proposition 4.7. *For $l \gg 0$ we have*

$$\frac{\varepsilon_{l+1}(e(l+1))}{\varepsilon_l(e(l))} = \pm \prod_{\text{codim}\,\Gamma = 1} E_{A\cap\Gamma}(f, K(\Gamma) \cap \mathbf{Z}^{k-1}, \mathbf{Z}^{k-1})^{(-1)^{k-1}\rho(\Gamma)},$$

where Γ runs through all facets of Q.

Proof. Consider the following diagram of quasi-isomorphic embeddings of complexes (the differential in each of them is given by the exterior multiplication with df):

$$V^\bullet(l) = (F_l \cap G^l)\Omega^\bullet \overset{\varphi_l}{\hookrightarrow} (F_{l+1} \cap G^l)\Omega^\bullet \overset{\psi_l}{\hookleftarrow} (F_{l+1} \cap G^{l+1})\Omega^\bullet = V^\bullet(l+1).$$

If $g : C^\bullet \to D^\bullet$ is a quasi-isomorphism of based complexes then we denote simply by $\det(g)$ the determinant of the based exact complex $\text{Cone}(g)$ (see Appendix A). In the particular case when $C^\bullet$ and $D^\bullet$ each consist of one vector space in degree 0, a quasi-isomorphism is just an isomorphism of vector spaces $g : C^0 \to D^0$ and $\det(g)$ is the ordinary determinant of the matrix of g in the chosen bases of C^0 and D^0.

The ratio $\psi_l^{-1} \circ \varphi_l$ represents a morphism in the derived category from $V^\bullet(l)$ to $V^\bullet(l+1)$ such that the induced morphism on the cohomology spaces is the identity map on $\bigoplus_{\alpha \in \text{Sing}(f)} \Omega^{k-1}_{T,\alpha}$. Therefore, $\varepsilon_{l+1}(e(l+1))/\varepsilon_l(e(l)) = \det(\varphi_l)/\det(\psi_l)$, and our assertion follows from the next lemma.

Lemma 4.8. *For $l \gg 0$ we have*

$$\det(\varphi_l) = \prod_{\Gamma \subset \Lambda_+} E_{A \cap \Gamma}(f, K(\Gamma) \cap \mathbf{Z}^{k-1}, \mathbf{Z}^{k-1})^{(-1)^{k-1}|\rho(\Gamma)|},$$

$$\det(\psi_l) = \prod_{\Gamma \subset \Lambda_-} E_{A \cap \Gamma}(f, K(\Gamma) \cap \mathbf{Z}^{k-1}, \mathbf{Z}^{k-1})^{(-1)^{k-1}|\rho(\Gamma)|}$$

where Γ runs over facets of Q and the determinants are taken with respect to the monomial bases.

Proof. Recall that we have a linear function h_Γ on $\mathbf{R}^{k-1}$ equal 1 on Γ (and 0 at ω_0 which is the zero of our group structure on $\mathbf{R}^{k-1}$). It follows from the definition of the number $|\rho(\Gamma)|$ (Section 1H) that in each region of the form $\{u \in \mathbf{R}^{k-1} : l \leq h_\Gamma(u) < l+1\}$, $l \in \mathbf{Z}$, there are exactly $|\rho(\Gamma)|$ translates of the hyperplane $\mathrm{Aff}_{\mathbf{R}}(\Gamma)$ which intersect $\mathbf{Z}^{k-1}$.

Let Σ, as before, denote the convex cone with apex ω_0 generated by Q and let $g : \Sigma \to \mathbf{R}$ be the function which is homogeneous of degree 1 and equal to 1 on Λ_+. Note that for any facet $\Gamma \subset Q$ such that $\Gamma \subset \Lambda_+$, the restriction of g on $K(\Gamma)$ (the convex cone with apex ω_0 generated by Γ) coincides with h_Γ.

Note that φ_l is a quasi-isomorphism and an embedding, and, moreover, it takes monomial bases in its domain into subsets of monomial bases in its range. This implies that $\det(\varphi_l)$, i.e., the determinant of the complex $\mathrm{Cone}(\varphi_l)$, is equal to the determinant of the complex $\mathrm{Coker}\,\varphi_l$ with respect to the natural system of bases there. This follows by considering a natural two-term filtration in $\mathrm{Cone}(\varphi_l)$ whose first quotient is the cone of the identity map and the second quotient is $\mathrm{Coker}(\varphi_l)$.

As a vector space,

$$(\mathrm{Coker}\,\varphi_l)^i = \bigoplus_{\substack{u \in \Sigma \cap \mathbf{Z}^{k-1}: \\ i+l \leq g(u) < i+l+1}} \bigwedge^i \mathbf{C}^{k-1}.$$

For any facet $\Gamma \subset \Lambda_+$ of Q, let $D^\bullet_{\Gamma,l} \subset \mathrm{Coker}\,\varphi_l$ be the subcomplex defined by

$$D^i_{\Gamma,l} = \bigoplus_{\substack{u \in \Sigma \cap \mathbf{Z}^{k-1}: \\ i+l \leq h_\Gamma(u) < i+l+1}} \bigwedge^i \mathbf{C}^{k-1}.$$

We have a surjection of complexes

$$\bigoplus_{\substack{\Gamma \subset \Lambda_+ \\ \mathrm{codim}_Q(\Gamma)=1}} D^\bullet_{\Gamma,l} \longrightarrow \mathrm{Coker}\,\varphi_l.$$

As in the proof of Lemma 3.3, we can extend this surjection to a Kouchnirenko-type resolution of Coker φ_l whose p-th term is the sum of certain complexes $D^{\bullet}_{\Gamma,l}$ associated to the faces $\Gamma \subset \Lambda_+$ of codimension $p+1$ in Q. If codim $\Gamma = 1$, the complex $D^{\bullet}_{\Gamma,l}$ has a decreasing filtration Φ with $|\rho(\Gamma)|$ layers given by

$$\Phi^m D^i_{\Gamma,l} = \bigoplus_{\substack{u \in \Sigma \cap \mathbf{Z}^{k-1}: \\ i+l+1-\frac{m}{|\rho(\Gamma)|} \leq h_\Gamma(u) < i+l+1}} \bigwedge^i \mathbf{C}^{k-1},$$

where $m = 1, 2, \ldots, |\rho(\Gamma)|$. Every quotient of this filtration is isomorphic to the complex

$$L^\bullet(A \cap \Gamma, K(\Gamma) \cap \mathbf{Z}^{k-1}, \mathbf{Z}^{k-1}). \tag{4.6}$$

For codim $\Gamma \geq 2$ the determinant of $D^{\bullet}_{\Gamma,l}$ is equal to ± 1 by the same reasoning as in the proof of Lemma 3.3 (dévissage into L-complexes corresponding to semigroups of smaller rank and application of Lemma 3.1 (b)). By multiplicativity of the determinants of complexes in filtrations (Appendix A, Proposition 17) we get the first equality claimed in Lemma 4.8. The second equality (for ψ_l) is similar. This concludes the proofs of Lemma 4.8 and Proposition 4.7.

We now consider the morphism of complexes $\chi_{f,l} : V^\bullet(l) \to V^\bullet(l+1)$ given by multiplication by f. Let Cone $(\chi_{f,l})$ be the cone of $\chi_{f,l}$. This is a complex with terms

$$\text{Cone}\,(\chi_{f,l})^i = V^i(l+1) \oplus V^{i+1}(l), \quad i = 0, 1, \ldots, k. \tag{4.7}$$

For $u_i \in V^i(l+1)$, $u_{i+1} \in V^{i+1}(l)$ the differential in Cone $(\chi_{f,l})$ acts by

$$d(u_i, u_{i+1}) = (df \wedge u_i + (-1)^{i+1} f \cdot u_{i+1}, df \wedge u_{i+1}) \in V^{i+1}(l+1) \oplus V^{i+2}(l),$$

(see formula (1) of Appendix A). For $l \gg 0$ and generic f, the morphism $\chi_{f,l}$ is a quasi-isomorphism. Indeed, each of these complexes has the unique cohomology space isomorphic to $\bigoplus_{\alpha \in \text{Sing}\,(f)} \Omega^{k-1}_{T,\alpha}$ and the action of f on this space is diagonal with diagonal entries $f(\alpha)$. So whenever all $f(\alpha) \neq 0$ $\chi_{f,l}$ is a quasi-isomorphism and so Cone $(\chi_{f,l})$ is acyclic. In addition, this cone is equipped with natural $\mathbf{Z}$-lattices, hence its determinant is uniquely determined up to sign.

Corollary 4.9. *For generic $f \in \mathbf{C}^A$ and for $l \gg 0$, the determinant of the acyclic based complex* Cone $(\chi_{f,l})$ *is equal to*

$$\left(\Pi_A(f) \cdot \prod_{\text{codim}\,\Gamma = 1} E_{A \cap \Gamma}(f, K(\Gamma) \cap \mathbf{Z}^{k-1}, \mathbf{Z}^{k-1})^{-\rho(\Gamma)} \right)^{(-1)^{k-1}}. \tag{4.8}$$

The exponent $(-1)^{k-1}$ appears because of the normalization of the grading of complexes we have chosen: the degrees go from 0 to $k-1$.

Proof. Consider the identification

$$\theta_l = \varepsilon_{l+1}^{-1} \circ \varepsilon_l : \mathrm{Det}(V^\bullet(l)) \xrightarrow{\sim} \mathrm{Det}(V^\bullet(l+1))$$

where the ε_l are given by (4.5). Let

$$f_* : \bigotimes_{\alpha \in \mathrm{Sing}\,(f)} \Omega_{T,\alpha}^{k-1} \longrightarrow \bigotimes_{\alpha \in \mathrm{Sing}\,(f)} \Omega_{T,\alpha}^{k-1}$$

be the map induced by the multiplication by f. Since the action of f on each $\Omega_{T,\alpha}^{k-1}$ is the multiplication by $f(\alpha)$, the map f_* is the multiplication by the product of $f(\alpha)$ over $\alpha \in \mathrm{Sing}\,(f)$, i.e., by $\Pi_A(f)$. Let

$$(\chi_{f,l})_* : \mathrm{Det}(V^\bullet(l)) \xrightarrow{\sim} \mathrm{Det}(V^\bullet(l+1))$$

be the identification induced by the quasi-isomorphism $\chi_{f,l}$. If we denote, as in Proposition 4.7, by $e(l)$ the basis vector in $\mathrm{Det}(V^\bullet(l))$ corresponding to the monomial basis in $V^\bullet(l)$ then, clearly,

$$\det(\chi_{f,l}) = \frac{e(l+1)}{(\chi_{f,l})_* e(l)}.$$

Note that

$$(\chi_{f,l})_* = \varepsilon_{l+1}^{-1} \circ (f_*)^{\otimes(-1)^{k-1}} \circ \varepsilon_l.$$

Thus

$$\det(\chi_{f,l}) = \frac{\varepsilon_{l+1}(e(l+1))}{\varepsilon_l(e(l))} \cdot \Pi_A(f)^{(-1)^{k-1}}$$

and our statement follows from Proposition 4.7.

Theorem 1.17 claims that the expression in (4.8), i.e., the determinant of Cone $(\chi_{f,l})$, coincides with a certain principal A-determinant, namely the determinant of the complex $L^\bullet(A, l, K(A) \cap \mathbf{Z}^k, \mathbf{Z}^k)$. Here $K(A) \subset \mathbf{R}^k$ is the cone with apex 0 generated by $A \times \{1\}$ (see Section 1H). Thus, the theorem will follow from the next proposition, which is a complete analog of Theorem 4.6 Chapter 2.

Proposition 4.10. *There is an isomorphism of complexes*

$$\left(L^\bullet(A, l, K(A) \cap \mathbf{Z}^k, \mathbf{Z}^k), \partial_f\right) \xrightarrow{\sim} \mathrm{Cone}\,(\chi_{f,l})[1], \qquad (4.9)$$

which induces an isomorphism of the integral lattices in all the terms.

Proof. We regard $\mathbf{Z}^{k-1}$ as the affine lattice $\{(m_1, \ldots, m_k) \in \mathbf{Z}^k : m_k = 1\}$ in $\mathbf{Z}^k$. We abbreviate the complex on the left hand side of (4.9) by $L^\bullet(l)$. Elements of $L^\bullet(l)$ can be viewed as discrete vector fields on $\mathbf{Z}^k$ and thus as differential forms on $(\mathbf{C}^*)^k$ (see Section 2A Chapter 9). Let $x_1, \ldots, x_k$ be the standard coordinates on $(\mathbf{C}^*)^k$. Similarly, elements of $V^\bullet(l)$ can be regarded as discrete vector fields on $\mathbf{Z}^{k-1} \subset \mathbf{Z}^k$ and hence as differential forms on $(\mathbf{C}^*)^{k-1} \subset (\mathbf{C}^*)^k$. The coordinates on $(\mathbf{C}^*)^{k-1}$ are $x_1, \ldots, x_{k-1}$.

Lemma 4.11. *For all i and l we have the decomposition*

$$L^{i+1}(l) = x_k^{i+l+1}\left(V^i(l+1) \wedge \frac{dx_k}{x_k} \quad \oplus \quad V^{i+1}(l)\right)$$

which induces an analogous decomposition of the integral lattices.

Proof. By definition, $L^{i+1}(l)$ is the direct sum of the copies of $\bigwedge^{i+1} \mathbf{C}^k$ over all points of $K(A) \cap \mathbf{Z}^k$ with last coordinate $i + l + 1$. Such points are in bijection with integral points in $(i + l + 1)Q \subset \mathbf{R}^{k-1}$, the dilation of Q with coefficient $i + l + 1$ and center ω_0. The space $V^i(l + 1)$ (resp. $V^{i+1}(l)$) is the sum of the copies of $\bigwedge^i \mathbf{C}^{k-1}$ (resp. $\bigwedge^{i+1} \mathbf{C}^{k-1}$) over the integral points of the above dilation. Under the correspondence between forms on a torus and "discrete vector fields" on a lattice (see Section 2A Chapter 9), the form dx_k/x_k corresponds to the vector field $(0, -e_k)$ where $e_k = (0, \ldots, 0, 1) \in \mathbf{Z}^k$ is the basis vector complementing $\mathbf{Z}^{k-1}$ to $\mathbf{Z}^k$. This makes the lemma obvious, since

$$\bigwedge^{i+1} \mathbf{C}^k = \left(\bigwedge^{i} \mathbf{C}^{k-1}\right) \wedge e_k \;\oplus\; \bigwedge^{i+1} \mathbf{C}^{k-1}.$$

To complete the proof of Proposition 4.10, we identify, for each i, l the space $V^i(l + 1) \oplus V^{i+1}(l)$ with $L^{i+1}(l)$ by the map

$$(u_i, u_{i+1}) \longmapsto x_k^{i+l+1}\left(u_i \wedge \frac{dx_k}{x_k} + u_{i+1}\right). \tag{4.10}$$

Then we write the action of the differential ∂_f on the image of (u_i, u_{i+1}) in $L^\bullet(l)$. This action is given by the multiplication by the 1-form $d\tilde{f}$, where $\tilde{f}(x_1, \ldots, x_k) = x_k f(x_1, \ldots, x_{k-1})$. (The replacement of f by $\tilde{f}$ is the consequence (on the level of Laurent polynomials) of regarding $\mathbf{Z}^{k-1}$ as an affine lattice in $\mathbf{Z}^k$ as above.) Therefore

$$d\tilde{f} = f dx_k + x_k df$$

so

$$d\tilde{f} \wedge x_k^{i+l+1}\left(u_i \wedge \frac{dx_k}{x_k} + u_{i+1}\right)$$

$$= x_k^{i+l+2}\left(df \wedge u_i \wedge \frac{dx_k}{x_k} + (-1)^{i+1} f u_{i+1} \wedge \frac{dx_k}{x_k} + df \wedge u_{i+1}\right).$$

If we map this element of $L^{i+2}(l)$ back into $V^{i+1}(l+1) \oplus V^{i+2}(l)$, by the map inverse to (4.10) with shifted i, we get the formula for the differential in the cone.

Proposition 4.10, and hence also Theorem 1.17, are proved.

CHAPTER 11

Regular A-Determinants and A-Discriminants

In the previous chapter we established some structural properties of the principal A-determinant E_A. Now we shall apply this information to the A-discriminant Δ_A. In the most important case when the toric variety X_A is smooth, we have

$$E_A(f) = \prod_{\Gamma \subset Q} \Delta_{A \cap \Gamma}(f)$$

where the product is taken over all the faces of the polytope $Q = \text{Conv}\,(A)$ (Theorem 1.2 Chapter 10). Since (in the case when X_A is smooth) a similar equality holds for each $E_{A\cap\Gamma}$, we have a system of equalities relating the polynomials $\Delta_{A\cap\Gamma}$ and $E_{A\cap\Gamma}$ that allows us to recover Δ_A as as an alternating product of the $E_{A\cap\Gamma}$. Consequently, alternating sums and products will appear in the expressions for the Newton polytope and coefficients of Δ_A.

1. Differential forms on a singular toric variety and the regular A-determinant

A. Definition of the regular A-determinant

Let $A \subset \mathbf{Z}^{k-1}$ be a finite set. If the toric variety X_A is smooth, we represented the A-discriminant Δ_A as the determinant of a combinatorially defined complex $C^\bullet(A, l)$ (see Section 2C Chapter 9). This combinatorial definition can in fact be given without any restrictions on A. Our first task will be to study this in general.

We make our standard assumptions that A affinely generates $\mathbf{Z}^{k-1}$ over $\mathbf{Z}$, and that $\mathbf{Z}^{k-1}$ is realized as $\{u \in \mathbf{Z}^k : h(u) = 1\}$ where $h(m_1, \ldots, m_k) = m_k$. As before, by S we denote the semigroup in $\mathbf{Z}^k$ generated by A and for any i we set $S_i = \{u \in S : h(u) = i\}$. We also denote by $Q \subset \mathbf{R}^{k-1}$ the convex hull of A, and by $K \subset \mathbf{R}^k$ the convex hull of S, i.e., the convex cone with apex 0 and base Q.

Let l be an integer. As in Section 2C Chapter 9, we consider the graded vector space $C^\bullet(A, l) = \oplus_i C^i(A, l)$ where

$$C^i(A, l) = \bigoplus_{u \in S, h(u) = l+i} \bigwedge^i \text{Lin}_{\mathbf{C}} \Gamma(u), \tag{1.1}$$

and $\Gamma(u)$ is the minimal face of the cone K containing u. For a polynomial $f(x) = \sum_{\omega \in A} a_\omega x^\omega \in \mathbf{C}^A$, we define the differential $\partial_f : C^i(A, l) \to C^{i+1}(A, l)$

by the same formula as (2.10) Chapter 9. It is immediately verified that $\partial_f^2 = 0$, i.e., we get a complex.

Let $Y = Y_A = \operatorname{Spec} \mathbf{C}[S]$ be the affine toric variety corresponding to S. We defined, for any i, the coherent sheaf $\tilde{\Omega}_Y^i$ of Danilov i-forms on Y in Section 2B Chapter 9. The restriction of $\tilde{\Omega}_Y^i$ to the smooth part of Y coincides with the usual sheaf of regular i-forms.

Proposition 1.1. *For $l \gg 0$, the space $C^i(A, l)$ is identified with the space of global sections of $\tilde{\Omega}_Y^i$ which are homogeneous of degree $i + l$ with respect to dilations of the conic variety Y. For $f \in \mathbf{C}^A$ the differential ∂_f is identified with the exterior multiplication by $df \in H^0(Y, \tilde{\Omega}^1)$.*

The proof is similar to that of Proposition 2.3 Chapter 10.

Proposition 1.2. *Let $f \in \mathbf{C}^A$ be such that $\Delta_{A \cap \Gamma}(f) \neq 0$ for any face $\Gamma \subset Q$. Then for $l \gg 0$ the complex $(C^\bullet(A, l), \partial_f)$ is exact.*

Proof. The proof is similar to that of Proposition 2.4 Chapter 10. We leave details to the reader.

Note that the terms of (1.1) are equipped with natural $\mathbf{Z}$-lattices

$$C_{\mathbf{Z}}^i(A, l) = \bigoplus_{u \in S, h(u) = l + i} \bigwedge^i \mathrm{Lin}_{\mathbf{Z}}(\Gamma(u) \cap A). \tag{1.2}$$

We take some system e of $\mathbf{Z}$-bases in these lattices and define the *regular A-determinant* of weight l as

$$D_{A,l}(f) = \det(C^\bullet(A, l), \partial_f, e)^{(-1)^k}. \tag{1.3}$$

This is a rational function on $\mathbf{C}^A$ defined uniquely up to a sign.

Theorem 1.3.

(a) *If A is such that X_A is smooth then, for $l \gg 0$, the regular A-determinant $D_{A,l}$ coincides with the A-discriminant Δ_A.*

(b) *For any A the rational function $D_{A,l}$ is, for $l \gg 0$, independent up to a sign of the choice of l; more specifically, we have*

$$D_{A,l}(f) = \pm \prod_{\Gamma \subset Q} E_{A \cap \Gamma}(f)^{(-1)^{\mathrm{codim}(\Gamma)}}. \tag{1.4}$$

Proof. Part (a) of this theorem is just Theorem 2.7, Chapter 9.

To prove part (b) we consider the complex $L^\bullet(A, l)$ whose determinant is E_A (see (1.2), (1.3) Chapter 10). Let us define the subspace $L_{\mathrm{int}}^i(A, l) \subset L^i(A, l)$ to be the sum of the summands in (1.2), Chapter 10, corresponding to those $u \in S_{i+l}$ which are interior points of the cone K. Clearly these subspaces form a subcomplex so we have the following embeddings of complexes:

$$L^{\bullet}_{\text{int}}(A, l) \subset C^{\bullet}(A, l) \subset L^{\bullet}(A, l).$$

Let us introduce in $C^{\bullet}(A, l)$ an increasing filtration F by setting

$$F_p C^i(A, l) = \bigoplus_{\substack{u \in S_{i+l} \\ \dim\, \Gamma(u) \leq p}} \bigwedge^i \mathrm{Lin}_{\mathbf{C}} \Gamma(u). \tag{1.5}$$

Then

$$\mathrm{gr}^F_p C^{\bullet}(A, l) = \bigoplus_{\substack{\Gamma \subset Q \\ \dim(\Gamma) = p-1}} L^{\bullet}_{\text{int}}(A \cap \Gamma, l). \tag{1.6}$$

(Note that Γ in (1.6) is a face of Q, and $\Gamma(u)$ in (1.5) is a face of K, the cone over Q, which explains the shift of dimension by 1.) For any face $\Gamma \subset Q$ we have

$$E_{A \cap \Gamma}(f) = \det(L^{\bullet}(A \cap \Gamma, l), \partial_f, e)^{(-1)^{\dim(\Gamma)+1}}, \tag{1.7}$$

where e is a system of $\mathbf{Z}$-bases in the natural $\mathbf{Z}$-lattices in the terms of the complex. Comparing (1.6) and (1.7) and using the multiplicativity of determinants with respect to filtrations (Proposition 17, Appendix A), we see that it is enough to prove the following.

Proposition 1.4. *For generic $f \in \mathbf{C}^A$ and $l \gg 0$, the complex $(L^{\bullet}_{\text{int}}(A, l), \partial_f)$ is exact and*

$$\det(L^{\bullet}_{\text{int}}(A, l), \partial_f) = \det(L^{\bullet}(A, l), \partial_f) = E_A(f).$$

More precisely, to prove part (b) of Theorem 1.3 we apply Proposition 1.4 to each $A \cap \Gamma$.

Proof of Proposition 1.4. We consider the quotient complex $L^{\bullet}(A, l)/L^{\bullet}_{\text{int}}(A, l)$. Its terms are direct sums of summands corresponding to the $u \in S$ which lie on the boundary of K. Hence our quotient has a Kouchnirenko-type resolution (see Section 3C Chapter 8) whose p-th term is the sum

$$\bigoplus_{\substack{\Gamma \subset Q \\ \dim(\Gamma) = k-2-p}} L^{\bullet}(A \cap \Gamma, l, S \cap \mathbf{R}_+\Gamma, \mathbf{Z}^k)$$

of complexes similar to those in Section 1E Chapter 10. The determinant of any of these complexes is equal to ± 1 by Theorem 1.11 (a) Chapter 10. Hence the quotient $L^\bullet(A, l)/L^\bullet_{\text{int}}(A, l)$ has determinant ± 1, which implies that the determinants of $L^\bullet(A, l)$ and $L^\bullet_{\text{int}}(A, l)$ coincide up to a sign. This completes the proofs of Proposition 1.4 and Theorem 1.3.

Using Theorem 1.3 (b), we shall write $D_A(f)$ for $D_{A,l}(f)$ with $l \gg 0$. It is important to notice that (in contrast to Δ_A and E_A) the function $D_A(f)$ in general is not a polynomial, but only a rational function (of course this may happen only when X_A is singular). We shall exhibit a concrete situation when D_A is not a polynomial in Example 2.5 below.

Nevertheless, we have the following corollary generalizing Theorem 2.8 Chapter 9.

Corollary 1.5. *The rational function $D_A(f)$ is homogeneous of degree*

$$\sum_{\Gamma \subset Q} (-1)^{\text{codim}\Gamma} (\dim\Gamma + 1)\text{Vol}_\Gamma(\Gamma). \tag{1.8}$$

Proof. For any set A, the A-resultant R_A is homogeneous of degree $\text{Vol}(Q)$ with respect to any of its arguments (Corollary 2.2 Chapter 8). Here $Q = \text{Conv}(A)$ and the volume is taken with respect to the lattice affinely generated by A. It follows from this and the definition of the principal A-determinant E_A (formula (1.1) Chapter 10) that E_A is homogeneous of degree $(\dim Q + 1)\text{Vol}(Q)$. Applying this to every set $A \cap \Gamma$ we find that $E_{A\cap\Gamma}$ is homogeneous of degree $(\dim \Gamma + 1)\text{Vol}_\Gamma(\Gamma)$. Now our statement follows from (1.4).

B. The case when X_A is quasi-smooth: the polynomiality of D_A

We have already mentioned that when X_A is not smooth then D_A is not necessarily a polynomial. However, we shall prove below that D_A is a polynomial provided that X_A is quasi-smooth. This means (Section 4D Chapter 5) that the polytope $Q = \text{Conv}\,(A)$ is simple and for any face $\Gamma \subset Q$, the index $i(\Gamma, A)$ is equal to 1.

Theorem 1.6. *If X_A is quasi-smooth then D_A is a polynomial. In particular, its degree (1.8) is a non-negative number.*

Note that we do not claim that $D_A = \Delta_A$.

The proof of Theorem 1.6 will occupy this and the next subsection. We start with the observation that by Theorem 1.3 and Theorem 1.2 of Chapter 10, the factorization of D_A into prime elements of the ring $\mathbf{Z}[(a_\omega)]$ is

$$D_A(f) = \pm \prod_{\Gamma \subset Q} \Delta_{A\cap\Gamma}(f)^{\mu(\Gamma)}, \tag{1.9}$$

where the $\mu(\Gamma) \in \mathbf{Z}$ are some exponents. We shall give an interpretation of the $\mu(\Gamma)$ in terms of constructible sheaves (as in Section 2 Chapter 10).

Let $V = \mathbf{C}^A$, so that the toric variety $Y = Y_A$ is embedded into V^*. Consider the constant sheaf $\underline{\mathbf{C}}_Y$ on Y. We can regard it as a sheaf on V^*. This sheaf is obviously constructible so we can consider the characteristic cycle $\mathbf{SS}(\underline{\mathbf{C}}_Y)$ (see Section 2C Chapter 10). We shall use the notation of the cited section, in particular the Lagrangian varieties $\mathrm{Con}\,(\Gamma) \subset V^* \times V$.

Theorem 1.7. *Let the set A be such that $i(\Gamma, A) = 1$ for any face $\Gamma \subset Q$. Then the exponent $\mu(\Gamma)$ in (1.9) is equal to $c(\mathrm{Con}\,(\Gamma), \underline{\mathbf{C}}_Y[n-k])$, the multiplicity of* $\mathrm{Con}\,(\Gamma)$ *in the characteristic cycle of the shifted sheaf $\underline{\mathbf{C}}_Y[n-k]$.*

The proof of Theorem 1.7 is quite similar to that of Theorem 2.15 Chapter 10. So we give only a sketch.

Let (a_ω) and (b_ω) be the dual coordinates in V and V^*, and $\mathbf{C}[a, b]$ be the polynomial ring in these coordinates, graded by $\deg\,(a_\omega) = 0$, $\deg\,(b_\omega) = 1$. We consider the $\mathbf{C}[S]$-modules $\mathrm{Vec}^i(S)$ (Section 2B Chapter 9). Note that $\mathrm{Vec}^i(S) = \bigoplus_l C^i(A, l)$. We regard each $\mathrm{Vec}^i(S)$ as a $\mathbf{C}[b]$-module by the homomorphism

$$\mathbf{C}[b] \to \mathbf{C}[S], \quad b_\omega \mapsto x^\omega.$$

We consider the complex of $\mathbf{C}[a, b]$-modules

$$\mathrm{Vec}^\bullet(S)[a] = \left\{\mathbf{C}[a] \otimes \mathrm{Vec}^0(S) \longrightarrow \mathbf{C}[a] \otimes \mathrm{Vec}^1(S) \longrightarrow \cdots\right\} \tag{1.10}$$

with the differential $\sum_{\omega \in A} a_\omega \otimes \partial_{x^\omega}$. Since $\mathbf{C}[a, b]$ is the coordinate ring of $V \times V^*$, any $\mathbf{C}[a, b]$-module has a well-defined multiplicity along any subvariety in $V \times V^*$.

Proposition 1.8. *We have*

$$\mu(\Gamma) = (-1)^k \mathrm{mult}_{\mathrm{Con}(\Gamma)} H^\bullet(\mathrm{Vec}^\bullet(S)[a]).$$

The proof is similar to that of Proposition 2.6 Chapter 10.

To establish the relation with constructible sheaves, we consider the analytic de Rham complex of Danilov differential forms on Y (cf. Theorem 2.5 Chapter 9):

$$\tilde{\Omega}^\bullet_{Y,\mathrm{an}} = \left\{\tilde{\Omega}^0_{Y,\mathrm{an}} \xrightarrow{d} \tilde{\Omega}^1_{Y,\mathrm{an}} \xrightarrow{d} \cdots\right\}. \tag{1.11}$$

Proposition 1.9. *Suppose that $i(\Gamma, A) = 1$ for any face $\Gamma \subset Q$. Then*
(a) *The complex $\tilde{\Omega}^\bullet_{Y,\mathrm{an}}$ is a right resolution of the constant sheaf $\underline{\mathbf{C}}_Y$.*

(b) *For any* $y \in Y$ *the formal completion* $\left(\tilde{\Omega}^{\bullet}_{Y,\mathrm{an}}\right)^{\wedge}_{y}$ *(considered as a complex of vector spaces) is a right resolution of the vector space* $\mathbf{C}$.

Proof. This is similar to Theorem 2.2 Chapter 10. As in that proof, we use the transversal slice to Y at y. Suppose that y lies in the orbit $Y^0(\Gamma) \subset Y$ where Γ is a face of Q. The condition $i(\Gamma, A) = 1$ means that the transversal slice is the toric variety $\mathrm{Spec}\, \mathbf{C}[S/\Gamma]$ (see Section 3A Chapter 5). So our statement is reduced to the following.

Lemma 1.10. *Let* Σ *be an admissible semigroup (Section 3C Chapter 5) embedded into some* $\mathbf{Z}^m$. *Let* $Y_\Sigma = \mathrm{Spec}\, \mathbf{C}[\Sigma]$ *be the corresponding affine toric variety and* $0 \in Y_\Sigma$ *be the unique* 0*-dimensional torus orbit. Then both the stalk and the formal completion at* 0 *of the complex* $\tilde{\Omega}^{\bullet}_{Y_\Sigma,\mathrm{an}}$ *are right resolutions of* $\mathbf{C}$.

Proof of the lemma. The space $\mathrm{Vec}^i(\Sigma) = H^0(Y_\Sigma, \tilde{\Omega}^i)$ consists, by definition, of finite sums

$$\sum_{u \in \Sigma} (u, \lambda_u), \quad \lambda_u \in \bigwedge^i \mathrm{Lin}_{\mathbf{C}} \Gamma(u), \tag{1.12}$$

where $\Gamma(u)$ is the minimal face of the cone $K = \mathrm{Conv}(\Sigma)$ containing u. The differential $d : \mathrm{Vec}^i(\Sigma) \to \mathrm{Vec}^{i+1}(\Sigma)$ acts by

$$(u, \lambda_u) \quad \longmapsto \quad (u, -u \wedge \lambda_u). \tag{1.13}$$

The stalk at 0 of $\tilde{\Omega}^i_{Y_\Sigma,\mathrm{an}}$ consists of (possibly infinite) sums of the form (1.12) with the condition that the λ_u grow at most polynomially. If $u \neq 0$ then the Koszul complex, given by the exterior multiplication with u, is exact, so from $u \wedge \lambda_u = 0$ it follows that $\lambda_u = u \wedge \mu_u$ for some $\mu_u \in \bigwedge^{i-1} \mathrm{Lin}_{\mathbf{C}} \Gamma(u)$, and the μ_u can be chosen to grow at most polynomially if the λ_u do. So the cohomology comes only from the summands in (1.12) with $u = 0$. The only non-trivial summand is the one with $i = 0$ and is generated by $(0, 1) \in \mathrm{Vec}^0(S)$. This gives the exactness of the analytic stalk. The case of formal completion is similar and amounts to considering the formal series of the form (1.12).

Now, to the complex $(\mathrm{Vec}^{\bullet}(S), d)$, we associate a complex of free right modules over the Weyl algebra $\mathrm{Dif} = \mathbf{C}[b_\omega, \partial/\partial b_\omega]$. This complex is

$$\left(\mathrm{Vec}^{\bullet}(S) \otimes_{\mathbf{C}[b]} \mathrm{Dif}, \ \tilde{\partial}\right) \tag{1.14}$$

with the differential given by (2.29) Chapter 10. The associated graded complex of this complex of Dif-modules is $\mathrm{Vec}^{\bullet}(S)[a]$ appearing in Proposition 1.8. The de Rham complex DR(Vec$^\bullet(S) \otimes$ Dif) is the usual de Rham complex (1.11). So by Theorem 2.13 (b) Chapter 10 we have

$$\begin{aligned}\mu(\Gamma) &= (-1)^k \mathrm{mult}_{\mathrm{Con}(\Gamma)} H^\bullet(\mathrm{Vec}^\bullet(S)[a]) \\ &= (-1)^{n-k}\ c(\mathrm{Con}(\Gamma),\ \mathrm{DR}(\mathrm{Vec}^\bullet(S) \otimes \mathrm{Dif})) = c(\mathrm{Con}(\Gamma), \underline{\mathbf{C}}_Y[n-k]).\end{aligned}$$

Theorem 1.7 is proved.

C. Perverse sheaves and positivity of $\mu(\Gamma)$

Let M be a complex algebraic variety. A constructible complex $\mathcal{F}^\bullet$ on M is called a *perverse sheaf* [BBD], [KS] if the following conditions hold:

(Perv$^+$). *For $i \geq 0$ the cohomology sheaf $\underline{H}^i(\mathcal{F}^\bullet)$ has support on an algebraic subvariety of codimension $\geq i$.*

(Perv$^+$). *For every smooth (not necessarily closed) subvariety $Z \subset M$ of codimension d, the sheaves $\underline{\mathbf{H}}^i_Z(\mathcal{F}^\bullet)$ of hypercohomology with support in Z are zero for $i < d$.*

The main property of perverse sheaves which we shall need is as follows (see [KS]).

Theorem 1.11. *If M is smooth then the characteristic cycle of any perverse sheaf $\mathcal{F}^\bullet$ is positive, i.e., $c(\Lambda, \mathcal{F}^\bullet) \geq 0$ for any conic Lagrangian variety $\Lambda \subset T^*M$.*

Returning to our situation, we have the following.

Theorem 1.12. *Let Y be a quasi-smooth toric variety. Then the constant sheaf $\underline{\mathbf{C}}_Y$ is perverse.*

Proof. In our case (Perv$^-$) is obvious, so it remains to show (Perv$^+$). Let $\pi : \tilde{Y} \to Y$ be the normalization morphism of Y. Since Y is quasi-smooth, it follows that π is a homeomorphism of topological spaces. Therefore it is enough to prove the perversity of the constant sheaf on $\tilde{Y}$. In other words, we can and will assume that Y is normal. Then $\underline{\mathbf{C}}_Y$ has the de Rham resolution $(\tilde{\Omega}^\bullet_Y, d)$ where the $\tilde{\Omega}^i$ are the sheaves of Danilov i-forms (see above). It was shown in [D] (Proposition 4.8) that, for a quasi-smooth normal Y, the sheaves $\tilde{\Omega}^i$ on Y satisfy the Cohen-Macaulay condition, i.e., $\underline{H}^j_Z(\tilde{\Omega}^i) = 0$ for $j \neq d$. This implies (Perv$^+$).

Now we can finish the proof of Theorem 1.6. We need to prove that all the exponents $\mu(\Gamma)$ are non-negative. Since X_A is assumed to be quasi-smooth, $Y - \{0\}$ is also quasi-smooth since this is the punctured cone over X_A. The perversity of the sheaf $\underline{\mathbf{C}}_{Y-\{0\}}$ on $Y-\{0\}$ implies the perversity of the shifted sheaf $\underline{\mathbf{C}}_{Y-\{0\}}[n-k]$ on $V^* - \{0\} \supset Y - \{0\}$. Since for any face $\Gamma \subset Q$ we have, by Theorem 1.7,

$$\mu(\Gamma) = c(\mathrm{Con}(\Gamma), \underline{\mathbf{C}}_{Y-\{0\}}[n-k]),$$

the non-negativity follows from the perversity.

2. Newton numbers and Newton functions

A. The Newton number of an admissible semigroup

The exponents $\mu(\Gamma)$ of prime factors in the decomposition (1.9) of the regular A-determinant D_A are, by virtue of Theorem 1.3 and Theorem 1.2 Chapter 10, certain alternating sums. In this section we study sums of this kind in a more systematic way and in a slightly more general context. This will allow us to apply them later (Section 4) in a different situation.

Let S be an admissible semigroup (see Section 3C, Chapter 5), Ξ its group completion (it is a free Abelian group), $k = \mathrm{rk}(\Xi)$, and $K \subset \Xi_{\mathbf{R}} = \mathbf{R}^k$ the convex hull of S (it is a polyhedral cone with apex 0). We define the *Newton number* of S as

$$\nu(S) = \sum_{\Gamma \subset K} (-1)^{\mathrm{codim}\,\Gamma} u(S \cap \Gamma) \tag{2.1}$$

where Γ runs over all faces of K (including 0 and K itself), and $u(\cdot)$ is the subdiagram volume of a semigroup (see Definition 3.8 Chapter 5). For the trivial semigroup $S = \{0\}$ we set $\nu(S) = 1$.

Example 2.1. For a free semigroup $S = \mathbf{Z}^k_+$, $k > 0$, we have $\nu(S) = 0$. Indeed, the subdiagram part $K_-(S)$ is an elementary simplex for the lattice $\Xi = \mathbf{Z}^k$, so $u(S) = \mathrm{Vol}_\Xi(K_-(S)) = 1$. Since each $S \cap \Gamma$ is also a free semigroup, we have $u(S \cap \Gamma) = 1$ for all faces $\Gamma \subset K = \mathbf{R}^k_+$. Since for each $i = 0, 1, \ldots, k$ there are $\binom{k}{i}$ faces of dimension i, we have

$$\nu(S) = \sum_{i=0}^{k} (-1)^{k-i} \binom{k}{i} = 0,$$

in view of the binomial formula.

The relationship between the Newton numbers and the exponents $\mu(\Gamma)$ in (1.9) is given by the following.

Proposition 2.2. *Suppose, in the situation and notation of Section 1B Chapter 10, that $i(\Gamma, A) = 1$ for any face $\Gamma \subset Q$ (this holds, for example when X_A is normal). Then $\mu(\Gamma) = \nu(S/\Gamma)$.*

Proof. Combining Theorem 1.2 Chapter 10 and Theorem 1.3 of this chapter, we obtain the formula which is valid for arbitrary A (without the assumption $i(\Gamma, A) = 1$):

$$\mu(\Gamma) = \sum_{\Theta:\Gamma\subset\Theta\subset Q} i(\Gamma, A\cap\Theta)u(S(A\cap\Theta)/\Gamma), \tag{2.2}$$

where $S(A\cap\Theta)$ is the semigroup in $\mathbf{Z}^k$ generated by $A\cap\Theta$. The assumption that all $i(\Theta, A)$ are equal to 1 implies that $i(\Gamma, A\cap\Theta) = 1$ as well. Replacing all these indices in (2.2) by 1, we obtain our statement.

In general, the Newton number $\nu(S)$ may be negative (and $D_A(f)$ may not be a polynomial in coefficients of f). There is, however, one important case when non-negativity holds.

Theorem 2.3. *Suppose an admissible semigroup S has the following two properties:*
(a) *The convex hull $K(S) \subset \Xi_{\mathbf{R}}$ is a simplicial cone.*
(b) $\mathrm{Lin}_{\mathbf{Z}}(S\cap\Gamma) = \Xi\cap\mathrm{Lin}_{\mathbf{R}}(\Gamma)$ *for any face $\Gamma\subset K(S)$.*
Then $\nu(S) \geq 0$.

This theorem is completely analogous to Theorem 1.6, but is slightly more general since it deals with a semigroup S not necessarily coming from a set A. In the case when $K(S)$ is strictly simplicial (i.e. $K(S)\cap\Xi$ is a free semigroup) it is due to Kouchnirenko [Kou].

The proof is based on the topological interpretation of Newton numbers similar to what was done in Section 1B. Let $Y_S = \mathrm{Spec}\ \mathbf{C}[S]$; note that the properties (a) and (b) in Theorem 2.3 mean exactly that Y_S is quasi-smooth, see Theorem 3.6 (b) Chapter 5. Let B be a finite set of generators of S. The variety Y_S is embedded into $(\mathbf{C}^B)^*$ and 0 is the unique 0-dimensional torus orbit on Y_S. We regard $\underline{\mathbf{C}}_{Y_S}$ as a sheaf on $(\mathbf{C}^B)^*$.

Theorem 2.4. *The Newton number of S has the following interpretation:*

$$\nu(S) = c(\mathrm{Con}(0), \underline{\mathbf{C}}_{Y_S}[\#(B) - k]).$$

This theorem is analogous to Theorem 1.7.

Proof. For every face $\Gamma\subset K(S)$ let $Y^0(\Gamma)_S \subset Y_S$ be the corresponding torus orbit and let $j_\Gamma : Y^0(\Gamma)_S \to Y_S$ be the embedding. Denote by $Y_S^r \subset Y_S$ the union of all torus orbits of codimension $\leq r$ and by j_r the embedding of this union into Y_S. Then $Y_S^k = Y_S$ and for every r we have an exact triangle

$$\to Rj_{r,*}\underline{\mathbf{C}}_{Y_S^r} \to Rj_{r-1,*}\underline{\mathbf{C}}_{Y_S^{r-1}} \to \bigoplus_{\mathrm{codim}\,\Gamma=r} Rj_{\Gamma,*}\underline{\mathbf{C}}_{Y^0(\Gamma)_S} \to$$

in the derived category of constructible complexes on $(\mathbf{C}^B)^*$ (see [KS] for background on derived categories and exact triangles). Theorem 2.4 follows from the additivity of multiplicities in exact triangles and from Theorem 2.12 Chapter 10 yielding

$$c(\mathrm{Con}(0), Rj_{\Gamma,*}\underline{\mathbf{C}}_{Y^0(\Gamma)_S}) = (-1)^{\mathrm{codim}\,\Gamma} u(S \cap \Gamma).$$

Theorem 2.3 follows now from the perversity of $\underline{\mathbf{C}}_{Y_S}[m-k]$ as a complex on $(\mathbf{C}^B)^*$ (Theorem 1.12) and from Theorem 1.11.

Example 2.5. Let us exhibit a semigroup S with $\nu(S) < 0$ and a set A such that D_A is not a polynomial. We consider $\mathbf{R}^2 \times \mathbf{R}^3$ with basis e_1, e_2, f_1, f_2, f_3 and the semigroup S in this space generated by the set B consisting of the six vectors $e_i + f_j$. The linear span $\mathrm{Lin}_{\mathbf{R}}(S)$ has dimension 4 and the cone $K = \mathrm{Conv}(S)$ has, as its base, the triangular prism Conv $(B) = \Delta^1 \times \Delta^2$. Note that

$$\nu(S) = \left(\sum_{\Gamma \subset \Delta^1 \times \Delta^2} (-1)^{\mathrm{codim}\,\Gamma} \mathrm{Vol}_\Gamma(\Gamma) \right) + 1.$$

Here the summation is taken over all non-empty faces Γ of $\Delta^1 \times \Delta^2$ and Vol_Γ is induced by the lattice $\mathrm{Aff}_{\mathbf{Z}}(B \cap \Gamma) \subset \mathrm{Aff}_{\mathbf{R}}(\Gamma)$. Indeed, for a face $\Gamma \subset \Delta^1 \times \Delta^2$ the corresponding face of K is $\mathbf{R}_+\Gamma$. The subdiagram volume $u(S \cap \mathbf{R}_+\Gamma)$ (i.e., the volume of $K_-(\mathbf{R}_+\Gamma)$ (see Section 3E Chapter 5)) coincides with $\mathrm{Vol}_\Gamma(\Gamma)$ since $K_-(\mathbf{R}_+\Gamma)$ is the pyramid with base Γ and height 1.

The volume of $\Delta^1 \times \Delta^2$ is equal to 3, that of a triangular face to 1, of a quadrangular face to 2, of any edge to 1. Hence

$$\nu(S) = 3 - (2 \cdot 1 + 3 \cdot 2) + 9 \cdot 1 - 6 \cdot 1 + 1 = -1.$$

Consider now A which is the set of vertices of the hypersimplex $\Delta(3, 6)$ (see Section 3A Chapter 6). Thus $A \subset \mathbf{R}^6$ consists of $\binom{6}{3}$ vectors $e_i + e_j + e_k$ where $e_1, \dots, e_6$ is the standard basis of $\mathbf{R}^6$. It is straightforward to see that, for any vertex $\Gamma = e_i + e_j + e_k$ of $Q = \Delta(3, 6)$, the semigroup $S(A)/\Gamma$ is isomorphic to the semigroup S described above (geometrically, near each vertex $\Delta(3, 6)$ looks like a cone over $\Delta^1 \times \Delta^2$). Thus $\nu(S_A/\Gamma) = -1$. Since for any face $\Gamma' \subset \Delta^1 \times \Delta^2$ we have $i(\Gamma', A) = 1$, we find by Proposition 2.2 that D_A is not a polynomial.

B. Combinatorial Newton numbers

Let $K \subset \mathbf{R}^m$ be a convex polyhedral cone. We shall assume that K is rational, i.e., it is given by a system of linear inequalities with rational coefficients. By a triangulation of K we mean a partition of K into a finite number of rational simplicial cones, any two of which intersect along a common face. The definition of a coherent triangulation is analogous to that in Section 1C Chapter 5.

Let Γ be a face of K, $\dim \Gamma = i$. By the *combinatorial volume* of Γ with respect to a triangulation Σ, we mean the number of i-dimensional cones in Σ into which Γ is partitioned. We let $CV_\Sigma(\Gamma)$ denote this number. If Γ is a vertex of K, then we set $CV_\Sigma(\Gamma) = 1$.

By the *combinatorial Newton number* of K with respect to Σ, we mean

$$CN_\Sigma(K) = \sum_\Gamma (-1)^{\operatorname{codim}\Gamma} CV_\Sigma(\Gamma), \tag{2.3}$$

where Γ runs through all (non-empty) faces of K, including K itself.

Theorem 2.6.

(a) *If K is a cone in $\mathbf{R}^3$, then for any triangulation Σ we have $CN_\Sigma(K) \geq 0$.*

(b) *If Σ is a coherent triangulation of the cone $\mathbf{R}^p \times \mathbf{R}^q_+ \subset \mathbf{R}^{p+q}$, then $CN_\Sigma(\mathbf{R}^p \times \mathbf{R}^q_+) \geq 0$.*

Proof. (a) Consider first the case when K is strictly convex, i.e., it does not contain straight lines. Let P be the plane polygon at the base of K. Then Σ induces a triangulation T of P, i.e., a decomposition of P into finitely many triangles. Assume that T has more than one triangle (otherwise $CN_\Sigma(K) = 0$ and there is nothing to prove). Let T_0 denote the set of vertices of P, T_1 denote the set of sides of triangles from T which lie on the boundary of P, and T_2 denote the set of triangles from T. By definition (2.5), we have

$$CN_\Sigma(K) = \#(T_2) - \#(T_1) + \#(T_0) - 1.$$

Clearly, each segment from T_1 is contained in exactly one triangle from T_2. We want to cancel out all the segments from T_1 with corresponding triangles. This is always possible except when two segments share the same triangle; however, such segments also share a vertex of P, and so they can be cancelled together with their common triangle and common vertex. Performing these cancellations, we see that

$$CN_\Sigma(K) = \#(T_2') + \#(T_0') - 1,$$

where T_2' is the set of interior triangles from T_2 (having no sides on the boundary), and T_0' is the set of vertices of P which belong to more than one triangle from T_2. Therefore, it remains to show that at least one of the sets T_2' and T_0' is non-empty.

Suppose, on the contrary, that $T_2' = T_0' = \emptyset$, i.e., T has no interior triangles, and each vertex of P belongs to exactly one triangle from T. Let us count how many triangles, edges and vertices can T have under such an assumption. First, T

has m triangles having a vertex of P as one of the vertices. Each of these triangles has two boundary edges and one interior edge. Suppose, T has p more triangles. Due to our assumption, each of them has one boundary edge and two interior edges. Hence, the number of boundary edges in T is $(2m+p)$, and the number of interior edges is $(\frac{m}{2}+p)$ (since each boundary edge belongs to one triangle, while each interior edge belongs to two triangles). Furthermore, it is easy to see that under our assumptions T has no interior vertices. So the number of vertices is equal to the number of boundary edges, i.e., to $(2m+p)$. It follows that

$$\#(\text{triangles of } T) - \#(\text{edges of } T) + \#(\text{vertices of } T)$$

$$= (m+p) - (\frac{m}{2}+p) = \frac{m}{2}.$$

On the other hand, by Euler's theorem, for a polygon the last expression is equal to 1. We conclude that $m=2$, which gives a desired contradiction.

The case when K is not strictly convex is actually easier, and we leave it to the reader.

(b) Suppose we are given a coherent triangulation Σ of $\mathbf{R}^p \times \mathbf{R}^q_+$. We regard Σ as a fan and consider the corresponding toric variety $X(\Sigma)$ (see Section 4A Chapter 5). This toric variety is glued from the affine charts $\operatorname{Spec} \mathbf{C}[\check{\sigma} \cap \mathbf{Z}^{p+q}]$, where $\sigma \in \Sigma$, and $\check{\sigma} \subset \mathbf{R}^{p+q} = (\mathbf{R}^{p+q})^*$ is the dual cone of σ. The r-dimensional torus orbits on $X(\Sigma)$ correspond to the cones of codimension r in Σ. We consider the map of fans $(\mathbf{R}^{p+q}, \Sigma) \to (\mathbf{R}^q, \mathbf{R}^q_+)$ extending the projection $\mathbf{R}^{p+q} \to \mathbf{R}^q$. This map corresponds to a proper morphism of toric varieties $\pi : X(\Sigma) \to \mathbf{C}^q = X(\mathbf{R}^q_+)$.

Let $\mathcal{F}^\bullet$ be a finite complex of sheaves on $\mathbf{C}^q$ which is constructible with respect to the stratification by coordinate subspaces (i.e., whose cohomology sheaves are locally constant on the interior of each coordinate subspace). For every subset $I \subset \{1, \ldots, q\}$, we let $\operatorname{rk}(\mathcal{F}^\bullet, \mathbf{C}^I)$ denote the alternating sum of the ranks of the cohomology sheaves $\underline{H}^i(\mathcal{F})$ at the generic point of the stratum $(\mathbf{C}^*)^I \times \{0\}^{\overline{I}}$, where $\overline{I} = \{1, \ldots, q\} - I$.

Lemma 2.7. *For every $I \subset \{1, \ldots, q\}$ the combinatorial volume $CV_\Sigma(\mathbf{R}^p \times \mathbf{R}^I_+)$ is equal to* $\operatorname{rk}(R\pi_* \underline{\mathbf{C}}_{X(\Sigma)}, \mathbf{C}^{\overline{I}})$. *Furthermore, $R^{2i+1}\pi_*\underline{\mathbf{C}}_{X(\Sigma)} = 0$ for all i.*

Proof. The cones in Σ of dimension $p + \#(I)$ into which $\mathbf{R}^p \times \mathbf{R}^I_+$ is partitioned are in one-to-one correspondence with the torus orbits on $X(\Sigma)$ which project isomorphically onto the orbit $(\mathbf{C}^*)^{\overline{I}} \subset \mathbf{C}^q$ corresponding to the face $\mathbf{R}^I_+ \subset \mathbf{R}^q_+$ (see Section 4A, Chapter 5). For all of the other orbits which project to $(\mathbf{C}^*)^{\overline{I}}$, the fibers of the projection are isomorphic to algebraic tori $(\mathbf{C}^*)^m$ for various $m \geq 1$.

Hence, they do not contribute to the Euler characteristic, i.e., to $\mathrm{rk}(R\pi_*\mathbf{C}_{X(\Sigma)}, \mathbf{C}^{\bar{I}})$. Thus, this number coincides with $CV_\Sigma(\mathbf{R}^p \times \mathbf{R}^I_+)$.

It remains to show that any higher direct image with odd index is 0. We shall prove that $R^{2i+1}\pi_*\underline{\mathbf{C}}_{X(\Sigma)}$ has the zero fiber at $0 \in \mathbf{C}^q$; the proof for other points is analogous, except that we reduce it to the case of lower dimensional varieties. Since π is a proper morphism, the fiber in question is $H^{2i+1}(\pi^{-1}(0), \mathbf{C})$. Furthermore, $\pi^{-1}(0)$ is a deformation retract of $X(\Sigma)$, so our fiber coincides with $H^{2i+1}(X(\Sigma), \mathbf{C})$.

Let us compactify $X(\Sigma)$. To do this we consider the diagonal subgroup $\mathbf{C}^* \subset (\mathbf{C}^*)^q$ which corresponds to homotheties, and we attach to $X(\Sigma)$ its set of infinite points, i.e., the set Z of orbits of $\mathbf{C}^*$ in $X(\Sigma)$ that are not contained in $\pi^{-1}(0)$. We let $\overline{X} = X \cup Z$ denote the resulting compactification. Both $\overline{X}$ and Z are quasi-smooth normal projective toric varieties. Hence (see [D], §10), the odd-dimensional cohomology groups of $\overline{X}$ and Z with complex coefficients are 0. In order to prove that $H^{2i+1}(X(\Sigma), \mathbf{C}) = 0$ for all i, it remains to use the Gysin exact sequence [BBD], [KS]. This sequence exists in our case because quasi-smooth toric varieties are rational homological manifolds (see again [D]). The vanishing of $H^1(X(\Sigma), \mathbf{C})$, which is situated at the very end of the Gysin sequence, follows from the fact that the fundamental group of $X(\Sigma)$ is finite ([D], Proposition 9.3). Lemma 2.7 is proved.

Lemma 2.8. *We have the following isomorphism in the derived category of constructive complexes on* $\mathbf{C}^q$:

$$R\pi_*\mathbf{C}_{X(\Sigma)} \cong \bigoplus_I \underline{\mathbf{C}}_{\mathbf{C}^I} \otimes M_I^\bullet,$$

where I runs through all subsets of $\{1, \dots, q\}$, $\underline{\mathbf{C}}_{\mathbf{C}^I}$ is the constant sheaf on the coordinate subspace $\mathbf{C}^I$, and the $M_I^\bullet$ are some graded vector spaces.

Proof. As noted above, $X(\Sigma)$ is a rational homological manifold. Hence, the constant sheaf $\underline{\mathbf{C}}_{X(\Sigma)}$ coincides with the intersection cohomology complex on $X(\Sigma)$ (see [BBD] for general background). As shown by Saito [Sa1], the intersection cohomology sheaf of any manifold extends to an object in the category of the pure polarizable Hodge modules. We let $H\mathbf{C}_{X(\Sigma)}$ denote the Hodge module corresponding to $\underline{\mathbf{C}}_{X(\Sigma)}$. Next, by the decomposition theorem in [Sa1], the direct image $R\pi_*(\underline{H\mathbf{C}}_{X(\Sigma)})$ is a direct sum of twists of pure polarizable Hodge modules. The category of such modules is semisimple, and its irreducible objects correspond to the intersection cohomology extensions of irreducible locally constant sheaves on the strata, equipped with a variation of the Hodge structure. In our case all of the strata are tori. In order to show that every irreducible summand is a shifted constant sheaf, it is sufficient to demonstrate that it has no monodromy on an open subset of its domain of definition. Notice that the action of $(\mathbf{C}^*)^q$ on $\mathbf{C}^q$ extends to an action on the sheaves $R^i\pi_*\underline{H\mathbf{C}}_{X(\Sigma)}$. Moreover, the holonomy (parallel translation) of any such sheaf from a point a to a point b is given by the action of any torus element taking a to b. Consequently, the monodromy around any closed loop is trivial, since it corresponds to the action of the identity element of the torus. Lemma 2.8 is proved.

We now conclude the proof of Theorem 2.6. Note that, if $J \subset \{1, \ldots, q\}$ is a non-empty subset, then

$$\sum_I (-1)^{\#(I)} \cdot \mathrm{rk}(\underline{\mathbf{C}}_{\mathbf{C}^J}, \mathbf{C}^I) = \sum_{I \subset J} (-1)^{\#(I)} = 0.$$

As for the case $J = \emptyset$, corresponding to the skyscraper sheaf at zero, the graded space of multiplicities $M^\bullet_\emptyset$ has only even graded components, since $R^{\mathrm{odd}}\pi_*\mathbf{C}_{X(\Sigma)} = 0$. Hence, $\sum_I (-1)^{|I|} \cdot \mathrm{rk}(\mathbf{C}^I, R\pi_*\underline{\mathbf{C}}_{X(\Sigma)}) = \sum_i \dim M^{2i}_\emptyset$ is a non-negative number, and Theorem 2.6 is proved.

In the process of the proof we have also obtained the following topological interpretation of the combinatorial Newton numbers.

Corollary 2.9. *Let Σ be a coherent triangulation of $\mathbf{R}^p \times \mathbf{R}^q_+$. Let $\pi : X(\Sigma) \to \mathbf{C}^q$ be the proper morphism corresponding to Σ. Then $CN_\Sigma(\mathbf{R}^p \times \mathbf{R}^q_+)$ is equal to the sum of the multiplicities in $R\pi_*\underline{\mathbf{C}}_{X(\Sigma)}$ of all the shifted sheaves $\underline{\mathbf{C}}_{\{0\}}[d]$ (here the shift parameter d is necessarily even).*

C. The Newton function of a triangulation

Let Q be a polytope in $\mathbf{R}^{k-1}$ with vertices in $\mathbf{Z}^{k-1}$. Let T be a coherent triangulation of Q into simplices with integral vertices.

Definition 2.10. A simplex σ of T (of any dimension $j \geq 0$) is called *massive* if it lies in a j-dimensional face of Q.

We define an integral-valued function CN_T on the set of simplices of T (of every dimension) by setting

$$CN_T(\sigma) = \sum_{\substack{\gamma \supset \sigma \\ \gamma \text{ massive}}} (-1)^{\dim Q - \dim \gamma}, \tag{2.4}$$

where γ runs through all of the massive simplices of T containing σ as a face. We call CN_T the *Newton function* of T.

The definition of the Newton function is quite similar to that of the combinatorial Newton number of a cone with respect to its triangulation, see (2.3). In fact, if we assume, as usual, that Q lies in an affine hyperplane in $\mathbf{R}^k$, and let K be the cone with apex 0 over Q, and Σ the triangulation of K formed by the cones over the simplices of T, then

$$CN_\Sigma(K) = CN_T(\emptyset)$$

(with the understanding that in (2.4) $\dim\emptyset = -1$).

On the other hand, the Newton function of any simplex σ can be expressed as a combinatorial Newton number by the following construction.

First, notice that for every affine subspace L in an affine space W we can define the quotient vector space W/L whose elements are parallel translations of L. In particular, this space has a distinguished vector 0 represented by L itself.

Now let σ be a simplex in T, and let $\Gamma(\sigma) \subset Q$ be the smallest face containing σ. Consider the image of Q in the quotient space $\mathrm{Aff}_{\mathbf{R}}(Q)/\mathrm{Aff}_{\mathbf{R}}(\sigma)$ (where $\mathrm{Aff}_{\mathbf{R}}$ stands for the affine span). Let $K(\sigma, Q)$ denote the cone in $\mathrm{Aff}_{\mathbf{R}}(Q)/\mathrm{Aff}_{\mathbf{R}}(\sigma)$ whose apex is the distinguished point 0 of this quotient space, and the base is the image of Q. For each simplex γ of T having σ as a face, let $K(\gamma)$ denote the cone in $\mathrm{Aff}_{\mathbf{R}}(Q)/\mathrm{Aff}_{\mathbf{R}}(\sigma)$ with apex 0 whose base is the image of γ. It is clear that the cones $K(\gamma)$ form a triangulation of $K(\sigma, Q)$, which we shall denote $\Sigma(T, \sigma)$ and call the triangulation induced by T. This triangulation is coherent if T is coherent.

It is implied at once that

$$CN_T(\sigma) = CN_{\Sigma(T,\sigma)}(K(\sigma, Q)). \tag{2.5}$$

Theorem 2.11. *Suppose that Q is simple or lies in $\mathbf{R}^3$. Let T be a coherent triangulation of Q as above. Then the Newton function CN_T takes only non-negative values.*

Proof. This follows from (2.5) and Theorem 2.6, once we observe that if Q is a simple polytope, σ is any simplex in T, and $\Gamma(\sigma) \subset Q$ is the smallest face containing σ then $K(\sigma, Q) \cong \mathbf{R}^p \times \mathbf{R}^q_+$, where $p = \dim\Gamma(\sigma) - \dim\sigma$, $q = \dim Q - \dim\Gamma(\sigma)$.

We conclude this subsection by a property of the Newton function, which will be used later.

Proposition 2.12. *For every simplex $\sigma \in T$ we have*

$$\sum_{\gamma:\gamma\supset\sigma} (-1)^{\dim\gamma-\dim\sigma} CN_T(\gamma) = \begin{cases} 0 & \text{if } \sigma \text{ is not massive,} \\ (-1)^{\dim Q-\dim\sigma} & \text{if } \sigma \text{ is massive,} \end{cases} \tag{2.6}$$

where the summation is over all simplices $\gamma \in T$ having σ as a face.

Proof. Expressing $CN_T(\gamma)$ in the left hand side of (2.6) with the help of (2.4) and changing the order of the summation, we obtain

$$\sum_{\substack{\delta\supset\sigma \\ \delta \text{ massive}}} (-1)^{\dim Q-\dim\delta} \sum_{\gamma:\sigma\subset\gamma\subset\delta} (-1)^{\dim\gamma-\dim\sigma}$$

If $\delta \neq \sigma$, the second sum has, up to sign, the form $\chi - 1$ where χ is the Euler characteristic of a certain polyhedral ball. Thus $\chi - 1 = 0$. When $\delta = \sigma$ we obtain the required expression.

Note that in [GZK3] what we now call the Newton function was called the Möbius function of a triangulation. The reason was that (2.6) is reminiscent of the property of the classical Möbius function. Still, we have finally decided that the present terminology is less ambiguous.

D. Non-negativity of Newton numbers and polynomiality of D_A for low-dimensional cases

Proposition 2.14.

(a) *Let $S \subset \mathbf{Z}^2$ be any admissible semigroup. Then $\nu(S) \geq 0$.*

(b) *Let $S \subset \mathbf{Z}^3$ be any admissible semigroup generating $\mathbf{Z}^3$. Suppose that for any face $\Gamma \subset K$ we have*

$$\mathrm{Lin}_{\mathbf{Z}}(S \cap \Gamma) = \mathbf{Z}^3 \cap \mathrm{Lin}_{\mathbf{R}}(\Gamma).$$

Then $\nu(S) \geq 0$.

Proof. (a) Consider an arbitrary lattice triangle ABC. Since the area and lengths of sides are normalized with respect to the corresponding lattices, the usual formulas of the plane geometry need some adjustment. Let $\langle ABC\rangle$ denote the normalized area $\mathrm{Vol}_{\mathbf{Z}^2}(ABC)$, and $|AB|$, $|AC|$, $|BC|$ denote the lengths of sides each normalized with respect to the induced one-dimensional lattice (so say $|AB|$ equals the number of integral points in the segment $[AB]$ minus 1). Using these definitions, we have the following "strange" inequality:

$$\langle ABC\rangle \;\geq\; |AB|\cdot|AC|. \tag{2.7}$$

(Indeed, if $B' \in [A, B]$, $C' \in [A, C]$ are lattice points such that $|AB'| = |AC'| = 1$ then $\langle ABC\rangle = |AB|\cdot\langle AB'C\rangle = |AB|\cdot|AC|\cdot\langle AB'C'\rangle \geq |AB|\cdot|AC|$.) It follows that

$$\langle ABC\rangle - |AB| - |AC| + 1 \geq (|AB| - 1)(|AC| - 1) \geq 0. \tag{2.8}$$

Let now $S \subset \mathbf{Z}^2$ be an admissible semigroup which we assume (without loss of generality) to generate $\mathbf{Z}^2$. Let $K_+(S) = \mathrm{Conv}(S - \{0\}) \subset \mathbf{R}^2$ (see Section 3E Chapter 5). The finite part of the boundary of $K_+(S)$ consists of finitely many segments, say, $[B_0, B_1], [B_1, B_2], \ldots, [B_{m-1}, B_m]$. By definition,

$$\nu(S) = \left(\sum_{i=0}^{m-1} \langle 0B_i B_{i+1} \rangle\right) - |0B_0| - |0B_m| + 1,$$

which can be rewritten as

$$\sum_{i=0}^{m-1} (\langle 0B_i B_{i+1} \rangle - |0B_i| - |0B_{i+1}| + 1) + \sum_{i=1}^{m-1} (2 \cdot |0B_i| - 1)\,.$$

Now the first sum in the last expression is ≥ 0 by (2.7), and the second sum is always positive if present at all (i.e., if $m \geq 2$). This proves part (a).

(b) Let $K_-(S)$ be the subdiagram part of $K(S)$ (Section 3E Chapter 5), i.e., the closure of $K(S) - K_+(S)$ where $K(S) = \mathrm{Conv}(S)$ and $K_+(S) = \mathrm{Conv}(S - \{0\})$. Thus the subdiagram volume of S is $u(S) = \mathrm{Vol}(K_-(S))$. Let Λ be the common boundary of $K_+(S)$ and $K_-(S)$. It is a polyhedral surface in $\mathbf{R}^3$. Consider the decomposition Υ of $K(S)$ into polyhedral cones obtained by taking a cone through every face of Λ. This decomposition is coherent since Λ is a part of the boundary of the convex polyhedron $K_+(S)$. Let us refine Υ to a coherent triangulation Σ of $K(S)$ by subdividing every 3-cone from Υ into simplicial cones without adding new edges. Then

$$u(S) = \sum_{\substack{L \in \Sigma \\ \dim L = 3}} u(S \cap L). \tag{2.9}$$

This is because, first, $K_-(S)$ is the union of $K_-(S \cap L)$ by construction, and, second, the lattice $\mathrm{Lin}_{\mathbf{Z}}(S \cap L)$ defining the volume of $K_-(S \cap L)$, coincides with $\mathrm{Lin}_{\mathbf{Z}}(S) = \mathbf{Z}^3$.

Let P be a (bounded) plane polygon (which is a section of $K(S)$) by a plane whose equation has rational coefficients. The triangulation Σ of $K(S)$ induces a triangulation T of P into triangles with rational vertices. For every simplex σ of T (of dimension 0,1 or 2) let $L(\sigma) \in \Sigma$ be the corresponding cone. We also add to T the empty set $\emptyset$ regarded as a simplex of dimension (-1) and set $L(\emptyset) = \{0\}$. Applying (2.9) and Proposition 2.12, we obtain

$$\nu(S) = \sum_{\sigma \in T} CN_T(\sigma) \cdot \nu(S \cap L(\sigma)),$$

where the sum is over all simplices including $\emptyset$, and $CN_T(\emptyset) = CN_\Sigma(K(S))$.

Every $\nu(S \cap L(\sigma))$ is non-negative by Theorem 2.3. Indeed, the $L(\sigma)$ are simplicial cones, so condition (1) of Theorem 2.3 is satisfied. As for condition (2), it also holds for any σ and any face $L(\tau) \subset L(\sigma)$. More precisely, if $L(\tau)$ has dimension 2 and lies on a 2-dimensional face Γ of $K(S)$, then the condition follows from our assumption

$$\mathrm{Lin}_{\mathbf{Z}}(S \cap \Gamma) = \mathbf{Z}^3 \cap \mathrm{Lin}_{\mathbf{R}}(\Gamma).$$

In other cases the condition follows because $S \cap L(\tau) = (S \cap L(\sigma)) \cap L(\tau)$. To finish the proof of Proposition 2.14 (b), it remains to note that the numbers $CN_T(\sigma)$ for all σ including $\sigma = \emptyset$ are non-negative by Theorem 2.6.

Corollary 2.15. *Let $A \subset \mathbf{Z}^{k-1}$ be a finite subset affinely generating $\mathbf{Z}^{k-1}$ over $\mathbf{Z}$. Suppose that $k-1 \leq 3$ and for any face $\Gamma \subset Q = \mathrm{Conv}(A)$ we have $i(\Gamma, A) = 1$. Then the regular A-determinant D_A is a polynomial.*

Proof. By (2.2), Δ_A always enters the factorization of D_A with exponent 1. As for other factors $\Delta_{A\cap\Gamma}$ corresponding to proper faces of Q, their exponents are the Newton numbers of the semigroups of the form S/Γ (Proposition 2.2), and so they are non-negative by Proposition 2.14.

3. The Newton polytope of the regular A-determinant and D-equivalence of triangulations

A. Vertices of the Newton polytope and D-equivalence

The regular A-determinant D_A is not always a polynomial (see Example 2.5). By Theorem 1.6 and Corollary 2.5, D_A is a polynomial provided the set $A \subset \mathbf{Z}^{k-1}$ satisfies the following two conditions:

(1) The polytope $Q = \mathrm{Conv}(A)$ is simple or has dimension ≤ 3.

(2) For every face $\Gamma \subseteq Q$ the index $i(\Gamma, A)$ (see Section 3A, Chapter 5) is equal to 1.

In this section we assume that A satisfies these conditions. In particular, they hold in the most important case when X_A is smooth. In this case D_A coincides with the A-discriminant Δ_A.

For $f = \sum_{\omega \in A} a_\omega x^\omega \in \mathbf{C}^A$ let us write $D_A(f)$ as the sum of monomials

$$D_A(f) = \sum_{\eta \in \mathbf{Z}_+^A} d_\eta \prod_{\omega \in A} a_\omega^{\eta(\omega)}. \tag{3.1}$$

The Newton polytope $N(D_A) \subset \mathbf{R}^A$ is the convex hull of the set of $\eta \in \mathbf{Z}_+^A$ for which $d_\eta \neq 0$. We are interested in describing this polytope.

Let $\Gamma \subseteq Q$ be a face, and let $\mathrm{Aff}_{\mathbf{R}}(\Gamma) \subset \mathbf{R}^{k-1}$ be its affine span. We introduce the volume form Vol_Γ on $\mathrm{Aff}_{\mathbf{R}}(\Gamma)$, induced by the lattice $\mathrm{Aff}_{\mathbf{Z}}(A \cap \Gamma)$.

Let T be a triangulation of (Q, A). According to Definition 2.10, we call a simplex σ of T of some dimension $j \geq 0$ *massive* if it lies in a j-dimensional face of Q. In that case the face is obviously unique; we shall denote it by $\Gamma(\sigma)$. In particular, all simplices of full dimension $k-1$ are massive, and for each of them $\Gamma(\sigma) = Q$.

For $\omega \in A$ we set $\eta_{T,j}(\omega) = \sum_\sigma \mathrm{Vol}_{\Gamma(\sigma)}(\sigma)$, where σ runs through all of the massive j-dimensional simplices in T for which ω is a vertex. We thus obtain functions $\eta_{T,j} : A \to \mathbf{Z}$; note that $\eta_{T,n-1}$ coincides with the function φ_T (Section 1D Chapter 7). Let

$$\eta_T = \sum_{j=0}^{k-1} (-1)^{k-1-j} \eta_{T,j} \in \mathbf{Z}^A. \tag{3.2}$$

Definition 3.1. We say that two triangulations T and T' of (Q, A) are *D-equivalent* if $\eta_T = \eta_{T'}$.

Theorem 3.2. *Under the above assumptions (1) and (2), we have*

(a) *The vertices of $N(D_A)$ are exactly the points η_T for all coherent triangulations T of (Q, A). Thus, they are in one-to-one correspondence with the D-equivalence classes of coherent triangulations of (Q, A).*

(b) *The monomial in D_A corresponding to (the class of) a coherent triangulation T has the form*

$$d_{\eta_T} \prod_{\omega \in A} a_\omega^{\eta_T(\omega)},$$

where

$$d_{\eta_T} = \pm \prod_{\sigma \in T-\text{massive}} \mathrm{Vol}_{\Gamma(\sigma)}(\sigma)^{(-1)^{k-1-\dim\sigma} \cdot \mathrm{Vol}_{\Gamma(\sigma)}(\sigma)}.$$

This theorem follows from the fact that $D_A = \prod_{\Gamma \subset Q} E_{A \cap \Gamma}^{(-1)^{\mathrm{codim}(\Gamma)}}$ (Theorem 1.3 above) and from Theorem 1.4 of Chapter 10.

Remarks 3.3. *(a)* Let T be a coherent triangulation of (Q, A). Theorem 3.2 implies, in particular, that the function $\eta_T : A \to \mathbf{Z}$ takes only non-negative values. Another non-trivial consequence is that the expression for d_{η_T} in part (b) is an integer. Furthermore, if T and T' are D-equivalent coherent triangulations then $d_{\eta_T} = d_{\eta_{T'}}$.

(b) Since the polynomial E_A is divisible by D_A, it follows that $N(E_A)$ is the Minkowski sum of $N(D_A)$ and some other polytope. Intuitively, $N(E_A)$ can be obtained from $N(D_A)$ by "cutting corners"; conversely, $N(D_A)$ can be obtained from $N(E_A)$ by a deformation merging together some vertices (see Figure 47).

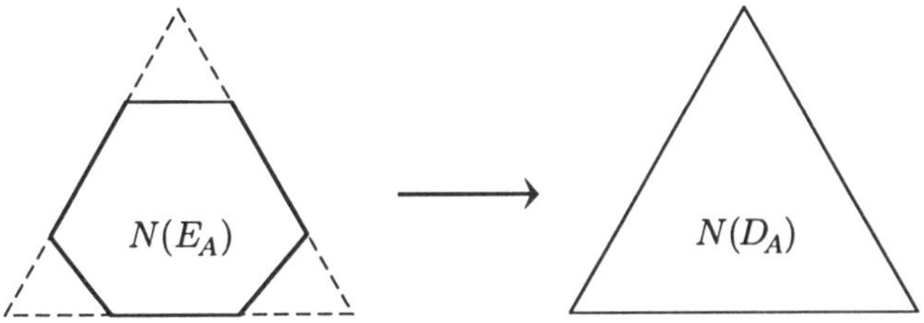

Figure 47.

The fact that D_A is a factor of E_A also implies some information about the face poset of $N(D_A)$. Recall that faces of $N(E_A)$ are in bijection with coherent subdivisions of (Q, A).

Theorem 3.4.

(a) *There is a monotone surjective map from the poset of coherent subdivisions of (Q, A) onto the poset of faces of $N(D_A)$.*

(b) *If $P = \{(Q_i, A_i)\}$ is a coherent subdivision of (Q, A) and $F \subset N(D_A)$ is the corresponding face then the coefficient restriction $D_A\|_F$ (see Section 1A Chapter 6) is equal to the product $\prod_i \prod_{\Sigma \subset Q_i} \Delta_{A_i \cap \Sigma}^{m_{i,\Sigma}}$ where Σ runs over all the faces of Q_i and the $m_{i,\Sigma}$ are some non-negative exponents.*

(c) *We have $m_{i,\Sigma} > 0$ whenever $A_i \cap \Sigma$ is not contained in a proper face of Q.*

Proof. (a) It follows from (1.4) that

$$E_A(f) = \prod_{\Gamma \subset Q} D_{A \cap \Gamma}(f). \tag{3.3}$$

By our assumptions, each $D_{A\cap\Gamma}$ is a polynomial. Thus E_A is divisible by D_A so the Newton polytope $N(D_A)$ is a Minkowski summand of $N(E_A)$. By Theorem 4.18 Chapter 5 the normal fan of $N(E_A)$ is a refinement of the normal fan of $N(D_A)$. However (Theorem 1.4 Chapter 10) the polytope $N(E_A)$ coincides with the secondary polytope $\Sigma(A)$ and its poset of faces is identified with the poset of coherent subdivisions of (Q, A) by Theorem 2.4 Chapter 7. This implies (a).

(b) Let $F(P)$ be the face of $N(E_A)$ corresponding to P. By Theorem 1.12′ Chapter 10, the coefficient restriction of E_A to F_P is the product of some positive powers of E_{A_i}. The prime factors of each E_{A_i} are the $\Delta_{A\cap\Sigma}$, $\Sigma \subset Q_i$ (Theorem 1.2 Chapter 10). Now expressing D_A by (3.3) we find the statement about the coefficient restriction of D_A as well.

(c) If $A_i \cap \Sigma$ is not contained in a proper face of Q, then $\Delta_{A_i \cap \Sigma}$ will occur in the factorization of $E_A \|_{F(P)}$ but not in the factorization of $E_{A \cap \Gamma} \|_{F(P)}$ for any proper face $\Gamma \subset Q$. This implies our statement.

B. The D-equivalence and modifications of triangulations

The notion of D-equivalence is rather implicit. In general, this is still an open problem to describe D-equivalence in combinatorial terms. Later in this chapter we shall reduce this problem to two classification problems: one from combinatorial geometry and the other from " the geometry of numbers." For small-dimensional polytopes, these latter problems can be solved explicitly thus providing a complete description of the D-equivalence.

Intuitively, the D-equivalent triangulations can differ only "near" the boundary of Q. To make such a statement more precise, we need a notion of integral distance between two subspaces.

Definition 3.5. Let W be a finite-dimensional affine space over $\mathbf{Q}$ equipped with an integral lattice Ξ, and let $U_1, U_2 \subset W$ be affine subspaces such that $\dim U_1 + \dim U_2 = \dim W - 1$. We call the *integral distance* between U_1 and U_2 and denote by $\rho(U_1, U_2)$ the volume $\mathrm{Vol}_\Xi(\sigma)$, where $\sigma \subset W \otimes \mathbf{R}$ is a simplex such that $\sigma \cap (U_1 \otimes \mathbf{R})$ and $\sigma \cap (U_2 \otimes \mathbf{R})$ are elementary simplices with vertices in $\Xi \cap U_1$ and $\Xi \cap U_2$, respectively (this number obviously does not depend on the choice of σ).

Now we can give some examples of D-equivalent triangulations. We shall use the notion of the modification of a triangulation along a circuit (see Section 2C Chapter 7).

Examples 3.6. *(a)* Let A be a subset of $\mathbf{Z}^2$ so $Q \subset \mathbf{R}^2$ is a plane polygon. Let Γ be an edge of Q. Suppose Γ contains at least three elements of A, say, a, b, c. Suppose also that there is a point $d \in A$ whose integral distance to the line $\mathrm{Aff}_{\mathbf{R}}(\Gamma)$ containing Γ is equal to 1 (this means that there are no lattice points inside the strip bounded by $\mathrm{Aff}_{\mathbf{R}}(\Gamma)$ and by the parallel line through d (see Figure 48).

Suppose that T is a triangulation of (Q, A) having the triangle acd as one of its simplices. Then T is supported on the circuit $Z = \{a, b, c\}$ (see Definition 2.9 Chapter 7). The modified triangulation $T' = s_Z(T)$ is obtained from T by subdividing the triangle acd into the two triangles abd and bcd. It follows at once from the definitions that T and T' are D-equivalent.

(b) More generally, let $A \subset \mathbf{Z}^{k-1}$ and let $\Gamma \subset Q$ be a facet. Suppose that there is a circuit $Z \subset A \cap \Gamma$ whose convex hull has the same dimension as Γ. Suppose also that there is a point $d \in A$ whose integral distance to $\mathrm{Aff}(\Gamma)$ equals 1. Since

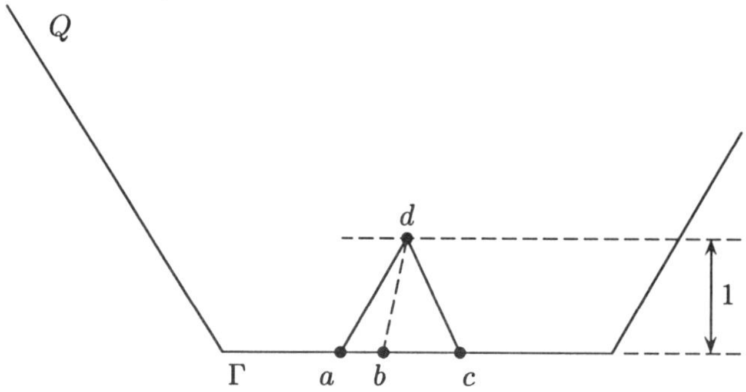

Figure 48. A D-equivalent modification of a triangulation of a plane polygon

the set $\{d\} \cup Z$ is weakly dependent, its convex hull has exactly two triangulations. Let T be a coherent triangulation of (Q, A) such that $\mathrm{Conv}(\{d\} \cup Z)$ is a union of simplices from T. Then the modification $T' = s_Z T$ is well-defined and T and T' are D-equivalent.

We let $\mathrm{Cl}(T)$ denote the D-equivalence class of a coherent triangulation T, and we let $\eta_{\mathrm{Cl}(T)}$ denote the corresponding vertex of $N(D_A)$.

Proposition 3.7. *The normal cone of $N(D_A)$ at $\eta_{\mathrm{Cl}(T)}$ is the union $\bigcup_{T' \in \mathrm{Cl}(T)} C(T')$ where $C(T')$ is the cone of Definition 1.4 Chapter 7.*

Proof. Since D_A divides E_A, the Newton polytope $N(D_A)$ is a Minkowski summand of $N(E_A)$. By Theorem 4.8 Chapter 5, the normal fan of $N(E_A)$ is a refinement of the normal fan of $N(D_A)$. The D-equivalence just describes this refinement, from which we obtain the statement.

Proposition 3.8.

(a) *Two vertices η and η' of $N(D_A)$ are joined by an edge if and only if there exist coherent triangulations T and T' of (Q, A) such that $\eta = \eta_T$, $\eta' = \eta_{T'}$, and $T' = s_Z(T)$ for some circuit Z.*

(b) *Any two D-equivalent coherent triangulations T and T' can be obtained from each other by a finite sequence of modifications such that all intermediate triangulations are coherent and belong to the same D-equivalence class.*

Proof. Both statements follow from Proposition 3.7 and from the description of the edges of $\Sigma(A) = N(E_A)$ (see Theorem 2.10 Chapter 7). Indeed, translating the cited theorem into the language of normal cones, we obtain that the cones $C(T)$ and $C(T')$ are separated by a "wall" of full dimension, if and only if $T' = s_Z(T)$ for some circuit Z. This implies both statements of our proposition.

In view of Proposition 3.8, to describe the D-equivalence relation, it is enough to describe which modifications lead to D-equivalent triangulations. We shall give a complete answer to this question for the case when $\dim Q \leq 3$. The proof of the following two propositions will be given in Section 4 below.

Proposition 3.9.

(a) *If Q has dimension 1 (i.e., is a segment), then no two different triangulations of (Q, A) are D-equivalent.*

(b) *Suppose Q has dimension 2 (i.e., is a plane polygon), and $Z \subset A$ is a circuit which modifies a triangulation T into a D-equivalent triangulation T'. Then Z consists of three points belonging to some side $\Gamma \subset Q$; furthermore, there is a unique separating subset $J \subset A - Z$, and J consists of a single point ω, having integral distance 1 from Γ. (In other words, we have the situation in Example 3.6 (a).)*

Proposition 3.9 can be applied, in particular, to the Newton polytope of the classical discriminant of a homogeneous polynomial of degree d in three variables. This case will be considered in more detail in Chapter 13.

Proposition 3.10. *Suppose* $\dim Q = 3$. *Then all of the situations when a modification along a circuit Z takes a coherent triangulation T to a D-equivalent triangulation are as follows.*

(a) *The circuit Z consists of four points and lies on a facet (= two-dimensional face) $\Gamma \subset Q$. The only separating subset consists of a single point $\omega \in A$ and the integral distance from ω to $\mathrm{Aff}(\Gamma)$ is equal to 1. In other words, we have the case of Example 3.6 (b).*

(b) *The circuit Z consists of three points and lies on an edge $R \subset Q$. Let Γ_1, Γ_2 be the facets of Q containing R. There is a unique separating subset $J = \{\omega_1, \omega_2\}$, and we have $\omega_1 \in \Gamma_1, \omega_2 \in \Gamma_2$. The integral distance between the affine spans of R and J is equal to 1.*

(c) *The circuit Z consists of three points and lies on a facet $\Gamma \subset Q$, but does not lie on an edge. There are two separating subsets: $\{\omega, \omega'\}$ and $\{\omega, \omega''\}$, where $\omega', \omega'' \in \Gamma$, $\omega \notin \Gamma$ and ω', ω'' lie on different sides of the line through Z. The integral distance from ω to $\mathrm{Aff}(\Gamma)$ is equal to 1.*

Figure 49 depicts various cases given by Proposition 3.10.

C. The Newton polytope of the A-discriminant

If the toric variety X_A is smooth then D_A coincides with the A-discriminant Δ_A so the previous analysis gives information about Δ_A. In general, the fact that Δ_A is a factor of E_A gives some information of the structure of the Newton polytope $N(\Delta_A)$. For example, the same argument as in Theorem 3.4 gives the following statement.

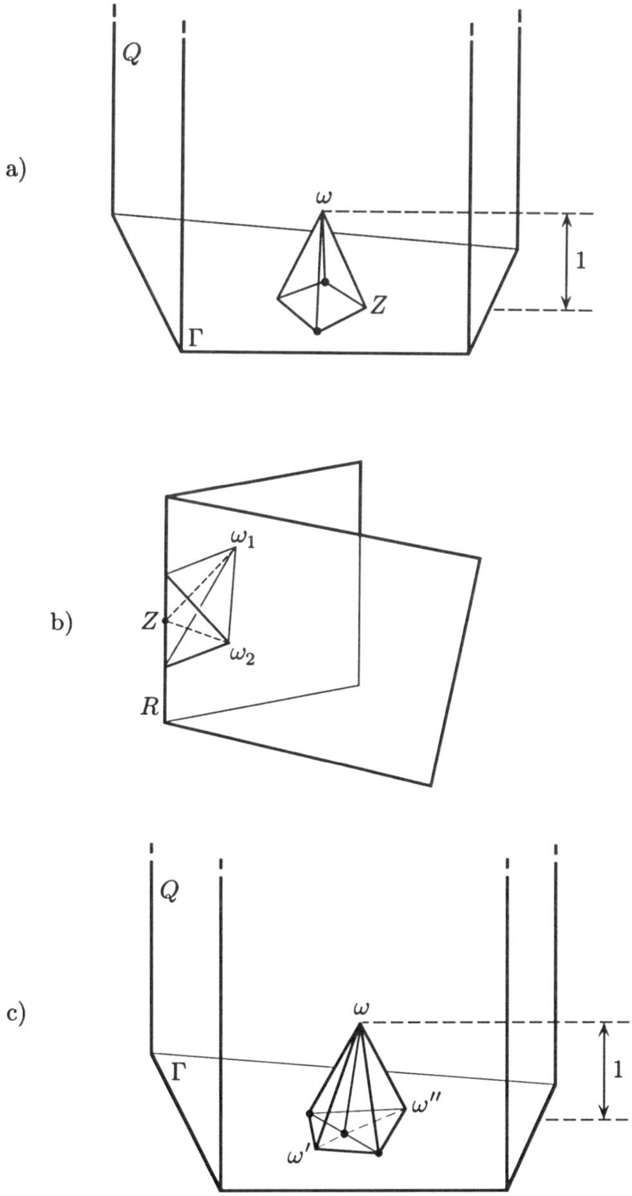

Figure 49. D-equivalent modifications of triangulations of 3-polytopes

Theorem 3.11. *All the statements of Theorem 3.4 remain valid if we replace D_A by Δ_A.*

Using Theorem 3.11, we can define a new relation on the set of coherent triangulations of (Q, A): "Δ-equivalence," which is stronger than the D-equivalence (when the latter is defined).

In order to describe the Δ-equivalence more explictly, we have to write explicitly the monomial in Δ_A corresponding to a coherent triangulation T. One way to do this is to express Δ_A in terms of the principal determinants $E_{A\cap\Gamma}$ for all faces $\Gamma \subseteq Q$. Namely, by Theorem 1.2 Chapter 10, each $E_{A\cap\Gamma}$ can be written as

$$E_{A\cap\Gamma}(f) = \pm \prod_{\Sigma\subset\Gamma} \Delta_{A\cap\Sigma}(f)^{m(\Sigma,\Gamma)}, \tag{3.4}$$

where Σ runs through the faces of Q contained in Γ, and the $m(\Sigma, \Gamma)$ are non-negative exponents. Under a suitable order of the faces of Q, these exponents form an upper-triangular matrix $M = \|m(\Sigma, \Gamma)\|$ with 1's on the diagonal. We then have

$$\Delta_A(f) = \pm \prod_{\Gamma\subset Q} E_{A\cap\Gamma}(f)^{(M^{-1})(\Gamma,Q)}, \tag{3.5}$$

where M^{-1} is the inverse matrix of M. This leads, in principle, to an explicit formula for the required monomial in Δ_A. The main difficulty of this approach is that we do not know an explicit formula for M^{-1}. Note that the analogs of (3.4) and (3.5) for D_A are the formulas (3.3) and (1.4), which are much simpler.

Let us finish the discussion of discriminants with an example.

Example 3.12. An elliptic curve in the Tate normal form. Consider the space $\mathbf{C}^A$ of polynomials

$$f(x, y) = a_{00} + a_{10}x + a_{20}x^2 + a_{30}x^3 + a_{01}y + a_{11}xy + a_{20}y^2. \tag{3.6}$$

So $A \subset \mathbf{Z}^2$ is the set of exponents of the monomials in (3.6). The polytope Q is a right triangle in $\mathbf{R}^2$ (see Figure 50). It easily follows from Proposition 2.2 that $D_A = a_{02}a_{30}\Delta_A$. Thus $N(D_A)$ is obtained from $N(\Delta_A)$ by a simple translation, and the Δ-equivalence in this case coincides with the D-equivalence. The polynomial Δ_A (i.e., the discriminant of the elliptic curve $\{f = 0\}$) can be

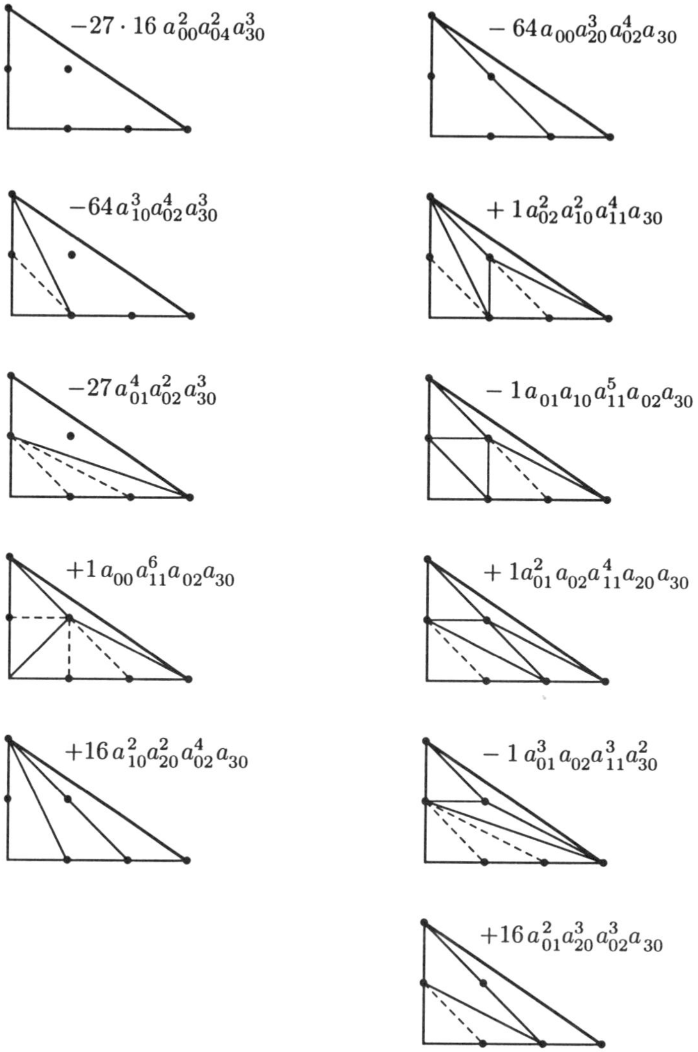

Figure 50.

found from formulas in [La] (Appendix 1). In expanded form it has 26 terms:

$$\Delta_A = -27\cdot 16a_{00}^2a_{02}^3a_{30}^2 - 64a_{00}a_{20}^3a_{02}^3 - 64a_{10}^3a_{02}^3a_{30} - 27a_{01}^4a_{02}a_{30}^2$$
$$+a_{00}a_{11}^6 + 16a_{10}^2a_{20}^2a_{02}^3 + 16a_{01}^2a_{20}^3a_{02}^2 + a_{02}a_{10}^2a_{11}^4 - a_{01}a_{10}a_{11}^5$$
$$+a_{01}^2a_{11}^4a_{20} - a_{01}^3a_{11}^3a_{30} + 32\cdot 9a_{00}a_{02}^3a_{10}a_{20}a_{30} + 48a_{00}a_{02}^2a_{11}^2a_{20}^2$$

$$+8\cdot 27a_{00}a_{01}^2a_{02}^2a_{30}^2 - 72a_{01}^2a_{02}^2a_{10}a_{20}a_{30} - 72a_{00}a_{02}^2a_{10}a_{11}^2a_{30}$$

$$-16a_{01}a_{02}^2a_{10}a_{11}a_{20}^2 - 8a_{02}^2a_{10}^2a_{11}^2a_{20} + 96a_{01}a_{02}^2a_{10}^2a_{11}a_{30}$$

$$-144a_{00}a_{01}a_{02}^2a_{11}a_{20}a_{30} - 12a_{00}a_{02}a_{11}^4a_{20} + 8a_{01}a_{02}a_{10}a_{11}^2a_{20}$$

$$-8a_{01}^2a_{02}a_{11}^2a_{20}^2 - 30a_{01}^2a_{02}a_{10}a_{11}^2a_{30} + 36a_{01}^3a_{02}a_{11}a_{20}a_{30}$$

$$+36a_{00}a_{01}a_{02}a_{11}^3a_{30}.$$

The first 11 terms in this formula correspond to the vertices of $N(\Delta_A)$. Figure 50 shows the 11 D-equivalence classes of triangulations of (Q, A) (all triangulations are easily seen to be coherent) and the terms which correspond to them by Theorem 3.2. The modifications inside a given D-equivalence class consist of taking away or adding back some of the dotted segments. Thus, $N(\Delta_A)$ has 11 vertices. Its dimension can be easily found to be 4 (cf. Theorem 1.7 Chapter 7).

4. More on D-equivalence

A. How does a modification affect η_T

In this section we keep assumptions (1) and (2) from Section 3A.

In view of Proposition 3.7, for a description of D-equivalence it is enough to describe which modifications preserve D-equivalence. By definition, two triangulations T, T' are D-equivalent if the functions $\eta_T, \eta_{T'} : A \to \mathbf{Z}$ coincide. It turns out that whenever T' is obtained from T by a modification, the differences $\eta_T(\omega) - \eta_{T'}(\omega)$ are subject to certain non-negativity conditions.

Theorem 4.1. *Let T be a coherent triangulation of (Q, A) which is supported on a circuit $Z \subset A$, and let $T' = s_Z(T)$ be the modification of T along Z. Let $T_+(Z) = T|_{\mathrm{Conv}(Z)}$ and $T_-(Z) = T'|_{\mathrm{Conv}(Z)}$ be the induced triangulations of* Conv(Z)*, and $Z_\pm = \{\omega \in Z : \mathrm{Conv}(Z - \omega) \in T_\pm(Z)\}$ be the corresponding subsets in Z. Then*

$$\eta_T(\omega) = \eta_{T'}(\omega) \quad \text{for} \quad \omega \notin Z,$$

$$\eta_T(\omega) \leq \eta_{T'}(\omega) \quad \text{for} \quad \omega \in Z_+,$$

$$\eta_T(\omega) \geq \eta_{T'}(\omega) \quad \text{for} \quad \omega \in Z_-.$$

For $\omega \notin Z$, this assertion follows immediately from (3.2). For $\omega \in Z_\pm$ we shall expand $\pm(\eta_T(\omega) - \eta_{T'}(\omega))$ as a sum of non-negative terms. This will be done using Newton functions (see Section 2C), and the following version of Newton numbers.

Let W be a real affine space equipped with a lattice Ξ. Let $\sigma \subset W$ be a simplex (of arbitrary dimension) with vertices in Ξ, and let τ be a face of σ (possibly the empty face). By the *relative Newton number* of (τ, σ) we mean

$$\nu_\Xi(\tau, \sigma) = \sum_{\gamma \in [\tau, \sigma]} (-1)^{\dim\sigma - \dim\gamma} \mathrm{Vol}(\gamma). \tag{4.1}$$

Here γ runs through all of the faces of σ containing τ (including σ and τ), and the volume of γ is taken with respect to the lattice $\mathrm{Aff}_{\mathbf{R}}(\gamma) \cap \Xi$ (we take the volume of the empty face to be 1). We shall adopt the abbreviation $\nu_\Xi(\sigma)$ for $\nu_\Xi(\emptyset, \sigma)$. We let σ/τ denote the simplex in the quotient space $W/\mathrm{Aff}_{\mathbf{R}}(\tau)$ whose vertices are the images of the vertices of σ. Let $\Xi/\mathrm{Aff}_{\mathbf{R}}(\tau) \subset W/\mathrm{Aff}_{\mathbf{R}}(\tau)$ be the image of the lattice Ξ.

Lemma 4.2.
(a) *We have* $\nu_\Xi(\tau, \sigma) = \mathrm{Vol}(\tau) \cdot \nu_{\Xi/\mathrm{Aff}_{\mathbf{R}}(\tau)}(\sigma/\tau)$.
(b) *The number* $\nu_\Xi(\tau, \sigma)$ *is always non-negative.*

Proof. Part (a) is obvious, since for every γ in (4.1) we have $\mathrm{Vol}(\gamma) = \mathrm{Vol}(\tau) \cdot \mathrm{Vol}(\gamma/\tau)$.

Using (a), we reduce the proof of (b) to the special case of $\nu_\Xi(\sigma) = \nu_\Xi(\emptyset, \sigma)$. To show that $\nu_\Xi(\sigma) \geq 0$, we remark that $\nu_\Xi(\sigma)$ is a particular case of Newton numbers (see Section 2A). Indeed, let us realize W as an affine hyperplane in a vector space V such that $0 \notin W$. Let $\tilde{\Xi}$ be the Abelian subgroup in V generated by Ξ, and let $K \subset V$ be the convex cone with apex 0 generated by σ. Let $S = \tilde{\Xi} \cap K$. Then $\nu_\Xi(\sigma)$ is easily seen to coincide with the Newton number $\nu(S)$ of the semigroup S. Since K is a simplicial cone, $\nu(S) \geq 0$ by Theorem 2.3 (note that condition (2) of that theorem is satisfied by the definition of S). Lemma 4.2 is proved.

Returning to Theorem 4.1, let CN_T be the Newton function of the triangulation T (see Section 2C). We shall say that a subset $J \subset A - Z$ is *subseparating* for T and T' if for some $\omega \in Z$ the set $(Z - \{\omega\}) \cup J$ is the set of vertices of a simplex of T; the only difference with separating subsets (Definition 2.11 Chapter 7) is that the simplex can now be of arbitrary (not necessarily maximal) dimension. It is easy to see that for $\omega \in Z$ and a subseparating subset $J \subset A - Z$, the number $CN_T(\mathrm{Conv}((Z - \{\omega\}) \cup J))$ does not depend on ω. We denote this number simply as $CN_{T,Z}(J)$.

We deduce Theorem 4.1 from the following more precise fact.

Proposition 4.3. *In the case of Theorem 4.1, for* $\omega \in Z_\pm$ *we have*

$$\pm(\eta_{T'}\omega) - \eta_T(\omega))$$
$$= \sum_{J-\text{ subsep.}} CN_{T,Z}(J) \cdot \nu_\Xi(\mathrm{Conv}(Z - \{\omega\}), \mathrm{Conv}((Z - \{\omega\}) \cup J)), \tag{4.2}$$

where the summation is taken over all subseparating subsets for T and T' (including $\emptyset$).

Proof. Let T_+ and T_- be the triangulations of $\mathrm{Conv}(Z)$ induced by T and T' respectively. Note that for every $\omega \in Z_+$ we have

$$\sum_{\substack{\tau \in T_- \\ \omega \in \mathrm{Vert}(\tau)}} \mathrm{Vol}(\tau) - \sum_{\substack{\tau \in T_+ \\ \omega \in \mathrm{Vert}(\tau)}} \mathrm{Vol}(\tau) = \mathrm{Vol}(\mathrm{Conv}(Z - \{\omega\})), \tag{4.3}$$

where the sums are taken over maximal simplices τ of $T_\pm$ having ω as a vertex. Recall also that by definition

$$\eta_T(\omega) = \sum_{\substack{\text{massive } \sigma \in T \\ \omega \in \mathrm{Vert}(\sigma)}} (-1)^{\mathrm{codim}\,\sigma} \mathrm{Vol}(\sigma)$$

and similarly for $\eta_{T'}(\omega)$. Since T' is obtained from T by modification along Z, both T and T' are supported on Z (Definition 2.9 Chapter 7). Thus simplices of T having ω as a vertex are $\mathrm{Conv}(\tau \cup J')$, where τ is a maximal simplex in T_+ and J' is subseparating. Similarly for simplices of T'. Thus, by (4.3), we can write

$$\begin{aligned}
&\eta_{T'}(\omega) - \eta_T(\omega) \\
&\quad = \sum_{\substack{\tau \in T_-, J' \text{ subsep.:} \\ \mathrm{Conv}(\tau \cup J') \text{ massive} \\ \omega \in \mathrm{Vert}(\tau)}} (-1)^{\mathrm{codim}\,\mathrm{Conv}(\tau \cup J')} \mathrm{Vol}\,\mathrm{Conv}(\tau \cup J') \\
&\qquad - \sum_{\substack{\tau \in T_+, J' \text{ subsep.:} \\ \mathrm{Conv}(\tau \cup J') \text{ massive} \\ \omega \in \mathrm{Vert}(\tau)}} (-1)^{\mathrm{codim}\,\mathrm{Conv}(\tau \cup J')} \mathrm{Vol}\,\mathrm{Conv}(\tau \cup J') \\
&\quad = \mathrm{Vol}(\mathrm{Conv}(Z - \{\omega\})) \\
&\qquad \times \sum_{\substack{\text{subsep.} J' : \\ \mathrm{Conv}(Z \cup J) \text{ massive}}} (-1)^{\mathrm{codim}\,\mathrm{Conv}(Z \cup J')} \rho(\mathrm{Aff}_{\mathbf{Q}}(J'), \mathrm{Aff}_{\mathbf{Q}}(Z)))
\end{aligned}$$

where ρ means the integral distance and the condition "$\mathrm{Conv}(Z \cup J)$ massive" is an abbreviation for "$\mathrm{Conv}(\sigma \cup J)$ massive for any maximal simplex $\sigma \in T_+$ or T_-" (if this holds for one such simplex it holds for any other). By Proposition

2.13, the last sum is equal to

$$\mathrm{Vol}(\mathrm{Conv}(Z - \{\omega\})) \times \sum_{\text{subsep. } J'} \sum_{\text{subsep. } J \supset J'} CN_{T,Z}(J)(-1)^{\#(J)-\#(J')}\rho(\mathrm{Aff}_{\mathbf{Q}}(J'), \mathrm{Aff}_{\mathbf{Q}}(Z))$$

$$= \sum_{\text{subsep. } J} CN_{T,Z}(J) \sum_{J' \subset J} (-1)^{\#(J)-\#(J')} \Big\{ \mathrm{Vol}(\mathrm{Conv}(Z - \{\omega\})) \times \rho(\mathrm{Aff}_{\mathbf{Q}}(J'), \mathrm{Aff}_{\mathbf{Q}}(Z)) \Big\}$$

$$= \sum_{\text{subsep. } J} CN_{T,Z}(J) \cdot \nu_{\Xi}(\mathrm{Conv}(Z - \{\omega\}), \mathrm{Conv}((Z - \{\omega\}) \cup J)),$$

as claimed. For $\omega \in Z_-$ the argument is completely similar.

To conclude the proof of Theorem 4.1, it remains to note that each summand in (4.2) is the product of two non-negative terms: the value of the Newton function (non-negative by Theorem 2.15) and the relative Newton number (non-negative by Lemma 4.2).

Proposition 4.3 implies a criterion for two triangulations T and T' obtained from each other by modification along a circuit Z, to be D-equivalent. To formulate it, for every subseparating subset $J \subset A - Z$, we let $\overline{J}$ denote the image of J in the quotient space $\mathbf{R}^{k-1}/\mathrm{Aff}(Z)$.

Corollary 4.4. *In the case of Theorem 4.1, T and T' are D-equivalent if and only if for every subseparating subset $J \subset A - Z$ at least one of the numbers $CN_{T,Z}(J)$ and $\nu_{\mathbf{Z}^{k-1}/\mathrm{Aff}_{\mathbf{Z}}(Z)}(\mathrm{Conv}(\overline{J}))$ is equal to* 0.

Corollary 4.5. *Suppose that T, T' are D-equivalent coherent triangulations of (Q, A) obtained from each other by modification along a circuit $Z \subset A$. Then Z lies on some proper face of Q.*

Proof. Corollary 4.4 implies, in particular, that $CN_{T,Z}(\emptyset) = 0$. Let p be the codimension of $\mathrm{Conv}(Z)$ in Q. Assume that Z does not lie in a proper face. Taking a transversal slice $\mathbf{R}^p$ to $\mathrm{Conv}(Z)$, we derive from T a decomposition of this transversal slice into simplicial cones, i.e., a triangulation Σ of the whole $\mathbf{R}^p$ (here we use our assumption: if Z lies in a proper face we get a triangulation of some cone in $\mathbf{R}^p$). Note that $CN_{T,Z}(\emptyset) = CN_{\Sigma}(\mathbf{R}^p)$. Since $\mathbf{R}^p$ has no proper faces, $CN_{\Sigma}(\mathbf{R}^p) = CV_{\Sigma}(\mathbf{R}^p) > 0$.

Intuitively, Corollary 4.5 means that a modification which leads to a D-equivalent triangulation must take place near the boundary of Q.

B. Thin lattice simplices

We shall study both of the possibilities in Corollary 4.4.

Definition 4.6. Let W be a real affine space equipped with an integral lattice Ξ. We say that a simplex $\sigma \subset W$ with vertices in Ξ is *thin* (with respect to Ξ) if $\nu_\Xi(\sigma) = 0$.

Examples 4.7. *(a)* If σ has a vertex ω of height 1 (i.e., the integral distance from ω to the opposite face is equal to 1), then σ is thin. Indeed, for every face $\tau \subset \sigma$ not containing ω, the face $\mathrm{Conv}(\tau \cup \{\omega\})$ has the same volume as τ under our normalization. So the volumes of these two faces cancel out in (4.1), and the sum in (4.1) is equal to 0.

(b) The triangle in $\mathbf{R}^2$ with vertices (2, 0), (0, 2), (0, 0) is thin with respect to $\mathbf{Z}^2$. Indeed, in this case (4.1) becomes $4 - 3 \cdot 2 + 3 \cdot 1 - 1 = 0$.

(c) A segment joining two points from Ξ is thin if and only if it has length 1 with respect to Ξ, i.e., it does not contain other lattice points.

In general, a classification of thin lattice simplices seems to be an interesting problem in the "geometry of numbers."

The following statement explains the term "thin simplex."

Proposition 4.8. *If $\sigma \subset \mathbf{R}^m$ is a thin simplex with respect to $\mathbf{Z}^m$, then there are no integral points in the interior of σ.*

Proof. Suppose there is such a point p. Let $k = \dim \sigma$. We divide σ into $(k+1)$ simplices of dimension k by forming the cones from p over the facets of σ. Let T denote the resulting triangulation of σ, and let CN_T denote the Newton function of T. It easily follows from Proposition 2.13 that

$$\nu(\sigma) = \sum_{\gamma \in T} CN_T(\gamma)\nu(\gamma),$$

where γ runs through the simplices of arbitrary dimension in T. All terms in this sum are non-negative, and we have $CN_T(\{p\}) = k + 1$, $\nu(\{p\}) = 1$ and so $\nu(\sigma) > 0$, a contradiction.

Proposition 4.9. *Up to integral affine equivalence, all thin triangles (i.e., thin two-dimensional simplices) are exhausted by the triangles in Example 4.7 (a), (b).*

Proof. Let $\sigma \subset \mathbf{R}^2$ be a thin triangle with vertices in $\mathbf{Z}^2$. If σ has a vertex of height 1, then we are in the situation of Example 4.7 (a), and there is nothing to prove. So we assume that all heights of σ are ≥ 2. We want to prove that such a thin simplex is essentially unique, i.e., coincides with the triangle in Example 4.7 (b) up to integral affine equivalence. This will be done in several steps.

First, we claim that all sides of σ have length (with respect to $\mathbf{Z}^2$) ≤ 2. Suppose this is not true, i.e., σ has a side of length $l \geq 3$. Let h be the corresponding height; by assumption, $h \geq 2$. Under our normalization, the area of σ is lh. It follows from the "strange" formula (2.7) that each of the other two sides has length $\leq h$. Hence, $\nu(\sigma) \geq lh - 2h - l + 2 = (l-2)(h-1) \geq 1$, a contradiction.

Now let us show that all sides of σ have length 2. Suppose the lengths are a, b, c with $a \leq b \leq c \leq 2$. Since σ is thin, its area is equal to $a+b+c-2$. On the other hand, since we assumed the heights of σ to be ≥ 2, the area is $\geq 2c$. We obtain the inequality $a+b+c-2 \geq 2c$, which can be rewritten as $a+b \geq c+2$. Obviously, this implies $a = b = c = 2$, as required.

Finally, we see that σ has area 4. If we join the midpoints of the sides of σ, we obtain a partition of σ into 4 integral triangles, each of which must have area 1. Hence, σ is equivalent to the triangle in Example 4.7 (b). This completes the proof of Proposition 4.9.

C. Thin triangulations

The second possibility for a subseparating set J given in Corollary 4.4 is the vanishing of $CN_{T,Z}(J)$, which is equal to $CN_T(\mathrm{Conv}((Z - \{\omega\}) \cup J))$ for any $\omega \in Z$. According to Lemma 2.16, this number is a "combinatorial analog" of the Newton number. More precisely, suppose that $\Gamma(Z \cup J)$ is the smallest face of Q containing $Z \cup J$, $p = \dim\Gamma(Z \cup J) - \dim\mathrm{Conv}(Z \cup J)$, and $q = \dim Q - \dim\Gamma(Z \cup J)$. Let $K(Z, J)$ be the cone in $\mathrm{Aff}_{\mathbf{R}}(Q)/\mathrm{Aff}_{\mathbf{R}}(Z \cup J)$ whose apex is the zero of this vector space, and the base is the image of Q. Since Q is simple, $K(Z, J)$ is isomorphic to $\mathbf{R}^p \times \mathbf{R}^q_+$. Let $\Sigma(T, Z, J)$ be the triangulation of $K(Z, J)$, induced by T. Lemma 2.15 says that $CN_{T,Z}(J)$ is equal to the combinatorial Newton number $CN_{\Sigma(T,Z,J)}(K(Z, J))$.

Lemma 4.10. *The j-dimensional cones in $\Sigma(T, Z, J)$ are in one-to-one correspondence with the subseparating subsets $J' \supset J$ for which $\#(J' - J) = j$. In particular, the cones of maximal dimension correspond to the separating subsets containing J.*

The proof follows from the definition of separating subsets (Definition 2.11 Chapter 7).

Definition 4.11. Let K be a cone isomorphic to $\mathbf{R}^p \times \mathbf{R}^q_+$. A coherent rational triangulation Σ of K is *thin* if $CN_\Sigma(K) = 0$.

Examples 4.12. *(a)* Let Σ be an arbitrary coherent rational triangulation of $\mathbf{R}^p \times \mathbf{R}^{q-1}_+$. We consider the new triangulation $\mathrm{Cone}(\Sigma)$ of $\mathbf{R}^p \times \mathbf{R}^q_+ =$

$(\mathbf{R}^p \times \mathbf{R}_+^{q-1}) \times \mathbf{R}_+$, which consists of the cones of the form $\Theta \times \{0\}$ and $\Theta \times \mathbf{R}_+$ for all $\Theta \in \Sigma$. We call this triangulation the cone over Σ. The argument similar to that in Example 4.7 (a) (the cones $\Theta \times \{0\}$ and $\Theta \times \mathbf{R}_+$ will give cancelling contributions) shows that Cone (Σ) is always thin.

(b) Let T be a triangulation of a plane triangle P into triangles all of whose vertices lie on the boundary of P. Let K be the cone over P. Let Σ be the triangulation of K corresponding to T. Then Σ is thin. Indeed, let B be the set of vertices of T, so T is a triangulation of (P, B). It is verified immediately that if T' is another triangulation of (P, B) obtained from T by a modification and Σ' is the corresponding triangulation of K then $CN_\Sigma(K) = CN_{\Sigma'}(K)$. By a chain of modifications we can obtain from T any other coherent triangulation of (P, B), in particular the triangulation consisting of just one triangle P, for which the statement is obvious.

(c) The triangulation of $\mathbf{R} \times \mathbf{R}_+^2 = \{(x, y, z) \in \mathbf{R}^3 : x \geq 0, y \geq 0\}$ whose slice by the plane $x + y = 1$ is the triangulation of the strip shown on Figure 51, is thin. Infinite regions τ_+ and τ_- are cut out at the left and right. The polygonal line $A_1B_1 \ldots A_rB_r$ contains at least two vertices, and it may begin and end on either of the lines bounding the strip. Each of the triangles bordered by this line can be further subdivided, as indicated by the dotted lines.

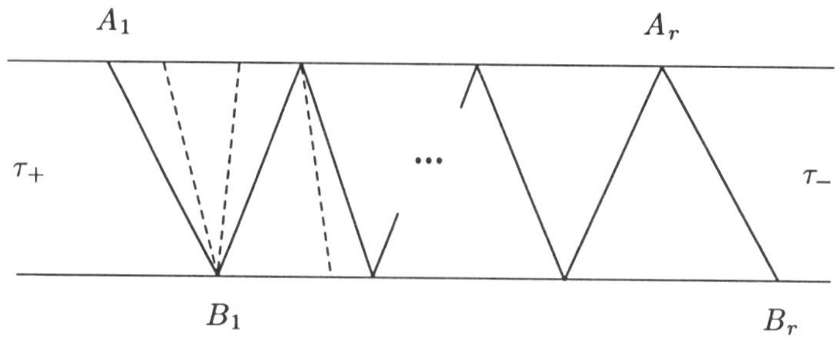

Figure 51.

It is clear that a triangulation of $\mathbf{R}_+^2$ is thin if and only if it consists of just $\mathbf{R}_+^2$. The classification of thin triangulations of $\mathbf{R}^p \times \mathbf{R}_+^q$ (when $p = 0$ we essentially have triangulations of a $(q-1)$-dimensional simplex) seems to be an interesting problem in combinatorial geometry.

Similar to the case of lattice simplices, the term "thin triangulation" is explained by the following statement.

Proposition 4.13. *Let Σ be a thin triangulation of $\mathbf{R}^p \times \mathbf{R}_+^q$. Then every ray*

$\Theta \in \Sigma$ (i.e., a one-dimensional cone from Σ) projects to the boundary of $\mathbf{R}_+^q$ under the natural projection onto $\mathbf{R}_+^q$.

Proof. Let $\pi : X(\Sigma) \to \mathbf{C}^q$ be the "desingularization of the affine space" corresponding to Σ (see the proof of Theorem 2.8 above). A ray Θ defines the orbit closure $X(\Theta)$ in $X(\Sigma)$ which is a hypersurface. If the projection of Θ does not lie on the boundary of $\mathbf{R}_+^q$, then $\pi(X(\Theta)) = \{0\}$. Since the fiber $\pi^{-1}(y)$ of a point $y \in \mathbf{C}^q - \{0\}$ has dimension less than $p+q-1$, the shifted sheaf $\underline{\mathbf{C}}_{\{0\}}[2p+2q-2]$ is a direct summand of the complex $R\pi_*\underline{\mathbf{C}}_{X(\Sigma)}$ in the derived category. This contradicts Corollary 2.9.

Proposition 4.14.

(a) *Every thin triangulation of $\mathbf{R}^p \times \mathbf{R}_+$ is linearly equivalent to the cone over some triangulation of $\mathbf{R}^p$.*

(b) *Every thin triangulation of $\mathbf{R} \times \mathbf{R}_+^2$ is combinatorially equivalent to a triangulation in Example 4.12 (c).*

(c) *Every thin triangulation of $\mathbf{R}_+^3$ is combinatorially equivalent to a triangulation in Example 4.12 (b).*

Proof. (a) The combinatorial Newton number is equal to the number of $(p+1)$-dimensional cones into which $\mathbf{R}^p \times \mathbf{R}_+$ is divided, minus the number of p-dimensional cones into which $\mathbf{R}^p \times \{0\}$ is divided. In order to obtain 0, each $(p+1)$-dimensional cone must be supported on $\mathbf{R}^p \times \{0\}$ by a p-dimensional face. This implies part (a).

Parts (b) and (c) follow from the fact that there are no interior rays (Proposition 4.13).

D. Proofs of Propositions 3.9 and 3.10

Now we are in a position to prove Propositions 3.9 and 3.10 which give a classification of modifications preserving the D-equivalence in dimensions ≤ 3. In the proofs below we make use of the following fact which is a direct consequence of Definition 3.5. If σ is a lattice simplex whose set of vertices is decomposed into a disjoint union $I \cup J$ and $\tau = \mathrm{Conv}(I)$, $\delta = \mathrm{Conv}(J)$ then

$$\mathrm{Vol}_\sigma(\sigma) = \mathrm{Vol}_\tau(\tau) \cdot \mathrm{Vol}_\delta(\delta) \cdot \rho, \tag{4.4}$$

where ρ is the integral distance between $\mathrm{Aff}_{\mathbf{Q}}(I)$ and $\mathrm{Aff}_{\mathbf{Q}}(J)$.

Proof of Proposition 3.9. If $\dim(Q) = 1$, i.e., Q is a segment, then E_A differs from D_A by multiplication with a monomial so the corresponding Newton polytopes are linearly isomorphic. Hence no two different triangulations are D-equivalent.

Now suppose that $\dim(Q) = 2$, i.e., Q is a plane polygon. Let T, T' be D-equivalent coherent triangulations of (Q, A) which are obtained from each other by modification along a circuit Z. First, by Corollary 4.5, Z should consist of 3 points, say $Z = \{a, b, c\}$, and lie on some side $\Gamma \subset Q$. We can assume that b lies inside the segment $[a, c]$. In this situation there is just one separating subset: it consists of a single point $\omega \in A$ lying inside Q. Comparing $\eta_T(b)$ and $\eta_{T'}(b)$, we find (by using (4.4) with $I = \{\omega\}$) that the integral distance from ω to Γ is 1. This proves Proposition 3.9.

Proof of Proposition 3.10. Now we consider the case $\dim(Q) = 3$. Let T, T', Z have the same meaning as above. Corollary 4.5 implies that Z lies on some proper face Γ of Q. Hence, Z consists of ≤ 4 points.

Suppose that $\#(Z) = 4$ so $\mathrm{Conv}(Z)$ is a polygon with 3 or 4 vertices lying on some 2-face $\Gamma \subset Q$. Then there is only one separating subset, and it consists of a single point $\omega \in A$ lying inside Q. Again a simple computation (by using (4.4) with $I = \{\omega\}$) shows that the integral distance from ω to Γ should be equal to 1, so we have the situation of Proposition 3.10 (a).

Suppose that $\#(Z) = 3$. There are two cases to consider: the minimal face $\Gamma \subset Q$ containing Z can be a 2-face or an edge. First, suppose Γ is a 2-face. By Corollary 4.4 and Lemma 2.15, the triangulation $\Sigma(T, Z, \emptyset)$ of the cone $K(Z, \emptyset) \simeq \mathbf{R} \times \mathbf{R}_+$ must be thin. So by Proposition 4.14 this triangulation consists of two cones, i.e., the separating subsets (which correspond to the maximal cones of the triangulation) are as in Proposition 3.10 (c). Let z be the middle point of Z. Then

$$\eta_T(z) - \eta_{T'}(z) = \mathrm{Vol}_Q\mathrm{Conv}(Z \cup \{\omega, \omega', \omega''\}) - \mathrm{Vol}_\Gamma\mathrm{Conv}(Z \cup \{\omega', \omega''\}).$$

Since $\mathrm{Conv}(Z \cup \{\omega, \omega', \omega''\})$ is a pyramid with base $\mathrm{Conv}(Z \cup \{\omega', \omega''\})$, it follows that the difference of volumes vanishes if and only if the height of the pyramid (i.e., the integral distance from ω to Γ) is equal to 1.

Now consider the last case when $\#(Z) = 3$ and Z lies on an edge $R \subset Q$. Then the cone $K(Z, \emptyset)$ is $\mathbf{R}^2_+$, and its induced triangulation $\Sigma(T, Z, \emptyset)$ must be thin. Consequently, this triangulation consists only of $\mathbf{R}^2_+$ itself, i.e., the separating subsets are as in Proposition 3.10 (b). The statement concerning the integral distance is proved in the same way as before (by using (4.4) with J as in Proposition 3.10 (b)).

Proposition 3.10 is proved.

5. Relations to real algebraic geometry

Results of previous sections provide some information about Newton polytopes of the A-discriminant Δ_A and related polynomials E_A and D_A. As we have seen (Section 1A Chapter 6), the structure of the Newton polytope of any (Laurent) polynomial is closely related to the geometry of the hypersurface defined by this polynomial. In particular, the Newton polytope of Δ_A sheds some light on the geometry of the discriminantal hypersurface ∇_A. In this section we shall be interested in the study of ∇_A from the point of view of real algebraic geometry.

A. Hilbert's sixteenth problem

The original problem posed by Hilbert, is as follows:

> *Investigate possible types of topological behavior of a non-singular real algebraic curve in* $\mathbf{R}P^2$ *of a given degree* d.

Let us make the terminology more precise. By a real plane curve we mean an algebraic curve $Z \subset \mathbf{C}P^2$ given by a homogeneous equation $f(x_0, x_1, x_2) = 0$ with real coefficients. For such a curve Z by $Z(\mathbf{R})$ we denote the set $Z \cap \mathbf{R}P^2$ of real points of Z. (It may happen that $Z(\mathbf{R}) = \emptyset$.) A curve is non-singular if it has no singular points in $\mathbf{C}P^2$. For such a curve Z, the set $Z(\mathbf{R})$ is a smooth one-dimensional real submanifold in $\mathbf{R}P^2$; so topologically $Z(\mathbf{R})$ is a disjoint union of circles ("ovals"). These ovals are situated in some way in $\mathbf{R}P^2$; for instance, some of them divide $\mathbf{R}P^2$ into two parts and some do not; some may lie inside others etc. The problem is to describe all possible patterns of behavior of ovals up to isotopy, i.e., continuous deformation of the ambient $\mathbf{R}P^2$.

This problem and its higher-dimensional generalizations have been the subject of many works, see surveys [Gud], [Wi]. The complete classification is by now known for curves up to degree 7. We cannot give here even a partial overview of main developments since this would lead us too far away. We shall concentrate only on one aspect of the problem related to the kind of questions studied elsewhere in this book.

Fix a natural number d. Let A be the set of monomials in x_0, x_1, x_2 of degree d, and $\mathbf{R}^A$ (resp. $\mathbf{C}^A$) the space of real (resp. complex) homogeneous polynomials of degree d. A real plane curve of degree d is just a curve with the equation $f = 0$ for $f \in \mathbf{R}^A$. Such a curve is non-singular if and only if the A-discriminant $\Delta_A(f)$ does not vanish. Let $\nabla_A \subset \mathbf{C}^A$ be the A-discriminantal hypersurface, i.e., the hypersurface $\{\Delta_A(f) = 0\}$, and $\nabla_A(\mathbf{R}) \subset \mathbf{R}^A$ the set of real points of ∇_A. The hypersurface $\nabla_A(\mathbf{R})$ divides $\mathbf{R}^A$ into several components. If f, g are two polynomials lying in the same connected component of the complement $\mathbf{R}^A - \nabla_A(\mathbf{R})$ then the curves $\{f = 0\}$ and $\{g = 0\}$ in $\mathbf{R}P^2$ are isotopic. Thus the classification problem of topological types of curves $\{f = 0\}$, $f \in \mathbf{R}^A - \nabla_A(\mathbf{R})$ can be reduced to two steps. First, we have to describe all the connected components of $\mathbf{R}^A - \nabla_A(\mathbf{R})$ and second, describe the topological behavior of the curve given by a polynomial from a given component.

It is convenient to consider a more general framework of hypersurfaces in toric varieties, not only in P^2. Let $A \subset \mathbf{Z}^{k-1}$ be a finite set of lattice points which we regard, as usual, as Laurent monomials in $k-1$ variables. Let X_A be the corresponding projective toric variety (Section 1B, Chapter 5). This variety is defined over $\mathbf{Q}$ and hence over $\mathbf{R}$ so we can consider the set of its real points $X_A(\mathbf{R})$. Let $\mathbf{R}^A$ (resp. $\mathbf{C}^A$) be the space of real (resp. complex) polynomials all of whose monomials are from A. Any $f \in \mathbf{C}^A$ defines a hypersurface $Z_f \subset X_A$. We are interested in the case when $f \in \mathbf{R}^A$ and we consider the set of real points $Z_f(\mathbf{R}) \subset X_A(\mathbf{R})$.

Let $\nabla_A \subset \mathbf{C}^A$ be the A-discriminantal variety (Section 1A, Chapter 9). If X_A is smooth then Z_f is non-singular if and only if $f \notin \nabla_A$. As in the particular case considered above, if $f, g \in \mathbf{R}^A$ lie in the same connected component of $\mathbf{R}^A - \nabla_A(\mathbf{R})$ then the hypersurfaces $Z_f(\mathbf{R})$ and $Z_g(\mathbf{R})$ are isotopic.

The natural "Hilbert problem" associated with A is to study the possible types, up to isotopy, of topological behavior of hypersurfaces $Z_f(\mathbf{R})$ for $f \in \mathbf{R}^A - \nabla_A$.

There is also a second setup for the problem. Namely, let $Q \subset \mathbf{R}^{k-1}$ be the convex hull of A. For any face $\Gamma \subset Q$, let $p_\Gamma : \mathbf{R}^A \to \mathbf{R}^{A\cap\Gamma}$ be the natural projection (the coefficient restriction of polynomials). Consider the open subset

$$\mathbf{R}^A_{\text{gen}} = \mathbf{R}^A - \bigcup_{\Gamma\subseteq Q} p_\Gamma^{-1}(\nabla_{A\cap\Gamma}). \tag{5.1}$$

We call $\mathbf{R}^A_{\text{gen}}$ the *generic stratum* in $\mathbf{R}^A$. The complex generic stratum $\mathbf{C}^A_{\text{gen}} \subset \mathbf{C}^A$ is defined similarly. The "refined Hilbert problem" for A is to study the components of $\mathbf{R}^A_{\text{gen}}$.

If X_A is smooth then the condition $f \in \mathbf{C}^A_{\text{gen}}$ means that not only the hypersurface Z_f is smooth but so are its intersections with all the orbit closures $X(\Gamma) \subset X_A$. Thus the study of connected components of $\mathbf{R}^A_{\text{gen}}$ amounts to the study of possible types of generic topological behavior of Z_f with respect to all $X(\Gamma)$.

Examples 5.1. *(a)* Let $A = \{1, x, x^2, \ldots, x^d\}$ so $\mathbf{R}^A$ consists of polynomials of degree $\leq d$ in one variable x. Two polynomials $f, g \in \mathbf{R}^A - \nabla_A$ lie in the same connected component if and only if they have the same number of real roots (including infinity; by our conventions, f has a root of multiplicity i at infinity if $\deg(f) = d-i$, so for $f \notin \nabla_A$ we have $\deg(f) \geq d-1$). In Figure 52 we consider the case $d = 4$ and we have depicted the section of the hypersurface $\nabla_A(\mathbf{R})$ by the 3-dimensional affine space $\{a_0 + \cdots + a_4x^4 : a_3 = 0, a_4 = 1\}$.

(b) Let A consist of bilinear monomials x_iy_j where the x_i and y_j ($i = 1, \ldots, m$; $j = 1, \ldots, n$) are two sets of variables. (see Example 1.1 (b)

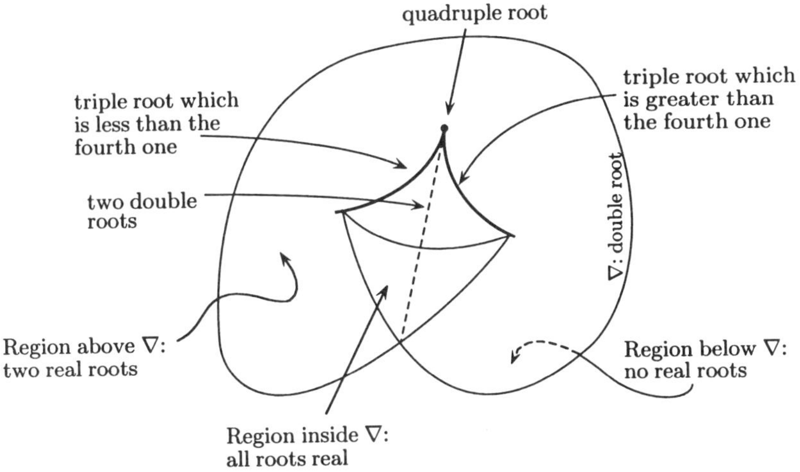

Figure 52. The space of polynomials $a_0 + a_1x + a_2x^2 + x^4$

Chapter 5). The space $\mathbf{R}^A$ consists of real $m \times n$ matrices $\|a_{ij}\|$. The generic stratum $\mathbf{R}^A_{\text{gen}}$ consists of matrices of which all the square minors, including 1×1, are non-zero (see Example 1.3 (b) Chapter 10). Denote by M the $(m+n) \times n$ matrix obtained from $\|a_{ij}\|$ by appending a unit $n \times n$ matrix. The fact that all the square minors of $\|a_{ij}\|$ are non-zero is equivalent (see (1.3) Chapter 3) to the fact that all the maximal minors of M are non-zero or, in other words, that the system of vectors in $\mathbf{R}^n$, formed by the standard basis vectors and the columns of $\|a_{ij}\|$, is in general position (no n vectors are linearly dependent). Therefore connected components of $\mathbf{R}^A_{\text{gen}}$ are in bijection with connected components of the configuration space of $(m+n)$-tuples of vectors in $\mathbf{R}^n$ in general position.

The relevance of the Newton polytope of Δ_A for the "A-Hilbert problem" considered above stems from Corollary 1.7 Chapter 10. More precisely, let T be a coherent triangulation of (Q, A). Let $C(T)$ be the normal cone to the secondary polytope $\Sigma(A)$ at the vertex corresponding to T, see Definition 1.4 Chapter 7. The above cited corollary guarantees the existence of a vector $b = (b_\omega)_{\omega\in A} \in C(T)$ such that $\Delta_A(f) \neq 0$ whenever $f = \sum_{\omega\in A} a_\omega x^\omega \in \mathbf{C}^A$ has all $a_\omega \neq 0$ and the logarithmic vector $(\log|a_\omega|)$ belongs to the translated cone $C(T) + b$.

We apply this to polynomials with real coefficients. Let $(\mathbf{R}^*)^A$ be the set of polynomials in $\mathbf{R}^A$ with all the coefficients non-zero. Consider the region

$$U(T, b) = \left\{ f = \sum_{\omega\in A} a_\omega x^\omega \in (\mathbf{R}^*)^A : (\log|a_\omega|) \in C(T) + b \right\}. \tag{5.2}$$

When b lies "deeply enough" inside $C(T)$, the region $U(T,b)$ does not intersect the real discriminantal variety $\nabla_A(\mathbf{R})$. On the other hand, $U(T,b)$ is the union of $2^{\#(A)}$ connected components $U(T,b,\varepsilon)$ labeled by functions $\varepsilon : A \to \{\pm 1\}$ which prescribe signs of a_ω. By definition,

$$U(T,b,\varepsilon) = \left\{ f = \sum_{\omega\in A} a_\omega x^\omega \ \in\ U(T) :\ \operatorname{sgn}(a_\omega) = \varepsilon(\omega) \right\}. \qquad (5.3)$$

The logarithm map identifies each $U(T,b,\varepsilon)$ with $C(T)+b$.

We arrive at the following conclusion.

Proposition 5.2. *Let T be any coherent triangulation of (Q,A), and let $\varepsilon : A \to \{\pm 1\}$ be any function. Then for any $f \in U(T,b,\varepsilon)$ with b as above the isotopy type of the real hypersurface $Z_f(\mathbf{R}) = \{f = 0\} \subset X_A(\mathbf{R})$ is the same, i.e., it is completely determined by T and ε.*

Clearly, if b lies "deeply enough" inside $C(T)$, then all the $\Delta_{A\cap\Gamma}(f) \neq 0$ for all $f \in U(T,b,\varepsilon)$. Hence, (T,ε) determine in fact the isotopy class of the hypersurface together with its intersections with the orbit closures $X(\Gamma)$.

We are interested in a direct description of the isotopy type of $Z_f(\mathbf{R})$ for $f \in U(T,b,\varepsilon)$. Such a description is equivalent to a result of Viro [V 1] that was obtained without using secondary polytopes. We shall present his construction.

B. The real part of a toric variety

The first step in a combinatorial description of Z_f is to give a direct construction of the set of real points of the toric variety X_A.

For simplicity of exposition we shall assume that X_A is normal.

Let $A = \{\omega^{(1)}, \ldots, \omega^{(n)}\}$. Clearly the real part $X_A(\mathbf{R})$ can be defined in $\mathbf{R}P^{n-1}$ as the closure (in the real topology, not Zariski) of

$$X_A^0(\mathbf{R}) = \left\{ (t^{\omega^{(1)}} : \ldots : t^{\omega^{(n)}}),\ t = (t_1, \ldots, t_{k-1}) \in (\mathbf{R}^*)^{k-1} \right\}. \qquad (5.4)$$

The set $X_A^0(\mathbf{R})$ is the union of 2^{k-1} connected components, according to the choices of signs of the t_i. All these components are isomorphic to each other. Consider one of them, namely

$$X_A^0(\mathbf{R}_+) = \left\{ (t^{\omega^{(1)}} : \ldots : t^{\omega^{(n)}}),\ t = (t_1, \ldots, t_{k-1}) \in (\mathbf{R}_+^*)^{k-1} \right\}, \qquad (5.5)$$

where $\mathbf{R}_+^*$ stands for the set of positive real numbers. Let $X_A(\mathbf{R}_+)$ be the closure of $X_A^0(\mathbf{R}_+)$; we shall call it the *positive part* of X_A. It is stratified by its intersections with subvarieties $X(\Gamma)$ where Γ runs through the faces of Q.

a) 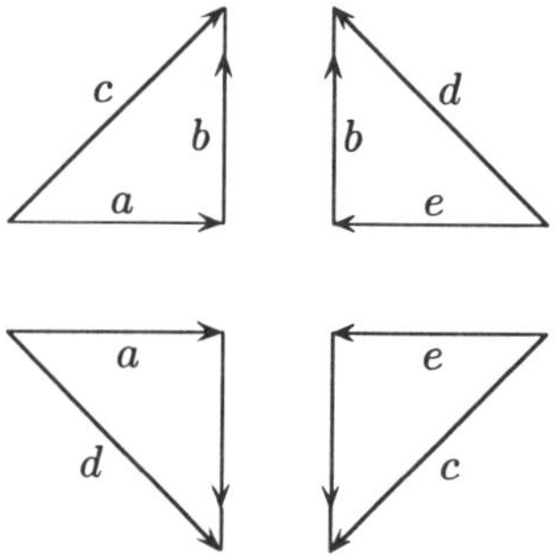

b) 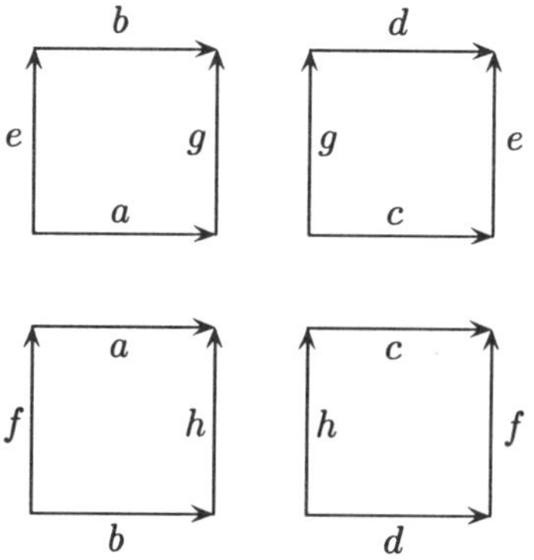

Figure 53. Gluing $\mathbf{R}P^2$ and $\mathbf{R}P^1 \times \mathbf{R}P^1$ out of 4 triangles (resp. squares).

Theorem 5.3. *The space $X_A(\mathbf{R}_+)$ stratified by its intersections with the $X(\Gamma)$ is homeomorphic to the polytope $Q = \mathrm{Conv}(A)$ stratified by its faces. The homeomorphism is given by the restriction of the moment map $\mu : X_A \to Q$ (Section 1C Chapter 6).*

This fact is well-known and we refer to [Fu2] for the proof.

Theorem 5.3 implies that $X_A(\mathbf{R})$ can be obtained by gluing together 2^{k-1} copies of Q by identifying some faces. More precisely, for any group homomorphism $\xi : \mathbf{Z}^{k-1} \to \{\pm 1\}$, we take a copy $Q(\xi)$ of Q. If $\Gamma \subset Q$ is a face then by $\Gamma(\xi)$ we shall denote the corresponding face of $Q(\xi)$.

Theorem 5.4. *The space $X_A(\mathbf{R})$ is identified with the result of the gluing of the polytopes $Q(\xi)$, $\xi \in \mathrm{Hom}(\mathbf{Z}^{k-1}, \{\pm 1\})$ by the following identifications: for any face $\Gamma \subset Q$ we identify $\Gamma(\xi) \subset Q(\xi)$ with $\Gamma(\xi') \subset Q(\xi')$ if and only if ξ and ξ' coincide on the affine sublattice $\mathrm{Aff}_{\mathbf{R}}(\Gamma) \cap \mathbf{Z}^{k-1}$.*

The proof, which is based on a straightforward interpretation of the moment map, is left to the reader.

Example 5.5. *(a)* Let $A \subset \mathbf{Z}^2$ consist of three points $(0,0)$, $(0,1)$, $(1,0)$. Then Q is a triangle, and X_A is the projective plane P^2. Theorem 5.4 represents $\mathbf{R}P^2$ as the result of the gluing of four triangles (copies of Q) as in Figure 53a.

(b) Let $A \subset \mathbf{Z}^2$ consist of four points $(0,0)$, $(0,1)$, $(1,0)$, $(1,1)$. Then Q is a square, and $X_A = P^1 \times P^1$. Theorem 5.4 represents $\mathbf{R}P^1 \times \mathbf{R}P^1$ (topologically a torus) as the result of the gluing of four squares as in Figure 53 b.

C. Viro's theorem

Let A satisfy the assumptions of subsection B. Denote by $\tilde{Q}$ the result of the gluing of 2^{k-1} copies of Q as in Theorem 5.4. So $\tilde{Q}$ is a CW-complex homeomorphic to $X_A(\mathbf{R})$.

Let T be a coherent triangulation of (Q, A) and $\varepsilon : A \to \{\pm 1\}$ any function. We shall construct a subspace $\mathcal{Z}(T, \varepsilon) \subset \tilde{Q}$ which will represent the isotopy type of any hypersurface Z_f for $f \in U(T, b, \varepsilon)$ (see subsection A).

For any group homomorphism $\xi : \mathbf{Z}^{k-1} \to \{\pm 1\}$, we denote by $A(\xi)$ the copy of A in $Q(\xi)$. Let $\tilde{A} \subset \tilde{Q}$ be the union of all $A(\xi)$. We extend any $\varepsilon : A \to \{\pm 1\}$ to a function $\tilde{\varepsilon} : \tilde{A} \to \{\pm 1\}$ by setting $\tilde{\varepsilon}(\omega(\xi)) = \varepsilon(\omega) \cdot \xi(\omega)$, where $\omega(\xi) \in A(\xi)$ is a copy of $\omega \in A$. Clearly, this is well-defined: if $\omega(\xi)$ is identified with $\omega(\xi')$ in $\tilde{Q}$ then $\xi(\omega) = \xi'(\omega)$ by definition of $\tilde{Q}$.

A triangulation T of (Q, A) obviously extends to a triangulation of $\tilde{Q}$ with vertices in $\tilde{A}$, denoted by $\tilde{T}$. It is convenient to work with the dual subdivision of $\tilde{Q}$ defined as follows. Let $\mathrm{Bar}\,(\tilde{T})$ be the first barycentric subdivision of the triangulation $\tilde{T}$ (see Figure 54).

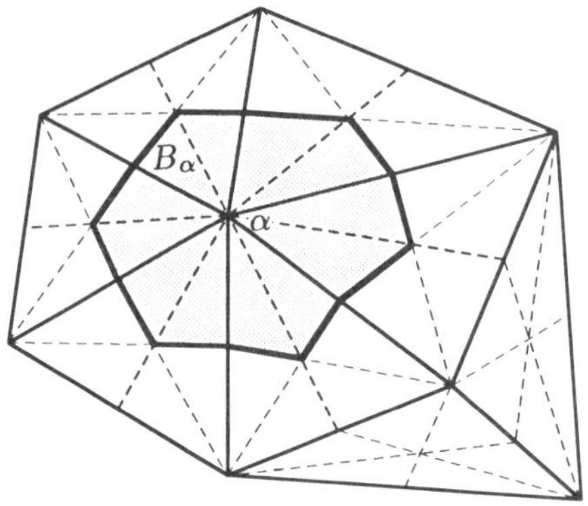

Figure 54.

For any $\alpha \in \tilde{A}$ let B_α be the union of all simplices of $\mathrm{Bar}\,(\tilde{T})$ having α as a vertex. (Note that $B_\alpha = \emptyset$ if α is not a vertex of $\tilde{T}$; otherwise $\dim\,(B_\alpha) = k - 1$.) The subdivision of $\tilde{Q}$ into the B_α will be called the *dual subdivision* to $\tilde{T}$ and denoted by $\mathrm{Dual}\,(\tilde{T})$. In the case when X_A is smooth, $X_A(\mathbf{R})$ is a smooth real manifold and each B_α is a cell so we have the dual CW-decomposition. In the general case the B_α may be more complicated, but we shall still call them "cells."

Finally, we define the subspace $\mathcal{Z}_+(T, \varepsilon)$ (resp. $\mathcal{Z}_-(T, \varepsilon)$) in $\tilde{Q}$ to be the union of "cells" $B_\alpha \in \mathrm{Dual}(\tilde{T})$ for which $\tilde{\varepsilon}(\alpha) = 1$ (resp. $\tilde{\varepsilon}(\alpha) = -1$). Let $\mathcal{Z}(T, \varepsilon) = \mathcal{Z}_+(T, \varepsilon) \cap \mathcal{Z}_-(T, \varepsilon)$.

The theorem of Viro is as follows.

Theorem 5.6. *Let T be a coherent triangulation of (Q, A), $\varepsilon : A \to \{\pm 1\}$ any function and $U(T, b, \varepsilon)$ the corresponding region (5.3) in $\mathbf{R}^A$. Then the hypersurface $Z_f(\mathbf{R}) = \{f = 0\} \subset X_A(\mathbf{R})$ for $f \in U(T, b, \varepsilon)$ has the same isotopy type as the subspace $\mathcal{Z}(T, \varepsilon) \subset \tilde{Q}$ (under the identification $X_A(\mathbf{R}) = \tilde{Q}$ given by Theorem 5.4). Moreover, the intersection of Z_f with any of the 2^{k-1} copies of Q constituting $X_A(\mathbf{R})$ has the same isotopy type as the intersection of $\mathcal{Z}(T, \varepsilon)$ with the corresponding copy of Q inside $\tilde{Q}$.*

The proof will be given in subsection D. We conclude this subsection with a couple of examples. The first one is an elementary particular case to be used in the general proof.

Example 5.7. Let A consist of monomials $1, x_1, \dots, x_{k-1}$, that is, of vectors $0, e_1, \dots, e_{k-1} \in \mathbf{Z}^{k-1}$, where the e_i are standard basis vectors. So $\mathbf{R}^A$ is the space of affine-linear functionals $f(x) = a_0 + \sum_{i=1}^{k-1} a_i x_i$. The polytope Q is the simplex

$$Q = \left\{ (t_1, \dots, t_{k-1}) : \; t_i \geq 0, \; \sum t_i \leq 1 \right\}$$

of dimension $k-1$. The vertices of Q are the basis vectors $e_1, \dots, e_{k-1}$ and the zero, denoted by e_0 for uniformity. The variety X_A is the projective space P^{k-1} and $x_1, \dots, x_{k-1}$ are coordinates in an affine chart of this space. In these coordinates the moment map $\mu : P^{k-1} \to Q$ is given by (cf. Section 1C Chapter 6):

$$\mu(x_1, \dots, x_{k-1}) = \left(\frac{|x_1|}{1 + \sum |x_i|}, \quad \dots, \quad \frac{|x_{k-1}|}{1 + \sum |x_i|} \right). \tag{5.6}$$

Fix some $f(x) = a_0 + \sum_{i=1}^{k-1} a_i x_i \in (\mathbf{R}^*)^A$ (so all the coefficients are non-zero). Then $Z_f(\mathbf{R})$ is a hyperplane in $X_A(\mathbf{R}) = \mathbf{R}P^{k-1}$ transversal to all the coordinate hyperplanes. Let us see what Theorem 5.6 means in this case. Consider first the intersection of Z_f with the positive part of X_A, i.e., the set of solutions of $f(x_1, \dots, x_{k-1}) = 0$ with all $x_i > 0$. The moment map (5.6) takes this set to the following hyperplane section of Q:

$$\mu(Z_f \cap X_A^0(\mathbf{R}_+)) = \left\{ (t_1, \dots, t_{k-1}) \in Q : a_0 \left(1 - \sum_{i=1}^{k-1} t_i \right) + \sum_{i=1}^{k-1} a_i t_i = 0 \right\}. \tag{5.7}$$

This section separates the vertices e_i for which $a_i > 0$ from those with $a_i < 0$. So topologically the same result will be obtained if we take the CW-decomposition of Q dual to that given by its faces, take the union $\mathcal{Z}_+$ of cells around the vertices

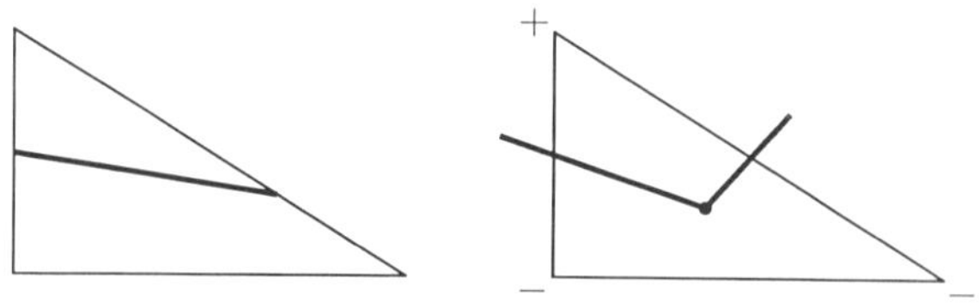

Figure 55.

e_i with $a_i > 0$ and a similar union $\mathcal{Z}_-$ of cells around the e_i with $a_i < 0$, and then consider the common boundary of these parts (see Figure 55).

We see that Theorem 5.6 gives a correct answer for the positive part of X_A. The study of other parts (correponding to other choices of signs of the x_i) can be reduced to the positive part by changing the signs of the a_i. Indeed, to study the equation $a_0 + \sum_{i=1}^{k-1} a_i x_i = 0$ in the region $\{\varepsilon_i x_i > 0\}$ for some fixed $\varepsilon_i = \pm 1$, is the same as to study the equation $a_0 + \sum_{i=1}^{k-1} \varepsilon_i a_i x_i = 0$ in the region $\{x_i > 0\}$. Thus the theorem is true in this case.

Example 5.8. We remarked (Section 2C Chapter 7) that the modification of a triangulation along a circuit is analogous to the surgery in the theory of manifolds. Viro's theorem provides a direct connection between these two notions.

More precisely, let M be a (real) m-dimensional topological manifold and let p, q be such that $p + q = m + 1$. A *surgery of type* (p, q) on M consists in finding a submanifold (of full dimension) in M homeomorphic to the product $S^{p-1} \times B^q$ of a sphere and a ball, deleting it from M (so the result will be a manifold with boundary $S^{p-1} \times S^{q-1}$) and then gluing in the product $B^p \times S^{q-1}$ along the boundary; see [Miln1] for more background information.

Now suppose that $Z \subset A$ is a circuit lying in the interior of Q such that $\mathrm{Conv}(Z)$ has full dimension. Let T_+, T_- be the only two triangulations of $\mathrm{Conv}(Z)$. Let $Z_\pm = \{\omega \in Z : \mathrm{Conv}(Z - \{\omega\}) \in T_\pm\}$. Then $\#(Z_+) + \#(Z_-) = \#(Z) = k+1$. Let $p = \#(Z_+) - 1$ and $q = \#(Z_-) - 1$. Suppose we are given a function $\varepsilon : A \to \{\pm 1\}$ such that $\varepsilon(\omega) = 1$ for $\omega \in Z_+$ and $\varepsilon(\omega) = -1$ for $\omega \in Z_-$. Let T, T' be two triangulations of (Q, A) obtained from each other by modification along Z. Then the parts of the hypersurfaces $\mathcal{Z}(T, \varepsilon)$ and $\mathcal{Z}(T', \varepsilon)$ lying inside the positive copy Q of $\tilde{Q}$ (the one corresponding to the trivial homomorphism $\mathbf{Z}^{k-1} \to \{\pm 1\}$), are obtained from each other by a surgery of type (p, q) (see Figure 56 a, b).

If a function ε is not compatible with the decomposition of Z into $Z_\pm$, i.e., if there are two elements ω, ω' lying in say Z_+ for which $\varepsilon(\omega) \neq \varepsilon(\omega')$ then $\mathcal{Z}(T, \varepsilon)$ and $\mathcal{Z}(T', \varepsilon)$ have the same isotopy type (see Figure 56c). For other parts of $\tilde{Q}$ the situation is similar.

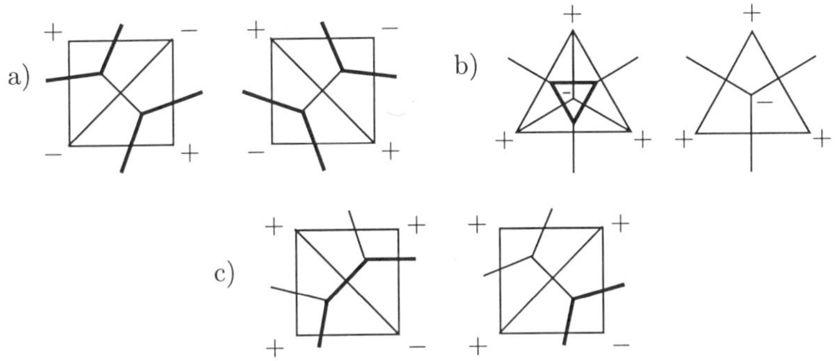

Figure 56. Modification of triangulations and Morse surgery

It is also possible to see, in some instances, the effect of a D-equivalent modification of a triangulation on the corresponding hypersurface. This amounts to moving the hypersurface across some stratum of $\tilde{Q}$ (or equivalently, across some torus orbit on $X_A(\mathbf{R})$), without changing the isotopy type of the hypersurface (see Figure 57).

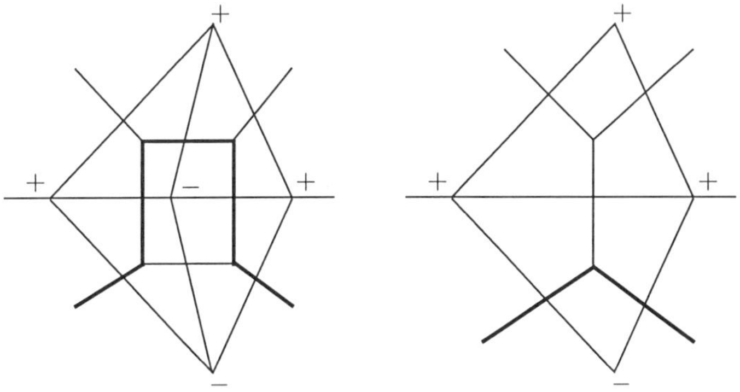

Figure 57.

D. Proof of Theorem 5.6

First we note that, similar to the end of Example 5.7, studying the hypersurface Z_f in any domain $Q(\xi) \cong Q$ of the variety $X_A(\mathbf{R})$ can be reduced to studying the positive domain $X_A(\mathbf{R}_+)$. So we shall restrict ourselves to this case. We denote by $\mathrm{Dual}(T)$ the CW-decomposition of Q dual to a triangulation T. It consists of cells B_ω, $\omega \in A$. (Now when we consider only Q, these are indeed cells.)

By $\mathcal{Z}_+$ (resp. $\mathcal{Z}_-$) we denote the union of cells B_ω such that $\mathrm{sgn}(a_\omega) = 1$ (resp. $\mathrm{sgn}(a_\omega) = -1$), and by $\mathcal{Z}$ we denote the common boundary of $\mathcal{Z}_+$ and $\mathcal{Z}_-$.

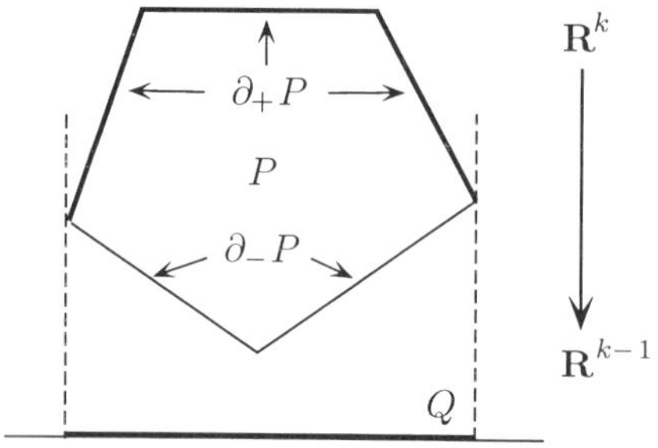

Figure 58.

We need to prove that the intersection of $Z_f(\mathbf{R})$ with $X_A(\mathbf{R}_+) \cong Q$ can be taken into $\mathcal{Z}$ by a self-homeomorphism of Q which takes each face into itself.

Since for any $f \in U(T, b, \varepsilon)$ the isotopy class of $Z_f(\mathbf{R})$ is the same, it suffices to consider some particular polynomial. We fix an integral vector $(\lambda_\omega) \in \mathbf{Z}^A$ lying in the interior of the cone $C(T)$ and consider the Laurent polynomial $F(x, t) = F(x_1, \ldots, x_{k-1}, t)$ in k variables $x_1, \ldots, x_{k-1}, t$ defined by

$$F(x, t) = \sum_{\omega \in A} \varepsilon(\omega) t^{\lambda_\omega} x^\omega. \tag{5.8}$$

We also regard $F(x, t)$ as a family of polynomials $F_t(x) := F(x, t)$ with a parameter t. Then for $t \gg 0$ we have $F_t \in U(T, b, \varepsilon)$, so we shall study the hypersurface defined by F_t with $t \gg 0$.

Consider the Newton polytope $P = N(F(x, t))$. By definition, P is the convex hull in $\mathbf{R}^k = \mathbf{R}^{k-1} \times \mathbf{R}$ of the set

$$B = \{(\omega, \lambda_\omega) \in \mathbf{Z}^k \,:\, \omega \in A\}.$$

Let $p : \mathbf{R}^k \to \mathbf{R}^{k-1}$ be the projection to the first factor. Then $p(P) = Q$. Let ∂P be the boundary of P. With respect to the projection p it splits into two parts: the top part $\partial_+ P$ and the bottom part $\partial_- P$ (see Figure 58).

The top part $\partial_+ P$ is the graph of the concave T-piecewise-linear function $g_\lambda : Q \to \mathbf{R}$ (see Lemma 1.8 Chapter 7); we have $g_\lambda(\omega) = \lambda_\omega$ for any $\omega \in A$ which is a vertex of some simplex in T. Since λ lies in the interior of $C(T)$, the faces of $\partial_+ P$ are simplices projecting bijectively under p to the simplices of T.

Let X_B be the toric variety corresponding to B and $\mu_B : X_B \to P$ the corresponding moment map. Consider the hypersurface $Z_F(\mathbf{R}) \subset X_B(\mathbf{R})$ given by $F(x,t)=0$. Let us study the intersection of Z_F with the positive part $X_B(\mathbf{R}_+)$. More precisely, we consider its image under the moment map

$$\mu_B : Z_F \cap X_B(\mathbf{R}_+) \to P. \tag{5.9}$$

Since the coefficient restriction of F to any face of P is essentially an affine-linear function, the analysis in Example 5.7 implies that if the intersection of $\mu_B(Z_F \cap X_B(\mathbf{R}_+))$ with a face of P (which is a simplex) is non-empty then it is a hyperplane section of this simplex of the form (5.7). We are interested only in the faces lying in $\partial_+ P$. Example 5.7 implies the following.

Lemma 5.9. *The projection $p(\mu_B(Z_F \cap X_B(\mathbf{R}_+)) \cap \partial_+ P) \subset Q$ is isotopic to the subcomplex $\mathcal{Z} = \mathcal{Z}_+ \cap \mathcal{Z}_-$ defined above. The isotopy may be chosen so as to preserve the faces of Q.*

It remains to compare the projection in Lemma 5.9 with the image of

$$\mu_A : Z_f \cap X_A(\mathbf{R}_+) \to Q, \tag{5.10}$$

where $f(x) = F_{t_0}(x)$ for some $t_0 \gg 0$. Consider the hypersurface $H_{t_0} = \{t = t_0\}$ in $X_B(\mathbf{R}_+)$. Let $\mu_B(H_{t_0}) \subset P$ be the image of H_{t_0} under the moment map.

Proposition 5.10. *For $t_0 \geq 0$ the hypersurface $\mu_B(H_{t_0})$ is the graph of a continuous function $\theta_{t_0} : Q \to \mathbf{R}$ which is smooth everywhere outside the boundary of Q. For $t_0 \to \infty$ this function converges uniformly to the function g_λ (whose graph is $\partial_+ P$).*

Assuming Proposition 5.10, Theorem 5.6 is proved as follows. By Lemma 5.9, the subcomplex

$$p(\mu_B(Z_F \cap X_B(\mathbf{R}_+)) \cap \partial_+ p) \subset Q$$

is isotopic to the polyhedral hypersurface $\mathcal{Z}$. By Proposition 5.10, for $t_0 \gg 0$ the intersection

$$\mu_B(Z_F \cap X_B(\mathbf{R}_+)) \cap H_{t_0} = \mu_A(Z_f \cap X_A(\mathbf{R}_+)) \subset Q$$

is also isotopic to the above subcomplex, hence to $\mathcal{Z}$, thus proving the theorem.

The behavior of H_t, $t \gg 0$ and Z_F inside the positive part $X_B(\mathbf{R}_+) \cong P$ is depicted in Figure 59.

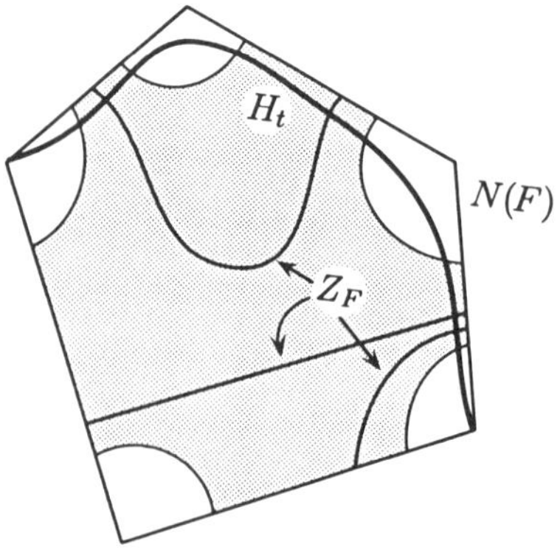

Figure 59. Intersection of H_t and $Z_F \cap X_B(\mathbf{R}_+)$.
The shadowed resgion is the compactified amoeba of F (Figure 19).

Proof of Proposition 5.10. We start by recalling the definition of the moment map μ_B (formula (1.3) Chapter 6). For clarity of notation we denote the system of variables $x_1, \dots, x_{k-1}, t$ by $y_1, \dots, y_k$. Thus $y_k = t$ and $y_i = x_i$, $i \le k-1$. Then

$$\mu_B(y) = \frac{\sum_{\gamma \in B} |y^\gamma| \cdot \gamma}{\sum_{\gamma \in B} |y^\gamma|}. \tag{5.11}$$

Thus $\mu_B(H_{t_0})$ is obtained by setting in (5.11) $y_k = t_0$ and letting the rest of the variables vary freely. Since (5.11) depends only on the absolute values of the y_i, we will assume in the sequel that all the y_i are positive. Note that

$$\left(y_i \frac{\partial}{\partial y_i} \right) y^\gamma = \gamma_i y^\gamma$$

where γ_i is the i-th component of γ. Thus for $y \in \mathbf{R}_+^k$ we have

$$\mu_B(y) = \left(y_1 \frac{\partial \Phi}{\partial y_1}, \ \dots, \ y_k \frac{\partial \Phi}{\partial y_k} \right), \tag{5.12}$$

where

$$\Phi(y) = \log\left(\sum_{\gamma \in B} y^\gamma \right).$$

Let us make a change of variables $y_i = e^{z_i}$ where $z_i \in \mathbf{R}, i = 1, \dots, k$. We regard μ_B as a function of the vector of new variables $z = (z_1, \dots, z_k)$. Then from (5.12) we get

$$\mu_B(z) = \left(\frac{\partial \Psi}{\partial z_1}, \ \dots \ , \frac{\partial \Psi}{\partial z_k} \right), \tag{5.13}$$

where

$$\Psi(z) = \log\left(\sum_{\gamma\in B} e^{(\gamma,z)}\right), \qquad (\gamma, z) = \sum \gamma_i z_i.$$

Suppose that $t_0 > 0$ and $q > 0$ are given. Consider the points of $\mu_B(Z_f)$ (where $f = F_{t_0}$) which project under p onto q. By (5.13), such points correspond to the points $z' \in \mathbf{R}^{k-1}$ where the gradient of the function

$$(z_1, \ldots, z_{k-1}) \longmapsto \Psi(z_1, \ldots, z_{k-1}, \log t_0)$$

is equal to $q \in \mathbf{R}^{k-1}$. The uniqueness of such z' (provided it exists) is implied by the next lemma

Lemma 5.11. *The function $\Psi : \mathbf{R}^k \to \mathbf{R}$ is strictly convex, i.e., the region*

$$\Gamma_+(\Psi) = \left\{(z, w) \in \mathbf{R}^k \times \mathbf{R} : w \geq \Psi(z)\right\}$$

is convex and its boundary does not contain line segments.

Proof. Let $\Gamma(\Psi)$ denote the graph of Ψ, i.e., the boundary of $\Gamma_+(\Psi)$. Consider the Laurent polynomial ζ in $k+1$ variables defined by

$$\zeta(y, u) = u - \sum_{\gamma\in A} y^\gamma, \qquad u \in \mathbf{C}^*,\ y \in (\mathbf{C}^*)^k. \tag{5.14}$$

We claim that $\Gamma_+(\Psi)$ is a component of the complement of the amoeba of this polynomial (see Section 1B Chapter 6) so our statement will follow from Corollary 1.6 Chapter 6. Indeed, by definition, the amoeba of ζ is the set of points $(\log|y_1|, \ldots, \log|y_k|, \log|u|)$ for all $y_1, \ldots, y_k, u \in \mathbf{C}^*$ such that $\zeta(y, u) = 0$. Thus, in view of the correspondence between the variables y_i and z_i the graph $\Gamma(\Psi)$ is a part of the amoeba (it corresponds to the choice of y_i, u in $\mathbf{R}_+$, not just in $\mathbf{C}^*$). If, however, a point (z, w) lies strictly above $\Gamma(\Psi)$, i.e., $w > \Psi(z)$ then there cannot exist $y_i, u \in \mathbf{C}^*$ such that

$$\zeta(y, u) = 0, \quad \log|u| = w, \log|y_i| = z_i.$$

Indeed, $w > \Psi(z)$ would mean that $|u| > \sum |y^\Gamma|$ which is inconsistent with $u = \sum y^\Gamma$, i.e., the vanishing of ζ. Therefore the graph $\Gamma(\Psi)$ is a component of the boundary of the amoeba so Ψ is convex. To see that it is strictly convex, it remains to note that the Hessian determinant $\det\|\partial^2\Psi/\partial z_i \partial z_j\|$ is the same as the Jacobian of the moment map hence non-zero. The lemma is proved.

Thus, to find the point of $\mu_B(Z_f)$ which projects under p onto $q \in Q$, we should take the minimum over $z_1, \ldots, z_{k-1}$ of

$$\Psi(z_1, \ldots, z_{k-1}, \log t_0) - \sum_{i=1}^{k-1} q_i z_i$$

and evaluate the gradient of Ψ at the point where the minimum is achieved. (The point of $\mu_B(Z_f)$ with the required property exists if and only if the minimum exists.) Since the first $k-1$ components of the gradient will be the components of q, we arrive at the following conclusion.

Lemma 5.12. *The set $\mu_B(H_{t_0})$ consists of the points $(q, \theta_{t_0}(q))$ where*

$$\theta_{t_0}(q) = \frac{\partial}{\partial s}\bigg|_{s=\log t_0} \min_{z_1,\ldots,z_{k-1}\in\mathbf{R}} \left(\Psi(z_1, \ldots, z_{k-1}, s) - \sum_{i=1}^{k-1} q_i z_i\right) \tag{5.15}$$

and $q \in Q$ is such that the minimum exists.

To show that the minimum in (5.15) exists for all $q \in Q$, note that the interpretation of the graph of Ψ in terms of the amoeba can be exploited further. Let Π be the Newton polytope of the polynomial ζ. This is a pyramid in $\mathbf{R}^{k+1}$ with base P and apex $e = (0, \ldots, 0, 1)$. Obvious estimates of the growth of Ψ imply that the region $\Gamma_+(\Psi)$ contains some affine cone with non-empty interior. By Corollary 1.8 Chapter 6, this region, being a component of the complement to the amoeba of ζ, contains a translation of the normal cone $N_v P$ for some vertex v of Π. This vertex is easily seen to be e, the apex of Π. Thus $\Gamma_+(\Psi)$ contains a translation of the cone $N_e(\Pi)$ and does not intersect some translation of all the other normal cones $N_v(\Pi)$, $v \neq e$. This means that asymptotically, as $\|z\| \to \infty$, the boundary of $\Gamma_+(\Psi)$, i.e., the graph of Ψ becomes parallel to the boundary of $N_e\Pi$. The latter boundary projects onto $\mathbf{R}^k$ in a one-to-one way so it is the graph of a piecewise-linear function $\mathbf{R}^k \to \mathbf{R}$ denoted by W. This function is homogeneous of degree 1. Now note that, for any $s \in \mathbf{R}$, the set of $q = (q_1, \ldots, q_{k-1}) \in \mathbf{R}^{k-1}$ for which the minimum

$$\min_{z_1,\ldots,z_{k-1}\in\mathbf{R}} \left(W(z_1, \ldots, z_{k-1}, s) - \sum_{i=1}^{k-1} q_i z_i\right) \tag{5.16}$$

exists, is precisely the polytope Q; this follows from the definition of the normal cone. This implies that for every $q \in Q$ the minimum in (5.15) indeed exists. This proves the first statement of Proposition 5.10.

Let us prove the second part of Proposition 5.10. Denote the minimum in (5.16) by $W^*(q, s)$. This is a linear function in s so its derivative in s depends only on q. It is straightforward to verify that

$$\frac{\partial}{\partial s} W^*(q, s) = g_\lambda(q),$$

i.e., it is the function whose graph is $\partial_+ P$. Thus when $t_0 \to \infty$, the graphs of $\Psi\Big|_{\mathbf{R}^{k-1} \times \{\log t_0\}}$ and $W\Big|_{\mathbf{R}^{k-1} \times \{\log t_0\}}$ become more and more parallel so the the derivative in (5.15) converges to $g_\lambda(q)$, as claimed. This concludes the proof of Proposition 5.10 together with Viro's theorem 5.6.

PART III

Classical Discriminants and Resultants

CHAPTER 12

Discriminants and Resultants for Polynomials in One Variable

In this chapter we consider the most classical case of the discriminant of a polynomial in one variable, and the resultant of two such polynomials. In the general language of Part II, we consider the A-discriminant Δ_A where A consists of monomials $1, x, x^2, \ldots, x^n$ and the (A_1, A_2)-resultant where $A_1 = \{1, x, \ldots, x^m\}$ and $A_2 = \{1, x, \ldots, x^n\}$.

1. An overview of classical formulas and properties

Here we collect some well-known formulas and properties of discriminants and resultants (for polynomials in one variable). Many of these formulas have already appeared as examples of the general theory developed earlier. We collect them in one place for the benefit of the practical reader. Whenever possible, we give short independent proofs of the formulas, not referring to the general theory.

A. The resultant of two polynomials in one variable

Let $m, n \geq 1$ and

$$f(x) = a_0 + a_1 x + \cdots + a_m x^m, \quad g(x) = b_0 + b_1 x + \cdots + b_n x^n \qquad (1.1)$$

be two polynomials in one variable of degrees less than or equal to m and n respectively. By $R(f, g)$ we denote their resultant. If necessary, we use the notation $R_{m,n}(f, g)$ to emphasize the dependency of m and n. For example, f is also a polynomial of degree $\leq m+1$ but $R_{m+1,n}(f, g)$ is not necessarily the same as $R_{m,n}(f, g)$ (see below).

Equivalently, let

$$F(x_0, x_1) = a_0 x_0^m + a_1 x_0^{m-1} x_1 + \cdots + a_m x_1^m,$$

$$G(x_0, x_1) = b_0 x_0^n + b_1 x_0^{n-1} x_1 + \cdots + b_n x_1^n \qquad (1.2)$$

be two binary forms (homogeneous polynomials) of degrees m and n. If the coefficients a_i, b_j in (1.2) are the same as in (1.1) then the resultant $R(F, G)$ is the same as $R(f, g)$. The use of binary forms is sometimes more convenient.

Product formula. If $a_m \neq 0$ and $b_n \neq 0$ then

$$R(f, g) = a_m^n b_n^m \prod_{i,j}(x_i - y_j), \tag{1.3}$$

where $x_1, \dots, x_m$ are roots of f and $y_1, \dots, y_n$ are roots of g.

This formula is often taken as the definition of the resultant. According to the general theory (Chapter 8), the resultant is defined only up to sign as the irreducible integral polynomial in the coefficients a_i, b_j which vanishes whenever f and g have a common root. Note that (1.3) does not have sign ambiguity. The connection between the two definitions is straightforward. To see that the right hand side of (1.3) is a polynomial in the a_i, b_j, we use the fundamental theorem on symmetric polynomials and the Vietae formulas which say that the a_i/a_m (respectively the b_j/b_n) are the elementary symmetric polynomials in $x_1, \dots, x_m$ (respectively, $y_1, \dots, y_n$). So the product in (1.3) is a polynomial in a_i, b_j which obviously vanishes whenever f and g have a common root. Its irreducibility follows from elementary Galois theory.

Vanishing of the resultant. For two concrete binary forms F, G as in (1.2) the vanishing of $R(F, G)$ is equivalent to the fact that F and G have a common root other than $(0, 0)$.

For two concrete polynomials f and g, the vanishing of $R(f, g) = R_{m,n}(f, g)$ is equivalent to the fact that f and g satisfy at least one of the following two conditions:

(a) f and g have a common root;

(b) $\deg f < m$ and $\deg g < n$, i.e., $a_m = b_n = 0$.

The case (b) means that the common root is at infinity.

The next three properties are immediate consequences of (1.3).

Symmetry.

$$R_{m,n}(f, g) = (-1)^{mn} R_{n,m}(g, f). \tag{1.4}$$

Multiplicativity. If f' is another polynomial of degree $\leq m'$ then

$$R_{m+m',n}(ff', g) = R_{m,n}(f, g) \cdot R_{m',n}(f', g). \tag{1.5}$$

The multiplicativity in the second argument follows from (1.4).

Quasi-homogeneity. The polynomial $R(f, g) = R(a_0, \dots, a_m, b_0, \dots, b_n)$ is homogeneous of degree n in the a_i and of degree m in the b_j. In addition, it has the following quasi-homogeneity:

$$R(\lambda^0 a_0, \lambda^1 a_1, \ldots, \lambda^m a_m, \lambda^0 b_0, \lambda^1 b_1, \ldots, \lambda^n b_n) = \lambda^{mn} R(a_0, \ldots, a_m, b_0, \ldots, b_n). \tag{1.6}$$

The homogeneity conditions can be written in differential form:

$$\sum_i a_i \frac{\partial R}{\partial a_i} = nR, \quad \sum_j b_j \frac{\partial R}{\partial b_j} = mR,$$

$$\sum_i i a_i \frac{\partial R}{\partial a_i} + \sum_j j b_j \frac{\partial R}{\partial b_j} = mnR. \tag{1.7}$$

PGL(2)-invariance. For any non-degenerate matrix $h = \begin{pmatrix} \alpha & \beta \\ \gamma & \delta \end{pmatrix} \in GL(2, \mathbf{C})$ let

$$(h^* f)(x) = (\gamma x + \delta)^m f\left(\frac{\alpha x + \beta}{\gamma x + \delta}\right), \quad (h^* g)(x) = (\gamma x + \delta)^n g\left(\frac{\alpha x + \beta}{\gamma x + \delta}\right).$$

Then

$$R(h^* f, h^* g) = \det(h)^{mn} R(f, g). \tag{1.8}$$

Proof. For any given h both sides of (1.8) are polynomials in coefficients of f and g which vanish for the same pairs (f, g). Indeed, f and g have a common root if and only if the same holds for $h^*(f)$ and $h^*(g)$. So $R(h^* f, h^* g)$ and $R(f, g)$ are proportional for any h. The coefficient of proportionality $\varepsilon(h)$ satisfies the property $\varepsilon(hh') = \varepsilon(h) \cdot \varepsilon(h')$. This implies that $\varepsilon(h)$ is a power of $\det(h)$. The exponent of the power is found from the homogeneity of R.

For the case of a diagonal matrix h, the equality (1.8) follows from the homogeneity conditions above. The validity of (1.8) for all matrices leads to two additional differential equations on R:

$$\sum_{i=0}^{m-1} (m - i) a_i \frac{\partial R}{\partial a_{i+1}} + \sum_{j=0}^{n-1} (n - j) b_j \frac{\partial R}{\partial b_{j+1}} = 0. \tag{1.9}$$

$$\sum_{i=1}^{m} i a_i \frac{\partial R}{\partial a_{i-1}} + \sum_{j=1}^{n} j b_i \frac{\partial R}{\partial b_{j-1}} = 0. \tag{1.10}$$

Finding the common root. If, for given f, g, we have $R(f, g) = 0$ but at least one first partial derivative of R at (f, g) is non-zero then f and g have a unique common root α (possibly $\alpha = \infty$), and it can be found from the proportions:

$$(1 : \alpha : \alpha^2 : \ldots : \alpha^m) = \left(\frac{\partial R}{\partial a_0}(f, g) : \frac{\partial R}{\partial a_1}(f, g) : \ldots : \frac{\partial R}{\partial a_m}(f, g)\right), \tag{1.11}$$

$$(1:\alpha:\alpha^2:\ldots:\alpha^n) = \left(\frac{\partial R}{\partial b_0}(f,g):\frac{\partial R}{\partial b_1}(f,g):\ldots:\frac{\partial R}{\partial b_n}(f,g)\right). \qquad (1.11')$$

This is a particular case of Corollary 3.7 Chapter 3. The general proof given there (reduction via the Cayley trick, to the biduality theorem for projective dual varieties), is probably the most transparent.

Sylvester formula. We have

$$R(f,g) = \begin{vmatrix} a_0 & a_1 & a_2 & \ldots & a_{m-1} & a_m & 0 & 0 & \ldots & 0 \\ 0 & a_0 & a_1 & \ldots & a_{m-2} & a_{m-1} & a_m & 0 & \ldots & 0 \\ \vdots & \vdots & \ddots & \ddots & \vdots & \vdots & \vdots & \ddots & \ddots & \vdots \\ 0 & 0 & 0 & \ldots & a_0 & a_1 & a_2 & a_3 & \ldots & a_m \\ b_0 & b_1 & b_2 & \ldots & b_{n-1} & b_n & 0 & 0 & \ldots & 0 \\ 0 & b_0 & b_1 & \ldots & b_{n-2} & b_{n-1} & b_n & 0 & \ldots & 0 \\ \vdots & \vdots & \ddots & \ddots & \vdots & \vdots & \vdots & \ddots & \ddots & \vdots \\ 0 & 0 & 0 & \ldots & b_0 & b_1 & b_2 & b_3 & \ldots & b_n \end{vmatrix}. \qquad (1.12)$$

This determinant is of order $m+n$; it has n rows involving the a's and m rows involving the b's.

Proof. The determinant in (1.12) is a polynomial in the a_i, b_j of the same degree as R. It is non-zero since, e.g., it contains the monomial $a_0^n b_n^m$ with coefficient 1. So it is enough to show that if f and g have a common root, say α, then the determinant vanishes. The matrix in (1.12) is the matrix of the linear operator

$$\partial : S_{n-1} \oplus S_{m-1} \to S_{m+n-1}, \quad \partial(u,v) = fu + gv,$$

where by S_d we denote the space of polynomials of degree $\leq d$. If f, g both vanish at α then any polynomial in the image of ∂ also vanishes; so ∂ is not of full rank and $\det(\partial) = 0$.

Dependence on m and n. If $m' \geq m$ then

$$R_{m',n}(f,g) = b_n^{m'-m} R_{m,n}(f,g).$$

Similarly, if $n' \geq n$ then

$$R_{m,n'}(f,g) = (-1)^{m(n'-n)} a_m^{n'-n} R_{m,n}(f,g).$$

This follows at once from the Sylvester formula.

Examples. For two linear polynomials

$$R_{1,1}(a_0 + a_1x, b_0 + b_1x) = a_0b_1 - a_1b_0. \qquad (1.13)$$

For two quadratic polynomials

$$R_{2,2}(a_0 + a_1 + a_2x^2, b_0 + b_1x + b_2x^2)$$
$$= a_0^2b_2^2 + a_0a_2b_1^2 - a_0a_1b_1b_2 + a_1^2b_0b_2 - a_1a_2b_0b_1 + a_2^2b_0^2 - 2a_0a_2b_0b_2. \quad (1.14)$$

Expression in terms of coefficients of $g(x)/f(x)$. Suppose that $a_0 = 1$ and

$$g(x)/f(x) = r_0 + r_1x + r_2x^2 + \cdots.$$

Then

$$R(f,g) = \begin{vmatrix} r_n & r_{n+1} & \ldots & r_{n+m-1} \\ r_{n-1} & r_n & \ldots & r_{n+m-2} \\ \vdots & & & \vdots \\ r_{n-m+1} & r_{n-m+2} & \ldots & r_n \end{vmatrix}. \quad (1.15)$$

Proof. Multiply the Sylvester matrix (1.12) on the right by

$$\begin{pmatrix} 1 & s_1 & s_2 & \ldots & s_{m+n-1} \\ 0 & 1 & s_1 & \ldots & s_{m+n-2} \\ 0 & 0 & 1 & \ldots & s_{m+n-3} \\ \vdots & & \ddots & \vdots & \vdots \\ 0 & 0 & 0 & \ldots & 1 \end{pmatrix},$$

where $1/f(x) = 1 + \sum_{i=1}^{\infty} s_ix^i$.

Bezout-Cayley formula. Suppose $m = n$. Consider the polynomial in two variables

$$\varphi(x,y) = \frac{f(x)g(y) - g(x)f(y)}{x - y} = \sum_{i,j=0}^{n-1} c_{ij}x^iy^j. \quad (1.16)$$

Then

$$R(f,g) = \det\|c_{ij}\|. \quad (1.17)$$

Proof. It is not hard to show that the determinant in (1.17) is a non-zero polynomial in the a_i, b_j having the same degree as R and the same coefficient 1 at the monomial $a_0^nb_n^n$. It remains to show that the determinant vanishes whenever f and g have a common root. We claim that if α is such a root then the vector $(1, \alpha, \alpha^2, \ldots, \alpha^{n-1})$ is annihilated by the matrix $C = \|c_{ij}\|$. Indeed, we have $\varphi(\alpha, y) = 0$ for any y. By (1.16), this means that the row vector $(1, \alpha, \alpha^2, \ldots, \alpha^{n-1})C$ is orthogonal to all vectors of the form $(1, y, y^2, \ldots, y^{n-1})$. Since vectors of the form $(1, y, y^2, \ldots, y^{n-1})$ generate $\mathbf{C}^n$, we conclude that $(1, \alpha, \alpha^2, \ldots, \alpha^{n-1})C = 0$, as claimed. Hence $\det(C) = 0$.

Let us denote by $[ij]$ the expression $a_i b_j - a_j b_i$. An easy computation shows that the coefficients c_{ij} in (1.16) are given by

$$c_{ij} = \sum_{p=0}^{\min(i,j)} [p, i+j+1-p].$$

Thus the explicit form of (1.17) is

$$R(f,g) = \begin{vmatrix} [01] & [02] & [03] & \dots & [0n] \\ [02] & [03]+[12] & [04]+[13] & \dots & [1n] \\ [03] & [04]+[13] & [05]+[14]+[23] & \dots & [2n] \\ \vdots & \vdots & \vdots & \vdots & \vdots \\ [0n] & [1n] & [2n] & \dots & [n-1,n] \end{vmatrix}. \quad (1.18)$$

Interpolating between Sylvester and Bezout. Suppose $m = n$. Let r be any integer such that $n \leq r \leq 2n$. Then

$$R(f,g) = \det \begin{pmatrix} \mathrm{Syl}(a)_{r-n,r} \\ \mathrm{Syl}(b)_{r-n,r} \\ \mathrm{Bez}_{2n-r,r} \end{pmatrix} \quad (1.19)$$

where $\mathrm{Syl}(a)_{r-n,n}$ is the left corner submatrix of the Sylvester matrix (1.12) on the first $(r-n)$ rows of the a's and first r columns; similarly, $\mathrm{Syl}(b)_{r-n,n}$ is the submatrix in (1.12) on first $(r-n)$ rows of the b's and first r columns. Finally, $\mathrm{Bez}_{2n-r,r}$ is the left corner submatrix in the Bezout matrix (1.18) on the first $2n-r$ rows and first n columns which is accompanied by the $(2n-r) \times (r-n)$-block of zeros. For $r = n$ the formula (1.19) gives the Bezout-Cayley formula (1.18) and for $r = 2n$ we get the Sylvester formula (1.12).

For $n = 4$ (i.e. for the resultant of two quartic polynomials) and $r = 5, 6, 7$ we get

$$R(f,g) = \begin{vmatrix} a_0 & a_1 & a_2 & a_3 & a_4 \\ b_0 & b_1 & b_2 & b_3 & b_4 \\ [0,1] & [0,2] & [0,3] & [0,4] & 0 \\ [0,2] & [0,3]+[1,2] & [0,4]+[1,3] & [1,4] & 0 \\ [0,3] & [0,4]+[1,3] & [1,4]+[2,3] & [2,4] & 0 \end{vmatrix}$$

$$= \begin{vmatrix} a_0 & a_1 & a_2 & a_3 & a_4 & 0 \\ 0 & a_0 & a_1 & a_2 & a_3 & a_4 \\ b_0 & b_1 & b_2 & b_3 & b_4 & 0 \\ 0 & b_0 & b_1 & b_2 & b_3 & b_4 \\ [0,1] & [0,2] & [0,3] & [0,4] & 0 & 0 \\ [0,2] & [0,3]+[1,2] & [0,4]+[1,3] & [1,4] & 0 & 0 \end{vmatrix}$$

$$= \begin{vmatrix} a_0 & a_1 & a_2 & a_3 & a_4 & 0 & 0 \\ 0 & a_0 & a_1 & a_2 & a_3 & a_4 & 0 \\ 0 & 0 & a_0 & a_1 & a_2 & a_3 & a_4 \\ b_0 & b_1 & b_2 & b_3 & b_4 & 0 & 0 \\ 0 & b_0 & b_1 & b_2 & b_3 & b_4 & 0 \\ 0 & 0 & b_0 & b_1 & b_2 & b_3 & b_4 \\ [0,1] & [0,2] & [0,3] & [0,4] & 0 & 0 & 0 \end{vmatrix}. \tag{1.20}$$

Formula (1.19) first appeared in [WZ1]; its proof can be obtained by unraveling the resultant spectral sequence $C^{\bullet\bullet}_{\bullet-}(\mathcal{O}(n), \mathcal{O}(n)|\mathcal{O}(r-1))$ (see Section 4C Chapter 3).

B. The discriminant of a polynomial in one variable

Let

$$f(x) = a_0 + a_1 x + \cdots + a_n x^n \tag{1.21}$$

be a polynomial in one variable of degree $\leq n$. We denote by $\Delta(f)$ its discriminant. If necessary, we shall use the notation $\Delta_n(f)$ to emphasize the dependence on n.

Equivalently, let

$$F(x_0, x_1) = a_0 x_0^n + a_1 x_0^{n-1} x_1 + \cdots + a_n x_1^n \tag{1.22}$$

be a binary form of degree n. If the a_i in (1.22) and (1.21) are the same then the discriminant $\Delta(F)$ is the same as $\Delta_n(f)$. As for the case of resultants the use of binary forms is sometimes more convenient.

Here are the main properties of discriminants. They are parallel to the properties of resultants discussed above, and we usually omit the proofs.

Product formula. If $a_n \neq 0$ then

$$\Delta(f) = (-1)^{\frac{n(n-1)}{2}} a_n^{2n-2} \prod_{i<j} (x_i - x_j)^2, \tag{1.23}$$

where $x_1, \ldots, x_n$ are roots of f.

As for the resultants, the equivalence of (1.23) to the general definition is proven by using symmetric polynomials and Galois theory. The formula (1.23) is used to normalize the discriminant uniquely, not just up to sign. Later we will use this normalization.

Vanishing of the discriminant. For a concrete binary form $F(x_0, x_1)$ of degree n, the vanishing of $\Delta(F)$ means that F is divisible by the square of a linear form.

For a concrete polynomial $f(x)$ of degree $\leq n$, we have $\Delta_n(f) = 0$ if and only if f satisfies at least one of two conditions:

(a) f has a double root;

(b) $\deg(f) \leq n-2$, i.e., $a_n = a_{n-1} = 0$.
Under the second condition, the double root is at infinity.

Quasi-homogeneity. The discriminant $\Delta(f) = \Delta(a_0, \dots, a_n)$ is a homogeneous polynomial in the a_i of degree $2n-2$. In addition, it satisfies the quasi-homogeneity condition:

$$\Delta(\lambda^0 a_0, \lambda^1 a_1, \dots, \lambda^n a_n) = \lambda^{n(n-1)} \Delta(a_0, \dots, a_n). \tag{1.24}$$

In the differential form the homogeneity conditions are

$$\sum a_i \frac{\partial \Delta}{\partial a_i} = (2n-2)\Delta, \quad \sum i a_i \frac{\partial \Delta}{\partial a_i} = n(n-1)\Delta. \tag{1.25}$$

PGL(2)-invariance. For a non-degenerate matrix $h = \begin{pmatrix} \alpha & \beta \\ \gamma & \delta \end{pmatrix}$, let $(h^* f)(x) = (\gamma x + \delta)^n f\left(\frac{\alpha x + \beta}{\gamma x + \delta}\right)$. Then

$$\Delta(h^* f) = \det(h)^{n(n-1)} \Delta(f). \tag{1.26}$$

The equality (1.26) is equivalent to the homogeneity conditions (1.25) together with two additional differential equations:

$$\sum_{i=0}^{n-1} (n-i) a_i \frac{\partial \Delta}{\partial a_{i+1}} = \sum_{i=1}^{n} i a_i \frac{\partial \Delta}{\partial a_{i-1}} = 0. \tag{1.27}$$

Finding the double root. If for a particular f we have $\Delta(f) = 0$ but at least one partial derivative of Δ at f is non-zero, then f has a unique double root α (possibly $\alpha = \infty$), and it can be found from the proportions:

$$(1 : \alpha : \alpha^2 : \dots : \alpha^n) = \left(\frac{\partial \Delta}{\partial a_0}(f) : \frac{\partial \Delta}{\partial a_1}(f) : \dots : \frac{\partial \Delta}{\partial a_n}(f) \right). \tag{1.28}$$

Relations with resultants. We have

$$\Delta(f) = \frac{1}{a_n} R_{n,n-1}(f, f') = n^n a_n^{n-1} \prod_{\alpha : f'(\alpha) = 0} f(\alpha). \tag{1.29}$$

In view of (1.29), every formula for the resultant gives rise to a formula for the discriminant. For instance, Sylvester formula (1.12) represents $\Delta(f)$ in terms of the determinant of order $2n-1$:

$$\Delta(f) = \frac{1}{a_n}\begin{vmatrix} a_0 & a_1 & a_2 & \dots & a_{n-1} & a_n & 0 & \dots \\ 0 & a_0 & a_1 & \dots & a_{n-2} & a_{n-1} & a_n & \dots \\ \vdots & \dots & & \dots & & \dots & & \vdots \\ 1\cdot a_1 & 2\cdot a_2 & \dots & (n-1)a_{n-1} & na_n & 0 & 0 & \dots \\ 0 & 1\cdot a_1 & \dots & (n-2)a_{n-2} & (n-1)a_{n-1} & na_n & 0 & \dots \\ \vdots & \dots & & \dots & & \dots & & \vdots \end{vmatrix} \tag{1.30}$$

The following two equalities are similar to (1.29) but are more conveniently formulated in terms of a binary form F in (1.22):

$$\Delta(F) = \frac{1}{n^{n-2}} R_{n-1,n-1}\left(\frac{\partial F}{\partial x_0}, \frac{\partial F}{\partial x_1}\right) = \frac{1}{n^n a_0 a_n} R_{n,n}\left(x_0\frac{\partial F}{\partial x_0}, x_1\frac{\partial F}{\partial x_1}\right). \tag{1.31}$$

Conversely, using (1.3) and (1.23), the resultant can be obtained as the "polarization" of the discriminant: If g is a polynomial of degree $\le m$ then

$$(R_{n,m}(f,g))^2 = (-1)^{mn}\frac{\Delta_{m+n}(f\cdot g)}{\Delta_n(f)\cdot\Delta_m(g)}. \tag{1.32}$$

Examples. For a quadratic polynomial

$$\Delta(a_0 + a_1x + a_2x^2) = 4a_0a_2 - a_1^2. \tag{1.33}$$

For a cubic polynomial

$$\Delta(a_0+a_1x+a_2x^2+a_3x^3) = 27a_0^2a_3^2+4a_0a_2^3+4a_1^3a_3-a_1^2a_2^2-18a_0a_1a_2a_3. \tag{1.34}$$

For a quartic polynomial

$$\begin{aligned}\Delta(a_0 &+ a_1x + a_2x^2 + a_3x^3 + a_4x^4)\\ &= 256a_0^3a_4^3 - 27a_0^2a_3^4 - 27a_1^4a_4^2 + 16a_0a_2^4a_4 - 4a_0a_2^3a_3^2 - 4a_1^2a_2^3a_4 - 4a_1^3a_3^3 + a_1^2a_2^2a_3^2\\ &\quad -192a_0^2a_1a_3a_4^2 - 128a_0^2a_2^2a_4^2 + 144a_0^2a_2a_3^2a_4 + 144a_0a_1^2a_2a_4^2 - 6a_0a_1^2a_3^2a_4\\ &\quad -80a_0a_1a_2^2a_3a_4 + 18a_0a_1a_2a_3^3 + 18a_1^3a_2a_3a_4. \end{aligned}\tag{1.35}$$

For a quintic polynomial

$$\begin{aligned}
&\Delta(a_0 + a_1 x + \cdots + a_5 x^5) \\
&= 3125 a_0^4 a_5^4 - 2500 a_0^3 a_1 a_4 a_5^3 - 3750 a_0^3 a_2 a_3 a_5^3 + 2000 a_0^3 a_2 a_4^2 a_5^2 + 2250 a_0^3 a_3^2 a_4 a_5^2 \\
&\quad -1600 a_0^3 a_3 a_4^3 a_5 + 256 a_0^3 a_4^5 + 2000 a_0^2 a_1^2 a_3 a_5^3 - 50 a_0^2 a_1^2 a_4^2 a_5^2 + 2250 a_0^2 a_1 a_2^2 a_5^3 \\
&\quad -2050 a_0^2 a_1 a_2 a_3 a_4 a_5^2 + 160 a_0^2 a_1 a_2 a_4^3 a_5 - 900 a_0^2 a_1 a_3^3 a_5^2 + 1020 a_0^2 a_1 a_3^2 a_4^2 a_5 \\
&\quad -192 a_0^2 a_1 a_3 a_4^4 - 900 a_0^2 a_2^3 a_4 a_5^2 + 825 a_0^2 a_2^2 a_3^2 a_5^2 + 560 a_0^2 a_2^2 a_3 a_4^2 a_5 - 128 a_0^2 a_2^2 a_4^4 \\
&\quad -630 a_0^2 a_2 a_3^3 a_4 a_5 + 144 a_0^2 a_2 a_3^2 a_4^3 + 108 a_0^2 a_3^5 a_5 - 27 a_0^2 a_3^4 a_4^2 - 1600 a_0 a_1^3 a_2 a_5^3 \\
&\quad +160 a_0 a_1^3 a_3 a_4 a_5^2 - 36 a_0 a_1^3 a_4^3 a_5 + 1020 a_0 a_1^2 a_2^2 a_4 a_5^2 + 560 a_0 a_1^2 a_2 a_3^2 a_5^2 \\
&\quad -746 a_0 a_1^2 a_2 a_3 a_4^2 a_5 + 144 a_0 a_1^2 a_2 a_4^4 + 24 a_0 a_1^2 a_3^3 a_4 a_5 - 6 a_0 a_1^2 a_3^2 a_4^3 \\
&\quad +356 a_0 a_1 a_2^2 a_3^2 a_4 a_5 - 80 a_0 a_1 a_2^2 a_3 a_4^3 - 630 a_0 a_1 a_2^3 a_3 a_5^2 + 24 a_0 a_1 a_2^3 a_3^2 a_5 \\
&\quad -72 a_0 a_1 a_2 a_3^4 a_5 + 18 a_0 a_1 a_2 a_3^3 a_4^2 + 108 a_0 a_2^5 a_5^2 - 72 a_0 a_2^4 a_3 a_4 a_5 + 16 a_0 a_2^4 a_4^3 \\
&\quad +16 a_0 a_2^3 a_3^3 a_5 - 4 a_0 a_2^3 a_3^2 a_4^2 + 256 a_1^5 a_5^3 - 192 a_1^4 a_2 a_4 a_5^2 - 128 a_1^4 a_3^2 a_5^2 \\
&\quad +144 a_1^4 a_3 a_4^2 a_5 - 27 a_1^4 a_4^4 + 144 a_1^3 a_2^2 a_3 a_5^2 - 6 a_1^3 a_2^2 a_4^2 a_5 - 80 a_1^3 a_2 a_3^2 a_4 a_5 \\
&\quad +18 a_1^3 a_2 a_3 a_4^3 + 16 a_1^3 a_3^4 a_5 - 4 a_1^3 a_3^3 a_4^2 - 27 a_1^2 a_2^4 a_5^2 + 18 a_1^2 a_2^3 a_3 a_4 a_5 \\
&\quad -4 a_1^2 a_2^3 a_4^3 - 4 a_1^2 a_2^2 a_3^3 a_5 + a_1^2 a_2^2 a_3^2 a_4^2. \qquad (1.36)
\end{aligned}$$

The discriminant of a binomial:

$$\Delta(a + bx^n) = n^n a^{n-1} b^{n-1}. \qquad (1.37)$$

This is a special case of the second equality in (1.29).

The discriminant of a trinomial:

For $0 < m < n$ and m and n relatively prime, we have

$$\Delta(a+bx^m+cx^n) = n^n a^{n-1} c^{n-1} + (-1)^{n-1} m^m (n-m)^{n-m} a^{m-1} b^n c^{n-m-1}. \qquad (1.38)$$

Proof. Using again the second inequality in (1.29), we obtain

$$\Delta(a + bx^m + cx^n) = n^n c^{n-1} a^{m-1} \prod_{\alpha} (a + b\alpha^m + c\alpha^n), \qquad (1.39)$$

the sum over all the roots of the equation $\alpha^{n-m} = -\frac{mb}{nc}$. For every such α, we have

$$a + b\alpha^m + c\alpha^n = a + \frac{n-m}{n} b\alpha^m.$$

Since m and n are relatively prime, we can make the change of variables $\beta = \alpha^m$ and rewrite (1.39) as

$$\Delta(a + bx^m + cx^n) = n^n c^{n-1} a^{m-1} \prod_{\beta} (a + \frac{n-m}{n} b\beta), \tag{1.40}$$

where β runs over the roots of the equation $\beta^{n-m} = (-\frac{mb}{nc})^m$. Introducing the polynomial $q(x) = x^{n-m} - (-\frac{mb}{nc})^m$, we see that the product in (1.40) equals

$$(-\frac{n-m}{n} b)^{n-m} \cdot q(-\frac{an}{b(n-m)}).$$

Substituting this into (1.40) yields (1.38).

It is also possible to prove (1.38) in a more conceptual way. Denote by Z and A the sets of monomials $\{1, x^m, x^n\}$ and $\{1, x, x^2, \ldots, x^n\}$ respectively. Then our discriminant Δ is, in the notation of Chapter 9, Δ_A. Proposition 1.8 of Chapter 9 gives that the discriminant Δ_Z coincides, up to a non-zero scalar factor, with the right hand side of (1.38). On the other hand, the restriction of Δ_A to $\mathbf{C}^Z$, the space of trinomials $a + bx^m + cx^n$, should have the form $\varepsilon \cdot \Delta_Z^l$ for some $\varepsilon \in \mathbf{C}^*, l \geq 1$, since the two discriminants define the same locus in $\mathbf{C}^Z$. By comparing the degrees we see that $l = 1$. This means that (1.38) holds up to a constant factor. To show that the constant factor is 1, we put $b = 0$ and use (1.37).

Dependence on n. We have

$$\Delta_{n+1}(f) = (-1)^n a_n^2 \Delta_n(f), \quad \Delta_{n'}(f) = 0 \ for \ n' \geq n+2. \tag{1.41}$$

Proof. If $\Delta_{n+1}(f)$ vanishes then either f has a double root (so $\Delta_n(f) = 0$), or f has degree $\leq n-1$ (so $a_n = 0$). This implies (1.41) up to a constant factor (for $n \geq 2$ this follows by comparing degrees; the case $n \leq 2$ is obvious). To show that the constant factor is 1, we take $f = a + bx^n$ and apply (1.38) to trinomials of the form $a + bx^n + cx^{n+1}$.

Hilbert's formula: Consider a *terminating hypergeometric polynomial*

$${}_2F_1(\alpha, \beta, \gamma; x) = 1 + \frac{\alpha\beta}{1 \cdot \gamma} x + \frac{\alpha(\alpha+1)\beta(\beta+1)}{1 \cdot 2 \cdot \gamma(\gamma+1)} x^2 + \cdots,$$

where $\alpha = -n$ is a negative integer. This is a polynomial in x of degree n. We have

$$\Delta({}_2F_1(-n, \beta, \gamma; x)) = \prod_{j=1}^{n-1} \frac{j^j (\beta + j - 1)^{n-j} (\beta - \gamma + j - 1)^{n-j}}{(\gamma + j)^{2n-1-j}}. \tag{1.42}$$

This formula was proved in [Hil2].

C. The monomial expansion of the resultant

We return to the resultant of two polynomials f and g. Let us write down the resultant as the sum of monomials:

$$R(f, g) = \sum_{p,q} c_{pq} a^p b^q, \tag{1.43}$$

where $p = (p_0, p_1, \dots, p_m) \in \mathbf{Z}_+^{m+1}$, $q = (q_0, q_1, \dots, q_m) \in \mathbf{Z}_+^{n+1}$. Despite many explicit formulas for $R(f, g)$ given above, there seems to be no simple formula for the coefficients c_{pq}. In fact, these coefficients can be put into the context of the classical theory of symmetric polynomials. We are now going to describe this relation.

Let $\Delta^m(n)$ denote the set of all $p = (p_0, p_1, \dots, p_m) \in \mathbf{Z}_+^{m+1}$ with $\sum p_i = n$. Then $c_{pq} = 0$ unless $p \in \Delta^m(n), q \in \Delta^n(m)$. We associate to $q \in \Delta^n(m)$ a symmetric polynomial $M_q(x_1, \dots, x_m)$ as follows: $M_q(x_1, \dots, x_m)$ is the sum of all monomials $x_1^{\alpha_1} x_2^{\alpha_2} \cdots x_m^{\alpha_m}$ such that q_0 of the exponents α_i are equal to n, q_1 of them are equal to $n-1$, and so on, q_n of them are equal to 0. Being symmetric, each M_q is uniquely represented as a polynomial in elementary symmetric polynomials

$$e_i(x) = e_i(x_1, \dots, x_m) = \sum_{1 \le j_1 < \dots < j_i \le m} x_{j_1} \cdots x_{j_i}.$$

Proposition 1.1. *For every $p \in \Delta^m(n)$, $q \in \Delta^n(m)$, the coefficient c_{pq} in the resultant is equal to $(-1)^{\sum_i ip_i}$ times the coefficient at $\prod_i e_i(x)^{p_i}$ in the expansion of M_q as a polynomial in elementary symmetric polynomials.*

For example, if $m = n = 2$, and $q = (1, 0, 1)$ then $M_q = x_1^2 + x_2^2 = e_1(x)^2 - 2e_2(x)$, which contributes $b_0 b_2 (a_1^2 - 2a_0 a_2)$ to (1.14).

Proof. Since $g(x) = b_n \prod_j (x - y_j)$, we can write (1.3) as

$$R(f, g) = a_m^n \prod_i g(x_i).$$

Expanding this product into the sum of monomials in $x_1, \dots, x_m$, we see that

$$R(f, g) = a_m^n \sum_q b^q M_q(x_1, \dots, x_m).$$

It remains to expand each $M_q(x_1, \dots, x_m)$ as a polynomial in elementary symmetric polynomials and notice that $e_i(x_1, \dots, x_m) = (-1)^{m-i} a_i / a_m$.

By a *partition* of length $\leq m$, we shall understand a collection $\lambda = (\lambda_1, \ldots, \lambda_m)$ of non-negative integers such that $\lambda_1 \geq \lambda_2 \geq \cdots \geq \lambda_m$. The numbers λ_i are called parts of λ. For any such λ we have the following *monomial symmetric function* in m variables:

$$m_\lambda(x_1, \ldots, x_m) = \sum_{\alpha \in S_m \lambda} x_1^{\alpha_1} \cdots x_m^{\alpha_m}, \tag{1.44}$$

where S_m is the group of permutations acting on $\mathbf{Z}^m$, and α runs through the S_m-orbit of λ (cf. Section 2C Chapter 4). For example, the function M_q above coincides with m_λ, where the partition λ has q_i parts equal to $n-i$ for $i = 0, \ldots, n$; the elementary symmetric polynomial e_i is also of the form m_λ, where λ has i parts equal to 1 and $m - i$ parts equal to 0.

Clearly, the functions m_λ form a basis in the vector space of all symmetric polynomials in $x_1, \ldots, x_m$. A different basis in the space of symmetric polynomials is provided by the products $\prod_{i=1}^m e_i^{p_i}$ of elementary symmetric polynomials. Such a product is also encoded by a partition μ having p_i parts equal to i for $i = 1, \ldots, m$; so we shall write

$$e_\mu(x) = \prod_{i=1}^m e_i(x)^{p_i}. \tag{1.45}$$

Thus, μ runs over the partitions all of whose parts are $\leq m$. However, such a partition μ can have more parts than m, so we obtain a different class of partitions.

The relationship between these two classes of partitions (with length $\leq m$, and with parts not exceeding m) is given by the *conjugation* of partitions. Recall [Macd] that the *diagram* of a partition λ is defined as the finite set

$$\{(i, j) \in \mathbf{Z}^2 : 1 \leq i \leq m, 1 \leq j \leq \lambda_i\}.$$

The *conjugate* partition λ^* is defined by the condition that its diagram is transpose of the diagram of λ. This means that if μ has p_i parts equal to i for $i = 1, \ldots, m$ then the parts of μ^* are $(p_1 + p_2 + \cdots + p_m, p_2 + \cdots + p_m, \ldots, p_m)$. Thus, the correspondence $\mu \mapsto \mu^*$ is a bijection between the partitions with parts $\leq m$ and those of length $\leq m$.

Proposition 1.1 says that the coefficients in the resultant are essentially the entries of the transition matrix between the two bases (m_λ) and (e_μ). Although there is known some combinatorial interpretation of these entries [ER], it is not good enough even for deciding what are the signs of the c_{pq}, or which of the coefficients are non-zero.

A much better interpretation is known for the entries of the inverse matrix $d_{\lambda\mu}$, i.e., the coefficients in the expansion

$$e_\mu = \sum_\lambda d_{\lambda\mu} m_\lambda.$$

Expanding the product of elementary symmetric polynomials, we see that $d_{\lambda\mu}$ is equal to the number of (0, 1)-matrices with row sums $\lambda_1, \ldots, \lambda_m$ and column sums $\mu_1, \ldots, \mu_n$. The quantities $d_{\lambda\mu}$ play an important role in combinatorics, representation theory of symmetric groups and classical theory of symmetric polynomials, see e.g. [Macd]. The description of all pairs (λ, μ) such that $d_{\lambda\mu} > 0$ is known as the Gale-Ryser theorem. To state it we need the so-called *dominance partial order* on partitions.

Let $\lambda = (\lambda_1, \ldots, \lambda_m)$, $\nu = (\nu_1, \ldots, \nu_m)$ be two partitions of length $\leq m$. We say that ν *dominates* λ and write $\lambda \leq \nu$ if $\lambda_1 + \cdots + \lambda_i \leq \nu_1 + \cdots + \nu_i$ for $i = 1, \ldots, m-1$, and $\lambda_1 + \cdots + \lambda_m = \nu_1 + \cdots + \nu_m$.

Proposition 1.2. (Gale-Ryser Theorem). *We have $d_{\lambda\mu} > 0$ if and only if $\lambda \leq \mu^*$. Furthermore, $d_{\mu^*\mu} = 1$.*

For the proof see [Ry], Theorem 1.1 Chapter 6.

Combining Propositions 1.1 and 1.2, we can find an upper bound for the set of all pairs (p, q) such that the corresponding coefficient c_{pq} in the resultant is non-zero.

Proposition 1.3. *Let $p \in \Delta^m(n)$, $q \in \Delta^n(m)$, and let λ, μ be the partitions such that λ has q_i parts equal to $n - i$ for $i = 0, \ldots, n$, and μ has p_i parts equal to i for $i = 1, \ldots, n$. Then $c_{pq} = 0$ unless $\mu^* \leq \lambda$. Furthermore, if $\mu^* = \lambda$ then $c_{pq} = (-1)^{\sum_i ip_i}$.*

We shall see in Section 2 that the pairs (p, q) such that $\mu^* = \lambda$ are exactly the vertices of the Newton polytope of the resultant. It is interesting to note that although the concept of the Newton polytope was not popular among the algebraists in the 19th century, the extreme monomials in the resultant were singled out by Gordan [Go2]. To explain his observation, we represent the resultant by the Sylvester formula (1.12). Expanding the Sylvester determinant into monomials, we get

$$R(f, g) = \sum_{\pi} \operatorname{sgn}(\pi) b_{\pi_1 - 1} b_{\pi_2 - 2} \cdots b_{\pi_m - m} a_{\pi_{m+1} - 1} a_{\pi_{m+2} - 2} \cdots a_{\pi_{m+n} - n}, \quad (1.46)$$

the sum over all permutations $\pi = (\pi_1, \ldots, \pi_{m+n})$ of the indices $1, \ldots, m+n$ (with the convention that $a_i = 0$ unless $0 \leq i \leq m$, $b_j = 0$ unless $0 \leq j \leq n$). We call a permutation π an (m, n)-*shuffle permutation* if

$$\pi_1 < \pi_2 < \cdots < \pi_m, \quad \pi_{m+1} < \pi_{m+2} < \cdots < \pi_{m+n}. \quad (1.47)$$

It is easy to see that the monomial in (1.46) corresponding to an (m, n)-shuffle permutation π has the form $(-1)^{\sum_i ip_i} a^p b^q$, where the exponents p_i and q_j are given by

$$p_0 = \pi_1 - 1,\ \ p_i = \pi_{i+1} - \pi_i - 1 \text{ for } i = 1, \ldots, m-1,\ \ p_m = m + n - \pi_m, \quad (1.48)$$

$$q_0 = \pi_{m+1} - 1,\ \ q_j = \pi_{m+j+1} - \pi_{m+j} - 1 \text{ for } j = 1, \ldots, n-1,$$

$$q_n = m + n - \pi_{m+n}. \quad (1.49)$$

It is easy to see that every such monomial appears exactly once in (1.46), and that (1.48) and (1.49) establish a bijection between (m, n)-shuffle permutations and pairs (p, q) from Proposition 1.3 such that the corresponding partitions λ and μ are conjugate to each other.

2. Newton polytopes of the classical discriminant and resultant

A. The Newton polytope of the discriminant

We write the discriminant of a polynomial $f(x) = a_0 + a_1x + \cdots + a_nx^n$ in one variable x as a linear combination of monomials

$$\Delta(f) = \sum_{\varphi_0, \ldots, \varphi_n} c_{\varphi_0, \ldots, \varphi_n} a_0^{\varphi_0} a_1^{\varphi_1} \cdots a_n^{\varphi_n}$$

or in abbreviated form as $\Delta(f) = \sum_\varphi c_\varphi a^\varphi$, where $\varphi = (\varphi_0, \ldots, \varphi_n)$ runs over some finite set in $\mathbf{Z}_+^{n+1}$. No simple explicit formula seems to be known for the coefficients c_φ. Nevertheless, it is possible to describe the Newton polytope $N(\Delta)$ of Δ.

By definition (see Chapter 6), $N(\Delta) \in R^{n+1}$ is the convex hull of the vectors φ such that $c_\varphi \neq 0$. Let $A = \{0, 1, \ldots, n\} \subset \mathbf{Z}$ be the set of exponents of monomials which may occur in f. By definition, our discriminant $\Delta(f)$ coincides with the A-discriminant $\Delta_A(f)$. In Theorem 1.4 Chapter 10 we have described the Newton polytope of the principal A-determinant E_A. On the other hand, in our situation Theorem 1.2 Chapter 10 means that $E_A(f) = a_0a_n\Delta_A(f)$. Combining these two facts we get the following description of $N(\Delta)$.

Theorem 2.1. *The Newton polytope of $\Delta(f)$ is the secondary polytope $\Sigma(A) \subset \mathbf{R}^{n+1}$ translated by the vector* $(-1, 0, 0, \ldots, 0, -1)$.

Here the secondary polytope is taken with respect to the measure on $\mathbf{R}$ normalized in a usual way (so that the segment $[0, 1]$ has length 1). This polytope was described in Section 3A Chapter 7: the "triangulations" which parametrize the vertices of the secondary polytope are, in this simple case, just subdivisions of the segment $Q = [0, n]$ into several subsegments. We summarize these results explicitly by taking into account the translation vector in Theorem 2.1.

Theorem 2.2. *The Newton polytope $N(\Delta)$ is combinatorially equivalent to an $(n-1)$-dimensional cube. It consists of all points $\varphi = (\varphi_0, \varphi_1, \dots, \varphi_n) \in \mathbf{R}^{n+1}$ satisfying two linear equations*

$$\sum_i \varphi_i = 2n-2, \ \sum_i (n-i)\varphi_i = n(n-1),$$

and $2n-2$ linear inequalities

$$\varphi_j \geq 0, \ \sum_{i=0}^{j} (j-i)\varphi_i \geq j(j-1) \quad \text{for} \quad j = 1, 2, \dots, n-1.$$

The vertices of $N(\Delta)$ are in a bijection with the subsets $I \subset \{1, 2, \dots, n-1\}$; the vertex $\varphi(I)$ corresponding to $I = \{i_1 < i_2 < \cdots < i_s\}$ has coordinates $\varphi_0 = i_1 - 1$, $\varphi_n = n - i_s - 1$, $\varphi_{i_p} = i_{p+1} - i_{p-1}$ for $p = 1, \dots, s$, and $\varphi_i = 0$ for $i \notin \{0, n\} \cup I$.

The absolute values of the coefficients of extreme monomials in Δ are given by specializing Theorem 1.4 (b) Chapter 10. Theorem 1.14 of Chapter 10 gives the ratio of signs for any two triangulations obtained from each other by a modification. By "integrating" this ratio, starting from the coefficient at the monomial $a_0^{n-1}a_n^{n-1}$ which is, by (1.37), equal to $+n^n$, we arrive at the following.

Theorem 2.3. *For every subset $I = \{i_1 < i_2 < \cdots < i_s\} \subset \{1, 2, \dots, n-1\}$ the monomial*

$$a^{\varphi(I)} = a_0^{i_1 - 1} a_{i_1}^{i_2} a_{i_2}^{i_3 - i_1} a_{i_3}^{i_4 - i_2} \dots a_{i_{s-1}}^{i_s - i_{s-2}} a_{i_s}^{n - i_{s-1}} a_n^{n - i_s - 1}$$

appears in Δ with the coefficient

$$c_{\varphi(I)} = (-1)^{n(n-1)/2} \prod_{p=0}^{s} (-1)^{l_p(l_p - 1)/2} l_p^{l_p},$$

where $l_p = i_{p+1} - i_p$ (with the convention $l_0 = i_1, l_s = n - i_s$).

The case of a cubic polynomial is illustrated by Figure 60.

B. The Newton polytope of the resultant: the vertices

We fix numbers $m, n \geq 1$ and consider the resultant $R(f, g) = R_{m,n}(f, g)$ where

$$f(x) = a_0 + a_1 x + \cdots + a_m x^m, \quad g(x) = b_0 + b_1 x + \cdots + b_n x^n$$

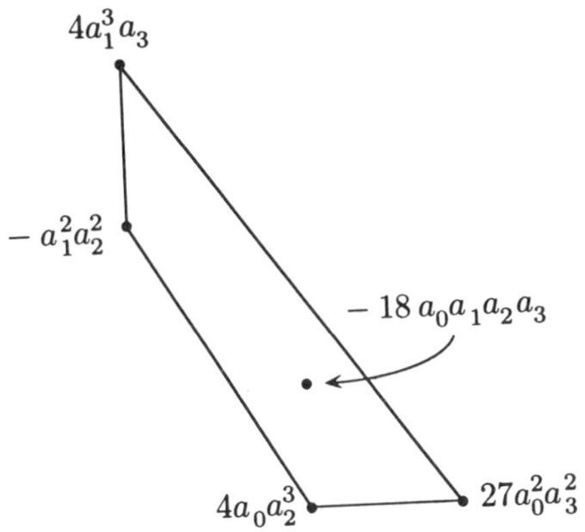

Figure 60. The discriminant of a cubic polynomial

are indeterminate polynomials. So $R(f, g)$ is a polynomial in the a_i and b_j and we write it as a sum of monomials, as in (1.43):

$$R(f, g) = \sum_{p,q} c_{pq} a^p b^q$$

where $p = (p_0, p_1, \ldots, p_m)$ and $q = (q_0, q_1, \ldots, q_n)$ are multi-indices. The Newton polytope of the resultant lies in the space $\mathbf{R}^{m+n+2}$ with coordinates $p_0, p_1, \ldots, p_m, q_0, q_1, \ldots, q_n$. By definition, it is the convex hull of all lattice points (p, q) such that $c_{pq} \neq 0$. We shall denote this polytope by $N_{m,n}$.

Let A be the set of points $(0, 0), (1, 0), \ldots, (m, 0), (0, 1), (1, 1), \ldots, (n, 1)$ in $\mathbf{Z}^2$. The convex hull Q of A is the trapezoid shown in Figure 43.

Using the Cayley trick (Proposition 1.7 Chapter 9) we see that $R(f, g)$ is equal to $\Delta_A(f(x) + yg(x))$. The toric variety X_A associated to A is smooth: this is the rational normal scroll of type (m, n) (see Examples 1.1 (d) of Chapter 5 and 3.6 of Chapter 3). Thus Δ_A coincides with the regular A-determinant D_A (see Theorem 1.3 Chapter 11). It follows that $N_{m,n} = N(D_A)$.

By Theorem 3.2 Chapter 11, the vertices of $N(D_A) = N_{m,n}$ are in bijection with the D-equivalence classes of coherent triangulations of (Q, A). We recall that every coherent triangulation T of (Q, A) gives rise to a vertex η_T of the Newton polytope $N(\Delta_A)$. The component of η_T corresponding to $\omega \in A$ is equal to

$$\eta_T(\omega) = \sum_{\sigma} (-1)^{\dim(Q)-\dim(\sigma)} \mathrm{Vol}\,(\sigma)$$

where the sum is over the simplices Δ of arbitrary dimension appearing in T that have ω as a vertex and are massive (i.e., affinely span a face of Q); the volume form on each face $\Gamma \subset Q$ is induced by the integral lattice spanned by $A \cap \Gamma$. Two triangulations T and T' are called D-equivalent if $\eta_T = \eta_{T'}$. In our situation when Q is a trapezoid it is easy to see that all triangulations of (Q, A) are coherent (this is proved in the same way as Proposition 3.4 of Chapter 7). Furthermore, the D-equivalence is generated by the following relation: $T \sim T'$ if T' is obtained from T by a subdivision of a triangle into the union of two smaller triangles. This follows from the description of the D-equivalence for plane polygons given in Proposition 3.9 Chapter 11.

We say that a triangulation T is *coarse* if no triangle of T is the union of two other triangles. Clearly, coarse triangulations form a set of representatives of D-equivalence classes.

The vertex η_T of $N_{m,n}$ corresponding to a coarse triangulation T can be described as follows. The coordinate $p_i = p_i(T)$ is equal to $|b-c|$ if there is a triangle in T with vertices $(i, 0)$, $(b, 1)$, $(c, 1)$, otherwise $p_i(T) = 0$; similarly, $q_j = q_j(T)$ is equal to $|b - c|$ if there is a triangle in T with vertices $(j, 1)$, $(b, 0)$, $(c, 0)$, otherwise $q_j(T) = 0$. Summarizing, we have

Theorem 2.4. *The correspondence $T \mapsto (p(T), q(T))$ is a bijection between the set of coarse triangulations of (Q, A) and the set of vertices of $N_{m,n}$.*

We illustrate this theorem for the case $m = n = 2$. By (1.14), the resultant of two quadratic forms contains 7 monomials, but only 6 of them are vertices of $N_{2,2}$. They correspond to 6 coarse triangulations of (Q, A) as shown in Figure 61.

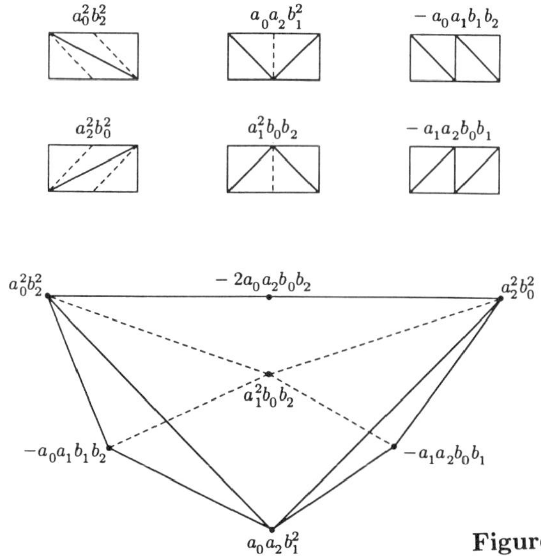

Figure 61. The polytope $N_{2,2}$.

We are going to study $N_{m,n}$ in detail. Unfortunately, we cannot apply the general theory of Chapter 7 since $N_{m,n}$ is not a secondary polytope. So we start describing $N_{m,n}$ from scratch using nothing but Theorem 2.4. First we present several convenient combinatorial interpretations of the vertices of $N_{m,n}$.

Let us label the horizontal unit segments on the boundary of Q as follows: we attach the symbol A_i to the segment $[(i-1,0),(i,0)]$ for $i=1,\dots,m$, and the symbol B_j to the segment $[(j-1,1),(j,1)]$ for $j=1,\dots,n$. Now let T be a coarse triangulation of (Q, A). We associate to T a shuffle $w(T)$ of two words $A_1A_2\dots A_m$ and $B_1B_2\dots B_n$. We recall that a shuffle is a word of length $m+n$ containing each of the $A_1A_2\dots A_m$ and $B_1B_2\dots B_n$ as a subsequence; the shuffles were introduced (Section 3D Chapter 7) in relation to lattice paths. To construct $w(T)$ we order the triangles of T into a sequence $\sigma_1,\sigma_2,\dots,\sigma_r$ so that σ_1 is the triangle containing the side $[(0,0),(0,1)]$, and each σ_i has a common side with σ_{i-1}. The shuffle $w(T)$ lists all the horizontal unit segments on the boundary of Q in the following order: first the segments on the horizontal side of σ_1, then those on the horizontal side of σ_2, etc. (see Figure 62). This construction makes clear all the statements in the next proposition.

Proposition 2.5. *The correspondence $T\mapsto w(T)$ is a bijection between the set of coarse triangulations of (Q, A) and the set of shuffles of the $A_1A_2\dots A_m$ and $B_1B_2\dots B_n$. Hence the number of coarse triangulations, i.e., the number of vertices of $N_{m,n}$ is $\binom{m+n}{m}$. The components of the vertex η_T are recovered from $w(T)$ as follows: $p_i(T)$ is the number of letters between A_i and A_{i+1}, and $q_j(T)$ is the number of letters between B_j and B_{j+1} (with the convention that A_0 and B_0 stand to the left of the whole shuffle, while A_{m+1} and B_{n+1} stand to the right).*

Motivated by Proposition 2.5, we call $N_{m,n}$ the *(m,n)-shuffle polytope.* It turns out that $N_{m,n}$ can be realized as a subpolytope of the *permutohedron* P_{m+n}. We recall (see Section 3C Chapter 7) that P_{m+n} is the convex hull in $\mathbf{R}^{m+n}$ of $(m+n)!$ points $\pi=(\pi_1,\dots,\pi_{m+n})$ representing permutations of $\{1,2,\dots,m+n\}$. Recall also that a permutation π is an (m,n)-shuffle permutation if it satisfies (1.47). There is a natural bijection between (m,n)-shuffle permutations and shuffles of $A_1A_2\dots A_m$ and $B_1B_2\dots B_n$: we put A_i on the π_i-th place in the shuffle, and B_j on the π_{m+j}-th place. Translating Proposition 2.5 into the language of shuffle permutations, we see that if π is a shuffle permutation corresponding to the shuffle $w(T)$ then the components of η_T are given by (1.48) and (1.49). This implies the following.

Proposition 2.6. *The shuffle polytope $N_{m,n}$ can be affinely embedded into P_{m+n} so that each vertex η_T goes to the vertex π corresponding to the shuffle $w(T)$.*

We know (from Section 3D Chapter 7) that the shuffles of $A_1A_2\dots A_m$ and

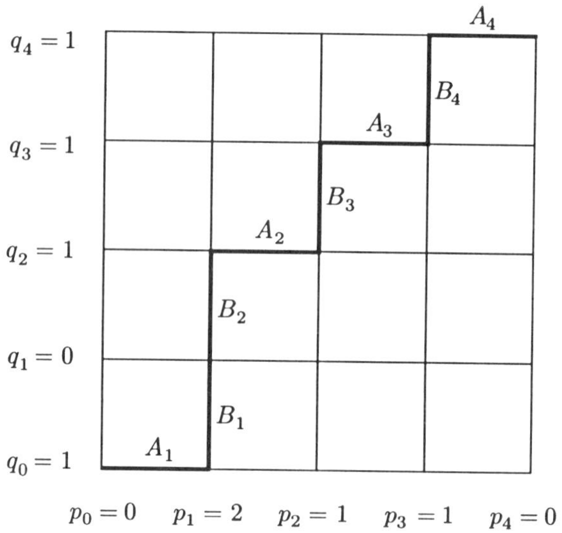

$$w = w(T) = A_1 B_1 B_2 A_2 B_3 A_3 B_4 A_4.$$

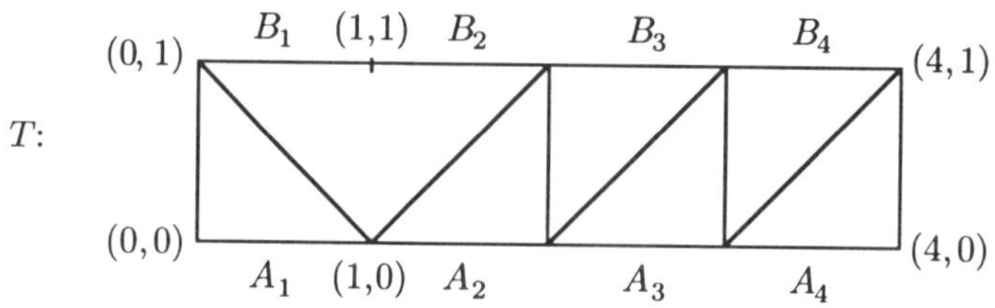

Figure 62.

$B_1 B_2 \ldots B_n$ are also in bijection with the *lattice paths* from (0, 0) to (m, n). This time the symbols in the shuffle are interpreted as unit moves, the symbol A_i meaning the horizontal move from the vertical line $x = i - 1$ to the line $x = i$, and B_j meaning the vertical move from the horizontal line $y = j - 1$ to the line $y = j$. For a lattice path L from (0, 0) to (m, n), we define the integers $p_i(L)$ $(i = 0, \ldots, m)$ and $q_j(L)$ $(j = 0, \ldots, n)$ as follows: $p_i(L)$ is the length of the part of L lying on the vertical line $x = i$, and $q_j(L)$ is the length of the part of L lying on the horizontal line $y = j$. In particular, $p_i(L) = 0$ if L just crosses the line $x = i$ in the horizontal direction, and similarly for $q_j(L)$. Translating Proposition 2.5 into the language of lattice paths, we obtain the following.

Proposition 2.7. *If L is the lattice path corresponding to the shuffle $w(T)$ then the vertex η_T has components $p_i(T) = p_i(L)$ and $q_j(T) = q_j(L)$ for all i, j.*

In view of this proposition, it will be convenient for us to use the notation $\eta_T = \eta_L$ for a lattice path L corresponding to $w(T)$.

The correspondence between coarse triangulations of the trapezoid, shuffles, and lattice paths is illustrated in Figure 62.

C. The Newton polytope of the resultant: further properties

Theorem 2.8. *The polytope $N_{m,n}$ has dimension $m+n-1$. Its affine span in $\mathbf{R}^{m+n+2}$ is given by three linear equations*

$$\sum_{i=0}^{m} p_i = n, \ \sum_{j=0}^{n} q_j = m, \ \sum_{i=0}^{m}(m-i)p_i + \sum_{j=0}^{n}(n-j)q_j = mn. \qquad (2.1)$$

Proof. First we show that every vertex of $N_{m,n}$ satisfies (2.1). From the point of view of the resultant, this is equivalent to the three homogeneity conditions (1.7). It will also be instructive for us to deduce the last equation in (2.1) in a purely combinatorial way (the first two equations are obvious from any of the above interpretations of vertices). In fact, we shall prove a stronger statement. For every integral point $(k, l) \in [0, m] \times [0, n]$, we define an affine-linear function h_{kl} on $\mathbf{R}^{m+n+2}$ by

$$h_{kl} = \sum_{i=0}^{k}(k-i)p_i + \sum_{j=0}^{l}(l-j)q_j - kl. \qquad (2.2)$$

Lemma 2.9. *Every vertex η_L of $N_{m,n}$ satisfies the inequalities $h_{kl}(\eta_L) \geq 0$ for all k, l. Furthermore, $h_{kl}(\eta_L) = 0$ if and only if a lattice path L passes through the point (k, l).*

The last equation in (2.1) is a special case of Lemma 2.9 corresponding to $(k, l) = (m, n)$.

Proof of Lemma 2.9. By Proposition 2.7, the component p_i of η_L is equal to $p_i(L)$, i.e., to the length of the intersection of the polygonal line L with the vertical line $x = i$. We write $\sum_{i=0}^{k}(k-i)p_i$ in the form $\sum_{i=0}^{k-1}(p_0 + p_1 + \cdots + p_i)$. Clearly, this sum is equal to the area of the part of the rectangle $[0, k] \times [0, n]$ lying below the polygonal line L. Similarly, $\sum_{j=0}^{l}(l-j)q_j$ is equal to the area of the part of the rectangle $[0, m] \times [0, l]$ lying above L. These two parts cover the rectangle $[0, k] \times [0, l]$ and do not overlap; moreover, their union is equal to $[0, k] \times [0, l]$ if

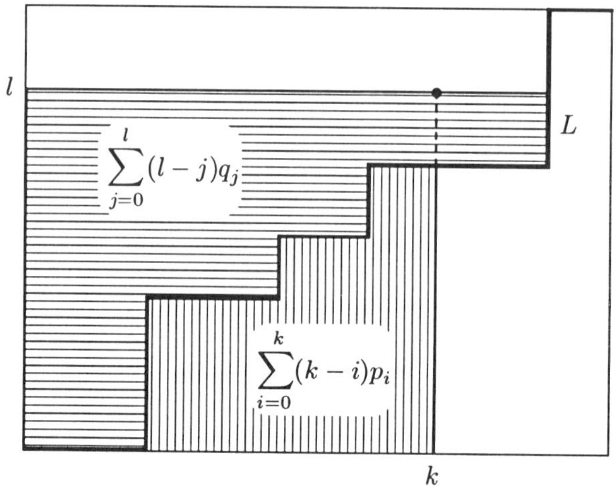

Figure 63. Proof that $h_{kl}(\eta_L) \geq 0$

and only if L passes through (k, l). Hence, the sum of their areas is $\geq kl$ with the equality if and only if L passes through (k, l) (see Figure 63). This proves Lemma 2.9 and the equations (2.1).

Clearly, the three linear equations in (2.1) are independent, i.e., define an affine subspace of dimension $m + n - 1$. To complete the proof of Theorem 2.8, it remains to construct $m + n$ affinely independent vertices of $N_{m,n}$. To do this, we consider all (m, n)-shuffles with the property that at least one of the words $A_1 A_2 \dots A_m$ and $B_1 B_2 \dots B_n$ is a subword of our shuffle, i.e., is not separated by other symbols. There are exactly $m + n$ such shuffles, and the corresponding vertices of $N_{m,n}$ given by Proposition 2.5 have the form

$$ne_0 + me'_n,\ ne_m + me'_0,\ ne_i + ie'_0 + (m-i)e'_n, \qquad i = 1, \dots, m-1,$$

$$je_0 + (n-j)e_m + me'_j, \qquad j = 1, \dots, n-1,$$

where $e_0, \dots, e_m$ is the standard basis in $\mathbf{R}^{m+1}$, and $e'_0, \dots, e'_n$ is the standard basis in $\mathbf{R}^{n+1}$. The proof that these points are affinely independent is straightforward, and we leave it to the reader.

Note that Lemma 2.9 is a lattice analog of the classical Young inequality (see, e.g., [HLP]), stated as follows. Let $y = \varphi(x)$ be a continuous, strictly increasing function of x for $x \geq 0$, with $\varphi(0) = 0$. Then for $a, b \geq 0$

$$ab \leq \int_0^a \varphi(x)dx + \int_0^b \varphi^{-1}(y)dy,$$

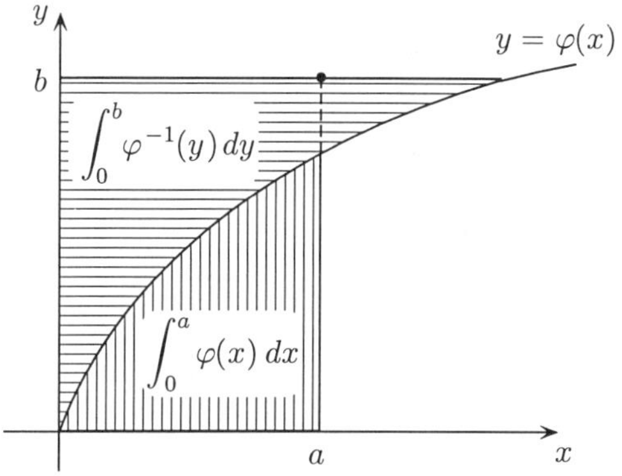

Figure 64. The Young inequality

where $\varphi^{-1}(y)$ is the function inverse to $\varphi(x)$. The inequality is strict unless $b = \varphi(a)$. The proof is the same as that of Lemma 2.9 (see Figure 64).

Proposition 2.10. *Each of the polytopes $N_{1,n}$ and $N_{n,1}$ is an n-dimensional simplex.*

This is clear since both $N_{1,n}$ and $N_{n,1}$ are of dimension n and have $n + 1$ vertices.

Now we describe the face lattice of $N_{m,n}$. We need some terminology. By a *block* in the rectangle $[0, m] \times [0, n]$, we mean a sub-rectangle $[k, k'] \times [l, l']$ with integral vertices, together with a collection of some vertical chords $x = k_1, x = k_2, \ldots, x = k_r$ and some horizontal chords $y = l_1, y = l_2, \ldots, y = l_s$. A block can be *degenerate*, i.e., a vertical or horizontal segment, but not a point. The chords can be taken only for non-degenerate blocks, but even then each of the collections of selected vertical or horizontal chords (or both) can be empty. The k_i's and l_j's are integers such that $k < k_1 < \cdots < k_r < k', l < l_1 < \cdots < l_s < l'$. For a block $\mathcal{B}$ we denote its south-west corner (k, l) by $\min(\mathcal{B})$ and its north-east corner (k', l') by $\max(\mathcal{B})$.

By a *(m, n)-labyrinth* we mean a (non-empty) collection Λ of blocks $\mathcal{B}_1, \mathcal{B}_2, \ldots, \mathcal{B}_r$ in $[0, m] \times [0, n]$ such that $\min(\mathcal{B}_1) = (0, 0)$, $\max(\mathcal{B}_r) = (m, n)$, and $\min(\mathcal{B}_i) = \max(\mathcal{B}_{i-1})$ for $i = 2, \ldots, r$. By a slight abuse of notation, we denote by the same symbol Λ the subset of the rectangle $[0, m] \times [0, n]$ formed by all the sides and chords of all blocks $\mathcal{B}_1, \mathcal{B}_2, \ldots, \mathcal{B}_r$. Figure 65 shows an example of a block and a labyrinth.

Theorem 2.11. *The faces of $N_{m,n}$ are in bijection with (m, n)-labyrinths: the face $F(\Lambda)$ corresponding to a labyrinth Λ has vertices η_L for all lattice paths L contained in Λ.*

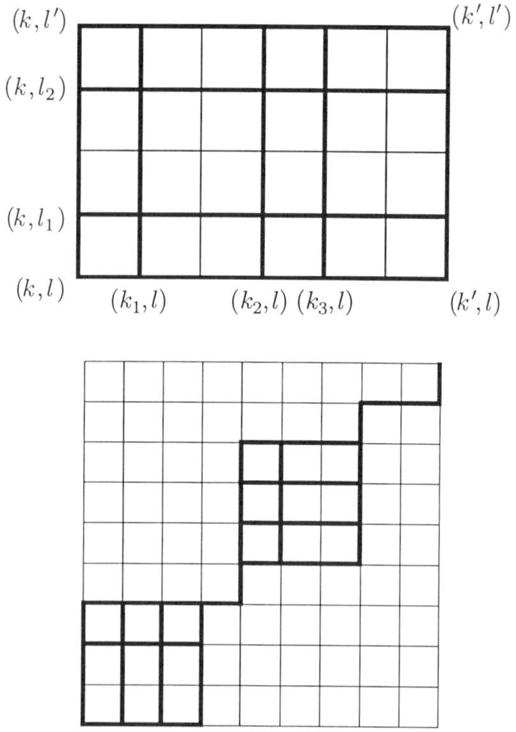

Figure 65. A block and a labyrinth

Corollary 2.12. *The bijection* $\Lambda \mapsto F(\Lambda)$ *is order-preserving:* $F(\Lambda) \subset F(\Lambda')$ *if and only if* $\Lambda \subset \Lambda'$, *where* Λ *and* Λ' *are considered as subsets of the rectangle* $[0, m] \times [0, n]$.

Corollary 2.13. *The edges of* $N_{m,n}$ *correspond to labyrinths having only one non-degenerate block, with no chords selected (see Figure 66).*

In the course of the proof of Theorem 2.11 we shall also prove the following.

Theorem 2.14. *The polytope* $N_{m,n}$ *is defined in* $\mathbf{R}^{m+n+2}$ *by linear equations (2.1) and linear constraints*

$$p_k \geq 0, \;\; q_l \geq 0, \;\; h_{kl} \geq 0 \quad \text{for} \quad (k, l) \in [0, m] \times [0, n]. \tag{2.3}$$

Proof of Theorems 2.11 and 2.14. We temporarily denote by $N'_{m,n}$ the set of points in $\mathbf{R}^{m+n+2}$ satisfying (2.1) and (2.3). Clearly, $N'_{m,n}$ is bounded, hence is a convex polytope. By Theorem 2.8 and Lemma 2.9, $N_{m,n} \subseteq N'_{m,n}$.

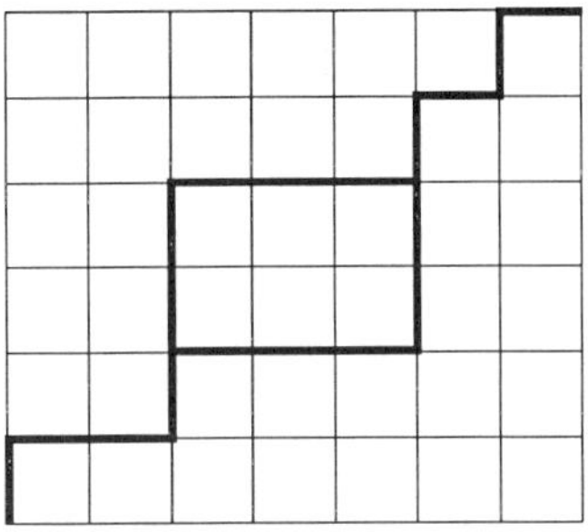

Figure 66. An edge of $N_{m,n}$

For every three subsets $I \subset [0,m]$, $J \subset [0,n]$, $K \subset [0,m] \times [0,n]$ we set

$$F(I,J,K) =$$

$$\{(p,q) \in N'_{m,n} : p_i = 0 \; for \; i \in I, \; q_j = 0 \; for \; j \in J, \; h_{k,l} = 0 \; for \; (k,l) \in K\}, \tag{2.4}$$

where the functions h_{kl} are defined by (2.2). By definition, the faces of $N'_{m,n}$ are exactly non-empty subsets of the type $F(I,J,K)$.

We say that a triple (I,J,K) is *related* to an (m,n)-labyrinth Λ if for a lattice path L the conditions $\eta_L \in F(I,J,K)$ and $L \subset \Lambda$ are equivalent to each other.

Lemma 2.15.

(a) *For every (m,n)-labyrinth Λ there exists a triple (I,J,K) related to Λ.*

(b) *Fix a (non-empty) face F of $N'_{m,n}$, and set*

$$I = \{i \in [0,m] : p_i = 0 \text{ on } F\}, \;\; J = \{j \in [0,n] : q_j = 0 \; on \; F\},$$

$$K = \{(k,l) \in [0,m] \times [0,n] : h_{kl} = 0 \text{ on } F\}.$$

Then the triple (I,J,K) is related to some labyrinth Λ.

Before proving this lemma we derive Theorems 2.11 and 2.14. Since every labyrinth contains some lattice path, it follows from Lemma 2.15 (b) that every face F of $N'_{m,n}$ contains some vertex η_L of $N_{m,n}$. In particular, every vertex of $N'_{m,n}$ is a vertex of $N_{m,n}$, which proves Theorem 2.14. Applying again Lemma 2.15 (b), we see that every face F of $N_{m,n}$ has as vertices the points η_L, where L runs over all lattice paths contained in some labyrinth Λ. Hence F has the form $F(\Lambda)$ from Theorem 2.11. Conversely, by Lemma 2.15 (a), each $F(\Lambda)$ is a face of $N_{m,n}$, which completes the proof of Theorem 2.11.

Proof of Lemma 2.15. (a) We can take K to be the set of all lattice points lying on degenerate blocks of Λ, and I (resp. J) to be the set of all $i \in [0, m]$ (resp. $j \in [0, n]$) such that the line $x = i$ (resp. $y = j$) is a chord of some non-degenerate block of Λ but not a selected one. The fact that (I, J, K) is related to Λ follows from Lemma 2.9.

The proof of Lemma 2.15 (b) is more elaborate. We deduce it from the next description of the triple (I, J, K) corresponding to a face F of $N'_{m,n}$.

Lemma 2.16. *Let (I, J, K) be a triple associated to a face F of $N'_{m,n}$ as in Lemma 2.15 (b).*

(a) *The set K is the disjoint union of several lattice paths $L_1, \ldots, L_r$ such that $\min(L_1) = (0,0)$, $\max(L_r) = (m,n)$, and $\min(L_i)$ lies to the northeast of $\max(L_{i-1})$ for $i = 2, \ldots, r$. In other words, if L_i is a path from $[k_i, l_i]$ to $[k'_i, l'_i]$ then $0 = k_1 \le k'_1 < k_2 \le k'_2 < \cdots < k_r \le k'_r = m$ and $0 = l_1 \le l'_1 < l_2 \le l'_2 < \cdots < l_r \le l'_r = n$.*

(b) *Suppose i is such that K crosses the vertical line $x = i$. Then $i \in I$ if and only if i is different from $k'_1, k_2, k'_2, \ldots, k'_{r-1}, k_r$ and the intersection $K \cap \{x = i\}$ consists of one point. Analogously, if K crosses a horizontal line $y = j$ then $j \in J$ if and only if j is different from $l'_1, l_2, l'_2, \ldots, l'_{r-1}, l_r$ and the intersection $K \cap \{y = j\}$ consists of one point.*

Using Lemma 2.16, we associate to (I, J, K) an (m, n)-labyrinth Λ in the following way. The degenerate blocks of Λ are horizontal and vertical segments of lattice paths $L_1, \ldots, L_r$ from Lemma 2.16 (a). The non-degenerate blocks of Λ are rectangles $\mathcal{B}_1, \mathcal{B}_2, \ldots, \mathcal{B}_{r-1}$ in $[0, m] \times [0, n]$ such that $\min(\mathcal{B}_i) = (k'_i, l'_i)$, $\max(\mathcal{B}_i) = (k_{i+1}, l_{i+1})$. Finally, in every non-degenerate block we select the chords $x = i$ for all $i \notin I$, and the chords $y = j$ for all $j \notin J$. The fact that (I, J, K) is related to Λ is clear.

It remains to prove Lemma 2.16.

Proof of Lemma 2.16. (a) It suffices to prove the following three statements:

(a1) K contains $(0, 0)$ and (m, n).

(a2) K cannot contain two lattice points (k, l), (k', l') such that $k < k'$, $l > l'$.

(a3) If K contains two lattice points u, v lying on the same horizontal or vertical line then K contains all the lattice points of the segment $[u, v]$.

The item (a1) is clear. To prove (a2) suppose that K contains (k, l), (k', l') such that $k < k'$, $l > l'$, i.e., $h_{kl} = h_{k'l'} = 0$ on F. To arrive at a contradiction, we consider the following identity:

$$h_{kl} + h_{k'l'} - h_{k'l} - h_{kl'} = (k' - k)(l - l'), \tag{2.5}$$

which follows directly from (2.2). The left hand side of (2.5) takes a non-positive value at each point of F, which contradicts the fact that the right hand side is positive.

To prove (a3) it is enough to consider the case when u and v lie on a horizontal line (the proof for a vertical line is the same). So we have to prove the following: if $k < k'' < k'$ and $(k, l), (k', l) \in K$ then $(k'', l) \in K$. To see this we use another identity, which is also a direct consequence of (2.2):

$$h_{kl} + h_{k'l} = h_{k+1,l} + h_{k'-1,l} + \sum_{k<k''<k'} p_{k''}. \tag{2.6}$$

Evaluating both sides of (2.6) at some point of F, we see that if $(k, l), (k', l) \in K$ then the left hand side vanishes, hence all summands on the right hand side also vanish. In particular, this shows that $(k + 1, l), (k' - 1, l) \in K$. Repeating this argument if necessary, we see that $(k'', l) \in K$ for $k < k'' < k'$, which completes the proof of (a3).

(b) We shall prove only the statement about I (the proof for J is completely analogous). It is easy to see that our statement is a consequence of the following two statements.

(b1) If $(0, j) \in K$ for some $j \geq 1$ then $0 \notin I$; if $(m, j) \in K$ for some $j \leq n - 1$ then $m \notin I$.

(b2) If $(i, j) \in K, i \in I$, and $|i' - i| = 1$ then $(i', j) \in K$.

The proof of (b1) and (b2) uses the same method as the above proof of (a2) and (a3). For (b1) we use the identities

$$h_{0j} + p_0 = h_{1j} + j, \;\; h_{mj} + p_m = h_{m-1,j} + (n - j) + (p_0 + \cdots + p_m - n). \tag{2.7}$$

If $i \in [1, m - 1]$ then (b2) follows from the identity

$$2h_{ij} + p_i = h_{i-1,j} + h_{i+1,j},$$

which is a special case of (2.6). If $i = 0$ then in view of (b1) we have $j = 0$, hence in this case (b2) follows from the first identity in (2.7) for $j = 0$. The case $i = m$ in (b2) is treated in the same way. This completes the proof of Lemma 2.16 and hence of Theorems 2.11 and 2.14.

Corollary 2.17. *If $m, n \geq 2$ then $N_{m,n}$ has exactly $mn + 3$ facets. They are given by linear equations $p_i = 0$ for $i = 0, 1, \ldots, m$, $q_j = 0$ for $j = 0, 1, \ldots, n$, and $h_{kl} = 0$ for $k = 1, 2, \ldots, m - 1$, $l = 1, 2, \ldots, n - 1$.*

Proof. By Theorem 2.14, every facet of $N_{m,n}$ is supported by one of the inequalities in (2.3). It remains to check which of these inequalities defines a facet, i.e, corresponds to a maximal proper labyrinth. The face $\{p_i = 0\}$ for $1 \le i \le m-1$ corresponds to the labyrinth having one block $[0, m] \times [0, n]$ with all the chords selected except the chord $\{x = i\}$. The face $\{p_0 = 0\}$ corresponds to the labyrinth having one non-degenerate block $[1, m] \times [0, n]$ with all the chords selected, and the unit horizontal segment from $(0, 0)$ to $(1, 0)$. The face $\{p_m = 0\}$ and all the faces $\{q_j = 0\}$ are described in a similar way. As for the face $\{h_{kl} = 0\}$ for $k = 1, 2, \ldots, m-1$, $l = 1, 2, \ldots, n-1$, it corresponds to the labyrinth with two blocks $[0, k] \times [0, l]$ and $[k, m] \times [l, n]$ and all chords selected. Each of these labyrinths is seen to be a maximal proper one.

It remains to check the faces $\{h_{0l} = 0\}$ and $\{h_{ml} = 0\}$ for $l = 1, 2, \ldots, n-1$, and $\{h_{k0} = 0\}$, $\{h_{kn} = 0\}$ for $k = 1, 2, \ldots, m-1$. Clearly, each of the faces $\{h_{0l} = 0\}$ is contained in the facet $\{q_0 = 0\}$ already taken into account. All the remaining cases are totally similar.

Theorem 2.18. *The face $F(\Lambda)$ of $N_{m,n}$ corresponding to an (m, n)-labyrinth Λ is combinatorially equivalent to $N_{m_1,n_1} \times \cdots \times N_{m_r,n_r}$, the product over all non-degenerate blocks $\mathcal{B}_1, \mathcal{B}_2, \ldots, \mathcal{B}_r$ in Λ, where $m_i - 1$ and $n_i - 1$ are numbers of vertical and horizontal chords selected in the block $\mathcal{B}_i$.*

Proof. Using induction on $m+n$, it is enough to prove the statement for the facets $F(\Lambda)$ of $N_{m,n}$. The cases $m = 1$ and $n = 1$ are clear from Proposition 2.10, so we assume that $m, n \ge 2$. Then the facets of $N_{m,n}$ are given by Corollary 2.17.

Consider first the facet $\{h_{kl} = 0\}$ for some $k = 1, 2, \ldots, m-1$, $l = 1, 2, \ldots, n-1$. We claim that this facet is not only combinatorially but affinely isomorphic to $N_{k,l} \times N_{m-k,n-l}$. To see this, consider the linear maps

$$\alpha : \mathbf{R}^{k+1} \times \mathbf{R}^{l+1} \times \mathbf{R}^{m-k+1} \times \mathbf{R}^{n-l+1} \to \mathbf{R}^{m+1} \times \mathbf{R}^{n+1},$$

$$\beta : \mathbf{R}^{m+1} \times \mathbf{R}^{n+1} \to \mathbf{R}^{k+1} \times \mathbf{R}^{l+1} \times \mathbf{R}^{m-k+1} \times \mathbf{R}^{n-l+1}$$

defined as follows. For

$$(p', q') \in \mathbf{R}^{k+1} \times \mathbf{R}^{l+1}, \ (p'', q'') \in \mathbf{R}^{m-k+1} \times \mathbf{R}^{n-l+1}$$

we set $\alpha(p', q', p'', q'') = (p, q)$, where

$$p_i = p'_i \quad \text{for} \quad 0 \le i < k; \quad p_k = p'_k + p''_0,$$

$$p_i = p''_{i-k} \quad \text{for} \quad k < i \le m,$$

and similarly,

$$q_j = q'_j \quad \text{for} \quad 0 \le j < l, \ q_l = q'_l + q''_0, \ q_j = q''_{j-l} \quad \text{for} \quad l < j \le n.$$

Conversely, for $(p, q) \in \mathbf{R}^{m+1} \times \mathbf{R}^{n+1}$ we set $\beta(p, q) = (p', q', p'', q'')$, where

$$p'_i = p_i \quad \text{for} \quad 0 \le i < k, \quad p'_k = l - p_0 - p_1 - \cdots - p_{k-1},$$
$$p''_0 = n - l - p_{k+1} - \cdots - p_m, \quad p''_i = p_{k+i} \quad \text{for} \quad 1 \le i \le m - k,$$

and similarly,

$$q'_j = q_j \quad \text{for} \quad 0 \le j < l, \quad q'_l = k - q_0 - q_1 - \cdots - q_{l-1},$$
$$q''_0 = m - k - q_{l+1} - \cdots - q_n, \quad q''_j = q_{l+j} \quad \text{for} \quad 1 \le j \le n - l.$$

It is straightforward to verify that α and β define mutually inverse isomorphisms between $N_{k,l} \times N_{m-k,n-l}$ and $N_{m,n} \cap \{h_{kl} = 0\}$. The same argument shows that each of the facets $\{p_0 = 0\}$ and $\{p_m = 0\}$ is affinely isomorphic to $N_{m-1,n}$, and each of $\{q_0 = 0\}$ and $\{q_n = 0\}$ is affinely isomorphic to $N_{m,n-1}$. It remains to show that a facet $\{p_i = 0\}$ for $i = 1, 2, \ldots, m - 1$ is combinatorially isomorphic to $N_{m-1,n}$, and a facet $\{q_j = 0\}$ for $j = 1, 2, \ldots, n - 1$ is combinatorially isomorphic to $N_{m,n-1}$. As shown in the proof of Corollary 2.17, each of these facets corresponds to a one-block labyrinth. More generally, consider a one-block labyrinth Λ on $[0, m] \times [0, n]$ with vertical chords $x = k_1, x = k_2, \ldots, x = k_{r-1}$ and horizontal chords $y = l_1, y = l_2, \ldots, y = l_{s-1}$. We claim that $F(\Lambda)$ is combinatorially equivalent to $N_{r,s}$. The vertices of $F(\Lambda)$ correspond to lattice paths from $(0, 0)$ to (m, n) lying in Λ. The only difference with $N_{r,s}$ is that the lattice paths are now taken on the rectangular lattice, where the sizes of all rectangles can be different. Going through the above arguments, we can show that Theorem 2.8, Lemma 2.9, and Theorem 2.11 after a suitable modification extend to the polytope $F(\Lambda)$, hence $F(\Lambda)$ is combinatorially equivalent to $N_{r,s}$.

Shuffle polytopes $N_{m,n}$ are reminiscent of another family of polytopes: the hypersimplices $\Delta_{m,n}$ (see Section 3A Chapter 6). The hypersimplex $\Delta_{m,n}$ has the same dimension $m + n - 1$ and the same number of vertices $\binom{m+n}{m}$ as $N_{m,n}$, but in general $\Delta_{m,n}$ and $N_{m,n}$ are not combinatorially equivalent.

We conclude this chapter with an illustration of how the algebraic properties of discriminants and resultants can provide geometric properties of the Newton polytopes. Consider the discriminant $\Delta(f)$ of a binary form f of degree $r + 1$. Denote its Newton polytope by N_r. The description of N_r is given by Theorem 2.2; it is a subpolytope in $\mathbf{R}^{r+2}$ combinatorially equivalent to an r-cube. According to (1.31), $\Delta(f)$ is equal to the resultant of two partial derivatives of the binary form $F(x_0, x_1)$ corresponding to f. Translating this statement into the language of Newton polytopes, we obtain the following.

Proposition 2.19. *Let $\pi : \mathbf{R}^{r+1} \times \mathbf{R}^{r+1} \to \mathbf{R}^{r+2}$ be the projection defined by*

$$\pi(p_0, \ldots, p_r, q_0, \ldots, q_r) = (p_0, p_1 + q_0, \ldots, p_r + q_{r-1}, q_r).$$

Then $\pi(N_{r,r}) = N_r$.

CHAPTER 13

Discriminants and Resultants for Forms in Several Variables

We now consider the most straightforward generalization of the resultants and discriminants treated in Chapter 12, namely the resultants and discriminants for homogeneous forms in several variables. Although they are very classical objects of study, many fundamental questions about them still remain open. To realize how little is known it is enough to mention that an explicit polynomial expression still remains unknown with the exception of some very special cases. The methods developed in Parts I and II help to put this and other questions in a general perspective. Without pretending to be complete, we present an overview of some fairly classical results together with more fresh developments.

1. Homogeneous forms in several variables

Here we collect some basic facts on discriminants and resultants for forms in several variables. To benefit the reader less interested in generalizations and more in the classical material, we have tried to make the exposition reasonably self-contained, at the price of some repetition.

A. The resultant: basic properties and Poisson formula

Let $f_0, f_1, \ldots, f_k$ be $k+1$ polynomials in k variables $x_1, x_2, \ldots, x_k$*. The *resultant* $R(f_0, \ldots, f_k)$ is an irreducible polynomial in the coefficients of $f_0, f_1, \ldots, f_k$ which vanishes whenever $f_0, f_1, \ldots, f_k$ have a common root. When $k = 1$ this amounts to the resultant of two polynomials in one variable, treated in Chapter 12. It is not so easy to describe what exactly the vanishing of the resultant means because we have to take care of roots "at infinity." The best way to do it is to "homogenize" polynomials $f_0, f_1, \ldots, f_k$, i.e., to replace each f_i by the homogeneous form $x_0^{d_i} f(x_1/x_0, \ldots, x_k/x_0)$, where d_i is the degree of f_i. With some abuse of notation we denote the homogeneous forms thus obtained also by $f_0, f_1, \ldots, f_k$. The advantage of this homogeneous setting is that now the vanishing of $R(f_0, \ldots, f_k)$ is equivalent to the fact that the forms $f_0, \ldots, f_k$ have a common root in $C^{k+1} - \{0\}$. It is also helpful to view each f_i as a section of the line bundle $\mathcal{O}(d_i)$ on P^k; then

---

* In this and in the next chapter we find it more convenient to switch from the notation in Parts I and II where we mostly considered k forms in $k-1$ variables.

the vanishing of $R(f_0, \ldots, f_k)$ means that these sections have a common root in P^k.

As usual, the resultant is defined up to sign by the requirement that it is irreducible over **Z**, i.e, it has integral coefficients with the greatest common divisor equal to 1. In the present case it is possible to fix the resultant uniquely by the requirement that $R(x_0^{d_0}, x_1^{d_1}, \ldots, x_k^{d_k}) = 1$.

The resultant $R(f_0, \ldots, f_k)$ is a special case of the mixed $(A_0, \ldots, A_k)$-resultant (Section 1 Chapter 8). Here A_i is the set of all monomials of degree d_i in $(k+1)$ variables or, equivalently, the set of all integral points in the simplex

$$d_i \Delta^k = \left\{ (\omega_0, \ldots, \omega_k) \in R^{k+1} : \omega_j \geq 0, \ \sum_j \omega_j = d_i \right\}.$$

The degree of the resultant can be found with the help of Corollary 1.6 Chapter 8. We arrive at the following well-known proposition.

Proposition 1.1. *The resultant $R(f_0, \ldots, f_k)$ is a homogeneous polynomial in the coefficients of each form f_i of degree $d_0 d_1 \cdots d_{i-1} d_{i+1} \cdots d_k$.*

In particular, in the case when all the forms $f_0, \ldots, f_k$ have the same degree d, the resultant has degree d^k in the coefficients of each f_i, i.e., has total degree $(k+1)d^k$.

Now we give an analog of the product formula (1.3) from Chapter 12. It is more convenient to state it in the non-homogeneous setting, so we consider again $k+1$ polynomials $f_0, f_1, \ldots, f_k$ in k variables $x_1, x_2, \ldots, x_k$. Let d_i be the degree of f_i for $i = 0, 1, \ldots, k$, so each f_i contains some monomials of degree d_i and possibly some monomials of smaller degree. For $i = 1, 2, \ldots, k$, let $\overline{f}_i$ denote the homogeneous component of degree d_i in f_i. By the Bezout theorem, if $f_1, f_2, \ldots, f_k$ are sufficiently generic then they have $d_1 d_2 \cdots d_k$ distinct common roots. In such a situation we denote by $\Pi(f_0; f_1, \ldots, f_k)$ the product of values of f_0 at all common roots of $f_1, f_2, \ldots, f_k$.

Theorem 1.2. *The resultant $R(f_0, \ldots, f_k)$ is given by*

$$R(f_0, \ldots, f_k) = R(\overline{f}_1, \ldots, \overline{f}_k)^{d_0} \Pi(f_0; f_1, \ldots, f_k). \tag{1.1}$$

Formula (1.1) is known as the *Poisson formula* (cf. [Jou], Proposition 2.7). The term $R(\overline{f}_1, \ldots, \overline{f}_k)$ appearing on the right hand side is the resultant of k *homogeneous* forms in k variables. We can prove (1.1) using the same approach as in the proof of Theorem 1.17 Chapter 10 (in fact, the present case is much easier).

The Poisson formula implies the following multiplicative property.

Proposition 1.3. *Let $f_0 = f_0' f_0''$ be a product of two homogeneous forms. Then*

$$R(f_0' f_0'', f_1, \dots, f_k) = R(f_0', f_1, \dots, f_k) R(f_0'', f_1, \dots, f_k).$$

B. The Cayley determinantal formula

We now present the classical formula due to Cayley for the resultant of $k+1$ homogeneous forms in $k+1$ variables, expressing it as the determinant of the *resultant complex.* This is a special case of the general formula given by Theorem 4.2 Chapter 3, but we wish to state it in the most "down-to-earth" manner. For simplicity we consider the case when all the forms $f_0, f_1, \dots, f_k$ have the same homogeneity degree d.

The resultant complex depends on m and is the *twist parameter* $m \in \mathbf{Z}$

$$C^\bullet = C^\bullet(m; f_0, \dots, f_k) = \left\{ C^{-k-1} \xrightarrow{\partial} C^{-k} \xrightarrow{\partial} \cdots \xrightarrow{\partial} C^{-1} \xrightarrow{\partial} C^0 \right\},$$

where

$$C^{-p} = S^{m-pd}\mathbf{C}^{k+1} \otimes \Lambda^p \mathbf{C}^{k+1}. \tag{1.2}$$

Here $S^{m-pd}\mathbf{C}^{k+1}$ is regarded as the space of forms of degree $m-pd$ in $x_0, x_1, \dots, x_k$ (so this term is absent if $m - pd < 0$). The differential $\partial : C^{-p} \to C^{-p+1}$ acts by $\partial = \sum_{j=0}^k f_j \otimes \partial_j$, where each f_j acts by multiplication, and ∂_j is the derivation of the exterior algebra $\Lambda^\bullet \mathbf{C}^{k+1}$ with respect to the j-th standard basis vector.

We choose a basis in each C^{-p} consisting of monomials of degree $m - pd$ in $x_0, x_1, \dots, x_k$, tensored with the wedge monomials of degree p in the standard basis vectors of $\mathbf{C}^{k+1}$.

Theorem 1.4. *For each $m > (k+1)(d-1)$ the resultant $R(f_0, \dots, f_k)$ is equal to the determinant of the complex $C^\bullet(m; f_0, \dots, f_k)$.*

Proof. The statement follows from Theorem 4.2 Chapter 3 (or from Theorem 2.5 Chapter 8) provided the resultant complex is *stably twisted.* This means that for each $p = 0, 1, \dots, k+1$ the sheaf $\mathcal{O}(m-pd)$ on P^k has no higher cohomology. By Serre's theorem (Theorem 2.12 Chapter 2), this happens exactly when $m - pd \geq -k$ for $p = 0, 1, \dots, k+1$, i.e., when $m > (k+1)(d-1)$.

Theorem 1.4 gives an explicit formula for the resultant. Unfortunately, for large k and d this formula is impractical for the purposes of actual computation. This is so because even in the most "economic" case $m = (k+1)(d-1)+1$ the resultant complex has many non-zero terms. Namely, $C^{-p} \neq 0$ for p such that $(k+1)(d-1)+1-pd \geq 0$, i.e., for $0 \leq p \leq k+1-\lceil \frac{k}{d} \rceil$, where $\lceil x \rceil$ is the smallest integer $\geq x$. Remembering the explicit formula for the determinant of a complex (see Theorem 13, Appendix A), we see that the Cayley formula expresses $R(f_0, \ldots, f_k)$ as the alternating product of $k + 1 - \lceil \frac{k}{d} \rceil$ ordinary determinants whose entries are linear combinations of the coefficients of the f_i.

The number $k - \lceil \frac{k}{d} \rceil$ can be thought of as the "determinantal complexity" of Cayley's formula. It is always non-negative and is equal to 0 only in two extreme cases when $k = 1$ or $d = 1$. For $k = 1$ Cayley's formula becomes the Sylvester formula (1.12), Chapter 12 for the resultant of two binary forms. For $d = 1$ Cayley's formula expresses the resultant of a system of linear forms as the determinant of their coefficient matrix.

The differential at the right in the resultant complex has the form

$$S^{k(d-1)}\mathbf{C}^{k+1} \otimes \mathbf{C}^{k+1} \xrightarrow{\partial_0} S^{(k+1)(d-1)+1}\mathbf{C}^{k+1},$$
$$\partial_0 = \partial_0(f_0, \ldots, f_k) : g_0 \otimes e_0 + g_1 \otimes e_1 + \cdots + g_k \otimes e_k \mapsto g_0 f_0 + g_1 f_1 + \cdots + g_k f_k,$$

where $e_0, e_1, \ldots, e_k$ are standard basis vectors in $\mathbf{C}^{k+1}$. The resultant $R(f_0, f_1, \ldots, f_k)$ is equal to the greatest common divisor of all the maximal minors of the matrix of $\partial_0 = \partial_0(f_0, \ldots, f_k)$ (see Appendix A, Theorem 34). (As above, we choose a basis in $S^{(k+1)(d-1)+1}\mathbf{C}^{k+1}$ consisting of all monomials of degree $(k+1)(d-1)+1$ in $x_0, x_1, \ldots, x_k$, and a basis in $S^{k(d-1)}\mathbf{C}^{k+1} \otimes \mathbf{C}^{k+1}$ consisting of all monomials of degree $k(d-1)$ tensored with $e_0, e_1, \ldots, e_k$.)

In fact, in the present case we do not need all the minors; it suffices to take only $k+1$ specific minors of the matrix of ∂_0. In order to see this for each $i = 0, 1, \ldots, k$, let M_i denote the set of all monomials $x_0^{\alpha_0} x_1^{\alpha_1} \cdots x_k^{\alpha_k}$ of degree $k(d-1)$ such that $\max(\alpha_0, \alpha_1, \ldots, \alpha_{i-1}) < d$ (so M_0 consists of all monomials of degree $k(d-1)$). We claim that the set $\bigcup_{i=0}^{k} (M_i \otimes \{e_i\})$ is in bijection with the set of all monomials of degree $(k+1)(d-1)+1$ in $x_0, x_1, \ldots, x_k$. Such a bijection can be given by sending $x^\alpha \otimes e_i \in M_i \otimes \{e_i\}$ to the monomial $x^\alpha x_i^d$ (the inverse bijection takes every monomial $x^\beta = x_0^{\beta_0} x_1^{\beta_1} \cdots x_k^{\beta_k}$ of degree $(k+1)(d-1)+1$ to x^β / x_i^d, where i is the minimal index such that $\beta_i \geq d$). It follows that the subspace of $S^{k(d-1)}\mathbf{C}^{k+1} \otimes \mathbf{C}^{k+1}$ spanned by $\bigcup_{i=0}^{k} (M_i \otimes \{e_i\})$ has the same dimension as $S^{(k+1)(d-1)+1}\mathbf{C}^{k+1}$. Let $D_k(f_0, \ldots, f_k)$ denote the maximal minor of the matrix of $\partial_0(f_0, \ldots, f_k)$ corresponding to its restriction to the span of $\bigcup_{i=0}^{k} (M_i \otimes \{e_i\})$. Likewise, for each $j = 0, 1, \ldots, k-1$, we denote by $D_j(f_0, \ldots, f_k)$ the maximal minor of the matrix of $\partial_0(f_0, \ldots, f_k)$ corresponding to its restriction to the span of $\bigcup_{i=0}^{k} (M_{\sigma(i)} \otimes \{e_i\})$, where σ is some permutation of indices $0, 1, \ldots, k$ such that $\sigma(j) = k$.

Theorem 1.5. *The resultant $R(f_0, \dots, f_k)$ is the greatest common divisor of the polynomials $D_0, D_1, \dots, D_k$.*

Proof. Clearly, the resultant being the g.c.d of all maximal minors of the matrix of ∂_0 divides the g.c.d $(D_0, D_1, \dots, D_k)$. On the other hand, by construction each polynomial D_j has degree d^k in the coefficients of f_j. Therefore, the g.c.d. $(D_0, D_1, \dots, D_k)$ has degree less than or equal to d^k in the coefficients of each of the forms $f_0, \dots, f_k$. Hence its total degree is less than or equal to that of the resultant. This implies our theorem.

Note that each minor D_j in Theorem 1.5 depends on the choice of a permutation σ of indices, but this dependence does not matter, i.e., Theorem 1.5 remains valid for all possible choices.

C. Polynomial expressions for the resultant via Weyman's complexes

It follows from the general theory developed in Section 4 Chapter 3 that even those values of the twist parameter $m \in \mathbf{Z}$, for which the resultant complex (1.2) is not stably twisted, still give rise to an explicit expression for the resultant as the determinant of a certain spectral sequence $(C_r^{\bullet\bullet}, \partial_r)$, the so-called resultant spectral sequence. Recall that this spectral sequence arises from the twisted Koszul complex

$$\bigwedge^{k+1}\left(\mathcal{O}(-d)^{\oplus k+1}\right) \otimes \mathcal{O}(m) \to \cdots \to \bigwedge^{1}\left(\mathcal{O}(-d)^{\oplus k+1}\right) \otimes \mathcal{O}(m) \to \mathcal{O}(m) \quad (1.3)$$

of sheaves on P^k. By Theorem 4.11 of Chapter 3, we can interpret the determinant of the spectral sequence $(C_r^{\bullet\bullet})$ as the determinant of some complex which incorporates, in a sense, all the higher differentials in $(C_r^{\bullet\bullet})$. Using Serre's description of the cohomology of the sheaves $\mathcal{O}(n)$ on a projective space (Theorem 2.12 Chapter 2), we can describe the complex in question as follows.

The complex corresponding to an arbitrary twist parameter $m \in \mathbf{Z}$ has the form

$$C^\bullet = C^\bullet(m; f_0, \dots, f_k) = \left\{C^{-k-1} \xrightarrow{\partial} \cdots \xrightarrow{\partial} C^{-1} \xrightarrow{\partial} C^0 \xrightarrow{\partial} \cdots \xrightarrow{\partial} C^k\right\},$$

where

$$C^{-p} = \left[S^{m-pd}\mathbf{C}^{k+1} \otimes \Lambda^p \mathbf{C}^{k+1}\right] \oplus \left[(S^{pd-m+k(d-1)-1}\mathbf{C}^{k+1})^* \otimes \Lambda^{p+l}\mathbf{C}^{k+1}\right]. \quad (1.4)$$

Here the first summand in (1.4) can appear only for p such that $0 \le p \le k+1$ and $m - pd \ge 0$; it arises as the space of global sections

$$H^0\left(P^k, \bigwedge^{p}\left(\mathcal{O}(-d)^{\oplus k+1}\right) \otimes \mathcal{O}(m)\right).$$

Similarly, the second summand can appear only for $-k \le p \le 1$ and $pd - m + k(d-1) - 1 \ge 0$; it arises as the top cohomology space

$$H^k\left(P^k, \bigwedge^{p+k}(\mathcal{O}(-d)^{\oplus k+1}) \otimes \mathcal{O}(m)\right).$$

In particular, in the stably twisted case when $m > (k+1)(d-1)$, the second summand never appears so that (1.4) reduces to (1.2).

The differential

$$\begin{aligned}\partial : &\left[S^{m-pd}\mathbf{C}^{k+1} \otimes \Lambda^p\mathbf{C}^{k+1}\right] \oplus \left[(S^{pd-m+k(d-1)-1}\mathbf{C}^{k+1})^* \otimes \Lambda^{p+k}\mathbf{C}^{k+1}\right]\\ &\longrightarrow \left[S^{m-(p-1)d}\mathbf{C}^{k+1} \otimes \Lambda^{p-1}\mathbf{C}^{k+1}\right] \oplus \left[(S^{(p-1)d-m+k(d-1)-1}\mathbf{C}^{k+1})^* \otimes \Lambda^{p+k-1}\mathbf{C}^{k+1}\right]\end{aligned}$$

is represented by a block 2×2 matrix

$$\begin{pmatrix} \partial_{11} & \partial_{12} \\ 0 & \partial_{22} \end{pmatrix}.$$

Here the operator

$$\partial_{11} : S^{m-pd}\mathbf{C}^{k+1} \otimes \Lambda^p\mathbf{C}^{k+1} \longrightarrow S^{m-(p-1)d}\mathbf{C}^{k+1} \otimes \Lambda^{p-1}\mathbf{C}^{k+1}$$

is given by the same formula $\partial_{11} = \sum_{j=0}^k f_j \otimes \partial_j$ as in subsection 1B. This is one of the components of the differential ∂_1 in the resultant spectral sequence, namely the map of the spaces of global sections of sheaves in (1.3) induced by the differential in (1.3). The operator

$$\begin{aligned}\partial_{22} : &(S^{pd-m+k(d-1)-1}\mathbf{C}^{k+1})^* \otimes \Lambda^{p+k}\mathbf{C}^{k+1}\\ &\longrightarrow (S^{(p-1)d-m+k(d-1)-1}\mathbf{C}^{k+1})^* \otimes \Lambda^{p+k-1}\mathbf{C}^{k+1}\end{aligned}$$

is seen to be given by $\partial_{22} = \sum_{j=0}^k f_j^* \otimes \partial_j$, where f_j^* stands for the operator adjoint of the multiplication by f_j. This is also one of the components of the differential ∂_1 in the resultant spectral sequence, this time the map on the top cohomology of the sheaves in (1.3) induced by the differential there. The operator

$$\partial_{12} : (S^{pd-m+k(d-1)-1}\mathbf{C}^{k+1})^* \otimes \Lambda^{p+k}\mathbf{C}^{k+1} \longrightarrow S^{m-(p-1)d}\mathbf{C}^{k+1} \otimes \Lambda^{p-1}\mathbf{C}^{k+1}$$

can appear only for $p = 1$ where it takes the form

$$\partial_{12} : (S^{(k+1)(d-1)-m}\mathbf{C}^{k+1})^* \otimes \Lambda^{k+1}\mathbf{C}^{k+1} \longrightarrow S^m\mathbf{C}^{k+1}.$$

This is a lifting of a component of the differential ∂_{k+1} of the resultant spectral sequence. In a more symmetric form ∂_{22} can be viewed as a bilinear form

$$(S^{(k+1)(d-1)-m}\mathbf{C}^{k+1})^* \times (S^m\mathbf{C}^{k+1})^* \longrightarrow \mathbf{C};$$

this form depends polynomially on the coefficients of $f_0, f_1, \ldots, f_k$ and is antisymmetric with respect to permutations of $f_0, f_1, \ldots, f_k$. No nice explicit expression of this form seems to be known at present.

For general values of k, d and m the complex (1.3) can have many non-zero terms, and its determinant is a quite complicated rational expression. But there are some special cases when the complex has only two non-zero terms C^{-1} and C^0, so its determinant is the ordinary determinant of the matrix of the differential $\partial : C^{-1} \longrightarrow C^0$. This leads to a *polynomial* expression for the resultant $R(f_0, \ldots, f_k)$. An example of such an expression is the Sylvester formula for the resultant of three ternary forms (Section 4D Chapter 3). In the present notation this corresponds to $k = 2$, $d \geq 2$ and $m = 2d - 1$ or $2d - 2$, so $R(f_0, f_1, f_2)$ is the determinant of the matrix of the differential

$$\partial : \left[S^{d-1-\varepsilon}\mathbf{C}^3 \otimes \Lambda^1\mathbf{C}^3\right] \oplus \left[(S^{d-2+\varepsilon}\mathbf{C}^3)^* \otimes \Lambda^3\mathbf{C}^3\right] \longrightarrow S^{2d-1-\varepsilon}\mathbf{C}^3,$$

where $\varepsilon = 0, 1$.

The following proposition lists all the values of k, d and m which give rise to polynomial formulas for the resultant.

Proposition 1.6. *The complex (1.3) has only two non-zero terms exactly for the following values of* k, d, m.

(a) **Linear forms:** $k \geq 1, d = 1, 1 \geq m \geq -1.$

(b) **Binary forms:** $k = 1, d \geq 2, 2d - 1 \geq m \geq -1.$

(c) **Ternary forms:** $k = 2, d \geq 2, 2d - 1 \geq m \geq d - 2.$

(d) **Quaternary forms:** $k = 3, d \geq 2, 2d - 1 \geq m \geq 2d - 3.$

(e) **Quadratic forms in five variables:** $k = 4, d = 2, 3 \geq m \geq 2.$

(f) **Cubic forms in five variables:** $k = 4, d = 3, m = 5.$

(g) **Quadratic forms in six variables:** $k = 5, d = 2, m = 3.$

In each of the cases the only two non-zero terms in the complex are

$$C^{-1} = \left[S^{m-d}\mathbf{C}^{k+1} \otimes \Lambda^1\mathbf{C}^{k+1}\right] \oplus \left[(S^{(k+1)(d-1)-m}\mathbf{C}^{k+1})^* \otimes \Lambda^{k+1}\mathbf{C}^{k+1}\right],$$

$$C^0 = S^m\mathbf{C}^{k+1} \oplus \left[(S^{(k+1)(d-1)-d-m}\mathbf{C}^{k+1})^* \otimes \Lambda^k\mathbf{C}^{k+1}\right],$$

so the resultant is equal to the determinant of the matrix of $\partial : C^{-1} \longrightarrow C^0$.

Proof. The classification follows directly from (1.3). First we see that all the terms C^{-p} with $p \geq 2$ vanish exactly when $2d - 1 \geq m$ (we have to care only about the first summand in (1.3) because the second summand involves $\Lambda^{k+p}\mathbf{C}^{k+1}$, hence is zero for all m). Similarly, all the terms C^{-p} with $p \leq -1$ become zero exactly when $m \geq kd - k - d$. Hence the triples (k, d, m) such that (1.3) has only two non-zero terms are exactly those satisfying the inequalities

$$2d - 1 \geq m \geq kd - k - d. \tag{1.5}$$

This is now an elementary exercise to check that the cases (a) to (g) provide a list of all integral solutions of (1.5) with $k, d \geq 1$. (One way to see this is to rewrite the inequality $2d - 1 \geq kd - k - d$ in the form $(k - 3)(d - 1) \leq 2$.)

D. The discriminant

For the convenience of the reader we collect here some basic properties of the discriminant of a polynomial in several variables. Most were already mentioned as examples in the preceding chapters.

We can consider the discriminant in two equivalent versions: for polynomials in k variables of degree less than or equal to d, or for (homogeneous) forms in $k + 1$ variables of degree d. In the homogeneous version, the discriminant is an irreducible polynomial $\Delta(f)$ in the coefficients of a form $f = f(x_0, x_1, \ldots, x_k)$ which vanishes if and only if all the partial derivatives $\partial f/\partial x_0, \partial f/\partial x_1, \ldots, \partial f/\partial x_k$ have a common zero in $\mathbf{C}^{k+1} - \{0\}$. The polynomial $\Delta(f)$ is defined uniquely up to sign by the requirement that it is irreducible over $\mathbf{Z}$, i.e., it has relatively prime integer coefficients.

Under the geometric approach of Part I, the discriminant corresponds to the Veronese embedding $P^k = P(\mathbf{C}^{k+1}) \longrightarrow P(S^d\mathbf{C}^{k+1})$ which associates to each $(x_0 : x_1 : \ldots : x_k)$ a point whose homogeneous coordinates are all monomials of degree d in $x_0, \ldots, x_k$. Since this embedding is toric, $\Delta(f)$ is also the A-discriminant in Part II, where A is the set of all homogeneous monomials in $x_0, \ldots, x_k$ of degree d, or, equivalently, the set of all integral points in the scaled simplex

$$d\Delta^k = \left\{(\omega_0, \omega_1, \ldots, \omega_k) : \omega_i \geq 0, \omega_0 + \omega_1 + \cdots + \omega_k = d\right\}.$$

The discriminant of a form f of degree d in $k + 1$ variables is a polynomial in the coefficients of f whose degree is equal to $(k + 1)(d - 1)^k$ (see (2.12) in Chapter 9). In particular the discriminant is non-trivial exactly when $d \geq 2$.

The relationships between the discriminants and the resultants are similar to the case of one variable (see formula (1.31) of Chapter 12), but more complicated as far as the numerical constants are concerned. To keep the notation straight, we denote by Δ_d the discriminant of a form of degree d and by R_d the resultant of $k+1$ forms of degree d in $k+1$ variables. We also use the notation $a \uparrow b$ for a^b.

Proposition 1.7. *For a form $f(x_0, \dots, x_k)$ of degree d, we have*

$$\Delta_d(f) = c_{d,k} \cdot R_{d-1}\left(\frac{\partial f}{\partial x_0}, \dots, \frac{\partial f}{\partial x_k}\right), \tag{1.6}$$

where

$$c_{d,k} = \pm d \Big\uparrow \left(\frac{(-1)^{k+1} - (d-1)^{k+1}}{d}\right). \tag{1.7}$$

Proof. By comparing degrees we find that (1.6) holds with some non-zero constant $c_{d,k}$, since both sides define the same locus in the space of polynomials. To find $c_{d,k}$ we take a special polynomial $f_0(x) = x_0^d + \dots + x_k^d$. Then

$$R_{d-1}\left(\frac{\partial f_0}{\partial x_0}, \dots, \frac{\partial f_0}{\partial x_k}\right) = R_{d-1}(dx_0^{d-1}, \dots, dx_k^{d-1}) = d^{(k+1)(d-1)^k} \tag{1.8}$$

by the homogeneity of the resultant. To calculate $\Delta_d(f_0)$, we first find the principal A-determinant $E_A(f_0)$ where A is the set of homogeneous monomials of degree d (so Δ_d is the same as Δ_A). Let $B \subset A$ be the set $\{x_0^d, \dots, x_k^d\}$, so that $f_0 \in \mathbf{C}^B$. We apply the formalism of Section 1E Chapter 10. Thus in the notation of this section, $E_A(f_0) = E_A(f_0, S, \Xi)$ where S and Ξ are, respectively, the subsemigroup and subgroup in $\mathbf{Z}^{k+1}$ generated by A. Clearly, Ξ consists of $\omega = (\omega_0, \dots, \omega_k) \in \mathbf{Z}^{k+1}$ such that $\sum \omega_i$ is divisible by d. The definition of the principal A-discriminant calls also for a function $h : \Xi \to \mathbf{Q}$ which can be chosen as $h(\omega) = \frac{1}{d}\sum \omega_i$. For our particular choice of f_0 we can rewrite $E_A(f_0) = E_B(f_0, S, \Xi)$, by using Theorem 1.11 (b) Chapter 10, as

$$[\Xi : \mathrm{Lin}_{\mathbf{Z}}(B)]^{\mathrm{Vol}_\Theta(Q)} \cdot E_B(f_0)^{[\mathrm{Lin}_{\mathbf{Z}}(S_{\mathbf{Z}}):\mathrm{Lin}_{\mathbf{Z}}(A)]},$$

where $Q = \mathrm{Conv}(A) = d\Delta^k$ is the scaled simplex and $\Theta = \mathrm{Aff}_{\mathbf{Z}}(A)$. Note that $E_B(f_0) = 1$ and both $[\Xi : \mathrm{Lin}_{\mathbf{Z}}(B)]$ and $\mathrm{Vol}_\Theta(Q)$ are equal to d^k. Thus

$$E_A(f_0) = \left(d^k\right)^{(d^k)} = d^{kd^k}. \tag{1.9}$$

The discriminant $\Delta_A(f_0)$ is obtained as the alternating product

$$\Delta_A(f_0) = \prod_{\Gamma \subset Q} E_{A\cap\Gamma}(f_0)^{(-1)^{\mathrm{codim}(\Gamma)}}$$

(see Theorem 1.3 Chapter 11). Since every face of $Q = d\Delta^k$ is a simplex, for each j-dimensional face Γ, we have $E_{A\cap\Gamma}(f_0) = d^{jd^j}$. Thus

$$\Delta_A(f_0) = d \Big\uparrow \left(\sum_{i=0}^{k} (-1)^i \binom{k+1}{i} (k-i) d^{k-i}\right). \tag{1.10}$$

Simplifying the exponent in (1.10) by the binomial formula and combining this with (1.8) we obtain (1.7). Proposition 1.7 is proved.

Proposition 1.8. *For A as above and any form $f = f(x_0, \ldots, x_k)$ of degree d, we have*

$$R_d\left(x_0\frac{\partial f}{\partial x_0}, \ldots, x_k\frac{\partial f}{\partial x_k}\right) = d^{d^k} E_A(f) = d^{d^k} \prod_{\Gamma \subset d\Delta^k} \Delta_{A\cap\Gamma}(f). \qquad (1.11)$$

Proof. Again, the equality up to a constant factor follows by comparing degrees, since both sides of (1.11) define the same locus. The value of the constant is found by considering the same polynomial $f_0(x) = \sum x_i^d$ as in the proof of Proposition 1.7 and using (1.9) and the following equality:

$$R_d\left(x_0\frac{\partial f_0}{\partial x_0}, \ldots, x_k\frac{\partial f_0}{\partial x_k}\right) = R_d(dx_0^d, \ldots, dx_k^d) = d^{(k+1)d^k}.$$

Propositions 1.7 and 1.8 are higher-dimensional analogs of the equalities (1.31) of Chapter 12.

By Theorem 1.4 Chapter 10, the Newton polytope of $E = E_A$ is the secondary polytope $\Sigma(A)$. Thus, the extreme monomials in $E_A(f)$ correspond to coherent triangulations of (Q, A) (here Q is the scaled simplex $d\Delta^k$, and A is the set of all integral points in Q). When d and k are increasing the number of these monomials seems to grow very fast. As for the Newton polytope of the discriminant $\Delta_d(f)$, Theorem 3.2 Chapter 11 implies that its vertices correspond to so-called D-equivalence classes of coherent triangulations of (Q, A).

E. Example: ternary quadrics

Consider the case $k = d = 2$ when A consists of the quadratic monomials in three variables; using the language of lattice points, we write A as

$$A = \{(2,0,0), (0,2,0), (0,0,2), (1,1,0), (1,0,1), (0,1,1)\}. \qquad (1.12)$$

Thus, A consists of three vertices and three midpoints of edges of a regular triangle, (see Figure 67).

The A-resultant is the resultant $R(f_1, f_2, f_3)$ of three quadratic forms in three variables; in accordance with the numeration of points of A in (1.12), we write these forms as

$$\begin{aligned} f_1(x,y,z) &= a_{11}x^2 + a_{12}y^2 + a_{13}z^2 + a_{14}xy + a_{15}xz + a_{16}yz,\\ f_2(x,y,z) &= a_{21}x^2 + a_{22}y^2 + a_{23}z^2 + a_{24}xy + a_{25}xz + a_{26}yz,\\ f_3(x,y,z) &= a_{31}x^2 + a_{32}y^2 + a_{33}z^2 + a_{34}xy + a_{35}xz + a_{36}yz. \end{aligned}$$

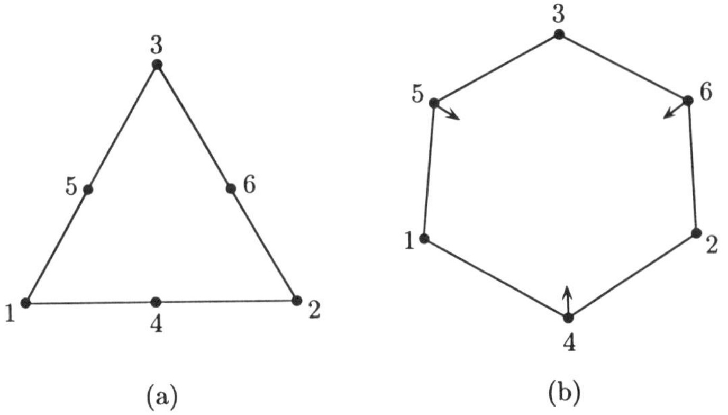

Figure 67.

The resultant $R(f_1, f_2, f_3)$ is a polynomial in the a_{ij} of degree 12. It can be also expressed as a polynomial of degree 4 in brackets

$$[j_1 j_2 j_3] = \det \begin{pmatrix} a_{1,j_1} & a_{1,j_2} & a_{1,j_3} \\ a_{2,j_1} & a_{2,j_2} & a_{2,j_3} \\ a_{3,j_1} & a_{3,j_2} & a_{3,j_3} \end{pmatrix},$$

where $1 \leq j_1 < j_2 < j_3 \leq 6$.

The polytope $\Sigma(A)$ has dimension 3. In fact, it is combinatorially equivalent to the *associahedron* $\Sigma(A')$, where A' is the set of vertices of a convex hexagon (the associahedron was described in detail in Section 3B Chapter 7). The combinatorial equivalence $\Sigma(A) \cong \Sigma(A')$ is induced by the deformation indicated in Figure 67. Thus, (Q, A) has fourteen coherent triangulations corresponding to the triangulations of the hexagon shown in Figure 34 in Chapter 7.

By Theorem 3.2 Chapter 8, the extreme terms in $R(f_1, f_2, f_3)$ are in bijection with the coherent triangulations of (Q, A); these extreme terms are the following fourteen bracket monomials:

$$-[145][246][356][456], \ [146][156][246][356], \ [145][245][256][356],$$

$$[145][246][346][345], \ [126]^2[156][356], \ [125]^2[256][356], \ [134]^2[246][346],$$

$$[136]^2[146][246], \ [145][245][235]^2, [145][345][234]^2, \ [136]^2[126]^2,$$

$$[125]^2[235]^2, \ [134]^2[234]^2, \ [123]^4.$$

(Here in each term the brackets correspond to the triangles of a coherent triangulation, and the exponent of a bracket is the volume of the corresponding triangle.)

The complete decomposition of the resultant into the sum of bracket monomials can be found in [KSZ2].

Now let

$$f(x, y, z) = a_1x^2 + a_2y^2 + a_3z^2 + a_4xy + a_5xz + a_6yz$$

be a quadratic form. The A-discriminant $\Delta(f)$ can be computed from Proposition 1.7. We obtain

$$\Delta(f) = a_1a_6^2 + a_5^2a_2 + a_4^2a_3 - a_4a_5a_6 - 4a_1a_2a_3. \tag{1.13}$$

According to Theorem 1.2 Chapter 10, the principal A-determinant $E_A(f)$ of a quadratic form is

$$E_A(f) = a_1a_2a_3(4a_1a_2 - a_4^2)(4a_1a_3 - a_5^2)(4a_2a_3 - a_6^2)\Delta(f), \tag{1.14}$$

where $\Delta(f)$ is given by (1.13), and the remaining six factors are discriminants corresponding to three vertices and three sides of the triangle $Q = \text{Conv}(A)$.

We leave to the reader an easy calculation of the lattice points φ_T and η_T corresponding to all coherent triangulations of (Q, A). The points φ_T (resp. η_T) are exponent vectors of the fourteen extreme monomials in $E_A(f)$ (resp. five extreme monomials in $\Delta(f)$). Thus, fourteen coherent triangulations of (Q, A) break into five D-equivalence classes.

2. Forms in several groups of variables

Here we extend the results of Section 1 to forms in several groups of variables which are homogeneous in the variables of each group.

A. The resultant: degree and Sylvester type formulas

We shall be dealing with the following version of the resultant. Fix natural numbers $k_1, k_2, \ldots, k_r$ and consider r sets of variables

$$x^{(1)} = (x_0^{(1)}, x_1^{(1)}, \ldots, x_{k_1}^{(1)}), \ldots, x^{(r)} = (x_0^{(r)}, x_1^{(r)}, \ldots, x_{k_r}^{(r)}),$$

each set $x^{(j)}$ consisting of $k_j + 1$ variables. For every sequence of non-negative integers $d_1, \ldots, d_r$, we denote by $S(d_1, \ldots, d_r)$ the space of polynomials in $x^{(1)}, \ldots, x^{(r)}$ homogeneous of degree d_j in $x^{(j)}$. We call the elements of $S(d_1, \ldots, d_r)$ *the forms of type* $(k_1, \ldots, k_r; d_1, \ldots, d_r)$. Such a form can also be regarded as a section of the line bundle $\mathcal{O}(d_1, \ldots, d_r)$ on the product of projective spaces $P^{k_1} \times \cdots \times P^{k_r}$, defined as:

$$\mathcal{O}(d_1, \ldots, d_r) = p_1^*\mathcal{O}(d_1) \otimes \cdots \otimes p_r^*\mathcal{O}(d_r),$$

where $p_j : P(V_1^*) \times \cdots \times P(V_r^*) \to P(V_j^*)$ is the projection.

Let $k = k_1 + \cdots + k_r$, and let $V = S(d_1, \ldots, d_r)^{k+1}$ be the space of $(k+1)$-tuples of forms $(f_0, f_1, \ldots, f_k)$ from $S(d_1, \ldots, d_r)$. The resultant $R(f_0, \ldots, f_k)$ is an irreducible polynomial function on V which vanishes at $(f_0, \ldots, f_k)$ if and only if a tuple $(f_0, \ldots, f_k)$ has a common root in $P^{k_1} \times \cdots \times P^{k_r}$. In the general framework of Part 1, this is the resultant corresponding to the embedding

$$P^{k_1} \times \cdots \times P^{k_r} = P(\mathbf{C}^{k_1+1}) \times \cdots \times P(\mathbf{C}^{k_r+1}) \longrightarrow P(S^{d_1}\mathbf{C}^{k_1+1} \otimes \cdots \otimes S^{d_r}\mathbf{C}^{k_r+1}),$$

which is a combination of classical Veronese and Segre embeddings. We call $R(f_0, \ldots, f_k)$ the resultant of type $(k_1, \ldots, k_r; d_1, \ldots, d_r)$.

Let $\Delta^k(d)$ denote the set of all non-negative integral vectors $(i_0, i_1, \ldots, i_k)$ with $i_0 + i_1 + \ldots + i_k = d$. The resultant of type $(k_1, \ldots, k_r; d_1, \ldots, d_r)$ coincides with the A-resultant (defined in Section 2 Chapter 8), where

$$A := \Delta^{k_1}(d_1) \times \cdots \times \Delta^{k_r}(d_r) \subset \mathbf{Z}^{k_1+1} \times \cdots \times \mathbf{Z}^{k_r+1}.$$

The degree of the resultant is given by the following.

Proposition 2.1. *The resultant $R(f_0, \ldots, f_k)$ of type $(k_1, \ldots, k_r; d_1, \ldots, d_r)$ is a polynomial of total degree*

$$(k+1)\binom{k}{k_1, k_2, \ldots, k_r} d_1^{k_1} d_2^{k_2} \cdots d_r^{k_r}.$$

Proof. By Corollary 2.2 Chapter 8, the resultant is an irreducible polynomial function on $S(d_1, d_2, \ldots, d_r)^{l+1}$ of degree $(k+1) \cdot \mathrm{Vol}\,(Q)$, where $Q = \mathrm{Conv}\,(A)$, and Vol is the normalized volume form with respect to the affine lattice spanned by A (so that an elementary lattice simplex has volume 1).

The polytope Q is a product of simplices. If $d_1 = d_2 = \ldots = d_r = 1$, then it is well-known that $\mathrm{Vol}\,(Q) = \binom{k_1+\cdots+k_r}{k_1, k_2, \ldots, k_r}$ (the most geometrically transparent way to see it is to construct a triangulation of Q into $\binom{k_1+\cdots+k_r}{k_1, k_2, \ldots, k_r}$ elementary simplices, by iterating the construction for the product of two simplices given in Section 3D Chapter 7). Clearly, the volume of any product of polytopes of dimensions

$k_1, \ldots, k_r$ is homogeneous of degree k_j with respect to scaling of the j-th factor. Hence the normalized volume of Q equals $\binom{k_1+\cdots+k_r}{k_1,k_2,\ldots,k_r} d_1^{k_1} d_2^{k_2} \cdots d_r^{k_r}$.

The construction of the resultant complex (1.2) and Theorem 1.4 can be extended to the resultant of type $(k_1, \ldots, k_r; d_1, \ldots, d_r)$ in a completely straightforward way. We leave the details to the reader. Note only that the resultant complex will now depend on r integers ("twist parameters") $m_1, \ldots, m_r$, which we abbreviate as a vector $m \in \mathbf{Z}^r$. Thus, whenever $\mathbf{m}$ is such that the complex is stably twisted, our resultant is the determinant of the resultant complex. The explicit formulas for the resultant obtained in this way are, understandably, more complicated than the corresponding formulas for the case of forms in only one group of variables (see Subsection B above). So it is natural to single out some "lucky" cases when the answer reduces to just one determinant of a square matrix. This will happen when the resultant complex has only two non-zero terms, at the far right.

The differential at the right of the complex has the form

$$S(m_1 - d_1, \ldots, m_r - d_r) \otimes \mathbf{C}^{k+1} \xrightarrow{\partial_0} S(m_1, \ldots, m_r), \tag{2.1}$$

$$\begin{aligned} \partial_0 = \partial_0(f_0, \ldots, f_k) : g_0 \otimes e_0 + g_1 \otimes e_1 + \cdots + g_k \otimes e_k \\ \mapsto g_0 f_0 + g_1 f_1 + \cdots + g_k f_k, \end{aligned} \tag{2.2}$$

where $e_0, e_1, \ldots, e_k$ are standard basis vectors in $\mathbf{C}^{k+1}$. So, if $\mathbf{m}$ is such that the resultant complex is stably twisted and reduces to (2.1), then, under the natural choices of the bases, we have

$$R(f_0, \ldots, f_k) \quad = \quad \det(\partial_0). \tag{2.3}$$

For instance, the Sylvester formula for the resultant of two binary forms of degree d (see (1.12) of Chapter 12) appears in this way when $r = 1$, $k = 1$ and $\mathbf{m} = m_1 = 2d - 1$. The following Proposition describes all such situations (for the proof see [SZ1]).

Proposition 2.2. *For given $k_1, \ldots, k_r, d_1, \ldots, d_r$, the following two conditions are equivalent:*

(a) There exists $\mathbf{m} \in \mathbf{Z}^r$ such that the corresponding resultant complex is stably twisted and has only two non-zero terms at the far right.

(b) $\min(k_j, d_j) = 1$ *for all* $j = 1, \ldots, r$.

Under these conditions, there are exactly $r!$ choices of the vector $\mathbf{m}$ satisfying (a). They correspond to the permutations π of $\{1, \ldots, r\}$. Given such π, the corresponding $\mathbf{m}$ has components

$$m_j = d_j\Bigl(1 + \sum_{i:\pi(i)\le\pi(j)} k_i\Bigr) - k_j. \tag{2.4}$$

We see that whenever the condition (a) of Proposition 2.2 is satisfied, we get $r!$ explicit polynomial formulas for the resultant. We call them *Sylvester type formulas* since they include the classical Sylvester formula as a particular case. Another particular case of Proposition 2.2 is the classical formula of Dixon [Di] for the resultant of three forms $f_i(x_0, x_1, y_0, y_1)$, $(i = 0, 1, 2)$ bihomogeneous of degree d_1 in x_0, x_1 and d_2 in y_0, y_1. Here $r = 2$, $k_1 = k_2 = 1$ and the two choices of $\mathbf{m}$ leading to a Sylvester type formula are

$$\mathbf{m} = (2d_1 - 1, 3d_2 - 1) \quad \text{and} \quad \mathbf{m} = (2d_2 - 1, 3d_1 - 1).$$

As a third example, let us mention the case of multilinear forms (all d_j are equal to 1). The resultant in this case will be discussed in Chapter 14 below.

The class of situations covered by Proposition 2.2 is not very large. There is, however, a possibility of "Sylvester type formulas" of a more general kind. We mean the situations when the resultant complex has more than two non-zero terms but the dimensions of the two last components coincide and are equal to the degree of the resultant. In other words, suppose that $\mathbf{m} = (m_1, \dots, m_r)$ satisfies the following two Diophantine equations:

$$(k+1) \cdot \prod_{j=1}^{r} \binom{k_j + m_j - d_j}{k_j} = \prod_{j=1}^{r} \binom{k_j + m_j}{k_j}$$

$$= (k+1) \binom{k}{k_1, k_2, \dots, k_r} d_1^{k_1} d_2^{k_2} \cdots d_r^{k_r}. \tag{2.5}$$

In this case the determinant of the corresponding matrix ∂_0 in (2.2) has the same degree as the resultant $R(f_0, \dots, f_k)$ and is divisible by it. Hence $\det(\partial_0)$ is either equal to the resultant or is identically zero.

Any vector $\mathbf{m}$ given by (2.4) satisfies (2.5) (this can be either shown directly or deduced from Proposition 2.2). However, (2.5) has other solutions as well. For instance, if $r = 4$ then the following values of the k_j, d_j and $\mathbf{m}$ (found with the help of MAPLE in [SZ1]) give a solution of (2.5) but do not come from (2.4):

$$(k_1, \dots, k_4) = (1, 1, 1, 1), \quad (d_1, \dots, d_4) = (1, 3, 5, 13), \quad \mathbf{m} = (3, 12, 17, 24).$$

We do not know at present whether such solutions can provide new formulas for the resultants.

The construction of Weyman's complexes (Section 1C) also extends to our present setting. We can extend Proposition 1.6 by giving a complete description of twist parameters $m_1, \dots, m_r$ such that the corresponding complex has only two non-zero terms. (We have seen that every such complex gives rise to a polynomial expression for the resultant.) It turns out that this method provides a polynomial expression for the resultant of type $(k_1, \ldots, k_r; d_1, \ldots, d_r)$ if and only if each pair (k_j, d_j) for $j = 1, \ldots, r$ belongs to the list in Proposition 1.6 (proved in [WZ1]).

B. The degree of the discriminant

Now we discuss the *discriminant* $\Delta(f)$ of a multihomogeneous form f on the product of projective spaces $P^{k_1} \times P^{k_2} \times \cdots \times P^{k_r}$. Recall that we denote by $S(d_1, \ldots, d_r)$ the space of forms on $P^{k_1} \times \cdots \times P^{k_r}$ which are homogeneous of degree d_j in the homogeneous coordinates on each P^{k_j}. We call the discriminant of a form from $S(d_1, \ldots, d_r)$ the *discriminant of type* $(k_1, \ldots, k_r; d_1, \ldots, d_r)$.

By definition, the discriminant of type $(k_1, \ldots, k_r; d_1, \ldots, d_r)$ is non-trivial if and only if the dual variety $(P^{k_1} \times \cdots \times P^{k_r})^\vee$ is a hypersurface, where $P^{k_1} \times \cdots \times P^{k_r}$ is taken in the embedding into $P(S^{d_1}\mathbf{C}^{k_1+1} \otimes \cdots \otimes S^{d_r}\mathbf{C}^{k_r+1})$. The criterion for this was given in Corollary 5.11 Chapter 1 which we reproduce here for the convenience of the reader.

Proposition 2.3. *The discriminant of type $(k_1, \ldots, k_r; d_1, \ldots, d_r)$ is non-trivial if and only if*

$$2k_j \leq k_1 + \cdots + k_r \tag{2.6}$$

for all j such that $d_j = 1$ (in particular, if all $d_j > 1$ then the discriminant is non-trivial for arbitrary dimensions $k_1, \ldots, k_r$).

Now we compute the *degree* $N(k_1, \ldots, k_r; d_1, \ldots, d_r)$ of the discriminant of type $(k_1, \ldots, k_r; d_1, \ldots, d_r)$. We fix positive integers $d_1, \ldots, d_r$, and consider the generating function

$$F_{d_1,\ldots,d_r}(z_1, \ldots, z_r) = \sum_{k_1,\ldots,k_r \geq 0} N(k_1, \ldots, k_r; d_1, \ldots, d_r) z_1^{k_1} \cdots z_r^{k_r}. \tag{2.7}$$

Theorem 2.4. *The generating function $F_{d_1,\ldots,d_r}(z_1, \ldots, z_r)$ is given by*

$$F_{d_1,\ldots,d_r}(z_1, \ldots, z_r) = \frac{1}{\left(\prod_j (1 + z_j) - \sum_j d_j z_j \prod_{i \neq j}(1 + z_i)\right)^2}. \tag{2.8}$$

Proof. We use the general formula for the degree of the A-discriminant given in Theorem 2.8 Chapter 9:

$$N(k_1, \ldots, k_r; d_1, \ldots, d_r) = \sum_{\Gamma \subset Q} (-1)^{\dim(Q)-\dim(\Gamma)} (\dim(\Gamma) + 1)\mathrm{Vol}\,(\Gamma), \quad (2.9)$$

where $Q = \mathrm{Conv}\,(A)$, the sum is over all faces $\Gamma \subset Q$, and the volume form on each face Γ is normalized so that an elementary simplex on the lattice affinely spanned by $A \cap \Gamma$ has volume 1.

In our situation Q is the product of scaled simplices, and its volume was computed in Proposition 2.1:

$$\mathrm{Vol}\,(Q) = \binom{k_1 + k_2 + \cdots + k_r}{k_1, k_2, \ldots, k_r} d_1^{k_1} d_2^{k_2} \cdots d_r^{k_r}.$$

Clearly, each face $\Gamma \subset Q$ has the form $\Gamma = d_1\Delta^{m_1} \times \cdots \times d_r\Delta^{m_r}$ for some $0 \leq m_j \leq k_j$; for given $m_1, \ldots, m_r$ there are $\prod_k \binom{k_j+1}{m_j+1}$ faces of this type. Furthermore, for a face $\Gamma = d_1\Delta^{m_1} \times \cdots \times d_r\Delta^{m_r}$ we have

$$\dim\,(\Gamma) = m_1 + \cdots + m_r, \ \mathrm{Vol}\,(\Gamma) = \binom{m_1 + \cdots + m_r}{m_1, \ldots, m_r} d_1^{m_1} d_2^{m_2} \cdots d_r^{m_r}.$$

Substituting all this into (2.9) we get

$$N(k_1, \ldots, k_r; d_1, \ldots, d_r) = \sum_{0 \leq m_j \leq k_j} (-1)^{\sum (k_j - m_j)} (m_1 + \cdots + m_r + 1)$$

$$\times \binom{m_1 + \cdots + m_r}{m_1, \ldots, m_r} \cdot \prod_j \binom{k_j + 1}{m_j + 1} d_1^{m_1} d_2^{m_2} \cdots d_r^{m_r}. \quad (2.10)$$

Setting $k_j - m_j = p_j$ we see that our generating function takes the form

$$F_{d_1,\ldots,d_r}(z_1, \ldots, z_r)$$

$$= \sum_{m_1,\ldots,m_r \geq 0} (m_1 + \cdots + m_r + 1) \binom{m_1 + \cdots + m_r}{m_1, \ldots, m_r} (d_1 z_1)^{m_1} \cdots (d_r z_r)^{m_r}$$

$$\times \prod_j \sum_{p_j \geq 0} (-1)^{p_j} \binom{m_j + p_j + 1}{m_j + 1} z_j^{p_j}. \quad (2.11)$$

By the binomial formula, the inner sum on the right hand side of (2.11) is equal to $(1 + z_j)^{-m_j-2}$. It follows that the generating function is equal to

$$\prod_j \frac{1}{(1 + z_j)^2} \sum_{s \geq 0} (s + 1) \sum_{m_1 + \cdots + m_r = s} \binom{s}{m_1, \ldots, m_r} \prod_j \left(\frac{d_j z_j}{1 + z_j}\right)^{m_j}. \quad (2.12)$$

By the multinomial formula, the inner sum in (2.12) is equal to $(\sum_j \frac{d_j z_j}{1+z_j})^s$, and then the summation over s gives

$$\frac{1}{\left(1-\sum_j \frac{d_j z_j}{1+z_j}\right)^2}.$$

Substituting this into (2.12) yields (2.8).

Now we derive a combinatorial formula for the degree $N(k_1, \dots, k_r; d_1, \dots, d_r)$, simply by expanding (2.8) as a power series in $z_1, \dots, z_r$. To state the formula we need some notation. Let $B = B_r$ denote the set of all non-empty subsets $\Omega \subset \{1, 2, \dots, r\}$ (so $\#(B) = 2^r - 1$). For each $\Omega \in B$ we set

$$d_\Omega = \sum_{j\in\Omega} d_j, \tag{2.13}$$

and let $\delta(\Omega) \in \mathbf{Z}_+^r$ be the characteristic vector of Ω (so $\delta(\Omega)_j = 1$ for $j \in \Omega$, and $\delta(\Omega)_j = 0$ for $j \notin \Omega$). For every $\kappa = (k_1, \dots, k_r) \in \mathbf{Z}_+^r$ let $\mathcal{P}(\kappa)$ denote the set of all *partitions* of κ into a sum of vectors $\delta(\Omega)$, i.e., $\mathcal{P}(\kappa)$ is the set of all non-negative integral vectors $(m_\Omega)_{\Omega\in B}$ such that $\sum_{\Omega\in B} m_\Omega \delta(\Omega) = \kappa$.

Theorem 2.5. *The degree $N(k_1, \dots, k_r; d_1, \dots, d_r)$ is given by*

$$N(k_1, \dots, k_r; d_1, \dots, d_r) = \sum_{(m_\Omega)\in\mathcal{P}(\kappa)} (1 + \sum_{\Omega\in B} m_\Omega)! \prod_{\Omega\in B} \frac{(d_\Omega - 1)^{m_\Omega}}{m_\Omega!}. \tag{2.14}$$

Proof. Expanding the products in the right hand side of (2.8), we can rewrite it as

$$\frac{1}{\left(1-\sum_{\Omega\in B}(d_\Omega - 1)z^{\delta(\Omega)}\right)^2}, \tag{2.15}$$

where $z^{\delta(\Omega)}$ means, according to our general conventions, the monomial $\prod_{j\in\Omega} z_j$. Expanding (2.15) in a power series with the help of the binomial formula, we obtain (2.14).

Comparing (2.10) and (2.14) we see that (2.14) has the advantage that all its terms are positive. This implies, in particular, another proof of the criterion in Proposition 2.3. Namely, the right hand side of (2.14) is non-zero if and only if $\kappa = (k_1, \dots, k_r)$ can be represented as a sum of vectors $\delta(\Omega)$ corresponding to subsets Ω such that $d_\Omega > 1$. But this condition is easily seen to be equivalent to the criterion in Proposition 2.3.

In the case when all d_j's are equal to 1, the formula (2.14) can be simplified further. This will be done in Chapter 14, where we also give some numerical examples.

CHAPTER 14

Hyperdeterminants

The goal of this chapter is to provide a natural "higher dimensional" generalization of the classical notion of the determinant of a square matrix. There were some attempts toward a rather straightforward definition of the "hyperdeterminant" for "hypercubic" matrices using alternating summations over the product of several symmetric groups (see e.g., [P], §54 and references therein). Here we systematically develop another approach under which the hyperdeterminant becomes a special case of the general discriminant studied in the previous chapters. As so many other ideas in the field, this approach is due to Cayley [Ca1].

1. Basic properties of the hyperdeterminant

A. Definitions and the non-triviality criterion

Let $r \geq 2$ be an integer, and $A = (a_{i_1,\dots,i_r})$, $0 \leq i_j \leq k_j$ be an r-dimensional complex matrix (or array) of format $(k_1 + 1) \times \cdots \times (k_r + 1)$.

The definition of the hyperdeterminant of A can be stated in geometric, analytic or algebraic terms. Let us give all three formulations.

Geometrically, consider the product $X = P^{k_1} \times \cdots \times P^{k_r}$ of several projective spaces in the Segre embedding into the projective space $P^{(k_1+1)\cdots(k_r+1)-1}$ (if P^{k_j} is the projectivization of a vector space $V_j^* = \mathbf{C}^{k_j+1}$ then the ambient projective space is $P(V_1^* \otimes \cdots \otimes V_r^*)$). The *hyperdeterminant of format* $(k_1 + 1) \times \cdots \times (k_r + 1)$ is the X-discriminant as defined in Chapter 1, i.e., a homogeneous polynomial function on $V_1 \otimes \cdots \otimes V_r$ which is a defining equation of the projectively dual variety $X^\vee$ (provided $X^\vee$ is a hypersurface in $(P^{(k_1+1)\cdots(k_r+1)-1})^*$). We denote the hyperdeterminant by Det. As usual, if $X^\vee$ is not a hypersurface, we set Det equal to 1, and refer to this case as *trivial.* If each V_j is equipped with a basis then an element $f \in V_1 \otimes \cdots \otimes V_r$ is represented by a matrix $A = (a_{i_1,\dots,i_r})$, $0 \leq i_j \leq k_j$ as above, and so Det (A) is a polynomial function of matrix entries. It is determined uniquely up to sign by the requirement that Det (A) has integral coefficients and is irreducible over $\mathbf{Z}$.

Analytically, the hyperplane $\{f = 0\}$ belongs to $X^\vee$ if and only if f vanishes at some point of X with all its first derivatives. If we choose a coordinate system $x^{(j)} = (x_0^{(j)}, x_1^{(j)}, \dots, x_{k_j}^{(j)})$ on each V_j^* then $f \in V_1 \otimes \cdots \otimes V_r$ is represented after restriction on X by a multilinear form

$$f(x^{(1)}, \dots, x^{(r)}) = \sum_{i_1,\dots,i_r} a_{i_1,\dots,i_r} x_{i_1}^{(1)} \cdots x_{i_r}^{(r)}. \tag{1.1}$$

Therefore, the condition Det $(A) = 0$ means that the system of equations

$$f(x) = \frac{\partial f(x)}{\partial x_i^{(j)}} = 0 \tag{1.2}$$

(for all i, j) has a solution $x = (x^{(1)}, \dots, x^{(r)})$ with all $x^{(j)} \neq 0$. We say that a multilinear form f (or a matrix A) satisfying this condition is *degenerate.*

Algebraically, the degeneracy of a form f can be easily characterized as follows. We denote by $\mathcal{K}(f)$ (or $\mathcal{K}(A)$) the set of points

$$x = (x^{(1)}, \dots, x^{(r)}) \in X = P^{k_1} \times \cdots \times P^{k_r}$$

such that

$$f(x^{(1)}, \dots, x^{(j-1)}, y, x^{(j+1)}, \dots, x^{(r)}) = 0$$

for every $j = 1, \dots, r$ and $y \in V_j^*$. We shall sometimes call $\mathcal{K}(A)$ the *kernel* of A. For a bilinear form $f(x, y)$, there is a notion of left and right kernels

$$K_l(f) = \{x : f(x, y) = 0, \forall y\}, \qquad K_r(f) = \{y : f(x, y) = 0, \ \forall x\}$$

and $\mathcal{K}(f) = K_l(f) \times K_r(f)$.

Proposition 1.1. *A form f is degenerate if and only if $\mathcal{K}(f)$ is non-empty.*

Proof. Computing the differential of f we see that $\mathcal{K}(f)$ is exactly the set of solutions of (1.2).

In particular, for $r = 2$ when f is a bilinear form with a matrix A, the degeneracy of f just defined coincides with the usual notion of degeneracy and means that A is not of maximal rank. Obviously, this condition is of codimension one if and only if A is a square matrix, and in this case Det (A) coincides with the ordinary determinant Det (A).

The following proposition is a special case of Theorem 1.4 Chapter 1.

Proposition 1.2. *Suppose that an r-dimensional matrix $A_0 = (a_{i_1,\dots,i_r})$ is of such a format that the hyperdeterminant is non-trivial, and that A_0 is a smooth point of the hypersurface of degenerate matrices. Then $\mathcal{K}(A_0)$ consists of the unique point $(x^{(1)}, \dots, x^{(r)})$. Furthermore, under a suitable normalization we have*

$$x_{i_1}^{(1)} \cdots x_{i_r}^{(r)} = \frac{\partial \mathrm{Det}(A)}{\partial a_{i_1,\dots,i_r}}\bigg|_{A=A_0} \tag{1.3}$$

for all $i_1, \dots, i_r$.

The first natural question about hyperdeterminants is to describe all matrix formats for which Det (A) is non-trivial, i.e., $X^\vee$ is a hypersurface, or in other words, the degeneracy of A is a codimension one condition. The matrices of such formats can be viewed as multidimensional generalizations of ordinary square matrices.

Theorem 1.3. *The hyperdeterminant of format* $(k_1 + 1) \times \cdots \times (k_r + 1)$ *is non-trivial if and only if*

$$k_j \leq \sum_{i \neq j} k_i \tag{1.4}$$

for all $j = 1, \ldots, r$.

Proof. By definition, the hyperdeterminant of format $(k_1+1) \times \cdots \times (k_r+1)$ is a special case of the discriminant of type $(k_1, \ldots, k_r; d_1, \ldots, d_r)$ (Section 2 Chapter 13), corresponding to $d_1 = \cdots = d_r = 1$. Hence Theorem 1.3 is a special case of Proposition 2.3 Chapter 13.

B. Invariance properties

Until the end of this section we assume that (1.4) holds, i.e., the hyperdeterminant of a matrix A is non-trivial. The next property of Det follows at once from any of the definitions.

Proposition 1.4. *The hyperdeterminant is relatively invariant under the action of the group* $GL(V_1) \times \cdots \times GL(V_r)$ *(and so invariant under the action of* $SL(V_1) \times \cdots \times SL(V_r)$*).*

To state explicitly the consequences of Proposition 1.4, we need some terminology. We shall identify the set of matrix (multi-)indices $I = \{(i_1, \ldots, i_r) : 0 \leq i_j \leq k_j\}$ of a matrix A with the set of vertices of the product $\Delta^{k_1} \times \cdots \times \Delta^{k_r}$ of r standard simplices. Thus, the submatrices of A correspond to faces of $\Delta^{k_1} \times \cdots \times \Delta^{k_r}$. By a *slice in the* j*-th direction* we mean the subset of all indices in I with the fixed j-th component, and also the corresponding submatrix of A. Two slices in the same direction are called *parallel.*

Corollary 1.5.

(a) *Interchanging two parallel slices leaves the hyperdeterminant invariant up to sign (which may equal 1).*

(b) *The hyperdeterminant is a homogeneous polynomial in the entries of each slice. The degree of homogeneity is the same for parallel slices.*

(c) *The hyperdeterminant does not change if we add to some slice a scalar multiple of a parallel slice.*

(d) *The hyperdeterminant of a matrix having two parallel slices proportional to each other is equal to* 0*. In particular,* Det $(A) = 0$ *if* A *has a zero slice.*

Proof. The properties (a) to (c) express the (relative) invariance of Det (A) under the action of various elements from $GL(V_j) = GL(k_j+1, \mathbf{C})$: permutation matrices, diagonal matrices, and unipotent matrices with only one non-zero off-diagonal entry. In fact, these matrices are known to generate the group $GL(k_j+1, \mathbf{C})$, and so the combination of these three properties is equivalent to Proposition 1.4. Part (d) follows at once from (b) and (c).

It is clear that a polynomial $P(a_{i_1,\dots,i_r})$ satisfies the condition (c) from Corollary 1.5 if and only if P is annihilated by the differential operators

$$D_{ij}^{(1)} = \sum_{i_2,\dots,i_r} a_{j,i_2,\dots,i_r} \frac{\partial}{\partial a_{i,i_2,\dots,i_r}}$$

(for all $i \neq j$), and by similar operators for the slices in other directions. This is probably the most practical way of verifying this condition.

Our next result will lead to a combinatorial characterization of the hyperdeterminant. This again requires some terminology. We define the *support* of a monomial $\prod_{i_1,\dots,i_r} a_{i_1,\dots,i_r}^{\alpha_{i_1,\dots,i_r}}$ as the set of all indices $(i_1, \dots, i_r)$ such that $\alpha_{i_1,\dots,i_r} \neq 0$. By the *star* of $(i_1, \dots, i_r)$ we mean the set of all indices which differ from $(i_1, \dots, i_r)$ in at most one place. In other words, if we represent the indices by vertices of the product of simplices then the star consists of a vertex itself and all the vertices connected to it by an edge. To give one more interpretation, we consider the set $\prod_{j=1}^{r}[0, k_j]$ with the *Hamming metric* used in the coding theory: the distance between two indices $(i_1, \dots, i_r)$ and $(i_1', \dots, i_r')$ is the number of positions j such that $i_j \neq i_j'$. Then the star of $\mathbf{i} = (i_1, \dots, i_r)$ is the Hamming ball of radius 1 with center at $\mathbf{i}$.

Proposition 1.6. *For a polynomial $P(a_{i_1,\dots,i_r})$ the following conditions are equivalent:*

(a) *P is relatively invariant under the group $GL(k_1+1, \mathbf{C}) \times \cdots \times GL(k_r+1, \mathbf{C})$ and is divisible by* Det (A).

(b) *P satisfies the conditions (a) to (c) of Corollary 1.5, and there exists an index $(i_1, \dots, i_r)$ such that the support of each monomial in $P(a_{i_1,\dots,i_r})$ meets the star of $(i_1, \dots, i_r)$.*

(c) *P satisfies the conditions (a) to (c) of Corollary 1.5, and the support of each monomial in $P(a_{i_1,\dots,i_r})$ meets the star of every index $(i_1, \dots, i_r)$.*

Proof. We already mentioned in the proof of Corollary 1.5 that P is relatively invariant if and only if it satisfies the conditions (a) to (c) of Corollary 1.5. Suppose this is the case. By definition, P is divisible by Det (A) if and only if P vanishes at all degenerate matrices A. By Proposition 1.1, A is degenerate if and only if $\mathcal{K}(A) \neq \emptyset$. Using the invariance of P we can assume that $\mathcal{K}(A)$ contains a point $(x^{(1)}, \ldots, x^{(r)})$ such that each $x^{(j)}$ is a basis vector $e_{i_j}^{(j)}$. Then the condition $(x^{(1)}, \ldots, x^{(r)}) \in \mathcal{K}(A)$ means that $a_{i'_1,\ldots,i'_r} = 0$ for all $(i'_1, \ldots, i'_r)$ in the star of $(i_1, \ldots, i_r)$. Clearly, P vanishes at all such matrices if and only if the support of each monomial in P meets the star of $(i_1, \ldots, i_r)$, and we are done.

Note that the second assertion in part (b) means that the support of any monomial from Det forms a 1-net in $\prod_{j=1}^{r}[0, k_j]$ with respect to the Hamming metric. Such nets are known as error-correcting codes. It would be interesting to study monomials in the hyperdeterminant from this point of view.

To illustrate the use of Proposition 1.6, we give an explicit formula for the three-dimensional hyperdeterminant of format $2 \times 2 \times 2$. The hyperdeterminant in this case was already known to Cayley (see [Ca1], p. 89).

Proposition 1.7. *The hyperdeterminant of a matrix* $A = (a_{ijk})$ $(i, j, k = 0, 1)$ *is given by*

$$\begin{aligned} \mathrm{Det}\,(A) = {} & (a_{000}^2 a_{111}^2 + a_{001}^2 a_{110}^2 + a_{010}^2 a_{101}^2 + a_{011}^2 a_{100}^2) \\ & -2(a_{000}a_{001}a_{110}a_{111} + a_{000}a_{010}a_{101}a_{111} + a_{000}a_{011}a_{100}a_{111} \\ & +a_{001}a_{010}a_{101}a_{110} + a_{001}a_{011}a_{110}a_{100} + a_{010}a_{011}a_{101}a_{100}) \\ & +4(a_{000}a_{011}a_{101}a_{110} + a_{001}a_{010}a_{100}a_{111}). \end{aligned} \tag{1.5}$$

Proof. First, we claim that the polynomial P defined by (1.5) satisfies conditions (a) to (c) of Corollary 1.5. This is verified directly (to check (c) we can use the remark after Corollary 1.5).

The monomials appearing in P can be visualized as follows. If we represent matrix entries by the vertices of the cube then the monomials in the first group correspond to four main diagonals of the cube; the monomials in the second group correspond to six rectangles formed by pairs of opposite edges, and the monomials in the third group to two tetrahedra whose edges are diagonals of the cube's faces. Obviously, the support of each of these monomials meets the star of every vertex of the cube. This shows that P satisfies the equivalent conditions of Proposition 1.6 and hence is divisible by Det (A).

There are several ways to complete the proof, i.e., to show that $P = \mathrm{Det}\,(A)$. For instance, it is not hard to see that P is irreducible. We can also refer to the results of Section 3 below which show that Det (A) in our case has degree 4. In fact, it is easy to see that in our case Det (A) is the $SL(2) \times SL(2) \times SL(2)$ invariant of minimal degree.

By definition, the vanishing of $\mathrm{Det}(A)$ for a $2\times2\times2$ matrix A (or, equivalently, degeneracy of A) means that the following system of 6 homogeneous equations with 6 unknowns has a non-trivial solution:

$$\begin{aligned}
a_{000}x_0y_0 + a_{010}x_0y_1 + a_{100}x_1y_0 + a_{110}x_1y_1 &= 0,\\
a_{001}x_0y_0 + a_{011}x_0y_1 + a_{101}x_1y_0 + a_{111}x_1y_1 &= 0,\\
a_{000}x_0z_0 + a_{001}x_0z_1 + a_{100}x_1z_0 + a_{101}x_1z_1 &= 0,\\
a_{010}x_0z_0 + a_{011}x_0z_1 + a_{110}x_1z_0 + a_{111}x_1z_1 &= 0,\\
a_{000}y_0z_0 + a_{001}y_0z_1 + a_{010}y_1z_0 + a_{011}y_1z_1 &= 0,\\
a_{100}y_0z_0 + a_{101}y_0z_1 + a_{110}y_1z_0 + a_{111}y_1z_1 &= 0.
\end{aligned} \tag{1.6}$$

It is not so easy (although possible) to prove directly that (1.6) has a non-trivial solution if and only if (1.5) vanishes.

C. Algebraic properties

We start with a multidimensional generalization of the fact that transposing a matrix preserves the determinant. For a matrix $A = (a_{i_1,\dots,i_r})$ of format $(k_1+1)\times\cdots\times(k_r+1)$ and a permutation σ of indices $1,\dots,r$, we denote by $\sigma(A)$ the matrix of format $(k_{\sigma^{-1}(1)}+1)\times\cdots\times(k_{\sigma^{-1}(r)}+1)$, whose $(i_1,\dots,i_r)$-th entry is equal to $a_{i_{\sigma(1)},\dots,i_{\sigma(r)}}$. The following result is an immediate consequence of definitions.

Proposition 1.8. *We have* $\mathrm{Det}\,(\sigma(A)) = \mathrm{Det}\,(A)$ *for every permutation σ. In particular, if A is degenerate then $\sigma(A)$ is degenerate.*

Now we discuss an analog of the multiplicative property of the ordinary determinant. Let $A = (a_{i_1,\dots,i_r})$ be a matrix of format $(k_1+1)\times\cdots\times(k_r+1)$ and $B = (b_{j_1,\dots,j_s})$ be a matrix of format $(l_1+1)\times\cdots\times(l_s+1)$. Suppose that $k_r = l_1$. We define the convolution (or product) $A * B$ to be the $(r+s-1)$-dimensional matrix C of the format

$$(k_1+1)\times\cdots\times(k_{r-1}+1)\times(l_2+1)\times\cdots\times(l_s+1)$$

with entries

$$c_{i_1,\dots,i_{r-1},j_2,\dots,j_s} = \sum_{h=0}^{k_r} a_{i_1,\dots,i_{r-1},h}\,b_{h,j_2,\dots,j_s}.$$

Similarly, we can define the convolution $A \underset{p,q}{*} B$ with respect to a pair of indices p, q such that $k_p = l_q$.

Proposition 1.9. *If A, B are degenerate then $A * B$ is also degenerate.*

Proof. We shall use Proposition 1.1. The definitions readily imply that if $(x^{(1)}, \ldots, x^{(r)}) \in \mathcal{K}(A)$ and $(y^{(1)}, \ldots, y^{(s)}) \in \mathcal{K}(B)$ then

$$(x^{(1)}, \ldots, x^{(r-1)}, y^{(2)}, \ldots, y^{(s)}) \in \mathcal{K}(A * B),$$

which implies our statement.

Corollary 1.10. *Let the formats of A, B be such that the hyperdeterminants of A, B and $A * B$ are non-trivial. Then there exist polynomials $P(A, B)$ and $Q(A, B)$ in entries of A and B such that*

$$\mathrm{Det}\,(A * B) = P(A, B)\mathrm{Det}\,(A) + Q(A, B)\mathrm{Det}\,(B). \tag{1.7}$$

Since $\mathrm{Det}\,(A)$ and $\mathrm{Det}\,(B)$ depend on disjoint sets of variables, it follows that P and Q in (1.7) are defined uniquely up to transformations

$$P \mapsto P + R(A, B)\mathrm{Det}\,(B), \quad Q \mapsto Q - R(A, B)\mathrm{Det}\,(A).$$

Multiplicative properties of hyperdeterminants deserve further study, but this goes beyond the aims of the present book.

2. The Cayley method and the degree

A. The discriminantal complex

We describe the discriminantal complex whose determinant is the hyperdeterminant of an r-dimensional matrix. We use the language of differential forms developed in Section 4 Chapter 2.

As in Section 1, we interpret A as the matrix of a multilinear form f on $V_1^* \times \cdots \times V_r^*$, where each V_j is a finite-dimensional vector space and $x_0^{(j)}, x_1^{(j)}, \ldots, x_{k_j}^{(j)}$ are coordinates in V_j^*. We often make use of the standard invertible sheaves $\mathcal{O}(m_1, \ldots, m_r)$ on $P(V_1^*) \times \cdots \times P(V_r^*)$; see Chapter 13, Section 2A.

Let Ω^p be the space of polynomial differential p-forms on $V_1^* \times \cdots \times V_r^*$. For each vector field ξ on $V_1^* \times \cdots \times V_r^*$, let $i_\xi : \Omega^p \to \Omega^{p-1}$ be the contraction with ξ, and $\mathrm{Lie}_\xi = di_\xi + i_\xi d$ the Lie derivative. For $j = 1, \ldots, r$ let

$$\xi_j = \sum_i x_i^{(j)} \frac{\partial}{\partial x_i^{(j)}}$$

be the Euler field on V_j^* regarded as a vector field on $V_1^* \times \cdots \times V_r^*$.

Now let $m_1, \ldots, m_r \in \mathbf{Z}$. We associate to $m_1, \ldots, m_r$ and f the complex

$$C^\bullet = C^\bullet(m_1, \ldots, m_r; f) = \left\{C^0 \xrightarrow{\partial_f} C^1 \xrightarrow{\partial_f} \cdots\right\}$$

in the following way. We define $C^p = C^p(m_1, \ldots, m_r)$ to be the space of all differential p-forms ω with polynomial coefficients on $V_1^* \times \cdots \times V_r^*$ satisfying conditions

$$\mathrm{Lie}_{\xi_j}(\omega) = (p + m_j)\omega \quad (j = 1, \ldots, r) \tag{2.1}$$

and

$$i_{\xi_1}(\omega) = i_{\xi_2}(\omega) = \cdots = i_{\xi_r}(\omega). \tag{2.2}$$

The differential $\partial_f : C^p \to C^{p+1}$ is defined as the exterior multiplication by the 1-form df.

The terms of $C^\bullet$ do not depend on f. In fact, (2.1) means simply that $\omega \in C^p(m_1, \ldots, m_r)$ is homogeneous of degree $(p + m_j)$ with respect to the variables $x_0^{(j)}, x_1^{(j)}, \ldots, x_{k_j}^{(j)}$, where we assume $\deg(x_i^{(j)}) = \deg(dx_i^{(j)}) = 1$.

Theorem 2.1. *Under a suitable choice of bases in the terms of $C^\bullet$, we have*

$$\mathrm{Det}\,(A) = \det\,(C^\bullet(m_1, \ldots, m_r; f))^{(-1)^{k_1+\cdots+k_r+1}} \tag{2.3}$$

for all non-negative integers $m_1, \ldots, m_r$.

Proof. We shall deduce the statement from Theorem 2.5 Chapter 2. To this end it is enough to prove the following.

(1) The complex $C^\bullet$ coincides with the discriminantal complex $C_+(X, \mathcal{O})$ where

$$X = \prod P^{k_j} = \prod P(V_j^*) \subset P\left(\bigotimes V_j^*\right)$$

is the Segre variety. In other words (see formula (2.1) Chapter 2), we have

$$C^p = H^0\left(\prod P^{k_j}, \bigwedge^p J(\mathcal{L})\right)$$

where $\mathcal{L} = \mathcal{O}_X(1, \ldots, 1)$ is the inverse image of the sheaf $\mathcal{O}(1)$ from the ambient projective space $P\left(\bigotimes V_j^*\right)$.

(2) The complex $C_+(X, \mathcal{O})$ is stably twisted, i.e.,

$$H^q\left(\prod P^{k_j}, \bigwedge^p J(\mathcal{L})\right) = 0, \qquad \forall p, \forall q > 0.$$

To prove (1), we denote the natural projection by

$$\pi : \prod (V_j^* - \{0\}) \longrightarrow X = \prod P(V_j^*).$$

All the tangent spaces to the fibers of π are spanned by the vector fields ξ_j. The conditions (2.1) and (2.2) define a certain subsheaf Ω_p^p in the sheaf of all regular forms on $\prod\left(V_j^* - \{0\}\right)$. The argument, similar to that in the proof of Propositions 4.1 and 1.3 of Chapter 2, shows that

$$\pi_*(\Omega_p^p) = \bigwedge^p J(\mathcal{L}),$$

which implies (1).

To prove the statement (2), we use Corollary 4.9 Chapter 2. In our situation, this corollary says that $C_+(X, \mathcal{O})$ is stably twisted if and only if the following sheaves on $X = \prod P(V_j^*)$ have no higher cohomology:

$$\Omega_X^p \otimes \mathcal{L}^{\otimes p}, \quad \Omega_X^{p-1} \otimes \mathcal{L}^{\otimes p}, \quad p \geq 0, \ \mathcal{L} = \mathcal{O}(1, \ldots, 1).$$

The vanishing of the higher cohomology of these sheaves follows, in a straightforward way, from the Künneth formula and the Bott theorem describing the cohomology of sheaves of the form $\Omega^i(j)$ on a single projective space (see [OSS], Chapter 1). This concludes the proof of Theorem 2.1.

The group $GL(V_1) \times \cdots \times GL(V_r)$ acts naturally on each term of $C^\bullet$. To describe this action we need the following notation. For a finite dimensional vector space V, denote by $S^{(p|q)}(V)$ the irreducible $GL(V)$-module corresponding to the *hook* Young diagram having one row of length q and p rows of length 1 (i.e., one column of length $p + 1$). We use the following well-known realization of this module.

Proposition 2.2. *The $GL(V)$-module $S^{(p|q)}(V)$ is isomorphic to the space $B^p(q)$ of polynomial differential p-forms ω on V^* such that* $\text{Lie}_\xi(\omega) = (p+q)\omega,\ i_\xi(\omega) = 0$, *where ξ is the Euler vector field on V^*.*

To prove Proposition 2.2 is is enough to show that $B^p(q)$ has a unique (up to multiple) highest vector, and that the corresponding highest weight is $(q, 1, 1, \ldots, 1, 0, \ldots, 0)$ (with p units and $k - p$ zeros, where $\dim V = k + 1$). We leave it to the reader to check that this is indeed the case; the highest vector in $B^p(q)$ has the form

$$x_0^{q-1} \sum_{i=0}^{p} (-1)^i x_i dx_0 \wedge \cdots \wedge dx_{i-1} \wedge dx_{i+1} \wedge \cdots \wedge dx_p,$$

where $x_0, \ldots, x_k$ are coordinates in V^*.

Proposition 2.3. *Each term $C^p(m_1, \dots, m_r)$ is a multiplicity free module over $GL(V_1) \times \dots \times GL(V_r)$ isomorphic to*

$$\bigoplus_{p_1,\dots,p_r} \left(S^{(p+m_1-p_1|p_1)}(V_1) \otimes \dots \otimes S^{(p+m_r-p_r|p_r)}(V_r) \right),$$

the sum over all $p_1, \dots, p_r$ with $p_1 + \dots + p_r = p$ or $p-1$.

Proof. Let $B^p(m_1, \dots, m_r)$ denote the subspace of $C^p(m_1, \dots, m_r)$ consisting of p-forms ω satisfying (2.1) and such that

$$i_{\xi_1}(\omega) = i_{\xi_2}(\omega) = \dots = i_{\xi_r}(\omega) = 0. \tag{2.2'}$$

Clearly, $B^p(m_1, \dots, m_r)$ is the kernel of the projection

$$C^p(m_1, \dots, m_r) \longrightarrow B^{p-1}(m_1 + 1, \dots, m_r + 1)$$

sending each $\omega \in C^p(m_1, \dots, m_r)$ to

$$\omega' = i_{\xi_1}(\omega) = i_{\xi_2}(\omega) = \dots = i_{\xi_r}(\omega).$$

Therefore, $C^p(m_1, \dots, m_r)$ is isomorphic as a $GL(V_1) \times \dots \times GL(V_r)$-module to

$$B^p(m_1, \dots, m_r) \oplus B^{p-1}(m_1 + 1, \dots, m_r + 1).$$

By definition, we have the decomposition

$$B^p(m_1, \dots, m_r) = \bigoplus_{p_1+\dots+p_r=p} \left(B^{p_1}(p + m_1 - p_1) \otimes \dots \otimes B^{p_r}(p + m_r - p_r) \right), \tag{2.4}$$

and the analogous decomposition for $B^{p-1}(m_1 + 1, \dots, m_r + 1)$. It remains to apply Proposition 2.2.

B. Degree of the hyperdeterminant

Fix $r \geq 2$ and let $N(k_1, \dots, k_r)$ be the degree of the hyperdeterminant of format $(k_1+1) \times \dots \times (k_r+1)$ (we assume that $N(k_1, \dots, k_r) = 0$ if the hyperdeterminant is trivial). The following theorem is essentially a special case of Theorem 2.4 of Chapter 13.

Theorem 2.4. *The generating function for the degrees $N(k_1, \ldots, k_r)$ is given by*

$$\sum_{k_1,\ldots,k_r\geq 0} N(k_1, \ldots, k_r) z_1^{k_1} \cdots z_r^{k_r} = \frac{1}{\left(1 - \sum_{i=2}^{r} (i-1) e_i(z_1, \ldots, z_r)\right)^2}, \quad (2.5)$$

where $e_i(z_1, \ldots, z_r)$ is the i-th elementary symmetric polynomial.

Proof. The generating function in question is obtained by setting all d_i equal to 1 in formula (2.8) of Chapter 13, i.e., is equal to

$$\frac{1}{\left(\prod_j (1+z_j) - \sum_j z_j \prod_{i\neq j} (1+z_i)\right)^2}. \quad (2.6)$$

It remains to observe that the polynomial

$$\prod_j (1+z_j) - \sum_j z_j \prod_{i\neq j} (1+z_i)$$

contains only square-free monomials in $z_1, \ldots, z_r$, and every such monomial of degree i occurs with the coefficient $(1-i)$.

A symmetric form (2.5) allows us to derive a combinatorial formula for $N(k_1, \ldots, k_r)$; this is more convenient for computations than the general formula (2.14) of Chapter 13. We shall use the terminology and notation of Section 1C Chapter 12 related to partitions and symmetric polynomials. Recall that a *partition* is a finite weakly decreasing sequence $\lambda = (\lambda_1, \ldots, \lambda_s)$ of non-negative integers; the numbers λ_i are called parts of λ. We shall denote the number of parts of λ equal to i by $m_i = m_i(\lambda)$, and write λ also as $\lambda = (1^{m_1}, 2^{m_2}, \ldots)$; for instance, the partition (4, 2, 2, 1) can also be written $(1, 2^2, 4)$.

The degree $N(k_1, \ldots, k_r)$ is obviously symmetric in $k_1, \ldots, k_r$ and so depends only on the partition κ obtained by rearranging $k_1, \ldots, k_r$ in the weakly decreasing order. Our expression will involve the quantities $d_{\kappa\lambda}$ (see Section 1C Chapter 12); recall that $d_{\kappa\lambda}$ is the number of (0, 1)-matrices whose row sums are the parts of κ and column sums are the parts of λ.

Theorem 2.5. *Let κ be the partition obtained by rearranging the numbers $k_1, \ldots, k_r$ in a weakly decreasing order. Then*

$$N(k_1, \ldots, k_r) = \sum_{\lambda} (m_2 + \cdots + m_p + 1)! \cdot d_{\kappa\lambda} \cdot \prod_{i=2}^{p} \frac{(i-1)^{m_i}}{m_i!}, \quad (2.7)$$

the sum over all partitions $\lambda = (1^{m_1}, 2^{m_2}, \ldots, p^{m_p})$ with $m_1 = 0$.

Proof. Expanding the right hand side of (2.5) by the binomial formula, we obtain

$$\sum_{\lambda}(m_2 + \cdots + m_p + 1)! \cdot e_\lambda \cdot \prod_{i=2}^{p} \frac{(i-1)^{m_i}}{m_i!}$$

(the same summation as in (2.7)), where

$$e_\lambda(z_1, \ldots, z_r) = \prod_{i=1}^{p} e_i(z_1, \ldots, z_r)^{m_i}.$$

It remains to notice that each monomial $z_1^{k_1} \cdots z_r^{k_r}$ appears in e_λ with the coefficient $d_{\kappa\lambda}$ (see Section 1C Chapter 12).

The behavior of $d_{\kappa\lambda}$ is controlled by the Gale-Ryser theorem (Proposition 1.2 Chapter 12). Using it, we can explicitly evaluate the $d_{\kappa\lambda}$ in many special cases which leads to a more explicit formula for the degree $N(k_1, \ldots, k_r)$. Probably, the most important special case is the following. We say that the matrix format $(k_1 + 1) \times \cdots \times (k_r + 1)$ is *boundary* if one of the numbers k_j equals the sum of all the others (i.e., one of the inequalities (1.4) becomes an equality); without loss of generality we can assume that $k_1 = k_2 + \ldots + k_r$.

Corollary 2.6. *The degree of the hyperdeterminant of the boundary format is*

$$N(k_2 + \cdots + k_r, k_2, \ldots, k_r) = (k_2 + \cdots + k_r + 1)\binom{k_2 + \cdots + k_r}{k_2, \ldots, k_r} = \frac{(k_1+1)!}{k_2! \cdots k_r!}. \tag{2.8}$$

Proof. We use the following obvious combinatorial statement.

Lemma 2.7. *We have*

$$d_{(k_2,\ldots,k_r),(1^{k_2+\cdots+k_r})} = \binom{k_2 + \cdots + k_r}{k_2, \ldots, k_r}.$$

It follows easily from the Gale-Ryser theorem that in the boundary case there is exactly one summand in (2.7), and it corresponds to $\lambda = (2^{k_2+\cdots+k_r})$. Clearly, every (0, 1)-matrix contributing to $d_{\kappa\lambda}$ has all the entries in the first row equal to 1. Hence $d_{\kappa\lambda}$ is given by Lemma 2.7. Substituting this into (2.7) yields our statement.

Note that for $r = 2$ the boundary format is just that of ordinary square matrices, and (2.8) expresses the fact that the (ordinary) determinant of a $(k+1) \times (k+1)$ matrix has degree $k + 1$.

We call the matrix format $(k_1+1)\times\cdots\times(k_r+1)$ with $k_1 \geq k_2 \geq \cdots \geq k_r$ *subboundary* if $k_1 = k_2+\cdots+k_r-1$.

Corollary 2.8. *The degree of the hyperdeterminant of the subboundary format is given by*

$$N(k_2+\cdots+k_r-1, k_2, \ldots, k_r) = 2\binom{k_2+\cdots+k_r}{k_2,\ldots,k_r}\cdot e_2(k_2,\ldots,k_r). \quad (2.9)$$

Here $e_2(k_2,\ldots,k_r) = \sum_{2\leq i<j\leq r} k_i k_j$ is the second elementary symmetric polynomial.

Proof. As for the boundary format, the Gale-Ryser theorem implies that (2.7) reduces to one summand corresponding to $\lambda = (2^{k_2+\cdots+k_r-2}, 3)$. Clearly, every (0, 1)-matrix contributing to $d_{\kappa\lambda}$ has all the entries in the first row equal to 1. Consider now the column with sum 3, i.e., containing three unit entries. We know that one of these units lies in the first row. Decomposing the set of our (0, 1)-matrices into the subsets according to the location of the remaining two units and using Lemma 2.7, we see that

$$\begin{aligned} d_{\kappa\lambda} &= \sum_{2\leq i<j\leq r}\binom{k_2+\cdots+k_r-2}{k_2,\ldots,k_i-1,\ldots,k_j-1,\ldots,k_r} \\ &= \frac{(k_2+\cdots+k_r-2)!}{k_2!\cdots k_r!}e_2(k_2,\ldots,k_r). \end{aligned}$$

Substituting this into (2.7) yields our statement.

Corollary 2.9. *The degree of the hyperdeterminant of the cubic format* $(k+1)\times(k+1)\times(k+1)$ *is given by*

$$N(k,k,k) = \sum_{0\leq j\leq k/2}\frac{(j+k+1)!}{j!^3(k-2j)!}\cdot 2^{k-2j}. \quad (2.10)$$

Proof. We have $\kappa = (k,k,k)$. By the Gale-Ryser theorem, the partitions λ contributing to (2.7) have the form $(2^{3j}, 3^{k-2j})$ for $0 \leq j \leq k/2$. Clearly, $d_{\kappa,(2^{3j},3^{k-2j})} = d_{((2j)^3),(2^{3j})}$ because every (0, 1)-matrix contributing to $d_{\kappa,(2^{3j},3^{k-2j})}$ has all the entries in the first $(k-2j)$ columns equal to 1. It is also easy to see that

$$d_{((2j)^3),(2^{3j})} = \binom{3j}{j,j,j} \quad (2.11)$$

because every $3\times 3j$ matrix contributing to $d_{((2j)^3),(2^{3j})}$ is determined by a disjoint decomposition of the set of columns into three j-element subsets C_{12}, C_{13}, C_{23}, where say C_{12} is the set of columns with unit entries in the first two rows. Substituting (2.11) into (2.7) yields (2.10).

For $k = 1, 2, 3$ the degree (2.10) is equal to 4, 36, and 272 respectively. It seems that the sum in (2.10) cannot be simplified.

Our last application of (2.7) is the expression for the degree of the hyperdeterminant of the r-dimensional format $2 \times 2 \times \cdots \times 2$. Denote this degree by $N((1^r))$.

Corollary 2.10. *The exponential generating function for the sequence $N((1^r))$ is given by*

$$\sum_{r \geq 0} N((1^r)) \frac{z^r}{r!} = \frac{e^{-2z}}{(1-z)^2}. \tag{2.12}$$

Proof. It follows from (2.7) and Lemma 2.7 that

$$N((1^r)) = r! \sum_{\lambda} \frac{(m_2 + \cdots + m_r + 1)!}{\prod_{i \geq 2} [(i-2)! i]^{m_i} m_i!}, \tag{2.13}$$

the sum over all partitions $\lambda = (2^{m_2}, 3^{m_3}, \ldots)$ with $2m_2 + 3m_3 + \cdots = r$. Therefore, we have

$$\sum_{r \geq 0} N((1^r)) \frac{z^r}{r!} = \sum \frac{(m_2 + \cdots + m_p + 1)! z^{2m_2 + 3m_3 + \cdots + pm_p}}{\prod_{i=2}^{p} [(i-2)! i]^{m_i} m_i!},$$

the sum over all sequences of non-negative integers $(m_2, \ldots, m_p)$ of arbitrary finite length p. The latter sum simplifies to

$$\frac{1}{\left(1 - \sum_{i \geq 2} \frac{z^i}{(i-2)! i}\right)^2}.$$

It remains to notice that

$$1 - \sum_{i \geq 2} \frac{z^i}{(i-2)! i} = (1-z) e^z$$

since

$$\frac{1}{(i-2)! i} = \frac{i-1}{i!} = \frac{1}{(i-1)!} - \frac{1}{i!}.$$

In particular, for $r = 2, 3, 4, 5, 6$ the degree is respectively 2, 4, 24, 128, 880 and then grows very fast.

3. Hyperdeterminant of the boundary format

In this section we study in detail the hyperdeterminant of the *boundary format*, i.e., the hyperdeterminant of an $(r+1)$-dimensional matrix $A = (a_{i_1,\dots,i_r,i_0})_{0\le i_j\le k_j}$ of format $(k_1+1)\times\cdots\times(k_r+1)\times(k_0+1)$ such that $k_0 = k_1+k_2+\cdots+k_r$. We will show that $\mathrm{Det}\,(A)$ in this case admits several different geometric interpretations, and has several nice explicit polynomial expressions.

A. Boundary hyperdeterminant as the resultant of a system of multilinear forms

We fix the matrix format $(k_1+1)\times\cdots\times(k_r+1)\times(k_0+1)$ such that $k_0 = k_1+k_2+\cdots+k_r$. We consider r groups of variables $x^{(j)} = (x_0^{(j)}, x_1^{(j)}, \dots, x_{k_j}^{(j)})$ for $1\le j\le r$. Let $S(m_1,\dots,m_r)$ denote the space of all polynomials in $x^{(1)},\dots,x^{(r)}$ which are homogeneous of degree m_j in the variables of each group $x^{(j)}$. We shall view a matrix A as a collection of (k_0+1) multilinear forms $f_0, f_1, \dots, f_{k_0} \in S(1,1,\dots,1)$ corresponding to the slices of A in the 0-th direction:

$$f_{i_0} = \sum_{i_1,\dots,i_r} a_{i_1,\dots,i_r,i_0} x_{i_1}^{(1)}\cdots x_{i_r}^{(r)}. \tag{3.1}$$

Theorem 3.1. *The hyperdeterminant* $\mathrm{Det}\,(A)$ *of the matrix of the boundary format is equal to the resultant of the system of multilinear forms* $f_0, f_1, \dots, f_{k_0}$. *In other words, A is degenerate if and only if the system of multilinear equations*

$$f_0(x) = f_1(x) = \cdots = f_{k_0}(x) = 0 \tag{3.2}$$

has a non-trivial solution.

Proof. By definition, $\mathrm{Det}\,(A)$ is the X-discriminant in the sense of Chapter 1, where X is the product of projective spaces $P^{k_1}\times\cdots\times P^{k_r}\times P^{k_0}$. Our statement is just a special case of the Cayley trick (Corollary 2.8 Chapter 3). In our case it is also very easy to give an independent proof. The "only if" part is obvious because the consistency of (3.2) is one of the conditions defining degeneracy (see Section 1). The "if" part follows from the fact that the consistency of (3.2) is a non-trivial condition on matrix entries.

We see that the hyperdeterminant of the boundary format is a special case of the resultant of type $(k_1,\dots,k_r; d_1,\dots,d_r)$ considered in Section 2 Chapter 13.

Analyzing the conditions of degeneracy (see Proposition 1.1), we can easily generalize Theorem 3.1 to the case when $k_0 > k_1+k_2+\cdots+k_r$. In this case we define the multilinear forms f_{i_0} by the same formula (3.1).

Theorem 3.1′. *Suppose that* $k_0 \ge k_1+k_2+\cdots+k_r$. *Then a matrix A of format* $(k_1+1)\times\cdots\times(k_r+1)\times(k_0+1)$ *is degenerate if and only if the system (3.2)* *has a non-trivial solution. The subvariety of degenerate matrices has codimension* $k_0 - (k_1 + k_2 + \ldots + k_r) + 1$.

B. Sylvester type formula and its applications

We assume again that $k_0 = k_1 + k_2 + \cdots + k_r$. Let

$$m_j = k_1 + k_2 + \cdots + k_{j-1}, \ j = 1, \ldots, r \tag{3.3}$$

(with the convention $m_1 = 0$). We associate to our matrix A the linear operator

$$\partial_A : S(m_1, m_2, \ldots, m_r)^{k_0+1} \to S(1 + m_1, 1 + m_2, \ldots, 1 + m_r)$$

given by $\partial_A(g_0, \ldots, g_{k_0}) = \sum_{i=0}^{k_0} f_i g_i$.

Proposition 3.2. *Each of the spaces* $S(m_1, m_2, \ldots, m_r)^{k_0+1}$ *and* $S(1 + m_1, 1 + m_2, \ldots, 1 + m_r)$ *has the same dimension* $N = \frac{(k_0+1)!}{k_1!k_2!\cdots k_r!}$.

Proof. This follows at once from the standard fact that $\dim(S^m(\mathbf{C}^{k+1})) = \binom{k+m}{k}$.

Let us choose in each of the spaces $S(m_1, \ldots, m_r)^{k_0+1}$ and $S(1+m_1, \ldots, 1+m_r)$ the basis consisting of monomials. We will denote by the same symbol ∂_A the matrix of the operator ∂_A in these bases. By Proposition 3.2, this matrix is square.

Theorem 3.3. *We have* $\mathrm{Det}(A) = \det(\partial_A)$.

Proof. Our statement is in fact a special case of Proposition 2.2 Chapter 13. Because of the significance of the result we give an independent "elementary" proof.

First suppose that A is degenerate. By Theorem 3.1, this means that the system (3.2) has a non-trivial solution x. This obviously implies that each polynomial $h \in \mathrm{Im}(\partial_A)$ vanishes at x. Therefore, ∂_A is not onto, and so $\det(\partial_A) = 0$. This implies that the polynomial $\det(\partial_A)$ is divisible by $\mathrm{Det}(A)$.

Clearly, each entry of the matrix ∂_A is a linear form in the matrix entries of A. Comparing Proposition 3.2 with Corollary 2.6 we see that the polynomials $\mathrm{Det}(A)$ and $\det(\partial_A)$ have the same degree. Therefore, Theorem 3.3 is a consequence of the following lemma.

Lemma 3.4. *The polynomial* $\det(\partial_A)$ *is non-zero, and it is irreducible over* $\mathbf{Z}$.

Proof of Lemma 3.4. It suffices to exhibit a matrix E with integral entries such that $\det(\partial_E) = \pm 1$ (recall that $\mathrm{Det}(E)$ is defined only up to sign). Let E be the matrix whose entry $a_{i_1,\ldots,i_r,i_0}$ is equal to 1 if $i_0 = i_1 + \cdots + i_r$, and is equal to 0 otherwise. To show that $\det(\partial_E) = \pm 1$ it is enough to establish the following.

Proposition 3.5. *The matrix ∂_E becomes triangular with ones along the main diagonal under a suitable ordering of its rows and columns.*

Proof. First we give an explicit description of the matrix ∂_A. We identify the set of all monomials of degree m in $(k+1)$ variables $x_0, \ldots, x_k$ with the set of their exponent vectors

$$\Delta^k(m) = \left\{ b = (b_0, \ldots, b_k) \in \mathbf{Z}^{k+1} : b_i \geq 0, \sum b_i = m \right\}.$$

Thus the set of all monomials in $S(m_1, \ldots, m_r)$ is identified with

$$D = D(k_1, \ldots, k_r) = \Delta^{k_1}(m_1) \times \cdots \times \Delta^{k_r}(m_r),$$

and the set of all monomials in $S(1+m_1, 1+m_2, \ldots, 1+m_r)$ with

$$R = R(k_1, \ldots, k_r) = \Delta^{k_1}(1+m_1) \times \cdots \times \Delta^{k_r}(1+m_r).$$

Now the rows of ∂_A are labeled by the set R, and the columns are labeled by $C = C(k_1, \ldots, k_r) = D \times [0, k_0]$, where $[0, k_0] = \{0, 1, \ldots, k_0\}$. We will denote a matrix entry of ∂_A by

$$\langle \mathbf{c}; i_0 | \mathbf{b} \rangle = \langle c^{(1)}, \ldots, c^{(r)}; i_0 | b^{(1)}, \ldots, b^{(r)} \rangle,$$

where $\mathbf{c} = (c^{(1)}, \ldots, c^{(r)}) \in D$, $i_0 \in [0, k_0]$, $\mathbf{b} = (b^{(1)}, \ldots, b^{(r)}) \in R$. We say that $\mathbf{b}$ *covers* $\mathbf{c}$ if $b^{(j)} - c^{(j)}$ has the form e_{i_j} for each $j = 1, \ldots, r$, where e_i is a vector with the i-th component 1 and zeros elsewhere; in this case we write $\mathbf{b} \longrightarrow \mathbf{c}$ or $\mathbf{b} \xrightarrow{i_1 \ldots i_r} \mathbf{c}$. By definition, $\langle \mathbf{c}; i_0 | \mathbf{b} \rangle = 0$ unless $\mathbf{b}$ covers $\mathbf{c}$; if $\mathbf{b} \xrightarrow{i_1 \ldots i_r} \mathbf{c}$ then $\langle \mathbf{c}; i_0 | \mathbf{b} \rangle = a_{i_1, \ldots, i_r, i_0}$.

In particular, we see that ∂_E is a $(0,1)$-matrix, and its entry $\langle \mathbf{c}; i_0 | \mathbf{b} \rangle$ is equal to 1 if and only if $\mathbf{b} \xrightarrow{i_1 \ldots i_r} \mathbf{c}$ for some $i_1, \ldots, i_r$ such that $i_1 + \cdots + i_r = i_0$. In this case we say that $\mathbf{b} \in R$ and $(\mathbf{c}; i_0) \in C$ are *incident* to each other.

For $0 \leq j \leq r$ we let

$$R_j = R_j(k_1, \ldots, k_r) = \{\mathbf{b} \in R : b_{k_p}^{(p)} > 0 \text{ for } j < p \leq r, b_{k_j}^{(j)} = 0\}$$

(for $j = 0$ the last condition is empty). For $1 \leq j \leq r$ we let

$$C_j = C_j(k_1, \ldots, k_r)$$

$$= \{(\mathbf{c}; i_0) \in C : c_{k_p}^{(p)} > 0 \text{ for } j < p \leq r, c_{k_j}^{(j)} = 0, \ i_0 < k_0\};$$

also let

$$C_0 = C_0(k_1, \dots, k_r) = \{(\mathbf{c}; i_0) \in C : i_0 = k_0, \ \mathbf{c} \ arbitrary\}.$$

Lemma 3.6.

(a) *We have*

$$R = \bigcup_{0 \le j \le r} R_j, \quad C = \bigcup_{0 \le j \le r} C_j,$$

both unions disjoint.

(b) *If* $\mathbf{b} \in R_j$ *is incident to* $(\mathbf{c}; i_0) \in C_p$ *then* $p \ge j$.

(c) *For every* $(\mathbf{c}; i_0) \in C_0$ *there is exactly one* $\mathbf{b} \in R$ *which is incident to* $(\mathbf{C}; i_0)$.

(d) *For every* $j = 1, \dots, r$ *there are natural bijections*

$$R_j(k_1, \dots, k_r) \longrightarrow R(k_1, \dots, k_{j-1}, k_j - 1, k_{j+1}, \dots, k_r),$$

$$C_j(k_1, \dots, k_r) \longrightarrow C(k_1, \dots, k_{j-1}, k_j - 1, k_{j+1}, \dots, k_r)$$

such that the elements $\mathbf{b} \in R_j(k_1, \dots, k_r)$ *and* $(\mathbf{c}; i_0) \in C_j(k_1, \dots, k_r)$ *are incident to each other if and only if their images are incident to each other.*

Proof. Statements (a) to (c) are immediate consequences of the definitions. The bijections in (d) are defined as follows. The image of $\mathbf{b} \in R_j(k_1, \dots, k_r)$ is obtained from $\mathbf{b}$ by forgetting the coordinate $b_{k_j}^{(j)}$ and subtracting 1 from $b_{k_p}^{(p)}$ for $j < p \le r$; the image of $(\mathbf{c}; i_0) \in C_j(k_1, \dots, k_r)$ is defined in exactly the same way (with i_0 remaining unchanged). Now (d) is also straightforward.

We can now easily complete the proof of Proposition 3.5. Choose an ordering of R so that elements of R_j will precede elements of R_p for $j < p$ (and similarly for C). By Lemma 3.6 (b), under such orderings the matrix ∂_E becomes block triangular with $(r+1)$ diagonal blocks, the j-th block being the incidence matrix for the incidence relation between R_j and C_j, where $0 \le j \le r$. By Lemma 3.6 (c), the 0-th block becomes the identity matrix under a suitable ordering of R_0 and C_0. But if $1 \le j \le r$ then by Lemma 3.6 (d), the j-th block of ∂_E coincides with the matrix of the same type ∂_E of the smaller format $(k_1+1) \times \cdots \times (k_{j-1}+1) \times k_j \times (k_{j+1}+1) \times \cdots \times (k_r+1) \times k_0$. Using induction on $k_0 = k_1 + \cdots + k_r$, we can assume that all the diagonal blocks of ∂_E can be made unitriangular by a permutation of rows and columns. Therefore, the same is true for ∂_E itself. This proves Proposition 3.5, and hence Lemma 3.4 and Theorem 3.3.

The matrix E can be viewed as a multidimensional analog of the identity matrix. In fact, we can show easily that the corresponding system (3.2) has only a trivial solution. To see this we represent each vector $x^{(j)} = (x_0^{(j)}, x_1^{(j)}, \dots, x_{k_j}^{(j)})$

by a "generating" polynomial $P^{(j)}(t) = \sum_{i=0}^{k_j} x_i^{(j)} t^i$. Then the system (3.2) for $A = E$ can be written

$$P^{(1)}(t)P^{(2)}(t)\cdots P^{(r)}(t) = 0,$$

which implies that some $P^{(j)}(t)$ is a zero polynomial, i.e., $x^{(j)} = 0$.

Proposition 3.5 has an interesting combinatorial corollary.

Corollary 3.7. *There exists exactly one bijection $\varphi : R \to C$ such that $\varphi(\mathbf{b})$ is incident to $\mathbf{b}$ for each $\mathbf{b} \in R$.*

The bijection φ from Corollary 3.7 and its inverse can be explicitly constructed as follows. For $\mathbf{b} = (b^{(1)}, \ldots, b^{(r)}) \in R$, we define the indices $i_1, \ldots, i_r$ successively: if $i_1, \ldots, i_{j-1}$ are already constructed, we define i_j as the minimal index such that

$$b_0^{(j)} + b_1^{(j)} + \cdots + b_{i_j}^{(j)} > i_1 + i_2 + \cdots + i_{j-1}$$

(the index i_1 is defined by $b^{(1)} = e_{i_1}$). We define then

$$\varphi(\mathbf{b}) = (b^{(1)} - e_{i_1}, \ldots, b^{(r)} - e_{i_r}; i_1 + \cdots + i_r). \tag{3.4}$$

It is easy to see that φ is a well-defined mapping from R to C such that $\varphi(\mathbf{b})$ is incident to $\mathbf{b}$ for each $\mathbf{b} \in R$.

To show that φ is a bijection we construct the inverse mapping $\psi : C \to R$. For $(c^{(1)}, \ldots, c^{(r)}; i_0) \in C$ we define the indices $i_r, i_{r-1}, \ldots, i_1$ successively: if $i_r, \ldots, i_{j+1}$ are already constructed, we define i_j as the minimal index such that

$$(1 + c_0^{(j)}) + (1 + c_1^{(j)}) + \cdots + (1 + c_{i_j}^{(j)}) > i_0 - i_r - \cdots - i_{k+1}.$$

Now we define

$$\psi(\mathbf{c}; i_0) = (c^{(1)} + e_{i_1}, \ldots, c^{(r)} + e_{i_r}). \tag{3.5}$$

It is straightforward to verify that ψ is well-defined, and both compositions $\psi \circ \varphi$ and $\varphi \circ \psi$ are identity mappings.

Example 3.8. Let $r = 3$ and $k_0 = 2$, $k_1 = k_2 = 1$, i.e., $A = (a_{ijk})$ $(0 \le i, j \le 1, 0 \le k \le 2)$ is a 3-dimensional matrix of format $2 \times 2 \times 3$. Let A_0 and A_1 be two slices of A in the second direction, i.e.,

$$A_j = \begin{pmatrix} a_{0j0} & a_{0j1} & a_{0j2} \\ a_{1j0} & a_{1j1} & a_{1j2} \end{pmatrix} \quad (j = 0, 1).$$

Then the 6×6 matrix ∂_A can be written as the block matrix

$$\partial_A = \begin{pmatrix} A_0 & 0 \\ A_1 & A_0 \\ 0 & A_1 \end{pmatrix}. \tag{3.6}$$

This corresponds to the following ordering of the sets R and C:

$$R = \{(0,0), (1,0), (0,1), (1,1), (0,2), (1,2)\},$$

$$C = \{(0;0), (0;1), (0;2), (1;0), (1;1), (1;2)\},$$

where (a, b) stands for $((1-a, a), (2-b, b)) \in \Delta^1(1) \times \Delta^1(2) = R$, and $(c; k)$ stands for $((1-c, c); k) \in \Delta^1(1) \times [0, 2] = C$. By Theorem 3.3, we have $\mathrm{Det}\,(A) = \det(\partial_A)$. This polynomial can be rewritten in many different ways. For instance, taking the Laplace expansion of $\det(\partial_A)$ in three first columns we see that

$$\mathrm{Det}\,(A) = \det \begin{pmatrix} a_{000} & a_{001} & a_{002} \\ a_{100} & a_{101} & a_{102} \\ a_{010} & a_{011} & a_{012} \end{pmatrix} \det \begin{pmatrix} a_{100} & a_{101} & a_{102} \\ a_{010} & a_{011} & a_{012} \\ a_{110} & a_{111} & a_{112} \end{pmatrix} \tag{3.7}$$

$$- \det \begin{pmatrix} a_{000} & a_{001} & a_{002} \\ a_{100} & a_{101} & a_{102} \\ a_{110} & a_{111} & a_{112} \end{pmatrix} \det \begin{pmatrix} a_{000} & a_{001} & a_{002} \\ a_{010} & a_{011} & a_{012} \\ a_{110} & a_{111} & a_{112} \end{pmatrix}.$$

More examples for three dimensional matrices will be given in subsection D below.

The expression (3.7) has a nice geometric interpretation. The 3×3 determinants appearing in (3.7) can be viewed as Plücker coordinates (or "brackets") on the variety of lines in P^3, and (3.7) is the Chow form of the Segre subvariety $P^1 \times P^1 \subset P^3$.

Returning to general boundary format, we can always interpret the hyperdeterminant as the Chow form of the subvariety $P^{k_1} \times \cdots \times P^{k_r}$ in the Segre embedding. Therefore, an expression of type (3.7) can be given for an arbitrary boundary format. To do this we introduce some terminology. For every subset $\sigma \subset [0, k_1] \times \cdots \times [0, k_r]$ of cardinality $k_0 + 1$ we define the polynomial $[\sigma] = [\sigma](A)$ to be

$$[\sigma] = \det(a_{\mathbf{i}, i_0}) \quad (\mathbf{i} \in \sigma, 0 \le i_0 \le k_0) \tag{3.8}$$

(note that if we do not specify an ordering of σ then $[\sigma]$ is defined only up to sign).

Let R and D have the same meaning as above. We say that a mapping $\pi : R \to D$ is a *covering* if it satisfies two conditions:

(1) An element $\mathbf{b}$ covers $\pi(\mathbf{b})$ for each $\mathbf{b} \in R$.

(2) For each $\mathbf{c} \in D$ the subset $\pi^{-1}(\mathbf{c}) \subset R$ has cardinality $k_0 + 1$.

To every covering $\pi : R \to D$ and every $\mathbf{c} = (c^{(1)}, \ldots, c^{(r)}) \in D$, we associate a subset $\sigma_\pi(\mathbf{c}) \subset [0, k_1] \times \cdots \times [0, k_r]$ of cardinality $k_0 + 1$ in the following way:

$$\sigma_\pi(\mathbf{c}) = \{(i_1, \ldots, i_r) : (c^{(1)} + e_{i_1}, \ldots, c^{(r)} + e_{i_r}) \in \pi^{-1}(\mathbf{c})\}.$$

Proposition 3.9. *The hyperdeterminant* Det (A) *can be written* $\sum_\pi \pm \prod_{\mathbf{c} \in D} [\sigma_\pi(\mathbf{c})]$, *the sum over all coverings* $\pi : R \to D$.

Proof. Consider the Laplace expansion of det (∂_A) corresponding to the following grouping of columns of ∂_A: we join together the columns $(\mathbf{c}; i_0)$ having the same component $\mathbf{c}$. By definition, summands in the Laplace expansion correspond to coverings $\pi : R \to D$, the summand corresponding to a covering π being just $\pm \prod_{\mathbf{c} \in D} [\sigma_\pi(\mathbf{c})]$.

Note that the signs in Proposition 3.9 are determined directly once we specify orderings of all subsets σ; since Det (A) itself is defined only up to sign, we can actually compute only the ratio of signs for every two summands.

Note also that even the existence of a covering $\pi : R \to D$ is a non-trivial combinatorial fact. Obviously, this implies (and in fact is equivalent to) the following property: for every subset $\Xi \subset D$ we have

$$\#\{\mathbf{b} \in R : \mathbf{b} \text{ covers } \mathbf{C} \text{ for some } \mathbf{c} \in \Xi\} \geq (k_0 + 1) \cdot \#(\Xi).$$

C. Another geometric interpretation

Now we give another geometric interpretation of Det (A) for the boundary format. Let V' be the space of all matrices of format $(k_2 + 1) \times \cdots \times (k_r + 1) \times (k_0 + 1)$. Let ∇' be the variety of all degenerate matrices in V', and let $X' \subset P(V')$ be the projectivization of ∇'. By Theorem 3.1′, X' has codimension $k_0 - (k_2 + \cdots + k_r) + 1 = k_1 + 1$ in $P(V')$. Consider its Chow form $R_{X'}$: by definition, this is an element of the coordinate ring of the Grassmann variety $G(k_1 + 1, V')$ which defines the hypersurface

$$\left\{\xi \in G(k_1 + 1, V') : \xi \cap X' \neq \emptyset\right\}.$$

We will show that this Chow form is given by the hyperdeterminant of the boundary format $(k_1 + 1) \times \cdots \times (k_r + 1) \times (k_0 + 1)$.

To be more precise we represent a matrix A of this boundary format as an ordinary (two-dimensional) matrix $(a_{i_1,j})$ of the format

$$(k_1+1)\times[(k_2+1)\cdots(k_r+1)(k_0+1)],$$

where $i_1 \in [0, k_1]$, and j is a multi-index $(i_2, \ldots, i_r, i_0) \in [0, k_2] \times \cdots \times [0, k_r] \times [0, k_0]$. Thus, A represents a linear operator $\tilde{A} : \mathbf{C}^{k_1+1} \longrightarrow V'$. Clearly, for all A except the subvariety of codimension more than 1, we have $\mathrm{rk}\,(\tilde{A}) = k_1 + 1$, i.e., $\mathrm{Im}\,(\tilde{A}) \in G(k_1+1, V')$.

Theorem 3.10. *The Chow form $R_{X'}$ evaluated at* $\mathrm{Im}\,(\tilde{A})$ *is equal to* $\mathrm{Det}\,(A)$.

Proof. Remembering all the definitions, we have only to prove that A is degenerate if and only if $\tilde{A}(x) \in \nabla'$ for some non-zero $x \in \mathbf{C}^{k_1+1}$. But this follows at once from characterizations of degenerate matrices given by Theorems 3.1 and 3.1′ (we apply Theorem 3.1 to A and Theorem 3.1′ to $\tilde{A}(x) \in V'$).

Note that the format $(k_2+1)\times\cdots\times(k_r+1)\times(k_0+1)$ is (up to permutation of its r directions) an arbitrary format not satisfying (1.4), i.e., such that the variety ∇' of all degenerate matrices of this format is of codimension ≥ 1. Since any subvariety in the projective space can be recovered from its Chow form (see Section 2C Chapter 3), it is possible to express the conditions of degeneracy in terms of the hyperdeterminant of boundary format. These conditions take the following form (cf. Corollary 2.6 Chapter 3).

Corollary 3.11. *A matrix A' of format $(k_2+1)\times\cdots\times(k_r+1)\times(k_0+1)$ is degenerate if and only if the hyperdeterminant* $\mathrm{Det}\,(A)$ *of the boundary format $(k_1+1)\times(k_2+1)\times\cdots\times(k_0+1)$ vanishes whenever A has A' as a slice in the first direction.*

To make Theorem 3.10 more explicit, we represent $R_{X'}$ as a polynomial in Plücker coordinates $[\Omega]$ on the Grassmannian $G(k_1+1, V')$, where Ω runs over the subsets of $[0, k_2]\times\cdots\times[0, k_r]\times[0, k_0]$ of cardinality (k_1+1). For such a subset we denote by $[\Omega] = [\Omega](A)$ the corresponding Plücker coordinate of $\mathrm{Im}\,(\tilde{A})$, i.e., the minor

$$[\Omega] = \det\,(a_{i_1,\mathbf{j}}),\ i_1 \in [0, k_1],\ \mathbf{j} \in \Omega \tag{3.9}$$

(as the polynomials $[\sigma]$ above, $[\Omega]$ is defined only up to sign). Then Theorem 3.10 means, in particular, that $\mathrm{Det}\,(A)$ can be expressed as a polynomial in these minors. Such an expression can be given quite explicitly in full analogy with Proposition 3.9. For this we need some more terminology.

Let $R' = \Delta^{k_2}(1+m_2)\times\cdots\times\Delta^{k_r}(1+m_r)$, where $m_2, \ldots, m_r$ are given by (3.3) (thus the set R of row indices for the matrix ∂_A in subsection B is equal to

$\Delta^{k_1}(1) \times R'$). Let $C = D \times [0, k_0]$ have the same meaning as in subsection B. We say that $(\mathbf{c}; i_0) = (c^{(1)}, c^{(2)}, \ldots, c^{(r)}; i_0) \in C$ *covers* $\mathbf{b}' = (b^{(2)}, \ldots, b^{(r)}) \in R'$ if $b^{(j)} - c^{(j)}$ has the form e_{i_j} for $j = 2, 3, \ldots, r$. We say that a mapping $\tau : C \longrightarrow R'$ is a *covering* if it satisfies two conditions:

(1) Every $(\mathbf{c}; i_0) \in C$ covers its image $\tau(\mathbf{C}; i_0)$.

(2) For each $\mathbf{b}' \in R'$ the fiber $\tau^{-1}(\mathbf{b}') \subset C$ has cardinality $k_1 + 1$.

To every covering $\tau : C \to R'$ and every $\mathbf{b}' = (b^{(2)}, \ldots, b^{(r)}) \in R'$ we associate the subset $\Omega_\tau(\mathbf{b}') \subset [0, k_2] \times \cdots \times [0, k_r] \times [0, k_0]$ of cardinality $k_1 + 1$ in the following way:

$$\Omega_\tau(\mathbf{b}') = \{(i_2, \ldots, i_r, i_0) : (0, b^{(2)} - e_{i_2}, \ldots, b^{(r)} - e_{i_r}; i_0) \in \tau^{-1}(\mathbf{b}')\}.$$

Proposition 3.12. *The hyperdeterminant* Det (A) *can be written*

$$\sum_{\tau} \pm \prod_{b' \in R'} [\Omega_\tau(\mathbf{b}')],$$

the sum over all coverings $\tau : C \to R'$.

Proof. Consider the Laplace expansion of $\det(\partial_A)$ corresponding to the following grouping of rows of ∂_A: we join together the rows $(b^{(1)}, \mathbf{b}')$ having the same component $\mathbf{b}'$. By definition, summands in the Laplace expansion correspond to coverings $\tau : C \to R'$, the summand corresponding to a covering τ being just $\pm \prod_{\mathbf{b}' \in R'} [\Omega_\tau(\mathbf{b}')]$.

Combining Theorem 3.3 with Proposition 1.8, we obtain $r!$ different determinantal formulas for the hyperdeterminant Det (A) of an $(r+1)$-dimensional boundary format. Namely, for each permutation σ of indices $1, \ldots, r, r+1$ leaving $(r+1)$ invariant we have

$$\operatorname{Det}(A) = \det(\partial_{\sigma(A)}), \tag{3.10}$$

where $\partial_{\sigma(A)}$ is a square matrix constructed as in the proof of Lemma 3.4 but with respect to the "transpose" matrix $\sigma(A)$ instead of A. All the matrices $\partial_{\sigma(A)}$ are of the same order and consist of zeros and matrix entries of A. But their block structures can differ substantially, so it is far from obvious that they have the same determinant. In particular, the expressions for Det (A) as Chow forms given by Propositions 3.9 and 3.12 depend upon the choice of σ. Of course, two different expressions for the same Chow form can be transformed one to another by means of Plücker relations on the corresponding Grassmannian. In order to avoid cumbersome indices we illustrate all these phenomena by treating in great details the case of three-dimensional matrices.

D. Three-dimensional boundary format

Now we deal with the matrices of a three-dimensional boundary format. We reproduce the results of previous subsections but also add some further results. It will be convenient to modify our notation as follows. We fix three positive integers $m, n, p \geq 2$ with $n = m + p - 1$. Let $\mathbf{C}^{m\times n\times p}$ denote the space of matrices of boundary format $m \times n \times p$. We write a matrix $A \in \mathbf{C}^{m\times n\times p}$ as $A = (a_{ijk})_{i\in[m], j\in[n], k\in[p]}$, where we use the notation $[m] = \{1, 2, \ldots, m\}$. By Corollary 2.6, the hyperdeterminant is a polynomial in matrix entries a_{ijk} of degree $\frac{n!}{(m-1)!(p-1)!}$.

A matrix $A \in \mathbf{C}^{m\times n\times p}$ can be represented by each of three ordinary matrices $A^{12} \in \mathbf{C}^{mn\times p}$, $A^{13} \in \mathbf{C}^{n\times mp}$, $A^{23} \in \mathbf{C}^{np\times m}$. Here A^{12} has the same entries as A but they are arranged so that the row index set is $[m] \times [n]$, and the column index set is $[p]$ (likewise for A^{13} and A^{23}). For example, if A has format $3 \times 4 \times 2$ then

$$A^{12} = \begin{pmatrix} a_{111} & a_{112} \\ a_{121} & a_{122} \\ a_{131} & a_{132} \\ a_{141} & a_{142} \\ a_{211} & a_{212} \\ a_{221} & a_{222} \\ a_{231} & a_{232} \\ a_{241} & a_{242} \\ a_{311} & a_{312} \\ a_{321} & a_{322} \\ a_{331} & a_{332} \\ a_{341} & a_{342} \end{pmatrix},$$

$$A^{13} = \begin{pmatrix} a_{111} & a_{112} & a_{211} & a_{212} & a_{311} & a_{312} \\ a_{121} & a_{122} & a_{221} & a_{222} & a_{321} & a_{322} \\ a_{131} & a_{132} & a_{231} & a_{232} & a_{331} & a_{332} \\ a_{141} & a_{142} & a_{241} & a_{242} & a_{341} & a_{342} \end{pmatrix},$$

$$A^{23} = \begin{pmatrix} a_{111} & a_{211} & a_{311} \\ a_{112} & a_{212} & a_{312} \\ a_{121} & a_{221} & a_{321} \\ a_{122} & a_{222} & a_{322} \\ a_{131} & a_{231} & a_{331} \\ a_{132} & a_{232} & a_{332} \\ a_{141} & a_{241} & a_{341} \\ a_{142} & a_{242} & a_{342} \end{pmatrix}.$$

The three "flattenings" of A give rise to three different geometric interpretations of the degeneracy of A and of $\mathrm{Det}(A)$.

1. Let $\alpha^{12} : \mathbf{C}^p \to \mathbf{C}^{m\times n}$ be the linear map with matrix A^{12}. We interpret the $p \times p$ minors of A^{12} as the Plücker coordinates on the Grassmannian of subspaces of dimension p in $\mathbf{C}^{m\times n}$. Now Theorem 3.10 takes the following form.

Proposition 3.13. *A matrix A is degenerate if and only if either* $\mathrm{rk}\,(A^{12}) < p$, *or* $\mathrm{rk}\,(A^{12}) = p$ *and* $\mathrm{Im}\,(\alpha^{12})$ *has non-zero intersection with the variety $\nabla_{m,n}$ of degenerate $m \times n$ matrices. The hyperdeterminant* $\mathrm{Det}\,(A)$ *is a polynomial of degree $\binom{n}{p}$ in the $p \times p$ minors of A^{12}, and this polynomial is the Chow form of $P(\nabla_{m,n})$.*

2. Let $\alpha^{13} : \mathbf{C}^{m\times p} \to \mathbf{C}^n$ be the linear map with matrix A^{13}. Now we interpret the $n \times n$ minors of A^{13} as the dual Plücker coordinates on the Grassmannian of subspaces of codimension n in $\mathbf{C}^{m\times p}$. The following result is a geometric reformulation of Theorem 3.1.

Proposition 3.14. *A matrix A is degenerate if and only if either* $\mathrm{rk}\,(A^{13}) < n$, *or* $\mathrm{rk}\,(A^{13}) = n$ *and* $\mathrm{Ker}\,(\alpha^{13})$ *has non-zero intersection with the variety $\nabla^1_{m,p}$ of $m \times p$ matrices of rank 1. The hyperdeterminant* $\mathrm{Det}\,(A)$ *is a polynomial of degree $\binom{n-1}{m-1}$ in the $n \times n$ minors of A^{13}, and this polynomial is the Chow form of $P^{m-1} \times P^{p-1}$ in the Segre embedding.*

3. Let $\alpha^{23} : \mathbf{C}^m \to \mathbf{C}^{n\times p}$ be the linear map with matrix A^{23}. We interpret the $m \times m$ minors of A^{23} as the Plücker coordinates on the Grassmannian of subspaces of dimension m in $\mathbf{C}^{n\times p}$. Interchanging the roles of m and p in Proposition 3.13, we obtain the following.

Proposition 3.15. *A matrix A is degenerate if and only if either* $\mathrm{rk}\,(A^{23}) < m$, *or* $\mathrm{rk}\,(A^{23}) = m$ *and* $\mathrm{Im}\,(\alpha^{23})$ *has non-zero intersection with the variety $\nabla_{n,p}$ of degenerate $n \times p$ matrices. The hyperdeterminant* $\mathrm{Det}\,(A)$ *is a polynomial of degree $\binom{n}{m}$ in the $m \times m$ minors of A^{23}, and this polynomial is the Chow form of $P(\nabla_{n,p})$.*

Now we present several explicit polynomial expressions for the hyperdeterminant of the boundary $m \times n \times p$ format. We start with Sylvester type formulas. Let $f_1, f_2, \ldots, f_n$ be linear forms on $\mathbf{C}^{m\times p}$ corresponding to the rows of A^{13}. We can also think of $f_1, f_2, \ldots, f_n$ as bilinear forms on $\mathbf{C}^m \times \mathbf{C}^p$ given by

$$f_j(x_1, \ldots, x_m, y_1, \ldots, y_p) = \sum_{i,k} a_{ijk} x_i y_k. \tag{3.11}$$

For every two integers $q, r \geq 0$ let $S(q, r)$ denote the space of bihomogeneous forms in $x_1, \ldots, x_m, y_1, \ldots, y_p$ having degree q in $x_1, \ldots, x_m$ and degree r in $y_1, \ldots, y_p$. In particular, $f_1, f_2, \ldots, f_n \in S(1, 1)$. We define a linear map $\Phi_{q,r} : S(q, r) \otimes \mathbf{C}^n \to S(q + 1, r + 1)$ by

$$\Phi_{q,r}(g_1 \otimes e_1 + \cdots + g_n \otimes e_n) = g_1 f_1 + \cdots + g_n f_n, \tag{3.12}$$

where $e_1, \ldots, e_n$ is the standard basis in $\mathbf{C}^n$. We denote by the same symbol $\Phi_{q,r}$ the matrix of this linear map in the bases consisting of monomials in $S(q+1, r+1)$ and of monomials tensored with the e_j's in $S(q, r) \otimes \mathbf{C}^n$. The following is a special case of Theorem 3.3 (and also of Proposition 2.2 Chapter 13).

Proposition 3.16. *Each of the matrices $\Phi_{0,m-1}$ and $\Phi_{p-1,0}$ is square of order $\frac{n!}{(m-1)!(p-1)!}$, and*

$$\mathrm{Det}\,(A) = \det\,(\Phi_{0,m-1}) = \det\,(\Phi_{p-1,0}). \tag{3.13}$$

To present the matrix $\Phi_{0,m-1}$ in a more explicit form, we use the notation

$$\Delta^{p-1}(m) = \left\{ \alpha = (\alpha_1, \alpha_2, \ldots, \alpha_p) \in \mathbf{Z}_+^p \;:\; \alpha_1 + \alpha_2 + \cdots + \alpha_p \;=\; m \right\}.$$

Then the columns of $\Phi_{0,m-1}$ are labeled by $\Delta^{p-1}(m-1) \times [n]$, and the rows are labeled by $[m] \times \Delta^{p-1}(m)$. The matrix entry $\langle i, \alpha \mid \beta, j \rangle$ of $\Phi_{0,m-1}$ is equal to a_{ijk} if $\alpha = \beta + e_k$ for some $k \in [p]$, and 0 otherwise. The matrix $\Phi_{0,m-1}$ is described in exactly the same way with m and p interchanged.

For the $3 \times 4 \times 2$ format these matrices are as follows:

$$\Phi_{0,2} = \begin{pmatrix} A_{\bullet\bullet 1} & 0 & 0 \\ A_{\bullet\bullet 2} & A_{\bullet\bullet 1} & 0 \\ 0 & A_{\bullet\bullet 2} & A_{\bullet\bullet 1} \\ 0 & 0 & A_{\bullet\bullet 2} \end{pmatrix},$$

where for $k = 1, 2$ we have

$$A_{\bullet\bullet k} = \begin{pmatrix} a_{11k} & a_{12k} & a_{13k} & a_{14k} \\ a_{21k} & a_{22k} & a_{23k} & a_{24k} \\ a_{31k} & a_{32k} & a_{33k} & a_{34k} \end{pmatrix};$$

$$\Phi_{1,0}^t = \begin{pmatrix} A_{1\bullet\bullet} & A_{2\bullet\bullet} & A_{3\bullet\bullet} & 0 & 0 & 0 \\ 0 & A_{1\bullet\bullet} & 0 & A_{2\bullet\bullet} & A_{3\bullet\bullet} & 0 \\ 0 & 0 & A_{1\bullet\bullet} & 0 & A_{2\bullet\bullet} & A_{3\bullet\bullet} \end{pmatrix},$$

where ^{t} stands for the transpose matrix, and for $i = 1, 2, 3$ we have

$$A_{i\bullet\bullet} = \begin{pmatrix} a_{i11} & a_{i12} \\ a_{i21} & a_{i22} \\ a_{i31} & a_{i32} \\ a_{i41} & a_{i42} \end{pmatrix}.$$

The above formulas represent $\mathrm{Det}\,(A)$ as the determinant of a sparse matrix whose non-zero entries are the entries a_{ijk} of A arranged into a non-trivial pattern.

Our next formula is essentially of the same kind although it does not fit into the general framework of Sylvester type formulas (see Section 2 Chapter 13). We consider again the bilinear forms $f_1, f_2, \ldots, f_n$ on $\mathbf{C}^m \times \mathbf{C}^p$ given by (3.11). Now we think of f_j as a linear map $\mathbf{C}^m \to (\mathbf{C}^p)^*$ which sends each vector $x \in \mathbf{C}^m$ to a linear form $f_j(x, \cdot)$ on $\mathbf{C}^p$. We define a linear map $\Phi : \mathbf{C}^m \otimes \Lambda^m \mathbf{C}^n \to (\mathbf{C}^p)^* \otimes \Lambda^{m-1}\mathbf{C}^n$ by

$$\Phi(x \otimes \xi) = \sum_j f_j(x, \cdot) \otimes \partial_j \xi, \tag{3.14}$$

where ∂_j for $j \in [n]$ is the j-th partial derivative in the exterior algebra $\Lambda^\bullet \mathbf{C}^n$. We denote by the same symbol Φ the matrix of this linear map in the bases formed by tensoring standard basis vectors in $\mathbf{C}^m$ and $(\mathbf{C}^p)^*$ with wedge monomials in the exterior algebra.

Theorem 3.17. *The matrix Φ is square of order $\frac{n!}{(m-1)!(p-1)!}$, and*

$$\mathrm{Det}\,(A) = \det\,(\Phi). \tag{3.15}$$

Proof. This is a consequence of Theorem 4.7 Chapter 3. More precisely, we consider the resultant spectral sequence

$$C^{\bullet\bullet}_{r-}\big(\mathcal{O}(1,1), \ldots, \mathcal{O}(1,1) | \, \mathcal{O}(-1,m)\big)$$

for the variety $X = P^{m-1} \times P^{p-1}$, whose determinant, according to the cited theorem, is equal to $R_X(f_1, \ldots, f_n) = \mathrm{Det}(A)$. An explicit calculation of the cohomology by the Künneth formula and Serre's theorem (Theorem 2.12 Chapter 2) shows that the only non-trivial differential in this spectral sequence is

$$\partial_1 : H^{m-1}\left(P^{m-1}, \bigwedge^{m}(\mathcal{O}(-1,-1)^{\oplus n}) \otimes \mathcal{O}(-1,m)\right)$$

$$\longrightarrow H^{m-1}\left(P^{m-1}, \bigwedge^{m-1}(\mathcal{O}(-1,-1)^{\oplus n}) \otimes \mathcal{O}(-1,m)\right)$$

and that this differential is canonically identified with Φ. This implies (3.15).

It is also possible to give an "elementary" proof of (3.15) similar to that of Theorem 3.3 above. We leave such a proof as an exercise for the reader.

The companion formula obtained from (3.15) by interchanging m and p appears to coincide with (3.15). To see this we represent the matrix Φ in the following form which exhibits the symmetry between m and p. We use the notation $\binom{[n]}{m}$ for the set of all m-element subsets of $[n]$. Then the columns of Φ are labeled by $[m] \times \binom{[n]}{m}$, and the rows can be labeled by $[p] \times \binom{[n]}{p}$ (we identify $\binom{[n]}{m-1}$ with $\binom{[n]}{p}$ by taking complements). The matrix entries of Φ are given by

$$\Phi_{(k,\rho),(i,\sigma)} = \begin{cases} \pm a_{ijk} & \text{if } \sigma \cap \rho = \{j\}, \\ 0 & \text{if } \#(\sigma \cap \rho) \geq 2. \end{cases} \tag{3.16}$$

Here $(k, \rho) \in [p] \times \binom{[n]}{p}$, $(i, \sigma) \in [m] \times \binom{[n]}{m}$, and the sign $\pm$ is equal to $(-1)^{r-1}$ if j is the r-th member of σ in the increasing order.

For the $3 \times 4 \times 2$ format this matrix is as follows:

$$\Phi^t = \begin{pmatrix} A_{\bullet 3 \bullet} & -A_{\bullet 2 \bullet} & 0 & A_{\bullet 1 \bullet} & 0 & 0 \\ A_{\bullet 4 \bullet} & 0 & -A_{\bullet 2 \bullet} & 0 & A_{\bullet 1 \bullet} & 0 \\ 0 & A_{\bullet 4 \bullet} & -A_{\bullet 3 \bullet} & 0 & 0 & A_{\bullet 1 \bullet} \\ 0 & 0 & 0 & A_{\bullet 4 \bullet} & -A_{\bullet 3 \bullet} & A_{\bullet 2 \bullet} \end{pmatrix},$$

where for $j = 1, 2, 3, 4$ we set

$$A_{\bullet j \bullet} = \begin{pmatrix} a_{1j1} & a_{1j2} \\ a_{2j1} & a_{2j2} \\ a_{3j1} & a_{3j2} \end{pmatrix}.$$

Our next goal is to present three more formulas for $\mathrm{Det}\,(A)$ explicitly expressing it in terms of maximal minors of each of the three matrices A^{12}, A^{13} and A^{23}. Consider the linear map $\Psi^{23} : \Lambda^m \mathbf{C}^n \to S^m(\mathbf{C}^p)^*$ given by

$$\Psi^{23}(e_{j_1} \wedge \cdots \wedge e_{j_m}) = \det \left\| \sum_k a_{i,j_q,k} y_k \right\|_{i,q \in [m]}. \tag{3.17}$$

Here $S^m(\mathbf{C}^p)^*$ is identified with the space of forms of degree m in $y_1, \ldots, y_p$. We denote by the same symbol Ψ^{23} the matrix of the map (3.17) in the monomial bases. Clearly, Ψ^{23} is square of order $\binom{n}{m}$, and its entries are polynomials of degree m in the (a_{ijk}). The following formula was communicated to the authors by A.I. Bondal.

Theorem 3.18. *We have* $\mathrm{Det}\,(A) = \det\,(\Psi^{23})$.

For an elementary proof of this theorem see [SZ2], Theorem 6.5. We leave it to the reader to put it into the context of the resultant spectral sequence or that of the Weyman complex (cf. [WZ1], Section 4).

In an explicit matrix form, the rows of Ψ^{23} are labeled by $\Delta^{p-1}(m)$, and the columns are labeled by $\binom{[n]}{m}$. For $\alpha = (\alpha_1, \ldots, \alpha_p) \in \Delta^{p-1}(m)$, $\sigma = \{j_1 < j_2 < \cdots < j_m\} \in \binom{[n]}{m}$ the matrix entry $\Psi^{23}_{\alpha\sigma}$ of Ψ^{23} is given by

$$\Psi^{23}_{\alpha\sigma} = \sum [j_1^{k_1}, j_2^{k_2}, \ldots, j_m^{k_m}], \tag{3.18}$$

where $[j_1^{k_1}, j_2^{k_2}, \ldots, j_m^{k_m}] = \det(a_{i,j_q,k_q})_{i,q\in[m]}$ is a maximal minor of A^{23}, and the sum in (3.18) is over all sequences $(k_1, \ldots, k_m)$ such that $y_{k_1} \cdots y_{k_m} = y_1^{\alpha_1} \cdots y_p^{\alpha_p}$.

For the $3 \times 4 \times 2$ format the matrix $(\Psi^{23})^t$ is as follows:

$$\begin{pmatrix} [1^1 2^1 3^1] & [1^1 2^1 3^2]+[1^1 2^2 3^1]+[1^2 2^1 3^1] & [1^1 2^2 3^2]+[1^2 2^1 3^2]+[1^2 2^2 3^1] & [1^2 2^2 3^2] \\ [1^1 2^1 4^1] & [1^1 2^1 4^2]+[1^1 2^2 4^1]+[1^2 2^1 4^1] & [1^1 2^2 4^2]+[1^2 2^1 4^2]+[1^2 2^2 4^1] & [1^2 2^2 4^2] \\ [1^1 3^1 4^1] & [1^1 3^1 4^2]+[1^1 3^2 4^1]+[1^2 3^1 4^1] & [1^1 3^2 4^2]+[1^2 3^1 4^2]+[1^2 3^2 4^1] & [1^2 3^2 4^2] \\ [2^1 3^1 4^1] & [2^1 3^1 4^2]+[2^1 3^2 4^1]+[2^2 3^1 4^1] & [2^1 3^2 4^2]+[2^2 3^1 4^2]+[2^2 3^2 4^1] & [2^2 3^2 4^2] \end{pmatrix}.$$

Let Ψ^{12} denote the linear map $\Lambda^p \mathbf{C}^n \to S^p(\mathbf{C}^m)^*$ and the corresponding matrix constructed in the same way as Ψ^{23} but with m and p interchanged. Thus Ψ^{12} is a square matrix of order $\binom{n}{p}$, and its entries are polynomials of degree p in the (a_{ijk}). The companion Bondal formula is

$$\mathrm{Det}(A) = \det(\Psi^{12}). \tag{3.19}$$

The matrix description of Ψ^{12} is totally similar to that of Ψ^{23}. For the $3\times4\times2$ format the matrix $(\Psi^{12})^t$ is as follows:

$$\begin{pmatrix} [^1 1^1 2] & [^1 1^2 2]+[^2 1^1 2] & [^1 1^3 2]+[^3 1^1 2] & [^2 1^2 2] & [^2 1^3 2]+[^3 1^2 2] & [^3 1^3 2] \\ [^1 1^1 3] & [^1 1^2 3]+[^2 1^1 3] & [^1 1^3 3]+[^3 1^1 3] & [^2 1^2 3] & [^2 1^3 3]+[^3 1^2 3] & [^3 1^3 3] \\ [^1 1^1 4] & [^1 1^2 4]+[^2 1^1 4] & [^1 1^3 4]+[^3 1^1 4] & [^2 1^2 4] & [^2 1^3 4]+[^3 1^2 4] & [^3 1^3 4] \\ [^1 2^1 3] & [^1 2^2 3]+[^2 2^1 3] & [^1 2^3 3]+[^3 2^1 3] & [^2 2^2 3] & [^2 2^3 3]+[^3 2^2 3] & [^3 2^3 3] \\ [^1 2^1 4] & [^1 2^2 4]+[^2 2^1 4] & [^1 2^3 4]+[^3 2^1 4] & [^2 2^2 4] & [^2 2^3 4]+[^3 2^2 4] & [^3 2^3 4] \\ [^1 3^1 4] & [^1 3^2 4]+[^2 3^1 4] & [^1 3^3 4]+[^3 3^1 4] & [^2 3^2 4] & [^2 3^3 4]+[^3 3^2 4] & [^3 3^3 4] \end{pmatrix}$$

where $[^{i_1} j_1^{i_2} j_2] := a_{i_1,j_1,1}a_{i_2,j_2,2} - a_{i_1,j_1,2}a_{i_2,j_2,1}$ is a maximal minor of A^{12}.

Our last formula for $\mathrm{Det}(A)$ resembles the classical Bezout formula. We associate to a family $f_1, \ldots, f_n$ of bilinear forms a polynomial function in $x_1, \ldots, x_m$, $y_1, \ldots, y_p$ which we call the *generalized Jacobian* of $f_1, \ldots, f_n$. To this end, we associate to every family $(g_1, \ldots, g_s)$ of functions on $\mathbf{C}^m \times \mathbf{C}^p$ a differential form of degree $s - 1$ given by

$$\omega(g_1, \ldots, g_s) = \sum_{i=1}^{s} (-1)^{i-1} g_i dg_1 \wedge \cdots \wedge dg_{i-1} \wedge dg_{i+1} \wedge \cdots \wedge g_s.$$

It is not hard to show that the form $\omega(f_1, \dots, f_n)$ is proportional to $\omega(x_1, \dots, x_m) \wedge \omega(y_1, \dots, y_p)$. We define the generalized Jacobian $J(f_1, \dots, f_n)$ as the coefficient of proportionality in

$$\omega(f_1, \dots, f_n) = J(f_1, \dots, f_n)(\omega(x_1, \dots, x_m) \wedge \omega(y_1, \dots, y_p)). \tag{3.20}$$

Clearly, $J(f_1, \dots, f_n)$ is a bihomogeneous form of degree $p-1$ in $x_1, \dots, x_m$ and of degree $m-1$ in $y_1, \dots, y_p$. We write it down as a linear combination of monomials:

$$J(f_1, \dots, f_n) = \sum_{\alpha,\beta} b_{\alpha,\beta} x^\alpha y^\beta, \tag{3.21}$$

where α runs over $\Delta^{m-1}(p-1)$, and β runs over $\Delta^{p-1}(m-1)$. Consider the matrix $B = (b_{\alpha,\beta})$. This is a square matrix of order $\binom{n-1}{m-1}$ whose entries are polynomials of degree n in the a_{ijk}.

Theorem 3.19. *We have* $\mathrm{Det}\,(A) = \det\,(B)$.

Sketch of the proof. We consider the variety $X = P^{m-1} \times P^{p-1}$ and apply Theorem 4.7 Chapter 3 to the resultant spectral sequence

$$C_{r-}^{\bullet\bullet}\big(\mathcal{O}(1,1), \dots, \mathcal{O}(1,1) | \, \mathcal{O}(0, m-1)\big).$$

Thus the determinant of this spectral sequence is $R_X(f_1, \dots, f_n) = \mathrm{Det}(A)$. An explicit calculation of the cohomology involved shows that the first term of our spectral sequence has only two non-trivial components, namely

$$H^{n-1}\left(P^{m-1} \times P^{p-1}, \bigwedge^{n}(\mathcal{O}(-1,-1)^{\oplus n}) \otimes \mathcal{O}(0, m-1)\right) = S^{p-1}\mathbf{C}^m,$$

$$H^{0}\left(P^{m-1} \times P^{p-1}, \bigwedge^{0}(\mathcal{O}(-1,-1)^{\oplus n}) \otimes \mathcal{O}(o, m-1)\right) = S^{m-1}(\mathbf{C}^p)^*$$

which are connected by the (unique) differential ∂_n. Thus $\mathrm{Det}(A) = \det(\partial_n)$. It can be seen by an explicit calculation (similar to the one in the proof of Proposition 5.4 Chapter 2) that ∂_n coincides with $J(f_1, \dots, f_n)$, if we regard the latter as a linear map $S^{p-1}\mathbf{C}^m \to S^{m-1}(\mathbf{C}^p)^*$. This implies our theorem.

It is also possible to give an elementary proof of Theorem 3.19, similar to what was done in the proof of Theorem 3.3.

Let us describe the matrix entries $b_{\alpha,\beta}$ in Theorem 3.19. We need some terminology. We shall identify a subset $\Omega \subset [m] \times [p]$ with a bipartite graph with the vertex set $[m] \cup [p]$ by representing each $(i, k) \in [m] \times [p]$ as an edge joining

$i \in [m]$ with $k \in [p]$. We say that Ω is a *base* if the corresponding graph is a spanning tree (in particular, $\#(\Omega) = m + p - 1 = n$). We say that a base Ω is of type (α, β) and write $\Omega \vdash (\alpha, \beta)$ if Ω has $\alpha_i + 1$ elements in the i-th row and $\beta_k + 1$ elements in the k-th column for $i \in [m]$, $k \in [p]$.

Lemma 3.20. *Every base Ω can be uniquely represented in the form*

$$\Omega = \{(1, k_1), (2, k_2), \dots, (m, k_m), (i_2, 2), (i_3, 3), \dots, (i_p, p)\} \tag{3.22}$$

for some $k_1, \dots, k_m \in [p]$, $i_2, \dots, i_p \in [m]$.

Proof. Let us orient the tree Ω so that all the edges point out from the vertex $1 \in [p]$. Clearly, every vertex in $[m] \cup ([p] \setminus \{1\})$ is an end-point of some edge, and we write an edge $(i, k) \in \Omega$ as (i, k_i) if its end-point is $i \in [m]$, and as (i_k, k) if its end-point is $k \in [p]$. This establishes the existence of (3.22). The proof of uniqueness is an easy combinatorial exercise (cf. [SZ2], Lemma 1.6 (b)).

For every sequence $(i_1, k_1), \dots, (i_n, k_n) \in [m] \times [p]$ we denote by $[i_1k_1, \dots, i_nk_n]$ the corresponding maximal minor of A^{13} given by

$$[i_1k_1, \dots, i_nk_n] = \det\,(a_{i_q, j, k_q})_{j,q \in [n]}. \tag{3.23}$$

If a base Ω is expressed as in (3.22) then we write

$$[\Omega] := [1k_1, 2k_2, \dots, mk_m, i_22, \dots, i_pp].$$

The definitions (3.20) and (3.21) readily imply the following.

Proposition 3.21. *The matrix entries of B are given by*

$$b_{\alpha,\beta} = \sum_{\Omega \vdash (\alpha,\beta)} [\Omega]. \tag{3.24}$$

For the $3 \times 4 \times 2$ format the matrix B is as follows:

$$B = \begin{pmatrix} [11, 21, 31, 12] & [11, 21, 32, 12] + [11, 22, 31, 12] & [11, 22, 32, 12] \\ [11, 21, 31, 22] & [11, 21, 32, 22] + [12, 21, 31, 22] & [12, 21, 32, 22] \\ [11, 21, 31, 32] & [11, 22, 31, 32] + [12, 21, 31, 32] & [12, 22, 31, 32] \end{pmatrix}. \tag{3.25}$$

The formula (3.25) can be found in [Sal], Art. 277. One can find there some ideas on the general $m \times n \times p$ case as well, but Theorem 3.19 and Proposition 3.21 seem to be new.

4. Schläfli's method

A. Statements and proofs

In this section we study Schläfli's method of computing hyperdeterminants ([Schl]). Although this method does not give an answer in general, it works in some important special cases and provides interesting additional information.

Let $A = (a_{i_0,i_1,\ldots,i_r})$ be an $(r+1)$-dimensional matrix of format $(k_0+1) \times (k_1+1) \times \cdots \times (k_r+1)$. We associate to A a family of r-dimensional matrices $\tilde{A}(x)$ linearly depending on the auxiliary variables $x = (x_0, \ldots, x_{k_0})$:

$$\tilde{A}(x)_{i_1,\ldots,i_r} = \sum_{i_0=0}^{k_0} a_{i_0,i_1,\ldots,i_r} x_{i_0}. \tag{4.1}$$

In other words, $\tilde{A}$ is the linear operator

$$\mathbf{C}^{k_0+1} \to \mathbf{C}^{k_1+1} \otimes \cdots \otimes \mathbf{C}^{k_r+1}$$

naturally associated to A.

Let us assume that $k_1, \ldots, k_r$ satisfy (1.4), i.e., the r-dimensional hyperdeterminant of format $(k_1+1) \times \cdots \times (k_r+1)$ is non-trivial. We associate to A a polynomial function $F_A(x) = \mathrm{Det}\,(\tilde{A}(x))$. This is a homogeneous form in $x_0, \ldots, x_{k_0}$ of degree $N(k_1, \ldots, k_r)$ (see Section 2). Denote by $\Delta(F_A)$ the discriminant of F_A. We consider $\Delta(F_A)$ as a polynomial in matrix entries of A. Using the known formula for the degree of the discriminant ((2.12) of Chapter 9) we see that $\Delta(F_A)$ has degree

$$\deg(\Delta(F_A)) = (k_0+1)(N(k_1, \ldots, k_r) - 1)^{k_0} N(k_1, \ldots, k_r). \tag{4.2}$$

Theorem 4.1. *The polynomial $\Delta(F_A)$ is divisible by the $(r+1)$-dimensional hyperdeterminant* $\mathrm{Det}\,(A)$.

Note that $\Delta(F_A)$ might be identically zero.

Proof. Suppose that $A = (a_{i_0,i_1,\ldots,i_r})$ is degenerate; we have to show that the corresponding form $F_A(x) = \mathrm{Det}\,(\tilde{A}(x))$ has a zero discriminant. Choose a point $(x^{(0)}, x^{(1)}, \ldots, x^{(r)}) \in \mathcal{K}(A)$ (see Proposition 1.1). It is enough to show that F_A vanishes at $x^{(0)}$ with all its first derivatives. Denote

$$A_0 = \tilde{A}(x^{(0)}), \quad b_{i_1,\ldots,i_r} = \left.\frac{\partial \mathrm{Det}}{\partial a_{i_1,\ldots,i_r}}\right|_{A_0}.$$

Then we have $F_A(x^{(0)}) = \mathrm{Det}\,(A_0)$, and

$$\left.\frac{\partial F_A(x)}{\partial x_{i_0}}\right|_{x=x^{(0)}} = \sum_{i_1,\ldots,i_r} a_{i_0,i_1,\ldots,i_r} b_{i_1,\ldots,i_r} \tag{4.3}$$

for all $i_0 = 0, 1, \ldots, k_0$. Clearly, $(x^{(1)}, \ldots, x^{(r)}) \in \mathcal{K}(A_0)$, hence $F_A(x^{(0)}) = 0$. If $b_{i_1,\ldots,i_r} = 0$ for all $i_1, \ldots, i_r$ then by (4.3) $\left.\frac{\partial F_A(x)}{\partial x_{i_0}}\right|_{x=x^{(0)}} = 0$ for all i_0, and we are done. So we can assume that some $b_{i_1,\ldots,i_r}$ is non-zero. But this means that A_0 is a smooth point of the variety of degenerate matrices, and we can apply Proposition 1.2. By this proposition, we can assume that $b_{i_1,\ldots,i_r} = x^{(1)}_{i_1} \cdots x^{(r)}_{i_r}$ for all $i_1, \ldots, i_r$. Substituting this into (4.3) and remembering the definition of $\mathcal{K}(A)$ we see that all the first partial derivatives of F_A vanish at $x^{(0)}$ which proves our theorem.

Denote by $\nabla = \nabla(k_1, \ldots, k_r)$ the variety of all degenerate matrices of format $(k_1+1) \times \cdots \times (k_r+1)$; by definition, the projectivization of ∇ is the projectively dual variety $X^\vee$ of $X = P^{k_1} \times \cdots \times P^{k_r}$. Let ∇_{sing} be the variety of singular points of ∇, and $X^\vee_{sing}$ the projectivization of ∇_{sing}. Let $c = c(k_1, \ldots, k_r)$ denote the codimension (i.e., the minimum of codimensions of irreducible components) of $X^\vee_{sing}$ in the projective space $P(\mathbf{C}^{(k_1+1)\cdots(k_r+1)})$.

Analyzing the proof of Theorem 4.1 we get the following refinement.

Theorem 4.2. *The ratio $G(A) = \Delta(F_A)/\mathrm{Det}\,(A)$ has the following form:*

(a) If $k_0 + 1 < c(k_1, \ldots, k_r)$ then G is a non-zero constant.

(b) If $k_0+1 = c(k_1, \ldots, k_r)$ then $G(A) = \prod R_Z^{m_Z}(\mathrm{Im}(\tilde{A}))$, where Z ranges over irreducible components of $X^\vee_{sing}$ having codimension $c(k_1, \ldots, k_r)$, R_Z is the Chow form of Z, and the m_Z are some multiplicities.

(c) If $k_0 + 1 > c(k_1, \ldots, k_r)$ then G (and hence $\Delta(F_A)$) is identically zero.

Proof. First we establish the following.

Lemma 4.3. *If $\Delta(F_A)$ is not identically zero then it is not divisible by* $\mathrm{Det}\,(A)^2$.

Proof. By definition, $X^\vee$ is the union of projective spaces P_x, $x \in X$ where P_x is formed by hyperplanes tangent to X at x. The codimension of P_x is equal to $dim\,(X) + 1 = k_1 + \cdots + k_r + 1$. The vanishing of $\Delta(F_A)$ means that the image $\mathrm{Im}\,(\tilde{A}) = \tilde{A}(\mathbf{C}^{k_0+1})$ is tangent to ∇ at some non-zero point. Suppose that $\Delta(F_A)$ is divisible by $\mathrm{Det}\,(A)^2$. Then for any one-parameter algebraic family of $(r+1)$-dimensional matrices A_t such that $\mathrm{Im}\,(\tilde{A}_0)$ is tangent to ∇, the function $\Delta(F_{A_t})$ is divisible by t^2. We will show that this is impossible by constructing a suitable "generic" family.

Let B be a generic point of ∇ and $\xi \in X^\vee$ be the projectivization of B. We can assume that ξ lies on exactly one P_x (if this were generically not so then $X^\vee$ would not be a hypersurface). Consider the variety Z of all k_0-dimensional projective subspaces in $P(\mathbf{C}^{(k_1+1)\cdots(k_r+1)})$, tangent to $X^\vee$ at ξ. Since $k_0 \leq k_1 + \cdots + k_r$ it follows that a dense open part of Z is formed by subspaces which meet P_x only at ξ. Let L be a generic element of Z. Then L has a simple tangency with $X^\vee$. Now take a generic one-parameter family of matrices A_t such that L is the projectivization of Im $(\tilde{A}_0)$. The simple tangency condition implies that the function $t \mapsto \Delta(F_{A_t})$ has a simple zero at $t = 0$. This completes the proof of Lemma 4.3.

Now we can easily complete the proof of Theorem 4.2. In the course of the proof of Theorem 4.1 we have actually shown that $\Delta(F_A) = 0$ if and only if either Det $(A) = 0$ or $\tilde{A}(x^{(0)}) \in \nabla_{sing}$ for some non-zero $x^{(0)} \in \mathbf{C}^{k_0+1}$. Denote by W the variety of all matrices A such that Im $(\tilde{A})$ meets ∇_{sing} at some non-zero point. Taking into account Lemma 4.3 we see that the ratio $G(A)$ may vanish only when $A \in W$.

Clearly, codim $(W) > 1$ for $k_0 + 1 < c(k_1, \ldots, k_r)$, codim $(W) = 1$ for $k_0 + 1 = c(k_1, \ldots, k_r)$, and W coincides with the whole matrix space for $k_0 + 1 > c(k_1, \ldots, k_r)$. Now all assertions of our theorem follow at once from the definition of the Chow form.

B. Examples

Example 4.4. Let $r = 2$, and ∇ be the space of degenerate $m \times m$ matrices, $m \geq 2$. The variety ∇_{sing} consists of matrices of rank $\leq (m-2)$ and has codimension four. Therefore, the hyperdeterminant of a three-dimensional matrix A of format $2 \times m \times m$ (resp. $3 \times m \times m$) is equal to the discriminant of the binary (resp. ternary) form Det $\tilde{A}(x)$. Note that for matrices of format $2 \times 2 \times 2$ or $3 \times 3 \times 3$, we obtain three different formulas for the hyperdeterminant corresponding to three different choices of a distinguished direction. For the format $2 \times 2 \times 2$ the hyperdeterminant is given by (1.5). For each of the formats $2 \times m \times m$ and $3 \times m \times m$ we obtain from (4.2) the formula for the degree of the hyperdeterminant:

$$N(1, m-1, m-1) = 2m(m-1), \quad N(2, m-1, m-1) = 3m(m-1)^2. \quad (4.4)$$

Note that $2 \times m \times m$ is a subboundary format, and the first of the formulas (4.4) is consistent with (2.9). It is an easy exercise to deduce the second formula in (4.4) from the general formula (2.7).

For the format $4 \times m \times m$ Theorem 4.2 (b) says that $\Delta(F_A)$ is equal to the product of Det (A) and some power R^ν of the Chow form R of $X^\vee_{sing}$. The value of ν can be obtained by calculation of degrees. By (4.2), the degree of $\Delta(F_A)$ is equal to $4m(m-1)^3$. It follows easily from (2.7) that the degree of Det (A) is equal to $\frac{2}{3}m(m-1)(m-2)(5m-3)$. On the other hand, the degree of the variety

$X^\vee_{sing}$ is known to be $m^2(m-1)(m+1)/12$ (see [ACGH], Chapter 2, formula (5.1)). Since $X^\vee_{sing}$ has codimension four, the degree of its Chow form R as a polynomial in matrix entries of A is four times the degree of $X^\vee_{sing}$, i.e., is equal to $m^2(m-1)(m+1)/3$. These three expressions imply that the exponent ν is equal to 2.

Example 4.5. Let $r = 3$ and let $V_1 = V_2 = V_3 = \mathbf{C}^2$ be three two-dimensional vector spaces. Let $V = V_1 \otimes V_2 \otimes V_3$ be the space of $2 \times 2 \times 2$ -matrices, and let e_{ijk} $(i, j, k \in \{0, 1\})$ be its standard basis vectors (matrix units). Let $\nabla \subset V$ be the variety of degenerate matrices. The group $G = GL(V_1) \times GL(V_2) \times GL(V_3)$ acts on the space V, leaving varieties ∇ and ∇_{sing} invariant. It is known and easy to check that G has only seven orbits on V, including $\{0\}$ (and hence six orbits on $P(V)$). The closures of six orbits in $P(V)$ and representatives of these orbits are the following:

dim = 7: $P(V)$ itself; a representative $e_{000} + e_{111}$.

dim = 6: The projectivization $X^\vee$ of ∇; a representative $e_{100} + e_{010} + e_{001}$.

dim = 4: Three varieties

$$P(V_1) \times P(V_2 \otimes V_3),\ P(V_2) \times P(V_1 \otimes V_3),\ P(V_1 \otimes V_2) \times P(V_3);$$

representatives $e_{010} + e_{001}$, $e_{100} + e_{001}$, $e_{010} + e_{100}$.

dim = 3: The product $P(V_1) \times P(V_2) \times P(V_3)$; a representative e_{000}.

The singular locus $X^\vee_{sing}$ has three irreducible components, namely the orbit closures of dimension four just described. This can be seen by calculating partial derivatives of the hyperdeterminant of a $2 \times 2 \times 2$ -matrix (given by (1.5)) at all the representatives listed above.

In particular, we see that $X^\vee_{sing}$ has codimension three. Hence for a $2\times2\times2\times2$ matrix A we have $\mathrm{Det}\,(A) = \Delta(F_A)$. This was already known to Schläfli ([Schl], §19). The degree of $\mathrm{Det}\,(A)$ is equal to 24.

For a $3 \times 2 \times 2 \times 2$ matrix A, it follows from Theorem 4.2 (b) and the obvious symmetry that for some $\nu \geq 0$

$$\Delta(F_A) = \mathrm{Det}\,(A) \cdot (R_{12}R_{13}R_{23})^\nu(\mathrm{Im}\,(\tilde{A})),$$

where R_{ab} is the Chow form of the component $P(V_c) \times P(V_a \otimes V_b) \subset X^\vee_{sing}$ for $\{a, b, c\} = \{1, 2, 3\}$. The exponent ν can be found as in the previous example. By (4.2), the degree of $\Delta(F_A)$ is equal to $3^3 \cdot 4 = 108$. By (2.9), the degree of $\mathrm{Det}\,(A)$ is equal to $N(2, 1, 1, 1) = 2 \cdot 3! \cdot 3 = 36$. Finally, each of the Chow forms

$R_{ab}(\mathrm{Im}(\tilde{A}))$ is easily seen to have degree 12 as a polynomial in matrix entries of A. It follows that $\nu = (108 - 36)/3 \cdot 12 = 2$.

It seems likely that in the general case of Theorem 4.2 (b), for any component Z of $X^\vee_{sing}$ the exponent with which R_Z enters $\Delta(F_A)$ is equal to the multiplicity of $X^\vee$ along Z (i.e., the degree of the normal cone, see [ACGH], Ch.2, §1). In both Examples 4.4 and 4.5 the normal cone at a generic point of $X^\vee_{sing}$ can be seen to be a quadratic cone, and the exponent is equal to two.

It follows from Theorem 4.2 that whenever the hyperdeterminant of format $(k_1 + 1) \times \cdots \times (k_r + 1)$ is non-trivial, we can apply Schläfli's method to matrices of format $2 \times (k_1 + 1) \times \cdots \times (k_r + 1)$, and conclude that $\Delta(F_A)$ is the product of $\mathrm{Det}\,(A)$ with some extra factors. In particular, this gives a method for calculating the hyperdeterminant of format 2^r by successive computations of discriminants of binary forms. However, the extra factors grow very fast with r.

We conjecture that formats $2 \times m \times m$, $3 \times m \times m$ and $2 \times 2 \times 2 \times 2$ are the only ones for which Schläfli's method gives the hyperdeterminant exactly (i.e., $\Delta(F_A)$ is not identically zero and does not contain extra factors). This is equivalent to the assertion that, for any formats other than $m \times m$ and $2 \times 2 \times 2$, the singular locus of ∇ has codimension two in the matrix space.

APPENDIX A

Determinants of Complexes

Complexes

Let k be a field. A *complex* of k-vector spaces is a graded vector space $W^\bullet = \bigoplus_{i\in\mathbf{Z}} W^i$ together with a family of linear operators

$$d = \{d_i : W^i \to W^{i+1}\}, \qquad d_{i+1} \circ d_i = 0.$$

The spaces W^i are called the *terms* of $W^\bullet$. Throughout this Appendix we shall assume that all but finitely many terms of a complex are zero. Cohomology spaces of a complex $W^\bullet$ are defined as $H^i(W^\bullet) = \mathrm{Ker}(d_i)/\mathrm{Im}(d_{i-1})$. A complex $W^\bullet$ is called *exact* if all $H^i(W^\bullet)$ are equal to zero.

Example 1. Any linear operator between two vector spaces, say, W^{-1} and W^0, can be regarded as a complex with only two non-zero terms:

$$\{\cdots \to 0 \to W^{-1} \xrightarrow{d_{-1}} W^0 \to 0 \to \cdots\}.$$

Such a complex is exact if and only if d_{-1} is invertible.

For any graded vector space $W^\bullet$ and any $m \in \mathbf{Z}$ we shall denote by $W^\bullet[m]$ the same vector space but with the grading shifted by m:

$$\left(W^\bullet[m]\right)^i = W^{m+i}.$$

The same notation will be used for complexes.

Let $V^\bullet$ and $W^\bullet$ be two complexes of vector spaces. A morphism of complexes $f : V^\bullet \to W^\bullet$ is a collection of linear operators $f_i : V^i \to W^i$ which commute with the differentials in $V^\bullet$ and $W^\bullet$. The cone of a morphism f is a new complex $\mathrm{Cone}(f)$ with terms $\mathrm{Cone}(f)^i = W^i \oplus V^{i+1}$ and the differential given by

$$d(w, v) = \big(d_W(w) + (-1)^{i+1} f(v),\ d_V(v)\big), \quad w \in W^i, v \in V^{i+1}. \tag{1}$$

Any morphism of complexes $f : V^\bullet \to W^\bullet$ induces a morphism of cohomology spaces $H^i(f) : H^i(V^\bullet) \to H^i(W^\bullet)$. A morphism f is called a *quasi-isomorphism* if all $H^i(f)$ are isomorphisms. The *derived category* of vector spaces $D(\mathrm{Vect})$ is obtained from the category of complexes of vector spaces by formally inverting the quasi-isomorphisms.

The proofs of several results in this Appendix and in the main body of the book require the use of more general derived categories involving, for example, complexes of modules over a ring R, complexes of sheaves etc. We refrain from giving more detailed background material on this subject referring the reader to first chapters in [Bo], [KS].

Determinantal vector spaces

Let W be a vector space of dimension n over a field k. Denote the one-dimensional vector space $\bigwedge^n(W)$ by $\mathrm{Det}(W)$. For $W = 0$, we set $\mathrm{Det}(0) = k$.

Let now $W^\bullet = \bigoplus W^i$ be a finite-dimensional graded vector space. We set

$$\mathrm{Det}(W^\bullet) = \bigotimes_i \mathrm{Det}(W^i)^{\otimes(-1)^i}, \tag{2}$$

where V^{-1} for a one-dimensional vector space V stands for the dual vector space V^*. The correspondence $W^\bullet \longmapsto \mathrm{Det}(W^\bullet)$ defines a functor Det from the category of finite-dimensional graded vector spaces and their isomorphisms to the category of one-dimensional vector spaces and isomorphisms. Let us list some basic properties of this functor.

Proposition 2.

(a) Det *takes multiplication by* $\lambda \in k^*$ *to the multiplication by* $\lambda^{\chi(W^\bullet)}$, *where* $\chi(W^\bullet) = \sum_i (-1)^i \dim(W^i)$.

(b) *Let* $V^\bullet$ *and* $W^\bullet$ *be two graded vector spaces. Then there is a natural isomorphism*
$$\mathrm{Det}(V^\bullet \oplus W^\bullet) \cong \mathrm{Det}(V^\bullet) \otimes \mathrm{Det}(W^\bullet).$$

(c) *There is a natural isomorphism*
$$\mathrm{Det}(W^\bullet[1]) \cong \mathrm{Det}(W^\bullet)^{-1}.$$

The proofs are obvious.

The Euler isomorphism

Let $(W^\bullet, d)$ be a finite-dimensional complex of vector spaces. Forgetting the differential, we can regard $W^\bullet$ simply as a graded vector space. The cohomology spaces of our complex form another graded vector space $H^\bullet(W^\bullet) = \bigoplus H^i(W^\bullet)$.

Proposition 3. *There is a natural isomorphism*

$$\mathrm{Eu}_d : \mathrm{Det}(W^\bullet) \longrightarrow \mathrm{Det}(H^\bullet(W^\bullet)).$$

We call Eu_d the *Euler isomorphism* because of the following corollary.

Corollary 4. *We have an equality*

$$\sum(-1)^i \dim(W^i) = \sum(-1)^i \dim(H^i(W^\bullet).$$

To deduce Corollary 4 from Proposition 3 it suffices to consider the action of $\lambda \in k^\star$ on $\mathrm{Det}(W^\bullet)$ and $\mathrm{Det}(H^\bullet(W^\bullet))$.

Proof of Proposition 3. As in the standard proof of the above corollary, we split our complex into several short exact sequences:

$$0 \to \mathrm{Ker}(d_i) \to W^i \to \mathrm{Im}(d_i) \to 0,$$

and

$$0 \to \mathrm{Im}(d_{i-1}) \to \mathrm{Ker}(d_i) \to H^i(W^\bullet) \to 0$$

for all i. The proof then reduces to the following particular case.

Lemma 5. *For an exact sequence of vector spaces*

$$0 \to A \to B \to C \to 0$$

there is a natural isomorphism

$$\mathrm{Det}(B) \cong \mathrm{Det}(A) \otimes \mathrm{Det}(C).$$

Proof. Let m, n, p be the dimensions of A, B, C respectively. We identify A with its image in B, and C with the quotient space B/A. Let

$$a_1 \wedge \cdots \wedge a_m \in \mathrm{Det}(A) = \bigwedge^m(A), \qquad c_1 \wedge \cdots \wedge c_p \in \mathrm{Det}(C) = \bigwedge^p(C).$$

We define a map $\mathrm{Det}(A) \otimes \mathrm{Det}(C) \to \mathrm{Det}(B)$ by sending

$$(a_1 \wedge \cdots \wedge a_m) \otimes (c_1 \wedge \cdots \wedge c_p) \quad \mapsto \quad a_1 \wedge \cdots \wedge a_m \wedge \hat{c}_1 \wedge \cdots \wedge \hat{c}_p,$$

where $\hat{c}_j \in B$ is some representative of $c_j \in C = B/A$. We leave it to the reader to check that this map is well-defined and establishes the desired isomorphism.

Corollary 6. *Let $(W^\bullet, d)$ be an exact complex of k-vector spaces. Then there is a natural identification*

$$\mathrm{Eu}_d : \mathrm{Det}(W^\bullet) \to k.$$

Indeed, $\mathrm{Det}(H^\bullet(W^\bullet))$ is canonically identified with k. (We can easily avoid appealing to the determinant of the zero vector space by splitting an exact complex into only one group of short exact sequences and then applying Lemma 5.)

Note that even if we fix the terms W^i, the isomorphism in Corollary 6 depends on d in a very essential way.

The determinant of a based exact complex

Let $e = \{e_\alpha,\ \alpha = 1, \dots, \dim(W)\}$ be a basis of a vector space W. Denote by $\mathrm{Det}(e) \in \mathrm{Det}(W)$ the wedge product $e_1 \wedge \cdots \wedge e_{\dim(W)}$. Clearly, $\mathrm{Det}(e)$ is a basis (i.e., a non-zero vector) in $\mathrm{Det}(W)$.

Now let $W^\bullet = \bigoplus W^i$ be a graded vector space and suppose that $e = \{e(i)\}$ is a system of bases in all W^i, so that each $e(i)$ is a basis $\{e(i)_1, e(i)_2, \dots, e(i)_{\dim(W^i)}\}$ in W^i. Let us define a non-zero vector $\mathrm{Det}(e) \in \mathrm{Det}(W^\bullet)$ as

$$\mathrm{Det}(e) = \bigotimes_i \mathrm{Det}(e(i))^{(-1)^i},$$

where for any non-zero element l of a 1-dimensional vector space L the element $l^{-1} \in L^{-1} = L^*$ is defined by $l^{-1}(l) = 1$.

We call a *based complex* a system $(W^\bullet, d, e)$, where $(W^\bullet, d)$ is a complex and $e = \{e(i)\}$ is a system of bases in all terms of $W^\bullet$.

Definition 7. Let $(W^\bullet, d, e)$ be a based exact complex. We call its *determinant* the number

$$\det(W^\bullet, d, e) = \mathrm{Eu}_d(\mathrm{Det}(e)) \in k^*.$$

Proposition 8.

(a) *Let*

$$W^{-1} \xrightarrow{d} W^0$$

be a two-term exact complex (i.e., just an isomorphism of vector spaces). Then $\det(W^\bullet, d, e)$ *is the usual determinant of the matrix of* d *with respect to the chosen bases.*

(b) *For the shifted complex we have*

$$\det(W^\bullet[1], d, e) = \det(W^\bullet, d, e)^{-1}.$$

Proposition 9. *Let* $e = \{e(i)\}$ *and* $f = \{f(i)\}$ *be two systems of bases in an exact complex* $(W^\bullet, d)$. *Let* $\{A(i)\}$ *be the transition matrices, i.e., we have*

$$f(i)_p = \sum_{q=1}^{\dim W_i} A(i)_{pq} e(i)_q.$$

Then we have

$$\det(W^\bullet, d, f) = \det(W^\bullet, d, e) \cdot \prod_i \det(A(i))^{(-1)^i}.$$

The proofs of Propositions 8 and 9 are obvious.

Example 10. Suppose we have a three-term exact complex

$$0 \to A \overset{d_{-1}}{\to} B \overset{d_0}{\to} C \to 0.$$

Let $\{a_1, \ldots, a_m\}, \{b_1, \ldots, b_n\}, \{c_1, \ldots, c_p\}$ be bases in A, B, C respectively, so $n = m + p$. Let us fix the grading of the complex so that B lies in degree 0. Let us calculate the determinant of this exact complex with respect to the chosen bases.

Denote by D_{-1} and D_0 the matrices of d_{-1} and d_0 in the chosen bases; thus, D_{-1} is an $n \times m$ matrix of rank m, and D_0 is a $p \times n$ matrix of rank p. Let $\overline{D_{-1}}$ be the submatrix of D_{-1} given by all the m columns and first m rows. Permuting if necessary the basis vectors b_i (i.e., the rows of D_{-1}) we can assume that $\det(\overline{D_{-1}}) \neq 0$. Denote by $\overline{D_0}$ the submatrix of D_0 formed by the last $p = n - m$ columns and all the p rows. It follows from our assumption that $\det(\overline{D_0}) \neq 0$ either.

Proposition 11. *In the above situation, the determinant of the complex* $0 \to A \overset{d_{-1}}{\to} B \overset{d_0}{\to} C \to 0$ *with respect to the chosen bases is equal to* $\frac{\det(\overline{D_{-1}})}{\det(\overline{D_0})}$.

Proof. Let $\hat{c}_1, \ldots, \hat{c}_p \in B$ have the same meaning as in the proof of Lemma 5, i.e., $d_0(\hat{c}_j) = c_j$. Using the construction in Lemma 5, we see that the determinant in question is the coefficient of proportionality

$$\frac{d_{-1}(a_1) \wedge \cdots \wedge d_{-1}(a_m) \wedge \hat{c}_1 \wedge \cdots \wedge \hat{c}_p}{b_1 \wedge \cdots \wedge b_n}.$$

Since $\overline{D_0}$ is invertible, we can choose all the $\hat{c}_j$ to be linear combinations of $b_{m+1}, \ldots, b_n$. Hence the transition matrix between the bases $\{d_{-1}(a_1), \ldots, d_{-1}(a_m), \hat{c}_1, \ldots, \hat{c}_p\}$ and $\{b_1, \ldots, b_n\}$ in B has the block form

$$\begin{pmatrix} \overline{D_{-1}} & 0 \\ * & \overline{D_0}^{-1} \end{pmatrix}.$$

The determinant of our complex is the determinant of this matrix, i.e., is equal to $\frac{\det(\overline{D_{-1}})}{\det(\overline{D_0})}$, as claimed.

Thus we have several different ways to calculate the determinant of a three-term exact complex: they correspond to different choices of a non-zero minor of order m of the matrix D_{-1}.

The Cayley formula for the determinant of a complex

Proposition 11 can be generalized to the case of an exact complex of arbitrary length. For simplicity we assume that the non-zero terms of the complex are situated in degrees between 0 and $r > 0$ (the general case can be reduced to this one by using Proposition 8 (b)). We shall write our complex in the explicit coordinate form

$$W^\bullet = \left\{0 \to k^{B_0} \xrightarrow{D_0} k^{B_1} \longrightarrow \dots \xrightarrow{D_{r-1}} k^{B_r} \to 0\right\} \tag{3}$$

so that a totally ordered set B_i is the indexing set for the chosen basis in the i-th term of the complex, and D_i is a matrix with columns B_i and rows B_{i+1}. We denote by B the system of standard bases in all the terms of $W^\bullet$. For any subsets $X \subset B_i$, $Y \subset B_{i+1}$ we shall denote by $(D_i)_{XY}$ the submatrix in D_i on columns from X and rows from Y. Let n_i be the cardinality of B_i.

Definition 12 . A collection of subsets $I_i \subset B_i$, $i = 0, \dots, r$ is called *admissible* if $I_0 = \emptyset$, $I_r = B_r$, and for any $i = 0, \dots, r-1$ we have $\#(B_i - I_i) = \#(I_{i+1})$ and the submatrix $(D_i)_{B_i - I_i, I_{i+1}}$ is invertible.

Proposition 13. *Admissible collections exist. For any admissible collection* (I_i), *we have*

$$\#(I_i) = \sum_{j=0}^{i-1} (-1)^{i-1-j} n_j.$$

Proof. Since $\#(I_{i+1}) = n_i - \#(I_i)$, the statement about the cardinality follows by induction on i. Denote by $K_i \subset k^{B_i}$ the kernel of D_i which coincides with the image of D_{i-1}. It follows from the definition that (I_i) is an admissible collection if and only if for each i we have $k^{B_i} = k^{B_i - I_i} \oplus K_i$. Thus, the existence of an admissible collection follows from the obvious fact that any vector subspace of a coordinatized vector space has a complementary coordinate subspace.

Theorem 14. *Let* (I_i) *be an admissible collection of subsets for a based exact complex* (3). *Denote by* Δ_i *the determinant of the matrix* $(D_i)_{B_i - I_i, I_{i+1}}$. *Then the determinant of* $W^\bullet$ *equals*

$$\det(W^\bullet, D, B) = \prod_{i=0}^{r-1} \Delta_i^{(-1)^{r-1-i}}.$$

In this form the determinant of a based exact complex was introduced by Cayley in [Ca4] (see Appendix B).

Proof. Let K_i be as in the proof of Proposition 13; in particular, $K_0 = 0$, $K_r = k^{B_r}$. Consider the exact sequences

$$S_i = \{0 \to K_i \to k^{B_i} \to K_{i+1} \to 0\}$$

for $i = 0, \ldots, r-1$. For each i we choose a basis $f(i)$ in K_i and denote by $e(i)$ the system of bases $\{f(i), B_i, f(i+1)\}$ in the terms of S_i. By Definition 7 and the construction of the Euler isomorphism (Proposition 3), we have

$$\det(W^\bullet, D, B) = \prod_i \det(S_i, D, e(i))^{(-1)^{r-1-i}},$$

where the grading in each S_i is specified so that the middle term k^{B_i} has degree 1. According to the proof of Proposition 13, we can choose a basis $f(i)$ so that its image under the coordinate projection $k^{B_i} \to k^{I_i}$ is the standard basis of k^{I_i}. Now we can calculate $\det(S_i, D, e(i))$ by Proposition 10. For S_i the matrix $\overline{D_{-1}}$ becomes the identity matrix, and $\overline{D_0}$ is just $(D_i)_{B_i - I_i, I_{i+1}}$. Thus in the chosen grading we have $\det(S_i, D, e(i)) = \Delta_i$ (see Proposition 8 (b)). The theorem is proved.

Corollary 15. *Let* $(W^\bullet, d, e)$ *be any based exact complex and let*

$$\zeta = \sum_i (-1)^{i+1} \cdot i \cdot \dim(W^i).$$

Then we have

$$\det(W^\bullet, \lambda d, e) = \lambda^\zeta \det(W^\bullet, d, e).$$

For a complex situated in non-negative degrees this follows at once from Theorem 14 and Proposition 13. For an arbitrary based exact complex this follows from Proposition 8 (b).

Filtrations and triangularity

The determinant of a block-triangular matrix is equal to the product of the determinants of the diagonal blocks. We want to generalize this fact to the determinants of complexes.

A *filtration* on a vector space W is just an increasing family of subspaces $F_0W \subset F_1W \subset \cdots \subset F_mW = W$. The successive quotients of a filtration F will be denoted by $\mathrm{gr}_j{}^F(W) = F_jW/F_{j-1}W$. A filtration on a graded vector space is, of course, an increasing family of graded vector subspaces.

Proposition 16. *If F is a filtration on a (graded) vector space W then there is a canonical isomorphism*

$$\operatorname{Det}(W) \cong \bigotimes_j \operatorname{Det}\big(\operatorname{gr}_j^F(W)\big).$$

Proof. It suffices to consider the case of a filtration on a vector space with only one intermediate space, i.e., an exact sequence of three terms. The statement in this case follows from Lemma 5.

Let W be a filtered vector space and let $e = \{e_i\}$ be a basis of W. Then the filtration F induces a filtration of e by subsets: $F_j(e) = e \cap F_j W\}$. A basis e is said to be *compatible* with a filtration F if for each j the images $\bar{e}_i$ of the vectors e_i from $F_j(e) - F_{j-1}(e)$ form a basis in $\operatorname{gr}_j{}^F(W)$.

A filtration on a complex $(W^\bullet, d)$ is just an increasing family of subcomplexes $F_j W^\bullet \subset W^\bullet$. The successive quotients $\operatorname{gr}_j{}^F(W^\bullet)$ of the filtration F in this case will be complexes themselves; we denote the differentials in these complexes by $\bar{d}$. Let $e = \{e(i)\}$ be a system of bases in all the terms of $W^\bullet$. We say that e is compatible with F, if each $e(i)$ is compatible. In this case we denote by $\bar{e}$ the systems of bases in the quotients $\operatorname{gr}_j{}^F(W^\bullet)$, formed by images of the vectors from e.

Proposition 17. *Let $(W^\bullet, F)$ be a filtered finite complex of finite-dimensional k-vector spaces such that all the quotients $\operatorname{gr}_j{}^F(W^\bullet)$ are exact complexes, and let $e = \{e(i)\}$ be a system of bases in the W^i compatible with F. Then $W^\bullet$ is exact and*

$$\det(W^\bullet, d, e) = \prod_j \det(\operatorname{gr}_j{}^F W^\bullet, \bar{d}, \bar{e}).$$

Proof. Under the identification of Proposition 16, the Euler isomorphism for $W^\bullet$ is identified with the product of Euler isomorphisms for the $\operatorname{gr}_j^F(W^\bullet)$. A basis $\operatorname{Det}(e)$ in $\operatorname{Det}(W^\bullet)$ is taken by this identification to the tensor product of the bases $\operatorname{Det}(\bar{e})$. Our proposition follows from these two facts.

Let $f : V^\bullet \to W^\bullet$ be a morphism of exact complexes and $\operatorname{Cone}(f)$ its cone. We have an exact sequence of complexes

$$0 \to W^\bullet \to \operatorname{Cone}(f) \to V^\bullet[1] \to 0 \tag{4}$$

which implies that $\operatorname{Cone}(f)$ is also an exact complex. Since the i-th term of $\operatorname{Cone}(f)$ is $V^{i+1} \oplus W^i$, systems of bases in the terms of $V^\bullet$ and $W^\bullet$ naturally produce a system of bases in the terms of $\operatorname{Cone}(f)$. The following proposition is a consequence of the previous one and the exact sequence (4).

Proposition 18. *For an arbitrary choice of bases in the terms of $V^\bullet$ and $W^\bullet$, and the corresponding bases in the terms of* Cone(f) *we have*

$$\det(\mathrm{Cone}(f)) = \det(W^\bullet)/\det(V^\bullet).$$

Determinants of complexes of modules over a ring

Let R be a Noetherian integral domain. We denote by $R^\star$ the group of invertible elements of R. By a based complex over R we shall mean a complex of free R-modules $(W^\bullet, d)$ equipped with a system e of R-bases in all the terms. We would like to associate to an exact based complex $(W^\bullet, d, e)$ over R an element $\mathrm{Det}(W^\bullet, d, e) \in R^\star$. To do this we can consider the field of fractions k and replace a complex of free R-modules by the corresponding complex of k-vector spaces. The determinant of the latter complex will lie a priori in $k^\star$; we have to prove that it lies in fact in $R^\star$. To do this we shall extend the concept of the determinantal vector space to the case of rings.

It is necessary to consider not only free R-modules but also the projective ones. Recall that a finitely generated R-module P is called *projective* if it is a direct summand of some free module R^m. In algebraic-geometric language projective modules correspond to vector bundles over the spectrum of the ring, while free modules correspond to trivial vector bundles. Let us summarize the main properties of the projective modules. All of them are well-known (see, e.g., [Bou2], Chapter II,§5).

Proposition 19.

(a) *For any exact sequence of R-modules*

$$0 \to M \to N \xrightarrow{f} P \to 0$$

with P projective there exists a homomorphism $g : P \to N$ such that $fg = \mathrm{Id}_P$ (and hence $N \cong M \oplus P$).

(b) *If P is a projective R-module, $\mathfrak{p}$ is a prime ideal in R, and $S_\mathfrak{p} = R - \mathfrak{p}$ then the localization $P(S_\mathfrak{p}^{-1})$ is a free module over the local ring $R(S_\mathfrak{p}^{-1})$. Its rank does not depend on $\mathfrak{p}$; it is called the rank of P and denoted* $\mathrm{rk}(P)$.

(c) *If P, Q are projective R-modules of ranks m, n respectively, then R-modules*

$$P \otimes_R Q,\ S^r_R P,\ \bigwedge\nolimits^r_R P,\ \mathrm{Hom}_R(P, R)$$

are projective of ranks mn, $\binom{m+r-1}{r}$, $\binom{m}{r}$, m respectively.

(d) *If P is a projective R-module of rank 1, then the natural map*

$$P \otimes_R (\mathrm{Hom}_R(P, R)) \to R$$

is an isomorphism of R-modules.

Part (b) of this proposition means that projective modules are "locally free." In view of part (d), we shall call projective R-modules of rank 1 *invertible*. We denote by $\mathrm{Inv}(R)$ the category of invertible R-modules and their isomorphisms. The module $\mathrm{Hom}_R(P, R)$ for an invertible P will also be denoted by P^{-1}.

For a projective R-module M of rank r, we denote $\mathrm{Det}_R(M) = \bigwedge_R^r(M) \in \mathrm{Inv}(R)$. This is an invertible R-module. The determinantal module $\mathrm{Det}_R(M^\bullet)$ for a projective graded R-module $M^\bullet$ (in particular, for a complex of projective R-modules) is defined similarly to the vector space case.

Theorem 20. *Let* $(M^\bullet, d)$ *be a finite exact complex of projective R-modules. Then there is a natural isomorphism* $\mathrm{Eu}_d : \mathrm{Det}_R(M^\bullet) \to R$ *of invertible R-modules.*

Proof. Denote by $K_i \subset M^i$ the submodule $\mathrm{Ker}(d_i) = \mathrm{Im}(d_{i-1})$. Consider the exact sequences

$$0 \to K_i \to M^i \to K_{i+1} \to 0.$$

The theorem follows from two lemmas:

Lemma 21. *All* K_i *are projective R-modules.*

Lemma 22. *For a short exact sequence*

$$0 \to K \to M \to L \to 0$$

of projective R-modules there is a natural isomorphism

$$\mathrm{Det}_R(M) \cong \mathrm{Det}_R(K) \otimes_R \mathrm{Det}_R(L)$$

in the category $\mathrm{Inv}(R)$.

Lemma 21 is proved by induction starting from the right end of the complex using Proposition 19 (a). Note that even if all M^i are free we cannot guarantee in general that the K_i will be free; this is why we have to consider projective modules.

Lemma 22 is proved by localization using Proposition 19 (b). We define first the required isomorphisms for each localization similarly to Lemma 5. Then we have to check that these isomorphisms fit together into an isomorphism of invertible R-modules. We leave these details to the reader.

If e is a system of R-bases in terms of a finite complex $M^\bullet$ of free R-modules then we define a basis element $\mathrm{Det}(e)$ of the free R-module $\mathrm{Det}_R(M^\bullet)$ of rank 1 as in the case of vector spaces. Note that any two basis elements of a free module of rank 1 can be obtained from each other by multiplication by an invertible element.

Definition 23. Let $(M^\bullet, d, e)$ be a based exact complex of free R-modules. Its *determinant* is defined

$$\det_R(M^\bullet, d, e) = \mathrm{Eu}_d\mathrm{Det}(e)) \in R^\star.$$

Proposition 24. *If k is any R-algebra and $(M^\bullet, d, e)$ is a based exact complex of free R-modules then*

$$\det_R(M^\bullet, d, e) = \det_k(M^\bullet \otimes_R k, d, e).$$

The proof is obvious. Thus we have proved, in particular, that if k is the field of fractions of R, then $\det_k(M^\bullet \otimes_R k, d, e) \in R^\star$.

The case of a discrete valuation ground field

Another possibility of using a "triangular" structure of a complex appears when the ground field k itself is filtered or, more precisely, is equipped with a discrete valuation (see, e.g., [vdW], Chapter 18). In this book we use only the case when $k = \mathbf{C}((t))$, the field of formal Laurent series with complex coefficients $\sum_{i=-N}^{\infty} a_i t^i$, $a_i \in \mathbf{C}$. (Another important example would be the field $\mathbf{Q}_p$ of p-adic numbers.) But it is notationally even easier to consider the case of an arbitrary discrete valuation field k.

We denote the valuation by $\mathrm{ord} : k^\star \to \mathbf{Z}$. For $\mathbf{C}((t))$ it is given by

$$\mathrm{ord}\left(\sum a_i t^i\right) = \min\{i : a_i \neq 0\}.$$

We define a decreasing filtration on k by

$$F^m k = \{f \in k : \mathrm{ord}(f) \geq m\}.$$

We have

$$(F^m k)(F^n k) \subset F^{m+n} k.$$

In particular, $F^0 k$ is a subring called the ring of integers of k. (For example, for $k = \mathbf{C}((t))$ we have $F^0 k = \mathbf{C}[[t]]$, the ring of formal power series $\sum_{i=0}^{\infty} a_i t^i$.)

The quotient $\bar{k} = F^0k/F^1k$ is a field called the residue field of k (for $k = \mathbf{C}((t))$ we have $\bar{k} = \mathbf{C}$). The associated graded ring

$$\mathrm{gr}_F^\bullet k = \bigoplus_i F^ik/F^{i+1}k$$

is isomorphic to the Laurent polynomial ring $\bar{k}[t, t^{-1}]$. In general this isomorphism is not canonical and requires a choice of an element $\pi \in k$ with $\mathrm{ord}(\pi) = 1$ which will correspond to $t \in \bar{k}[t, t^{-1}]$. (For $k = \mathbf{C}((t))$ the obvious choice is $\pi = t$.) For any $f \in k$ we define the of f as the image of f in $\mathrm{gr}_F^{\mathrm{ord} f} k$. We denote this leading term by $\mathrm{Lt}(f) \in \mathrm{gr}_F^\bullet k$. (For $k = \mathbf{C}((t))$ the leading term of a series is just its first non-vanishing term; so it is "leading" when $t \to 0$.)

Definition 25. Let W be a finite-dimensional vector space over a discrete valuation field k. A decreasing filtration $\{F^m W\}$ of W by additive subgroups is said to be *compatible with the valuation* if the following conditions hold:

(a) $(F^m k)(F^n W) \subset F^{m+n} W$.

(b) If $\pi \in k^\star$ is any element of order 1, then the multiplication by π gives an isomorphism $F^i(W) \to F^{i+1}(W)$.

(c) The F^0k-module F^0W is free of finite rank and generates W over k.

Such a filtration is uniquely determined by the "lattice" $F^0W \subset W$. For any $w \in W$ we denote by $\mathrm{ord}_F(w)$ the minimal m such that $w \in F^m(W)$ and by $\mathrm{Lt}(w)$ the image of w in $\mathrm{gr}_F^{\mathrm{ord}(w)}(W)$. The following proposition is obvious.

Proposition 26. *If F is a filtration on W compatible with the valuation then the $\mathrm{gr}_F^\bullet k$-module $\mathrm{gr}_F^\bullet W$ is free. If the e_i form a basis of W, then the* $\mathrm{Lt}(e_i)$ *are free generators of* $\mathrm{gr}_F^\bullet W$.

Proposition 27. *Let $W^\bullet$ be a complex of k-vector spaces equipped with a decreasing filtration F compatible with the differentials and valuation. Suppose that the complex $\mathrm{gr}_F^\bullet(W^\bullet)$ of modules over $\mathrm{gr}_F^\bullet k = \bar{k}[t, t^{-1}]$ is exact. Then*

(a) *$W^\bullet$ is an exact complex.*

(b) *For any system e of bases in the terms of $W^\bullet$ we have*

$$\mathrm{Lt}(\det(W^\bullet, d, e)) = \det(\mathrm{gr}_F^\bullet(W^\bullet), \bar{d}, \mathrm{Lt}(e)).$$

Proof. (a) Consider the spectral sequence of the filtered complex

$$E_1^{pq} = H^{p+q}(\mathrm{gr}^{-q}(W^\bullet)) \Longrightarrow H^{p+q}(W^\bullet).$$

Each term of this spectral sequence is a finitely generated module over $\mathrm{gr}_\bullet^F k = \bar{k}[t, t^{-1}]$. Since the ring $\bar{k}[t, t^{-1}]$ is Noetherian, the spectral sequence converges (even although the filtration has infinitely many terms). Hence $W^\bullet$ is exact.

(b) By our assumptions, we have a filtration on the one-dimensional vector space $\mathrm{Det}_k(W^\bullet)$ which is compatible with the filtration on k. The associated graded module of this filtration is identified with $\mathrm{Det}_{\bar{k}[t,t^{-1}]}(\mathrm{gr}_F^\bullet(W^\bullet))$. Similarly, in $\mathrm{Det}_k(W^\bullet)$ and $\mathrm{Det}_{\bar{k}[t,t^{-1}]}(\mathrm{gr}_F^\bullet(W^\bullet))$ we have bases $\mathrm{Det}(e)$ and $\mathrm{Det}(\mathrm{Lt}(e))$. Our assertion follows from the fact that $\mathrm{Det}(\mathrm{Lt}(e)) = \mathrm{Lt}(\mathrm{Det}(e))$.

Note that in particular we have calculated the order of the determinant (this will be used below). Apart from this, Proposition 27 is a generalization of the following obvious fact.

Proposition 28. *Let D be an $n \times n$ matrix with entries in the ring $\mathbf{C}[[t]]$. Then the constant term of the series* $\det(D)$ *equals the determinant of the numerical matrix formed by constant terms of entries of D.*

Prime divisors of the determinant and cohomology

Let $D : \mathbf{Z}^n \to \mathbf{Z}^n$ be an integral $n \times n$ matrix. It is known that if $\mathrm{Det}(D)$ is non-zero then its absolute value coincides with the order of the quotient group $\mathbf{Z}^n/\mathrm{Im}(D)$. We give a generalization of this fact to the case of determinants of complexes over more general rings.

Let R be a Noetherian integral domain (in our applications we need the cases when $R = \mathbf{C}[x_1, \dots, x_n]$ or $\mathbf{Z}[x_1, \dots, x_n]$ is a polynomial ring). We shall impose on R the following conditions (which hold in the above examples):

(a) R is *regular*, i.e., any finitely generated R-module has a finite projective resolution.

(b) R is *factorial*, i.e., any non-zero element $a \in R$ has a prime decomposition

$$a = \varepsilon \prod \pi_\nu^{n_\nu}, \tag{5}$$

(where $\varepsilon \in R^*$ and the π_ν are prime elements) which is unique up to a permutation of the π_ν and multiplication of them by elements from R^*.

Condition (b) is not strictly necessary and is imposed to simplify an exposition. In the case of polynomial rings, in (a) we can replace "projective" by "free."

Let k be the field of fractions of R. Any non-zero $a \in k$ has a prime decomposition of the type (5) but with possibly negative exponents n_ν. We call n_ν the π_ν-adic order of a and denote it by $\mathrm{ord}_{\pi_\nu}(a)$.

Let $(M^\bullet, d, e)$ be a based complex of free R-modules. Suppose that the complex $M^\bullet \otimes_R k$ of k-vector spaces is exact but $M^\bullet$ is not necessarily exact. (Geometrically we can view $M^\bullet$ as a complex of vector bundles and our assumption means that it is exact "at a generic point", but not necessarily everywhere so we shall use the term "generically exact" for such complexes.) We are interested in the prime decomposition of $\det(M^\bullet \otimes_R k, d, e) \in k^\star$. Note that the problem discussed in the previous subsection is a particular case of this problem, corresponding to the case when $R = F^0 k$ is a discrete valuation ring and $M^\bullet = F^0 W^\bullet$.

Proposition 29. *For any prime element $\pi \in R$ the π-adic order of*

$$\det(M^\bullet \otimes_R k, d, e)$$

does not depend on the choice of R-bases in terms of $M^\bullet$.

Indeed, another choice of bases leads, by Proposition 9, to the multiplication of the determinant by an element from $R^\star$.

So π-adic orders of the determinant depend only on the complex $M^\bullet$ itself. If $M^\bullet$ is exact then its determinant lies in $R^\star$ and all its π-adic orders are zero. So in the general case, it is natural to relate these orders to the cohomology modules of $M^\bullet$.

We shall need the notion of *multiplicity* (or length) of a finitely generated R-module along a prime ideal $\mathfrak{p} \subset R$. It is defined as follows (cf. [Bou 2], Chapter 4 §5). Let $S_\mathfrak{p} = R - \mathfrak{p}$ and $R_\mathfrak{p} = R(S_\mathfrak{p}^{-1})$ be the localization of R at $\mathfrak{p}$. This is a local ring and we denote by $\mathbf{m}_\mathfrak{p} = \mathfrak{p} \cdot R_\mathfrak{p}$ its maximal ideal. The quotient $k_\mathfrak{p} = R_\mathfrak{p}/\mathbf{m}_\mathfrak{p}$ is a field which is the field of fractions of the domain $R/\mathfrak{p}$.

Let M' be a finitely generated module over $R_\mathfrak{p}$. Every quotient $\mathbf{m}_\mathfrak{p}^i M'/\mathbf{m}_\mathfrak{p}^{i+1} M'$ is a finite-dimensional vector space over $k_\mathfrak{p}$. We say that M' has *finite length* if $\mathbf{m}_\mathfrak{p}^i M' = 0$ for $i \gg 0$. For such M', we define the *multiplicity* of M'

$$\operatorname{mult}_{\mathbf{m}_\mathfrak{p}}(M') = \sum_i \dim_{k_\mathfrak{p}} \mathbf{m}_\mathfrak{p}^i M'/\mathbf{m}_\mathfrak{p}^{i+1} M'.$$

Now let M be a finitely generated R-module. We denote by $M_\mathfrak{p} = M \otimes_R R_\mathfrak{p}$ the localization of M at $\mathfrak{p}$. We define the *multiplicity of M at $\mathfrak{p}$* to be

$$\operatorname{mult}_\mathfrak{p}(M) = \begin{cases} \operatorname{mult}_{\mathbf{m}_\mathfrak{p}}(M_\mathfrak{p}), & \text{if } M_\mathfrak{p} \text{ has finite length;} \\ 0 & \text{otherwise.} \end{cases}$$

Theorem 30. *For any prime element $\pi \in R$ and any generically exact based complex $(M^\bullet, d, e)$ of free R-modules we have*

$$\operatorname{ord}_\pi(\det(M^\bullet \otimes k, d, e)) = \sum_i (-1)^i \operatorname{mult}_{(\pi)}(H^i M^\bullet),$$

where (π) is the principal ideal generated by π.

Note that the localization $R_{(\pi)}$ is a discrete valuation ring with the maximal ideal generated by π.

To prove Theorem 30 we shall define, following [Knud-Mum], the determinantal modules for R-modules and complexes which are no longer projective. The main idea is to use a projective resolution

$$0 \to P_m \to \cdots \to P_0 \to M \to 0$$

of an arbitrary module M and to define $\mathrm{Det}_R M$ as $\bigotimes_i \mathrm{Det}_R(P_i)^{\otimes(-1)^i}$, postulating the Euler isomorphism. This requires checking a lot of compatibility properties most of which we will refer the reader to *loc. cit.*

Let $D^b(R)$ be the derived category of finite complexes of finitely generated R-modules. Recall [Bo], [KS] that this category is *triangulated*, i.e., equipped with a class of diagrams

$$A \to B \to C \to A[1] \tag{6}$$

called *exact triangles*. They satisfy the four axioms of Verdier, see *loc. cit.*. For any triangulated category $\mathcal{A}$, we denote by $K_0(\mathcal{A})$ the Grothendieck group of $\mathcal{A}$, i.e., the Abelian group generated by symbols $[A]$, $A \in \mathrm{Ob}(\mathcal{A})$ which are subject to relations $[B] = [A] + [C]$ for any exact triangle (6). Recall that by $\mathrm{Inv}(R)$ we have denoted the category of invertible R-modules, i.e., of projective R-modules of rank 1.

Proposition 31. *There exists a functor* $\mathrm{Det}_R : D^b(R) \to \mathrm{Inv}(R)$ *which coincides with the above defined determinant functor on complexes of projective modules, and has the following property: for any exact triangle (6) there is a natural isomorphism*

$$\mathrm{Det}_R(B) \to \mathrm{Det}_R(A) \otimes_R \mathrm{Det}_R C.$$

The proof is based on the fact that any object of $D^b(R)$ is isomorphic to a complex of projective R-modules (existence of projective resolutions) and on Theorem 20 which ensures that Det_R behaves as expected on exact complexes (see [Knud-Mum]).

Let us mention also the following corollary of this proposition, to be used later. Let $M^\bullet$ be any finitely generated graded R-module. Regarding $M^\bullet$ as a complex with zero differential, we get a well-defined invertible R-module $\mathrm{Det}_R(M^\bullet)$.

Corollary 32. *If d is a differential making a graded R-module $M^\bullet$ into a complex then there is a natural isomorphism of invertible R-modules*

$$\mathrm{Eu}_d : \mathrm{Det}_R(M^\bullet) \to \mathrm{Det}_R(H^\bullet(M^\bullet))$$

which is compatible with the extension of scalars from R to k, its field of fractions.

Now let us prove Theorem 30. First note that both sides of the required equality are of local nature. Thus it suffices to consider the case when R is a local discrete valuation ring. Then projective R-modules are free and every submodule of a free R-module is free so every finitely generated R-module has a free resolution of length 2. Let $\mathcal{D}_1 \subset D^b(R)$ be the subcategory of generically exact complexes. For every complex $M^\bullet \in \mathcal{D}_1$, every cohomology module $H^i(M^\bullet)$ is torsion, i.e., is annihilated by some non-zero element of R.

Consider the function $M^\bullet \mapsto \mathrm{ord}_\pi \det(M^\bullet)$ on objects of $\mathcal{D}_1$. It is additive in exact triangles by Proposition 31 and so defines a homomorphism $K_0(\mathcal{D}_1) \to \mathbf{Z}$.

Note that the right hand side of the equality in Theorem 30 also defines a homomorphism of the same groups.

Proposition 33. *Let R be a discrete valuation ring and let $\pi \in R$ be an element of order 1. Then the Grothendieck group $K_0(\mathcal{D}_1)$ is isomorphic to $\mathbf{Z}$. The generator can be identified with the class of the R-module $R/\pi R$.*

Proof. This follows by considering the filtration of every R-module M by submodules $\pi^a M$. If M is torsion then $\pi^a M = 0$ for $a \gg 0$ so the class $[M]$ of M in $K_0(\mathcal{D}_1)$ is equal to $\sum[\pi^a M/\pi^{a+1} M]$; also every $\pi^a M/\pi^{a+1} M$ is a vector space over the field $R/\pi R$ so its class in K_0 is a multiple of the class of $R/\pi R$. This proves our proposition.

In order to complete the proof of Theorem 30, it remains to compare how the two homomorphisms in question act on the generator of the Grothendieck group. By definition, $\mathrm{mult}_\pi(R/\pi R) = 1$. On the other hand, this module has a two-term free resolution

$$0 \to R \xrightarrow{\pi} R \to R/\pi R \to 0.$$

The determinant of the two-term complex given by multiplication by π is π itself and its π-adic order is also 1. Theorem 30 is proved.

The determinant of a complex as a g.c.d. of maximal minors

Theorem 13 expresses the determinant of a complex as a rather complicated alternating product of certain minors of the differential matrices. Here we show that this determinant can be sometimes expressed as the greatest common divisor of the maximal minors of the rightmost differential.

Let R be a Noetherian integral domain satisfying conditions (a) and (b) above, and k be the field of fractions of R. Consider a generically exact based complex of free R-modules of the form

$$M^\bullet = \{0 \to R^{B_{-m}} \xrightarrow{D_{-m}} R^{B_{-m+1}} \xrightarrow{D_{-m+1}} \cdots \xrightarrow{D_{-2}} R^{B_{-1}} \xrightarrow{D_{-1}} R^{B_0} \to 0\}. \quad (7)$$

We assume that R^{B_i} is situated in degree i for $i = -m, \dots, 0$.

Theorem 34. *Suppose that* $\mathrm{mult}_{(\pi)} H^i(M^\bullet) = 0$ *for* $i < 0$ *and any prime element* $\pi \in R$. *Then* $\det(M^\bullet)$ *is equal, up to an invertible element* $\varepsilon \in R^\star$, *to the greatest common divisor of the maximal minors of the matrix* D_{-1}.

Proof. By considering localizations of R at various ideals (π), this reduces to the following lemma.

Lemma 35. *Let k be a discrete valuation field with the ring of integers R and maximal ideal m. Let $D : R^{B_{-1}} \to R^{B_0}$ be a homomorphism of free R-modules such that $D_k : k^{B_{-1}} \to k^{B_0}$ is surjective. Then*

$$\mathrm{mult}_{\mathbf{m}}(\mathrm{Coker}(D)) = \min_I \mathrm{ord}(\Delta_{I,B_0}), \tag{8}$$

where I runs over all subsets of B_{-1} of size $\#(B_0)$, and Δ_{I,B_0} is the minor of D on the columns from I and rows from B_0.

Proof of the lemma. Both sides of (8) remain unchanged if we subject D to elementary transformations of rows and columns (with coefficients in R) or multiply any row or column by an element from $R^\star$. Using a version of the Euclid algorithm for the ring R, we can bring, by transformations of the described type, any matrix over R to a matrix of the form $\|a_{ij}\|$ where $a_{ij} = 0$ for $i \neq j$ and each a_{ii} is either 0 or π^{m_i} where $\pi \in R$ is some fixed element with $\mathrm{ord}(\pi) = 1$. For such a matrix the statement is obvious.

Determinants of spectral sequences

Consider a spectral sequence (E_r^{pq}, d_r), $r \geq 1$ of k-vector spaces. We assume that the E_r^{pq} for $p, q \in \mathbf{Z}$ are finite-dimensional, and, for any given r, only a finite number of them are non-zero. Each d_r is thus a differential in $E_r^{\bullet\bullet} = \bigoplus_{p,q} E_r^{pq}$ of bi-degree $(r, r-1)$ such that the cohomology of $E_r^{\bullet\bullet}$ with respect to d_r is $E_{r+1}^{\bullet\bullet}$.

A spectral sequence is called *convergent* if for $r \gg 0$ all the differentials d_r are 0 so that $E_r^{pq} = E_{r+1}^{pq}$; this common vector space is denoted by E_∞^{pq}. We call a spectral sequence *exact* if $E_r^{\bullet\bullet} = 0$ for $r \gg 0$.

It is possible to generalize the notion of the determinant of a complex to based exact spectral sequences. Namely, for any spectral sequence we define the determinantal vector spaces

$$\mathrm{Det}(E_r^{\bullet\bullet}) = \bigotimes_{p,q} \mathrm{Det}(E_r^{pq})^{(-1)^{p+q}}.$$

The Euler isomorphism identifies, for each r, the space $\mathrm{Det}(E_r^{\bullet\bullet})$ with $\mathrm{Det}(E_{r+1}^{\bullet\bullet})$. Therefore, if the spectral sequence is exact, we obtain an isomorphism $\mathrm{Eu} : \mathrm{Det}(E_1^{\bullet\bullet}) \to k$.

Suppose now that we have chosen a system of bases in all the graded components of the first term E_1^{pq} of an exact spectral sequence. Denote this system of bases by e. Then we obtain a basis vector $\mathrm{Det}(e) \in \mathrm{Det}(E_1^{\bullet\bullet})$ as before. We call the *determinant* of a based exact spectral sequence $(E_r^{\bullet\bullet}, d_r, e)$ the image $\mathrm{Eu}(\mathrm{Det}(e)) \in k^\star$. Any exact complex $W^\bullet$ of vector spaces can be considered as an exact spectral sequence: $E_1^{p0} = W^p$, $E_1^{pq} = 0$ for $q \neq 0$, and $E_r^{pq} = 0$ for $r > 1$. The determinant of this spectral sequence coincides with the determinant of a complex defined earlier.

Now we extend Theorem 30 to describe the prime decomposition of the determinant of an exact spectral sequence. Let R be a Noetherian integral domain with the field of fractions k. As before, we assume that R is regular and factorial. Let $(E_r^{\bullet\bullet})$ be a spectral sequence of finitely generated R-modules such that the components E_1^{pq} of its first term are projective. We call $(E_r^{\bullet\bullet})$ *generically exact* if it becomes exact after tensor multiplication with k. Suppose that $(E_r^{\bullet\bullet})$ is convergent (over R) and generically exact. Then, as for usual complexes, we have an element $\det(E) \in k$ defined up to multiplication with elements of $R^\star$. We want to find π-adic orders of $\det(E)$ in terms of the limit R-modules E_∞^{pq}.

Theorem 36. *Let $\pi \in R$ be a prime element. Then*

$$\mathrm{ord}_\pi \det(E^{\bullet\bullet}) = \sum_{p,q} (-1)^{p+q} \mathrm{mult}_\pi E_\infty^{pq}.$$

Proof. Let $M^\bullet = \bigoplus M^i$ be a finitely generated graded R-module. The construction of the determinant R-module $\mathrm{Det}_R(M^\bullet)$ above may be reformulated as follows.

(1) The determinant k-vector space $\mathrm{Det}_k(M^\bullet \otimes k)$ contains a distinguished vector $\mathrm{Det}(M^\bullet)$ defined up to multiplication by elements of $R^\star$.

(2) If $M^\bullet$ is a complex with differential d then the Euler isomorphism Eu_d takes $\mathrm{Det}(M^\bullet)$ to $\mathrm{Det}(H^\bullet(M^\bullet))$.

(3) If each M^i is torsion, then

$$\det(M^\bullet) = \varepsilon \cdot \prod_{\pi \in R \text{ prime}} \prod_i \left(\mathrm{mult}_{(\pi)} H^i(M^\bullet)\right)^{(-1)^i}.$$

All these statements apply as well to doubly graded complexes like $E_r^{\bullet\bullet}$. By applying (2) to each $E_r^{\bullet\bullet}$ and (3) to $E_\infty^{\bullet\bullet}$ we get our theorem.

APPENDIX B

On the Theory of Elimination

A. Cayley

[*Cambridge and Dublin Math. Journal*, **3** (1848), 116–120]

Suppose the variables $X_1, X_2, \ldots, g$ in number, are connected by the h linear equations

$$\begin{aligned}\Theta_1 &= \alpha_1 X_1 + \alpha_2 X_2 + \cdots = 0,\\ \Theta_2 &= \beta_1 X_1 + \beta_2 X_2 + \cdots = 0,\\ &\vdots\end{aligned}$$

these equations not being all independent but connected by k linear equations

$$\begin{aligned}\Phi_1 &= \alpha_1' \Theta_1 + \alpha_2' \Theta_2 + \cdots = 0\\ \Phi_2 &= \beta_1' \Theta_1 + \beta_2' \Theta_2 + \cdots = 0\\ &\vdots\end{aligned}$$

these last equations not being independent but connected by the l linear equations

$$\begin{aligned}\Psi_1 &= \alpha_1'' \Phi_1 + \alpha_2'' \Phi_2 + \cdots = 0,\\ \Psi_2 &= \beta_1'' \Phi_1 + \beta_2'' \Phi_2 + \cdots = 0,\\ &\vdots\end{aligned}$$

and so on for any number of systems of equations.

Suppose also that $g - h + k - l + \cdots = 0$; in which case the number of quantities $X_1, X_2, \ldots$ will be equal to the number of really independent equations connecting them, and we may obtain by elimination of these quantities a result $\nabla = 0$.

To explain the formation of this final result, write

$$\nabla = \begin{array}{|ccc|ccc|ccc} \hline \alpha_1 & \beta_1 & \cdots & & & & & & \\ \alpha_2 & \beta_2 & & & & & & & \\ \vdots & & & & & & & & \\ \hline \alpha'_1 & \beta'_1 & \cdots & \alpha''_1 & \beta''_1 & \cdots & & & \\ \alpha'_2 & \beta'_2 & & \alpha''_2 & \beta''_2 & & & & \\ \vdots & & & \vdots & & & & & \\ \hline & & & \alpha'''_1 & \beta'''_1 & \cdots & & & \\ & & & \alpha'''_2 & \beta'''_2 & & & & \\ & & & \vdots & & & & & \\ \hline \end{array}$$

which for shortness may be thus represented,

$$\nabla = \begin{array}{|c|c|c} \hline \Omega & & \\ \hline \Omega' & \Omega'' & \\ \hline & \Omega''' & \\ \hline \end{array}$$

where Ω, Ω', Ω'', Ω''', Ω'''', ... contain respectively h, h, l, l, n, n, ... vertical rows and g, k, k, m, m, p, ... horizontal rows.

It is obvious from the form in which these systems have been arranged, what is meant by speaking of a certain number of the horizontal rows of Ω' and the *supplementary* vertical rows of Ω; or of a certain number of the horizontal rows of Ω'' and the supplementary vertical rows of Ω', &c.

Suppose that there is only one set of equations, or $g = h$: we have here only a single system Ω, which contains h vertical and h horizontal rows, and ∇ is simply the determinant formed with the system of quantities Ω. We may write in this case $\nabla = Q$.

Suppose that there are two sets of equations, or $g = h - k$: we have here two systems Ω, Ω', of which Ω contains h vertical and $h - k$ horizontal rows, Ω' contains h vertical and k horizontal rows. From any k of the h vertical rows of Ω' form a determinant, and call this Q'; from the supplementary $h - k$ vertical rows of Ω form a determinant, and call this Q: then Q' divides Q, and we have $\nabla = Q \div Q'$.

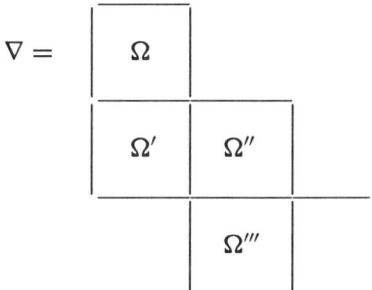

Suppose that there are three sets of equations, or $g = h - k + l$: we have here three systems, Ω, Ω', Ω'' of which Ω contains h vertical and $h - k + l$ horizontal rows, Ω' contains h vertical and k horizontal rows, Ω'' contains l vertical and k horizontal rows. From any l of the k horizontal rows of Ω'' form a determinant, and call this Q''; from the $k - l$ supplementary horizontal rows of Ω', choosing the vertical rows at pleasure, form a determinant, and call this Q'; from the $h - k + l$ supplementary vertical rows of Ω form a determinant, and call this Q: then Q'' divides Q', this quotient divides Q, and we have $\nabla = Q \div (Q' \div Q'')$.

Suppose that there are four systems of equations, or $g = h - k + l - m$: we have here four systems, Ω, Ω', Ω'' and Ω''', of which Ω contains h vertical and $h - k + l - m$ horizontal rows, Ω' contains h vertical and k horizontal rows, Ω'' contains l vertical and k horizontal rows, and Ω''' contains l vertical and m horizontal rows. From any m of the l vertical rows of Ω''' form a determinant, and call this Q'''; from the $l - m$ supplementary vertical rows of Ω'', choosing the horizontal rows at pleasure, form a determinant, and call this Q''; from the $k - l + m$ supplementary horizontal rows of Ω', choosing the vertical rows at pleasure, form a determinant, and call this Q'; from the $h - k + l - m$ supplementary vertical rows of Q form a determinant, and call this Q: then Q''' divides Q'', this quotient divides Q', this quotient divides Q, and $\nabla = Q \div \{Q' \div (Q'' \div Q''')\}$. The mode of proceeding is obvious.

It is clear that if all the coefficients $\alpha, \beta, \ldots$ be considered of the order unity, ∇ is of the order $h - 2k + 3l -$ &c.

What has preceded constitutes the theory of elimination alluded to in my memoir "On the Theory of Involution in Geometry", Journal *, vol. II, p. 52 - 61. And thus the problem of eliminating any number of variables x, y, ... from the same number of equations $U = 0$, $V = 0$, ... (where U, V, ... are homogeneous functions of any order whatever) is completely solved; though, as before remarked, I am not in posession of any method of arriving *at once* at the final result in its most simplified form; my process, on the contrary, leads me to a result encumbered by an extraneous factor which is only got rid of by a number of successive divisions less by two that the number of variables to be eliminated.

To illustrate the preceding method, consider the three equations of the second order,

$$U = ax^2 + by^2 + cz^2 + lyz + mzx + nxy = 0,$$

$$V = a'x^2 + b'y^2 + c'z^2 + l'yz + m'zx + n'xy = 0,$$

---

* Cambridge and Dublin Math. Journal

$$W = a''x^2 + b''y^2 + c''z^2 + l''yz + m''zx + n''xy = 0.$$

Here, to eliminate the fifteen quantities x^4, y^4, z^4, y^3z, z^3x, x^3y, yz^3, zx^3, xy^3, y^2z^2, z^2x^2, x^2y^2, x^2yz, y^2zx, z^2xy, we have the eighteen equations

$$x^2U = 0, \quad y^2U = 0, \quad z^2U = 0, \quad yzU = 0, \quad zxU = 0, \quad xyU = 0,$$

$$x^2V = 0, \quad y^2V = 0, \quad z^2V = 0, \quad yzV = 0, \quad zxV = 0, \quad xyV = 0,$$

$$x^2W = 0, \quad y^2W = 0, \quad z^2W = 0, \quad yzW = 0, \quad zxW = 0, \quad xyW = 0,$$

equations, however, which are not independent, but are connected by

$$a''x^2V + b''y^2V + c''z^2V + l''yzV + m''zxV + n''xyV$$
$$-(a'x^2W + b'y^2W + c'z^2W + l'yzW + m'zxW + n'xyW) = 0,$$
$$ax^2W + by^2W + cz^2W + lyzW + mzxW + nxyW$$
$$-(a''x^2U + b''y^2U + c''z^2U + l''yzU + m''zxU + n''xyU) = 0,$$
$$a'x^2V + b'y^2V + c'z^2V + l'yzV + m'zxV + n'xyV$$
$$-(ax^2V + by^2V + cz^2V + lyzUV + mzxV + nxyV) = 0.$$

Arranging these coefficients in the required form, we have the following value of ∇.

a						a'						a''					
	b						b'						b''				
		c						c'						c''			
	l		b				l'		b'				l''		b''		
		m		c				m'		c'				m''		c''	
n					a	n'					a'	n''					a''
		l	c					l'	c'					l''	c''		
m				a		m'				a'		m''				a''	
	n				b		n'				b'		n''				b''
	c	b	l				c'	b'	l'				c''	b''	l''		
c		a		m		c'		a'		m'		c''		a''		m''	
b	a				n	b'	a'				n'	b''	a''				n''
l			a	n	m	l'			a'	n'	m'	l''			a''	n''	m''
	m		n	b	l		m'		n'	b'	l'		m''		n''	b''	l''
		n	m	l	c			n'	m'	l'	c'			n''	m''	l''	c''
						a''	b''	c''	l''	m''	n''	$-a-$	$-b'$	$-c'$	$-l'$	$-m'$	$-n'$
$-a''$	$-b''$	$-c''$	$-l''$	$-n''$							a	b	c	l	m	n	
a'	b'	c'	l'	m'	n'	$-a$	$-b$	$-c$	$-l$	$-m$	$-n$						

which may be represented as before by

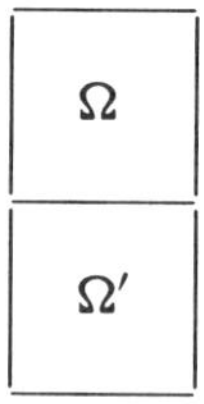

Thus, for instance, selecting the first, second and sixth lines of Ω' to form the determinant Q', we have $Q' = a''(a'b'' - a''b')$; and then Q must be formed from the third, fourth, fifth, seventh, &c. ... eighteenth lines of Ω. (It is obvious that if Q' had been formed from the first, second, and third lines of Ω', we should have had $Q' = 0$; the corresponding value of Q would also have vanished, and an illusory result be obtained; and similarly for several other combinations of lines.)

Bibliography

[Ad] J.F. Adams, *Infinite loop spaces*, Ann. Math. Studies **90**, Princeton Univ. Press, 1978.

[ACGH] E. Arbarello, M. Cornalba, P. Griffiths, J. Harris, *Geometry of algebraic curves, Vol.1* Grund. Math. Wiss, **267**, Springer-Verlag, 1990.

[Ar] V.I. Arnold, *Mathematical Methods of Classical Mechanics*, Springer-Verlag, 1989.

[AVG] V.I. Arnold, A.N. Varchenko, S.M. Gusein-Zade, *Singularities of differentiable maps, Vol. I - II,* Birkhäuser, Basel, 1985–1988.

[At1] M.F. Atiyah, *Angular momentum, convex polyhedra and algebraic geometry,* Proc. Edinburgh Math. Soc. **26** (1983), 121–138; *Collected Works, Vol. 5,* Cambridge Univ. Press, 1987, p. 379–391.

[At2] M.F. Atiyah, *Convexity and commuting Hamiltonians*, Bull. Lond. Math. Soc. **14** (1982), 1–15; *Collected Works, Vol. 5,* Cambridge Univ. Press 1987, p. 361–375.

[Au] R.J. Aumann, *Integrals of set-valued functions*, J. Math. Anal. Appl. **12** (1965), 1–12.

[Bar] D. Barlet, *Espace analytique réduit des cycles analytiques complexes compacts*, Lecture Notes in Math. **482** (1975), Springer-Verlag, 1–158,

[BE] H. Bateman, A. Erdelyi et al., *Higher Transcendental Functions Vol. 1*, McGraw-Hill, New York, 1953.

[BBD] A.A. Beilinson, I.N. Bernstein, P. Deligne, *Faisceaux Pervers*, Astérisque **100**, Soc. Math. France, 1981.

[BZ] A. D. Berenstein, A. V. Zelevinsky, *Triple multiplicities for $sl(r+1)$ and the spectrum of the exterior algebra of the adjoint representation*, J. Alg. Comb. **1** (1992), 7 –22.

[Ber] D. N. Bernstein, *The number of roots of a system of equations,* Funkt. Anal. Appl. **9** (1975), 183–185.

[BerZ] D. N. Bernstein, A. V. Zelevinsky, *Combinatorics of maximal minors*, J. Alg. Comb. **2** (1993), 111–121.

[BGI] P. Berthelot, A. Grothendieck, L. Illusie, *Théorie des Intersections et Théorème de Riemann-Roch* (SGA 6, 1966 - 67), Lecture Notes In Math. **225** (1971), Springer-Verlag.

[BFS] L.J. Billera, P. Filliman, B. Sturmfels, *Constructions and complexity of secondary polytopes*, Adv. Math. **83** (1990), 155 –179.

[BGS] L.J. Billera, I.M. Gelfand, B. Sturmfels, *Duality and minors of secondary polyhedra*, J. Comb. Theory B, **57** (1993), 258–268.

[BS1] L.J. Billera, B. Sturmfels, *Fiber Polytopes*, Ann. Math., **135** (1992), 527–549.

[BS2] L.J. Billera, B. Sturmfels, *Iterated fiber polytopes*, Preprint 1993.

[Bir] R. Birkeland, *Une proposition générale sur les fonctions hypergéometriques de plusieurs variables*, C. R. Acad. Sci. Paris, **185** (1927), 923–925.

[BF] J.-M. Bismut, D. Freed. *The analysis of elliptic families I. Metrics and connections on determinant bundles*, Comm. Math. Phys., **106** (1986), 139–176.

[Bj] J.-E. Björk, *Rings of differential operators*, North-Holland, Amsterdam, 1979.

[Bo] A. Borel et al., Algebraic *D-modules*, Perspectives in Math. **2** (1987), Academic Press, Boston.

[Bou1] N. Bourbaki, *Groupes et Algèbres de Lie*, Ch. VII-VIII, Hermann, Paris, 1975.

[Bou2] N. Bourbaki, *Commutative Algebra*, Springer -Verlag, 1989.

[Br1] A. Brill, *Das Zerfallen der Kurven in gerade Linien*, Math. Ann., **45** (1894), 410–427.

[Br2] A. Brill, *Vorlesungen über ebene algebraische Kurven und algebraische Funktionen*, F. Vieweg Verlag. Braunschweig 1925.

[BCG^3] R.L. Bryant, S.S. Chern, R.B. Gardner, H.L. Goldschmidt, P.A. Griffiths, *Exterior Differential Systems*, Math. Sci. Res. Inst. Publications **13** (1991), Springer-Verlag.

[Bry] J.L. Brylinski, *Loop spaces, Characteristic Classes and Geometric Quantization*, Progress in Math. **107**, Birkhäuser, Boston, 1993.

[Can] J. Canny, *Generalized characteristic polynomials*, J. Symb. Comp. **9** (1990), 241–250.

[Cat] F. Catanese, *Chow varieties, Hilbert schemes and moduli spaces of surfaces of general type*, J. Alg. Geometry, **1** (1992), 561–596.

[Ca1] A. Cayley, *On the theory of linear transformations*, Cambridge Math. J. **4** (1845), 1–16; *Collected Papers, Vol. 1*, p. 80–94, Cambridge Univ. Press, 1889.

[Ca2] A. Cayley, *On linear transformations,* Cambridge and Dublin Math. J. **1** (1846), 104–122; *Collected Papers, Vol. 1*, Cambridge Univ. Press, 1889, p. 95 -112.

[Ca3] A. Cayley, *On the theory of involution in geometry*, Cambridge and Dublin Math. J. **2** (1847), 52–61; *Collected Papers, Vol 1*, Cambridge Univ. Press 1889, p. 259–267,

[Ca4] A. Cayley, On the theory of elimination, *Cambridge and Dublin Math. J.* **3** (1848), 116–120; *Collected papers Vol. 1*, Cambridge Univ. Press, 1889, p. 370–374.

[Ca5] A. Cayley, *On a new analytical representation of curves in space*, Quarterly J. Pure and Appl. Math. **3** (1860), 225–236; *Collected Papers, Vol. 4*, Cambridge Univ. Press , 1889, p. 446–455.

[Ca6] A. Cayley, *On the partition of a polygon*, Proc. Lond. Math. Soc. (1) **22** (1890–91), 237–262; *Collected Papers, Vol. 13*, Cambridge Univ. Press, 1897, p. 93–113.

[C-vdW] W.L. Chow, B.L. van der Waerden, *Über zugeordnere Formen und algebraische Systeme von algebraishen Mannifaltigkeiten (Zur Algebraischen Geometrie IX)*, Math. Ann., **113** (1937), 692–704.

[Co] D.I.A. Cohen, *Basic techniques of combinatorial theory*, Wiley-Interscience, New York, 1978.

[D] V.I. Danilov, *The geometry of toric varieties*, Russian Math. Surveys, **33** (1987), 97–154.

[DEP] C. De Concini, D. Eisenbud, C. Procesi, *Hodge Algebras*, Astérisque **91**, Soc. Math. France.

[De1] P. Deligne, *Théorie de Hodge II*, Publ. Math. IHES, **40** (1971), 5–58.

[De2] P. Deligne, *Le formalisme des cycles évansecents*, Lecture Notes in Math., **340** (1973), Springer-Verlag, 82–115.

[De3] P.Deligne, "Le déterminant de la cohomologie", in: *Current trends in arithmetic algebraic geometry*, Contemporary Math., N. 67, Amer. Math. Soc., 1987, 93–177.

[Di] A.L. Dixon, *The eliminant of three quantics in two independent variables*, Proc. Lond. Math. Soc. **7** (1908), 49–69.

[ER] O. Egecioglu, J.B. Remmel, *The combinatorics of the inverse of the Kostka matrix*, Lin. and Multilin. Alg. **26** (1990), 59–84.

[E] H.G. Eggleston, *Convexity*, Cambridge Univ. Press, 1969.

[Fi] E.Fisher, *Ueber die Cayleyshe Eliminationsmethode*, Math. Zeit. **26** (1927), 497–550.

[Fr] W. Franz, *Die Torsion einer Überdeckung*, J. Reine u. angew. Math. **173** (1935), 245–254.

[Fu1] W. Fulton, *Intersection Theory*, Ergebnisse der Math. 3 Folge **2**, Springer-Verlag, 1984.

[Fu2] W. Fulton, *Introduction to Toric Varieties*, Princeton Univ. Press 1993.

[FH] W. Fulton. J. Harris, *Representation Theory. A first course*, Grad. Texts in Math. **129**, Springer-Verlag, 1991.

[Ga] F. Gaeta, "Associate forms, joins, multiplicities and intrinsic elimination theory", in: *Topics in Algebra, Banach Center Publications*, Vol. 26, part 2, PWN Polish Scientific Publishers, Warsaw, 1990.

[Ge] I. M. Gelfand, *General theory of hypergeometric functions*, Sov. Math. Dokl., **33** (1986), 573–577; *Collected Papers, Vol.3*, Springer-Verlag, 1989, p. 877–881.

[GGMS] I.M. Gelfand, R.M. Goresky, R.D. MacPherson, V.V. Serganova, *Combinatorial geometries convex polyhedra and Schubert cells*, Adv. Math., **63** (1988), 301–316; *Collected papers of I. M.Gelfand, Vol.3*, Springer-Verlag, 1989, p. 906–921.

[GGS] I.M. Gelfand, M.I. Graev, Z.Ya. Shapiro, *Integral geometry in projective space*, Funkt. Anal. Appl., **4** (1970), 12–28; *Collected Papers of I. M. Gelfand, Vol.3*, Springer-Verlag, 1988, p. 51–67.

[GGZ] I.M. Gelfand, M.I. Graev, A.V. Zelevinsky, *Holonomic systems of equations and series of hypergeometric type*, Sov. Math. Dokl., **36** (1988), 5–10; *Collected Papers of I.M. Gelfand, Vol. 3*, Springer-Verlag, 1989, p. 977–982.

[GK] I.M. Gelfand, M.M. Kapranov, "On the dimension and degree of the projective dual variety: a q-analog of the Katz-Kleiman formula", in: *Gelfand Mathematical Seminars 1990–92*, Birkhaüser 1993, p. 27–34.

[GKZ1] I.M. Gelfand, M.M. Kapranov, A.V. Zelevinsky, *Newton polytopes of the classical discriminant and resultant*, Adv. Math., **84** (1990), 237–254.

[GKZ2] I.M. Gelfand, M.M. Kapranov, A.V. Zelevinsky, *Generalized Euler integrals and A-hypergeometric functions*, Adv. Math., **84** (1990), 255–271.

[GKZ3] I.M. Gelfand, M.M. Kapranov, A.V. Zelevinsky, *Hyperdeterminants*, Adv. Math. **96** (1992), 226–263.

[GKZ4] I.M. Gelfand, M.M. Kapranov, A.V. Zelevinsky, "Hypergeometric functions, toric varieties and Newton polyhedra", in: M. Kashiwara, T. Miwa (Eds.), *Special functions (ICM - 90 Satellite Conference Proceedings)*, Springer-Verlag, 1991.

[GM] I.M.Gelfand, R.D.MacPherson, *Geometry in Grassmannians and a generalization of the dilogarithm*, Adv. Math., **44** (1982), 279–312; *Collected papers of I.M. Gelfand, Vol.3*, Springer-Verlag, 1989, p. 492 - 525.

[GZK1] I.M. Gelfand, A.V. Zelevinsky, M.M. Kapranov, *Hypergeometric functions and toric varieties*, Funct. Anal. and Appl. **23** (1989), 84–106.

[GZK2] I.M. Gelfand, A.V. Zelevinsky, M.M. Kapranov, *Projectively dual varieties and hyperdetermninants*, Sov. Math. Dokl., **39** (1989), 385–389.

[GZK3] I.M. Gelfand, A.V. Zelevinsky, M.M. Kapranov, *Discriminants of polynomials in several variables and triangulations of Newton polytopes*, Leningrad Math. J., **2** (1991), 449–505.

[GS] I.M. Gelfand, V.V. Serganova, *Combinatorial geometries and torus strata on homogeneous compact manifolds*, Russian Math. Surveys, **42** (1987), No. 2, 133–168; *Collected Papers of I.M. Gelfand, Vol. 3*, Springer -Verlag, 1989, p. 926–958.

[Gi] V. Ginzburg, *Characteristic varieties and vanishing cycles*, Invent. Math., **84** (1986), 327–402.

[Go1] P. Gordan, *Über die Bildung der Resultante zweier Gleichungen*, Math. Ann., **3** (1871), 355–414.

[Go2] P. Gordan, *Über die Resultante*, Math. Ann., **45** (1894), 405–409.

[Gr M] M. Green, I. Morrison, *The equations defining Chow varieties*, Duke Math. J., **53** (1986), 733–747.

[GH] P. Griffiths, J. Harris, *Principles of Algebraic Geometry*, Wiley-Interscience, London-New York, 1979.

[Gud] D.A. Gudkov, *The topology of real projective algebraic manifolds*, Russian Math. Surveys **29** (1974), 1–79.

[HLP] G.H. Hardy, J.E. Littlewood, G. Polya, *Inequalities*, Cambridge University Press, London 1951.

[Harr] J. Harris, *Algebraic Geometry*, Grad. Texts in Math. **133**, Springer-Verlag, 1992.

[Hart] R. Hartshorne, *Algebraic Geometry*, Grad. Texts in Math. **52**, Springer-Verlag, 1977.

[Hau] F. Hausdorff, *Set Theory*, Chelsea Publ. Co., New York, 1957.

[Hil1] D. Hilbert, *Über die Singularitäten der Discriminantalflache*, Math. Ann. **30** (1887), 437–441; *Gesammelte Abhandlungen, Bd.2*, S. 117–120, Springer-Verlag, 1933

[Hil2] D. Hilbert, *Über die Discriminante der im Endlichen abbrechenden hypergeometrischen Reihe*, J. Reine und angew. Math. **103** (1888) 337–345; *Gesammelte Abhandlungen, Bd.2*, S. 141–147, Springer-Verlag, 1933.

[Hilt] H. Hilton, *Plane Algebraic Curves*, Clarendon Press, Oxford, 1920.

[Hir] F. Hirzebruch, *Topological Methods in Algebraic Geometry*, Grund. Math. Wiss. **131** (1966), Springer-Verlag.

[Hoch] M. Hochster, *Rings of invariants of tori, Cohen-Macaulay rings generated by monomials and polytopes*, Ann. Math. **96** (1972), 318–337.

[HP] W.V.D. Hodge, D. Pedoe, *Methods of Algebraic Geometry, Vol.1*, Cambridge Univ. Press, Cambridge, 1953.

[Hol1] A. Holme, *On the dual of a smooth variety*, Lect. Notes in Math. **732** (1979), Springer-Verlag, 144 –156.

[Hol2] A. Holme, *The geometric and numerical properties of duality in projective algebraic geometry*, Manuscripta Math. **61** (1988), 145–162.

[Hop] H. Hopf, *Eine Verallgemeinerung der Euler-Poincaréschen Formel*, Nachr. der Gesell. der Wiss. zu Göttingen, Math. Wiss. Classe, 1928; *Selecta*, Springer-Verlag, 1964, p. 5–13.

[Hor] J. Horn, *Über die Konvergenz hypergeometrischer Reihen zweier und dreier Veränderlichen*, Math. Ann. **34** (1889), 544–600.

[I] A. Iarrobino, *Punctual Hilbert schemes*, Memoirs AMS, **10** (1977).

[Jo] J.P. Jouanolou. *Le formalisme du résultant*, Adv. Math. **90** (1991), 117–263.

[Ju1] F. Junker, *Über symmetrische Funktionen von mehreren Reihen von Veränderlichen*, Math. Ann., **43** (1893), 255–270.

[Ju2] F. Junker, *Die symmetrischen Funktionen und die Relationen zwischen der Elementarfunktionen derselben*, Math. Ann., **45** (1895), 1–84.

[Ka1] M.M. Kapranov, *A characterization of A-discriminantal hypersurfaces in terms of the logarithmic Gauss map*, Math. Ann. **290** (1991), 277–285.

[Ka2] M.M. Kapranov, *Veronese curves and Grothendieck-Knudsen moduli space* $\overline{M_{0,n}}$, J. Alg. Geometry, **2** (1993), 239–262.

[Ka3] M.M. Kapranov, *Chow quotients of Grassmannians I,* Adv. Soviet Math. **16**, Part 2, Amer. Math. Soc., 1993, 29–110.

[KSZ1] M.M. Kapranov, B. Sturmfels and A. Zelevinsky, *Quotients of toric varieties*, Math. Ann., **290** (1991), 643–655.

[KSZ2] M.M. Kapranov, B. Sturmfels and A. Zelevinsky, *Chow polytopes and general resultants*, Duke Math. J. **67** (1992), 189–218.

[KV] M.M. Kapranov, V.A. Voevodsky, *2-categories and Zamolodchikov tetrahedra equations*, Proc. Symp. Pure Math., to appear.

[Kas] M. Kashiwara, *Systems of microdifferential equations*, Progress in Math. **34**, Birkhäuser, Boston, 1983.

[KK] M. Kashiwara, T. Kawai, *On holonomic systems of microdifferential equations III: Systems with regular singularities*, Publ. RIMS, Kyoto Univ., **17** (1981), 813–979.

[KS] M. Kashiwara, P. Schapira, *Sheaves on Manifolds*, Grund. Math. Wiss. **292**, Springer-Verlag, 1990.

[Kat] N. Katz, *Pinceaux de Lefschetz: théorème d'existence,* Lecture notes in Math. **340** (1973), Springer -Verlag, 212–253.

[Kh] A.G. Khovansky, *Newton polyhedra and toral varieties*, Fuct. Anal. Appl. **11** (1977), 289–296.

[Kl] S. Kleiman, *Enumerative theory of singularities*, in: " Real and Complex Singularities", Proceedings of the Ninth Nordic Summer School /NAVF Sympos. Math. Oslo 1976, Sijthoff & Noordhoff, Alphen aan den Rijn 1977, 297–396.

[KP] S. Kleiman, R. Piene, "On the inseparability of the Gauss map", in: *Contemporary Math.* , **123**, Amer. Math. Soc., 1991, p. 107–130.

[Knop-Me] F. Knop, G. Menzel, *Duale Varietäten von Fahnenvarietäten*, Comm. Math. Helv., **62** (1987), 38–61.

[Knud-Mum] F. Knudsen, D. Mumford, *Projectivity of moduli space of stable curves I. Preliminaries on* Det *and Div*, Math. Scand., 39 (1976), 19–55.

[Kou] A.G. Kouchnirenko, *Polyèdres de Newton et nombres de Milnor*, Invent. Math., **32** (1976), 1–31.

[Kr] S. Kranz, *Function Theory of Several Complex Variables*, Wiley-Interscience, New York, 1982.

[La] S. Lang, *Elliptic functions*, Graduate Texts in Math. 112, Springer-Verlag, 1987

[Law] H.B. Lawson, *Algebraic cycles and homotopy theory, Ann. Math.*, **129** (1989), 253–291.

[Liu] L.A. Liusternik, *Convex Figures and Polyhedra*, Dover Publ. New York, 1963.

[Lee] C. Lee, *The associahedron and the triangulations of the n-gon,* European J. of Combinatorics, **10** (1989), 551–560.

[Lo] F. Loeser, *Polytopes secondaires et discriminants*, Séminaire Bourbaki 1990/91, Éxp.742, June 1991.

[Macd] I. Macdonald, *Symmetric functions and Hall polynomials*, Clarendon Press, Oxford, 1979.

[Macaul1] F.S. Macaulay, *Some formulae in elimination*, Proc. London Math. Soc. **35** (1902), 3–27.

[Macaul2] F.S. Macaulay, *Algebraic Theory of Modular Systems*, Cambridge Univ. Press, 1917.

[Macm] P.A. MacMahon, *Memoir on symmetric functions of the roots of systems of equations*, Phil. Trans., **181** (1890); *Collected papers, Vol. 2*, MIT Press, Cambridge MA, 1986, p. 32–84.

[Man] Y.I. Manin, *Lectures on the K-functor in algebraic geometry*, Russian Math. Surveys, **24**:5 (1969), 1–90.

[Mat] A. Mattuck, *The field of multisymmetric functions*, Proc. AMS **19** (1968), 764–765.

[Meb1] Z. Mebkhout, *Une equivalence des catégories*, Compos. Math. **51** (1984), 55–62.

[Meb2] Z. Mebkhout, *Une autre equivalence des catégories*, Compos. Math., **51** (1984), 63–69.

[Milg] R.J. Milgram, *Iterated loop spaces*, Ann. Math. **84** (1966), 386–403.

[Miln1] J. Milnor, *Morse Theory*, Ann. Math. Studies **51**, Princeton Univ. Press, 1963.

[Miln2] J. Milnor, *Whitehead torsion*, Bull. AMS, **72** (1966), 358–426.

[MS] J. Milnor, J.D. Stasheff, *Characteristic classes*, Ann. Math. Studies **76**, Princeton University Press, 1974.

[Min] F. Minding, *Über die Bestimmung des Grades der durch Elimination hervorgehenden Gleichung*, J. Reine und Angew. Math. **22** (1841), 178–183.

[Mum] D. Mumford, *Algebraic geometry I. Complex projective varieties*, Grund. Math. Wiss. **221**, Springer-Verlag, 1976.

[Nee] A. Neeman, *0-cycles in* P^n, Adv. Math. **89** (1991), 217–227.

[Net] E. Netto, *Vorlesungen über Algebra*, Bd. I, II, Teubner-Verlag, Leipzig, 1896.

[O] T. Oda, *Convex Bodies and Algebraic Geometry*, Erg. der Math., 3 Folge, **15**, Springer-Verlag, 1988.

[OSS] K. Okonek, P. Schneider, M. Spindler, *Vector bundles on complex projective spaces*, Progress in Math. **2**, Birkhäuser, Boston, 1982.

[Ore] O. Ore, *Sur la forme de fonctions hypergéometriques de plusieurs variables*, J. Math. Pure Appl. **9** (1930), 311–327.

[Q] D.Quillen, *Determinants of Cauchy-Riemann operators on Riemann surfaces*, Funct. anal. appl. **19** (1985), 31–34.

[P] E. Pascal, *Die Determinanten*, Teubner-Verlag, Leipzig, 1900.

[PS] P. Pedersen, B. Sturmfels, *Product formulas for sparse resultants*, to appear.

[Pen] R.C. Penner, *Universal constructions in Teichmüller theory*, Adv. Math. **98** (1993), 143–215.

[RSS] M. Rapoport, N. Shappacher, P. Schneider (Eds.), *Beilinson Conjectures on Special Values of L-Functions*, Perspectives in Math. **4**, Academic Press, 1988.

[RS] N. Ray, I. Singer, *R-torsion and the Laplacian for Riemannian manifolds*, Adv. in Math. **7** (1971), 145–210.

[Ris] J.-J. Risler, *Constructions d'hypersurfaces réelles (d'après Viro)*, Séminaire N. Bourbaki, Exp. 763, Novembre 1992.

[Ru] W. Rudin, *Real and Complex Analysis*, Mc Graw Hill, 1987.

[Ry] H.J. Ryser, *Combinatorial Mathematics*, Quinn & Boden, New Jersey, 1963.

[Sa1] M. Saito, *Modules de Hodge polarisables*, Publ. Math. RIMS, Kyoto Univ. **24** (1988), 849–995.

[Sa2] M. Saito, *Mixed Hodge modules*, Publ. Math. RIMS, Kyoto Univ. **26** (1990), 221–333.

[Sal] G. Salmon, *Lectures Introductory to the Modern Higher Algebra*, Dublin 1885; reprinted by Chelsea Publ., New York, 1969.

[Sau] D.J. Saunders, *The Geometry of Jet Bundles*, LMS Lect. Notes Series **142**, Cambridge Univ. Press, 1989.

[Schl] L. Schläfli, *Über die Resultante eines Systemes mehrerer algebraischen Gleichungen*, Denkschr. der Kaiserlicher Akad. der Wiss, math-naturwiss. Klasse, **4** Band, 1852; *Gesammelte Abhandlungen*, Band 2, S. 9-112, Birkhäuser Verlag, Basel, 1953).

[Se] B. Segre, *Bertini forms and Hessian matrices*, J. Lond. Math. Soc., **26** (1951), 164–176.

[Ser] J.P. Serre, *Faisceaux algébriques cohérents*, Ann. Math. **61** (1955), 197–278.

[Sh] I.R. Shafarevich, *Basic Algebraic Geometry*, Grund. Math. Wiss. **213**, Springer-Verlag, 1977.

[Shu] E. Shustin, *Critical points of real polynomials, subdivisions of Newton polyhedra and topology of real algebraic hypersurfaces*, Preprint 93-1, Tel-Aviv University, 1993.

[STT] D.D. Sleator, R.E. Tarjan and W.P. Thurston, *Rotation distance, triangulations and hyperbolic geometry*, J. AMS, **1** (1988), 647–681.

[Sm] Z. Smilansky, *Decomposability of polytopes and polyhedra*, Geom. Dedicata, **24** (1987), 29–49.

[So] N.P. Sokolov, *Space matrices and their applications*, Moscow, 1961 (in Russian).

[Stan] R.P. Stanley, *Subdivisions and local h -vectors*, J. AMS, **5** (1992), 805–852.

[St] J.D. Stasheff, *Homotopy associativity of H-spaces*, Trans. AMS, **108** (1963), 275–292.

[Stu1] B. Sturmfels, *Combinatorics of sparse resultants*, preprint 1992.

[Stu2] B. Sturmfels, *Viro's theorem for complete intersections*, preprint 1992.

[SZ1] B. Sturmfels, A. Zelevinsky, *Multigraded resultants of Sylvester type*, to appear in *J. Alg.*

[SZ2] B. Sturmfels, A. Zelevinsky, *Maximal minors and their leading terms*, Adv. in Math., **98** (1993), 65–112.

[V1] O.Y. Viro, *Gluing of algebraic hypersurfaces, removings of singularities and constructions of curves*, in: Proceedings of the International Topology Conference at Leningrad, Leningrad 1983 (in Russian), p. 149–197.

[V2] O.Y. Viro, *Gluing of plane algebraic curves and construction of curves of degrees 6 and 7*, in: Lecture notes in Math. **1060**, Springer-Verlag, 1984, 187–200.

[vdW1] B.L. van der Waerden, *Einführung in die algebraische Geometrie*, Springer-Verlag, 1955.

[vdW2] B.L. van der Waerden, *Algebra*, Springer-Verlag, 1991.

[Wal] R.J. Walker, *Algebraic curves*, Springer-Verlag, 1978.

[We1] J. Weyman, The equations of conjugacy classes of nilpotent matrices, Inv. Math. **98** (1989), 229–245.

[We2] J. Weyman, Calculating discriminants by higher direct images, to appear: Transactions of AMS.

[WZ1] J. Weyman, A. Zelevinsky, *Determinantal formulas for multigraded resultants*, to appear : J. Alg. Geom.

[WZ2] J. Weyman, A. Zelevinsky, *Multiplicative properties of projectively dual varieties*, to appear: Manuscripta Mathematica.

[Wi] G. Wilson, *Hilbert's sixteenth problem*, Topology **17** (1978), 53–73.

Notes and References

Here we collect some additional bibliographic comments and references.

Part I

Chapter 1. The concept of projective duality is very classical, going back at least to Poncelet. Duality for plane curves is discussed in many classical books, see, e.g., [Br 2], [Hilt], [vdW 1]. For a modern survey of projective duality in higher dimensions see [Kl], [Hol2]. A detailed discussion of subtleties appearing in finite characteristic (e.g., the violation of the biduality theorem) can be found in [KP], see also references therein. The exposition in Section 5 follows [WZ2].

Chapter 2. A better understanding of the origins of Cayley's "homological elimination" method can be obtained from his paper [Ca3] studying systems of hypersurfaces "in involution" (in modern terminology, linear systems of hypersurfaces). This paper contains a discussion of the idea of higher syzygies as well as an interesting bibliograpgy of still older works on the subject.

The material of Sections 2-4 is taken mostly from [GZK2]. Theorem 3.3 appeared first in [GK].

Chapter 3. More information on Grassmannians can be found in [GH], [HP], see also [DEP] for a modern treatment of the questions related to the coordinate ring. A classical reference for associated hypersurfaces and Chow forms is [vdW 1]; a more modern exposition is given in [Ga]. Theorem 2.7 appeared in [KSZ2].

In our approach the mixed $(\mathcal{L}_1, \ldots, \mathcal{L}_k)$ resultant is defined only up to a constant. This can be reformulated in a more "invariant" way by saying that the resultant takes values in a certain one-dimensional vector space which is not canonically identified with **C**. For a detailed discussion of this space as a functor of the bundles $\mathcal{L}_i$ see [De3], where this vector space is denoted $\langle \mathcal{L}_1, \ldots, \mathcal{L}_k \rangle$.

Chapter 4. As in Chapter 3, the standard reference on Chow varieties is [vdW1], §37. In the complex analytic situation the analogs of Chow varieties were constructed in [Bar]. The topology of Chow varieties $G(k, d, n)$ in the limit $d \to \infty$ was studied in [Law]. Theorem 1.6 was brought to our attention by L. Ein.

Symmetric functions in vector variables were studied at the turn of the century by F. Junker [Ju 1], [Ju 2], following earlier work of MacMahon [Macm].

The differential-geometric structure on Grassmannians (Section 3) is very important in integral geometry, see, e.g, [GGS].

Part II

Chapter 5. This material is mostly well-known (see e.g., [Fu2], [O]).

Chapter 6. The Newton polytopes (or rather polygons) go back to Newton who used them to construct power series expansions of algebraic functions in one variable, see [Wal] and also [Hilt], [vdW1]. In classical times mainly the lower part of the boundary of the Newton polygon (the so-called Newton diagram) was taken into account. See, however, [Br2] for a surprisingly modern treatment of the full Newton polygon and its application to the study of curves with singularities. Applications of higher-dimensional Newton polytopes to the study of singularities of functions can be found in [AVG].

The moment map can be defined for any Hamiltonian action of a Lie group G on a symplectic manifold M, see [Ar]. Here we consider the case when $G = (S^1)^k$ is the compact torus, and M is a toric variety with Kähler metric.

The material in Section 3 is mostly taken from [KSZ2].

Chapter 7. The material is mostly taken from [GZK3]. Sections 1E and 1F are based, respectively, on [BS] and [BFS]. The results in Section 3D were obtained by two of the authors (M.K. and A.Z) together with I.Pak and A. Postnikov.

Chapter 8. The results in Sections 2, 3 are mostly based on [KSZ2].

Chapter 9. The exposition in Sections 1, 2 is based on [GZK3]; that in Section 3 is based on [Ka1].

Chapter 10. The principal A-determinant was originally introduced in [GZK 3] as the determinant of the complex $L^\bullet(A, l)$; the interpretation (1.1) was found in [KSZ 2]. The exposition here is a modified and expanded version of [GZK 3].

Chapter 11. The material of Sections 1-4 is taken from [GZK 3] (with some corrections and modifications). The combinatorial Newton numbers and the Newton function of a triangulation were rediscovered and studied by R. Stanley in [Stan] under the name of "local h-vectors." The results and methods in [Stan] are similar to ours although the motivations are different. R. Stanley represents the Newton function (in our terminology) as the value at 1 of some polynomial in one variable with non-negative coefficients. The coefficients of these polynomials come from the multiplicities of sheaves $\underline{\mathbf{C}}_{\{0\}}[d]$ in Corollary 2.11.

An exposition of Viro's theorem (Section 5) and related questions was recently given in [Ris]. This theorem influenced many further developments, see, e.g., [Shu], [Stu 2]. In fact, the result of Viro is more general: in some cases it makes possible to decribe the isotopy type of a hypersurface by "gluing" isotopy types of simpler hypersurfaces.

Part III

Chapter 12. Discriminants and resultants for polynomials in one variable are discussed in many classical books on algebra, see, e.g., [Net], [Sal]; the latter reference contains extensive tables for discriminants and resultants written in terms of monomials, and also using the formalism of the symbolic method of the invariant theory. The paper [Go1] treats the problem of writing the resultant in the symbolic form. Some numerical invariants of singularities of discriminantal hypersurfaces were calculated by Hilbert [Hil1].

The material in Section 2 is mostly taken from [GKZ1]. For applications of the polytopes $N_{m,n}$ in the category theory see [KV]. The interpretation of the faces of $N_{m,n}$ in terms of labyrinths (equivalent to that of [GKZ1]) is due to M. Kapranov and V. Voevodsky.

Chapter 13. For a discussion of discriminants and resultants of forms in several variables from the classical point of view, see [Net], [Macaul2]. A modern treatment of resultants is given in [Jou], see also references therein. The Cayley formula is discussed in [Net], [Sal], [Fi]. Section 1C is based on [WZ1]. The discussion of Sylvester type formulas in Section 2A is based on [SZ1], [WZ1].

Chapter 14. The exposition is based mostly on [GKZ 3]. Section 3D is motivated by [WZ1].

Appendix A. A systematic early treatment of determinants of complexes can be found in [Fi] whose aim was to give a rigorous proof of Cayley's formula for the resultant [Ca 4]. (Fischer called complexes "Ketten der korrespondierenden Matrizen"; the present-day terminology goes back to [Hop].)

In topology determinants of complexes were introduced in 1935 by Reidemeister and Franz [Fr], who seemed to be unaware of the work of Cayley and Fischer. They used the word "torsion" for the determinant-type invariant they constructed. The 1950 paper of Whitehead [Wh] (see also [Miln 2]) contains a generalization of the Reidemeister-Franz torsion and applications to the theory of simple homotopy type.

There are several more recent works on the subject which influenced our exposition: [BF], [De 3], [Knud-Mum], [Q], [RS].

List of Notations

A: a finite subset of $\mathbf{Z}^{k-1}$ or $\mathbf{R}^{k-1}$
$\mathrm{Aff}_{\mathbf{R}}(I)$: the real affine subspace (in an affine space) spanned by a set I
$\mathrm{Aff}_{\mathbf{Z}}(I)$: the affine $\mathbf{Z}$-lattice (in an affine space) generated by a set I

$\mathcal{B} = \mathcal{B}(k, n)$: the coordinate ring of the Grassmannian $G(k, n)$
$\mathcal{B}_d$: degree d graded component of $\mathcal{B}$
$B^\bullet(l)$: one of two complexes whose cone is the discriminantal complex
$\mathbf{B}^\bullet(l)$: the complex of sheaves whose complex of global sections is $B^\bullet(l)$

$\mathbf{C}^A$: the space of Laurent polynomials on monomials
with exponents in $A \subset \mathbf{Z}^{k-1}$
$c(\Lambda, \mathcal{F}^\bullet)$: the multiplicity of a Lagrangian variety Λ in the characteristic cycle of a constructible complex $\mathcal{F}^\bullet$
$\mathrm{Char}(\mathcal{M})$: the characteristic variety of a $\mathcal{D}$-module $\mathcal{M}$
$\mathbf{Char}(\mathcal{M})$: the characteristic cycle of a $\mathcal{D}$-module $\mathcal{M}$
$c_i(E)$: the i-th Chern class of a bundle E
$c_X(q)$: the Chern polynomial of a projective variety X
$\mathrm{Con}(Z)$: the conormal space of a subvariety Z
$\mathrm{Conv}(I)$: the convex hull of a set I
$C^\bullet_\pm(X, \mathcal{M})$: the discriminantal complexes of a projective variety X
with respect to a twisting sheaf $\mathcal{M}$
$C^\bullet(X, l)$: the same as $C^\bullet_+(X, \mathcal{O}(l))$
$C^\bullet_\pm(\mathcal{L}_1, \ldots, \mathcal{L}_k|\mathcal{M})$: the mixed resultant complex
$C^\bullet(A, l)$: the discriminantal complex of the toric variety X_A;
the complex calculating the regular A-determinant D_A
$\mathrm{CV}_\Sigma(K)$: the combinatorial volume of a cone K with respect to
a triangulation Σ
$\mathrm{CN}_\Sigma(K)$: the combinatorial Newton number of a cone K with respect
to a triangulation Σ

D_A: the regular A-determinant
$\mathcal{D}_M$: the sheaf of differential operators on a manifold M

e: a basis vector, a basis or a system of bases in terms of a complex
E_A: the principal A-determinant

$E_A(f, S, \Xi)$: the generalized principal determinant

$\mathcal{F}$: a sheaf; a fan
$\mathcal{F}^\bullet$: a complex of sheaves
f: a polynomial from $\mathbf{C}^A$; a section of a line bundle.

$G(k, n)$: the Grassmann variety
$G(k, d, n)$: the Chow variety
g_ψ: the concave function corresponding to $\psi : A \to \mathbf{R}$

$\underline{H}^i(\mathcal{F}^\bullet)$: the cohomology sheaf of a complex of sheaves $\mathcal{F}^\bullet$
$\mathbf{H}^i(M, \mathcal{F}^\bullet)$: the hypercohomology of a space M with coefficients in a complex of sheaves $\mathcal{F}^\bullet$

$[i_1, \ldots, i_k]$: the bracket (Plücker coordinate)

$J(\mathcal{L})$: the first jet bundle of a line bundle $\mathcal{L}$
$j(f)$: the first jet of a section f

K: a convex cone
$K(I)$: the convex cone generated by a set I
$K_+(S)$: the convex hull of $S - \{0\}$ where S is a semigroup
$K_-(S)$: the subdiagram part of a semigroup S
$\mathcal{K}_\pm(E, s)$: the Koszul complexes associated to a section s of a vector bundle E

$L^i(\Xi)$: the space of discrete i-vector fields on a lattice Ξ
$L^i(S)$: the space of discrete i-vector fields on a semigroup S
$L^\bullet(A, l))$: the complex calculating the principal A-determinant E_A
$L^\bullet(A, l, S, \Xi)$: the complex calculating $E_A(f, S, \Xi)$
$\mathrm{Lin}_{\mathbf{R}}(I)$: the $\mathbf{R}$-vector subspace (in a vector space) generated by a set I
$\mathrm{Lin}_{\mathbf{Z}}(I)$: the abelian subgroup (in a vector space) generated by a set I

$\mathrm{mult}_Z(M)$: the multiplicity of a module M along a component Z of its support
$\mathrm{mult}_Z(Y)$: the multiplicity of a variety Y at the generic point of a subvariety Z

$N(f)$: the Newton polytope of a Laurent polynomial f
$N_\Gamma(Q)$: the normal cone to a polytope Q at a face Γ

$N(Q)$: the normal fan of a polytope Q

P: a projective space
P^*: the dual projective space
$P(V)$: the projectivization of a vector space V
$\mathrm{Proj}(R)$: the projective spectrum of a graded ring R
$p_{i_1,\dots,i_k}$: the Plücker coordinate
$p_\rho(\mathbf{x}_1, \dots, \mathbf{x}_d)$: the power sum symmetric function

Q: a convex polytope, typically $\mathrm{Conv}(A)$

R_X: the Chow form of a projective variety X
$\tilde{R}_X$: the X-resultant, i.e., the Chow form R_X written as a polynomial in k linear forms
$R_{\mathcal{L}_1,\dots,\mathcal{L}_k}$: the mixed $(\mathcal{L}_1, \dots, \mathcal{L}_k)$-resultant
$\mathbf{R}^A$: the space of real polynomials on monomials from A; the space of functions $A \to R$
$\mathbf{R}^A_{\mathrm{gen}}$: the generic stratum in the space of real polynomials

S: a semigroup
$\mathbf{SS}(\mathcal{F}^\bullet)$: the characteristic cycle of a constructible complex $\mathcal{F}^\bullet$

T: a triangulation; a torus

$u(S)$: the subdiagram volume of a semigroup S

$\mathrm{Vec}^i(S)$: the space of discrete i-vector fields on a semigroup S with some boundary conditions
Vol_Ξ: the volume form induced by a lattice Ξ

$\mathrm{Wt}(x)$: the weight polytope

X: a projective variety
X_A: the toric projective variety associated with $A \subset Z^{k-1}$
$X^\vee$: the projective dual variety

Y: an affine variety, typically the cone over X
Y_A: the affine cone over X_A
$Y^\vee$: the conic affine variety projectively dual to a conic affine variety Y

Z: a circuit; a subvariety

$\mathcal{Z}_X$: the associated hypersurface of a projective variety X

Δ^m: the m-dimensional simplex
Δ_X: the X- discriminant
Δ_A: the A-discriminant

η_T: the vector of exponents of the monomial in D_A corresponding to a triangulation T

Λ: a Lagrangian subvariety
Λ_+, Λ_-: parts of the boundary of the Newton polytope

μ: the moment map

$\nu(S)$: the Newton number of a semigroup S

Σ: a triangulation of a cone
$\Sigma(A)$: the secondary polytope of a set A

Ξ: a lattice

φ_T: the characteristic function of a triangulation T
Φ_φ: the complex of vanishing cycles

Ψ_φ: the complex of nearby cycles

χ: the Euler characteristic

ω: a typical element of A
Ω^i_X: the sheaf of differential forms on a smooth variety X
$\tilde{\Omega}^i_X$: the sheaf of Danilov differential forms on a (possibly singular) toric variety X
$\Omega^i_X(\log X_1)$: the sheaf of logarithmic differential forms

∇_A: the A-discriminantal variety
$\nabla_{\mathcal{L}_1,\dots,\mathcal{L}_k}$: the mixed resultant variety

Index